CURRENT TRENDS IN ORGANIC SYNTHESIS

Some Other IUPAC Titles of Interest from Pergamon Press

IUPAC Symposium Series

BENOIT & REMPP: Macromolecules

BRITTON & GOODWIN: Carotenoid Chemistry and Biochemistry

BROWN & DAVIES: Organ-Directed Toxicity — Chemical Indices and Mechanisms

CIARDELLI & GIUSTI: Structural Order in Polymers

FREIDLINA & SKOROVA: Organic Sulfur Chemistry

LAIDLER: Frontiers of Chemistry (Proceedings of the 28th IUPAC Congress)

MIYAMOTO: Pesticide Chemistry — Human Welfare and the Environment

ST-PIERRE & BROWN: Future Sources of Organic Raw Materials (CHEMRAWN I)

SHEMILT: Chemistry & World Food Supplies (CHEMRAWN II)

STEC: Phosphorus Chemistry Directed Towards Biology

TROST & HUTCHINSON: Organic Synthesis — Today and Tomorrow

IUPAC Nomenclature Guides

IRVING, FREISER & WEST: Compendium of Analytical Nomenclature

IUPAC: Nomenclature of Inorganic Chemistry & How to Name an Inorganic Substance (2-part set)

RIGAUDY & KLESNEY: Nomenclature of Organic Chemistry

WHIFFEN: Manual of Symbols & Terminology for Physicochemical Quantities and Units

Journals

CHEMISTRY INTERNATIONAL — IUPAC's international news magazine.

PURE AND APPLIED CHEMISTRY — IUPAC's official journal, featuring proceedings of IUPAC conferences, nomenclature rules and technical reports.

INTERNATIONAL UNION OF PURE AND APPLIED CHEMISTRY
(Organic Chemistry Division)

in conjunction with

The Science Council of Japan
The Chemical Society of Japan

CURRENT TRENDS IN ORGANIC SYNTHESIS

Proceedings of the Fourth International Conference on Organic Synthesis,
Tokyo, Japan, 22-27 August 1982

Edited by

HITOSI NOZAKI

*Department of Industrial Chemistry,
Kyoto University, Japan*

PERGAMON PRESS

OXFORD · NEW YORK · TORONTO · SYDNEY · PARIS · FRANKFURT

U.K.	Pergamon Press Ltd., Headington Hill Hall, Oxford OX3 0BW, England
U.S.A.	Pergamon Press Inc., Maxwell House, Fairview Park, Elmsford, New York 10523, U.S.A.
CANADA	Pergamon Press Canada Ltd., Suite 104, 150 Consumers Road, Willowdale, Ontario M2J 1P9, Canada
AUSTRALIA	Pergamon Press (Aust.) Pty. Ltd., P.O. Box 544, Potts Point, N.S.W. 2011, Australia
FRANCE	Pergamon Press SARL, 24 rue des Ecoles, 75240 Paris, Cedex 05, France
FEDERAL REPUBLIC OF GERMANY	Pergamon Press GmbH, Hammerweg 6, D-6242 Kronberg-Taunus, Federal Republic of Germany

Copyright © 1983 International Union of Pure and Applied Chemistry

All Rights Reserved. No part of this publication may be reproduced, stored in a retrieval system or transmitted in any form or by any means: electronic, electrostatic, magnetic tape, mechanical, photocopying, recording or otherwise, without permission in writing from the copyright holders.

First edition 1983

Library of Congress Cataloging in Publication Data
International Conference on Organic Synthesis (4th : 1982 : Tokyo, Japan)
Current trends in organic synthesis.
(IUPAC symposium series)
At head of title: International Union of Pure and Applied Chemistry (Organic Chemistry Division) in conjunction with the Science Council of Japan [and] the Chemical Society of Japan.
Includes bibliographical references and index.
1. Chemistry, Organic—Synthesis—Congresses.
I. Nozaki, Hitosi. II. International Union of Pure and Applied Chemistry. Organic Chemistry Division.
III. Nihon Gakujutsu Kaigi. IV. Nihon Kagakkai.
V. Title. VI. Series.
QD262.I55 1982   547'.2   82-22445

ISBN: 0-08-029217-8

In order to make this volume available as economically and as rapidly as possible the authors' typescripts have been reproduced in their original forms. This method unfortunately has its typographical limitations but it is hoped that they in no way distract the reader.

Printed in Great Britain by A. Wheaton & Co. Ltd., Exeter

CONTENTS

Organizing Committee	vii
Preface	ix
Acknowledgements	xi
Leukotrienes and Other Eicosanoids: Syntheses and Synthetic Methods E. J. COREY	1
Synthetic Approaches to Biologically Active Prostacyclin-Analogues W. BARTMANN, G. BECK, J. KNOLLE and R. H. RUPP	15
The Aldol Condensation as a Tool for Stereoselective Organic Synthesis C. H. HEATHCOCK	27
Studies Directed Towards Verrucarins: A Synthesis of Verucarol. An Approach to Verrucarin A B. M. TROST, P. G. McDOUGAL and J. H. RIGBY	45
New Approaches to the Total Synthesis of Biologically Active Natural Products M. E. JUNG	61
Studies Directed Towards the Chemical Synthesis of Yeast Alanine tRNA S. S. JONES, C. B. REESE and S. SIBANDA	71
Synthetic Aspects of 1-Oxacephem Antibiotics W. NAGATA	83
Studies Related to Beta-Lactam Compounds S. WOLFE	101
Stereochemistry of Palytoxin Y. KISHI	115
Diastereo- and Enantio-Selective Cycloaddition and Ene Reactions in Organic Synthesis W. OPPOLZER	131
Regio-, Diastereo-, and Enantioselective C-C Coupling Reactions Using Metalated Hydrazones, Formamides, Allylamines and Aminonitriles D. ENDERS	151
A New Methodology for the Generation of o-Quinodimethanes and Related Intermediates — An Approach to Asymmetric Synthesis of Polycycles Y. ITO	169
Asymmetric Synthesis Using Enantiomerically Pure Sulfoxides G. H. POSNER, K. MIURA, J. P. MALLAMO, M. HULCE and T. P. KOGAN	177
Stereocontrol via Cyclization Reactions P. A. BARTLETT	181
Stereocontrolled Synthesis of 25S,26-Dihydroxy and 1α,25S,26-Trihydroxy-cholecalciferol Side Chains A. D. BATCHO, J. SERENO, N. K. CHADHA, J. J. PARTRIDGE, E. G. BAGGIOLINI and M. R. USKOKOVIĆ	193
Carbohydrate Derivatives in the Asymmetric Synthesis of Natural Products B. FRASER-REID, L. MAGDZINSKI and B. MOLINO	197
Carbohydrates as "Chiral Templates" in Organic Synthesis — Target: Boromycin S. HANESSIAN, D. DELORME, P. C. TYLER, G. DEMAILLY and Y. CHAPLEUR	205
Stereochemical Control in Macrocycle Synthesis E. VEDEJS, J. M. DOLPHIN, D. M. GAPINSKI and H. MASTALERZ	221

Stereochemical Control in Macrocyclic Compounds W. C. STILL	233
New Hydroborating Agents H. C. BROWN	247
Palladium- or Nickel-Catalyzed Cross Coupling Involving Proximally Heterofunctional Reagents E. NEGISHI	269
Selective Reactions Using Organoaluminum Reagents H. YAMAMOTO and K. MARUOKA	281
New Access to Conjugated Dienes Via Carbocupration of Alkynes J. F. NORMANT and A. ALEXAKIS	291
Quinone Synthesis with Organometallic Reagents M. F. SEMMELHACK, T. SATO, J. BOZELL, L. KELLER, W. WULFF, A. ZASK and E. SPIESS	303
Organo-Silicon Mediated Synthetic Reactions I. KUWAJIMA	311
Trimethylsilanol as a Leaving Group in Preparative Organic Chemistry H. VORBRÜGGEN	323
Reaction of Silyl Enol Ethers with Carbenes and Carbenoids. New Syntheses of Various Carbonyl Compounds J.-M. CONIA and L. BLANCO	331
A New Synthetic Approach to 1,4-Dicarbonyl Systems and Functionalized Cyclopentenones Based on the Horner-Wittig Reaction of Phosphonates Containing Sulfur M. MIKOLAJCZYK	347
Vinyl and Beta-Alkoxy Radicals in Organic Synthesis G. STORK	359
Selective Hydrogenolysis of C-C Bonds in Small Ring Compounds H. MUSSO	371
Recent Synthetic Developments in Annulene Chemistry E. VOGEL	379
Vicarious Nucleophilic Substitution of Hydrogen. A New Method of Nucleophilic Alkylation of Nitroarenes M. MAKOSZA	401
Dichlorine Monoxide. A Powerful and Selective Chlorinating Reagent D. J. SAM, F. D. MARSH, W. B. FARNHAM and B. E. SMART	413
Index	423

Organizing Committee

Chairman:
T. Mukaiyama

Secretary General:
G. Tsuchihashi

Members:

T. Asahara, Y. Ban, M. Fujimaki, T. Goto, H. Hamada, M. Hasegawa,
T. Hata, Y. Inubushi, Y. Ishido, K. Ito, S. Itô, M. Kakudo, T. Kametani,
O. Kammori, M. Kobayashi, A. Kotera, K. Mori, K. Morita, R. Noyori,
H. Nozaki, S. Oae, M. Ohno, M. Ōki, S. Onogi, T. Saegusa, T. Sasaki,
T. Sato, H. Takei, Y. Takeuchi, J. Tsuji, H. Yajima, Y. Yoneda, Z. Yoshida

INTERNATIONAL UNION OF PURE AND APPLIED CHEMISTRY

IUPAC Secretariat: Bank Court Chambers, 2-3 Pound Way,
Cowley Centre, Oxford OX4 3YF, UK

PREFACE

The task of experts on organic synthesis lies in predicting accurately the actions of many invisible chemical compounds that undergo chemical changes in flasks, conducting discussion thereupon, and continuing their work, all on the basis of the experiences accumulated through experiment. Synthetic organic chemistry, which has so far created new potentials based on thinking on molecular levels, has now reached a stage at which it has become possible to synthesize extremely complicated molecules in flasks while controlling their three-dimensional structures.

The International Union of Pure and Applied Chemistry's (IUPAC) International Conference on Organic Synthesis is designed to promote the greater development on synthetic organic chemistry at the international level. The IUPAC conferences have been held in the past in Belgium, Israel and the United States. Four years ago when we were approached to organize this 4th IUPAC Symposium on Organic Synthesis we, as hosts, were both delighted and honoured to be given this immense task. We felt sure that the marvellous achievements which would be presented by the many chemists most actively engaged in the diversified frontier fields of organic synthesis would reflect future challenges for the synthetic organic chemistry. Furthermore, we could foresee considerable impact on such fields as biochemistry, inorganic chemistry, material science, natural resources, and energy problems. And, therefore, we were extremely enthusiastic to translate the newer frontier of organic chemistry into a stimulating and thought-provoking week.

We felt that this opportunity was an exciting challenge to display not only international progress but also a chance to show a new generation of vital and strengthening expansion of our own progress in organic synthesis while still retaining the aesthetics of Japan. This is reflected by the Conference Stamp, designed from four very old Chinese Characters. The upper two characters, YŪKI, mean Organic, and the lower two characters, GŌSEI, mean Synthesis. The meaning of this stamp is Organic Synthesis, and it is read as YŪKI GŌSEI.

The present volume contains thirty three lectures given by the principal participants of the Symposium in Tokyo during the period of 23 through 27 August, 1982, which have well rewarded our expectations. All the lectures have arbitarily been grouped into four divisions. The first one mainly focuses on Synthesis of Biologically Active Natural Compounds. This distinction applies to the lectures given by E. J. Corey, W. Bartmann, C. H. Heathcock, B. M. Trost, M. E. Jung, C. B. Reese, W. Nagata, S. Wolfe, and Y. Kishi. Topics are associated with metabolites of arachidonic acid, erythromycin A, verrucarins, steroids, anthracyclines, terpenes, yeast alanine t-RNA, beta-lactam antibiotics, and palitoxin. These lectures clearly indicate the state of the arts in organic synthesis, which obviously contribute much through achievements of constructing complex and unusual molecules.

One of the central problems in such efforts is Stereoselective and Chiral Synthesis. This particular field will be detailed in the following second group of lectures of W. Oppolzer, D. Enders, Y. Ito, G. H. Posner, P. A. Bartlett, M. R. Uskoković, B. Fraser-Reid, S. Hanessian, E. Vedejs, and W. C. Still. New processes of high degree of stereochemical control and asymmetric induction will be discussed. Chiral pool synthesis by means of carbohydrate precursors will follow and the section will be closed by discussions in large ring stereocontrol.

The subsequent sections are mainly concerned with Methodologies in Organic Synthesis, which now occupy the pivotal part of synthetic chemists' concern. New Reagents will be discussed in the third section of lectures given by H. C. Brown, E. Negishi, H. Yamamoto, J. F. Normant, M. F. Semmelhack, I. Kuwajima, H. Vorbrüggen, J.-M. Conia, and M. Mikołajczyk. The key atoms of the reagents are boron, aluminium, transition metals, silicon, phosphorus, and sulphur.

The final fourth division is devoted to New Reactions and contains lectures of G. Stork, H. Musso, E. Vogel, M. Makosza, and D. J. Sam. Discussions will involve radical initiated ring closure, small ring hydrogenolysis, annulene synthesis, vicarious nucleophilic substitution of aromatic hydrogen, and dichlorine monoxide mediated powerful chlorination.

The 4th Symposium in Tokyo is what we believe to be a continuing series of stimulating International Conferences on Organic Synthesis sponsored by the IUPAC. We were pleased and honoured to have been given the opportunity of transforming our enthusiasm into an outstanding symposium week, and are particularly appreciative of the tireless efforts and diligent work of all who participated in assisting with the many tasks that had to be done for the symposium and preparation of this manuscript.

October, 1982

Japan

Teruaki Mukaiyama

Hitosi Nozaki

ACKNOWLEDGEMENTS

We are grateful to the following organizations and companies for their sponsorship and/or generous financial assistance.

Sponsors:

The International Union of Pure and Applied Chemistry
The Science Council of Japan
The Chemical Society of Japan

Co-sponsors:

The Pharmaceutical Society of Japan
The Agricultural Chemical Society of Japan
The Society of Polymer Science, Japan
The Society of Synthetic Organic Chemistry, Japan

Financial Contributors:

Ajinomoto Co., Inc.
Association of Tokyo Stock Exchange Regular Members
Central Glass Co., Ltd.
Chugai Pharmaceutical Co., Ltd.
Commemorative Association for the Japan World Exposition 1970
Daicel Chemical Industries, Ltd.
Daiichi Seiyaku Co., Ltd.
Dainippon Ink & Chemicals, Inc.
Denki Kagaku Kogyo Kabushiki Kaisha
Earth Chemical Co., Ltd.
Fuji Photo Film Co., Ltd.
Godo Steel, Ltd.
Hoechst Japan, Ltd.
Itoman & Co., Ltd.
Japan Federation of Construction Contracts Inc.
Kanebo, Ltd., Pharmaceutical Research Center
Kao Corp.
Kikkoman Corp.
Konishiroku Photo Industry Co., Ltd.
Kumiai Chemical Industry Co., Ltd.
Kureha Chemical Industry Co., Ltd.
Lion Corp.
Meiji Seika Kaisha, Ltd.
Mitsubishi Corp.
Mitsubishi Petrochemical Co., Ltd.
Mitsubishi Steel Mfg. Co., Ltd.
Mitsui Petrochemical Industries, Ltd.
Mochida Pharmaceutical Co., Ltd.
Nagase & Co., Ltd.
Nichimen Corp.
Nippon Chemiphar Co., Ltd.
Nippon Kokan K.K.
Nippon Soda Co., Ltd.
Nippon Zeon Co., Ltd.
Nissin Flour Milling Co., Ltd. Central Research Laboratory
Nisshin Steel Co., Ltd.
Nitto Electric Industrial Co., Ltd.
Okura & Co., Ltd.
Regional Banks Association of Japan
Sankyo Co., Ltd.

Asahi Chemical Industry Co., Ltd.
Chori Co., Ltd.
C. Itoh & Co., Ltd.
Daido Steel Co., Ltd.
Daikin Industries, Ltd.
Dainippon Pharmaceutical Co., Ltd.
Du Pont Far East Inc., Japan
Eisai Co., Ltd.
Fujisawa Pharmaceutical Co., Ltd.
Grelan Pharmaceutical Co., Ltd.
Ihara Chemical Industry Co., Ltd.
Japan Automobile Manufactures Association Inc.
Japan Synthetic Rubber Co., Ltd.
Kanematsu-Gosho, Ltd.
Kawasaki Steel Corp., Ltd.
Kobe Steel, Ltd.
Kubota, Ltd.
Kuraray Co., Ltd.
Kyowa Hakko Kogyo Co., Ltd.
Marubeni Corp.
Mitsubishi Chemical Industries Ltd.
Mitsubishi Oil Co., Ltd.
Mitsubishi Rayon Co., Ltd.
Mitsui & Co., Ltd.
Mitsui Toatsu Chemicals Inc.
Monsanto Japan, Ltd.
Nakayama Steel Works, Ltd.
Nihon Tokushu Noyaku Seizo K.K.
Nippon Kayaku Co., Ltd.
Nippon Shinyaku Co., Ltd.
Nippon Steel Corp.
Nippon Zoki Pharmaceutical Co., Ltd.
Nissho Iwai Corp.
Nozaki & Co., Ltd.
Otsuka Pharmaceutical Co., Ltd.
Research Laboratories Morishita Pharmaceutical Co.
Sanyo-Kokusaku Pulp Co., Ltd.

Shin-Etsu Chemical Co., Ltd.
Shionogi Research Laboratories, Shionogi & Co., Ltd.
Showa Denko K.K.
Sumitomo Chemical Co., Ltd.
Sumitomo Metal Industries, Ltd.
Suntory Institute for Bio-Organic Research
Takasago Perfumery Co. Ltd.
Tanabe Seiyaku Co., Ltd.
T. Hasegawa Co., Ltd.
The Asahi Glass Foundation for Industrial Technology
The Federation of Electric Power Co.
The Japan Steel Works, Ltd.
The Life Insurance Association of Japan
The Marine and Fire Insurance Association of Japan
The Trust Company Association of Japan
Toa Nenryo Kogyo Kabushiki Kaisha
Tokyo Tanabe Co., Ltd.
Toray Industries, Inc.
Toshoku, Ltd.
Toyo Soda Manufacturing Co., Ltd.
Ube Industries, Ltd.
Unitika, Ltd. Research & Development Center
Yamada Science Foundation
Yodogawa Steel Works, Ltd.

Shiono Koryo Kaisha, Ltd.
Snow Brand Milk Products Co., Ltd.
Sumitomo Corp.
Suntory Central Research Institute
Syntex Corp.
Takeda Chemical Industries, Ltd.
Teijin, Ltd.
The Japan Gas Association
The Kajima Foundation
The Tokyo Bankers Association, Inc.
Toa Gosei Chemical Industry Co., Ltd.
Tokyo Ohka Kogyo Co., Ltd.
Topy Industries, Ltd.
Toshin Steel Co., Ltd.
Toyo Menka Kaisha, Ltd.
Toyobo Co., Ltd.
Ueno Fine Chemicals Industry, Ltd.
Wako Pure Chemical Industries, Ltd.
Yamanouchi Pharmaceutical Co., Ltd.
Yoshida Foundation for Science and Technology

LEUKOTRIENES AND OTHER EICOSANOIDS: SYNTHESES AND SYNTHETIC METHODS

E. J. Corey

*Department of Chemistry, Harvard University, Cambridge,
MA 02138, USA*

Abstract - Biologically active metabolites of arachidonic acid (eicosanoids), including prostaglandins and leukotrienes, are best obtained by total synthesis based on new synthetic methods. This article illustrates this point with a variety of examples selected to illustrate new methodology. Included are:

1. A biomimetic synthesis of leukotrienes.
2. A stereospecific synthesis of Z, E, E-trienes as applied to leukotriene B.
3. New methods for carbon-carbon single and triple bond formation as applied to the synthesis of mono acetylenic dehydroarachidonic acids, potent irreversible inhibitors of lipoxygenase enzymes.
4. A synthesis of 5-desoxyleukotrienes utilizing a sulfinyl-stabilized pentadienyl anion and a novel 1,5-rearrangement of sulfur.
5. A synthesis of trans-homoallylic alcohols from the new reagent lithiomethylenetriphenylphosphorane as a hyper-activated Wittig ylide.
6. An efficient synthesis of a medically interesting prostaglandin D metabolite employing new methods for functional group protection and reduction.
7. Finally, from a different area, new methodology which permits a most direct and effective synthesis of gibberellic acid.

Most of the new synthetic processes and reagents developed in our laboratory have resulted in no small measure from the perception of a methodological need which emerged either during the analysis or execution of specific projects on the synthesis of complex molecules. Stimuli for this work derive from the view that the rational discovery of new methods of synthesis is intrinsically linked to the activity of achieving new syntheses at the frontiers of organic chemistry and the conviction that new opportunity lies wherever the tools of contemporary synthesis are unequal to the task. More often than not the particular synthetic targets which generate a program of research are biologically active substances of unusual structure. These elements underlie the studies on synthetic methodology described herein.

During the late 1960's and early 1970's research at Harvard on the prostaglandins (PG's) resulted in the development of effective methods of total synthesis which made these rare and biologically important substances (and their analogs) readily available. Subsequent developments included the synthesis of stable analogs of PG endoperoxides and thromboxanes, prostacyclines, and various oxygenation products of arachidonic acid. Over the past five years our studies in the eicosanoid area have dealt largely with the leukotrienes (LT's) or "slow reacting substances" (SRS's), a newly recognized family of arachidonic acid-derived metabolites which are significantly involved in immunoregulation and which appear to be mediators in a variety of disease states including asthma, inflammation, immediate hypersensitivity, and allergy. As a result of this program, and partly in collaboration with Professor Bengt Samuelsson, the structures of the leukotrienes have been elucidated in all detail, their biosynthesis has been delineated and the compounds have been made available by synthesis in sufficient quantity so that biological and medical studies can advance unimpeded by problems of supply.[1]

The biosynthesis of leukotrienes A-E is summarized in Scheme 1. The pathway from arachidonic acid to leukotriene A, first proposed by us in March 1977, has been successfully duplicated chemically as outlined in Scheme 2. 5-Hydroperoxyeicosa-trans-6-cis-8,11,14-tetraenoic acid (5-HPETE) has been chemically synthesized (by selective reactions of arachidonic acid) in racemic form and enzymically synthesized as the 5-(S)-antipode.[2,3] Conversion of the methyl ester of 5-HPETE into LTA methyl ester, which finds precedent in a reaction discovered more than two decades ago in our laboratory,[4] could be accomplished under carefully selected and controlled

BIOSYNTHESIS OF LEUKOTRIENES

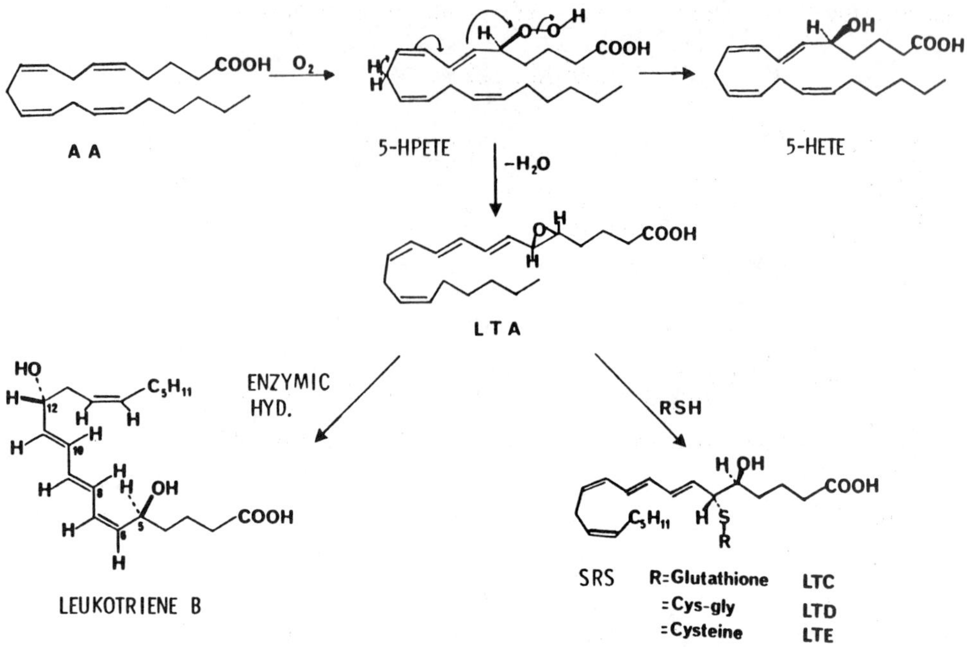

Scheme 1

CONVERSION OF ARACHIDONIC ACID TO 5-HPETE, LTA AND LTC-1

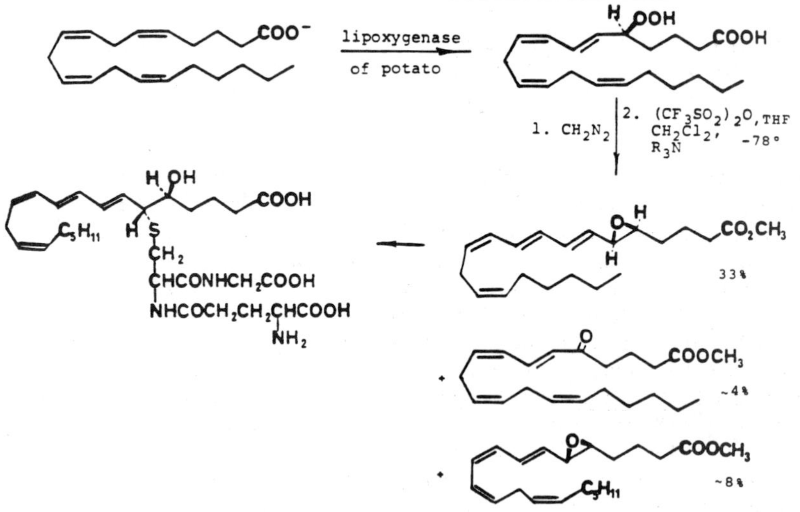

Scheme 2

conditions.[5,6] Because of the great sensitivity of LTA methyl ester to acid-catalyzed decomposition, an excess of tertiary amine and low temperature (-78°) must be maintained during the activation of the hydroperoxy function. In order to effect activation rapidly at -78° the highly reactive triflic anhydride is used. The amount of enone produced as a by-product may be minimized by use of the hindered tertiary amine pentamethyl piperidine in tetrahydrofuran (THF) as solvent. This new method, though only moderately efficient, provides a convenient and serviceable route to leukotrienes in amounts sufficient for biochemical experiments. The use of peroxytriflates in organic synthesis, of which this is the first example, is expected to be a valuable approach in the development of new reactions of hydroperoxides.

Leukotriene B, though formed enzymatically from LTA, has not thus far been obtainable in appreciable yield by chemical means from LTA. Total synthesis has both defined the stereochemistry of LTB and made this potent neutrophil chemotactic factor available for biological studies. A particularly effective method of synthesis of LTB which utilizes a new internal elimination process to establish the critical conjugated triene unit stereospecifically and in high yield is shown in Scheme 3.[7]

SYNTHESIS OF LTB

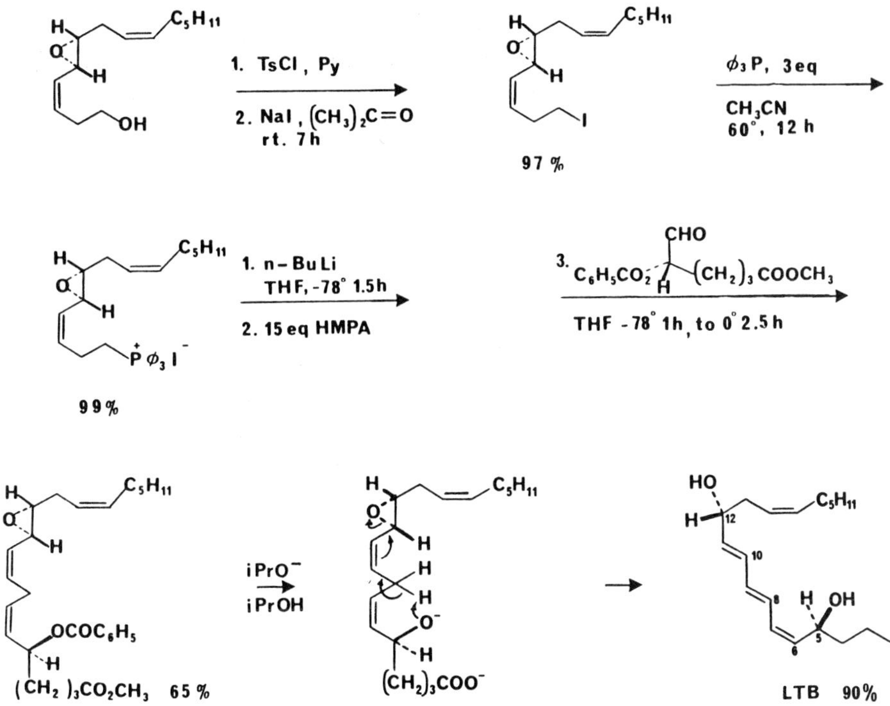

Scheme 3

Since the discovery of the leukotrienes and their possible role in disease, there has been enormous interest in the development of selective inhibitors of biosynthesis. Recently we have succeeded in developing by rational processes a family of irreversible and potent inhibitors of the lipoxygenation (LO) of arachidonic acid, including the 5-LO process which is the initial step in the formation of leukotrienes from arachidonic acid. The discovery was based on the proposition that if a triple bond were substituted for a cis HC=CH unit of arachidonic acid at the site of lipoxygenation, the occurrence of a LO reaction would produce a highly unstable vinylic hydroperoxide which might undergo very rapid ($t_{1/2} < 10^{-2}$ sec.) homolysis to products capable of deactivating the LO enzyme. The hypothesis is illustrated in Scheme 4 for the 15-LO reaction pathway characteristic of the well known lipoxygenase of soy bean. All four possible mono acetylenic dehydroarachidonic acids (DHA's) (Scheme 5) have been synthesized and tested.[8-11] It has been shown that 5,6-DHA, 11,12-DHA, and 14,15-DHA irreversibly inhibit 5-LO, 11-LO (prostaglandin synthetase) and 15-LO pathways, respectively. The effectiveness and utility of these DHA's make them valuable research tools, and in consequence a number of synthetic routes to these substances have been developed. Scheme 6 shows the first synthesis of 5,6-DHA, carried out starting from arachidonic acid, which employed an acetylene-forming elimination from the 2,4-dinitrobenzenesulfonylhydrazone of the 5,6-bromoketone.[8] The same elimination has been employed for the synthesis of 14,15-DHA[9] via the 14,15-epoxide of arachidonic acid, available in 98% yield by remarkably specific internal oxygen transfer reaction of peroxyarachidonic acid (Scheme 7).[12,13]

5,6-DHA has also been produced by a direct total synthesis which has several features of interest including the use of a terminally differentiated equivalent of 1,5-dilithio-cis,cis-1,4-pentadiene, a new C-C single/C-C triple bond forming coupling process of high efficiency, and a convergent overall plan as outlined in Scheme 8.[11] The vinylcopper-allenic iodide coupling reaction is exceptionally clean leading to acetylenic (S_N2') product in ca. 90% yield.[11] In contrast, the reaction of a Z-vinylcuprate reagent with the isomeric propargylic halides does not provide even modest yields of

DEVELOPMENT OF AN IRREVERSIBLE INHIBITOR OF SOYBEAN LIPOXYGENASE (15-LO)

Is 14, 15-DHA a substrate for 15-LO?

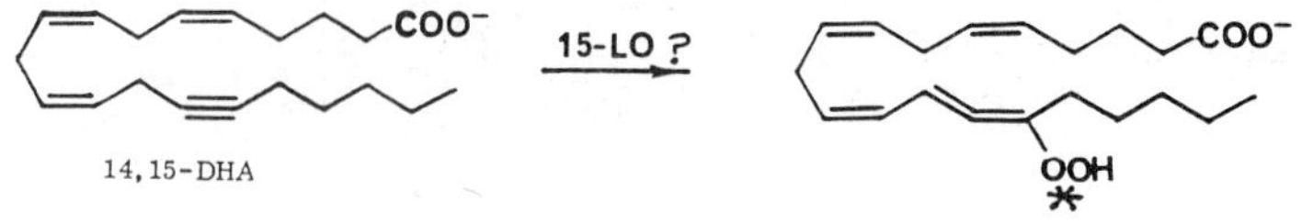

14, 15-DHA

Accelerated homolysis of a vinyl hydroperoxide.

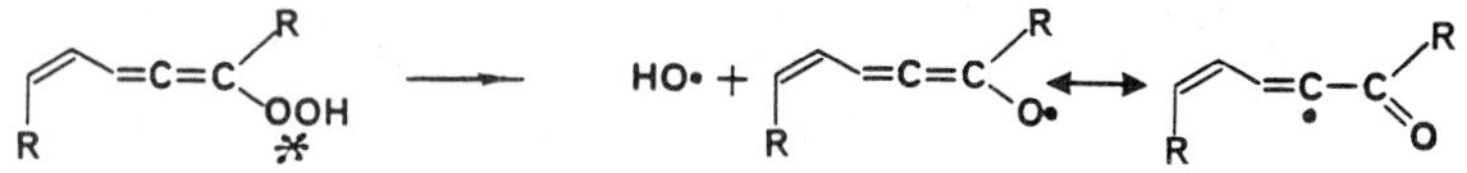

*O-O Bond energy estimated at 13 kcal/mole (33-20)

Scheme 4

FOUR MONOACETYLENIC DEHYDROARACHIDONIC ACIDS (DHA's)

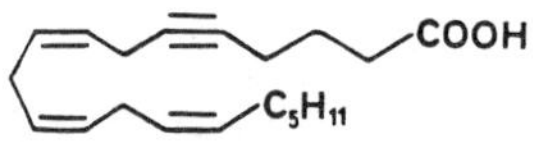

5, 6-DHA

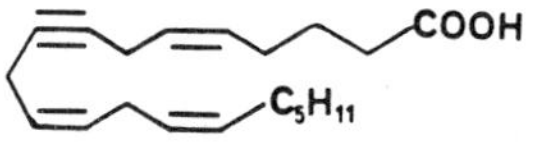

8, 9-DHA

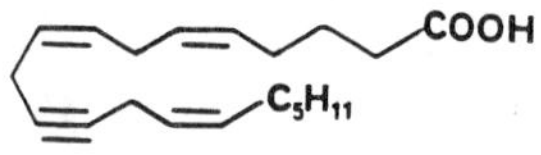

11, 12-DHA

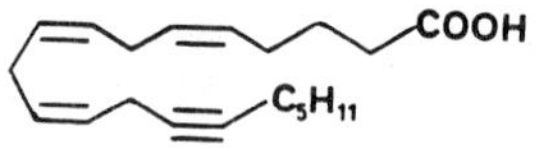

14, 15-DHA

Scheme 5

SYNTHESIS OF DHA'S

5, 6-DHA:

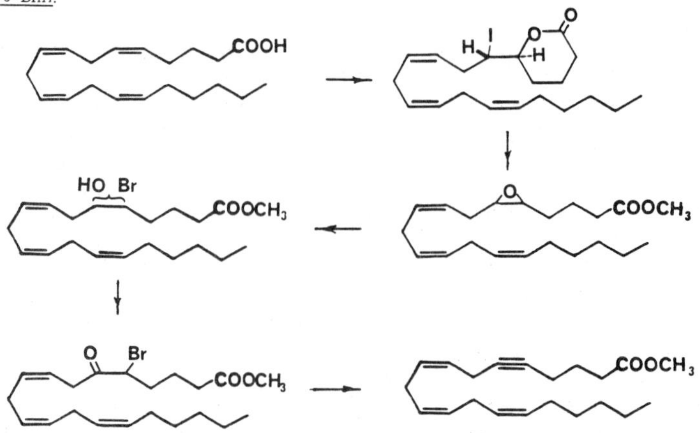

Scheme 6

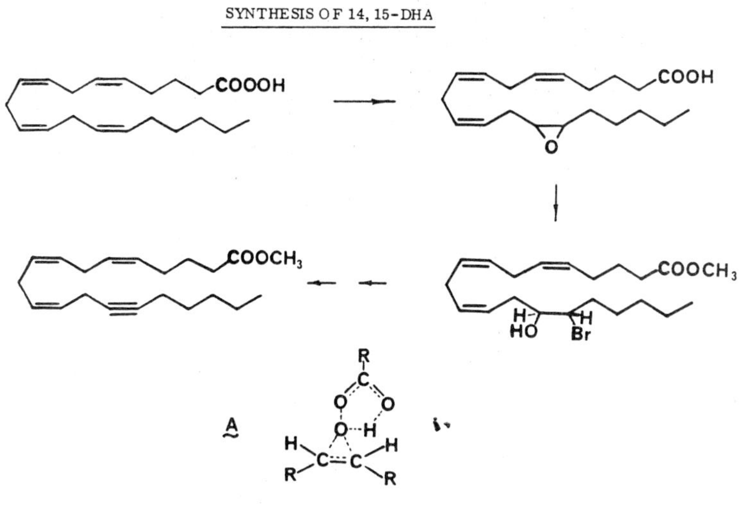

SYNTHESIS OF 14,15-DHA

Scheme 7

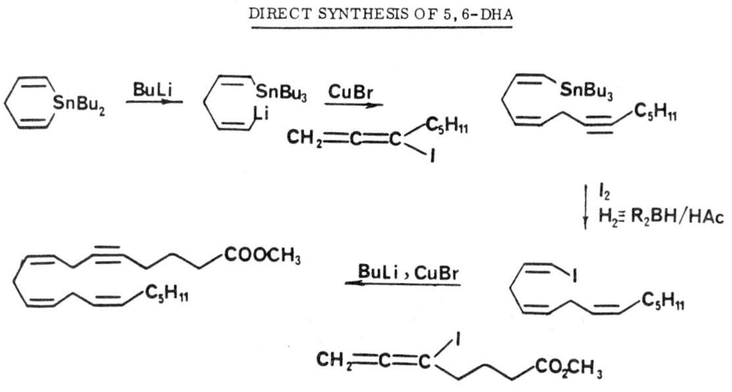

DIRECT SYNTHESIS OF 5,6-DHA

Scheme 8

acetylenic (S_N2) coupling product under the usual conditions. The same coupling methodology has been applied to simple syntheses of 8,9- and 11,12-DHA (Scheme 9)[11] and also to the first total synthesis of chiral 11-HETE (Scheme 10).[14]

Convergent routes to DHA's using the more conventional Wittig cis olefination method have also been demonstrated, as indicated for the case of 11,12-DHA in Scheme 11.[10] Of more than passing interest in this sequence is the use of a terminally differentiated building block corresponding to 3-hexyn-1,6-diol and the successful olefination of a Z-β,γ-unsaturated aldehyde, the latter not having been achieved previously.

The relatively slow onset (several minutes) and prolonged duration (30 min. to 2 hr.) of action of spasmogenic leukotrienes (LTC, LTD, and LTE) on the airway, suggest the possibility that these agents might attach covalently to the sites of recognition, perhaps by prior enzymic formation of a δ-lactone (Scheme 12). To test this question, and specifically to assess the importance of the 5-hydroxyl group to activity, 5-desoxy LTD was required. The retrosynthetic plan which was utilized for the synthesis is outlined in Scheme 13 and the successful execution is summarized in Schemes 14 and 15.[15] Of note here is the use of the conjugate base of trans-1-phenylsulfinylmethyl-1,3-butadiene as a synthetic equivalent of the 4-formyl-trans,trans-1,3-butadienyl anion in a novel five-carbon chain extension involving double [2,3] sigmatropic rearrangement. The rearrangement, which amounts to an interconversion of two position isomeric sulfoxides, is driven thermodynamically by the internalization of a conjugated diene unit. This facile transformation (occurring readily below 20°C), a Pummerer rearrangement, a Wittig condensation and an S_N2 displacement involving the thiolate of a protected form of cysteinyl glycine constitute the sequence of operations involved in this unusually simple and short route to 5-desoxy-LTD, which substance was in fact found to

SYNTHESIS OF 11,12-DHA

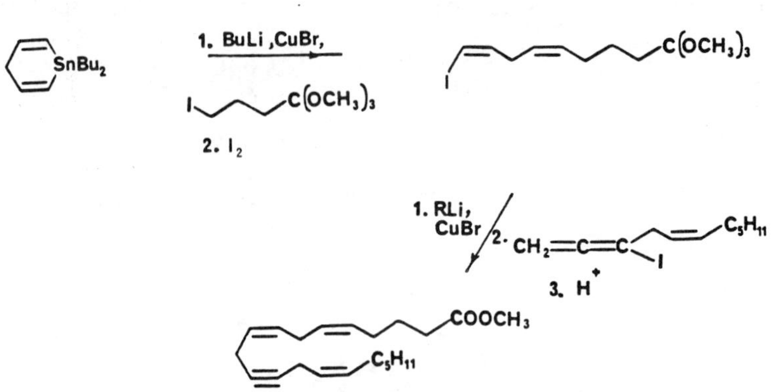

Scheme 9

SYNTHESIS OF 11(R)-HETE

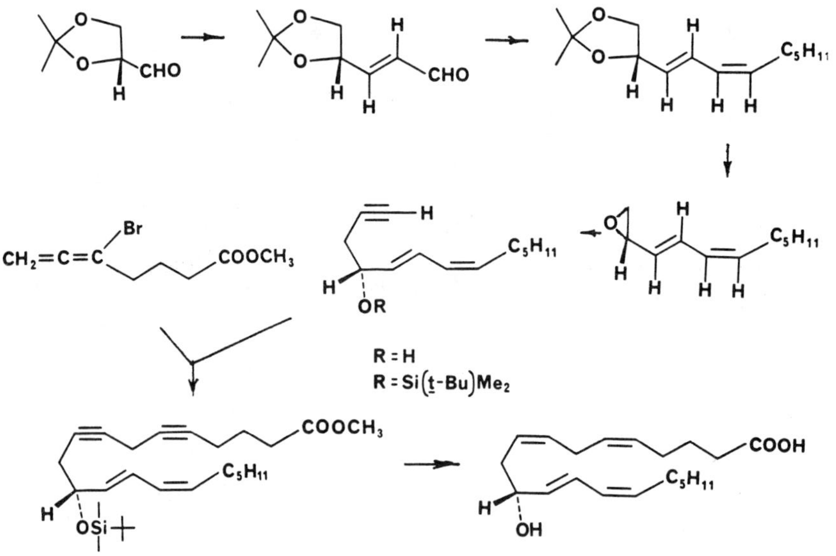

Scheme 10

WITTIG ROUTE TO 11,12-DHA

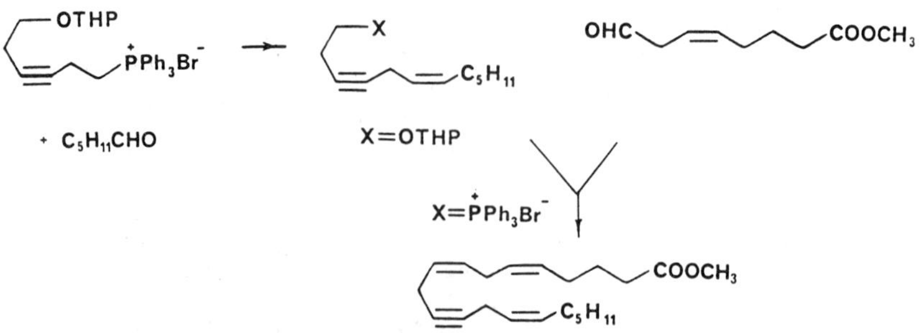

Scheme 11

SCHEME FOR POSSIBLE COVALENT ATTACHMENT
OF LT'S AT RECEPTOR(S)

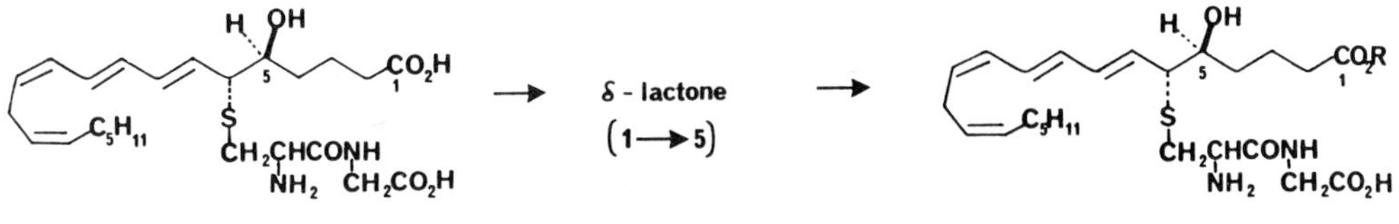

Scheme 12

RETROSYNTHETIC SCHEME FOR 5-DESOXY LTD

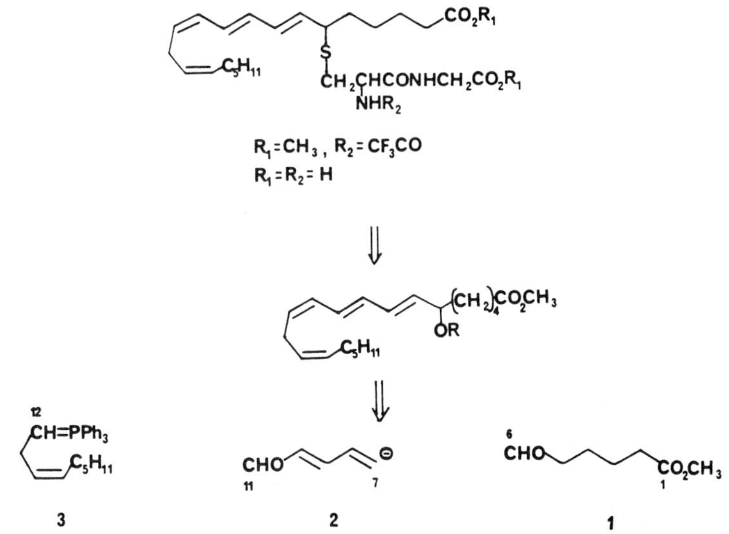

$R_1 = CH_3$, $R_2 = CF_3CO$
$R_1 = R_2 = H$

Scheme 13

SYNTHESIS OF 5-DESOXY LTD

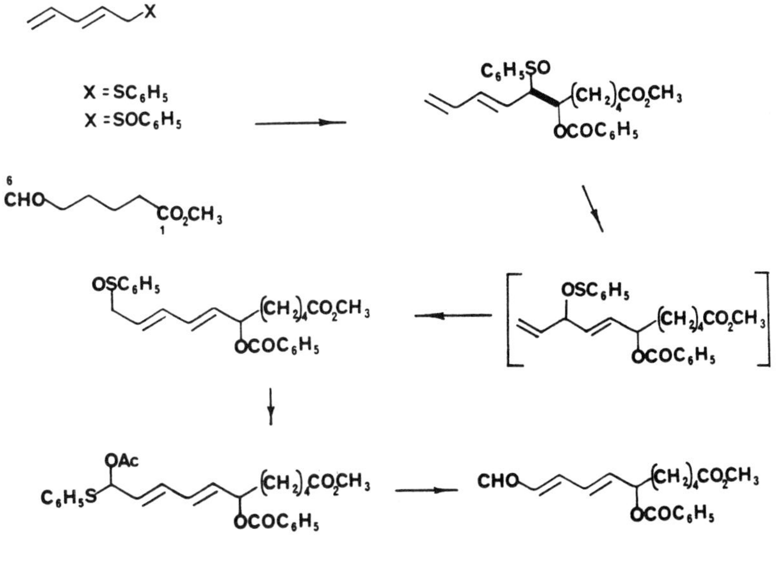

$X = SC_6H_5$
$X = SOC_6H_5$

Scheme 14

SYNTHESIS OF 5-DESOXY LTD

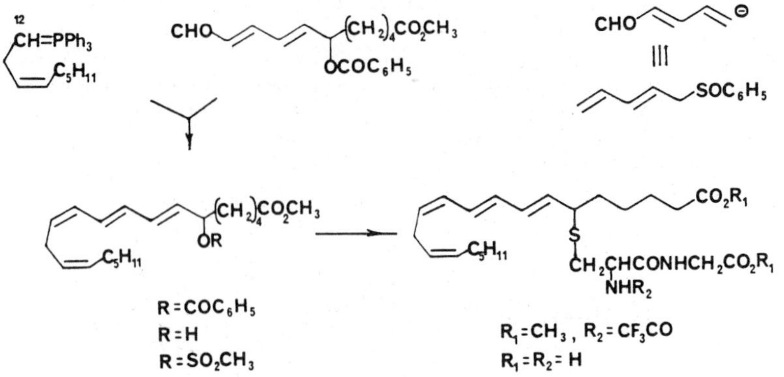

Scheme 15

to possess less than 0.1% of the biological activity of LTD itself.

In principle, a very useful synthetic route to homoallylic alcohols with concomitant formation of two carbon-carbon bonds could be effected by the sequence: (1) displacement reaction of a 1,2-epoxide with methylenetriphenylphosphorane; (2) proton abstraction to form a γ-oxido phosphonium ylide; and (3) condensation with a carbonyl compound. Unfortunately, this proposal suffers from the unreactivity of Wittig ylides towards all but the most reactive epoxides. It was therefore considered of interest to develop a method for the hyperactivation of phosphonium ylides, α-lithiation being the most plausible approach. Using such activation it might be possible, for example, to realize an exceedingly simple synthesis of 8-trans-11-HETE (a substance of interest in connection with the biomimetic synthesis of PG endoperoxides), as illustrated in Scheme 16.

Although methylenetriphenylphosphorane is commonly prepared by deprotonation of methyltriphenyl-phosphonium ion with alkyllithium reagents, the literature contains no indication of the possibility of further deprotonation. However, it has been discovered that this can be effected using tert-butyllithium in tetrahydrofuran (THF) solution (-78° to -40°C over 1 hr. and -40°C for 1 hr.) which produces a red to red-orange solution of the lithiated ylide (Scheme 17) at concentrations up to 0.2 M. The lithiated ylide can also be generated as an orange suspension in ether by the reaction of methyltriphenylphosphonium bromide with 2 equiv. of sec-butyllithium (-78° to -40°C for 1 hr. and 20°C for 3 hr.), or 1 equiv. of tert-butyllithium (-78° to 20°C over 2 hr. and 20°C for 3 hr.) in ether. Although fenchone is unaffected by treatment with methylenetriphenylphosphorane at temperatures up to 50°C in a variety of solvents (e.g., THF, THF-hexamethylphosphorictriamide (HMPA), or dimethyl sulfoxide), it reacts with the lithiated ylide in the presence of 20 equiv. of HMPA (-50° to 20°C for 1 hr.) to form an adduct which decomposes to the exo methylene derivative 4 (87% yield) in the presence of excess tert-butyl alcohol at 20°C (Scheme 17).[16]

The utility of the lithiated ylide as a synthetic reagent is also demonstrated by the reaction with epoxides, for example, with cyclopentene oxide (0°C for 2 hr. and 20°C for 20 hr.) to generate the γ-oxido ylide which reacts with benzaldehyde (-78°C for 1.5 hr. and 20°C for 8 hr.) to form the trans,trans-homoallylic alcohol (Scheme 17).[16]

PROJECTED SYNTHESIS OF 8-TRANS-11-HETE

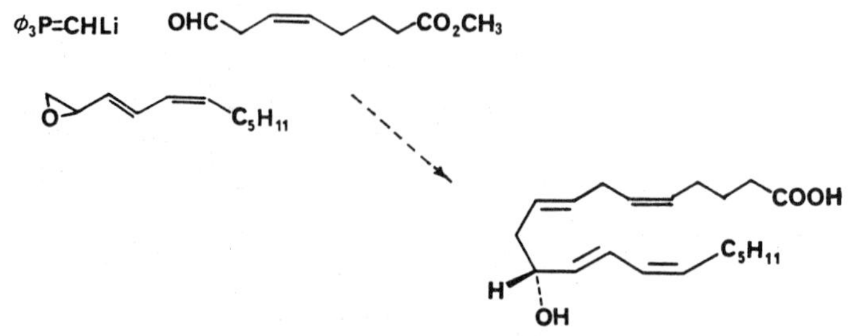

Scheme 16

METHOD FOR ACTIVATION OF WITTIG REAGENTS

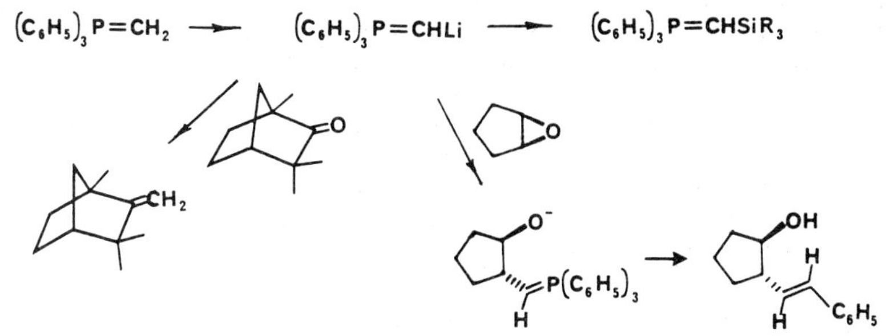

Scheme 17

Scheme 18 provides other examples of the application of lithiomethylenetriphenylphosphorane for the formation of two carbon-carbon bonds in one operation. Unfortunately, the projected synthesis of 8-trans-11-HETE using the lithiated ylide could not be achieved because of the occurrence of displacement at the wrong carbon atom, as illustrated by the reaction of a model system shown in Scheme 19.

LITHIATED WITTIG REAGENTS

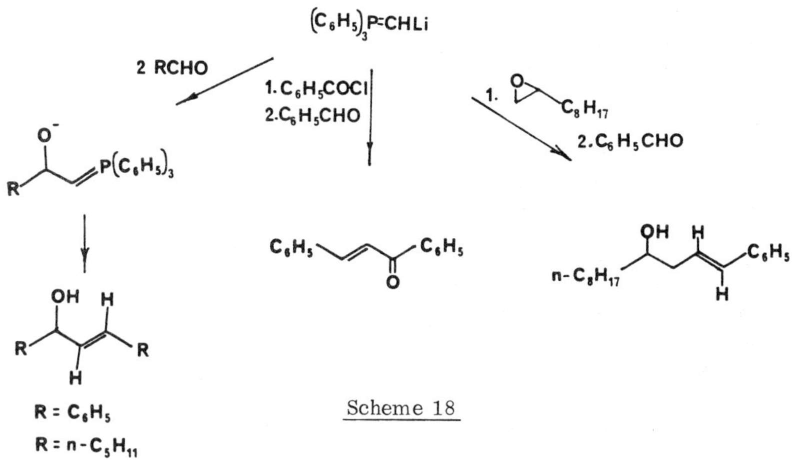

Scheme 18

MODEL EXPERIMENT FOR THE SYNTHESIS OF 8-TRANS-11-HETE

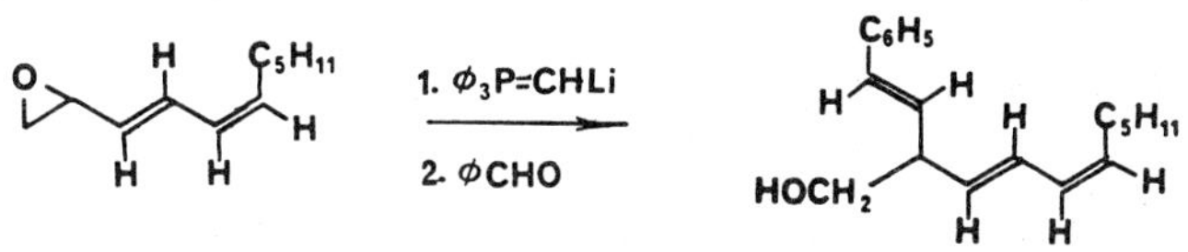

Scheme 19

Prostaglandin D_2 is produced in abnormally high amounts in humans suffering from the potentially fatal disorder, systemic mastocytosis, a disease in which tissue mast cells proliferate excessively.[17] Diagnosis of this and other diseases in which PGD_2 is overproduced could in principle be facilitated by a simple immunoassay for the major urinary metabolite of PGD_2 (substance 1 in Scheme 20).

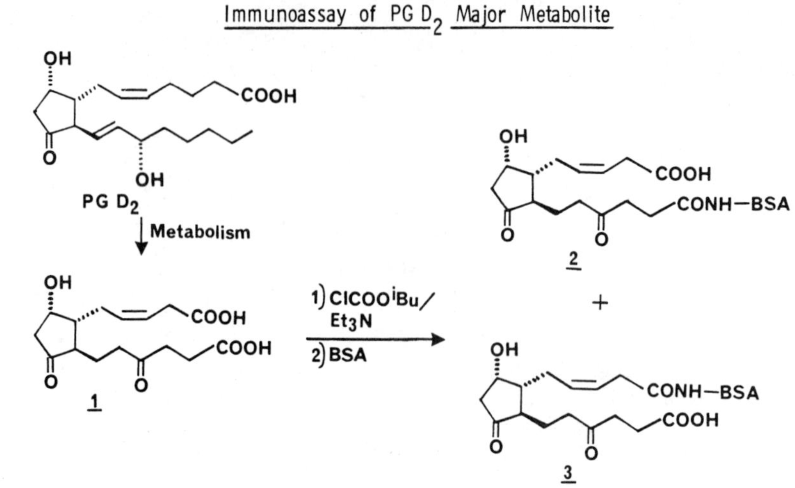

Scheme 20

For this reason we have developed a total synthesis of 1, a substance which is not available in significant amount from any natural source. The synthesis of this sensitive and highly functionalized prostanoid utilizes new reagents and techniques for functional group protection, a new method for the selective reduction of α,β-unsaturated ketones, and two useful new phosphonium ylides. Scheme 21 outlines the first stage of the synthesis, the conversion of the readily available prostaglandin precursor 4 to the thioketal lactone 9. Noteworthy here is the highly effective cleavage of the methyl ether 4, the mild oxidation of 7 to give the very acid- or base-sensitive ketone 8, and a new thioketalization method using zinc triflate as catalyst. Attempted thioketalization using many other catalysts failed completely.

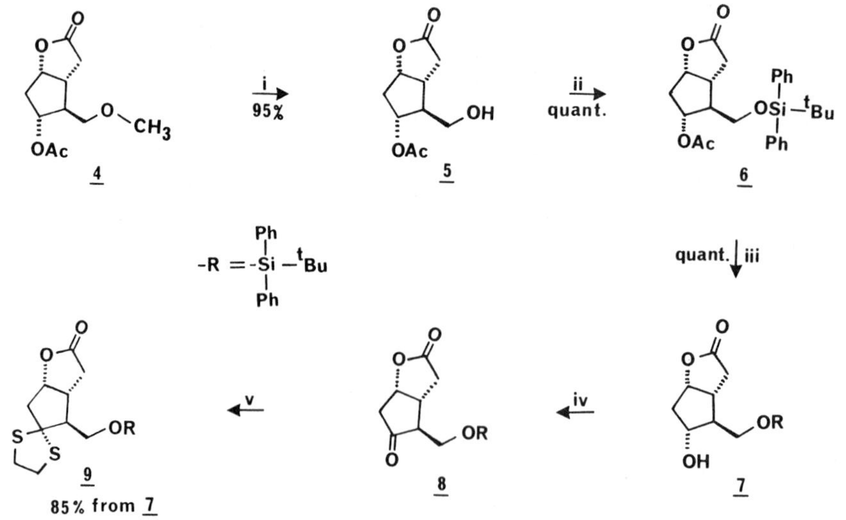

i) Me_3SiCl (2 equiv), NaI (2 equiv), CH_3CN, 50°C, 4 hr; ii) t-Bu(Ph)$_2$SiCl (1.5 equiv), DMAP (2 equiv), CH_2Cl_2, 0°C, 30 min; iii) K_2CO_3 (1 equiv), MeOH, then p-TsOH-H$_2$O (Cat.), CH_2Cl_2, 30 min.; iv) i-Pr-N=C=N-i-Pr (2 equiv), $Cl_2CHCOOH$ (0.5 equiv), DMSO-PhH (1:1), R.T., 30 min.; v) $HSCH_2CH_2SH$ (2 equiv), $Zn(OTf)_2$ (1.2 equiv), CH_2Cl_2, R.T., 5 hr.

Scheme 21

The second stage of the synthesis (Scheme 22) utilized the new ortho ester phosphonium reagent 19 for the introduction of the Z-β,γ-unsaturated acid unit corresponding to C(1) to C(3) in protected ortho ester form (11). The ortho ester allowed a unique internal protection of the cyclopentanol function (12) as well as desilylation and conversion of CH_2OH to CHO. The completion of the synthesis of 1 is shown in Scheme 23. Wittig chain extension from 14 produced the α,β-enone 15. Published methods for selective reduction of the α,β-double bond either failed or were a priori inapplicable. The problem was solved using a new two-step process: (1) conjugate addition of hydrogen sulfide and (2) phosphine-mediated[18] free radical chain desulfurization.

Thioketalization using zinc or magnesium triflate was found to be a mild and highly efficient process with application to a wide variety of ketones, including camphor, menthone, norcamphor, p-methoxyacetophenone and Δ^4-cholesten-3-one (yields generally >95% with ethane-1,2-dithiol).

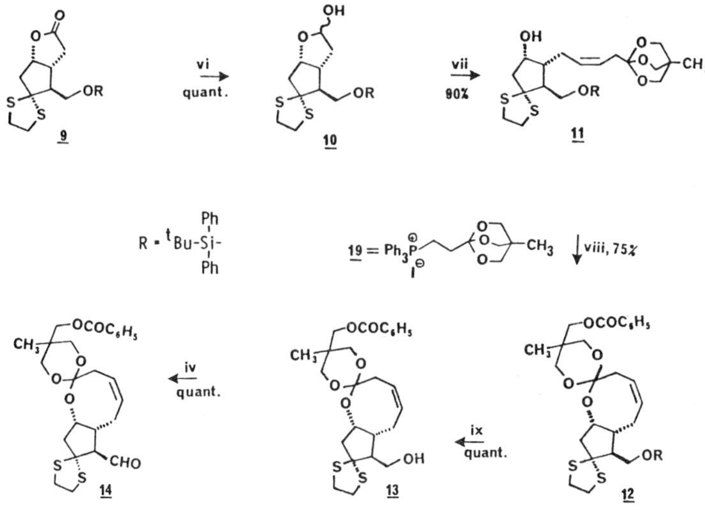

vi) DIBAL (1.0 equiv), Toluene, -78°C, 30 min; vii) 19, $NaCH_2S(O)CH_3$, DMSO, R.T., 4 hr; viii) PhCOCl (3 equiv), Pyridine (6 equiv), CH_2Cl_2, R.T., 4 hr; ix) n-$Bu_4N^⊕F^⊖$ (4 equiv), THF, 50°C, 9 hr.

Scheme 22

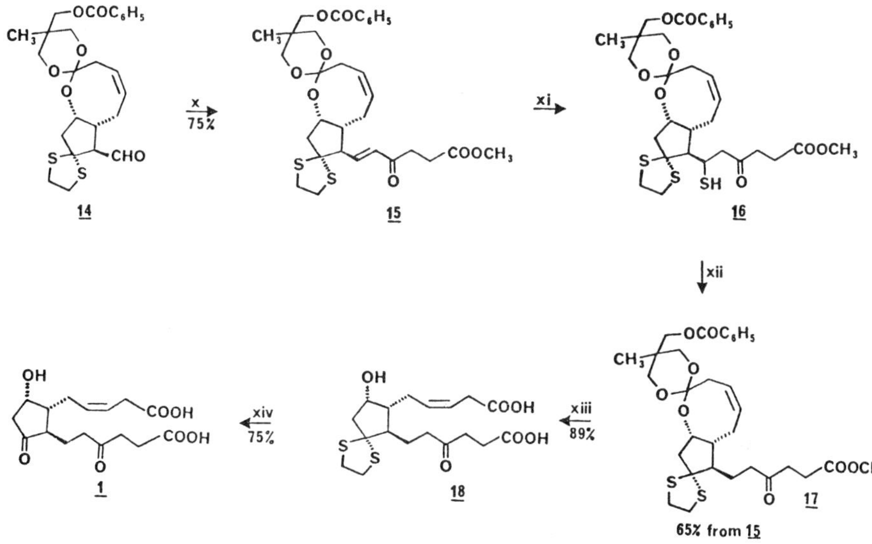

x) $(MeO)_2P(O)CH_2C(O)CH_2CH_2COOMe$, NaH, THF, R.T., 2 hr; xi) H_2S, K_2CO_3 (1.0 equiv), DMSO, R.T., 1 hr; xii) n-Bu_3P (10 equiv), hv, PhH, R.T., 5 hr; xiii) $2N-H_2SO_4$ (0.3 equiv), MeOH, R.T., 5 min., then 1N-NaOH (10 equiv), MeOH-H_2O, R.T., 10 hr; xiv) $HgCl_2$ (4 equiv), $CaCO_3$ (4 equiv), CH_3CN-H_2O (4:1), R.T., 8 hr.

Scheme 23

We close with an example of new synthetic methodology deriving from studies outside the eicosanoid field, specifically from work on the total synthesis of gibberellic acid. The total synthesis of gibberellic acid (GA$_3$, 1 in Scheme 24) was first achieved via the key intermediate 2 by a route which was unambiguous with regard to structure and stereochemistry.[19,20] In addition to the original synthesis of 2 from the Diels-Alder adduct 3, a second process was developed via the tricyclic ketone 4.[21] Both routes to 2 are fairly lengthy and although the individual steps are efficient, a simpler and more direct synthesis was clearly desirable. In this note we describe a novel nine-step synthesis of 4 which allows access to 2 in just 16 steps (vs. 25 in the original route).

Scheme 25 outlines the first five steps in the new synthesis of the tricyclic ketone 4, including two key steps: (1) a highly specific Diels-Alder addition and (2) a Cope rearrangement to produce a cis fused, angularly substituted hexahydro indene. It is noteworthy that the carbomethoxy substituent is essential for realization of the Cope rearrangement; without it, retro Diels-Alder reaction is the preferred thermolysis reaction. The completion of the synthesis (Scheme 26) used functional- and position-selective hydroboration (made possible by the deactivating effect of bromine on one double bond and steric screening by an angular group on one end of the other). Finally, functionally selective cyclization via copper-bromine exchange[22-24] selectively involved the cyclohexanone carbonyl over the cyclopentanone carbonyl (factor of ca. 20) to generate the required tricyclic structure. It is surmised that this selectivity arises partly from the greater reactivity of cyclohexanones over cyclopentanones in nucleophilic addition to carbonyl and partly from the lesser strain energy of the bicyclo[3,2,1]octane product.[25]

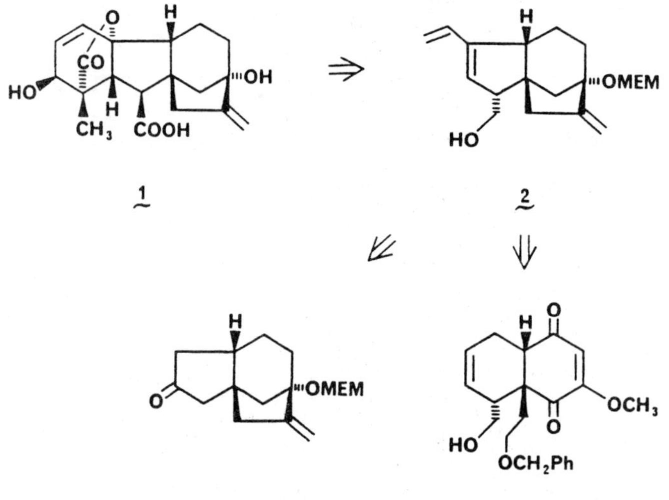

Scheme 24

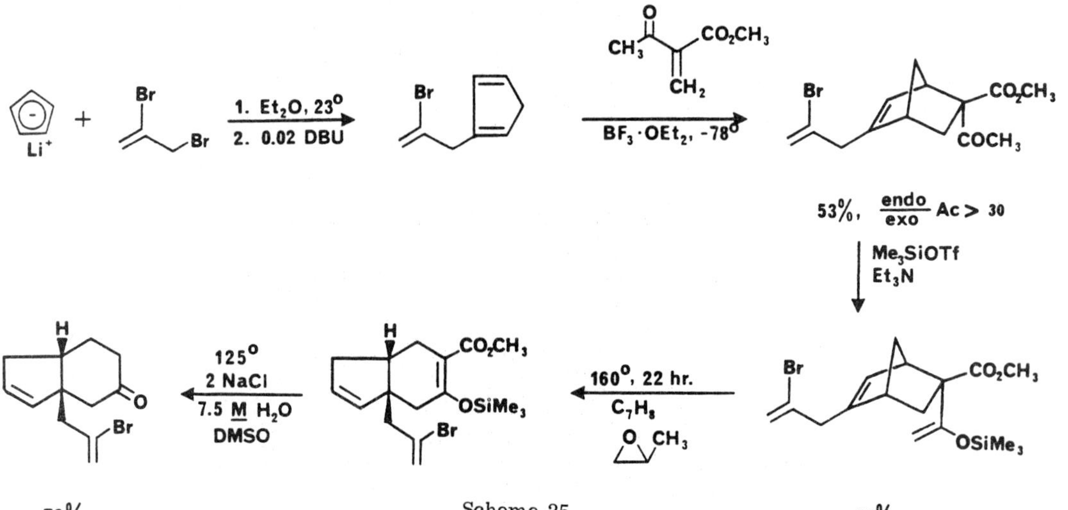

Scheme 25

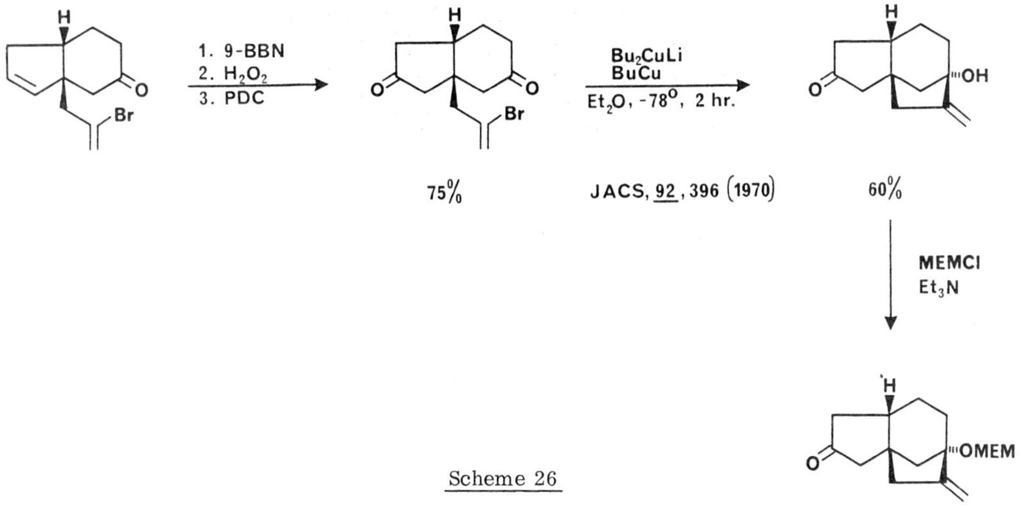

Scheme 26

Reagents above arrows:
1. 9-BBN
2. H_2O_2
3. PDC

→ (75%)

Bu_2CuLi / BuCu, Et_2O, $-78°$, 2 hr.
JACS, <u>92</u>, 396 (1970)

→ (60%)

MEMCl, Et_3N

Although the methods outlined above have been developed, as indicated in the introductory paragraph, for specific tasks, it is likely that they can be applied broadly with considerable advantage.

Acknowledgement – A substantial group of collaborators were responsible for the various developments outlined in this paper, for which they deserve major credit. Specifically, it is with great delight that I express appreciation to the following outstanding young chemists: David A. Clark, Alan E. Barton, Anthony Marfat, John E. Munroe, Paul B. Hopkins, Shun-ichi Hashimoto, John O. Albright, Francis Brion, Hokoon Park, Jahyo Kang, Dennis J. Hoover, Katsuichi Shimoji, and Kwan S. Kim. Other members of the group who provided help and advice at various stages include Giichi Goto, Hunseung Oh, Yasushi Nii, and Haruki Niwa.

This research was assisted financially by grants from the National Institutes of Health and the National Science Foundation.

REFERENCES AND NOTES

1. For a review of these developments see E. J. Corey, Experientia, in press (1982).
2. E. J. Corey, J. O. Albright, A. E. Barton and S.-i. Hashimoto, J. Am. Chem. Soc., <u>102</u>, 1435 (1980).
3. E. J. Corey and S.-i. Hashimoto, Tetrahedron Letters, <u>22</u>, 299 (1981).
4. I. Agata, E. J. Corey, A. G. Hortmann, J. Klein, S. Proskow and J. J. Ursprung, J. Org. Chem., <u>30</u>, 1698 (1965).
5. E. J. Corey, A. E. Barton, and D. A. Clark, J. Am. Chem. Soc., <u>102</u>, 4748 (1980).
6. E. J. Corey and A. E. Barton, Tetrahedron Letters, <u>23</u>, 2351 (1982).
7. E. J. Corey, A. Marfat, J. E. Munroe, K. S. Kim, P. B. Hopkins and F. Brion, Tetrahedron Letters, <u>22</u>, 1077 (1981).
8. E. J. Corey, H. Park, A. Barton and Y. Nii, Tetrahedron Letters, <u>21</u>, 4243 (1980).
9. E. J. Corey and H. Park, J. Am. Chem. Soc., <u>104</u>, 1750 (1982).
10. E. J. Corey and J. E. Munroe, J. Am. Chem. Soc., <u>104</u>, 1752 (1982).
11. E. J. Corey and J. Kang, Tetrahedron Letters, <u>23</u>, 1651 (1982).
12. E. J. Corey, H. Niwa, and J. R. Falck, J. Am. Chem. Soc., <u>101</u>, 1586 (1979).

13. E. J. Corey, H. Niwa, J. R. Falck, C. Mioskowski, Y. Arai, and A. Marfat in <u>Advances in Prostaglandin and Thromboxane Research</u>, Raven Press, New York, Vol. 6, p. 19 (1980).
14. E. J. Corey and J. Kang, <u>J. Am. Chem. Soc.</u>, <u>103</u>, 4618 (1981).
15. E. J. Corey and D. J. Hoover, <u>Tetrahedron Letters</u>, in press (1982).
16. E. J. Corey and J. Kang, <u>J. Am. Chem. Soc.</u>, in press (1982).
17. L. J. Roberts, B. J. Sweetman, R. A. Lewis, K. F. Austen, and J. A. Oates, <u>New Eng. J. Med.</u>, <u>303</u>, 1400 (1980).
18. F. W. Hoffmann, R. J. Ess, T. C. Simmons, and R. S. Hanzel, <u>J. Am. Chem. Soc.</u>, <u>78</u>, 6414 (1956).
19. E. J. Corey, R. L. Danheiser, S. Chandrasekaran, P. Siret, G. E. Keck, and J.-L. Gras, <u>J. Am. Chem. Soc.</u>, <u>100</u>, 8031 (1978).
20. E. J. Corey, R. L. Danheiser, S. Chandrasekaran, G. E. Keck, B. Gopalan, S. D. Larsen, P. Siret, and J.-L. Gras, <u>J. Am. Chem. Soc.</u>, <u>100</u>, 8034 (1978).
21. E. J. Corey and J. G. Smith, <u>J. Am. Chem. Soc.</u>, <u>101</u>, 1038 (1979).
22. E. J. Corey and I. Kuwajima, <u>J. Am. Chem. Soc.</u>, <u>92</u>, 395 (1970).
23. E. J. Corey, M. Narisada, T. Hiraoka, and R. A. Ellison, <u>J. Am. Chem. Soc.</u>, <u>92</u>, 396 (1970).
24. W. M. Grootaert and P. J. deClercq, <u>Tetrahedron Letters</u>, <u>23</u>, 3291 (1982).
25. Molecular mechanics calculations (Allinger MM-2) performed by Mr. Jay W. Ponder in these laboratories indicate that the desired bridged system is favored energetically over the alternative (cyclopentane-bridged) product by 2.5 kcal./mole.

SYNTHETIC APPROACHES TO BIOLOGICALLY ACTIVE PROSTACYCLIN-ANALOGUES

W. Bartmann, G. Beck, J. Knolle and R. H. Rupp

Hoechst Aktiengesellschaft, 6230 Frankfurt/Main 80, Federal Republic of Germany

Introduction

Prostacyclin (PGI_2) $\underline{2}$ the most active platelet-antiaggregatory prostanoid so far known was discovered by Vane and coworkers (1) in 1976 during intensive investigations of the arachidonic acid metabolism. In addition to its interaction with blood platelets PGI_2 $\underline{2}$ is a vasodilating antihypertensive and therefore an interesting objective of cardiovascular research.

PGI_2 $\underline{2}$ possesses general structural elements of a prostanoid - 20 carbon atoms, five of which are members of the bis oxygen-substituted cyclopentane ring, which itself bears an eight-membered so-called ω-sidechain; however, in PGI_2 $\underline{2}$ the seven-membered α-sidechain originally present in prostaglandin $F_{2\alpha}$ ($P\bar{G}F_{2\alpha}^-$) $\underline{1}$ is part of a tetrahydrofuran ring attached to the cyclopentane ring. Due to unsaturation in position 5 (prostaglandin nomenclature) this tetrahydrofuran ring is highly acid-labile, giving rise to the main hydrolysis product of PGI_2 $\underline{2}$ 6-oxo-prostaglandin $F_{1\alpha}$ $\underline{3}$ (2).

SCHEME 1

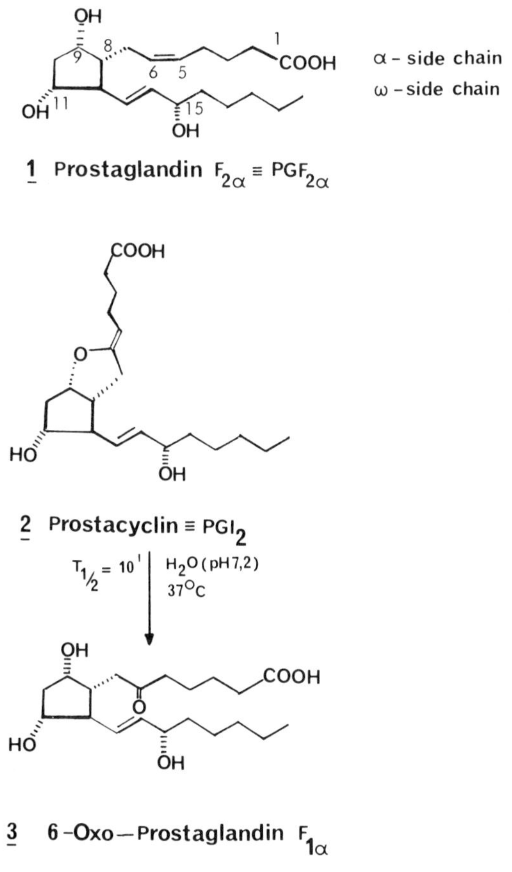

$\underline{1}$ Prostaglandin $F_{2\alpha} \equiv PGF_{2\alpha}$

α - side chain
ω - side chain

$\underline{2}$ Prostacyclin $\equiv PGI_2$

$T_{1/2} = 10'$ | H_2O (pH 7,2), $37°C$

$\underline{3}$ 6-Oxo—Prostaglandin $F_{1\alpha}$

The cis-configurated enolether moiety is essential for the biological activities of PGI$_2$ 2 which have been confirmed in human trials (3).

Two major objectives directed towards a broader clinical application of PGI$_2$ 2 or a corresponding analogue stimulated synthetic efforts worldwide. One goal was the stabilisation of the biologically essential acid-labile enolether function or its replacement by a biological equivalent isoelectronic and/or isosteric substructure opening the possibility of oral application. The other objective was the separation of the blood platelet antiaggregatory properties from the antihypertensive activities.

Up to now some 34 differently modified prostacyclin structures have been reported in the literature demonstrating the high interest in this rather unusual type of cardiovascularly active molecule.

The main task of a pharmaceutical chemist consists in the construction of a biological active molecule. At least in an early phase of his work methods and yields to achieve this goal will have less priority.

Therefore, this presentation tries to reproduce parts of the necessary dialogue between the chemist and the biologist.

Synthetic strategy

Our synthetic strategy was determined by formal considerations as well as by experimental observations. In order to mimic the essential enolether function 4 of PGI$_2$ 2 by a bioisosteric equivalent one could consider substructures 5, 6 and 7 in a very qualitative way as isoelectronic with the enolether function 4.

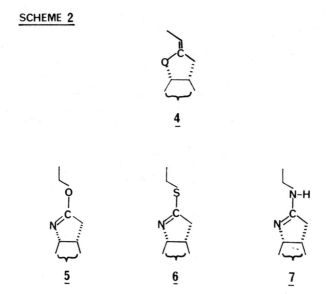

Furthermore, the sidechain attached to the heteroatoms in 5, 6 and 7 could adopt conformations close to the cis-configuration of the α-sidechain in PGI 2.

Model studies

From literature precedent the γ-oxocarboxylic acidamide 8a was expected to be in equiblibrium with its corresponding cyclic hydroxylactam 8b (4). This then gave the experimental entry into the planned series according to scheme 3. The hydroxyl-group in 8b could be removed either by hydrogenation over palladium/barium-sulfate in the presence of acid or by transformation to its corresponding thiophenylether 9 and subsequent desulfuration with Raney-Nickel leading to the lactam 10, which was smoothly converted into the corresponding thiolactam 11 by reaction with P_4S_{10} in pyridine. The assigned thermodynamically more stable cis-structure of 10 could be confirmed through NMR-studies.

Alkylation either of 10 or 11 with ethyl ω-bromo-butyrate or methyliodide proceeded normally giving rise to the desired structures 13 and 14 as well as to the intermediate 12. Aminolysis of 12 with γ-aminobutyric acid resulted in the formation of amidines 15a and 15b.

SCHEME 3

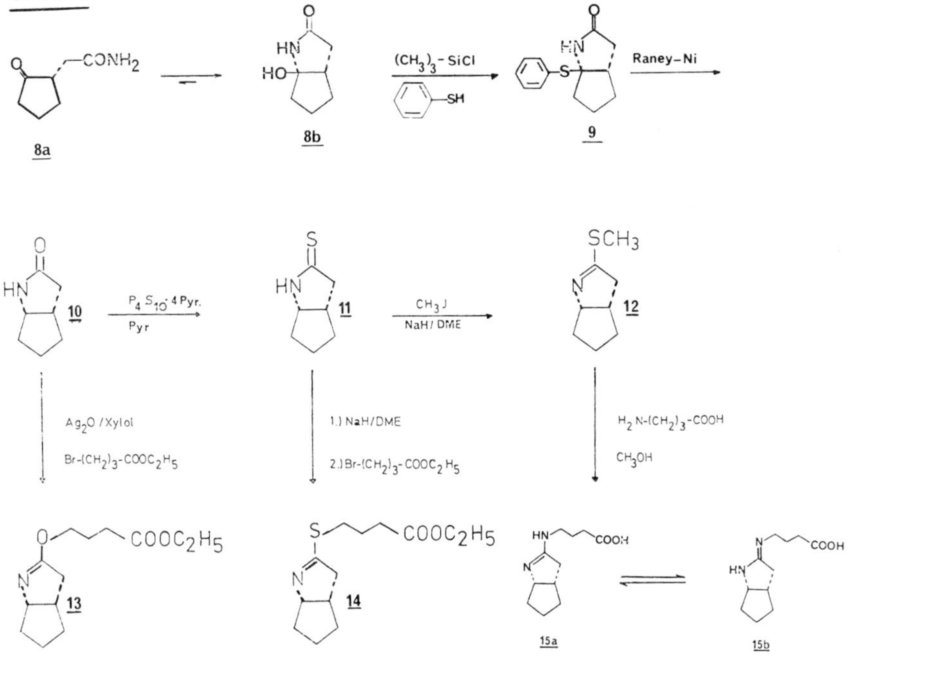

14 exhibited interesting biological activities. On i.v. application in anesthesized animals 14 lowered blood pressure and inhibited blood platelet aggregation in "in vitro" experiments, thus supporting our hypothesis that the enolether function of PGI_2 2 could be mimiced by at least one of the envisaged substructures. However, 14 lacked sufficient oral activity.

The fact that a comparatively simple molecule showed promising biological activity encouraged further studies with model compounds. Thus starting with the readily available lactam 16, conversion to the thiolactam 17 and boronate-reduction of the methoxycarbonyl function to the alcohol 18 provided a synthon to which both an α- and a ω-sidechain could be attached.

After alkylation with ethyl ω-bromo-butyrate yielding 19, the alcohol function in 19 was converted to the corresponding aldehyde 20 via a Moffat-oxidation in methylenchloride at -20°C with DMSO and oxalylchloride in the presence of triethylamine.

Applying the Corey methodology (6) for the synthesis of the ω-sidechain, aldehyde 20 could be used without further purification in the subsequent Horner-Emmons-Wittig-reaction leading to the enone 21, which on reduction with di-isobornyloxyaluminiumisopropylate led to a mixture of isomeric alcohols 22a and 22b separable by column chromatography.

SCHEME 4

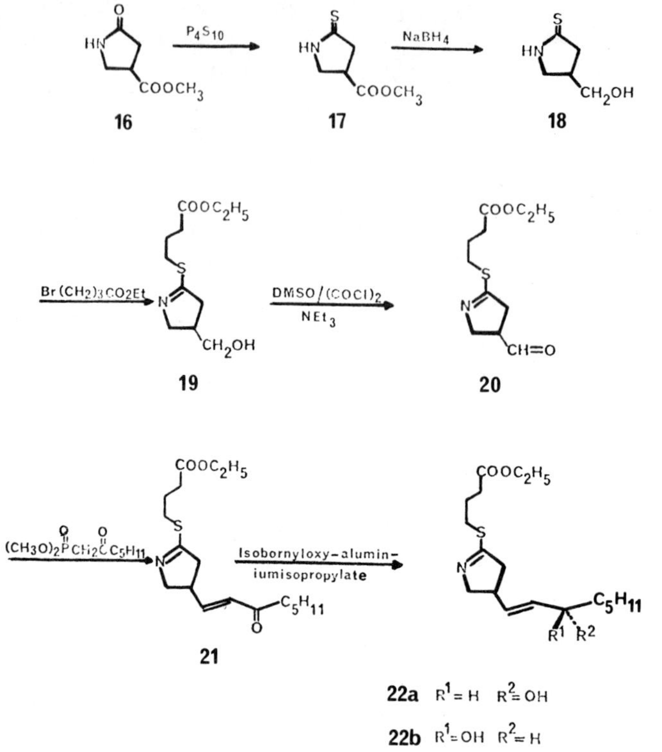

On i.v. application to anaesthesized rats 22a (5) displayed antihypertensive activity which was, however, weaker than that of 14.

ß-Hetero-imino-prostacyclin-analogues

At this stage it was necessary to prove our working hypothesis on racemic 40, a molecule identical with natural PGI_2 2 except that the enolether substructure 4 is replaced by the thioiminoether grouping 6.
For the synthesis of this compound we directed our investigations towards the adaption of the Corey-synthesis (6) that has become the basis for a commercially available $PGF_{2\alpha}$-derivative at Hoechst (7).

Although all necessary reaction steps to convert the tetrahydropyranylether of the Corey lactone 23 (6) to the target molecule 40 where either investigated during the synthesis of 14 or principally known from the literature, the preparation of the thiolactimether 40 starting from 23 proved to be a somewhat puzzling game with protecting groups.

Opening of the Corey lactone derivatives 24a or 24b proceeded in liquid ammonia and gave the carbonamides 25a or 25b which on Jones oxydation led to a mixture of the tautomeres 26a and 27a or 26b and 27b. Due to the presence of acid sensitive protecting groups the hydroxy group in 27a or 27b could only be removed via its transformation to the corresponding thiophenolethers 28a or 28b and their desulfuration with Raney-Nickel to 29a and 29b. The formation of the thiophenolethers 28a and 28b influenced favourably the tautomerisation of the amides 26a and 26b to the hydroxylactams 27a and 27b thus allowing the use of a crude mixture of amides 26a, 26b and hydroxylactams 27a, 27b for the reaction with thiophenol using trimethylchlorosilane as an auxiliary reagent.

The remaining lengthy route to target compounds 39, 40 and 41 followed reaction pathways which have been discussed above and are outlined in scheme 5.

Synthesis of Biologically Active Prostacyclin-Analogues

SCHEME 5

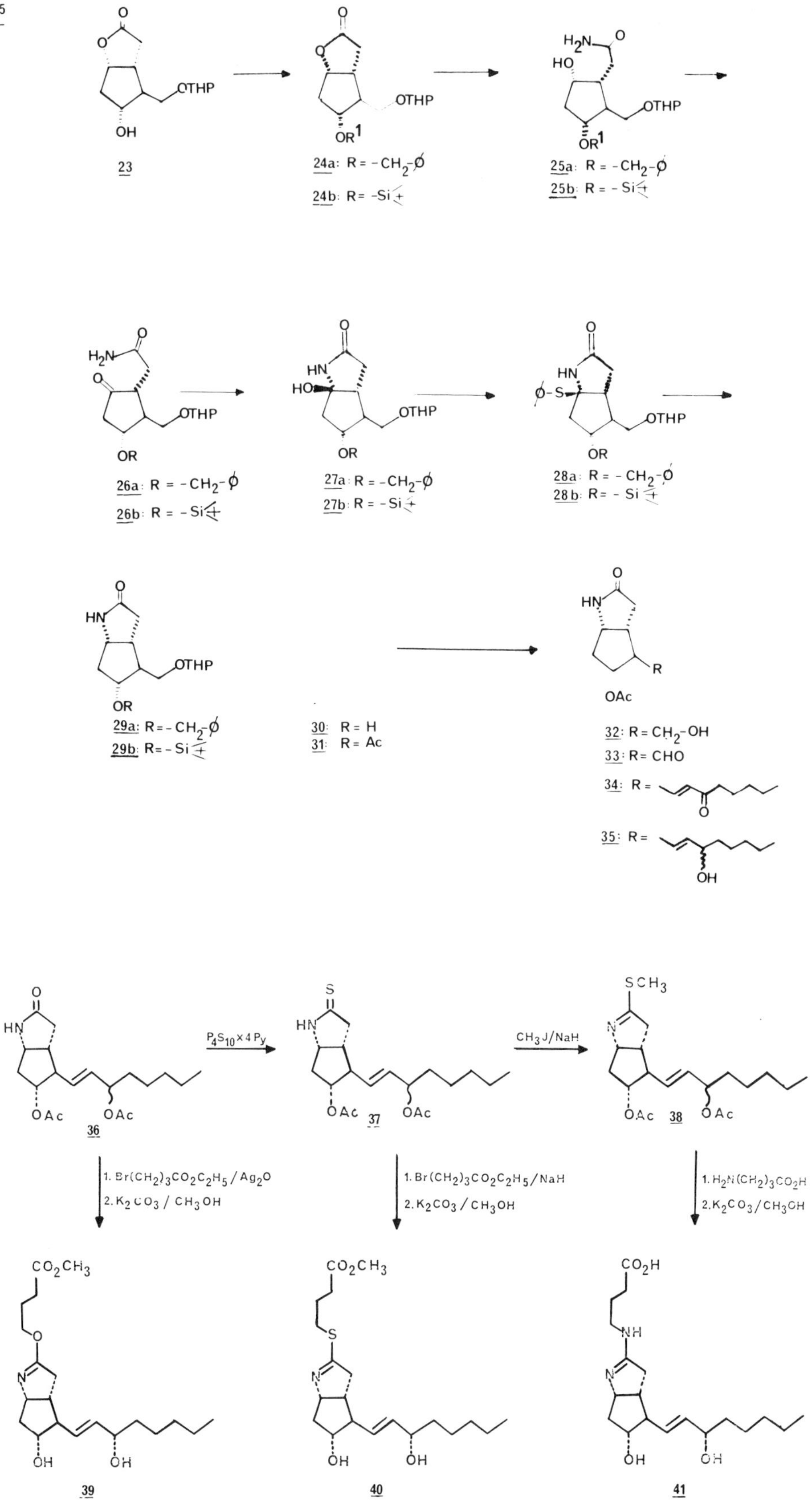

Again the assignment of the thermodynamically favoured cis-structure to the ring junction of the bicyclic system present in 40 (8) is in accordance with the NMR spectroscopical data; the cis-configuration was established by application of the Nuclear-Overhauser-Effect (NOE).

With regard to the antiaggregatory properties 40 proved to be at least as active as PGI_2 2 in a number of different biological models. Furthermore, 40 was sufficiently stable for oral application and showed biological activity. Finally a trend could be observed towards the desirable separation of antiaggregatory and antihypertensive activities, e.g. in the rat on intravenous application 40 was a ten times weaker antihypertensive than PGI_2 2.

Regarding structure-activity-relationships an open question remained: Is the 11-hydroxyl-group essential for a pronounced biological effect?

During the extensive studies towards the removal of the hydroxyl group in 27a, 27b via Raney-Nickel desulfuration of its thiophenolether derivative 28a, 28b partial hydrogenolysis of the benzyl-protected alcohol function was observed. The resulting deoxygenated lactone 42 could be used as a building block for a sequence ending with 45; the single reactions leading to 45 are practically identical with those described in the parallel series (e.g. scheme 3 or scheme 5).

SCHEME 6

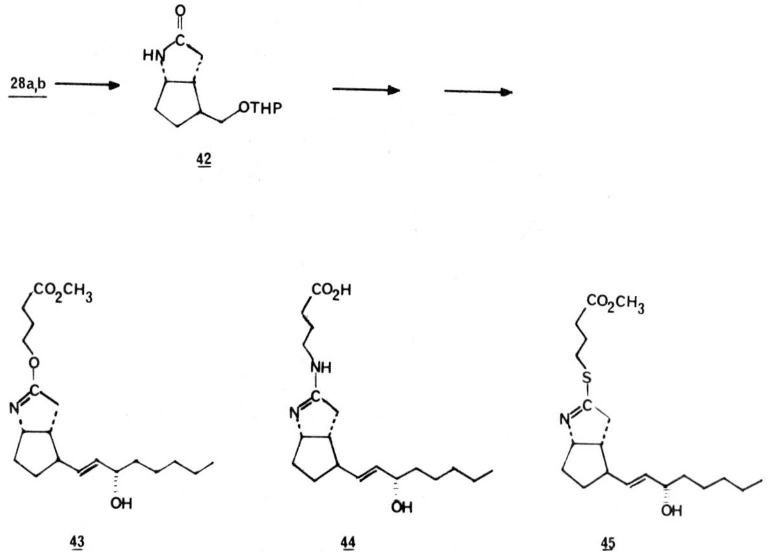

Again the less complicated structure 45 was biologically inferior to compound 40. Due to lack of time I cannot present the total catalogue of structure-activity-relationships worked out with the help of the four synthesized routes just described; the essential outcome was that contrary to our expectations the thioimino substructure has to be combined with all other structural elements of PGI_2.

Having reached our objective we were frightened by the length of the synthetic pathway.

In order to shorten the sequence and to improve yields in critical steps several approaches were followed almost simultaneously.

The product of a regioselective 1,2-addition of dichloroketene to cyclopentadiene 47 had been previously used as a cheap starting material for prostaglandin synthesis; e.g. the regio- and stereoselective ring opening of epoxide 49a derived from 47 through dechlorination and epoxidation had been thoroughly investigated by Roberts and coworkers (9).

SCHEME 7

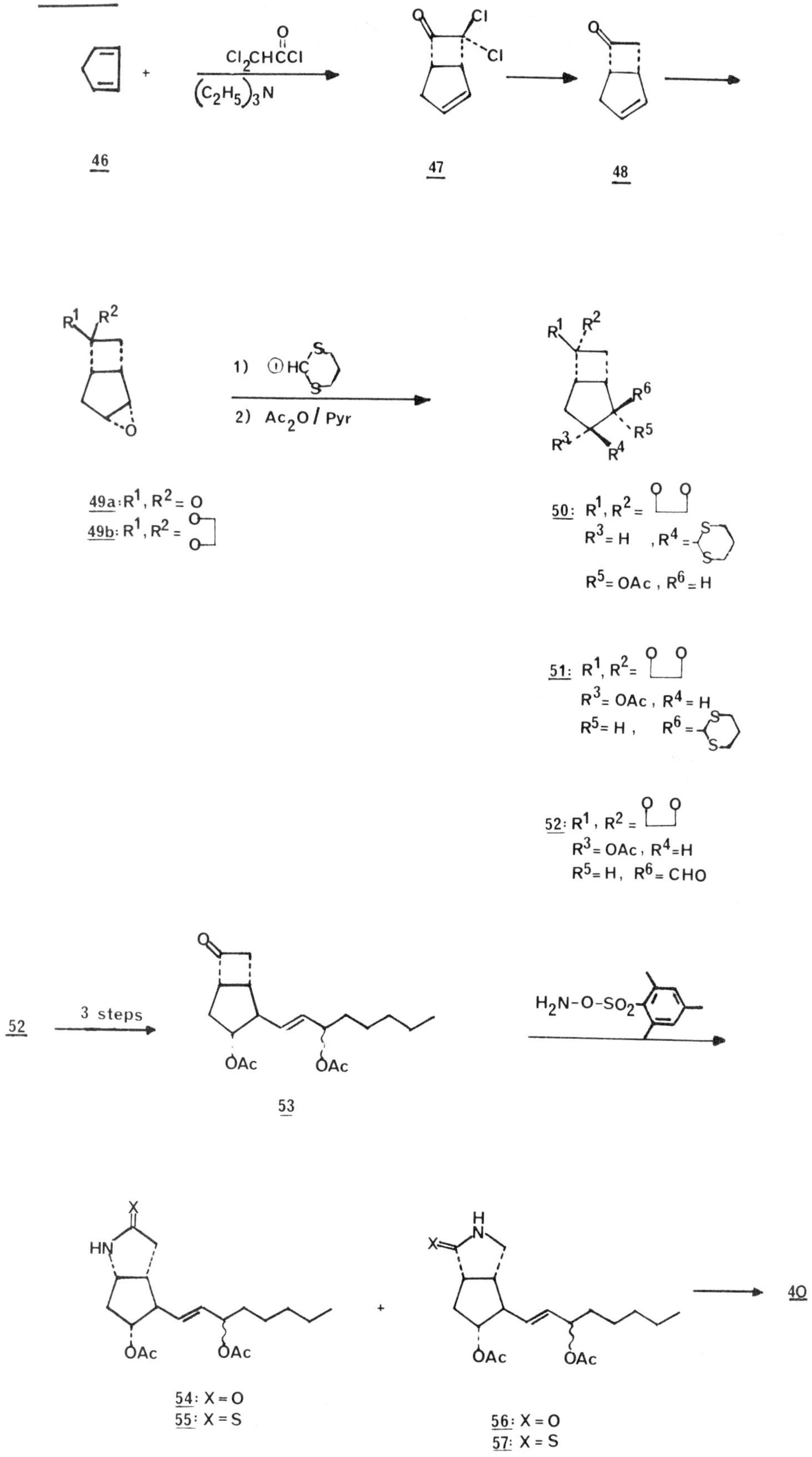

46 + Cl₂CHCCl / (C₂H₅)₃N → **47** → **48** →

49a: R¹, R² = O
49b: R¹, R² = O-O (dioxolane)

1) HC(S-S) (dithiane)
2) Ac₂O / Pyr

50: R¹, R² = O-O; R³ = H, R⁴ = dithianyl; R⁵ = OAc, R⁶ = H

51: R¹, R² = O-O; R³ = OAc, R⁴ = H; R⁵ = H, R⁶ = dithianyl

52: R¹, R² = O-O; R³ = OAc, R⁴ = H; R⁵ = H, R⁶ = CHO

52 →(3 steps)→ **53** (OAc, OAc)

H₂N-O-SO₂-mesityl →

54: X = O
55: X = S

+

56: X = O
57: X = S

→ **40**

Based on this experiments 48 was transformed to the corresponding epoxy-ketal 49b using standard methodology. Treatment of 49b with 2-lithiodithiane at - 50°C furnished after acetylation the masked aldehydes 50 and 51 (ratio 20 : 80) as a cristalline mixture which could be separated by fractional cristallisation. Selective hydrolysis of the dithiane moiety in acetate 51 to the corresponding aldehyde 52 was achieved by treatment with mercuric oxide/ bortrifluoride-etherate in moist THF at 50°C.

The crude aldehyde 52 could be transformed with standard methods to the ketone-diacetate 53 using a sequence of 3 steps already described above (Horner-Emmons-Wittig, boronate reduction, acetylation) followed by hydrolysis of the ketal function at room temperature with 0.2 N H_2SO_4. The crucial step in this sequence was the Beckmann-rearrangement of 53 to 54 which was expected to proceed with the same regio- and stereoselectivity as the corresponding Baeyer-Villiger oxidation (10). Normal Beckmann-rearrangement conditions failed. However, upon treatment of 53 with O-mesitylensulfonylhydroxylamine (11) (CH_2Cl_2,0°C, 30 min) the cristalline lactam 54 was obtained in 55 % yield along with 15 % of the isomeric lactam 56. The ratio of lactams 54 to 56 was easily determined by 270 MHZ^{1}H-NMR spectroscopy. Except for the identification of the pure compound and the determination of the yield the reaction sequence was carried on with the mixture of 54 and 56 which was converted to the mixture of thiolactams 55 and 57. (Compounds 54 and 55 proved to be identical in all respects with 36 and 37 synthesized according scheme 5.) Alkylation of 55 and 57 with ethyl bromobutyrate followed by hydrolysis of the acetate groups furnished after one single chromatography 40.

Both approaches for the synthesis of 40 described so far postponed the formation of the lactam moiety towards the end of the reaction sequence. A desirable starting material for an earlier construction of the lactam moiety is 58 (12).

SCHEME 8

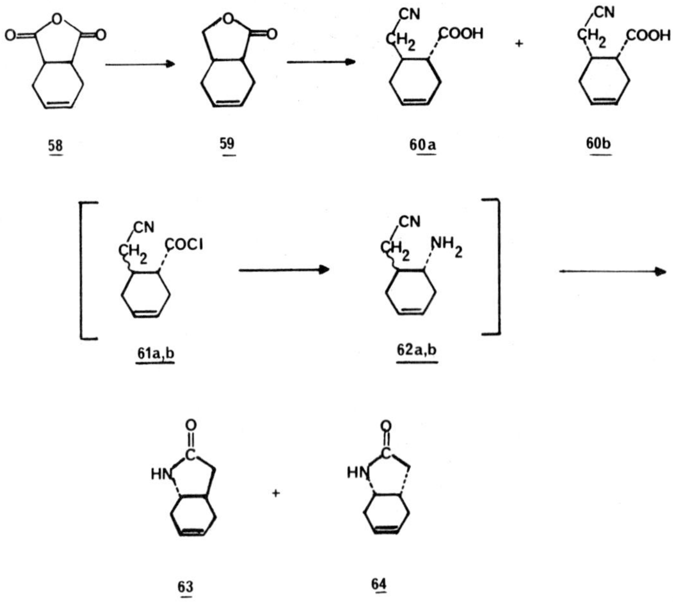

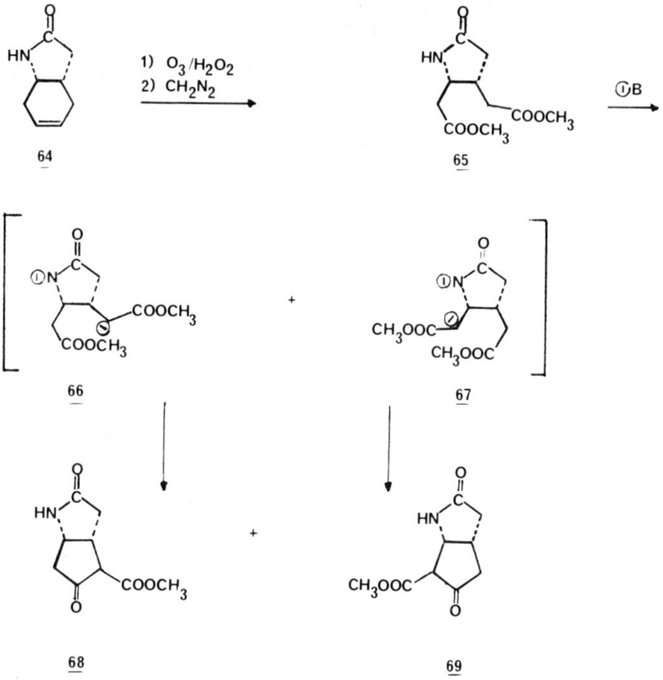

Starting from tetrahydrophthalic anhydride 58 and using known procedures the nitrilo-acids 60a and 60b can be obtained by two steps (12). Curtius degradation of 60a,b gave 61a,b and consequently 62a,b. Alkaline saponification of the crude mixture 62a,b yielded after acidification and refluxing in toluene preferentially the cis-isomer 64 which could be isolated by fractional cristallisation. Its synthesis was not optimized. In order to exclude any impurity caused by the formation of 63, 64 was purified by chromatography. The NMR spectra of cristalline 63 and 64 showed distinct differences with regard to the protons positioned at the cis- and trans-ring junction.

Ozonolysis of 64 followed by oxydative work-up and esterification of the resulting diacid yielded the diester 65 which on deprotonation could give rise to a mixture of dianions 66 and 67, but due to electrostatic repulsion one would expect the preferential formation of 66.

The equilibrium between both dianions - as expected - depended on the base used for deprotonation, the time and the temperature of the reaction. Two examples of a series of experiments may be mentioned. Using as a base lithiumdiisopropylamide in THF 4 hours at 0°C a mixture of 68 and 69 was obtained in 50 % yield; the ratio of 68 to 69 was 8 : 2 as determined by HPLC and NMR spectroscopy,whereas sodiummethoxide in toluene for 24 hours at room temperature produced a mixture of 68 and 69 with a ratio 2 : 8 in practically quantitative yield. None of these runs were satisfactory enough to bring up a larger amount of material necessary for toxicological trials. Due to this reason a third approach was tried which has proved to be the most practicable approach so far (scheme 9).

The chlorinated lactone 71 is easily accessible in pilot plant scale starting with norbornadiene 70 employing standard procedures (13). The nucleophilic replacement of chlorine in 71 causes difficulties because 1,3 elimination is favoured. Transformation of 71 to its corresponding diester 72, and protection of the OH-function in 72 with dihydropyran gave 73. The nucleophilic displacement(SN_2) of chlorine in this compound by sodium azide in DMSO at 100°C would ensure the stereochemistry of 74.

SCHEME 9

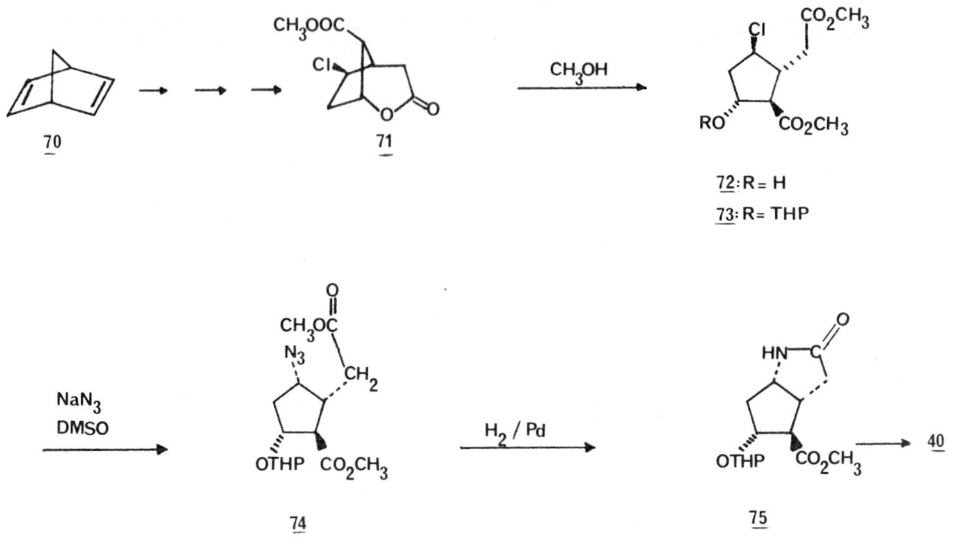

Except for its first identification 74 was usually not isolated but transformed as a crude product by hydrogenation to the lactam 75 which readily cristallised. Its configuration was confirmed by the transformation of 75 to 40.

Further shortening of the sequence could be accomplished at least on a laboratory scale.

SCHEME 10

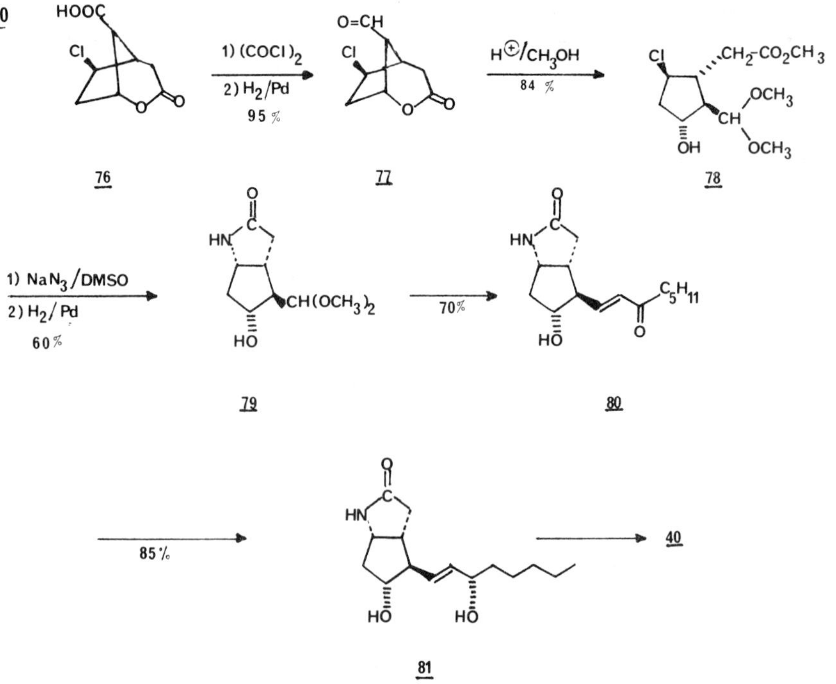

Rosenmund-reduction of the acid chloride of 76 yielded aldehyde 77 (14).
Acid catalysed methanolisis and acetalisation resulted in chloroacetal 78.
Again the chlorine atom in 78 could be smoothly exchanged against azide and
immediately hydrogenated to lactam 79. Removal of the protecting group in 79
gave the free aldehyde which was used without further purification as a
starting material for the Horner-Emmons-Wittig-reaction. The free hydroxyl
group of 80 enabled the stereoselective reduction of the ketone function in
80 to 81 (15). The final two steps from 81 to 40 have been described above.
By this synthesis the initial 24 steps could be reduced to 11, and more
importantly several of the intermediates cristallised readily without chromatography.

Acknowledgement

The authors wish to thank Dr. Schölkens and Dr. Weithmann for biological
experiments, Dr. Cavagna and Dr. Fehlhaber for measurements and helpful discussions concerning the NMR spectra. We also gratefully acknowledge the help
of Dr. Fickert for reviewing, Mrs. Böhm and Mrs. Pöllath for writing the
manuscript.

References

(1) S. Moncada, R. Gryglewski, S. Bunting and J.R. Vane, Nature 263, 663
    (1976)
(2) G.J. Dusting, S. Moncada and J.R. Vane, Brit. J. Pharmacol. 62, 414 P
    (1978)
(3) A. Szczeklik, R.J. Gryglewski, R. Nizankowski, J. Musial, R. Pieton and
    J. Mruk, Pharmacol. Res. Commun. Vol. 10, 545 (1978)
(4) W. Flitsch
    Chem. Ber. 103, 3205 (1970)
(5) The assignment of structure 22a to the biologically more active epimer
    followed prostaglandin practice but was not confirmed by spectroscopy.
(6) E.J. Corey, Th. K. Schaaf, W. Huber, U. Koelliker and N.M. Weinshenker
    J. Amer. Chem. Soc. 92, 397 (1970)
(7) W. Bartmann, G. Beck, U. Lerch, H. Teufel, B. Schölkens
    Prostaglandins 17, No. 2, 301 (1979)
(8) The assignment of the 15α-configuration to 40, R_f=0.11(TLC solvent system:
    ethyl acetate/methanol 8 : 1) was based on its biological activity and in
    comparison to the TCL behavior of $PGF_{2\alpha}$ and its corresponding 15ß-epimer,
    but not confirmed by spectroscopic measurements.
(9) R.F. Newton and S.M. Roberts, Tetrahedron 1980, 2163
(10) C.C. Howard, R.F. Newton, D.P. Reynolds, A.H. Wadsworth, D.R. Kelly,
     S.M. Roberts
     J. Chem. Soc. Perkin I, 852 (1979)
(11) Y. Tamura, J. Minamikawa and M. Ikeda, Synthesis 1, (1977)
(12) B. Belleau and J. Puranen
     Can. J. Chem. 43, 2551 (1965)
(13) R. Peel, J.K. Sutherland, J.C.S. Chem. Commun.151, (1974)
(14) N.R.A. Beeley, R. Peel, J.K. Sutherland
     Tetrahedron 37, 411 (1981)
(15) S. Iguchi, H. Nakai, H. Hayashi, H. Yamamoto
     J. Org. Chem. 44, 1363 (1979)

THE ALDOL CONDENSATION AS A TOOL FOR STEREOSELECTIVE ORGANIC SYNTHESIS

Clayton H. Heathcock

Department of Chemistry, University of California, Berkeley, California 94720, USA

"Erythromycin, with all our advantages, looks at present quite hopelessly complex, particularly in view of its plethora of asymmetric centers..."
R. B. Woodward, 1956

During the last half century, organic chemists have confronted and conquered a succession of synthetic obstacles of ever-increasing complexity.[1] Such formidable natural products as cholesterol, cortisone, prostaglandin, penicillin, terramycin, strychnine, lysergic acid, rifamycin, monensin, and vitamin B_{12} have all yielded to total synthesis. In spite of all these successes, one synthetic problem -- effective control of relative and absolute chirality when new asymmetric centers are created -- has continued to be especially troublesome. The problem is particularly acute when the asymmetric centers do not reside in a rigid cyclic or polycyclic framework. The quote which opens this account, taken from a well-known article by the foremost master of organic synthesis,[2] suggests the magnitude of the problem twenty-five years ago. However, since that time, and especially in the last decade, we have witnessed major advances in the general area of stereoselective synthesis, especially in the construction of acyclic and other conformationally-flexible molecules.[3] For the last five years my coworkers and I at Berkeley have investigated one of the oldest organic reactions, the aldol condensation,[4] as a method for achieving stereocontrol in the synthesis of such compounds.

It is appropriate that Woodward's last major scientific accomplishment was the conquest of erythromycin A (**1**).[5]

It is also appropriate that I use this fascinating molecule as the theme for an account of recent developments in aldol stereoselection, since it has provided the stimulus for most of the research we have carried out in this area. The alternating polyoxo backbone of erythronolide A (**2**), the aglycone of antibiotic **1**, strongly invites our attention to the aldol reaction as a possible method of synthesis. Our ultimate goal in this project is to learn enough about the aldol reaction that we can eventually employ it as the primary method for carbon-carbon bond construction and control of stereochemistry in a synthesis of **2**. While this goal has not quite been realized at the current time, much progress has been made, and some of the methodology developed *has* been employed in a total synthesis 6-deoxyerythronolide B (**3**), a biosynthetic precursor of **2**.[6]

2: X = OH
3: X = H

When an ethyl carbonyl compound reacts with an aldehyde, two diastereomeric aldols, referred to as erythro and threo isomers,[7] may be produced (eq 1). As is generally the case with diastereomeric reaction products, the erythro

$$\text{\~~COOMe} \xrightarrow[\text{THF}]{\text{LDA}} \xrightarrow[-78°C]{\text{PhCHO}} \text{Ph}\underset{\text{erythro (62\%)}}{\overset{\text{OH}}{\diagup}}\text{COOMe} + \text{Ph}\underset{\text{threo (38\%)}}{\overset{\text{OH}}{\diagup}}\text{COOMe} \quad (1)$$

and threo isomers are produced in unequal amounts. However, with simple ketones and esters, they are often formed in comparable amounts. In addition, there is a second stereochemical problem we must consider. In a chiral aldehyde, the two faces of the carbonyl group are diastereotopic, and addition of a nucleophile to them gives rise to a pair of diastereomers (eq 2). In such a case, the major and minor isomers may generally be predicted by Cram's

$$\text{(acetone)} \xrightarrow[\text{THF}]{\text{LDA}} \xrightarrow[-78°C]{\text{Ph-CHO}} \underset{\text{Cram (75\%)}}{\text{Ph}} + \underset{\text{anti-Cram (25\%)}}{\text{Ph}} \quad (2)$$

empirical rule for asymmetric induction;[8] thus, they are called the "Cram" and "anti-Cram" products. In the reaction of an ethyl carbonyl compound with a chiral aldehyde, there are four diastereomeric products, two erythro and two threo. An example is seen in the zinc-mediated reaction of ethyl α-bromopropionate with α-phenylpropionaldehyde (eq 3).[9] If we are effectively to use aldol technology for the synthesis of compounds such as 2 and 3, we must be able to control these two stereochemical features of the reaction. In fact, of the four stereochemical relationships embodied in the four products of eq 3, three appear in compounds 2 or 3.

$$\text{Ph-CHO} + \underset{\text{COOEt}}{\overset{\text{Br}}{\diagup}} \xrightarrow{\text{Zn}} \begin{Bmatrix} \text{erythro, Cram (55\%)} + \text{threo, Cram (24\%)} \\ \text{erythro, anti-Cram (16\%)} + \text{threo, anti-Cram (5\%)} \end{Bmatrix} \quad (3)$$

Simple Diastereoselection; Erythro-Threo Selectivity

The available evidence indicates that two factors determine the erythro-threo ratio in an aldol condensation. The first important factor in this regard is the enolate geometry, which can be either cis or trans (eq 4).[10] In an

$$\underset{\text{O}}{\overset{\text{R}}{\diagup}} \xrightarrow[\text{THF}]{\text{LDA}} \underset{\text{cis}}{\overset{\text{R}}{\diagup}}\text{O}^-\text{Li}^+ + \underset{\text{trans}}{\overset{\text{O}^-\text{Li}^+}{\diagup}}\text{R} \quad (4)$$

important series of papers published in 1967-69 the French chemist J.-E. Dubois showed that the enolates of cyclic ketones, which must be trans for geometric reasons, give rise mainly to threo aldols (eq 5).[11] Later, Dubois and Fellman found that ethyl t-butyl ketone, which affords a cis enolate, reacts with aldehydes to give erythro aldols (eq 6).[12]

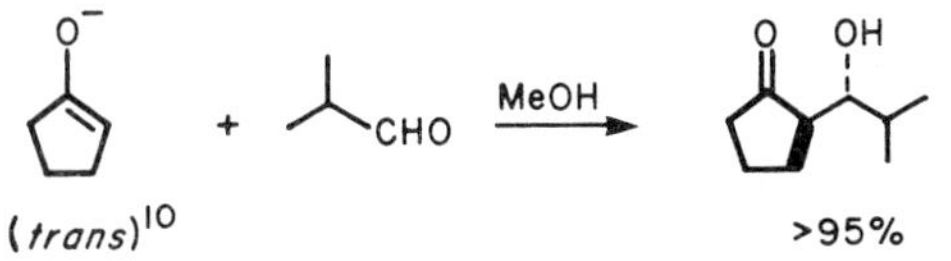

[Structure for eq (5): cyclopentanone enolate (trans) + iPrCHO, MeOH → aldol product, >95%]   (5)

[Structure for eq (6): enolate (cis) + PhCHO, THF → aldol product, >95%]   (6)

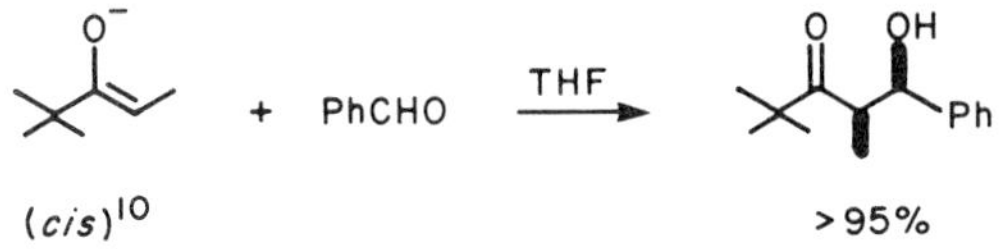

This work established a potentially useful relationship: trans enolates → threo aldols and cis enolates → erythro aldols. However, a subsequent investigation by Dubois and Fellman cast doubt on the scope of this generalization.[13] Although the cis enolate of 3-pentanone shows reasonably good erythro selectivity in its reaction with pivaldehyde (eq 7), the isomeric trans enolate is stereochemically indiscriminate (eq 8).[13] This brings us to the second important fac-

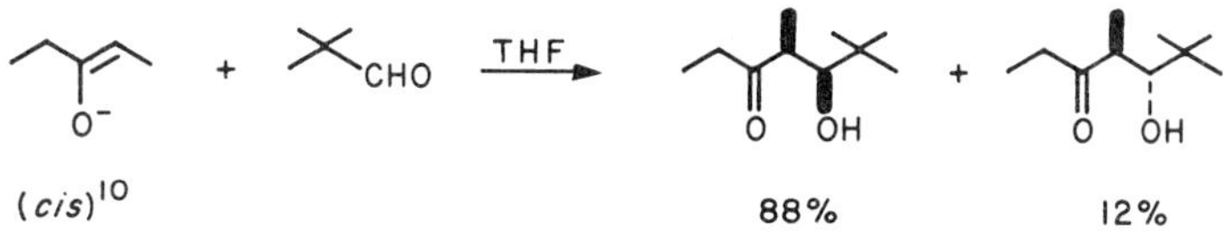

[Structure for eq (7): cis enolate + tBuCHO, THF → erythro 88% + threo 12%]   (7)

[Structure for eq (8): trans enolate + tBuCHO, THF → erythro 48% + threo 52%]   (8)

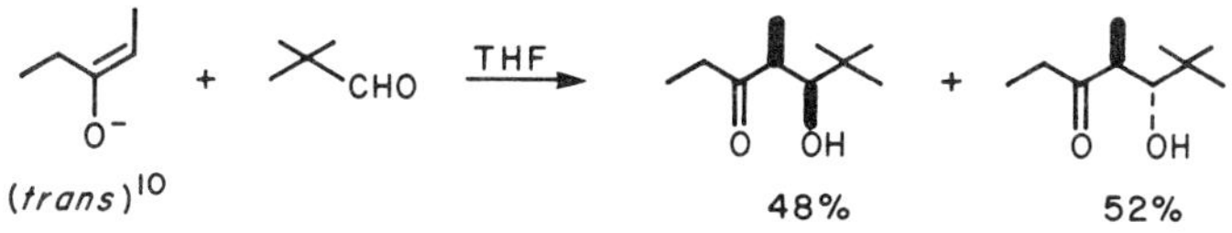

tor in determining erythro-threo selectivity -- the steric requirements of the substituent attached to the carbonyl group.[14,15] For example, whereas the cis enolate derived from ethyl t-butyl ketone shows excellent erythro-selectivity (eq 6), that derived from propionaldehyde gives erythro and threo products in equal amount (eq 9).[15] In a similar

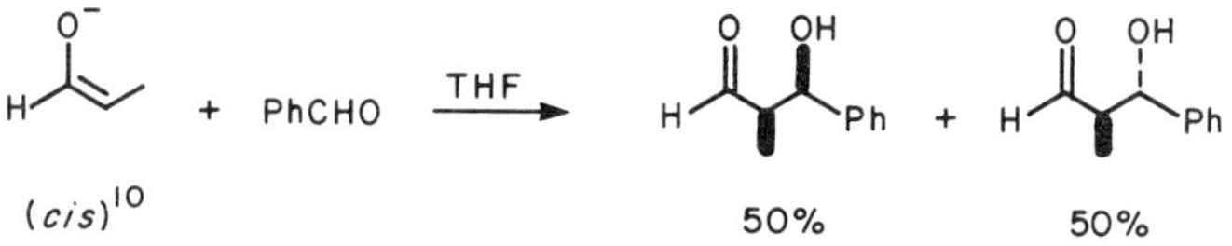

[Structure for eq (9): cis enolate + PhCHO, THF → 50% + 50%]   (9)

manner, the trans enolate derived from methyl propionate gives erythro and threo aldols in comparable amounts (eq 10)[15] but the trans enolate of ethyl mesityl ketone gives only the threo product (eq 11).[14,15]

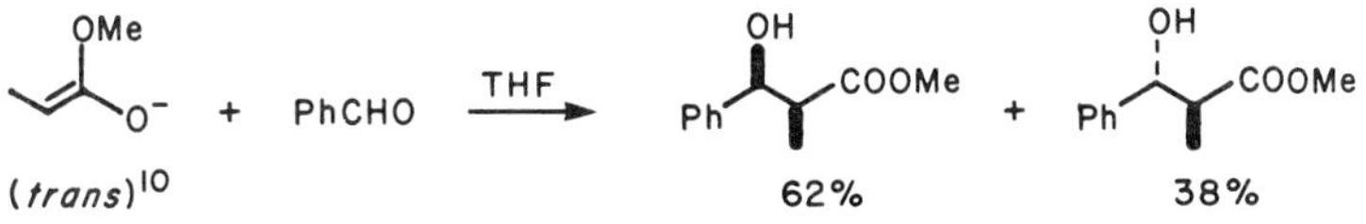

[Structure for eq (10): trans OMe enolate + PhCHO, THF → 62% + 38%]   (10)

[Structure for eq (11): mesityl ketone enolate + PhCHO, THF → >98%]   (11)

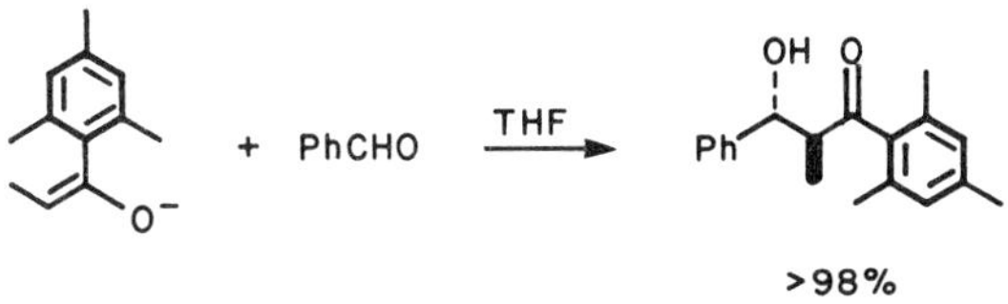

The data presented in the previous paragraph is nicely rationalized by a six-center transition state model first proposed in 1957 by Zimmerman and Traxler for the Ivanov condensation of phenylacetic acid with benzaldehyde (Fig. 1).[16] Applied to the aldol condensation, the Zimmerman-Traxler hypothesis predicts that a cis enolate will give

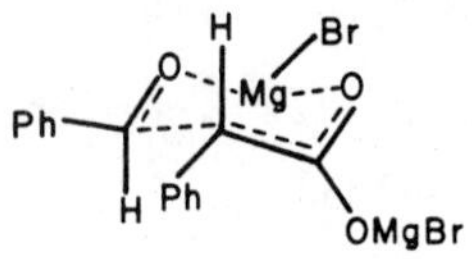

Figure I. Zimmerman-Traxler Transition State

rise to an erythro aldol if the R' group of the enolate is large (Fig. 2). As R' becomes smaller, the R:R' interaction becomes less important, and stereoselectivity decreases. It is a corollary of this hypothesis that trans enolates in which R' is large should give threo aldols, as is the case.

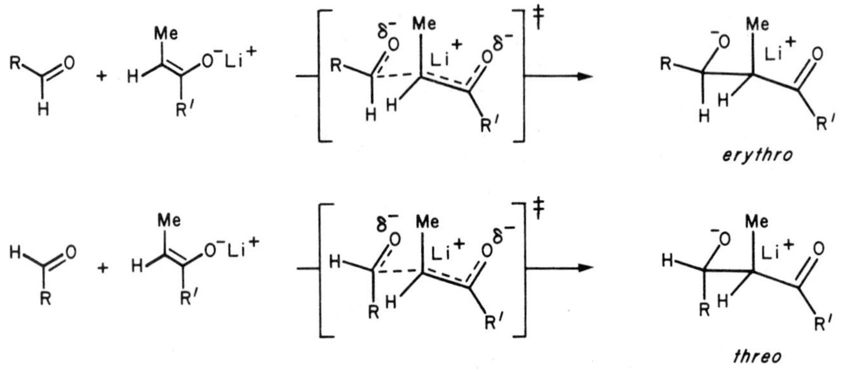

Figure 2. Alternative Transition States for the Aldol Condensation: cis Enolate giving *Erythro* or *Threo* Aldols

The transition states depicted in Figure 2 are certainly oversimplifications of reality. It is likely that lithium enolates exist and react in ethereal solutions as aggregates.[17a] Indeed, the lithium enolate of methyl *t*-butyl ketone has recently been crystallized from tetrahydrofuran and found by X-ray analysis to have the tetrameric structure shown in Figure 3.[17b] However, in this structure, each lithium is still ligated to one solvent molecule. Replacement of this weakly-bound ligand by the aldehyde oxygen would still allow the aldol reaction to proceed through a six-center transition state.

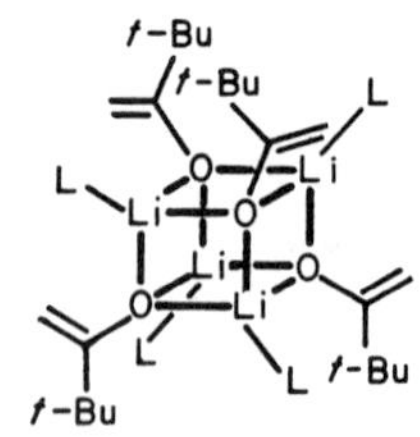

Figure 3. Schematic Representation of the Crystal Structure of the Lithium Enolate of Methyl *t*-Butyl Ketone (L = Tetrahydrofuran)

Another problem with the idealized representations of Figure 2 is that they do not explain the interesting observation that cis enolates are often more stereoselective than their trans isomers (eq 7 and 8). It has been suggested that this difference in stereoselectivity stems from a distortion in the transition state.[15,18] Thus, if we view the reacting array along the newly-forming bond, we see the Newman projection shown in Figure 4. In this transition state

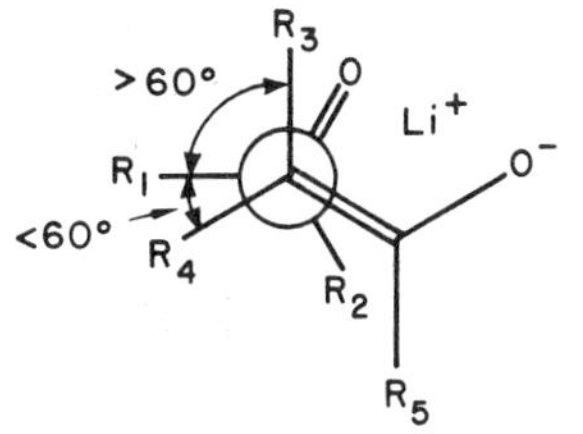

Figure 4. Transition State for the Aldol Condensation

model, there are two important interactions, between R_1 and R_4 and between R_2 and R_5. For a cis enolate, R_4 is hydrogen and therefore the R_2:R_5 interaction controls the stereochemistry of the reaction. However, in a trans enolate, the R_1:R_4 interaction must also be considered. If R_5 is of only modest size, the R_1:R_4 interaction can be as important as the R_2:R_5 interaction. Only when R_5 is very large does the trans enolate show good threo selectivity. This postulate is in accord with the observation that cis enolates react somewhat more rapidly than trans enolates.[13] Seebach has pointed out that the idealized Zimmerman-Traxler transition state indicated in Scheme 1 neglects the fact that Li-O bonds are rather long (approximately 1.95 Å).[19] This face may account for the distortion from a perfectly staggered arrangement that is suggested in Figure 4.

The reaction of a lithium enolate with an aldehyde is exceedingly facile. In ether solvents reaction is often complete in seconds at -78°C at concentrations of 0.5 M. Furthermore, evidence has been advanced that, in the absence of a chelatable cation, the addition is actually endothermic. Thus, the tris(diethylamino)sulfonium enolate of phenylpropanone appears to undergo no reaction with benzaldehyde at -78°C (eq 12).[20] However, if trimethylsilyl fluoride is added to the solution, the silylated aldol is produced immediately. If chelation of the cation is providing the driving

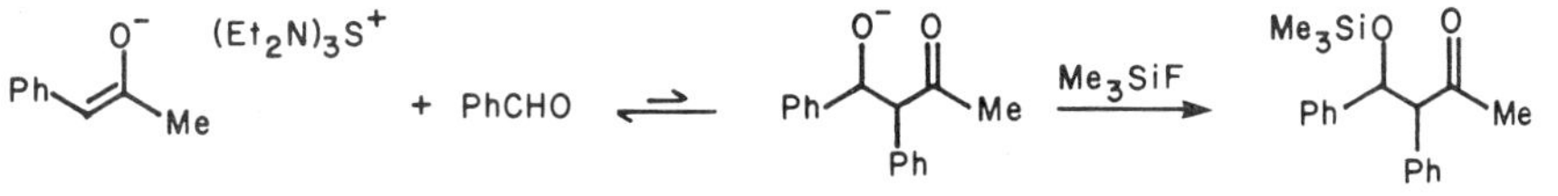

force that allows the aldol reaction to be thermodynamically favorable, the reaction must be energetically finely balanced, and just barely exothermic in many cases. Thus, it is no surprise that the reaction is freely reversible, and that this reversibility constitutes a mechanism for erythro-threo equilibration.

Several generalizations may be made in regard to the ease of erythro-threo equilibration. First, equilibration may be much slower than actual reverse aldolization if the enolate is highly stereoselective. For example, the half-life for erythro-threo equilibration of the lithium aldolate derived from benzaldehyde and the ethyl t-butyl ketone enolate is approximately eight hours at 25°C (eq 13).[15] However, the enolate produced by this aldol reversion has kinetic

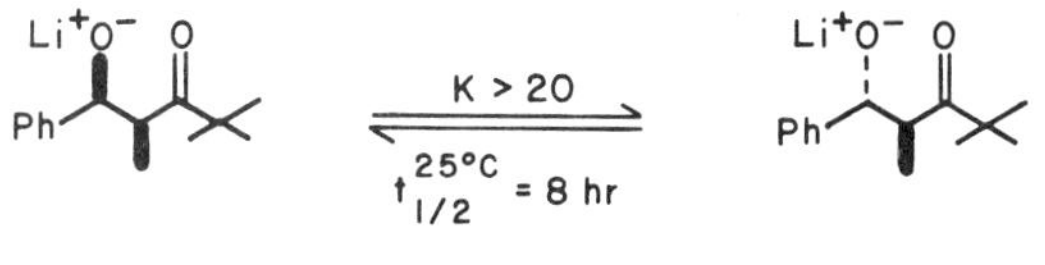

erythro:threo selectivity of 80:1.[15] Aldol reversion must occur, on the average, 80 times for each erythro molecule that is transformed into a threo molecule. Thus, the half-life for the reverse aldol reaction must be only 6 minutes at 25°C. With aldolates derived from less selective enolates, erythro-threo equilibration is much more rapid. For example, the lithium aldolates derived from 3-pentanone and benzaldehyde reach equilibrium within an hour at room temperature (eq 14).[15]

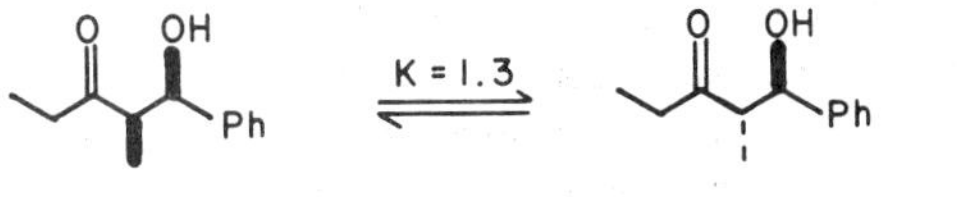

A second generalization regarding erythro-threo equilibration is that the rate of equilibration increases with decreasing basicity of the enolate. Thus, lithium aldolates of propiophenone equilibrate extremely easily (eq 15), presumably because the propiophenone enolate is less basic and more stable than enolates derived from dialkyl ketones. This trend is also consistent with the observation that aldols derived from esters are much less prone to undergo reverse aldolization and erythro-threo equilibration than are those derived from ketones.

$$\text{(15)}$$

Erythro-threo equilibration is also more rapid for aldolates that are substituted by bulky substituents at the position between the hydroxy and carbonyl functions. For example, the erythro and threo aldolates shown in eq 16 undergo equilibration at 25°C when R = t-butyl but not when R = methyl.[21]

$$\text{(16)}$$

Equilibration is also influenced by the nature of the cation associated with the aldolate. Boron aldolates are quite stable and do not undergo erythro-threo equilibration even at elevated temperatures (eq 17).[22] For the alkali metal

$$\text{(17)}$$

aldolates, the lithium compounds equilibrate most slowly and potassium compounds most rapidly. The rate of equilibration of potassium aldolates is further increased by addition of 18-crown-6.[15] Erythro-threo equilibration has also been observed with magnesium and zinc aldolates.[15,23] Zinc aldolates equilibrate sometimes more rapidly and sometimes more slowly than their lithium counterparts.[15]

Threo aldolates are often more stable than the erythro isomers, especially is associated with a zinc cation.[15,23] For example, the equilibrium constant for the erythro-threo equilibrium shown in eq 13 is greater than 20. That in eq 15 is only 1.1 for the lithium aldolate, but 3 for the zinc aldolate. This property of zinc aldolates has been used advantageously in several natural products syntheses. An example is seen in Evans' synthesis of ionophore A-23187, which involves the coupling of an aldehyde and a ketone by the House procedure using the zinc enolate (eq 18).[24]

$$\text{(18)}$$

The net threo:erythro ratio in this reaction is 70:30; the indicated diastereomer (the threo, Cram isomer) is obtained as the major product.

Erythro- and Threo-Selective Synthetic Reagents

The useful ketone **4** has been developed at Berkeley for use in the preparation of erythro β-hydroxy acids, aldehydes, and ketones.[15] Like ethyl t-butyl ketone, compound **4** gives the cis enolate, which reacts with various aldehydes to afford erythro aldols (e.g., eq 19). The resulting aldols may be oxidized with periodic acid to give the

$$\text{(19)}$$

corresponding erythro β-hydroxy acids (eq 20).[15] If the erythro aldol derived from the reaction of **4** with an aldehyde is first reduced with lithium aluminum hydride, desilylated, and then oxidized with sodium periodate, an erythro β-hydroxy aldehyde results (eq 21).[25] Finally, the hydroxy group can be protected, an organolithium reagent added to the ketone, and the product then cleaved with periodic acid to produce an erythro β-hydroxy ketone (eq 22).[26]

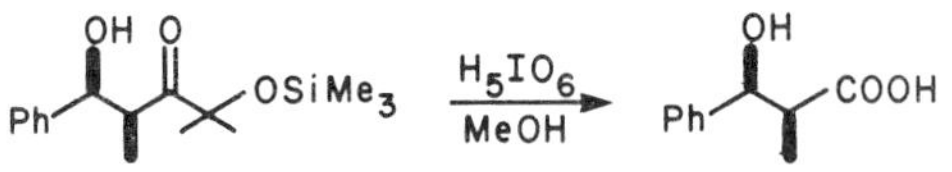

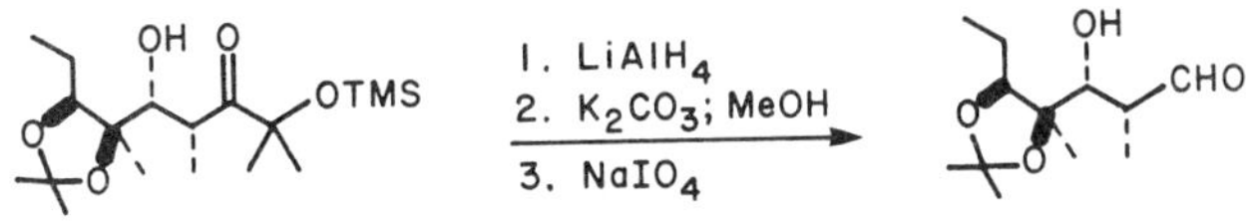

Another erythro-selective reagent is the boron enolate of S-phenyl propanethioate, which reacts with aldehydes to give erythro aldols of greater than 97% diastereomeric purity (eq 23).[27] Evans and Yamamoto and their respective coworkers have made the interesting discovery that zirconium enolates are erythro-selective regardless of enolate configuration.[28,29] Thus, N,N-diisopropylpropionamide forms a 4:1 mixture of cis and trans lithium enolates upon deprotonation with lithium diisopropylamide, and this mixture of enolates reacts with benzaldehyde to give a 61:39 mixture of erythro and threo aldols (eq 24).[15] However, when the lithium enolate mixture is treated with bis(cyclopentadienyl)zirconium dichloride prior to addition of the benzaldehyde, the erythro:threo ratio is >98:2.[28]

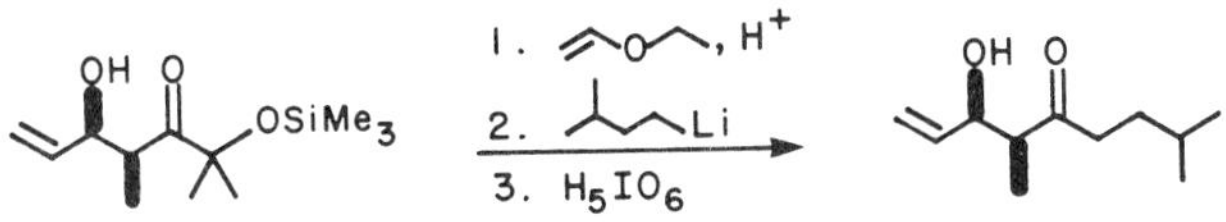

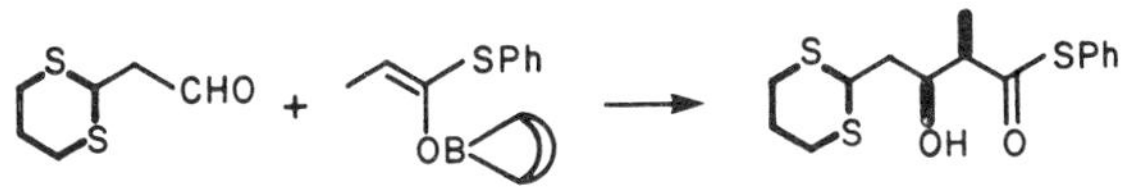

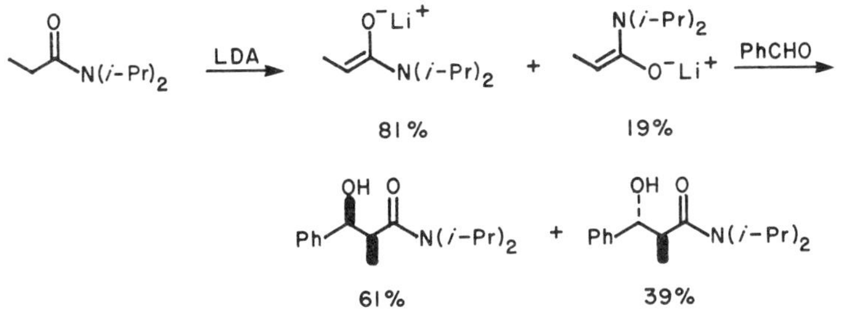

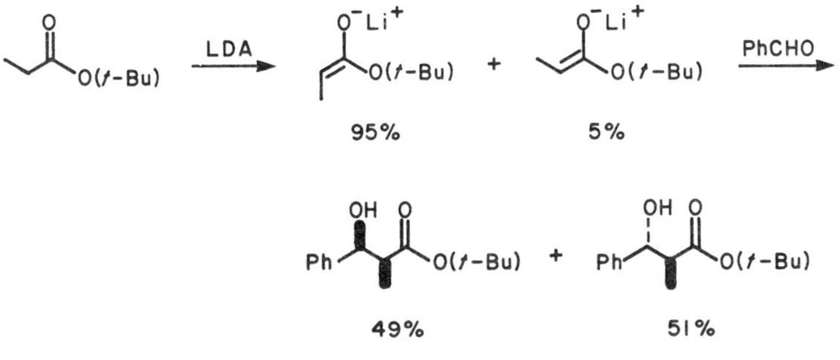

All of the highly threo-selective reagents are esters or thiol esters. With lithium diisopropylamide, esters give nearly pure trans enolates. However, with normal alkyl esters, kinetic stereoselectivity is low. For example, *t*-butyl propionate gives a 95:5 mixture of trans and cis enolates, but this enolate mixture reacts with benzaldehyde to produce an equimolar mixture of erythro and threo aldols (eq 25).[15] Reasoning that esters of considerably more bulky alcohols or phenols might show acceptable threo selectivity, we prepared and examined the chemistry of aryl propionates 5-7.[30] These compounds indeed show good threo-selectivity. For example, the enolate derived from DMP

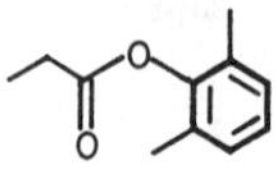

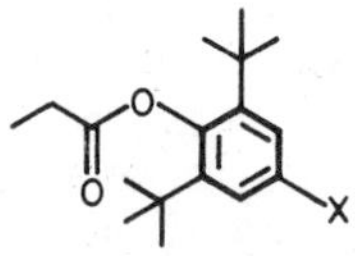

5
DMP propionate

6: X = Me, BHT propionate
7: X = OMe, DBHA propionate

propionate reacts with isobutyraldehyde to give essentially pure threo β-hydroxy ester (eq 26). If the aldehyde is aryl or an α-unbranched aliphatic one, ester **5** gives threo:erythro ratios of only 6:1. In such cases the more selective reagents **6** or **7** may be used to achieve threo:erythro ratios of greater than 30:1.

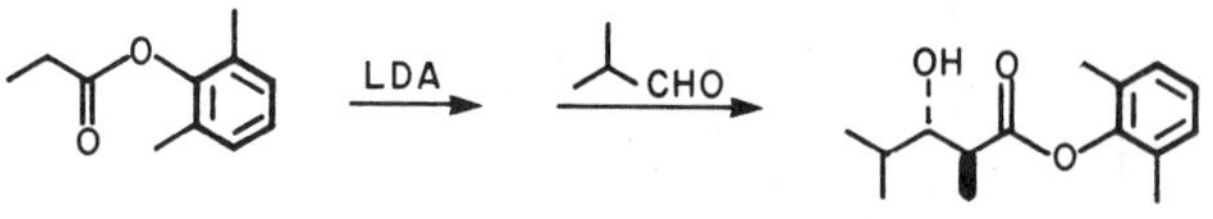

(26)

Recently we have found that certain O-alkyllactic acid esters show good simple diastereoselection.[31] For example, the BHT ester of O-benzyllactic acid (**8**) reacts with benzaldehyde to give solely the threo β-hydroxy ester (eq 27).[7] This is a useful result, since one of the major obstacles to application of aldol technology for the synthesis of erythronolide A (**2**) is the presence of vicinal diol units at C_5-C_6 and C_{11}-C_{12}. Not only does reagent **8** lead to an α,β-dioxy ester, but the relative configuration at the two new asymmetric carbons corresponds to that at the relevant positions in **2**.

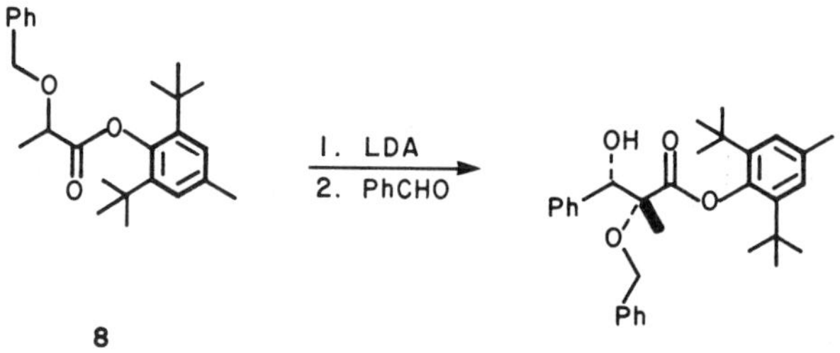

(27)

We do not know the configuration of the enolate ion derived from **8**. However, we assume that it is as shown below since it gives rise to threo aldols. The supposition that the lithium cation is coordinated to the α-alkoxy group

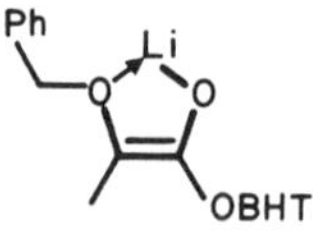

is based on recent observations we have made on the stereochemistry of some related α-phenoxy esters. Compound **9** reacts with benzaldehyde to provide a 2:1 mixture of threo and erythro esters (eq 28).[32] We believe that this

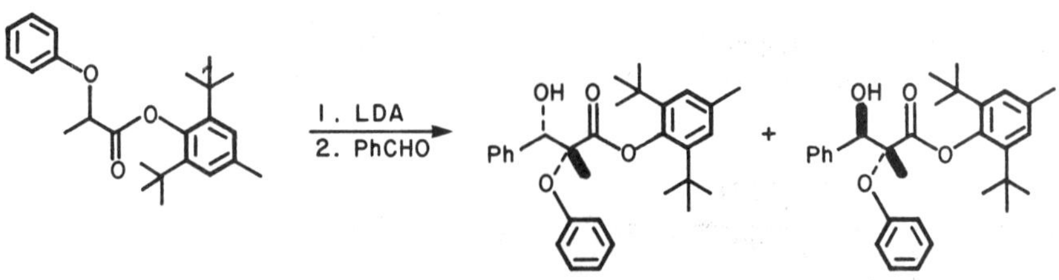

(28)

9   65%   35%

decrease in stereoselectivity results from the fact that ester **9** gives a mixture of cis and trans enolates:

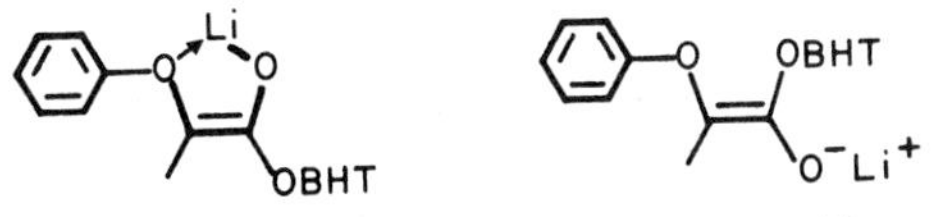

$$(\text{trans})^{10} \qquad (\text{cis})^{10}$$

We have been able to correct this situation by employing the *p*-anisyl, rather than the phenyl, compound. Thus, compound **10** reacts with benzaldehyde to afford only the threo diastereomer (eq 29).[32] The *p*-methoxy group in **10** renders the phenoxy oxygen more basic, and better able to internally coordinate the lithium cation. This gives rise to more trans enolate and higher threo-selectivity.

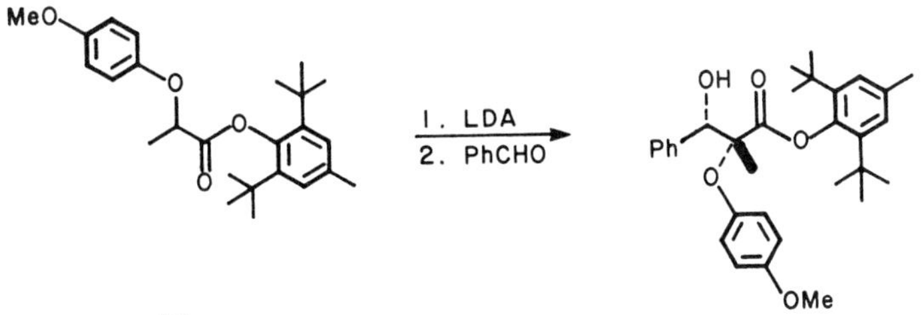

(29)

10

We have also made the surprising discovery that methyl α-methoxypropionate (**11**) also shows high simple diastereoselection, but in the opposite sense to that shown by esters **8** and **10**.[31] For example, the lithium enolate of **11** reacts with isobutyraldehyde to provide the erythro aldol (eq 30). Addition of compound **11** to the *t*-

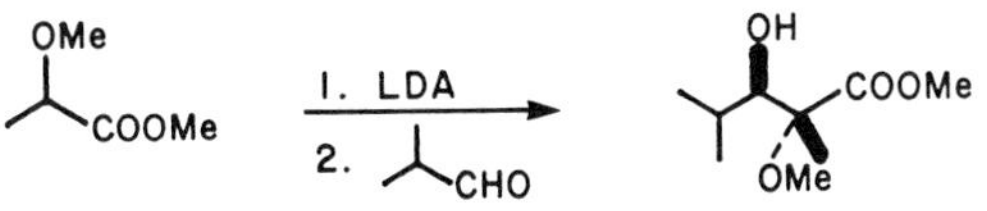

(30)

11

butyldiphenylsilyl ether of lactaldehyde gives a mixture of aldols. Treatment of this mixture with KF in DMF catalyzes migration of the silyl group from one oxygen to another and allows conversion to a 3:1 mixture of lactones (eq 31).[31] The major isomer of this pair has the three chiral centers of cladinose, the unusual carbohydrate that is attached to the C_3-hydroxy group in erythromycin A (**1**).

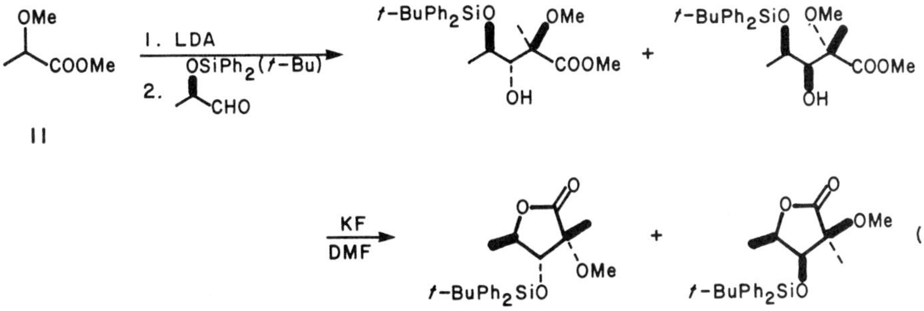

11

(31)

75%

The other effective threo-selective reagent is the dicyclopentylboron enolate of S-*t*-butyl propanethioate (**12**), developed independently by Evans and Masamune.[33,34] An example of the use of reagent **12** is seen in eq 32.[34]

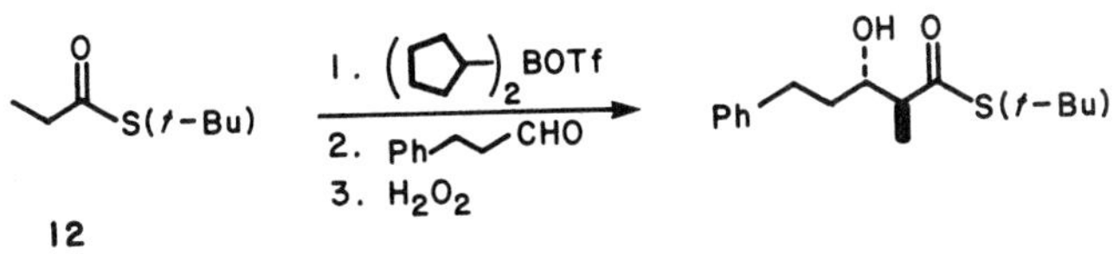

(32)

12

The Problem of Diastereoface Preference of Chiral Aldehydes

In the reaction of an ethyl carbonyl compound with a chiral aldehyde, four diastereomeric aldols may be produced (eq 3). This problem is partly alleviated by the use of one of the erythro- or threo-selective reagents discussed in the previous section (eq 33).[15] The 80:20 ratio obtained in eq 33 is similar to the net Cram:anti-Cram ratio of 79:21 seen in eq 3, and is presumably a manifestation of the inherent diastereoface preference of α-phenylpropionaldehyde.

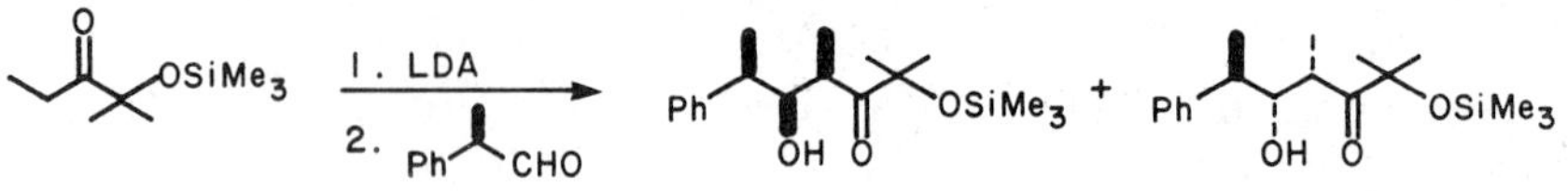

$$\text{4} \xrightarrow[\text{2. PhCHO}]{\text{1. LDA}} \text{erythro, "Cram" (80\%)} + \text{erythro, "anti-Cram" (20\%)} \quad (33)$$

To influence the Cram:anti-Cram ratio in additions to chiral aldehydes, we have applied the logical principle of double stereodifferentiation[35] (also known as double asymmetric induction).[36] An example is seen in the reaction of the fructose-derived ketone **13** with the enantiomeric glyceraldehyde acetonides (**14** and **15**). Ketone **13** reacts with

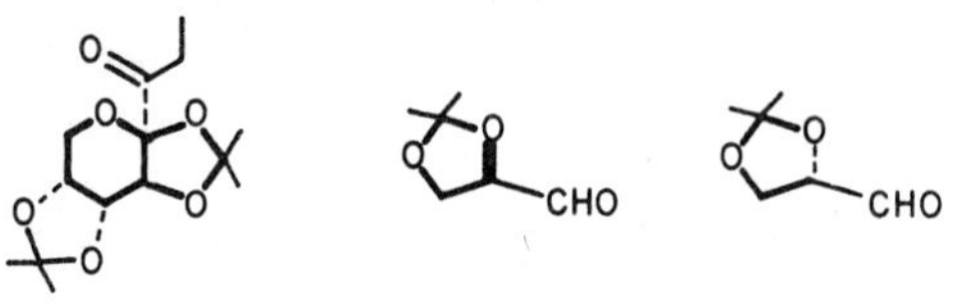

13 **14** **15**

benzaldehyde to give two erythro aldols in a ratio of 4:1 (eq 34). Aldehydes **14** and **15** have an inherent diastereoface

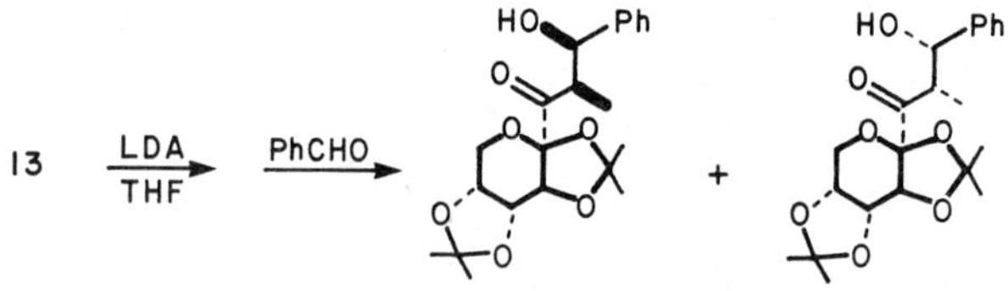

$$\text{13} \xrightarrow[\text{THF}]{\text{LDA, PhCHO}} (80\%) + (20\%) \quad (34)$$

preference of slightly greater than 4:1 (e.g., eq 35).[37] In the reaction of ketone **13** with aldehyde **14**, the two reactants

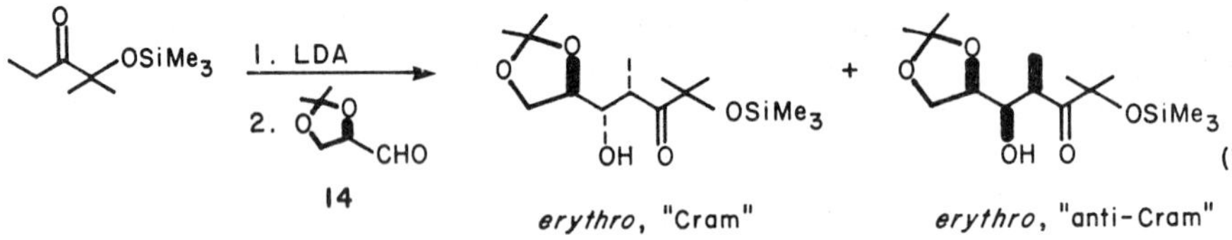

$$\xrightarrow[\text{2.}\ \text{14}\ \text{CHO}]{\text{1. LDA}} \text{erythro, "Cram" (81\%)} + \text{erythro, "anti-Cram" (19\%)} \quad (35)$$

are working at cross-purposes; the two erythro aldols are produced in a ratio of only 2:1 (eq 36). However, in the

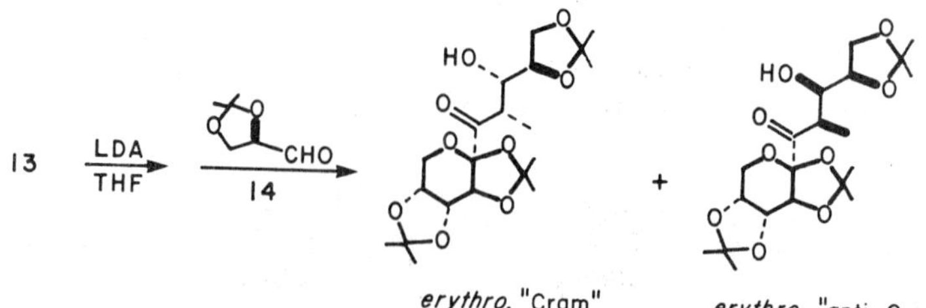

$$\text{13} \xrightarrow[\text{THF}]{\text{LDA}} \xrightarrow{\text{14 CHO}} \text{erythro, "Cram" (61\%)} + \text{erythro, "anti-Cram" (28\%)} \quad (36)$$

reaction of **13** with the enantiomeric aldehyde **15**, both the ketone and the aldehyde promote the same sense of chirality at the two new centers; in this case only the erythro, "Cram" isomer is produced (eq 37). Similar experiments have been reported by Masamune and coworkers.[38]

$$13 \xrightarrow{\text{LDA}}_{\text{THF}} \xrightarrow{\text{15}} \text{erythro, "Cram" (>97\%)} \tag{37}$$

Double stereodifferentiation is also observed when one of the chiral elements is the solvent.[35] Aldehyde **14** shows diastereoface preference of 4.3:1 in its reaction with ketone **4** in THF (eq 35). If the reaction is carried out in the enantiomeric tetramethoxybutanes **16** and **17**[39] the ratio of the "Cram" and "anti-Cram" diastereomers is 5.0:1 and 3.6:1, respectively.[40,41]

16   **17**

Mutual Kinetic Resolution

If the double stereodifferentiation effect is large enough, it can lead to "mutual kinetic resolution" in reactions of racemic enolates with racemic aldehydes. Such a case is seen in the reaction of the enolate derived from racemic ketone **18** with racemic α-phenylpropionaldehyde, which gives a single racemic aldol of greater than 97.7% diastereomeric purity (eq 38).[42] Thus, the S enantiomer of ketone **18** must react much faster with (S)- than with

$$(S)\text{-}18 \xrightarrow[\text{2. Ph-CHO}]{\text{1. LDA}} \tag{38}$$

(R)-α-phenylpropionaldehyde; i.e., kinetic resolution is operating. Since both reactants are racemic, the kinetic resolution is mutual, each enantiomer of **18** selecting the enantiomer of α-phenylpropionaldehyde with which it reacts more rapidly.

This interesting behavior turns out to be a consequence of double stereodifferentiation. Control experiments show that the enolate of ketone **18** shows a high inherent diastereoface preference in its reactions with achiral aldehydes having rather bulky groups attached to the carbonyl group (e.g., eq 39).[42] If we ignore the four threo

$$18 \xrightarrow[\text{2.}]{\text{1. LDA}} \quad >95\% \quad + \quad <5\% \tag{39}$$

diastereomers that might have been formed from the reaction of racemic **18** with racemic α-phenylpropionaldehyde, we are left with the four erythro isomers **19-22**. The first two result from reaction of (S)-enolate with (S)-aldehyde. The **19**:**20** ratio is expected to be large because of the effects of productive double stereodifferentiation. Erythro isomers **21** and **22** result from reaction of (R)-enolate with (S)-aldehyde. In this combination there are no erythro pro-

19 (S + S)   **20** (S + S)   **21** (S + R)   **22** (S + R)

ducts which satisfy the inherent diastereofacial preferences of both the enolate and aldehyde. Thus, the **21:22** ratio is expected to be modest. Furthermore, it can readily be shown that the **(19 + 20):(21 + 22)** ratio should be greater than unity. For example, suppose we take the inherent diastereofacial preference of the enolate of **18** and of α-phenylpropionaldehyde each to be 20:1. The relative rates of the four competing reactions leading to isomers **19-22** would then be (20 x 20), (1 x 1), (20 x 1), and (1 x 20) or 400, 1, 20, and 20. Thus, the **(19 + 20):(21 + 22)** ratio would be 10:1. This simple analysis shows that the phenomenon of mutual kinetic resolution may be observed in any reaction between two racemic reactants *if* both reactants show inherent diastereofacial preference in reactions with achiral reaction partners.

We have encountered several other cases of double stereodifferentiation with mutual kinetic resolution in our work at Berkeley. An example is the reaction of the racemic ketone **23** with racemic O-benzyllactaldehyde, which gives the two erythro aldols shown in a ratio of 10:1 (eq 40).[42b] The major isomer has been converted by a straightforward three-step sequence into (±)-blastmycinone, a degradation product of the antibiotic antimycin-A_3.[42b]

Reagents for Control of the "Cram's Rule Problem"

An ultimate goal of much of the research which has been carried out on aldol stereoselection is to be able to select either an erythro or threo product and to be able to determine which diastereotopic face of a chiral aldehyde reacts. That is, we would like to be able to produce any one of the four products of eq 3 as the sole product of an aldol reaction. The erythro-threo problem is under rather good control. As shown by the preceding two sections, diastereofacial control requires chiral enolate precursors. The first such reagent was ketone **18** and its relatives (e.g., **23**).[42a] Masamune and coworkers have recently employed similar reagents in a synthesis of 6-deoxyerythronolide B (**3**).[6] Compound **24**, which is obtained in three steps from (*R*)-mandelic acid, reacts as its dicyclopentylboron enolate with propionaldehyde to give a single erythro aldol, estimated to be of greater than 99% diastereomeric purity. After the normal periodic acid cleavage the corresponding β-hydroxy acid is obtained with high optical purity.

Evans and coworkers have introduced imides **25** and **26**, which are prepared in two steps from valinol and norephedrine, respectively.[43] The dibutylboron enolate of **25** reacts with benzaldehyde to give a single erythro aldol (eq 42); the diastereomeric purity of the product has been estimated to be greater than 99.96%. Compounds **24-26** have

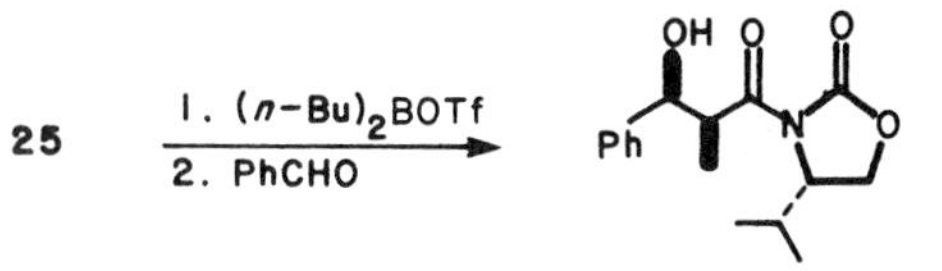

25 →[1. (n-Bu)₂BOTf][2. PhCHO] [product] (42)

such high inherent diastereoface preference that they overwhelm the modest inherent diastereoface preference of most chiral aldehydes. For example, the zirconium enolate of compound 25 reacts with both enantiomers of 3-benzyloxy-2-methylpropanal to give products having the same absolute configuration at the two newly-created asymmetric carbons (eq 43).[28b]

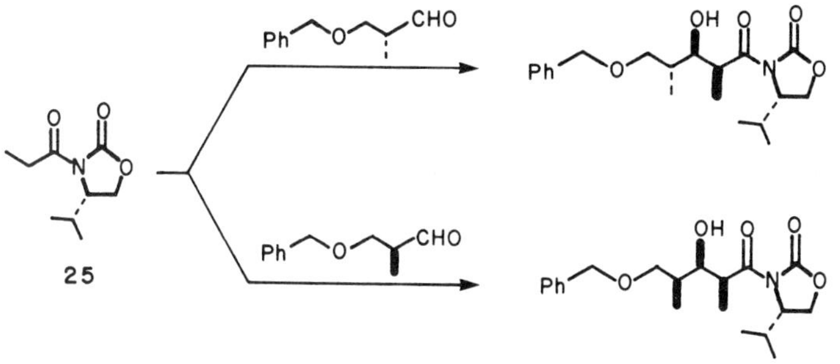

(43)

An Aldol Approach to the Synthesis of Erythromycin A

For about two years, my coworkers and I have been applying some of the methodology we have developed to an actual total synthesis of compound 1.[44] The starting point is aldehyde 27, which is prepared in racemic form as shown in eq 44 and in optically active form by the Evans procedure[43] as shown in eq 45. Compound 27 reacts smoothly with

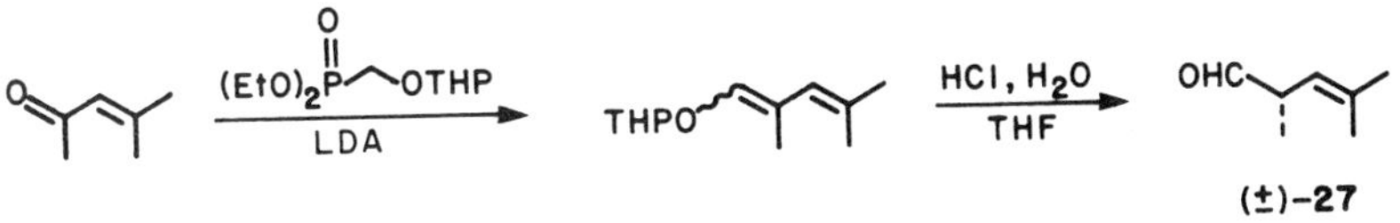

(44)

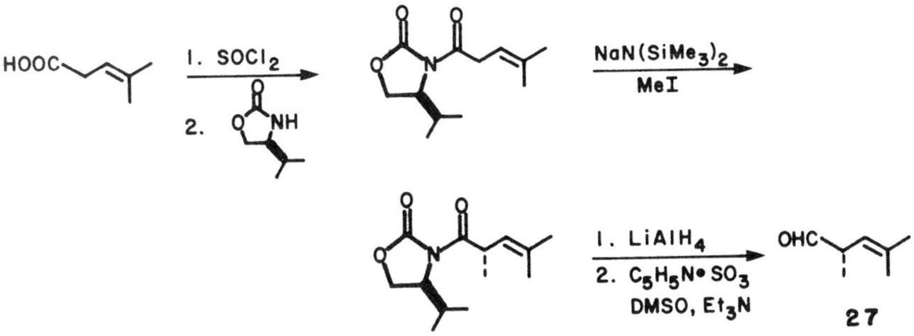

(45)

the enolate derived from ketone 4 to give two erythro aldols in a ratio of 94:6 (eq 46). After reduction of the ketone

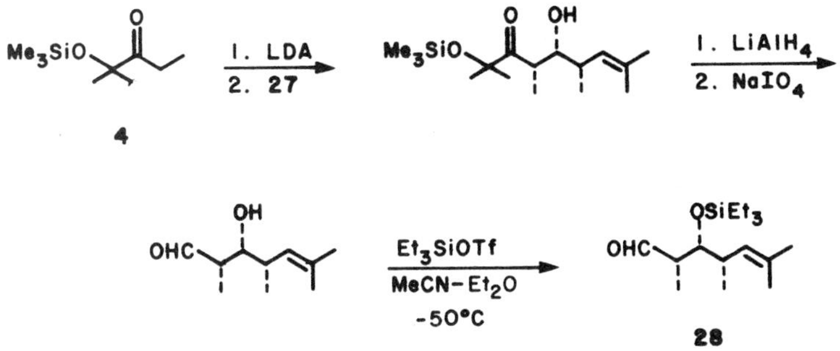

(46)

and cleavage with buffered sodium periodate, the β-hydroxy aldehyde is protected as its triethylsilyl ether in preparation for the second aldol condensation. This aldehyde (**28**) is allowed to react with the lithium enolate of BHT O-benzyllactate (**8**) to obtain two threo aldols in a ratio of 85:15. After protection of the future C_3 and C_5 hydroxy groups by formation of the acetonide the crystalline major isomer is reduced and the resulting primary alcohol is oxidized to provide aldehyde **29**, which contains five of the ten asymmetric carbons of erythronolide A.

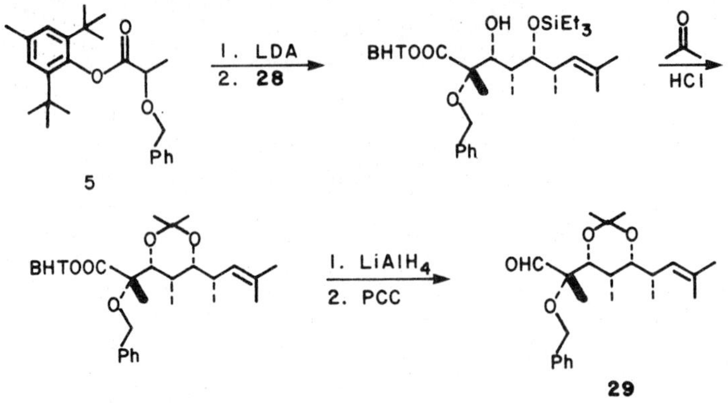

(47)

The β, γ-unsaturated aldehyde that serves as the starting point for synthesis of the C_8-C_{15} segment is also prepared by the Evans method, as described in eq 48. Aldehyde **30** reacts with the lithium enolate of BHT O-benzyl-

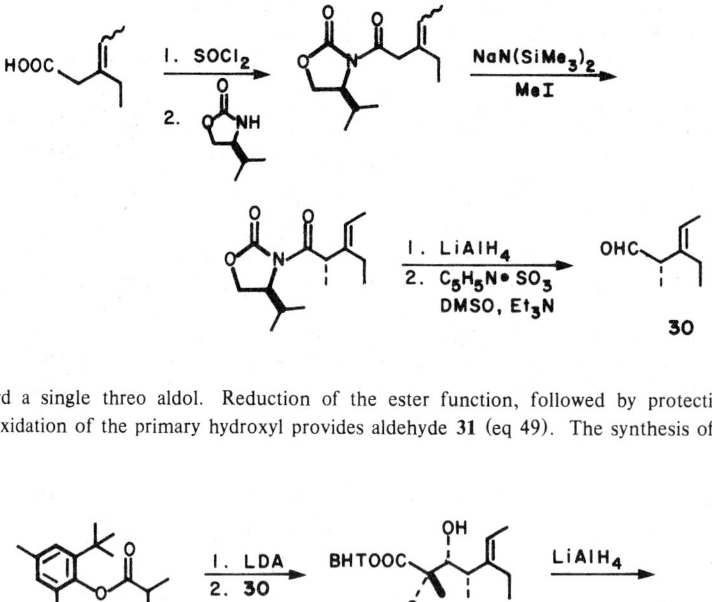

(48)

lactate to afford a single threo aldol. Reduction of the ester function, followed by protection of the secondary hydroxyl and oxidation of the primary hydroxyl provides aldehyde **31** (eq 49). The synthesis of the left-hand unit is

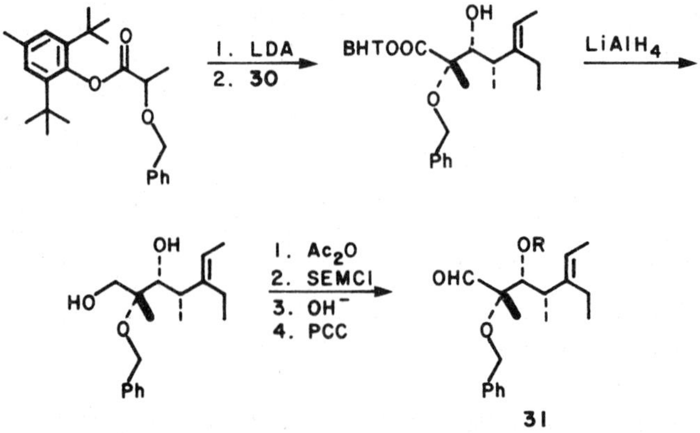

(49)

completed by addition of ethyllithium to aldehyde **31**; a 3:1 mixture of alcohol **32** and its diastereomer is produced. After protection of the new secondary hydroxyl, the alkene is cleaved to acquire ketone **33** (eq 50). Completion of the erythronolide A assembly will require coupling of ketone **33** with aldehyde **29**, followed by a number of redox reactions and formation of the macrocycle.

Concluding Remarks

A decade ago, organic chemists approached the problem of control of stereochemistry in the course of a synthesis in one of two ways. The timid simply ignored the problem and separated mixtures of stereoisomers if possible. The more courageous devised clever indirect methods to circumvent the problem. A classical method for stereocontrol in the synthesis of acyclic compounds has been to capitalize on the special properties of small-ring compounds. Thus, relative stereochemistry can be established at two or more asymmetric carbons in a six-membered ring by the use of sound principles of kinetic (steric hindrance) or thermodynamic (axial → equatorial) control. The ring can then be cleaved to obtain an acyclic compound having the several asymmetric carbons in the correct stereochemical relationship. The Woodward erythronolide A synthesis provides a beautiful example of this strategy. However, one should recognize that, for all its elegance, this classical approach to stereocontrol is indirect and must eventually give way to more direct methods as they become available.

At the beginning of this account I pointed out that the aldol condensation is one of the oldest organic reactions.[4] Its importance in organic chemistry has certainly not gone unrecognized; indeed, it is the *only* method to which an entire volume of *Organic Reactions* has been devoted.[45] Yet, the casual reader of the chemical literature cannot help but notice that the aldol condensation is undergoing a renaissance. The special attention that has been given to the reaction during the last five years has been inspired in part by the hypothesis that it might provide a simple, *direct* solution to the stereochemical problem in the synthesis of molecules such as **1**. I think the success we have had so far in our approach to erythromycin A suggests that this hypothesis is valid.

Acknowledgements

I wish to give special acknowledgement to the talented students and postdoctoral coworkers who have carried out the experiments upon which this account is based: Charles Buse, William Kleschick, John Sohn, Michael Pirrung, Charles White, John Lampe, Steven Young, Stephen Montgomery, James Hagen, Kenneth Kees, Ulrich Badertscher, Bai Dong-lu and Esa Jarvi. I also thank the National Institutes of Health for support of the project (AI-15027) and Professors David Evans and Dieter Seebach for valuable exchanges of information prior to publication.

References

1. See *inter alia*, "The Total Synthesis of Natural Products," vol. 1-4, J. ApSimon, ed., Wiley-Interscience, New York, 1973-81; "Natural Products Chemistry," vol. 1-2, K. Nakanishi, T. Goto, S. Itō, S. Natori and S. Nozoe, eds., Academic Press, Inc., New York, 1974; N. Anand, J.S. Bindra and S. Raganathan, "Art in Organic Synthesis," Holden-Day, Inc., San Francisco, 1970; J.S. Bindra and R. Bindra, "Creativity in Organic Synthesis," Academic Press, Inc., New York, 1975.

2. R.B. Woodward in "Perspectives in Organic Chemistry," A. Todd, ed., Interscience, New York, 1956.

3. For a good general review see P.A. Bartlett, *Tetrahedron*, **36**, 2 (1980).

4. R. Kane, *Ann. Physik. Chem.*, [2], **44**, 475 (1838); *J. Prakt. Chem.*, **15**, 129 (1838).

5. R.B. Woodward, *et. al*, *J. Am. Chem. Soc.*, **103**, 3210, 3213, 3215 (1981).

6. (a) S. Masamune, W. Choy, F.A.J. Kerdesky and B. Imperiali, *J. Am. Chem. Soc.*, **103**, 1566 (1981); (b) S. Masamune, M. Hirama, S. Mori, S.A. Ali, and D.S. Garvey, *ibid.*, **103**, 1568 (1981).

7. The convention used for describing the stereostructures of aldols is as follows: When the aldol is written with its backbone in an extended (zig-zag) manner, an erythro diastereomer is one in which the alkyl group at the α position and the hydroxy group at the β position both project either toward the viewer (bold bonds) or away from the viewer (dashed bonds). If one of these substituents projects toward the viewer and the other away from the viewer, this is a threo isomer. This convention is used whether or not there is also an additional hydroxy or alkoxy group at the α carbon.

8. D.J. Cram and F.A. Abd Elhafez, *J. Am. Chem. Soc.*, **74**, 5828 (1952).

9. T. Matsumoto, Y. Hosoda, K. Mori and K. Fukui, *Bull. Chem. Soc. Japan*, **45**, 3256 (1972).

10. In discussing stereoisomeric enolates, we use cis and trans to describe the relationship between the -O⁻ and alkyl group attached to the α-carbon. For most ketone enolates, cis is the same as *Z* and trans is the same as *E*. It is convenient to use the cis-trans, rather than the *E, Z* terminology, so that enolates of a given relative

configuration will always have the same stereochemical name. Otherwise, there is ambiguity which leads to confusion in discussions:

cis, (E)   cis, (Z)   cis, (?)

11. (a) J.-E. Dubois and M. Dubois, *Tetrahedron Lett.*, 4215 (1967); (b) J.-E. Dubois and M. Dubois, *Chem. Commun.*, 1567 (1968); (c) J.-E. Dubois and M. Dubois, *Bull. Soc. Chim. Fr.*, 3120 (1969); (d) J.-E. Dubois and M. Dubois, *ibid*, 3553 (1969).
12. J.-E. Dubois and P. Fellman, *C.R. Acad. Sci. Ser. C.*, **274**, 1207 (1972).
13. J.-E. Dubois and P. Fellman, *Tetrahedron Lett.*, 1225 (1975).
14. W.A. Kleschick, C.T. Buse, and C.H. Heathcock, *J. Am. Chem. Soc.*, **99**, 247 (1977).
15. C.H. Heathcock, C.T. Buse, W.A. Kleschick, M.C. Pirrung, J.E. Sohn, and J. Lampe, *J. Org. Chem.*, **45**, 1066 (1980).
16. H. Zimmerman and M. Traxler, *J. Am. Chem. Soc.*, **79**, 1920 (1957).
17. (a) L.M. Jackman and B.C. Lange, *Tetrahedron*, **33**, 2737 (1977); (b) R. Amstutz, W. B. Schweizer, D. Seebach, and J.D. Dunitz, *Helv. Chim. Acta*, **64**, 2617 (1981).
18. P. Fellman and J.-E. Dubois, *Tetrahedron*, **34**, 1249 (1978).
19. D. Seebach, R. Amstutz, and J. D. Dunitz, *Helv. Chim. Acta*, **64**, 2617 (1981); see also D. Seebach and J. Goliński, *ibid*, **64**, 1413 (1981) for an alternative proposal regarding the question of the greater intrinsic value of cis enolates.
20. R. Noyori, I. Nishida, J. Sakata and M. Nishizawa, *J. Am. Chem. Soc.*, **102**, 1223 (1980).
21. J. Mulzer, J. Segner, and G. Bruntrup, *Tetrahedron Lett.*, 4561 (1977).
22. D.A. Evans, J.V. Nelson, E. Vogel, and T.R. Taber, *J. Am. Chem. Soc.*, **103**, 3099 (1981).
23. H.O. House, D.S. Crumrine, A.Y. Teranishi, and H.D. Olmstead, *ibid.*, **95**, 3310 (1973).
24. D.A. Evans, C.E. Sacks, W.A. Kleschick, and T.A. Taber, *ibid.*, **101**, 6789 (1979).
25. C.H. Heathcock, S.D. Young, J.P. Hagen, M.C. Pirrung, C.T. White, and D. VanDerveer, *J. Org. Chem.*, **45**, 3846 (1980).
26. (a) C.T. White and C.H. Heathcock, *ibid.*, **46**, 91 (1981). (b) C.H. Heathcock and E.T. Jarvi, unpublished results.
27. M. Hirama, D.S. Garvey, L.D.-L. Lu and S. Masamune, *Tetrahedron Lett.*, 3937 (1979).
28. (a) D.A. Evans and L.R. McGee, *ibid.*, 3975 (1980). (b) D.A. Evans and L.R. McGee, *J. Am. Chem. Soc.*, **103**, 2876 (1981).
29. Y. Yamamoto and K. Maruyama, *Tetrahedron Lett.*, 4607 (1980).
30. (a) M.C. Pirrung and C.H. Heathcock, *J. Org. Chem.*, **45**, 1727 (1980). (b) C.H. Heathcock, M.C. Pirrung, S.H. Montgomery, and John Lampe, *Tetrahedron*, **37**, 4087 (1981).
31. C.H. Heathcock, J.P. Hagen, E.T. Jarvi, M.C. Pirrung, and S.D. Young, *J. Am. Chem. Soc.*, **103**, 4972 (1981).
32. C.H. Heathcock and U. Badertscher, unpublished results.
33. (a) D.A. Evans, E. Vogel, and J.V. Nelson, *J. Am. Chem. Soc.*, **101**, 6120 (1979). (b) D.A. Evans, J.V. Nelson, E. Vogel, and T. R. Taber, *ibid.*, **103**, 3099 (1981).
34. S. Masamune, S. Mori, D. Van Horn, and D.W. Brooks, *Tetrahedron Lett.*, 1665 (1978).
35. (a) C.H. Heathcock and C.T. White, *J. Am. Chem. Soc.*, **101**, 7076 (1979). (b) C.H. Heathcock, C.T. White, J.J. Morrison, and D. VanDerveer, *J. Org. Chem.*, **46**, 1296 (1981).
36. A. Horeau, H.-B. Kagan, and J.-P. Vigneron, *Bull. Soc. Chim. Fr.*, 3795 (1968).
37. C.H. Heathcock, S.D. Young, J.P. Hagen, M.C. Pirrung, C.T. White, and D. VanDerveer, *J. Org. Chem.*, **45**, 3846 (1980).
38. S. Masamune, S.A. Ali, D.L. Snitman, and D.S. Garvey, *Angew. Chem. Internat. Edn. Engl.*, **19**, 557 (1980).
39. D. Seebach, H.-O. Kalinowsky, B. Batsani, G. Cross, H. Daum, N. DuPreez, V. Ehrig, W. Langer, C. Nussler, H.-A. Oei, and M. Schmidt, *Helv. Chim. Acta.*, **60**, 301 (1977).
40. Seebach and Langer have examined aldol condensations between achiral reactants in chiral solvents and found observed low asymmetric induction (2-22% enantiomeric excess); D. Seebach and W. Langer, *Helv. Chim. Acta*, **62**, 1701 (1979).

41. Seebach and coworkers have observed double stereodifferentiation in the reduction of prochiral ketones by chiral reducing agents in chiral solvents; M. Schmidt, R. Amstutz, G. Grass, and D. Seebach, *Chem. Ber.*, **113**, 1691 (1980).

42. (a) C.H. Heathcock, M.C. Pirrung, C.T. Buse, J.P. Hagen, S.D. Young, and J.E. Sohn, *J. Am. Chem. Soc.*, **101**, 7077 (1979). (b) C.H. Heathcock, M.C. Pirrung, J. Lampe, C.T. Buse, and S.D. Young, *J. Org. Chem.*, **46**, 2290 (1981).

43. D.A. Evans, J. Bartoli, and T.L. Shih, *J. Am. Chem. Soc.*, **103**, 2127 (1981).

44. C.H. Heathcock, S.D. Young, J.P. Hagen, U. Badertscher, Bai Dong-lu and K. Kees, unpublished work.

45. D.A. Evans, M.D. Ennis, and D.J. Mathre, *J. Am. Chem. Soc.*, in press.

46. A.T. Nielsen and W.J. Houlihan, *Organic Reactions, Volume 16*, John Wiley & Sons, Inc., New York, 1968.

STUDIES DIRECTED TOWARDS VERRUCARINS: A SYNTHESIS OF VERRUCAROL. AN APPROACH TO VERRUCARIN A

Barry M. Trost, Patrick G. McDougal and James H. Rigby

McElvain Laboratories of Organic Chemistry, Department of Chemistry, University of Wisconsin, 1101 University Avenue, Madison, WI 53706, USA

Abstract - An approach to verrucarin A envisions dissection into verrucarol, verrucarinic acid, and a differentiated E,Z-muconate. A strategy embodying the concepts of geminal alkylation via cyclobutanones and modification of oxidation level via sulfenylation reactions generates a seco derivative of verrucarol. A synthesis of verrucarol highlights the use of a Diels-Alder strategy and a diastereotopic differentiation of two carbonyl groups that relies on an ene-retroene sequence. A completely stereocontrolled synthesis of the verrucarinic acid portion and the utilization of preferential conrotatory ring opening of a cyclobutene have set the stage for completion of the intact molecule.

INTRODUCTION

The trichothecanes are a group of sesquiterpenoid mycotoxins which exhibit a wide array of biological activity including significant cytotoxicity. Several members (such as verrucarins A and B and/or roridin A) show <u>in vivo</u> activity against sarcoma 37, Ehrlich ascites tumors in mice, Walker carcinoma in rats, and Yoshida sarcoma.[1] In addition to being among the most cytostatic compounds known, they are also among the most toxic substances lacking nitrogen. These compounds can be structurally divided into two major fragments, the sesquiterpene tricycle and a macrocyclic ring. For example, verrucarin A (**1**) and baccharin can be envisioned to derive from verrucarol (**2**) and an acyclic fragment that bridges the diol unit of verrucarol (in the case of baccharin, a further epoxidation of verrucarol is required). Other related sesquiterpenoid units include anguidine and calonectrin with the former compound being taken through phase II clinical trials for its cancer-fighting potential.

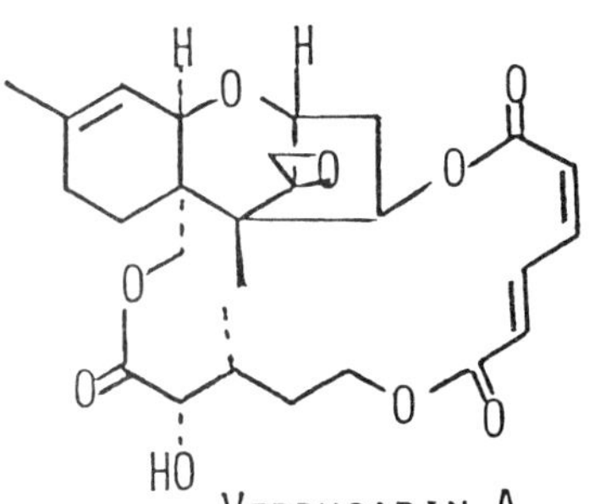

VERRUCARIN A
1

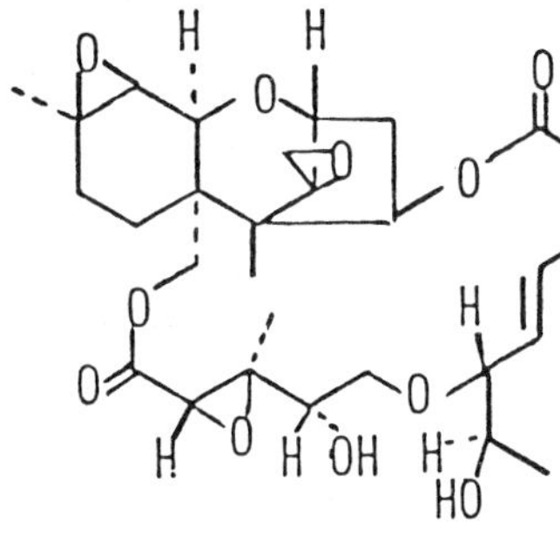

BACCHARIN

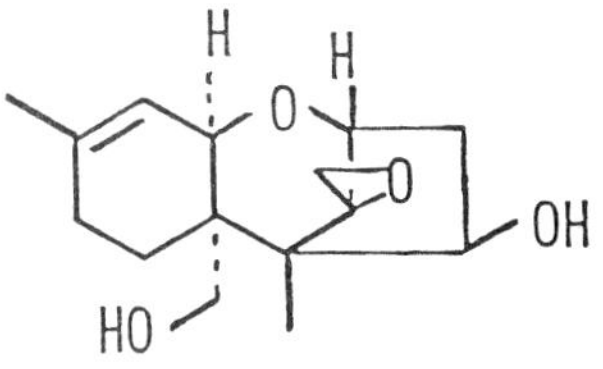

VERRUCAROL
2

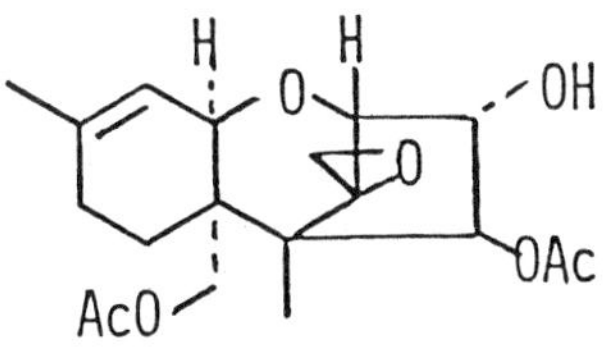

ANGUIDINE

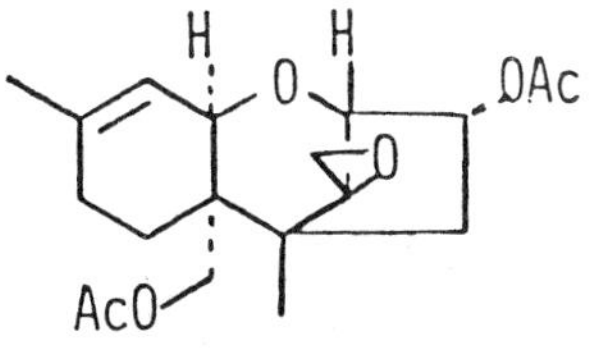

CALONECTRIN

In focusing on verrucarin A as a target, the natural and commonly employed strategy dissects it into verrucarol, verrucarinic acid (3) and a differentiated E,Z-muconic acid derivative 4.[2-6] Focusing initially on verrucarol, the common retrosynthetic analysis identified a

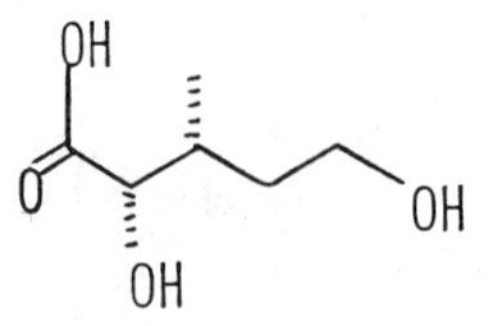

cis-fused tetrahydrochromanone 5 as a logical precursor. An approach to this ring system envisioned capitalizing upon the bias of a bicyclo[4.3.0]nonyl ring system to prefer a cis fusion to fix the requisite stereochemistry and then to effect a ring expansion (eq 1 and 2). Such a ring expansion could be of a pinacol type (eq 1)[5] or migration of a potential nucleophile to a Michael acceptor system as in 6 (eq 2). The latter approach has the advantage that the requisite acetic acid side chain of 5 is an integral part of the synthetic approach.

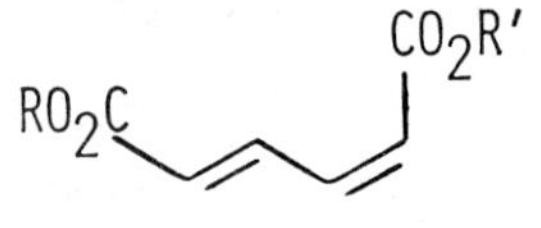

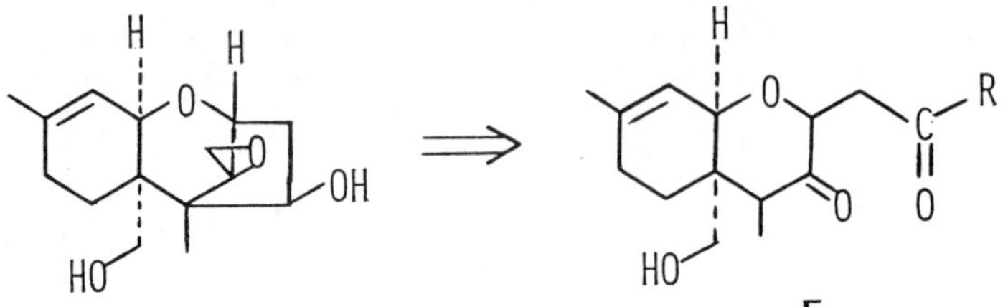

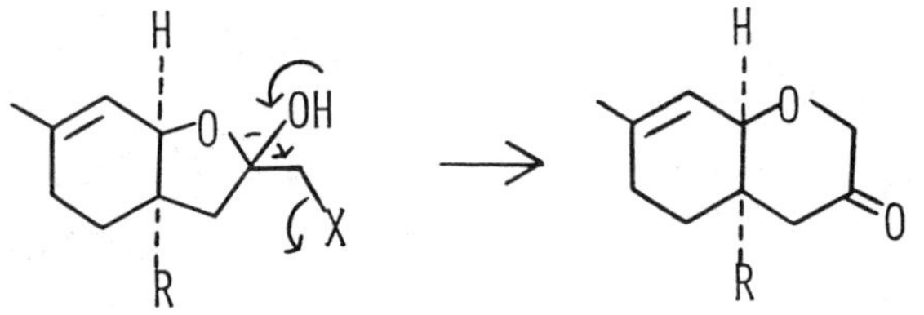

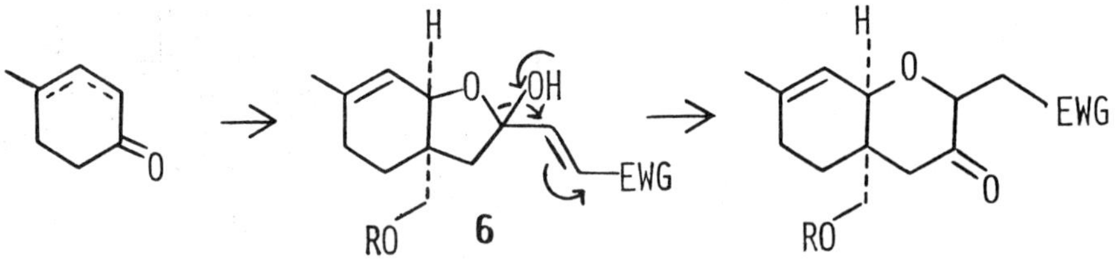

A SYNTHESIS OF A TETRAHYDROCHROMANONE

The bridgehead carbon bearing the hydroxymethyl substituent of the tetrahydrochromanone translates into a problem of geminal alkylation.[5b,7,8] With this perspective in mind, 4-methylcyclohex-2-en-1-one undergoes spiroannulation using 1-lithiocyclopropylphenyl sulfide[9] followed by fluoroboric acid induced rearrangement of the initial adduct to give cyclobutanone 7. Secosulfenylation[7e] of the cyclobutanone via in situ generation of the α,α-bis-sulfenylated cyclobutanone 8 [NaOCH$_3$, PhSSPh, CH$_3$OH] generates the chemodifferentiated geminal substituents as in 9. Lithium aluminum hydride reduction of the ester followed by standard methyl ether formation and aqueous hydrochloric acid hydrolysis of the thioacetal produces 10 which, save for the bridgehead methyl group of verrucarol, requires introduction of a synthon for ⁻CH$_2$CH$_2$CO$_2$R to complete the carbons of verrucarol. The γ-butyrolactone synthesis via cyclobutanones again offers a simple solution to this problem (eq 4).[10] Cyclobutanone annulation to 1 in identical fashion as before followed by chemoseletive Baeyer-Villiger oxidation with basic hydrogen peroxide generates lactone 12. With all of the

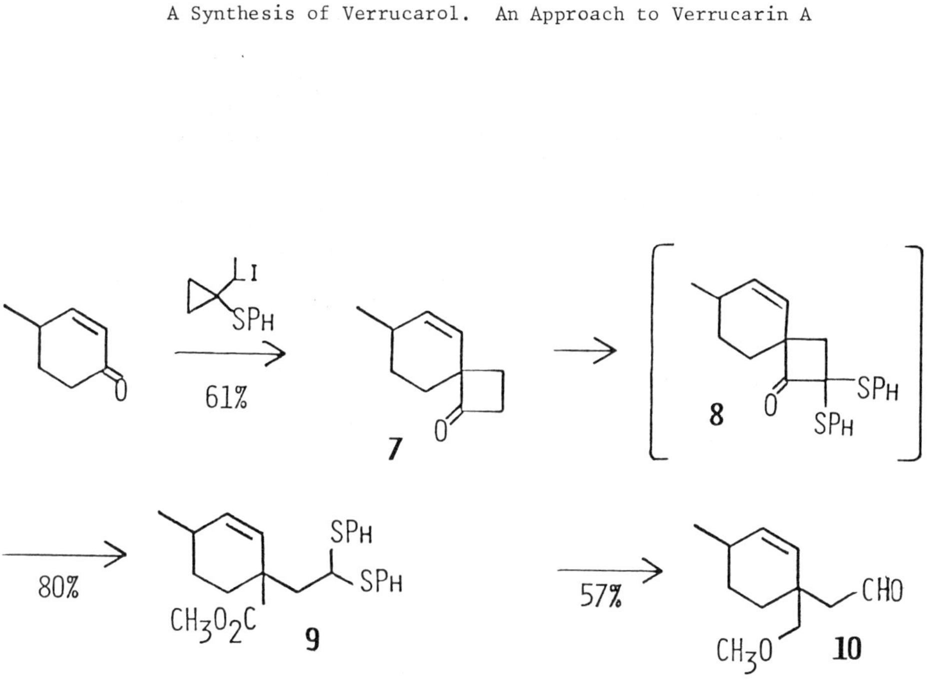

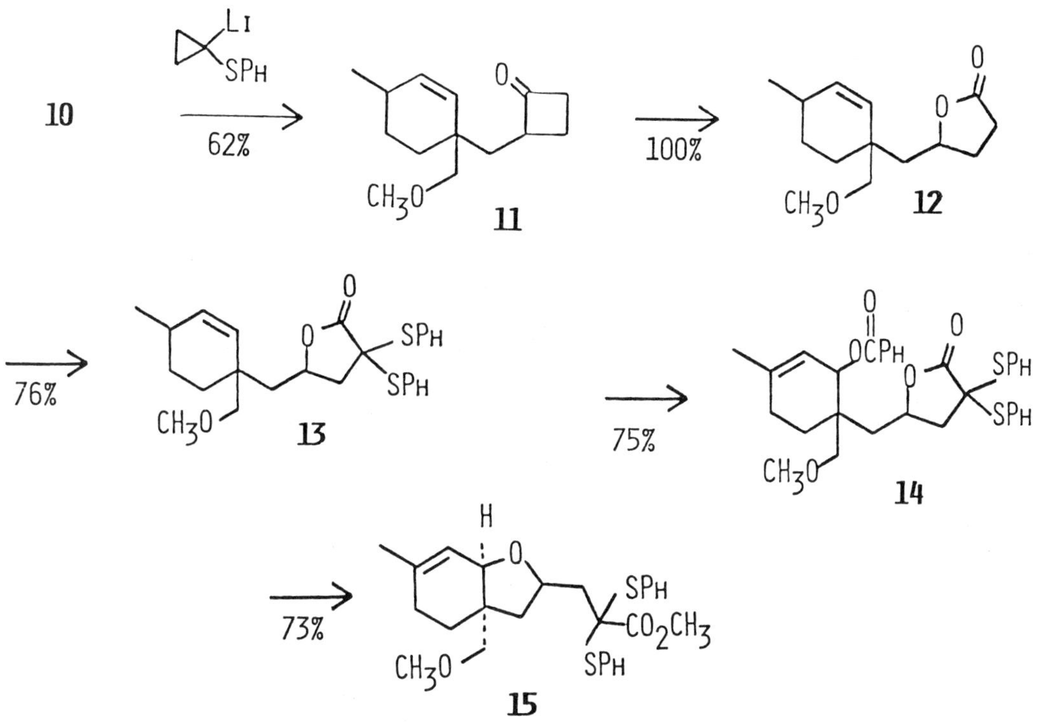

carbons in place, adjustment of the oxidation pattern remains to complete the synthesis of 5 - i.e., allylic oxidation and creation of a γ-hydroxy-α,β unsaturated ester. The latter envisioned taking advantage of the versatility of sulfenylation-dehydrosulfenylation (eq 5) and 2,3-sigmatropic rearrangements of allylic sulfoxides (eq 6). Both reactions were put in place by bis-sulfenylation of the lactone 12 [LDA, PhSSO$_2$Ph, THF]. t-Butyl perbenzoate with catalysis by cuprous bromide effects the chemoselective allylic oxidation of this bis-sulfenylated lactone 13 to allylic benzoate 14 which smoothly solvolyzes to the critical fused bicyclic ether 15 [i) NaOH, H$_2$O, THF; ii) HCl, H$_2$O; iii) CH$_2$N$_2$, ether].

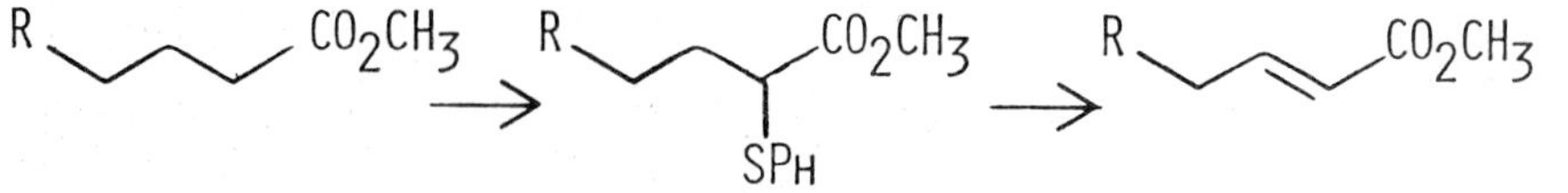

(5)

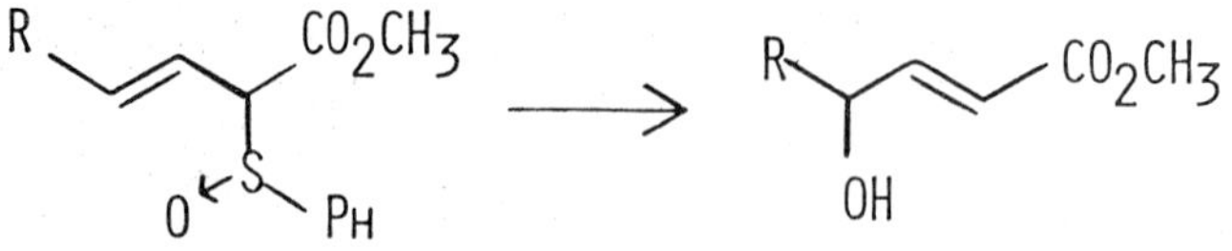

(6)

Timed release of the sulfur substituents sets in motion the sequence of transformations embodied in eq 5 and 6. MCPBA oxidation and warming to 46° in the presence of trimethyl phosphite produces the α-sulfenylated-α,β-unsaturated ester **16**.[11] A second MCPBA oxidation and then warming in the presence of base [DBU, CH_3OH] creates the requisite allylic sulfoxide **17** which, upon its genesis, suffers 2,3-sigmatropic rearrangement to the γ-hydroxylated enoate **18**. Delightfully, the conditions for these transformations are also conducive to the 1,2- oxygen shift such that the product directly isolated from **16** is indeed the sought after tetrahydrochromanone **19**. Most gratifyingly, all of the framework carbons of **19** derive from two sources - 4-methylcyclohex-2-en-1-one and cyclopropylphenyl sulfide.

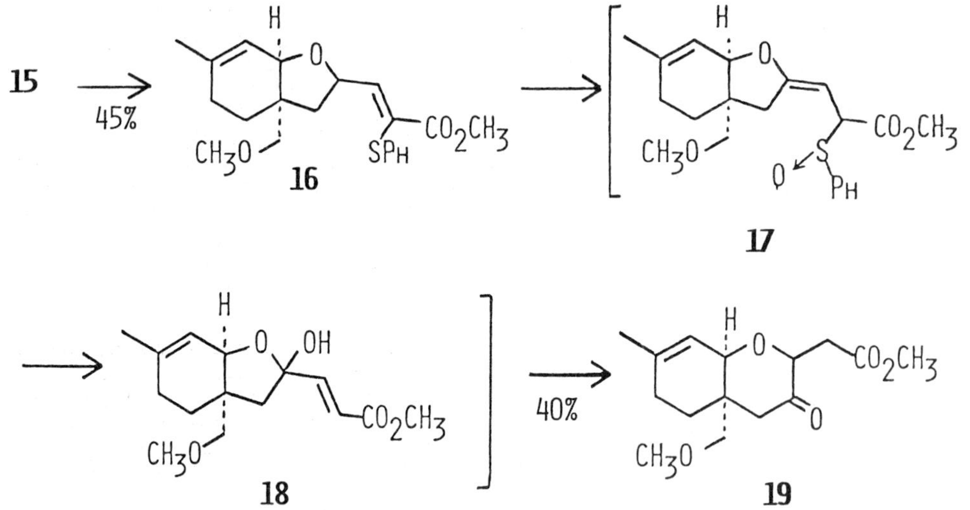

(7)

While we handsomely met our objectives utilizing the small ring and sulfur methodology under development in our laboratories, we were distressed by increasing informal reports regarding the closure of the five membered ring via aldol or related reactions. Our own experience with pilot reactions on **19** supported this concern. Thus, we turned to consideration of utilization of the same intrinsic approach to the tetrahydrochromanone ring system as embodied in eq 1 and 2 but one in which the five membered ring was already present.

AN ALTERNATIVE STRATEGY

The facility of the 1,2-shift as demonstrated by the direct isolation of **19** from **16** validated the use of a pinacol type of approach. Embodying the principle of eq 1 into a precursor bearing a five membered ring leads to **20** as a possibility; whereas, a direct analogy for the sequence represented in eq 2 and demonstrated by the example of eq 7 would invoke **21** as a possibility. Either **20** or **21** would easily derive from the cyclopentan-1,2-dione **22** in which differentiation of the diastereotopic carbonyl groups (labelled A and B in **22**) must be achieved. In **22**, which bears the secondary hydroxyl group

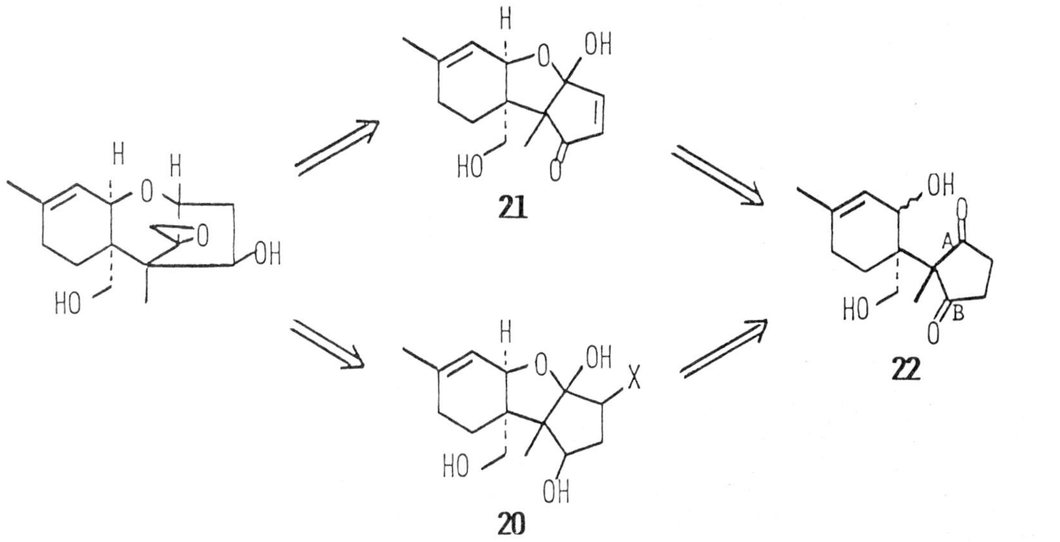

cis to the cyclopentan-1,3-dione unit, a high propensity to exist in a lactol form as
represented by **23** or **24** is anticipated. To the extent that a bias for **23** or **24** occurs, such
a cyclization provides the necessary differentiation since **23** represents selective blocking
of carbonyl group A of **22**, and **24** represents selective blocking of carbonyl group B of **22**.

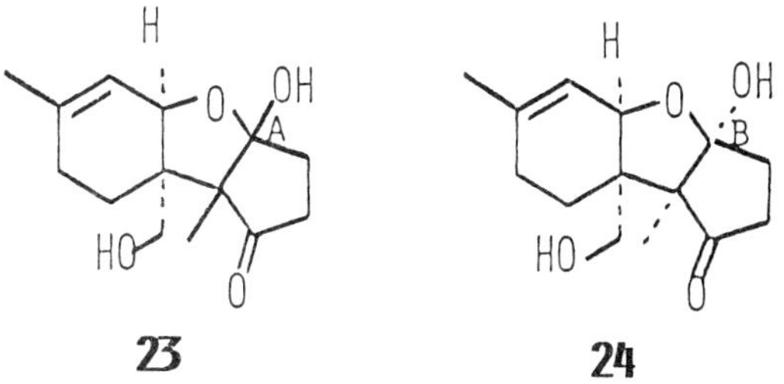

A prejudice in favor of the anti stereochemistry of **23** derives from perceived minimization
of non-bonded interactions. Thus, this approach benefits from resolving the problem of
stereochemistry of the ring fusion of the tetrahydrochromanone system and simultaneously
offering a simple solution to chemodifferentiation. Assuming that the stereochemistry of
the secondary alcohol will derive from preferential solvolysis to the cis fused
hexahydrobenzofuran, a natural strategy for **22** emerges - the Diels-Alder reaction (eq 8).[6]

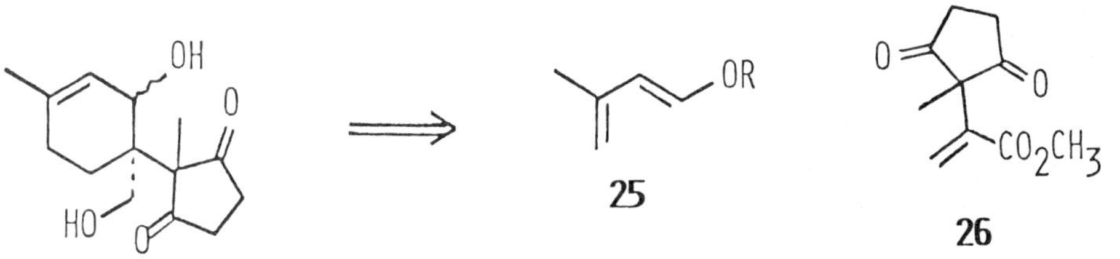

While such a reaction between diene **25** and dienophile **26** is predicted to produce the epi
series, this slight defect can be readily adjusted during the subsequent solvolysis. A
heartening aspect of this strategy is that all of the chirality derives from the Diels-Alder
reaction - a type of reaction that responds well to asymmetric induction.[12]

A VERRUCAROL SYNTHESIS

The synthesis of the unusual acrylate **26** utilizes the versatile 2-methylcyclopentan-1,3-dione, a building block of great importance in steroid synthesis. <u>In situ</u> enol

(9)

ether formation to **27** [TsOH, PhCH$_2$] followed by Claisen rearrangement [mesitylene, reflux] generates the C-allylated product **28**. Permanganate cleavage of the olefin directly to the carboxylic acid [KMnO$_4$, C$_{16}$H$_{33}$N$^+$(CH$_3$)$_3$Br$^-$, CH$_2$Cl$_2$, HOAc, H$_2$O][13] followed immediately by diazomethane esterification and DBU initiated elimination of the chloride [PhH, rt] completes the synthesis of the dienophile **26**, a low melting point solid (mp. 30-32°) that is available on large scale in 48% overall yield from starting dione.

Cycloaddition of **26** with the diene **25** R=Ac fails. Repetition of the reaction with the diene **25** R=TMS,[14] available from the acetate by acyl cleavage with <u>n</u>-butyllithium followed by silylation, in refluxing mesitylene produces the desired adduct **28** in 57% yield in addition to an isomeric by-product (vide infra).

Desilylation with fluoride ion generates the uninverted alcohol **29** which upon trifluoroacetolysis forms a new alcohol **30** whose isomeric nature with **29** was proven by oxidation of both **29** and **30** to the same triketone **31**. Methanolysis of **30** produces a >9:1 mixture of two lactol ethers; the identical mixture also derives by direct methanolysis of **28**. Surprisingly, the major isomer appears to be the <u>cis,syn,cis</u> stereochemistry in **32** rather than the <u>a priori</u> more logical <u>cis,anti,cis</u> depicted in **33**. Support for this assignment derived from the sodium borohydride reduction of **32** to a hydroxy ester which would not lactonize. While lactonization would be favored for the <u>cis,anti,cis</u> isomer, it is not possible for the <u>cis,syn,cis</u> isomer. While this preferential formation of **32** does provide the necessary chemodifferentiation, application of a pinacol type of rearrangement as represented by the conversion of **21** to **22** would produce the incorrect stereochemistry. To obviate the latter, the synthesis of the free lactol is required to permit, at a minimum, an interconversion of the <u>cis,anti,cis</u> and <u>cis,syn,cis</u> isomers such as **34** and **35** so that a preferential migration could be hoped for from **34**.

The lactol corresponding to **30**, <u>i.e.</u> **36**, which is available in 76% yield from the Diels-Alder adduct, sulfenylates (LDA, THF, HMPA, PhSSPh)[15] and dehydrosulfenylates (DBU, HgCl$_2$, THF) to generate what appears from NMR spectroscopy to be a single compound, either **34** or **35**. Unfortunately, all attempts to effect rearrangement of **34** and/or **35** fail.

Assessing the failure of the rearrangement as due to poor overlap between the migrating group and the migration terminus as depicted in **37**, we anticipate that the alternate alignment available in **38** should overcome this deficiency.

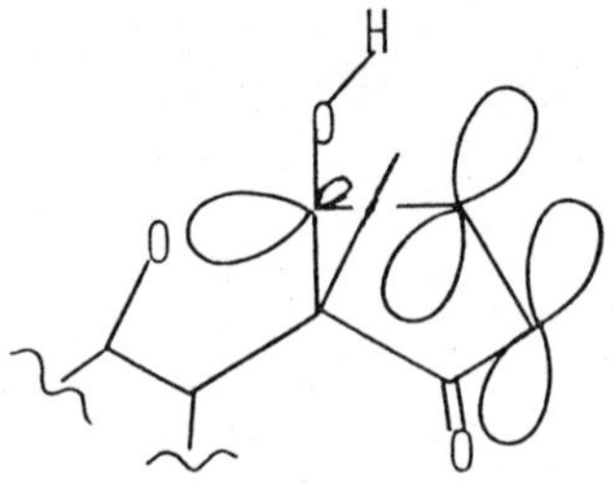

37

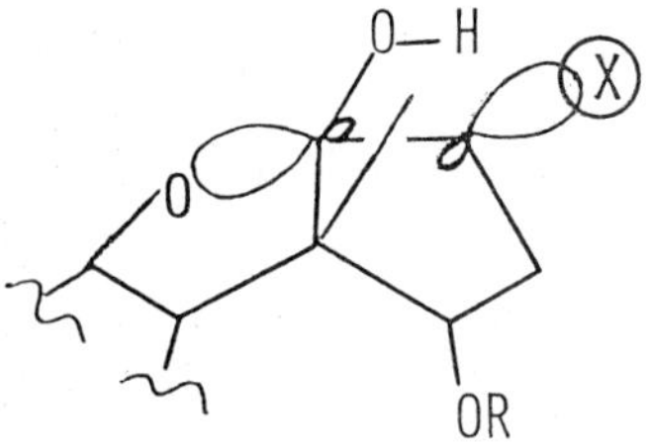

38

Juxtaposing the appropriate functionality correctly for **38** from **32** appeared somewhat cumbersome. At this point, we returned to the isomeric by-product obtained in the initial Diels-Alder reaction. Based upon the spectral data, this structure was deduced to be the tricyclic ketone **39**. The retrogression of **39** to the starting Diels-Alder adduct provides

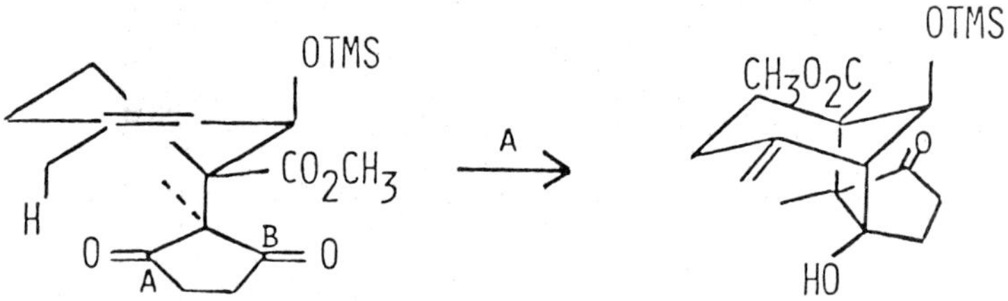

(10)

39

further structural support. This compound formally derives from an unusual intramolecular ene reaction[16] of the carbonyl group labelled A in eq 10. Such an adduct provides the chemodifferentiation of the two carbonyl groups of the Diels-Alder adduct required. Since this process can be reversed thermally, the ene-retroene represents a novel blocking method.

Reduction of **39** (NaBH$_4$, CH$_3$OH, rt) is accompanied by lactonization; cycloreversion subsequently unmasks the cyclopentanone **40** that is needed to generate the correct intermediate for the pinacol-type rearrangement. Introduction of the nucleofugal group was

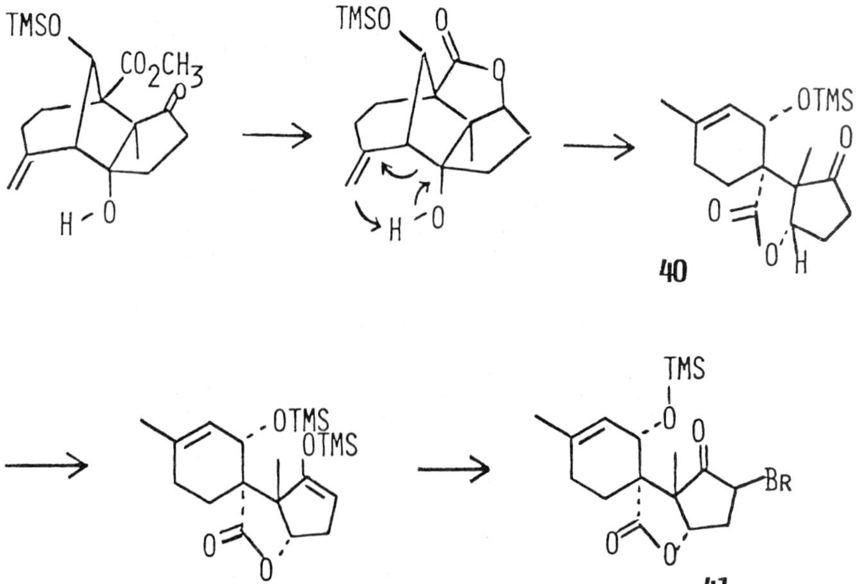

40

41

best achieved by bromination (Br$_2$·dioxane, CH$_2$Cl$_2$) of the enol silyl ether derived from **49** which produces the key intermediate **41** bearing the requisite leaving group in the correct stereochemistry.

Deblocking the alcohol with trifluoroacetic acid at 0° followed by fluoride ion [(C$_4$H$_9$)$_4$NF, THF] leads directly to the tricyclic ring skeleton of isoverrucarol in 62% yield from **40**. That epimerization of the alcohol did not accompany the trifluoroacetolysis was suggested by the lack of observable coupling between the proton at C(11) and the vinyl proton in either the lactone **42** or its further transformation product, the diol **43**. Confirmation that these

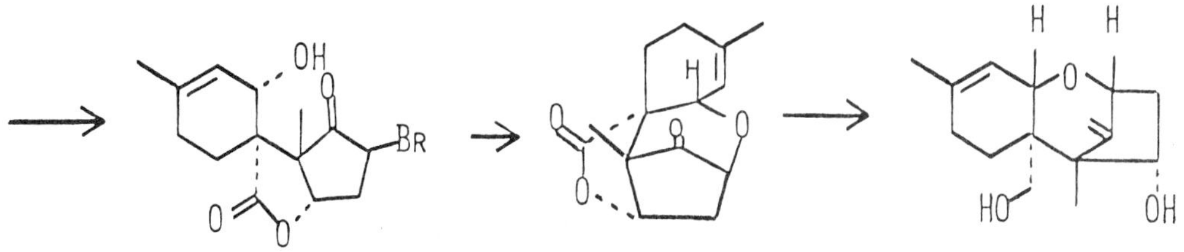

products derived from the 11-epi series arose in an x-ray structure determination of **42**. The fact that forcing the tetrahydropyran ring into a boat does not deter the cyclization attests to the facility of this approach to the verrucarol system.

Nevertheless, total stereochemical control can be exerted. Trifluoroacetolysis for a prolonged time [32° to 55°] leads to a lactol assigned structure **44**. While typical bases failed to effect the desired pinacol process, tetra-n-butylammonium fluoride treatment of **44** does indeed generate the sought-after ring system embodied in **45** in 70% overall yield from **41**. In this case, while no observable coupling could be seen between the proton at C(11)

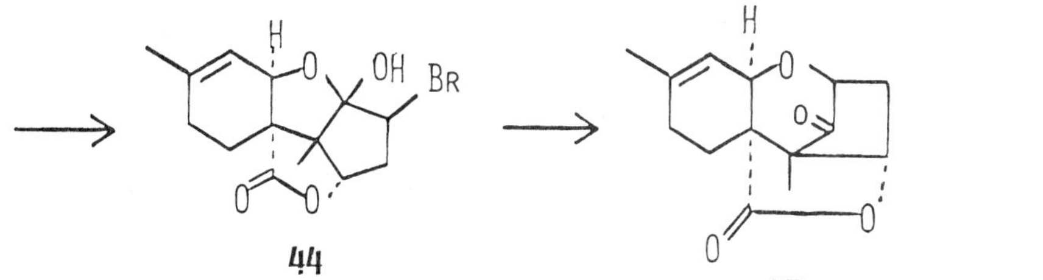

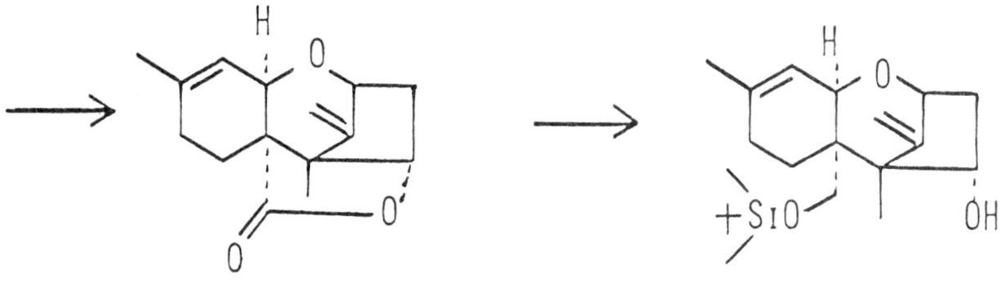

and the vinyl proton, the further transformation product **47** does show a 5.5 Hz coupling in complete accord with natural verrucarol. The transformation of the keto ester **45** to the monosilyl ether of the diol **47** followed standard methylenation [Ph$_3$PCH$_2$, LiBr, THF, 60°, 95%], reduction [DIBAL-H, PhCH$_3$, rt, 95%], and silylation [t-C$_4$H$_9$Si(CH$_3$)$_2$Cl, C$_3$H$_4$N$_2$, DMF, 42°, 82%].

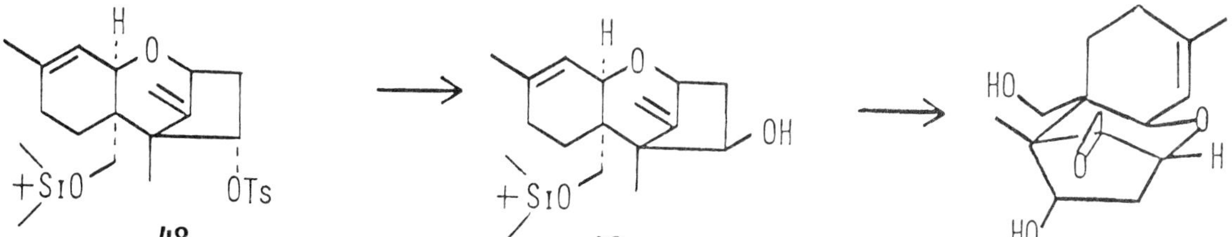

The final leg of this journey requires epimerization at C(4) and chemoselective epoxidation. The resistance of the alcohol to normal inversion procedures proved frustrating. It finally succumbed by converting **47** to the tosylate **48** [TsCl, C$_5$H$_5$N, 34°] followed by displacement with propionate anion [CsO$_2$CC$_2$H$_5$, 1,3-dimethyl-2-imidazolidinone, 150°][17] and hydrolysis [K$_2$CO$_3$, CH$_3$OH, rt] to give **49** with the product of elimination **50** as a co-equal by-product.

In the only extant synthesis of the verrucarol family, the trisubstituted double bond required protection to achieve selective epoxidation of the exocyclic double bond.[4] Such proved unnecessary in the case of **49** since it stereo- and chemoseletively epoxidized using t-butylhydroperoxide catalyzed by molybdenum hexacarbonyl to produce the mono-silyl ether of verrucarol in 85% yield and identical to an authentic sample prepared by chemoselective silylation of verrucarol (except for optical activity).[18] Fluoride initiated desilylation completes the synthesis and formation of racemic verrucarol, mp. 165.5-167°, identical spectrally and chromatographically with natural verrucarol, except for rotation.

This approach provides the trichothecane and the 11-epi-trichothecane ring systems in 11 and 10 steps respectively from the commercially available 2-methylcyclopentan-1,3-dione. In addition, the triene **50** offers an opportunity into other members of this family, the most important being anguidine. The use of the silyl ether derivative of the primary alcohol not only provides a basis for a highly chemoselective epoxidation but also for the ultimate introduction of the macrocycle.

SYNTHESIS OF VERRUCARINIC ACID AND DERIVATIVES

Having available the differentiated verrucarol piece, attention now shifted to the macrocycle, whose components are verrucarinic acid **3** and E,Z-muconic acid **4** (R=R'=H). In designing the synthesis of verrucarinic acid, its availability in optically active form was deemed essential in a convergent synthesis of the type underway. The first strategy recognized the possible advantage of a meso compound such as 1-cyclopenten-3,5-diol since mono-acylation of alcohol A generates the enantiomer that derives from acylation of alcohol B.[19] If a chiral ester is employed **51a** and **b** become diastereomers which could be separated, re-equilibrated, and separated – the process repeated or operated continuously so that a single enantiomeric series would ultimately result – an enantioconvergence that avoids the inefficiency of simple resolution.

With this concept in mind, the monobenzoate **52** was subjected to coupling with lithium dimethylcuprate (ether, -20°) to **53** contaminated by the product derived from allyl inversion and then to **54** [MCPBA, CH$_2$Cl$_2$, 0° then NaH, PhCOCl, ether, rt] in 53% overall yield from **52**.

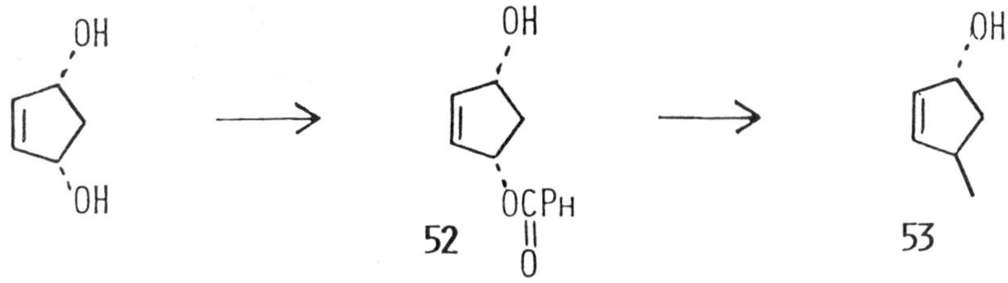

Regioselective opening of the epoxide with lithium thiophenoxide (CH$_3$CN, ether) proceeded exclusively in 75% yield to a single hydroxysulfide which was silylated [t-C$_4$H$_9$(CH$_3$)$_2$SiCl, C$_3$H$_4$N$_2$, DMF] and debenzoylated [DIBAL-H, ether, -78°] to give the hydroxy sulfide 55 in 90% yield.

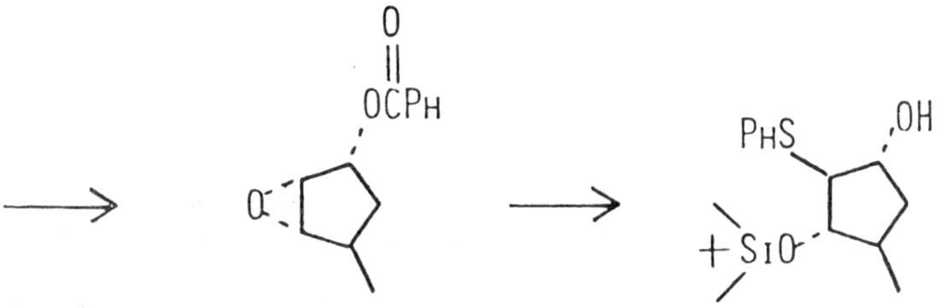

Oxidative cleavage of hydroxysulfides with lead tetraacetate offers an approach to C-C bond cleavage in which the resultant α,ω-dialdehyde remains differentiated (eq 11).[20] Combining

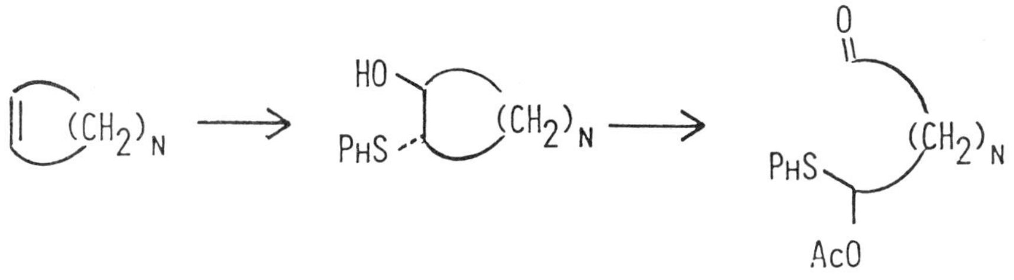

N = 1, 4, 8

this cleavage reaction with the direct hydroxysulfenylation of olefins provides a useful olefin cleavage in which chemodifferentiation is maintained.[20b]

Lead tetraacetate [C$_5$H$_5$N, PhH, reflux, 10 min] effects cleavage of the hydroxysulfide 55 to the mono-protected dialdehyde 56 in 90% yield. DIBAL-H [PhCH$_3$, THF, -78°] selectively reduces the free aldehyde which, upon work-up, deblocks the remaining aldehyde to produce the lactol 57 in 85% yield. Oxidation with the pyridine complex of chromium trioxide generates the mono-silyl ether of verrucarinolactone 58 (90% yield) whose spectral properties compare favorably to verrucarinolactone[21] itself.

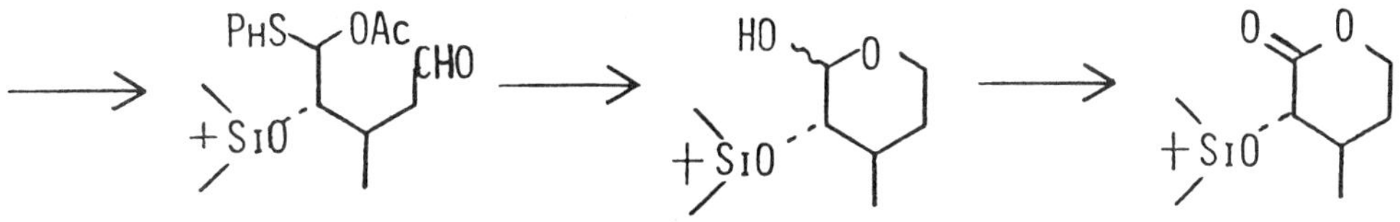

The key oxidative sequence 55 to 58 proceeded exceptionally well (overall 69%); nevertheless, the cuprate coupling **52 → 53** proved tedious due to the formation of the regioisomeric coupling product. To obviate this regioselectivity problem, the coupling product of diethyltrimethylsilylethynylalane with methyl E-2,3-epoxybutanoate **59**, obtained in 63% yield, was envisioned as the source of the verrucarinic acid portion, provided the absolute stereochemistry could be resolved. Subjection of the O-methylmandelate esters **60** and **61** to

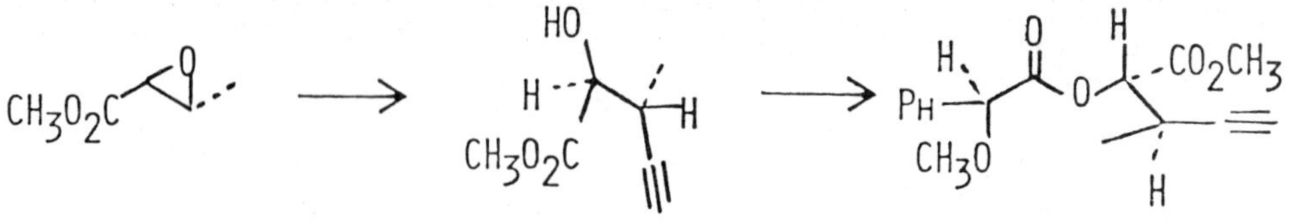

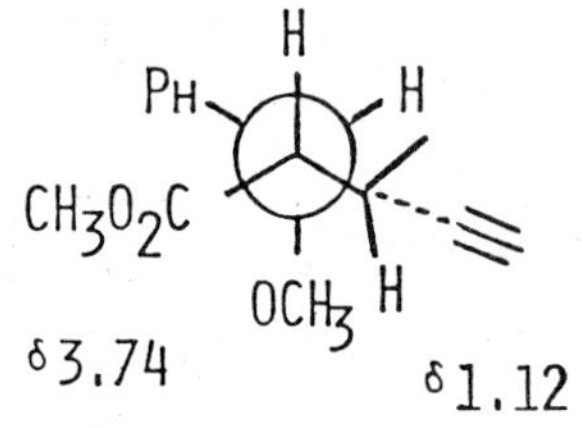

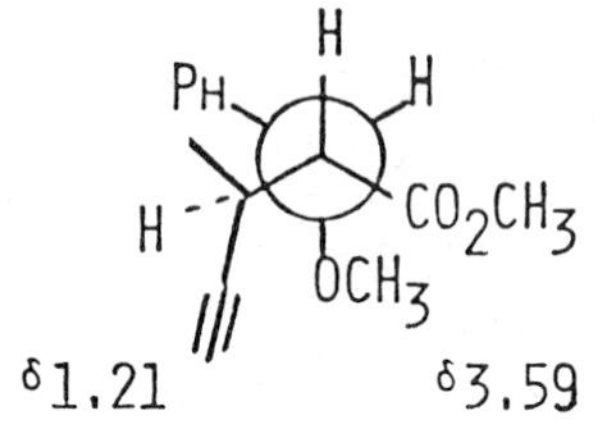

preparative HPLC not only provides the physical resolution but also a method for determination of the absolute configuration.[23] Using the conformations depicted in the Newman projections, the anisotropy of the phenyl group provides a differential deshielding of the ester group in **60** and the secondary methyl group in **61**.[24] This model has proven general and provides an excellent general solution to resolution of secondary alcohols.[25] Conversion of the terminal acetylene to the primary alcohol involved first catalytic reduction using Lindlar catalyst (96% yield) followed by hydroboration [disiamylborane,THF] - oxidation [MCPBA, THF][26] and silylation which required slightly forcing conditions [t-C$_4$H$_9$(CH$_3$)$_2$SiCl, DMAP, CH$_2$Cl$_2$] to give **62** in 89% overall yield. Replacement of the mandelate unit by a silyl group involved methanolysis in the presence of methoxide followed

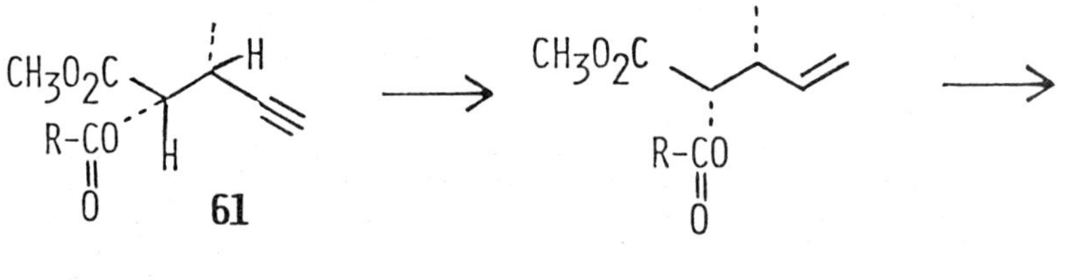

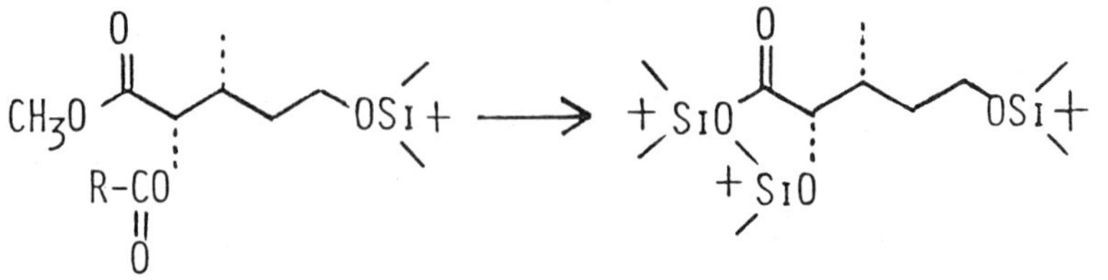

by standard silylation. Since the free carboxylic acid corresponding to **63** made the adjacent silyl ether labile, the corresponding silyl ester which can be directly employed in acylations was prepared by dealkylation [LiSCH$_3$, HMPA] and standard silylation in quantitative yield from **62**. The rotations for optically pure **62** and the methyl ester corresponding to **63** are +5.66 (c 1.13, acetone) and -13.09 (c 1.10, acetone) respectively.

SYNTHESIS OF DIFFERENTIATED E,Z-MUCONIC ACIDS AND MODEL MACROTRIOLIDES

Orienting the E,Z-muconic ester represents the key aspect for this portion of the molecule. While most approaches focussed upon specific synthesis of the differentiated unit,[2,27] the symmetry embodied in cyclobutene-3,4-dicarboxylic ester offers an alternative[28] - the key question being which of the two conrotatory processes illustrated in eq 12 will be realized.

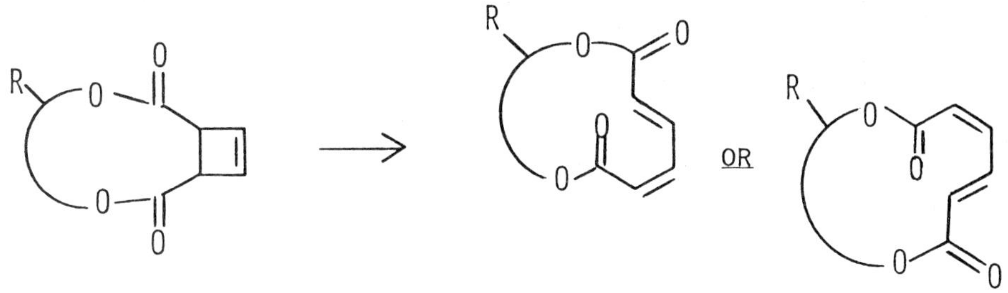

(12)

Selectivity in the opening of the mono ester of cis-cyclobutene-3,4-dicarboxylic acid has a remarkable, and, at this time, inexplicable solvent dependency (eq 13). Subjecting

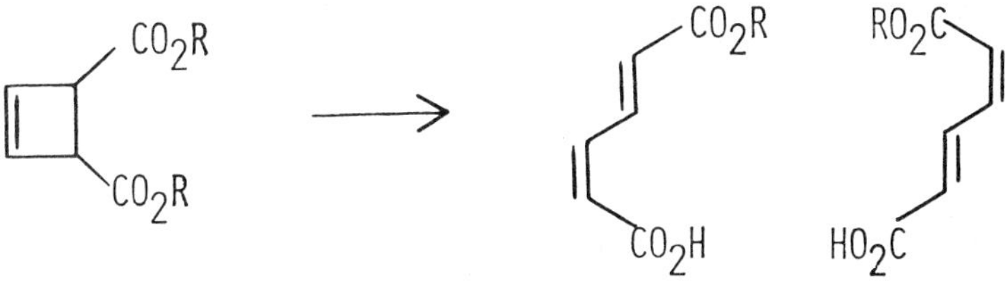

(13)

R = C_4H_9	DMSO	110°	1 : 1
	DME	83°	1.2 : 1
	CL$\sim$CL	84°	3 : 1
R = CH_2CH_2 TMS	CL$\sim$CL	84°	3 : 1

monoester **64** to tetra-n-butylammonium fluoride [DME, 72°] effects desilylation and simultaneous cyclobutene ring opening to give only the desired E,Z-muconate **65**, mp. 125.5-129° (eq 14) albeit in 39% yield. This result stems from selective destruction of the alternative isomer. To complete one strategy in this model system, the acid was esterified using alkylation techniques [I(CH$_2$)$_4$COSC$_2$H$_5$, DBU, CH$_3$CN][29] and cyclized [NBS, CH$_2$Cl$_2$, rt] to give the model macrotriolide **66**.

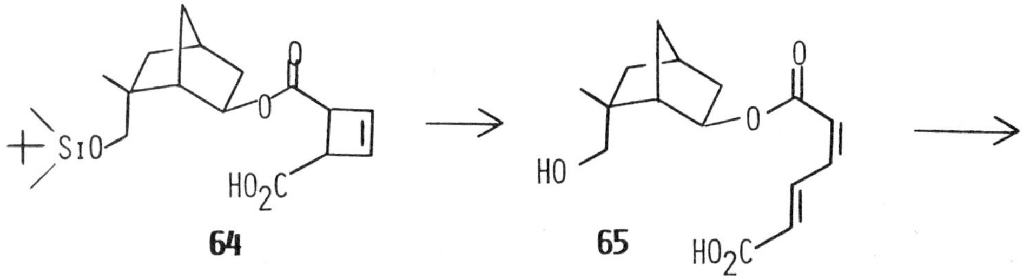

64   **65**

(14)

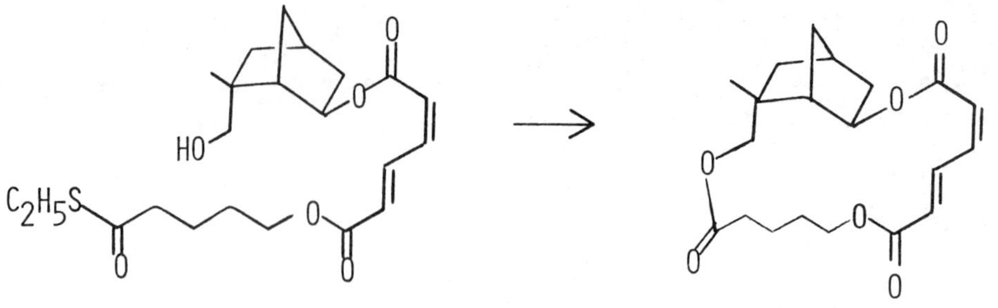

A more fascinating approach invokes the selective opening of triolide **69**. Selective esterification of the primary alcohol of **67** with the t-butyldimethylsilyl ester of 4-hydroxypentanoic acid [DCC, DMAP] followed by esterification of the secondary alcohol by cyclobutene-3,4-dicarboxylic anhydride [DME, DMAP] and desilylation [$(C_4H_9)_4NF$, THF] gives an 80% overall yield of the triolide precursor **68**. Efficient esterification occurs with

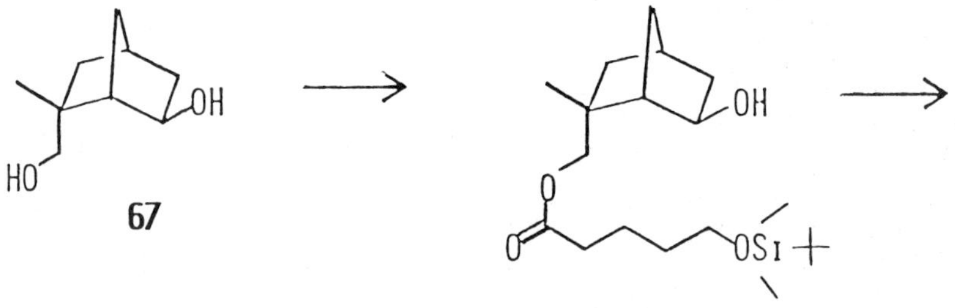

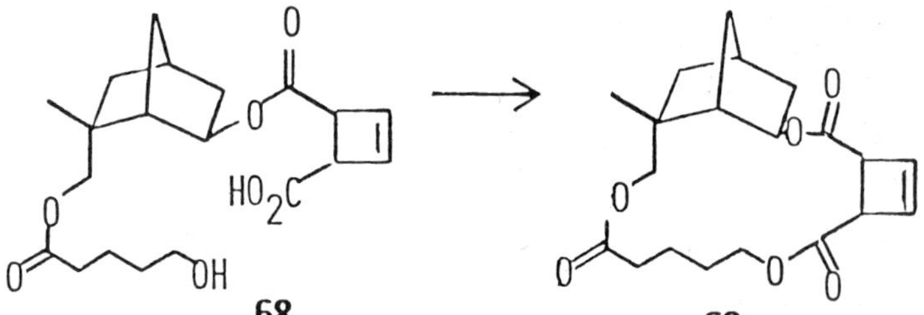

diethyl azodicarboxylate and triphenylphosphine [PhH, rt, 71%] to the triolide **69**. Thermolysis of **69** at 106° gives a 2:1 ratio of the two macrotriolide dienes in which the major isomer was identical to the previously obtained macrocycle. Most encouraging, a bias does exist for opening to the correct E,Z-muconate. The question of the ability of the additional substituents to enhance that bias remains to be answered.

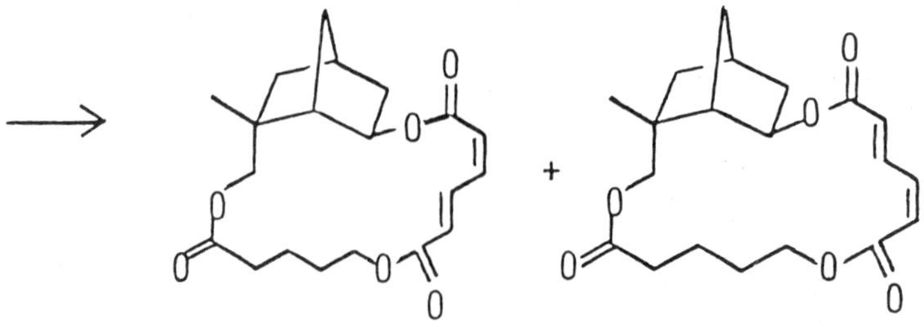

CONCLUSIONS

While the synthesis of verrucarin A must await the application of these later concepts to verrucarol, the pieces are in place. The trichothecane synthesis also offers further opportunities for development. For example, the asymmetric induction in the Diels-Alder reaction remains to be examined. The successful short synthesis of verrucarol not only provides a practical approach to this important family but also highlights the novel use of an intramolecular ene reaction as a diastereotopic differentiation and temporary blocking group - a strategy that might find wider applicability in organic synthesis.

ACKNOWLEDGMENT

We are grateful for the continuing support of the National Cancer Institute of the National Institutes of Health for their generous support of our programs. We are especially grateful for collaboration with Dr. Ken Haller on the x-ray structural determination of the 11-epi series.

REFERENCES

1. Reviews: C. Tamm, Fortsch. Chem. Org. Naturst. 31, 63 (1974). J.R. Bamberg, F.M. Strong, in "Microbial Toxins;" S. Kadis, Ed.; Academic Press: New York, 1973; Vol. 3 pp 207-292. T.W. Doyle, W.T. Bradner, in "Anticancer Agents Based on Natural Product Models;" J.M. Cassidy, J. Douros, Eds.; Academic Press: New York, 1980; Chapter 2.
2. W.C. Still, H. Ohmizu, J. Org. Chem. 46, 5242 (1981).
3. For some recent approaches see: (a) E.W. Colvin, S. Malchenko, R.A. Raphael, J.S. Roberts, J. Chem. Soc., Perkin Trans. 1, 658 (1978). (b) B.M. Trost, J.H. Rigby, J. Org. Chem. 43, 2938 (1978). (c) W.R. Roush, T.E. D'Ambra, Ibid. 45, 3927 (1980). (d) J.D. White, T. Matsui, J.A. Thomas, Ibid. 46, 3376 (1981). (e) A.J. Pearson, C.W. Ong, J. Am. Chem. Soc. 103, 6686 (1981), and references cited therein.
4. (a) E.W. Colvin, S. Malchenko, R.A. Raphael, J.S. Roberts, J. Chem. Soc., Perkin Trans. 1, 1989 (1973). (b) W.C. Still, M.-Y. Tasi, J. Am. Chem. Soc. 102, 3654 (1980). (c) R.H. Schlessinger, R.A. Nugent, J. Am. Chem. Soc. 104, 1116 (1982). (d) G.A. Kraus, B. Roth, K. Frazier, M. Shimagaki, J. Am. Chem. Soc. 104, 1114 (1982).
5. The 1-oxa[3.2.1]bicyclooctane skeleton has been formed by such a rearrangement. However, the migrating oxygen in both cases was a phenol and not a simple alcohol. W.K. Anderson, G.E. Lee, J. Org. Chem. 45, 501 (1980). D.J. Goldsmith, T.K. John, C.D. Kwong, G.R. Painter III, Ibid. 45, 3989 (1980).
6. The Diels-Alder has become a popular entry to the A-ring of the trichothecanes, see: (a) B.B. Snider, S.G. Amin, Synth. Commun. 8, 117 (1978). (b) Y. Nakahara, T. Tatsuno, Chem. Pharm. Bull. 28, 1981 (1980). (c) G.A. Kraus, K.J. Frazier, J. Org. Chem. 45, 4820 (1980). (d) G.A. Kraus, B.J. Roth, Ibid. 45, 4825 (1980). (e) R.E. Banks, J.A. Miller, M.J. Nunn, P. Stanley, T.J.R. Weakly, Z. Ullah, J. Chem. Soc., Perkin Trans. 1, 1096 (1981).
7. (a) B.M. Trost, M.J. Bogdanowicz, J. Am. Chem. Soc. 95, 2038 (1973). (b) B.M. Trost, M. Preckel, ibid. 95, 7862 (1973). (c) B.M. Trost, M.J. Bogdanowicz, J. Kern, ibid. 97, 2218 (1975). (d) B.M. Trost, M. Preckel, L. Leichter, ibid. 97, 2224 (1975). (e) B.M. Trost, J. Rigby, J. Org. Chem. 41, 3217 (1976).
8. For a review see: B.M. Trost, Pure Appl. Chem. 43, 563 (1975); Accounts Chem. Res. 7, 85 (1974).
9. B.M. Trost, D.E. Keeley, H.C. Arndt, J. Rigby, M.J. Bogdanowicz, J. Am. Chem. Soc. 99, 3080 (1977). B.M. Trost, D.E. Keeley, H.C. Arndt, M.J. Bodganowicz, J. Am. Chem. Soc. 99, 3088 (1977).
10. B.M. Trost, M.J. Bogdanowicz, J. Am. Chem. Soc. 95, 5321 (1973).
11. B.M. Trost, Accounts Chem. Res. 11, 453 (1978).
12. For use of chiral dienes see B.M. Trost, D. O'Krongly, J.L. Belletire, J. Am. Chem. Soc. 102, 7595 (1980). S. David, A. Lubineau, A. Thieffry, Tetrahedron, 34, 299 (1978). S. David, J. Eustache, J. Chem. Soc. Perkin Trans. 1, 2230 (1979). S. David, J. Eustache, A. Lubineau, Ibid. 1795 (1979). A. Korolev, V. Mur, Dokl. Akad. Nauk. S.S.S.R. 59, 251 (1948). Chem. Abstr. 42, 6776 (1949). Most work involves chiral dienophiles. See: R.K. Boeckmann, Jr. P.C. Naegeby, S.D. Arthur, J. Org. Chem. 45, 754 (1980). E.J. Corey, H.E. Ensley, J. Am. Chem. Soc. 97, 6908 (1979). J. Jurczak, M. Tracy, J. Org. Chem. 44, 3347 (1979). S. Hashimoto, N. Komeshima, K. Koga, Chem. Commun. 437 (1979). R.F. Farmer, J. Hamer, J. Org. Chem. 6359 (1966). H.M. Walborsky, L. Barash, T.C. Davis, Tetrahedron, 19, 2333 (1963).
13. A.P. Krapcho, J.R. Larson, J.R. Eldridge, J. Org. Chem. 42, 3749 (1977). D.G. Lee, V.S. Chang, Ibid. 43, 1532 (1978).
14. A. Rosner, K. Tolkiehn, K. Krohn, J. Chem. Res. (M), 3831 (1978). We have found that the siloxydiene 5b is most easily obtained by treating the acetoxydiene 4a with 2.1 eq of nBuLi at $-78°$ and quenching the yellow solution with trimethylsilyl chloride at $0°$.

15. B.M. Trost, T.N. Salzmann, K. Hiroi, <u>J. Am. Chem. Soc</u>. <u>98</u>, 4887 (1976). For a review see: B.M. Trost, <u>Chem. Rev</u>. <u>78</u>, 363 (1978).
16. For a recent review of intramolecular ene reactions see: W. Oppolzer, V. Snieckus, <u>Angew. Chem. Int. Ed. Eng</u>. <u>7</u>, 476 (1978). Intramolecular ene reactions where a ketone serves as the enophile are rare; see: M. Niva, M. Iguchi, S. Yamamura, <u>Bull. Chem. Soc. Japan</u>, <u>49</u>, 3148 (1976); P.A. Wender, J.C. Hubbs, <u>J. Org. Chem</u>. <u>45</u>, 365 (1980). B.B. Snider, E.A. Deutsch, <u>J. Org. Chem</u>. <u>47</u>, 745 (1982).
17. W.H. Kruizinga, B. Strytveen, R.M. Kellog, <u>J. Org. Chem</u>. <u>46</u>, 4321 (1981).
18. D.B. Tulshian, B. Fraser-Reid, <u>Tetrahedron Lett</u>. <u>21</u>, 4549 (1981).
19. Cf. S. Terashima, S. Yamada, <u>Tetrahedron Lett</u>. 1001, (1977). S. Miura, S. Kurozumi, T. Torie, T. Tanaka, M. Kobayashi, S. Matsubara, S. Ishimoto, <u>Tetrahedron</u>, <u>32</u>, 1893 (1976).
20. (a) B.M. Trost, K. Hiroi, <u>J. Am. Chem. Soc</u>. <u>97</u>, 6911 (1975). (b) B.M. Trost, M. Ochiai, P.G. McDougal, <u>Ibid</u>. <u>100</u>, 7103 (1978).
21. J. Gutzwiller, Ch. Tamm, <u>Helv. Chim. Acta</u>, <u>48</u>, 157 (1965).
22. Cf. P.A. Bartlett, J. Myerson, <u>J. Am. Chem. Soc</u>. <u>100</u>, 3950 (1978).
23. (a) J.A. Dale, H.S. Mosher, <u>J. Am. Chem. Soc</u>. <u>95</u>, 512 (1973). (b) B.M. Trost, D. Curran, <u>Tetrahedron Lett</u>. <u>22</u>, 4929 (1981).
24. A different conformation leading to predictable shielding effects by the phenyl ring leads to the same conclusions. See ref 23a.
25. B.M. Trost, <u>ACS Symposium Series</u>, <u>185, 1</u> (1982).
26. J.R. Johnson, M.G. VanCampen Jr, <u>J. Am. Chem. Soc</u>. <u>60</u>, 121 (1938).
27. D.B. Tulshian, B. Fraser-Reid, <u>J. Am. Chem. Soc</u>. <u>103</u>, 474 (1981). D.J. White, J.P. Carter, H.S. Kezar III, <u>J. Org. Chem</u>. <u>47</u>, 929 (1982).
28. W. Hartmann, <u>Chem. Ber</u>. <u>102</u>, 3974 (1969). G. Kollzenburg, P.G. Fuss, J. Leitich, <u>Tetrahedron Lett</u>. 3409 (1966). H. Scharf, J. Mattay, <u>Annalen</u>, 772 (1972).
29. C.G. Rao, <u>Org. Prep. Proc. Int</u>. <u>12</u>, 225 (1980).

NEW APPROACHES TO THE TOTAL SYNTHESIS OF BIOLOGICALLY ACTIVE NATURAL PRODUCTS

Michael E. Jung

Department of Chemistry, University of California, Los Angeles, California 90024, USA

<u>Abstract</u> - New concepts will be presented which employ intra- and intermolecular Diels-Alder reactions and electrocyclic rearrangements in the synthesis of several biologically active natural products of great structural diversity. Among the various target molecules are the steroids (cortisone and estrone), the anthracyclines and their analogues (aclacinomycin, collinemycin, etc.), and terpenes such as β-cuparenone and coronafacic acid. In particular, a new approach to estrone and cortisone via an intramolecular Diels-Alder reaction is described.

INTRODUCTION

Of all of the varied ways to construct organic molecules, few methods of carbon-carbon bond formation can match cycloadditions and electrocyclic rearrangements for their efficiency and their regiochemical and stereochemical control in the formation of new relative asymmetric centers in the molecule. For example, the Diels-Alder reaction not only results in the formation of two new carbon-carbon bonds, but also allows one the opportunity to control the stereochemistry of all of the new asymmetric centers. The Cope and Claisen rearrangements and their variants offer the possibility of new carbon-carbon bond formation with transfer of asymmetry at one center to another. The inherent elegance of these basic methods led us about five years ago to initiate a program concerned with the development of new general methodology for the facile construction of natural products of wide structural diversity utilizing as the key step both inter- and, in particular, intramolecular cycloadditions and facile electrocyclic rearrangements. As a special goal, we hoped to develop new dienes for the Diels-Alder reactions which would have higher reactivity or offer better regiochemical control than those available.

PERCHLORINATED DIENES

One of our initial goals in this broad program was the development of a general method for a process we termed functionalized <u>three-carbon annulation</u> (1), namely the attachment of a three-carbon unit to two adjacent carbons of a cyclic or acyclic precursor to form a functionalized cyclopentane ring. No good general method existed for this transformation, although the two corresponding ones - [2 + 2 → 4] (photochemical cycloaddition of two olefins) and [2 + 4 → 6] (Diels-Alder reaction) - were quite well known and of great synthetic utility. To give the greatest generality to the method, we desired to place as few restrictions as possible on the nature of the olefinic substrate so that not only electron-rich and electron-poor olefins but even simple unsubstituted olefins would afford good yields of the final cyclopentanone products. This problem was solved admirably by the use of the very highly reactive compound dimethoxytetrachlorocyclopentadiene <u>1</u> (2) as the diene component. Cycloaddition of <u>1</u> with representative olefins <u>2</u> afforded the adducts <u>3</u> in very good yields. Replacement of all of the chlorine atoms by hydrogen was smoothly effected by reduction with sodium in liquid ammonia/ethanol to furnish <u>4</u> in good yield. These are the products of a formal Diels-Alder reaction of dimethoxycyclopentadiene with the simple olefins, a reaction that is quite unlikely to produce especially in the case of the simple olefins. This technique of using a perchlorinated diene as the very reactive diene component, i.e., capable of cycloadding to "unreactive" dienophiles, followed by a high-yielding reductive dechlorination should permit the preparation of many Diels-Alder adducts that are unavailable at present. Final oxidation, hydrolysis, and decarboxylation of <u>4</u> produced the desired cyclopentanones <u>5</u> in moderate overall yields (34-40%).

This new synthetic approach allowed the facile preparation of other small highly functionalized molecules which are of great value in synthesis and are unavailable or available only with difficulty today. For example, the four step sequence shown for the preparation of the enone ketal <u>6</u> can be carried out on large scale in good overall yield (~60%) and is now an undergraduate laboratory preparation at UCLA. The dechlorination of <u>3e</u> is an exciting reaction to run since as one adds each drop of the ethereal solution of <u>3e</u> and ethanol to

		2		3	4	5
a	R = R' = $(CH_2)_4$	77%		87%	56%	37% overall
b	R = R' = $(CH_2)_5$	75%		86%	60%	38% overall
c	R = Ph R' = H	97%		76%	54%	40% overall
d	R = COOMe R' = H	95%		74%	50%	34% overall

the solution of sodium in ammonia at -78°C, a bright green light is produced! This occurs only for the endo alcohol **3e** and not for any of the other tetrachloronorbornene derivatives **3a-d**. We have no definite explanation for this unusual chemiluminescence. The enone ketal **6** proved to be a very valuable compound for natural products synthesis as described below in the section on anionic oxy-Cope rearrangements.

We have utilized this new concept of the use of perchlorinated dienes followed by dechlorination in the synthesis of the natural product β-cuparenone **7** and have begun an investigation of its use in heterocyclic systems. Whenever a new synthetic method is developed, a responsibility falls on the developer to determine its drawbacks as well as its advantages. We decided to do this for the diene **1** and have found that it is quite sensitive to steric hindrance. Therefore, although disubstituted olefins react quite well with **1**, tri- and tetrasubstituted olefins react very slowly, if at all. This was illustrated in a total synthesis of the sesquiterpene β-cuparenone **7** (3). The tetrasubstituted olefin **8** did not react with **1** under forcing conditions while the trisubstituted olefin isobutenyl acetate **9** required very vigorous conditions to give a fair yield of a mixture of the exo and endo adducts **10xn**. These adducts were then converted by a straightforward route into the natural product **7**.

This steric impediment to intermolecular Diels-Alder reactions of **1** with trisubstituted olefins could be readily overcome by making the cycloaddition intramolecular (4). Thus, when a solution of hexachlorocyclopentadiene **11** in dimethylallyl alcohol **12b** was treated with 2.3 equiv of potassium hydroxide at 25°C for 9h, the two intramolecular Diels-Alder adducts **15b** and **16b** were produced in yields of 12% and 31% respectively, along with 15% of the uncyclized dialkoxycyclopentadiene **14b**. It is interesting that this enormous difference in reaction rates between the intramolecular cycloaddition of **11** and **12** (9h, 25°C, 43%) and the intermolecular cycloaddition of **1** and **9** (3-4 wks., 131°C, 38%) is observed even though there is considerable ring strain in the products of the intramolecular case.

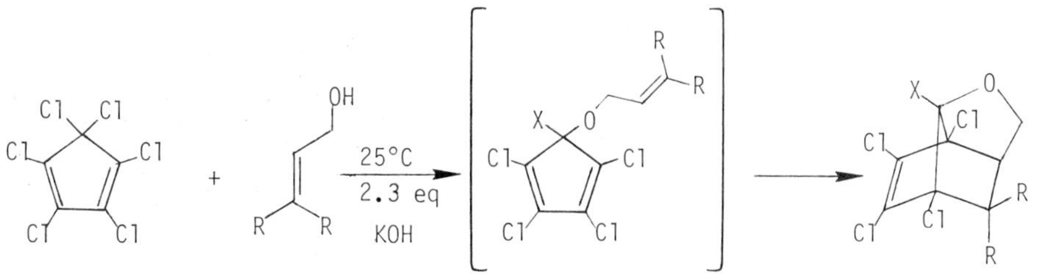

[Scheme showing reaction of 11 + 12a/12b → 13a/13b/14a/14b → 15a/15b/16a/16b]

11

12a R = H
12b R = Me

13a R = H X = Cl
14a R = H X = OR'
13b R = Me X = Cl
14b R = Me X = OR' (15%)

15a 44.5%
16a 15%
15b 12%
16b 31%

A similar reaction of 11 with allyl alcohol 12a produces the analogous products 15a and 16a in good yield.

We have begun investigating the extension of this methodology to heterocyclic systems. The 1-azapentachlorocyclopentadiene 17 rearranges to the 2-aza isomer 18 before undergoing cycloaddition with vinyl acetate to give 19 (5), in agreement with Wong's work (6). More exciting are our preliminary results with tetrachlorofuran 20 as a diene in Diels-Alder

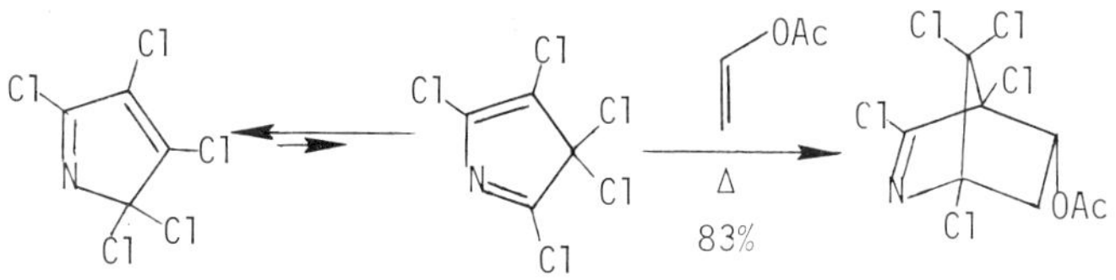

[Scheme: 17 ⇌ 18 + OAc-vinyl → 19, 83%]

17   18   19

reactions (7). Reaction of 20 with acrylic acid and a small amount of hydroquinone at 150°C for 15 sec produced a 90% yield of a mixture of exo and endo isomers, 21xn. This is to be contrasted with the cycloaddition of furan with methyl acrylate which requires several weeks at room temperature. Reductive dechlorination of 21 afforded the reduced products in fair yield (~50%). We also have preliminary results which show that 20 reacts with "unreactive" dienophiles such as allyl alcohol 12a, i.e., reaction of 12a and 20 at 150°C for 1d gave after chromatography a 75% yield of a mixture of products whose spectral characteristics are consistent with the isomeric Diels-Alder adducts 22xn.

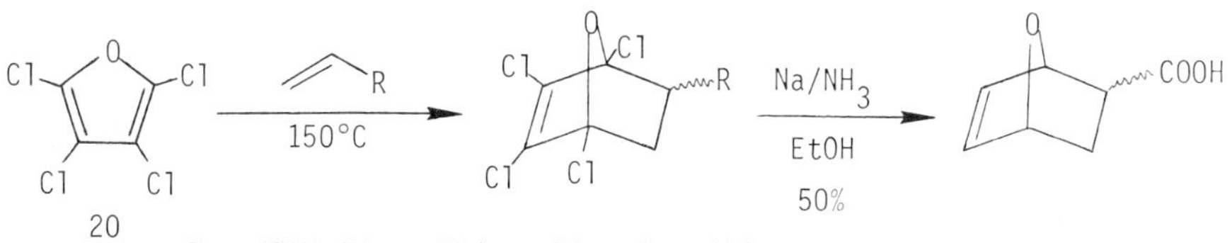

[Scheme: 20 + CH2=CHR → 21xn/22xn → Na/NH3, EtOH, 50% → COOH product]

20

R = COOH 15sec 90%
12a R = CH_2OH 1day 75%

21xn R = COOH
22xn R = CH_2OH

We are presently attempting to extend this general methodology of perchlorodiene cycloaddition-dechlorination to other systems where the Diels-Alder reaction of the unchlorinated diene does not occur or is troublesome.

SUBSTITUTED 2-PYRONES

As a new general approach to the synthesis of anthracyclines and other anthraquinone natural products, we have investigated the use of substituted 2-pyrones as dienes in Diels-Alder cycloadditions. Although 3-methoxy- and 3-hydroxy-2-pyrone (8) had been used in Diels-Alder cycloadditions, no reports on the use of 6-alkoxy-2-pyrones had been published. We reasoned that the reaction of 6-alkoxy-2-pyrones with substituted naphthoquinones such as juglone 23 would proceed with loss of carbon dioxide to give after oxidation a 1,8-dialkoxy-anthraquinone, the integral structural unit of the aclacinomycin class of anthracycline antitumor agents (9). The necessary pyrone 25 was readily available from β-methylglutaconic acid 24 in two steps, dehydrative cyclization and methylation (10). Reaction of 6-methoxy-4-methyl-2-pyrone 25 with juglone 23 followed by oxidation and demethylation furnished the natural anthraquinone chrysophanol 26 regiospecifically in 62% overall yield (10). This new methodology has been extended to the preparation of tetracyclic intermediates for anthracycline synthesis (11). For this application, a new method of regiospecific pyrone formation was developed, which involved directed cyclization of a specific glutaconic half ester. Thus, the acid ester 27 cyclized to the desired pyrone 28 on treatment with acetic anhydride in 96% yield. Cycloaddition of 28 with juglone 23 followed by oxidation produced the tetracyclic material 29 in 63% yield.

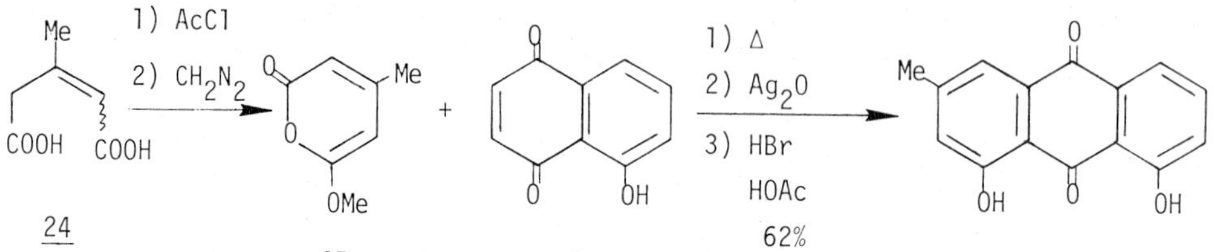

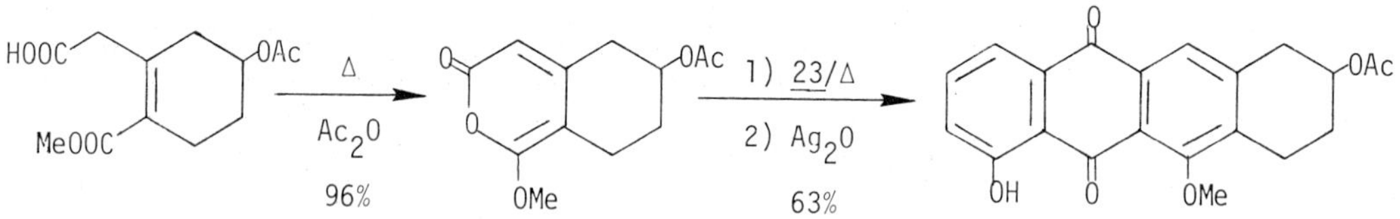

Other extensions of this method involved the regiospecific acetylation of 6-methoxy-4-alkyl-2-pyrones with acetic anhydride in trifluoroacetic acid (12). In this manner, 25 was converted regiospecifically into the 5-acetylpyrone 30 which could be reacted with 5,7-dihydroxynaphthoquinone to give the natural product 2-acetylemodin 31.

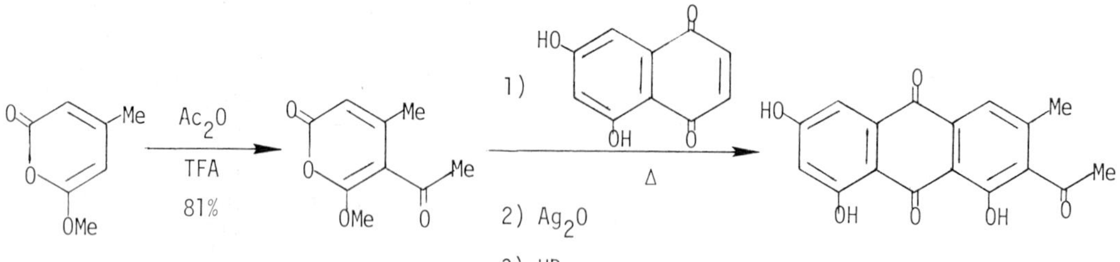

Further extensions using 2-benzopyran-3-ones and 2-pyrone-5-carboxylates are also under investigation currently in our laboratories (13).

ANIONIC OXY-COPE REARRANGEMENTS

Evans' discovery that the rate of the oxy-Cope rearrangement (14) was enhanced by up to a factor of 10^{17} by reaction of the anion of the allylic alcohol rather than the neutral compound (15) has greatly increased the usefulness of this process. We reasoned that an application of this reaction to organometallic adducts of the bicyclic enone ketal 6 would permit a rapid access to substituted cis-hydrindenones. This proved to be the case.

Addition of vinylmagnesium bromide to 6 occurred exclusively from the endo face, due to the steric hindrance of the syn-7-methoxy group toward exo attack, to give the exo alcohol 32 (16). Treatment of 32 with NaH in THF at 66°C for 1h produced a 72% yield of the hydrindenone 33. With the demonstration of the feasibility of this type of anionic rearrangement

in simple norbornenyl systems, attention was then directed to the possibility of utilizing aromatic rings as the olefinic components. At the outset of this research, there was only one example of the Cope rearrangement on an aromatic system, namely the pioneering work of Doering (17), i.e., 34a → 36a, which has recently been claimed to be in error (18). The difference in the reactivities of the two analogous systems toward [3,3]sigmatropic rearrangements, namely, the unreactivity of 4-phenyl-1-butene 34a toward Cope rearrangement under vigorous conditions (18) versus the high reactivity of allyl phenyl ether 34b toward Claisen rearrangement, has never been adequately explained. Lambert has shown that it is the first step of the Cope rearrangement that is the highly unfavorable step (19). We believe that the reasons for this can be easily understood by an examination of the overall thermodynamic changes of the two systems in the first step. For 34a, the loss of aromatic resonance energy is not compensated for in any way and thus the activation energy for this step should be very high. However, for 34b the loss of resonance energy is greatly compensated for by the thermodynamic driving force of forming a carbonyl group and two

34a X = CH$_2$ 35a X = CH$_2$ 36a X = CH$_2$
34b X = O   35b X = O   36b X = O

C-C bonds at the expense of a C=C bond and 2 C-O bonds. Thus, the starting material and product are of more nearly equal energy and one might expect a corresponding lowering of the activation energy of the pathway connecting them. In any event, we reasoned that by using the great rate enhancements offered by the anionic variation of the oxy-Cope rearrangement (15), one might be able to overcome the activation energy barrier and effect the Cope rearrangement on aromatic substrates. The substrates 37abc were prepared by addition of the corresponding aryl organometallic reagents to the enone 6. Rearrangement of the naphthyl and furyl adducts, 37ab, occurred quite readily (NaH, THF, 66°C, 1h) to give the rearranged products 38ab in high yields (75% and 72%, respectively) (16). However, the phenyl adduct 38c could not be induced to undergo [3,3]sigmatropic rearrangement, but rather underwent carbon-carbon bond cleavage to 39 instead (20). Thus, the driving force of the anionic oxy-Cope rearrangement does not seem to compensate for the loss of aromaticity of the phenyl system but does for the less stabilized naphthyl and furyl systems. The naphthyl systems could be taken on to steroid analogues by reduction of the two non-aromatic olefins and epimerization of the C9 hydrogen to the α-epimer.

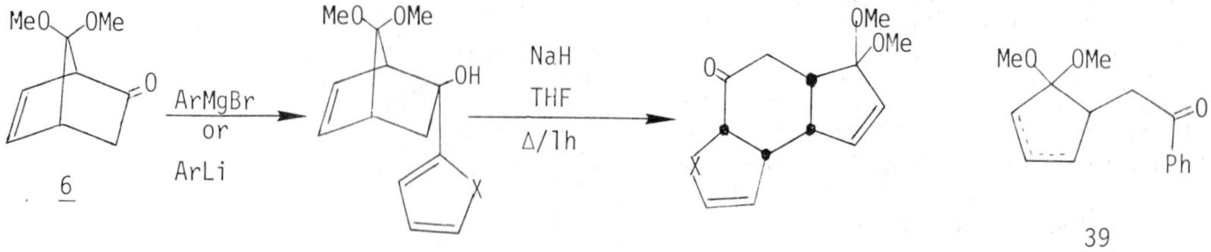

90%	37a	X = C₆H₄	75%	38a X = C₆H₄
94%	37b	X = O	72%	38b X = O
92%	37c	X = CH=CH		

This new synthetic concept of an anionic oxy-Cope rearrangement on an aromatic substrate has been used for the synthesis of the natural product coronafacic acid 43 (21). The benzofuryl adduct of the enone 40 was cleanly rearranged to 41 which was converted into the natural product 43 via the intermediate 42 (22).

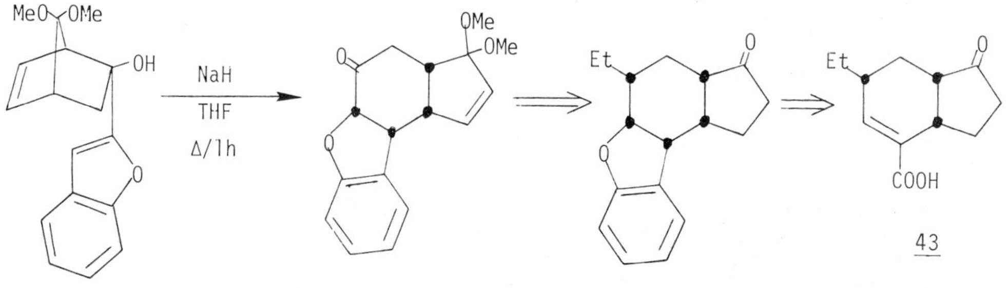

Recently we have extended this work to a synthesis of the angularly methyl-substituted hydrindenones necessary for the synthesis of natural steroids. The bicyclic enone 46 could be prepared in 8% overall yield in an 8-step synthesis starting from 2-methylcyclopentenone 44 via 45 as shown. Addition of vinylmagnesium bromide and rearrangement afforded a good yield of the desired methylhydrindenone 47 (23). The addition of groups other than vinyl and the conversion of these intermediates into steroids and other natural products, i.e., tricothecanes, is currently under investigation.

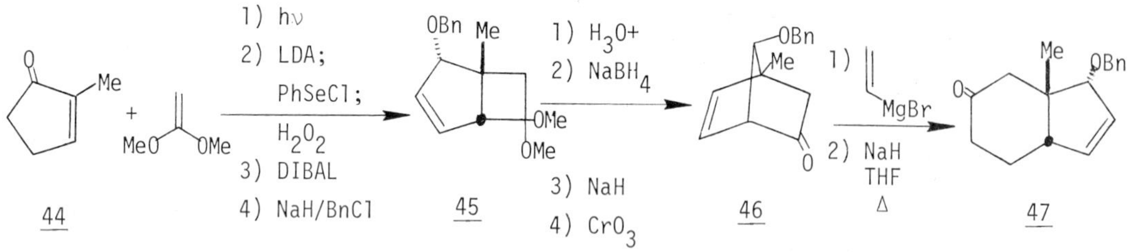

INTRAMOLECULAR DIELS-ALDER REACTIONS

In recent years, an enormous amount of research effort has been aimed at understanding the reactivity, stereo- and regioselectivity, and energetics of the intramolecular Diels-Alder reaction (24). By requiring the diene and dienophile to be part of the same molecule, one can change some of the normal reactivity requirements and stereo- and regiochemical preferences to give products that one would not expect from an intermolecular Diels-Alder reaction.

The acid component of the natural phytotoxin coronatine, coronafacic acid, 43, again was a model for our study of intramolecular cycloadditions (25). An intramolecular Lewis-acid promoted [2 + 2] cycloaddition of the ester 48 (prepared in quantitative yield from the corresponding acid and alcohol) afforded the cyclobutene 49. This is the first example of an intramolecular acid-promoted [2 + 2] cycloaddition of this type (26) and is a quite efficient way of constructing such molecules albeit in only fair yield. Conversion of 49 into the desired enone 50 was straightforward. Heating 50 to 100°C produced the trienone 51 with the necessary stereochemistry about the diene system to eventually produce 43. Heating of 51 to 180°C furnished the adduct 52 in 96% yield as a ~60:40 mixture of cis and trans ring junction isomers. Hydrolysis of this mixture produced coronafacic acid 43 in good yield, ending a short and efficient synthesis of 43.

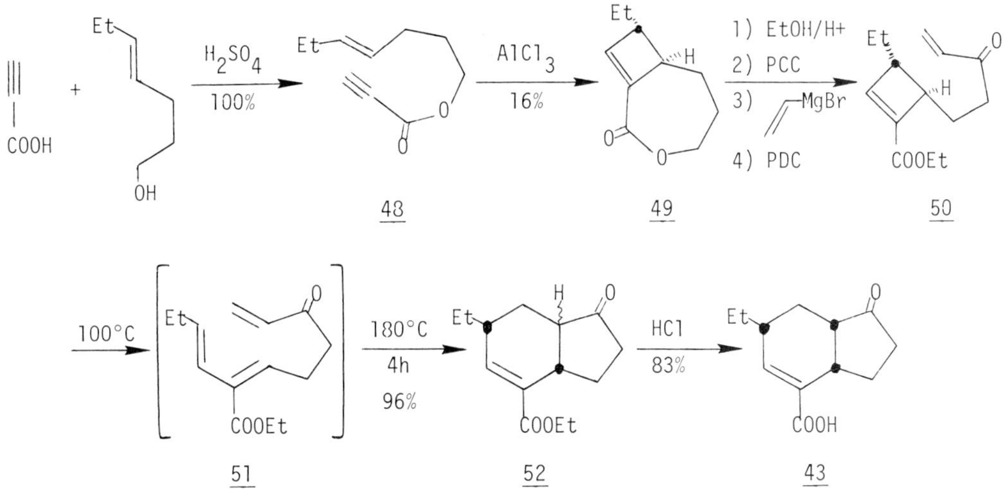

The fact that both cis and trans isomers of the hydrindenone 52 were formed in the intramolecular cycloaddition prompted us to examine this technique as a general method for the synthesis of the CD-ring portion of steroids. Sutherland (27) had already investigated the cyclization of the trienone 53 and reported that the predominant product was the cis hydrindenone 54c (the major component of a 30:4:3 mixture). We repeated this work and found that a quantitative yield of the hydrindenones were produced as a 70:30 mixture of cis and trans isomers, 54ct (28). However, close examination of molecular models of the transition states 55nx leading to the isomers (endo → cis, exo → trans) indicated that the endo transition state 55n could not attain the necessary geometry for the favorable secondary orbital overlap, which causes most endo transition states to be stabilized, because of the restrictions of the short methylene chain joining the two components. More importantly, it was

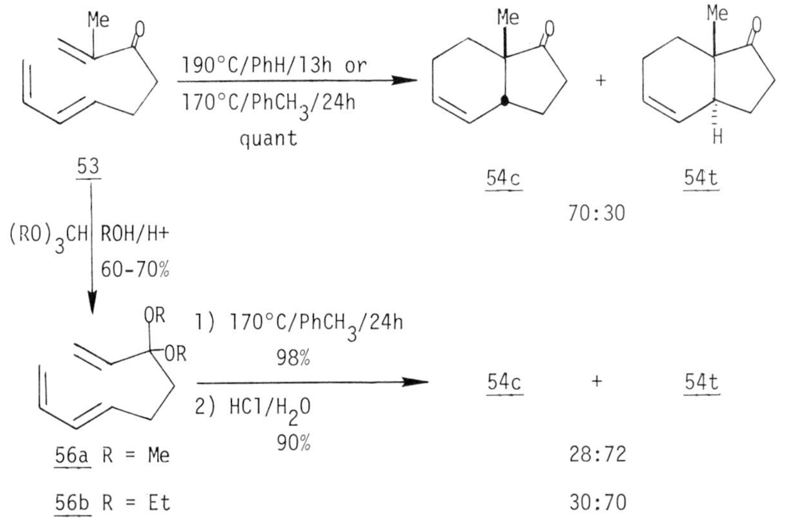

predicted that simple ketals would cause the cycloaddition to occur preferentially via the exo transition state, 55x. The endo transition state 55n for the ketals [X = (OR)$_2$] should now experience significant steric interference between one of the two alkoxy groups and the C-H bond of the butadiene system (marked with an asterisk). The corresponding exo transition state 55x should therefore be more stable with the methyl group in the endo position in place of the dialkoxymethylene group. This prediction proved to be true. Formation of the dimethyl or diethyl ketal 56 was straightforward. Heating 56a to 170°C for 1 day

                    55n   54c   55x   54t

produced a mixture of ketal and enol ether which could be hydrolyzed to a mixture of 54ct in which the trans isomer now predominated in a 28:72 ratio. The diethyl ketal 56b gave almost identical results. Thus, by a simple modification of the readily available trienone, one can prepare stereoselectively trans hydrindenones as potential CD-ring intermediates for steroid synthesis (29).

Unfortunately, all attempts at increasing the ratio of trans: cis to more than 3:1 have been unsuccessful, e.g., using lower temperatures, bulkier alkoxy groups, and Lewis and Bronsted acid catalysts. However, an examination of the transition state models 55nx indicates that if the starred H were replaced by an alkyl group (i.e., methyl or methylene), the difference between the two transition states would become much greater (due to the large interaction of the alkyl group with one of the alkoxy groups of the ketal) and should cause the reaction to occur totally via the exo transition state to give the desired trans isomer. This has been shown to be the case in octalone systems (30) but never in hydrindenones. To test this concept we have undertaken a very short total synthesis of steroids and, in particular, estrone 58 from commercially available 6-methoxy-α-tetralone 57 via the scheme shown.

                    57   58

Before preparing the fully substituted substrate, we synthesized model compounds to determine reaction conditions. Condensation of the lithium anion of 57 with butanal followed by addition of acetic anhydride produced a good yield (63%) of the E-enone 59 in addition to recovered starting ketone (~30%). The geometry of the olefinic unit was determined by a europium shift reagent study. Normal Wittig olefination of 59 proved unsuccessful as did the silyl-Wittig modification due presumably to the unreactivity of the carbonyl group. However, methyllithium added very well (92%) to the ketone to produce the allylic benzylic tertiary alcohol 60. Thermolysis of this alcohol absorbed on alumina or in a solution in HMPA caused dehydration to give the desired exocyclic methylene compound 61 which could be trapped by added N-phenylmaleimide to give 62 in good yield. If no trapping agent was present, the diene underwent a Diels-Alder dimerization to give 63. In the des-methoxy series, the diene corresponding to 61 could be isolated by reaction of the ketone corresponding to 59 with the silyl Wittig reagent followed by silica gel chromatography. This pure diene dimerized to the analogue of 63 overnight at 25°C in quantitative yield.

The information gained in the model study was then applied to the total synthesis of estrone. Alkylation of the lithium anion of the known isopropylidenedithiane 64 with the bromo acetal from acrolein followed by acidic hydrolysis produced the aldehyde 65. Aldol condensation as described above gave the dienone 66 in a one-pot reaction in 48% yield with 44% of the ketone recovered. Addition of methyllithium to 66 gave the alcohol 67 which was dehydrated by heating in HMPA to give the triene 68 which cyclized directly to the intramolecular Diels-Alder reaction product 69 as a mixture of isomers in which the trans greatly predominated. The final conversion of 69 into (±)-estrone 58 was accomplished in three straightforward steps (hydrolysis, reduction, demethylation) to end a very short, stereoselective synthesis of estrone which demonstates the power of the intramolecular Diels-Alder reaction in total synthesis.

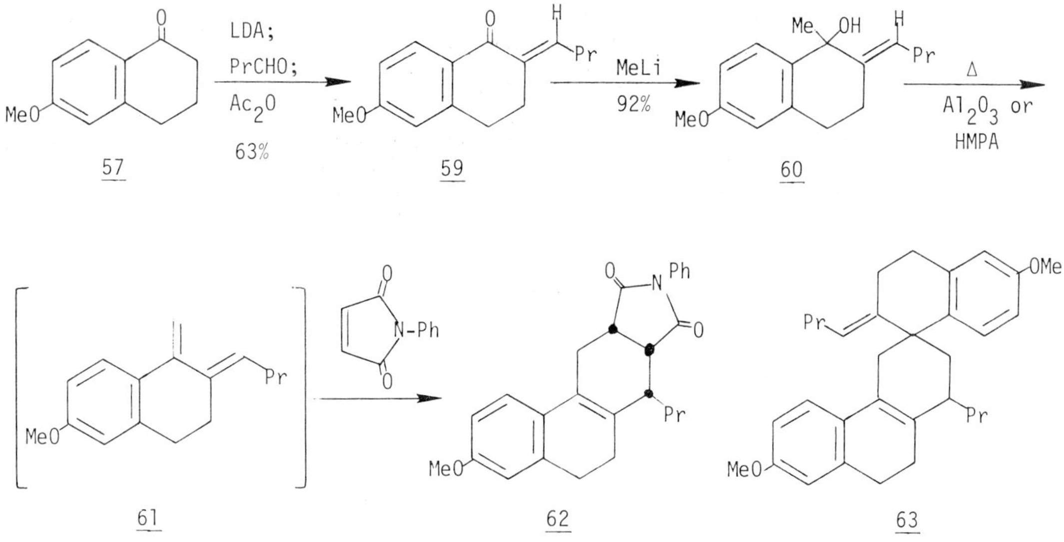

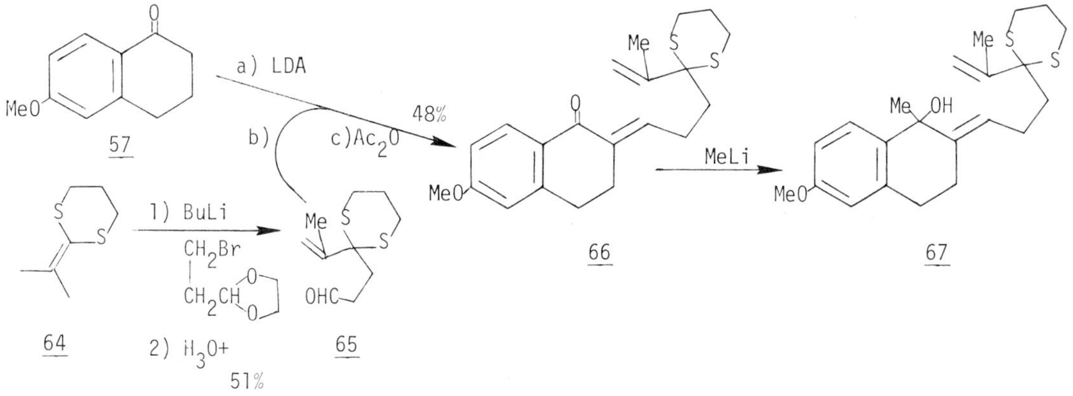

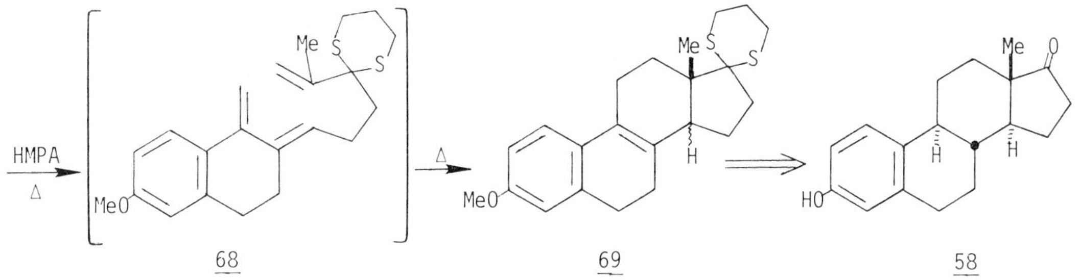

CONCLUSION

The foregoing examples help to illustrate the high utility of the new dienes for intermolecular Diels-Alder reactions, the anionic oxy-Cope rearrangement, and intramolecular cycloadditions for the total synthesis of biologically active natural products. We have shown that these methods can be applied in an elegant manner to complex molecules. Further developments and extensions of these methods are currently underway and will be reported in due course.

Acknowledgement - I would like to express my heartfelt thanks to my numerous coworkers who are mentioned individually in the references. They are directly responsible for the results described and the evolution of our research program. I also wish to thank the National Institutes of Health, the National Science Foundation, and the Research Corporation for financial support.

REFERENCES

1. M.E. Jung and J.P. Hudspeth, J. Am. Chem. Soc., 99, 5508 (1977).
2. E.T. McBee, W.R. Diveley, and J.E. Burch, Ibid., 77, 385 (1954).
3. M.E. Jung and C.D. Radcliffe, Tetrahedron Lett., 4397 (1980).
4. M.E. Jung and L.A. Light, J. Org. Chem., 47, 1084 (1982).
5. M.E. Jung and J.J. Shapiro, J. Am. Chem. Soc., 102, 7862 (1980).
6. P.H. Daniels, J.L. Wong, J.L. Atwood, L.G. Canada, and R.D. Rogers, J. Org. Chem., 45, 435 (1980).
7. M.E. Jung and P.D. Michna, unpublished results.
8. a) P. Bosshard, S. Fumagilli, R. Good, W. Treub, W.V. Philipsborn, and C.H. Eugster, Helv. Chim. Acta, 47, 769 (1964).
    b) E.J. Corey and A.P. Kozikowski, Tetrahedron Lett., 2389 (1975).
9. a) H. Umezawa, et al., J. Antibiot., 28, 830 (1975), 30, 616 (1977); b) D.E. Nettleton, Jr., et al., Ibid., 30, 525 (1977); J. Am. Chem. Soc., 101, 7041 (1979); c) F. Arcamone, et al., Ibid., 102, 1462 (1980).
10. M.E. Jung and J.A. Lowe, J. Chem. Soc. Chem. Commun., 95 (1978).
11. M.E. Jung, M. Node, R.W. Pfluger, M.A. Lyster, and J.A. Lowe, III, J. Org. Chem., 47, 1150 (1982).
12. M.E. Jung and R.W. Brown, Tetrahedron Lett., 3355 (1981).
13. M.E. Jung, R.W. Brown, and J.A. Hagenah, unpublished results.
14. a) J.A. Berson, et al., J. Am. Chem. Soc., 86, 5017, 5019 (1964); 90, 4729, 4730 (1968); b) A. Viola, et al., Ibid., 87, 1150 (1965); 89, 3462 (1967); 92, 2404 (1970).
15. D.A. Evans and A.M. Golob, Ibid., 97, 4765 (1975).
16. M.E. Jung and J.P. Hudspeth, Ibid., 100, 4309 (1978).
17. W. von E. Doering and R.A. Brazole, Tetrahedron, 22, 385 (1966).
18. M. Newcomb and R.S. Vieta, J. Org. Chem., 45, 4793 (1980).
19. J.B. Lambert, et al., Ibid., 44, 1480 (1979); J. Am. Chem. Soc., 101, 1793 (1979).
20. This cleavage reaction has now been used synthetically by Snowden. R.L. Snowden, et al., Helv. Chim. Acta, 64, 2193 (1981); Tetrahedron Lett., 335 (1982).
21. A. Ichihara, S. Sakamura, et al., Ibid., 269 (1977); 365 (1979); 4331 (1979); J. Am. Chem. Soc., 99, 636 (1977).
22. M.E. Jung and J.P. Hudspeth, Ibid., 102, 2463 (1980).
23. M.E. Jung and G.L. Hatfield, manuscript submitted.
24. G. Brieger and J.N. Bennett, Chem. Rev., 80, 63 (1980).
25. M.E. Jung and K.M. Halweg, Tetrahedron Lett., 2735 (1981).
26. B.B. Snider, Acc. Chem. Res., 13, 426 (1980).
27. J.K. Sutherland, J. Chem. Soc. Perkin Trans. 1, 1559 (1975).
28. M.E. Jung and K.M. Halweg, Tetrahedron Lett., 3929 (1981).
29. S.A. Bal and P. Helquist, Ibid., 3933 (1981) report nearly identical observations.
30. a) O.P. Vig, et al., Ind. J. Chem., 15B, 319 (1977); 16B, 449 (1978); b) J.-L. Gras and M. Bertrand, Tetrahedron Lett., 4549 (1979); c) D.F. Taber and B.P. Gunn, J. Am. Chem. Soc., 101, 3992 (1979).
31. M.E. Jung and K.M. Halweg, manuscript submitted.

STUDIES DIRECTED TOWARDS THE CHEMICAL SYNTHESIS OF YEAST ALANINE tRNA

Simon S. Jones, Colin B. Reese and Samson Sibanda

Department of Chemistry, King's College, Strand, London WC2R 2LS, UK

Abstract - The chemical synthesis of the 3'-terminal nonadecaribonucleoside octadecaphosphate sequence of yeast alanine tRNA, by the phosphotriester approach, is described. Adenine and cytosine residues are protected by N-acylation, guanine residues are protected as their 2-N-acyl-6-O-aryl derivatives, and uracil residues as their 4-O-aryl derivatives. The 2'-hydroxy functions of the ribose residues are protected by methoxytetrahydropyranyl groups and the terminal 2',3'-diol system is protected by a methoxymethylene group; the 2-dibromomethylbenzoyl group is used for the protection of 5'-hydroxy functions. Internucleotide linkages are protected by 2-chlorophenyl groups. The first phosphorylation step is effected with 2-chlorophenyl phosphorodi-(1,2,4-triazolide), and 1-(mesitylene-2-sulphonyl)-3-nitro-1,2,4-triazole is used as the condensing agent in the second phosphorylation step. The final unblocking procedure involves treatment with (i) N^1,N^1,N^3,N^3-tetramethylguanidinium 2-nitrobenzaldoximate in dioxan-water, (ii) concentrated aqueous ammonia, and (iii) 0.01 M - hydrochloric acid.

INTRODUCTION

In the past few years, much progress has been made in the chemical synthesis of oligo- and poly-deoxyribonucleotides of defined sequence (Refs. 1-6). This advance has been stimulated by the considerable need for synthetic oligodeoxyribonucleotides in biology - especially in the area of recombinant deoxyribonucleic acid (DNA) research - and has only been made possible by the development of the phosphotriester approach (Ref. 7) to oligonucleotide synthesis. Progress in the chemical synthesis of oligoribonucleotides has been less dramatic. This is perhaps due partly to the fact that the requirements in biological research, at the present time, are largely for synthetic oligodeoxyribo- rather than for oligoribo-nucleotides; it is also due to the additional complexity resulting from the need for all the 2'-hydroxy functions to be protected throughout the synthesis of oligoribonucleotides. We have long been interested in the particular problems associated with oligoribonucleotide synthesis (Ref. 8) and, in the past few years, have undertaken studies (Ref. 9) directed towards the chemical synthesis of yeast alanine transfer ribonucleic acid (tRNA^{Ala}). The latter polyribonucleotide is composed of 76 nucleotide residues and, like other tRNA molecules, contains some modified ribonucleosides in addition to adenosine, cytidine, guanosine and uridine. This article is concerned with the chemical synthesis of [ApUpUpCpCpGpGpApCpUpCpGpUpCpCpApCpCpA], the 3'-terminal nonadecaribonucleoside octadecaphosphate of yeast tRNA^{Ala}.

PROTECTION OF 2'- AND OTHER HYDROXY FUNCTIONS

Possibly the most crucial decision which has to be made in the synthesis of oligoribonucleotides (e.g. 1) is the choice of the protecting group (R) for the 2'-hydroxy functions (Ref. 10). This protecting group has to remain intact until the very last step of the synthesis and must then be removed under conditions which are mild enough to prevent subsequent attack of the released 2'-hydroxy functions on vicinal phosphodiester groups with consequent cleavage or migration of the internucleotide linkages.

The 4-methoxytetrahydro-2H-pyran-4-yl (Mthp) protecting group [as in ribonucleoside building block (2)] has been found (Ref. 11) to be particularly suitable for this purpose. It has the advantage of being achiral and is removable under very mild conditions of acidic hydrolysis such that the above requirements are met. In any oligoribonucleotide synthesis, it is also essential that the 2',3'-cis-diol system of the 3'-terminal nucleoside residue should, like the isolated 2'-hydroxy functions, remain protected until the very last step. The methoxymethylene (Mm) protecting group (Ref. 12) [as in ribonucleoside building block (3)], which undergoes acidic hydrolysis at ca. twice the rate of the Mthp and ca. 100 times the rate of the conventional isopropylidene protecting group, has been found to be particularly suitable for this purpose.

A third protecting group for hydroxy functions is required in oligoribonucleotide synthesis. In both stepwise (involving the addition of only one nucleotide residue at a time) and block (involving the addition of more than one nucleotide residue at a time) synthesis, a protecting group for 5'-hydroxy functions, which may be removed selectively without concomitant removal of the acid-labile [i.e. Mthp and Mm] and the base-labile 2-chlorophenyl protecting groups on the internucleotide linkages [see (4) and below], is required. Conventional (such as acetyl and benzoyl) and even much more sensitive acyl protecting groups (such as methoxy- and phenoxyacetyl) (Ref. 13) cannot be used to protect the 5'-hydroxy functions as even very mild conditions of alkaline hydrolysis or ammonolysis can lead to appreciable unblocking of the internucleotide linkages. We have designed the 2-dibromomethylbenzoyl (Dbmb) "protected" protecting group (Ref. 14) specifically for this purpose. The Dbmb protecting group can generally be removed from a fully-protected oligonucleotide (4) in two steps [(i) silver perchlorate-2,4,6-collidine/acetone-water (98:2 v/v); (ii) morpholine] to give the desired product (5) in good yield.

PROTECTION OF BASE RESIDUES

Three of the four principal bases of RNA have exocyclic amino groups. In order to avoid the possibility of phosphoramidate formation during phosphorylation and the possible occurrence of other side reactions, it has generally been considered desirable to protect the amino groups of adenine, cytosine and guanine residues by N-acylation [as in (6), (7) and (8), respectively] (Ref. 15). Prior to our very latest work (see below), uracil residues [as in (9)] had not been protected in oligoribonucleotide synthesis. However, following our recent observation (Ref. 16) that N-acyl guanine and uracil residues [as in (8) and (9), respectively] were susceptible to attack at their 6- and 4-positions, respectively, during the second phosphorylation step (see below) of the phosphotriester approach, we now protect (Ref. 17) guanine residues additionally on O-6 [as in (10)] and uracil residues on O-4 [as in (11)]. In the work described later in this article, adenine and cytosine residues are protected as their 6-N-(4-t-butylbenzoyl) derivatives [as in (6; R = 4-(Me$_3$C)C$_6$H$_4$) and (7; R = 4-(Me$_3$C)-C$_6$H$_4$), respectively], guanine residues as their 2-N-(4-t-butylphenylacetyl)-6-O-(2-nitrophenyl) derivatives [as in (10; R^1 = 4-(Me$_3$C)C$_6$H$_4$CH$_2$, R^2 = 2-(O$_2$N)C$_6$H$_4$)] and uracil residues as their 4-O-(2,4-dimethylphenyl) derivatives [as in (11; R = 2,4-Me$_2$C$_6$H$_3$)]. The use of lipophilic protecting groups increases the solubility of oligonucleotide phosphotriester intermediates in organic solvents and facilitates their purification (Ref. 18) by short column chromatography (Ref. 19). At the end of an oligoribonucleotide synthesis, the O-aryl protecting groups are removed from the guanine and uracil residues at the same time (Ref. 17) as the 2-chlorophenyl protecting groups (see below) are removed from the internucleotide linkages by treatment with N^1,N^1,N^3,N^3-tetramethylguanidinium 2-nitrobenzaldoximate (Ref. 20). The N-acyl protecting groups are then removed from adenine, cytosine and guanine residues by treatment with concentrated aqueous ammonia.

RIBONUCLEOSIDE BUILDING BLOCKS

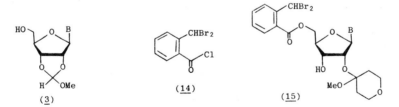

Three different types of building block are required for each ribonucleoside in our approach to the synthesis of oligoribonucleotides. First, 2'-O-methoxytetrahydropyranyl ribonucleoside derivatives (2) (Ref. 11) in which the base residues are appropriately protected [see above section on protection of base residues], are required. The latter building blocks (2) were originally prepared by treating corresponding 3',5'-di-O-acyl derivatives (Ref. 21) with 5,6-dihydro-4-methoxy-2H-pyran (12) (Refs. 11 and 22) in the presence of an acid catalyst and then removing the acyl groups. However, Markiewicz later found that ribonucleosides react with 1,3-dichloro-1,1,3,3-tetra-isopropyldisiloxane (Ref. 23) in the presence of imidazole to give 3',5'-protected derivatives (13) directly. If the latter compounds, which are generally more accessible than the corresponding 3',5'-diesters, are allowed to react with (12) in the presence of acid and the products then treated with tetra-n-butylammonium fluoride in tetrahydrofuran, the desired building blocks (2) are obtained, usually in satisfactory yields. We now use Markiewicz's procedure (Ref. 23) routinely except in the preparation of 2'-O-methoxytetrahydropyranyluridine (2; B = uracil-1-yl) (Ref. 11).

Secondly, 2',3'-O-methoxymethylene ribonucleoside derivatives (3) are required as 3'-terminal building blocks. The latter compounds (3) are readily prepared (Ref. 12) by treating ribonucleosides [appropriately protected, in the ways indicated above, on their base residues] with trimethyl orthoformate in the presence of an acid catalyst. Finally, 5'-O-(2-dibromomethylbenzoyl)-2'-O-methoxytetrahydropyranyl ribonucleoside derivatives (15) are required. These building blocks (15) may usually be prepared (Ref. 9) in satisfactory to good yields by treating the corresponding 2'-O-methoxytetrahydropyranyl derivatives (2) with 2-dibromomethylbenzoyl chloride (14) (Ref. 2) in acetonitrile-pyridine solution. It is our aim that all ribonucleoside building blocks should be obtained as pure crystalline solids and this aim has usually been achieved.

PROTECTION OF INTERNUCLEOTIDE LINKAGES

In almost all of the early work on oligonucleotide synthesis, the internucleotide linkages were left unprotected during the phosphorylation steps. While some success was achieved in the deoxyribose series (Ref. 24) using this so-called phosphodiester approach, it proved to be of very limited use in the ribose series (Ref. 25). It was for this reason that we originally turned our attention to the alternative phosphotriester approach in which the internucleotide linkages are all protected during the phosphorylation steps. The considerable advantages of the phosphotriester approach have been discussed elsewhere (Ref. 7) and it is now used virtually exclusively in the synthesis both of oligoribo- and oligodeoxyribonucleotides.

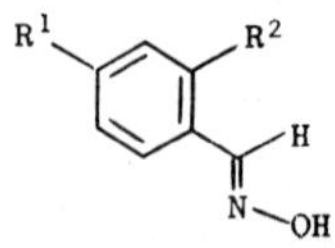

(16) a; $R^1 = NO_2$, $R^2 = H$
      b; $R^1 = H$, $R^2 = NO_2$

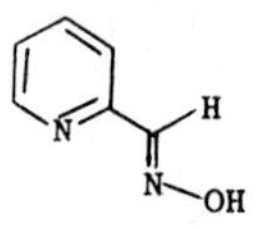

(17)

The success of the phosphotriester approach must clearly depend on the choice of the protecting group for the internucleotide linkages. We originally suggested (Ref. 26) that the phenyl group should be used for this purpose and shortly afterwards showed that substituted aryl groups (such as 2-chlorophenyl, 2-fluorophenyl, 4-chlorophenyl and p-cresyl) (Ref. 8) could also be used. We first carried out the unblocking of aryl (principally phenyl and 2-chlorophenyl) protected internucleotide linkages by treatment with sodium hydroxide in aqueous dioxan and found that the use of this procedure and other procedures involving hydroxide ion (Ref. 1) leads to an appreciable amount of concomitant internucleotide cleavage. However, we later found that 2-chlorophenyl protecting groups can readily be removed from the internucleotide linkages of fully-protected oligonucleotides [e.g. (4)] by treatment with the N^1,N^1,N^3,N^3-tetramethylguanidinium (TMG) salt of syn-4-nitrobenzaldoxime (16a) or syn-pyridine-2-carboxaldoxime (17) in dioxan-water (Ref. 27) and that, under these conditions, the extent of internucleotide cleavage is at most very slight. We have very recently found (Ref. 20) that the TMG salt of syn-2-nitrobenzaldoxime (16b) is an even more useful reagent for the removal of 2-chlorophenyl protecting groups from internucleotide linkages in that it effects unblocking more rapidly and without detectable internucleotide cleavage. Phosphotriester intermediates with 4-chlorophenyl protecting groups can also be unblocked at a convenient rate with the TMG salt of syn-2-nitrobenzaldoxime (16b), but the unblocking of phenyl-protected internucleotide linkages occurs at an inconveniently slow rate with this reagent (Ref. 20). We now invariably use 2-chlorophenyl groups to protect internucleotide linkages both in oligoribo- and oligodeoxyribo-nucleotide synthesis.

PHOSPHORYLATION METHODS

Scheme 1

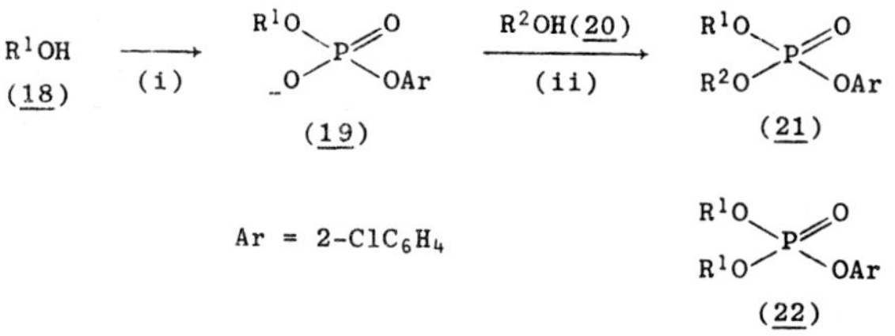

R^1OH (18) ⟶ (i) ⟶ (19) ⟶ R^2OH (20) / (ii) ⟶ (21)

Ar = $2\text{-}ClC_6H_4$

(22)

Two separate phosphorylation steps [(i) and (ii), Scheme 1] are required in the phosphotriester approach. Step (i) involves the reaction between the 3'-hydroxy function of a protected nucleoside or oligonucleotide (R^1OH, 18) and a suitable phosphorylating agent to give the corresponding 3'-(2-chlorophenyl) phosphate (19). Step (ii) involves the reaction between the latter phosphodiester intermediate (19) and the 5'-hydroxy function of a protected nucleoside or oligonucleotide (20) in the presence of a condensing agent to give the desired unsymmetrical phosphotriester (21).

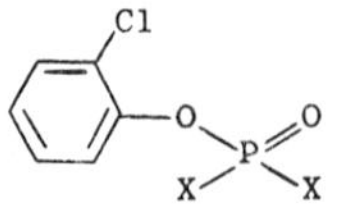

(23) a; X = Cl
     b; X = (24)

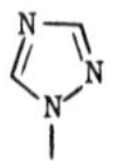

(24)

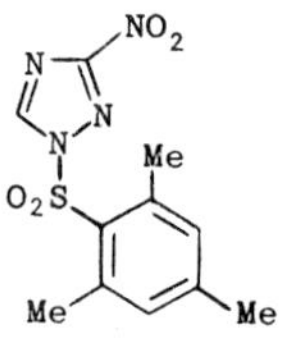

(25)

It is essential that the phosphorylating agent used in step (i) (Scheme 1) should be monofunctional and not react with two molecular equivalents of (18) to give a symmetrical phosphotriester (22). If this happens, the stoicheiometry of the reaction will be upset, mixtures of products will be obtained and the overall yield of the required unsymmetrical phosphotriester (21) will inevitably be lowered. We have found that when a 2',5'-protected ribonucleoside derivative (such as 15) or an oligoribonucleotide with a free 3'-hydroxy function is treated with a two- to three-fold excess of 2-chlorophenyl phosphorodi-(1,2,4-triazolide) (23b) for ca. 15 min at room temperature and the corresponding intermediate mono-(1,2,4-

triazolide) is then hydrolyzed by treatment with water and triethylamine, the desired 3'-(2-chlorophenyl) phosphate (19) is obtained (Refs. 2, 9 and 28) in virtually quantitative yield and may be isolated as a pure solid triethylammonium salt. The phosphorylating agent, 2-chlorophenyl phosphorodi-(1,2,4-triazolide) (23b), is readily generated in situ by treating 2-chlorophenyl phosphorodichloridate (23a) with two molecular equivalents each of 1,2,4-triazole and triethylamine. Despite the fact that the phosphorylating agent (23b) is potentially bifunctional, no symmetrical phosphotriester (22) is formed, providing that an excess of (23b) is used.

In the second phosphorylation step [Scheme 1, (ii)], it is essential that the condensing agent should not itself react directly with (20) but that it should activate the intermediate phosphodiester (19) so that it may then phosphorylate (20) to give (21) in high yield. We have found that 1-(mesitylene-2-sulphonyl)-3-nitro-1,2,4-triazole (MSNT, 25) (Refs. 9 and 27) meets this requirement and that it is a very satisfactory condensing agent indeed for the second step of the phosphotriester approach. MSNT (25) is a stable crystalline solid which effects condensation reactions [between monomers and oligomers corresponding to (19) and (20)] usually within 15-30 min in pyridine solution at room temperature.

SYNTHESIS OF DINUCLEOTIDE AND TRINUCLEOTIDE BLOCKS

In our approach to oligoribonucleotide synthesis, we first prepare a number of oligomer (usually trimer) blocks and then join these blocks together to give the fully-protected oligonucleotides (Ref. 9). The method used in the preparation of the oligomer blocks is outlined in Scheme 2.

Scheme 2

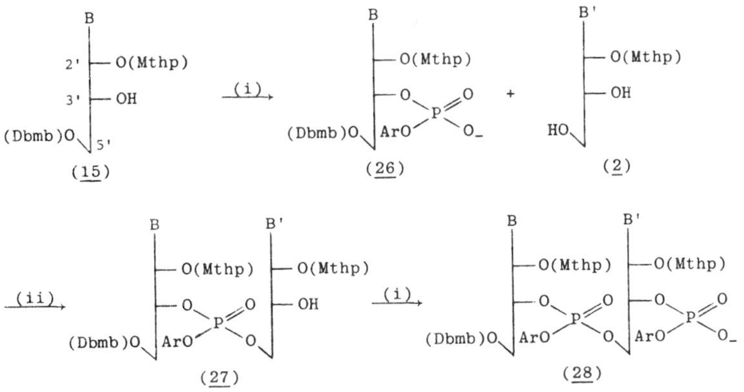

Reagents: (i)(a) 2-chlorophenyl phosphorodi-(1,2,4-triazolide)(23b)-1-methylimidazole/tetrahydrofuran, (b) Et_3N-H_2O; (ii) MSNT(25)/pyridine.

The 2',5'-protected ribonucleoside building blocks (15) with appropriately protected basic residues (see above) are first treated with 2-3 molecular equivalents of 2-chlorophenyl phosphorodi-(1,2,4-triazolide) (23b) in the presence of 1-methylimidazole in tetrahydrofuran solution (Ref. 29) at room temperature. After ca. 15 min, the putative corresponding intermediate phosphoromono-(1,2,4-triazolides) are hydrolyzed with water and triethylamine to give the desired phosphodiesters (26). The latter (26) (Ref. 9) are then isolated by precipitation as pure solid triethylammonium salts in high (usually 90% or above) yields.

The 3'-phosphodiesters (26) are then allowed to react with 2'-O-methoxytetrahydropyranyl ribonucleosides (2), again appropriately protected on their base residues, in the presence of a two- to three-fold excess of MSNT (25) in pyridine solution at room temperature. As indicated above, such condensation reactions are usually complete in 15-30 min and good yields (see Table 1) of the partially-protected dinucleoside phosphates (27) are often obtained (Ref. 9). It is generally advantageous if the phosphodiester components (26) are used in slight excess with respect to the nucleoside building blocks (2). Although the latter building blocks (2) are unprotected on their 3'-hydroxy functions, the condensation reactions appear to be virtually regioselective in that the isomeric dinucleoside phosphates with 3'→3'-internucleotide linkages cannot be detected in the products (Ref. 9). Furthermore, the extent of sulphonation of the nucleoside building blocks (2) is negligible despite the fact that MSNT (25) is used in excess.

As indicated in Scheme 2, the partially-protected dinucleoside phosphates (27) are then treated with 2-chlorophenyl phosphorodi-(1,2,4-triazolide) (23b) and are thereby converted into the corresponding dinucleotides (28). The latter, like the partially-protected mononucleotides (26), are isolated by precipitation as their solid triethylammonium salts, usually in very high yields (Ref. 9). The desired partially-protected trinucleoside diphosphate (29) and trinucleotide (30) blocks may be prepared by the same general procedures (Ref. 9). The dinucleotides (28) are allowed to react with nucleoside building blocks (2) in the presence of an excess of MSNT (25) to give trinucleoside diphosphates (29) which may be isolated, following short column chromatography, as pure precipitated solids in satisfactory to good yields (see Table 1). Again there is no evidence for the formation of isomeric products with 3'→3'-internucleotide linkages. Finally, the latter trinucleoside diphosphates (29) are converted into the corresponding trinucleotides (30) in high yields by the 2-chlorophenyl phosphorodi-(1,2,4-triazolide) procedure.

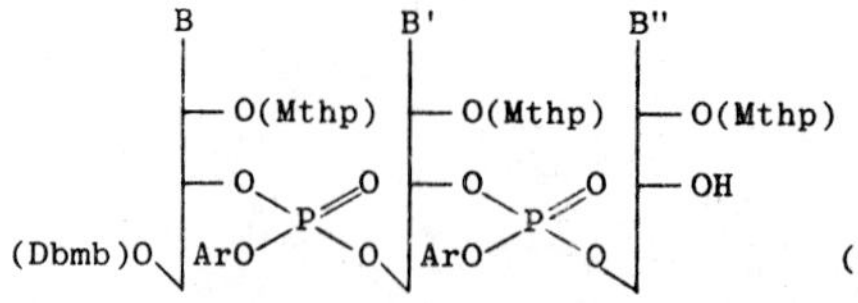

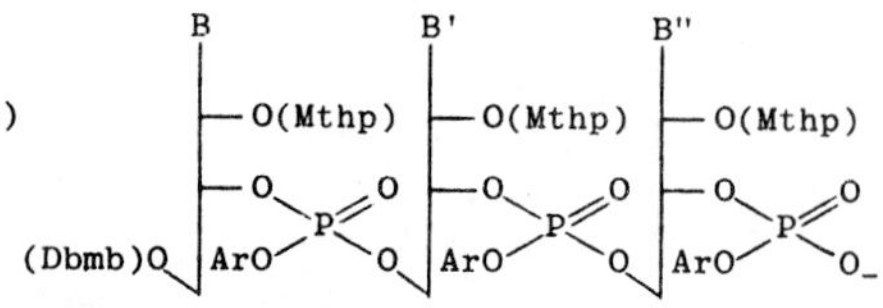

SYSTEM OF ABBREVIATIONS FOR OLIGORIBONUCLEOTIDE INTERMEDIATES

Scheme 3

$(5')$
Dbmb-$U'pC'pC'p$

(31)

$(5')$
Dbmb-$U'pC'pG'p$

(32)

$(5')$ $\qquad$ $(3')$
HO-$A'pC'pC'pA$-Mm

(33)

Dbmb = 2-dibromomethylbenzoyl; Mm = methoxymethylene; p = 2-ClC$_6$H$_4$O—P(=O)(O$^-$)—O—

A = 6-$\underline{N}$-(4-t-butylbenzoyl)adenosine; A' = 2'-$\underline{O}$-methoxytetrahydropyranyl-6-$\underline{N}$-(4-t-butylbenzoyl)adenosine; C' = 2'-$\underline{O}$-methoxytetrahydropyranyl-4-$\underline{N}$-(4-t-butylbenzoyl)cytidine; G' = 2'-$\underline{O}$-methoxytetrahydropyranyl-2-$\underline{N}$-(4-t-butylphenylacetyl)-6-$\underline{O}$-(2-nitrophenyl)guanosine; U' = 2'-$\underline{O}$-methoxytetrahydropyranyl-4-$\underline{O}$-(2,4-dimethylphenyl)uridine.

While the abbreviations suggested by the 1970 IUPAC-IUB Commission on Biochemical Nomenclature (Ref. 30) are quite clear for unprotected oligonucleotides, they are much more complex and hence less satisfactory for protected oligonucleotides. This is particularly true in the case of phosphotriester intermediates. For this reason, we recently introduced a simplified system (Ref. 2) of nomenclature for oligodeoxyribonucleotides in which base-protected nucleoside residues and protected internucleotide linkages are indicated simply by the appropriate italicized letters (i.e. A, C, G and p, respectively). This system is not entirely satisfactory for oligoribonucleotides as the terminal and the non-terminal nucleoside residues of the latter are likely to be protected both on their base and sugar moieties. However, the system can be suitably modified by the addition of a prime after the capital letter representing a particular ribonucleoside when the 2'-hydroxy function of the latter is protected by a methoxytetrahydropyranyl group (see Scheme 3). The usual convention of placing the 5'-group on the left-hand and the 3'-group on the right-hand side of a particular nucleoside residue has been observed.

When the above system is used, it is essential to include a key in which the precise meanings of the abbreviations are indicated. Such a key has been provided above. As indicated in the key, p represents a 2-chlorophenyl protected phosphate residue; if p occurs between two nucleoside residues [e.g. U' and C' in the partially-protected trinucleotide (31)], it represents a phosphotriester group but if it is attached to only one nucleoside residue [e.g. the p at the right-hand (i.e. 3'-) end of either (31) or (32)], it represents a phosphodiester group. In the partially-protected tetranucleoside triphosphate (33), both A' and A are italicized as the adenine moiety of each of these adenosine residues is protected by $\underline{N}$-acylation; however, the right-hand adenosine residue is abbreviated to A rather than A' as its 2'-hydroxy function is (as indicated) protected by part of a methoxymethylene and not by a methoxytetrahydropyranyl group. Unprotected adenine, cytosine, guanine and uracil moieties and unprotected internucleotide linkages are represented by non-italicized letters (A, C, G, U and p, respectively). The three partially-protected oligoribonucleotides (31-33) in Scheme 3 are the blocks required in the synthesis (see below) of the 3'-terminal decaribonucleoside nonaphosphate of yeast tRNA^{Ala}.

SYNTHESIS OF 3'-TERMINAL DECARIBONUCLEOSIDE NONAPHOSPHATE SEQUENCE

Our strategy for the synthesis of the 3'-terminal nonadecanucleoside octadecaphosphate sequence of yeast tRNA^{Ala} initially requires the synthesis of the fully-protected 3'-terminal decaribonucleoside nonaphosphate sequence (34). We had successfully completed a synthesis of this sequence [UpCpGpUpCpCpApCpCpA] earlier (Ref. 9) but had used different protecting groups on cytosine, guanine and uracil residues. When we undertook the first synthesis, we were unaware of the fact that side-reactions, involving 2-N-acyl guanine and uracil residues, can occur during the course of MSNT(25)-promoted condensation reactions. Following this discovery (Refs. 9 and 16), we now protect guanine residues on O-6 and uracil residues on O-4 with 2-nitrophenyl and 2,4-dimethylphenyl groups [as in 10; R^2 = 2-$(O_2N)C_6H_4$ and 11; R = 2,4-$Me_2C_6H_3$], respectively (Ref. 17). Guanine residues are also protected on N-2 with 4-t-butyl-phenylacetyl groups [as in 10; R = 4-$(Me_3C)C_6H_4CH_2CO$].

Dbmb-$U'pC'pG'pU'pC'pC'pA'pC'pC'pA$-Mm   Dbmb-$A'pC'pC'p$
(34)   (35)

Dbmb-$A'pC'pC'$-OH   Dbmb-$U'pC'pC'$-OH   Dbmb-$U'pC'pG'$-OH
(36)   (37)   (38)

Dbmb-$A'pC'$-OH   Dbmb-$U'pC'$-OH
(39)   (40)

TABLE 1. Preparation of Trinucleoside Diphosphate Blocks (29)

Expt No	5'-Protected Component (mmol)	5'-Hydroxy Component (mmol)	MSNT (25) (mmol)	Reaction Time (min)	Product (% isolated yield)
1	Dbmb-$A'p$ (1.73)	HO-C'-OH (1.6)	4.80	20	Dbmb-$A'pC'$-OH (80)
2	Dbmb-$A'p$ (0.99)	HO-U'-OH (0.90)	2.70	20	Dbmb-$A'pU'$-OH (80)
3	Dbmb-$C'p$ (0.98)	HO-C'-OH (0.89)	2.68	20	Dbmb-$C'pC'$-OH (82)
4	Dbmb-$G'p$ (0.95)	HO-A'-OH (0.87)	2.60	20	Dbmb-$G'pA'$-OH (79)
5	Dbmb-$U'p$ (1.56)	HO-C'-OH (1.30)	3.88	20	Dbmb-$U'pC'$-OH (83)
6	Dbmb-$A'pC'p$ (1.25)	HO-C'-OH (1.14)	3.42	20	Dbmb-$A'pC'pC'$-OH (62)
7	Dbmb-$A'pU'p$ (0.82)	HO-U'-OH (0.75)	2.25	20	Dbmb-$A'pU'pU'$-OH (64)
8	Dbmb-$C'pC'p$ (0.67)	HO-G'-OH (0.61)	1.83	20	Dbmb-$C'pC'pG'$-OH (84)
9	Dbmb-$G'pA'p$ (0.18)	HO-C'-OH (0.15)	0.55	20	Dbmb-$G'pA'pC'$-OH (81)
10	Dbmb-$U'pC'p$ (1.24)	HO-C'-OH (1.13)	3.72	20	Dbmb-$U'pC'pC'$-OH (78)
11	Dbmb-$U'pC'p$ (0.64)	HO-G'-OH (0.59)	1.93	25	Dbmb-$U'pC'pG'$-OH (69)

The synthesis of the fully-protected decaribonucleoside nonaphosphate (34) was based on three trinucleotide building blocks [(31), (32) and (35)]. The latter were prepared from the three corresponding trinucleoside diphosphates [(37), (38) and (36), respectively] by the 2-chlorophenyl phosphorodi-(1,2,4-triazolide) procedure (see above, Scheme 2). Average yields of over 90% were obtained. The three trinucleoside diphosphates [(37), (38) and (36)] were themselves prepared in satisfactory to good yields [Table 1, experiments nos. 10, 11 and 6, respectively] by MSNT(25)-promoted condensations between the appropriate dinucleotides [Dbmb-$U'pC'p$ and Dbmb-$A'pC'p$] and nucleoside building blocks [HO-C'-OH and HO-G'-OH]. The dinucleotides were prepared, in yields of over 90%, from the corresponding dinucleoside phosphates [Dbmb-$U'pC'$-OH (40) and Dbmb-$A'pC'$-OH (39)]. As indicated in Table 1 (experiments nos. 1 and 5, respectively), (39) and (40) were obtained in good yields by the MSNT(25)-promoted condensation procedure indicated in Scheme 2.

The strategy for the assembly of the three trinucleotide blocks [(31), (32) and (35)] into the fully-protected decaribonucleoside nonaphosphate (34) is outlined in Scheme 4. First, trinucleotide (35) (0.65 mmol) and the 3'-terminal building block, 6-N-(4-t-butylbenzoyl)-2',3'-O-methoxymethyleneadenosine (41) (0.62 mmol) were condensed together in the presence of MSNT (25) (1.85 mmol) to give the fully-protected tetraribonucleoside triphosphate [Dbmb-$A'pC'pC'pA$-Mm] in 86% isolated yield. The 5'-Dbmb protecting group was then removed by the usual two step procedure (Refs. 9 and 14) to give (33) in 93% isolated yield. The overall yield for the conversion of (41) into (33) [Scheme 4, (I)] was therefore 80%. Secondly, trinucleotide (31) (0.24 mmol) and partially-protected tetraribonucleoside triphosphate (33) (0.20 mmol) were condensed together in the presence of MSNT (25) (1.20 mmol) to give the fully-protected heptaribonucleoside hexaphosphate [Dbmb-$U'pC'pC'pA'pC'pC'pA$-Mm] in 86%

Scheme 4

(I)   Dbmb-$A'pC'pC'p$ + HO-A-Mm $\xrightarrow{\text{(i),(ii)}}$ HO-$A'pC'pC'pA$-Mm
         (35)   (41)   (33)

(II) Dbmb-$U'pC'pC'p$ + (33) $\xrightarrow{\text{(i),(ii)}}$ HO-$U'pC'pC'pA'pC'pC'pA$-Mm
         (31)   (42)

(III) Dbmb-$U'pC'pG'p$ + (42) $\xrightarrow{\text{(i)}}$ Dbmb-$U'pC'pG'pU'pC'pC'pA'pC'pC'pA$-Mm
         (32)   (34)

Reagents: (i) MSNT(25)/pyridine; (ii)(a) AgClO$_4$/acetone(or tetrahydrofuran)-water (98:2 v/v), (b) morpholine.

isolated yield. The 5'-Dbmb protecting group was then removed to give (42) in 83% isolated yield. The overall yield for the conversion of (33) into (42) [Scheme 4, (II)] was therefore ca. 71%. Finally, the trinucleotide (32) (0.175 mmol) and the partially-protected heptaribonucleoside hexaphosphate (42) (0.134 mmol) were condensed together [Scheme 4, (III)] in the presence of MSNT (25) (1.75 mmol, 50 min) to give the fully-protected decaribonucleoside nonaphosphate (34) in 63% isolated yield.

A portion of the latter fully-protected intermediate (34) was successfully unblocked by the standard three-step procedure (Ref. 9 and see below) to give the completely unprotected decaribonucleoside nonaphosphate ([UpCpGpUpCpCpApCpCpA]). The remainder of this material (34) was used (see below) in the synthesis of the 3'-terminal nonadecaribonucleoside octadecaphosphate sequence [ApUpUpCpCpGpGpApCpUpCpGpUpCpCpApCpCpA] of yeast tRNA[Ala].

SYNTHESIS OF 3'-TERMINAL NONADECARIBONUCLEOSIDE OCTADECAPHOSPHATE SEQUENCE

Dbmb-$A'pU'pU'pC'pC'pG'pG'pA'pC'$-OH   Dbmb-$A'pU'pU'$-OH
              (43)   (44)

Dbmb-$C'pC'pG'$-OH   Dbmb-$G'pA'pC'$-OH   Dbmb-$A'pU'pU'p$
      (45)   (46)   (47)

      HO-$C'pC'pG'$-OH   HO-$G'pA'pC'$-OH
           (48)   (49)

Our strategy for the synthesis of the 3'-terminal nonadecanucleoside octadecaphosphate of yeast tRNA[Ala] also required the partially-protected nonaribonucleoside octaphosphate (43), the preparation of which was based on three triribonucleoside diphosphates [(44), (45) and (46)]. The latter were all prepared by the procedure described above in connection with the synthesis of the three corresponding trimer blocks [(36), (37) and (38)] used in the assembly of the fully-protected decaribonucleoside nonaphosphate (34). Three partially-protected dinucleoside phosphates [Dbmb-$A'pU'$-OH, Dbmb-$C'pC'$-OH and Dbmb-$G'pA'$-OH] were first prepared in good yields [Table 1, experiments nos. 2, 3 and 4] by the method indicated in Scheme 2, and then converted into their 3'-(2-chlorophenyl) phosphates [Dbmb-$A'pU'p$, Dbmb-$C'pC'p$ and Dbmb-$G'pA'p$, respectively], in very high yields, by the 2-chlorophenyl phosphorodi-(1,2,4-triazolide) procedure (Scheme 2). The latter partially-protected dinucleotides were condensed with the appropriate nucleoside building blocks [HO-U'-OH, HO-G'-OH and HO-C'-OH, respectively] in the presence of MSNT (25) in pyridine solution to give the trinucleoside diphosphates [(44), (45) and (46)] in 64, 84 and 81% isolated yields, respectively [Table 1, experiments nos. 7, 8 and 9]. While Dbmb-$A'pU'pU'$-OH (44) was converted into Dbmb-$A'pU'pU'p$ (47) in very high yield by the 2-chlorophenyl phosphorodi-(1,2,4-triazolide) procedure, the 5'-Dbmb protecting groups were removed from Dbmb-$C'pC'pG'$-OH (45) and Dbmb-$G'pA'pC'$-OH (46) in the usual way (Scheme 4) to give HO-$C'pC'pG'$-OH (48) and HO-$G'pA'pC'$-OH (49) in 83 and 72% isolated yields, respectively.

Unlike the above fully-protected decaribonucleoside nonaphosphate (34), the partially-protected nonaribonucleoside octaphosphate (43) was assembled (Scheme 5) from its 5'-end. Assembly from the 3'-end, which is the more usual and generally preferred approach, was ruled out as we had not then succeeded in devising a suitable temporary protecting group for the 3'-end of the growing oligonucleotide. We are currently putting much effort into the development of such a protecting group. The trinucleotide (47) (0.20 mmol) and trinucleoside diphosphate (48) (0.16 mmol) blocks were condensed together [Scheme 5, (I)] in the presence of MSNT (1.04 mmol) to give the partially-protected hexaribonucleoside pentaphosphate [Dbmb-$A'pU'pU'pC'pC'pG'$-OH] in 62% isolated yield. As far as could be ascertained, phosphorylation of (48) occurred virtually exclusively on its 5'-hydroxy function. The condensation product was converted, in very high yield, into its 3'-(2-chlorophenyl) phosphate (50) by the usual

Scheme 5

(I)   Dbmb-A'pU'pU'p + HO-C'pC'pG'-OH $\xrightarrow{(i),(ii)}$ Dbmb-A'pU'pU'pC'pC'pG'p

             (47)   (48)   (50)

(II) (50) + HO-G'pA'pC'-OH $\xrightarrow{(i)}$ Dbmb-A'pU'pU'pC'pC'pG'pG'pA'pC'-OH

                  (49)   (43)

Reagents: (i) MSNT(25)/pyridine; (ii)(a) 2-chlorophenyl phosphorodi-(1,2,4-triazolide)(23b)-1-methylimidazole/tetrahydrofuran, (b) Et_3N-H_2O.

procedure. Finally, the latter hexanucleotide block (50) (0.073 mmol) was condensed together [Scheme 5, (II)] with the partially-protected trinucleoside diphosphate (49) (0.056 mmol) in the presence of MSNT (25) (0.37 mmol, 60 min) to give the desired partially-protected nona-ribonucleoside octaphosphate (43) in 47% isolated yield. The relatively low yield of (43) obtained possibly reveals the limitations of the approach (see above) adopted.

Scheme 6

Dbmb-A'pU'pU'pC'pC'pG'pG'pA'pC'p + HO-U'pC'pG'pU'pC'pC'pA'pC'pC'pA-Mm

              (51)   (52)

$\xrightarrow[C_5H_5N]{MSNT(25)}$ Dbmb-A'pU'pU'pC'pC'pG'pG'pA'pC'pU'pC'pG'pU'pC'pC'pA'pC'pC'pA-Mm

                                                (53)

Before the final condensation leading to the fully-protected nonadecaribonucleoside octadeca-phosphate (53), it was necessary to phosphorylate the partially-protected nonaribonucleoside octaphosphate (43) to give the nonanucleotide block (51) and remove the 5'-Dbmb protecting group from the fully-protected decaribonucleoside nonaphosphate (34) to give (52). These transformations were carried out by the usual procedures (see above), and (51) and (52) were obtained in yields of >90 and 73%, respectively. The condensation [Scheme 6] between (51) (0.02 mmol) and (52) (0.016 mmol) in the presence of MSNT (25) (0.40 mmol) in pyridine solution was allowed to proceed for 90 min. Following short column chromatography of the products, the crude isolated yield of the putative nonadecaribonucleoside octadecaphosphate (53) was ca. 70%. However, further chromatography of this material on Sephadex LH-60 (Ref. 31) suggested that it was only ca. 60% pure and that it contained lower molecular weight impurities. Therefore the yield of (53) obtained in the above condensation reaction [Scheme 6] may not have exceeded 40%. It should be noted that the molecular weight of (53) is over 13,500.

In our approach to oligoribonucleotide synthesis, the complete unblocking of a fully-protected oligoribonucleotide involves a three step procedure (Ref. 9). The unblocking of the nona-decaribonucleoside octadecaphosphate (53) by this procedure is illustrated in Scheme 7. In

Scheme 7

Dbmb-A'pU'pU'pC'pC'pG'pG'pA'pC'pU'pU'pG'pU'pC'pC'pA'pC'pC'pA-Mm

(53)

$\downarrow$ (i),(ii)

HO-A'pU'pU'pC'pC'pG'pG'pA'pC'pU'pU'pG'pU'pC'pC'pA'pC'pC'pA-Mm

(54)

$\downarrow$ (iii)

[ApUpUpCpCpGpGpApCpUpUpGpUpCpCpApCpCpA]

(55)

Reagents: (i) 0.3 M - N^1,N^1,N^3,N^3-tetramethylguanidinium (TMG) 2-nitrobenzal-doximate/dioxan-water (9:1 v/v), room temperature, 24 hr;
(ii) aqueous ammonia (d 0.88), room temperature, 72 hr;
(iii) 0.01 M - hydrochloric acid, room temperature, 6 hr.

step (i), the 2-chlorophenyl protecting groups were removed from the internucleotide linkages by treatment (see above) with the TMG salt of syn-2-nitrobenzaldoxime (16b) (Ref. 20) in dioxan-water. Treatment with 2-nitrobenzaldoximate ion under these conditions is known also to lead (Ref. 17) to the removal of the 6-O-(2-nitrophenyl) protecting groups from the 2-N-(4-t-butylphenylacetyl)-6-O-(2-nitrophenyl)guanine residues [as in (10; R^1 = 4-$(Me_3C)C_6H_4\overline{C}H_2$, R^2 = 2-$(O_2N)C_6H_4)$] and the 4-O-(2,4-dimethylphenyl) groups from the protected uracil residues [as in (11; R = 2,4-$Me_2C_6H_3$)]. It is likely that all of these oximate ion promoted unblocking processes were complete within, at most, 6-8 hr (Refs. 17 and 20). However, as they could not be monitored, oximate ion treatment was allowed to proceed for 24 hr. In step (ii), the 5'-terminal Dbmb and N-acyl protecting groups were removed by treatment with concentrated aqueous ammonia. In order to be reasonably certain that deacylation was complete, the reaction was again allowed to proceed for much longer than what was believed to be the minimum time necessary.

The final unblocking step [Scheme 7, step (iii)] involves the removal of the remaining 2'-Mthp and 3'-terminal Mm protecting groups by treatment with 0.01 M - hydrochloric acid (pH 2) (Ref. 9) at room temperature for 6 hr. However, as fully-unprotected oligo- and polyribonucleotides are susceptible to ribonuclease-promoted digestion, it was decided to purify the partially-protected nonadecaribonucleoside octadecaphosphate (54) [see above for system of abbreviations] before removing the acid-labile (i.e. Mthp and Mm) protecting groups. A sample of (53), which had been purified by chromatography both on silica gel and Sephadex LH-60, was therefore partially unblocked [by steps (i) and (ii), Scheme 7]. The resulting product (54) was chromatographed on DEAE Sephadex A25, using a linear gradient [from 0.001-1.5 M] triethylammonium bicarbonate buffer (pH 7.5). The elution profile obtained is illustrated in Fig. 1. The major peak, which related to material eluted from the column in the buffer concentration range 0.67 - 1.05 M, contained ca. 190 A_{260} units of material. The central fractions (41-50), which accounted for ca. 80% of the latter A_{260} units, were found to be virtually homogeneous by liquid chromatography on µBondapak C_{18}.

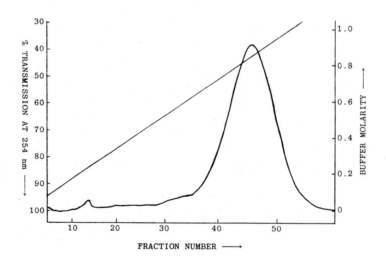

Fig. 1. DEAE-Sephadex A25 chromatography of the products obtained following the partial unblocking [by steps (i) and (ii) of the procedure indicated in Scheme 7] of the fully-protected nonadecaribonucleoside octadecaphosphate (53). A linear gradient of triethylammonium bicarbonate buffer (pH 7.5) was used.

The completely unprotected nonadecaribonucleoside octadecaphosphate (55) was generated when required by the action of aqueous acid [step (iii), Scheme 7] on (54); this material was completely degraded by treatment with aqueous sodium hydroxide, ribonuclease A, calf spleen phosphodiesterase and Crotalus adamanteus snake venom phosphodiesterase. Satisfactory qualitative and quantitative data were obtained by analysis of the hydrolysates.

We now intend to proceed to the synthesis of longer sequences of yeast tRNA^{Ala}. When the Dbmb protecting group has been removed from the 5'-end of the fully-protected nonadecaribonucleotide octadecaphosphate (53), further chain extension may be effected by a block condensation reaction. It is envisaged that such condensation reactions will involve relatively large (at least nonanucleotide) blocks. This should facilitate the isolation of the desired condensation products as purification by both gel filtration methods (e.g. on Sephadex LH-60) and conventional silica gel chromatography should then be possible. We have noted that the protection of guanine residues on O-6 [as in (10; R^2 = 2-$(O_2N)C_6H_4$)] and uracil residues on O-4 [as in (11; R = 2,4-$Me_2C_6H_3$)] has led to improved condensation reactions and, perhaps more importantly, has facilitated short column chromatographic separations of fully- and partially-protected phosphotriester intermediates. Indeed, it seems unlikely that the synthesis of the nonadecaribonucleoside octadecaphosphate (55) would have been possible without such protection, particularly of the guanine residues. Finally, as suggested above, we believe that it would have been better to have undertaken the synthesis of the partially-protected nonaribonucleoside octaphosphate (43) and the derived nonanucleotide (51) blocks from the 3'- rather than from the 5'-end, by making use of a suitable 3'-terminal protecting group. We are now in the process of repeating the synthesis of (51) in this way, and intend to synthesize other blocks of similar size by such an approach in the future.

Acknowledgements - We thank the Science and Engineering Research Council for generous support of this project. One of us (S.S.) also wishes to thank the British Council for the award of a Scholarship.

REFERENCES

1. N.J. Cusack, C.B. Reese, and J.H. van Boom, Tetrahedron Lett. 2209-2212 (1973); R. Arentzen and C.B. Reese, J. Chem. Soc. Perkin I 445-460 (1977).
2. J.B. Chattopadhyaya and C.B. Reese, Nucleic Acids Res. 8, 2039-2053 (1980).
3. K. Itakura, C.P. Bahl, N. Katagin, J.J. Michniewicz, R.H. Wightman, and S.A. Narang, Can. J. Chem. 51, 3649-3651 (1973); K. Itakura, N. Katagin, S.A. Narang, C.P. Bahl, K.J. Marians, and R. Wu, J. Biol. Chem. 250, 4592-4600 (1975).
4. K. Itakura, T. Hirose, R. Crea, A.D. Riggs, H.C. Heyneker, F. Bolivar, and H.W. Boyer, Science 198, 1056-1063 (1977).
5. K.L. Agarwal and F. Riftina, Nucleic Acids Res. 5, 2809-2823 (1978).
6. G.R. Gough, C.K. Singleton, H.L. Weith, and P.T. Gilham, Nucleic Acids Res. 6, 1557-1570 (1979).
7. C.B. Reese, Tetrahedron 34, 3143-3179 (1978).
8. C.B. Reese, Colloques Internationaux du C.N.R.S. 182, 319-328 (1970).
9. S.S. Jones, B. Rayner, C.B. Reese, A. Ubasawa, and M. Ubasawa, Tetrahedron 36, 3075-3085 (1980).
10. See Ref. 7, p.3146.
11. C.B. Reese, R. Saffhill, and J.E. Sulston, J. Am. Chem. Soc. 89, 3366-3368 (1967); Tetrahedron 26, 1023-1030 (1970).
12. M. Jarman and C.B. Reese, Chem. and Ind. 1493-1494 (1964); B.E. Griffin, M. Jarman, C.B. Reese, and J.E. Sulston, Tetrahedron 23, 2301-2313 (1967).
13. C.B. Reese and J.C.M. Stewart, Tetrahedron Lett. 4273-4276 (1968); C.B. Reese, J.C.M. Stewart, J.H. van Boom, H.P.M. de Leeuw, J. Nagel, and J.F.M. de Rooy, J. Chem. Soc. Perkin I 934-942 (1975).
14. J.B. Chattopadhyaya, C.B. Reese, and A.H. Todd, J. Chem. Soc. Chem. Commun. 987-988 (1979).
15. P.T. Gilham and H.G. Khorana, J. Am. Chem. Soc. 80, 6212-6222 (1958); R.K. Ralph, W.J. Connors, H. Schaller, and H.G. Khorana, ibid. 85, 1983-1988 (1963).
16. C.B. Reese and A. Ubasawa, Tetrahedron Lett. 21, 2265-2268 (1980); Nucleic Acids Res., Symposium Series No. 7, 5-21 (1980).
17. S.S. Jones, C.B. Reese, S. Sibanda, and A. Ubasawa, Tetrahedron Lett. 22, 4755-4758 (1981).
18. R.W. Adamiak, R. Arentzen, and C.B. Reese, Tetrahedron Lett. 1431-1434 (1977).
19. B.J. Hunt and W. Rigby, Chem. and Ind. 1868-1869 (1967).
20. C.B. Reese and L. Zard, Nucleic Acids Res. 9, 4611-4626 (1981).
21. C.B. Reese and J.E. Sulston, Proc. Chem. Soc. 214 (1964); H.P.M. Fromageot, B.E. Griffin, C.B. Reese, and J.E. Sulston, Tetrahedron 23, 2315-2331 (1967).
22. R. Arentzen, Y.T. Yan Kui, and C.B. Reese, Synthesis 509-510 (1975).
23. W.T. Markiewicz, J. Chem. Research (S) 24-25 (1979).
24. H.G. Khorana, Pure Appl. Chem. 17, 349-381 (1968); K.L. Agarwal, A. Yamazaki, P.J. Cashion, and H.G. Khorana, Angew. Chem. internat. Edit. 11, 451-459 (1972).
25. B.E. Griffin and C.B. Reese, Tetrahedron 25, 4057-4069 (1969).
26. C.B. Reese and R. Saffhill, Chem. Commun. 767-768 (1968).
27. C.B. Reese, R.C. Titmas, and L. Yau, Tetrahedron Lett. 2727-2730 (1978).
28. J.B. Chattopadhyaya and C.B. Reese, Tetrahedron Lett. 5059-5062 (1979).
29. C.B. Reese and M. Ubasawa, unpublished observations.
30. Biochemistry 9, 4022-4027 (1970).
31. J.F.M. de Rooij, R. Arentzen, J.A.J. den Hartog, G. van der Marel, and J.H. van Boom, J. Chromatog. 171, 453-459 (1979).

SYNTHETIC ASPECTS OF 1-OXACEPHEM ANTIBIOTICS

Wataru Nagata

Shionogi Research Laboratories, Shionogi & Co., Ltd., Fukushima-ku, Osaka 553, Japan

Abstract — Various aspects of the 1-oxa-1-dethia-cephem syntheses starting from penicillins are discussed, especially, from the viewpoint of industrial production. Consideration of several approaches showed that the approach via the intermediate a (Scheme 5) was the most efficient and industrially feasible. In this route, stereocontrol is effected by intramolecular alcoholysis of the enantio-oxazolinoazetidinone derivative 68 giving the 7-epi-3-methylene-1-oxacepham nucleus 56. Some improvements for conversion of the allylic iodide 67 into the allylic alcohol 68 are described, among which peracid oxidation of 67 to 68 was found to be the most advantageous. Interestingly, a new [2,3]-sigmatropic rearrangement of the allylic iodoso compound is involved in this oxidation reaction, which was developed into a new, general method for preparing the rearranged allylic alcohols from the allylic iodides. New methods are also introduced for converting the 7-epi-3-methylene-1-oxacepham nucleus 56 into various 1-oxacephem antibiotics including latamoxef (moxalactam, 6059-S) 12. One of these methods offers the advantage of forming the 3(4)-double bond at the final step after completing the necessary functionalization at positions 3' and 7 on 1-oxacepham intermediates which are generally more stable to nucleophiles than 1-oxa-3-cephem analogs.

INTRODUCTION

A principal role that organic syntheses play, especially in industrial laboratories, is to provide an efficient way to produce, in large quantities, a biologically valuable substance which either occurs in nature in small amounts or can only be prepared artificially. Many such examples can be found in the chemistry of β-lactam antibiotics. Clavulanic acid 1 (Ref. 1), a representative β-lactamase inhibitor, thienamycin 2 (Ref. 2), a representative of carbapenem antibiotics, and monobactam antibiotics 3 (Ref. 3 & 4), a novel class of antibiotics with a unique single β-lactam structure, have recently been isolated from fermentation broths in small amounts. In view of their unique antibacterial activities and structures, these antibiotics have become attractive synthetic targets for many β-lactam chemists.

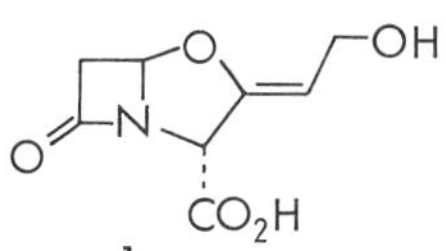

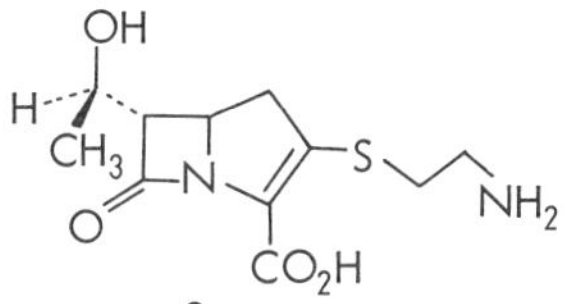

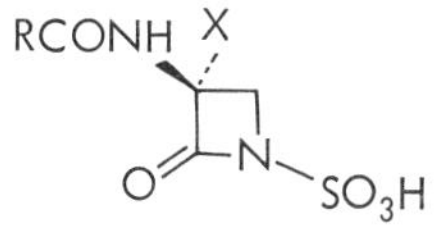

Scheme 1

On the other hand, recent advances in knowledge of the action mechanism and the structure-activity relationships in the field of β-lactam antibiotics have guided synthetic chemists to the challenge of modifying even the nuclei of penicillins and cephalosporins. Efforts in this direction are exemplified by synthetic studies of penems 5 (Ref. 5), which seem to be a structural hybrid of penicillin 4 and cephalosporin 6, 1-carba-1-dethia-cephems 7 (Ref. 6), and 1-oxa-1-dethia-cephems 8 (Ref. 7 & 8). The dream pursued in the field of organic syntheses is the discovery of a β-lactam compound with superior antibacterial activity, that is clinically useful for treating pathogenic infections and can be manufactured on an industrial scale. Synthesis of penems 5 was undertaken first by Woodward and his associates (Ref. 5). Interesting antibacterial activity of this class of compounds has been reported (Ref. 5b). 1-Carbacephems 7 were synthesized by Christensen and his associates at Merck in 1974 (Ref. 6). Syntheses of 1-oxacephems 8 have attracted much attention from many β-lactam chemists because of their anticipated antibacterial activity.

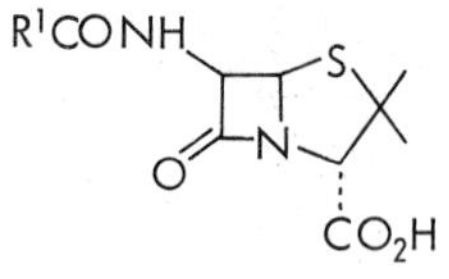

Scheme 2

4

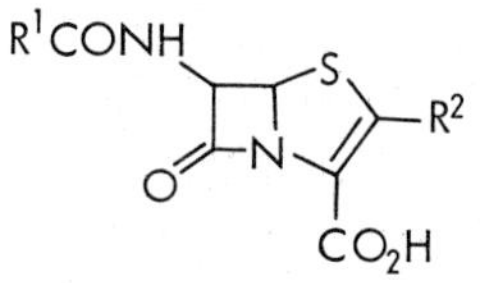

5

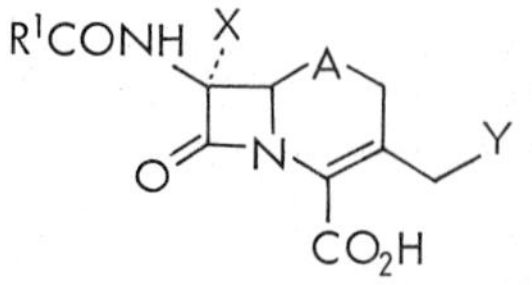

6 A = S
7 A = CH$_2$
8 A = O

Reviewed here are contributions from our laboratories to advances in the field of 1-oxa-cephem syntheses.

HISTORY

In 1974, Wolfe et al. (Ref. 7) reported the successful transformation of penicillin into methyl 1-oxacephem-4-carboxylate 9, most likely representing the first synthesis of this class of β-lactam compounds. However, this compound was not converted into its free acid and thus the antibacterial activity of this nuclear analog was not revealed. In the same year, Cama and Christensen (Ref. 8) succeeded in synthesizing the 1-oxa analog 10 of cephalothin as its racemic form and showed for the first time that 1-oxacephalothin possesses an antibacterial spectrum and potency parallel to those of the 1-thia congener. This result clearly demonstrated that the 1-sulfur atom is not essential for the antibiotic activity of cephalosporins. This important finding aroused the intense interest of many β-lactam chemists in synthesizing other nuclear analogs of cephalosporins. As for 1-oxa analogs, Beecham chemists (Ref. 9) reported the successful synthesis of the simplest but fundamental 1-oxa analog 11a in optically active form in 1977 and in the same year, we independently succeeded in synthesizing optically active 1-oxa analog 11b (Ref. 10) and then 1-oxacephalothin 10 (Ref. 11). We have found that, in contrast with the observation of the Merck group, optically active 10 exhibits antibacterial activity 4-8 times as active as that of cephalothin. To our surprise, the phenylmalonamido 1-oxacephem 11c was 16 times as potent as the corresponding 1-thia congener. Encouraged by this finding, we carried out the 1-oxacephem study more extensively and discovered a new potent antibacterial agent, latamoxef 12 (Ref. 11) (Note a) [moxalactam (Note b), 6059-S (Note c)]. This synthetic antibiotic has been found to be strongly active against not only a wide range of gram-negative bacteria including the pseudomonas species but also pathogenic anaerobes such as bacteroides fragilis. It is completely stable to various β-lactamases and thus is active against various resistant strains. In addition, it has several favorable therapeutic properties such as very low toxicity, high plasma peak level, and long duration.

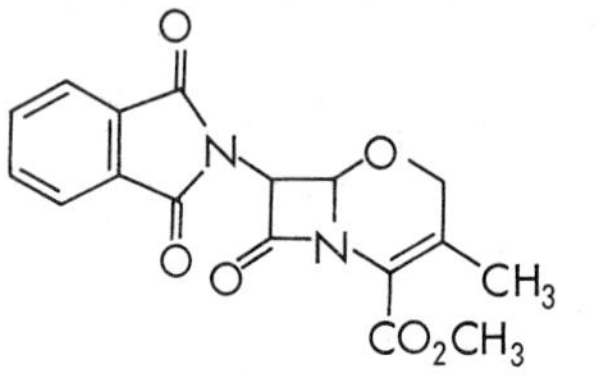

9

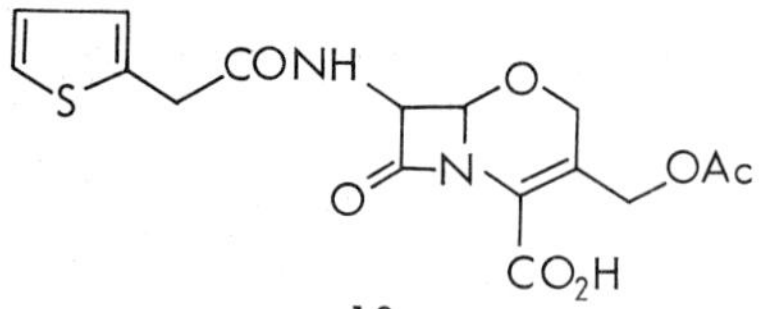

10

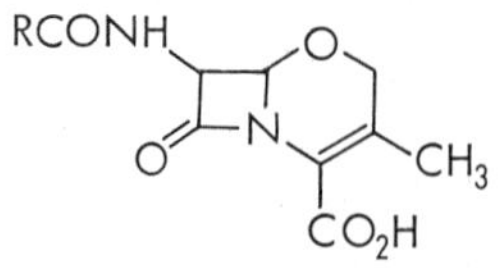

11a R = C$_6$H$_5$OCH$_2$
11b R = C$_6$H$_5$CH$_2$
11c R = C$_6$H$_5$CH
            |
            CO$_2$H

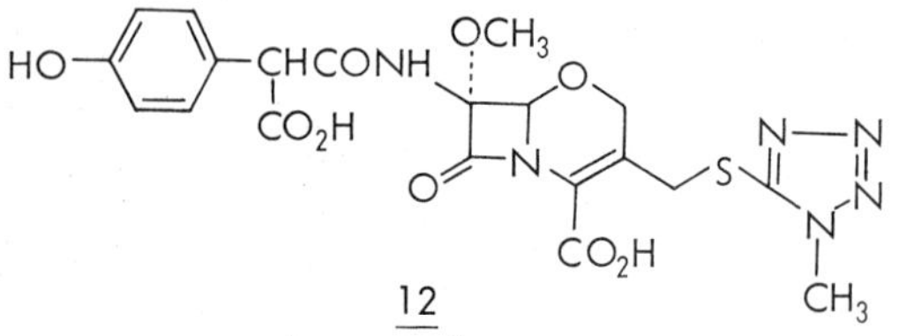

12
latamoxef
(moxalactam, 6059-S)

Scheme 3

Note a. International Nonproprietary Name (INN).
Note b. United States Adopted Name (USAN).
Note c. Code No. of Shionogi Research Laboratories.

Nevertheless, there appeared to be little prospect of commercial success, because the complexity of the molecule made it seem impossible to find a synthetic route suitable for industrial production. However, we succeeded in establishing a synthetic method leading to its production, and discussed here are the results of our synthetic studies with emphasis on those carried out in recent years.

STRATEGY

From the strategical point of view, penicillin was considered to be the most suitable starting material for the synthesis of 1-oxacephalosporin analog 15. First, as can be seen from Scheme 4, this compound can be obtained from penicillin 13 simply by removing the 1-sulfur atom followed by bridging with an oxygen atom between $C_{2\beta}$ and C_5 in the resulting seco-intermediate 14. The azetidinone part, which itself can not be easily prepared, is useful as a building block furnished with a chiral center. The second reason for its suitability as the starting material is the five carbon skeleton in the dihydrooxazine ring can be constructed by using the five carbon atoms of the penicillin thiazolidine ring to provide the most direct approach. And third, penicillin is inexpensive and easily available in large quantities.

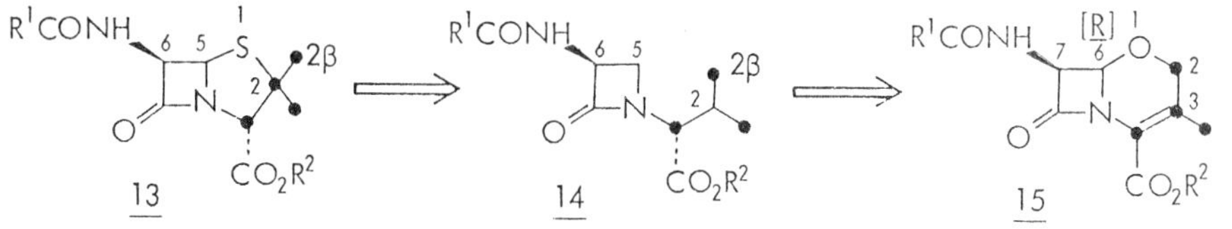

Scheme 4

Taking the known cephalosporin and 1-oxacephalosporin syntheses into consideration, we planned the following four feasible synthetic approaches a-d to construct the 1-oxacephem nucleus 16 on the basis of which bond must be formed for dihydrooxazine ring closure. In each approach, stereocontrol of C_6-O bond formation and ease of the ring closure are critical and thus some a priori arguments may be instructive. As discussed above, approaches via a and b provide the most direct and rational routes, whereas approaches via c and d are indirect and require several steps because the 3-methyl-3-butenoate-side chain must be partly or totally cut off and an appropriate alcoholic side chain must be added before closing the ring. Presumably, the C_6-O bond formation can be effected easily in the

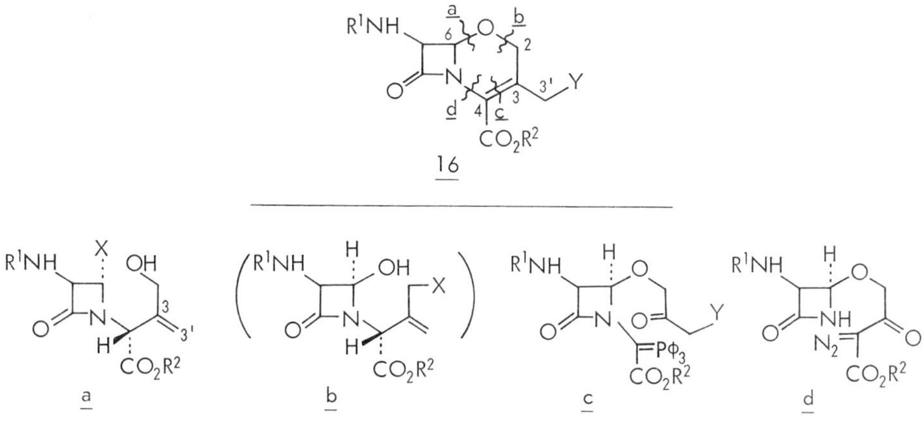

X = a leaving group
Y = a functional group

Scheme 5

approach via a under acetalization conditions, but stereocontrol to create the [R] chiral center at C_6 may cause a serious problem. This situation is well illustrated in Wolfe's pioneering work (Ref. 7). In his synthesis, the intermediate 17a or 17b was subjected to ring closure reaction using stannous chloride ($SnCl_2$). Whereas use of 1.33 molar equivalents of the reagent gave only the undesired trans isomer 18, use of 1.6 equivalents of the reagent gave the desired cis product 19 but only as a 1:1 mixture with the trans isomer 18. Such unsuccessful stereocontrol is also the case when preparing the intermediates c and d in Scheme 5, as discussed later. In contrast with the difficulty in stereocontrol at C_6, ring closure should be easy in the approaches via c and d, since the known intramolecular Wittig condensation (Ref. 12) and the intramolecular carbene insertion

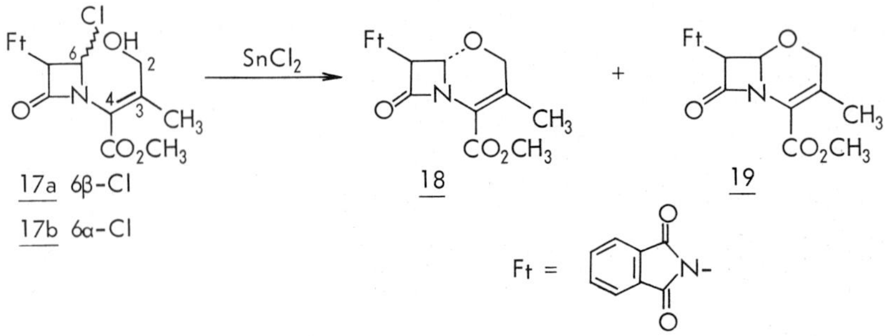

Scheme 6

reaction (Ref. 13), respectively, can be applied successfully. In the approach via b, ring closure between 1-oxygen and C_2 may be difficult, since intermediate b may be too unstable under the conditions for ring closure. In fact, we found by n.m.r. that an intermediate of type b is stable only below -30°C. Therefore this approach was not a promising one (Ref. 14) and is not discussed further. On the other hand, syntheses along other approaches were more or less successful and are discussed here in the order of c, d, and a.

In examining these approaches, various constraints required for industrial production were taken into consideration. The reactions had to be practical and easy to scale up. The reagents and the solvents needed to be inexpensive and, of course, nontoxic or non-noxious. Reactions which might give rise to explosions or fires were not feasible.

APPROACH VIA INTERMEDIATE c

Syntheses of azetidinone intermediates (stereochemical problems)
The principal problem in the approach via c is how to control the stereochemistry at C_6 (1-oxacephem numbering) in acetalization of the azetidinone fragment with the alcoholic side chain to form intermediate c. As is well known in glycosidation of aminosugars, the neighboring 7β acylamino substituent in 20 is considered to direct α-orientation of the entering alcoholic side chain at C_6 by participating in the formation of the carbonium ion 21 which is vulnerable only to the α-side attack. Thus, the trans product 23 formes exclusively. Similarly, acid-catalyzed alcoholysis of oxazolinoazetidinone 22 would proceed via the same carbonium ion 21 to give trans product 23 exclusively. In fact, the alcoholysis of 22 gave only the trans product 23 with none of the desired cis product 24 (Ref. 15). For this reason, we did not use the 7β-acylaminopenicillin fragment of type 20 in our first 1-oxacephem synthesis. Instead, we chose 7β-aminoazetidinone chloride 25 as the starting material, which was expected to give more or less a cis product. In fact, cis product 26a was formed in 26% yield together with 13% of the trans product 26b when 25 dissolved in propargyl alcohol was treated with zinc chloride and N-methylmorpholine (Ref. 10). The cis product, after separation by silica gel chromatography, was acylated and then hydrated to give the cis acetonyloxy azetidinone derivative 27. This result almost

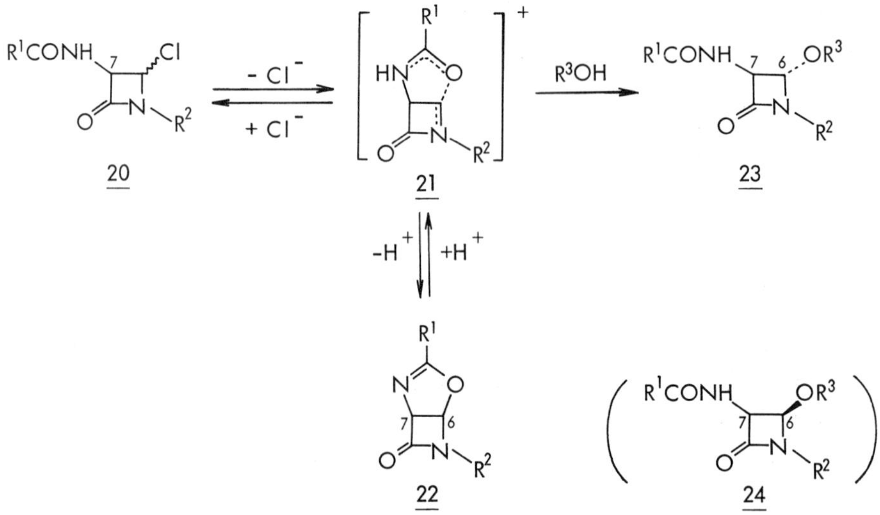

Scheme 7

Scheme 8

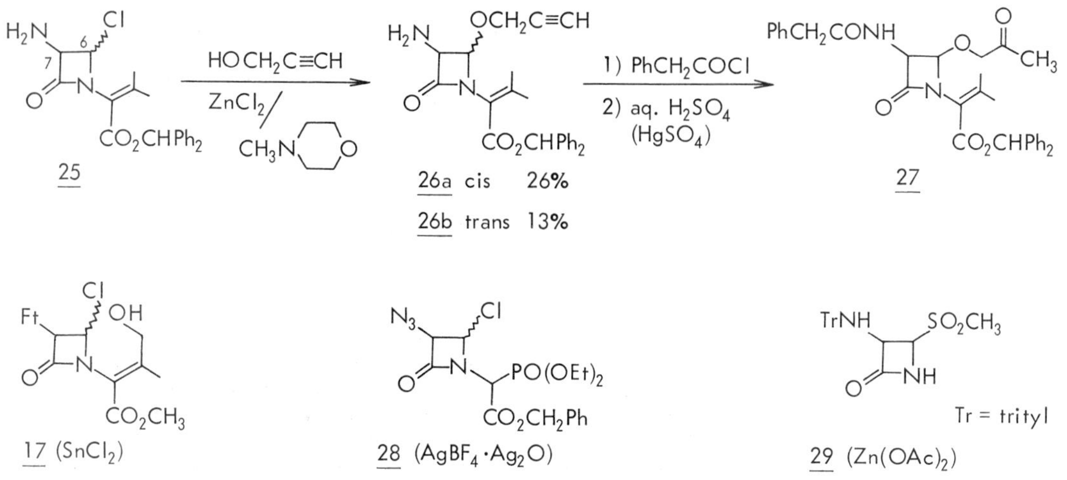

parallels the results on acetalization of the azetidinone derivatives 17 (Ref. 7), 28 (Ref. 8), and 29 (Ref. 9) substituted with the phthalimido, azido, and tritylamino group catalyzed by $SnCl_2$, $AgBF_4 \cdot Ag_2O$, and $Zn(OAc)_2$, respectively. The 7β substituents in these instances can not participate in complete exclusion of the cis product formation. Thus, the formation of the cis and trans mixture in an approximately 1:1 ratio in each case is reasonable.

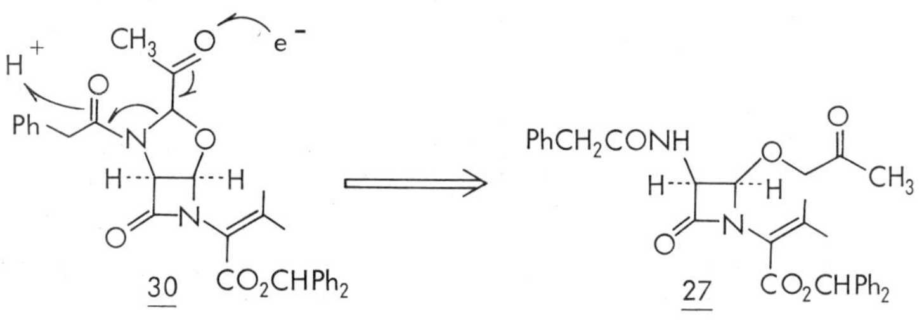

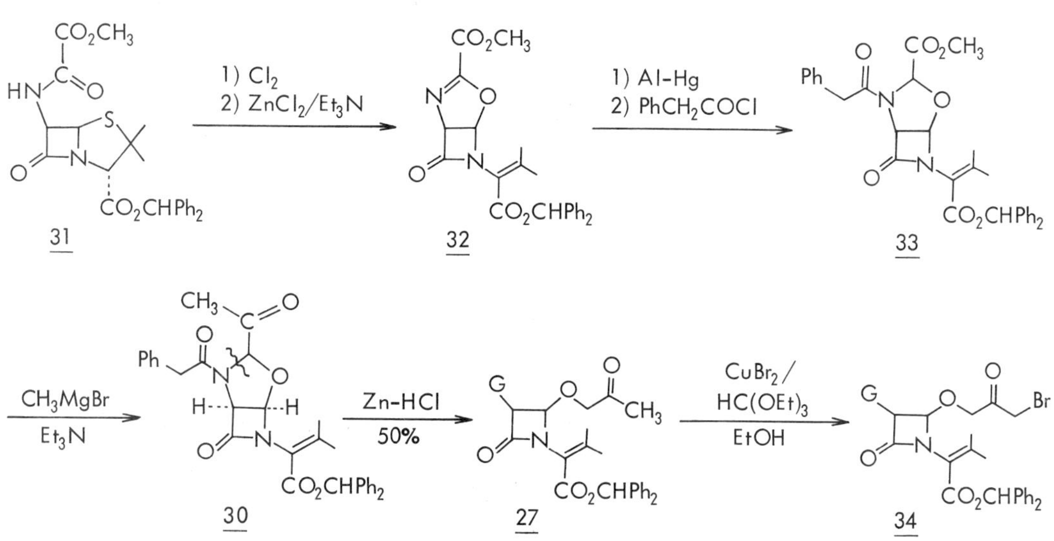

Needless to say, such a non-stereoselective conversion is unsatisfactory. Thus, we decided that the most reliable way to form the cis product exclusively would be to selectively

Scheme 9

Scheme 10 G = $PhCH_2CONH$

cleave the C-N bond of an appropriately designed oxazolidinoazetidinone such as 30. Reductive cleavage of compound 30 with a metal in the presence of a proton source should give the cis acetonyloxy azetidinone derivative 27 as a sole stereoisomer. Thus, the synthesis of the key compound 30 was undertaken (Ref. 16). The oxalylamino penicillin 31 was converted into the oxazoline 32, which on aluminum amalgam reduction followed by phenylacetylation gave 33 in high overall yield. Selective conversion of the oxazolidine ester group in 33 into the methyl ketone was effected with methylmagnesium bromide in the presence of triethylamine to give 30 in 70% yield. This key intermediate 30 underwent reductive cleavage with zinc and hydrogen chloride in tert-butanol and gave the expected acetonyloxyazetidinone 27 in an acceptable yield of 50% with recovery of 30% of 30. Terminal bromination at the acetonyl methyl in 27 was effected with cupric bromide in ethyl orthoformate and ethanol to give bromo ketone 34. Although exclusive formation of 27 was thus realized as expected, the synthesis involved too many steps and was not suitable for industrial production.

We finally investigated the acid-catalyzed cleavage of enantio-oxazolinoazetidinones such as 36 with an alcohol. According to the argument described above (Scheme 7), this reaction should exclusively give a trans compound in which the C_6-alkoxy group has the correct β-configuration, although the C_7-acylamino group is α. Since the 7α-acylamino group can be changed to β by either 7α-methoxylation or epimerization (see below), we thought this sequence of reactions might provide a useful method for 1-oxacephem synthesis. The

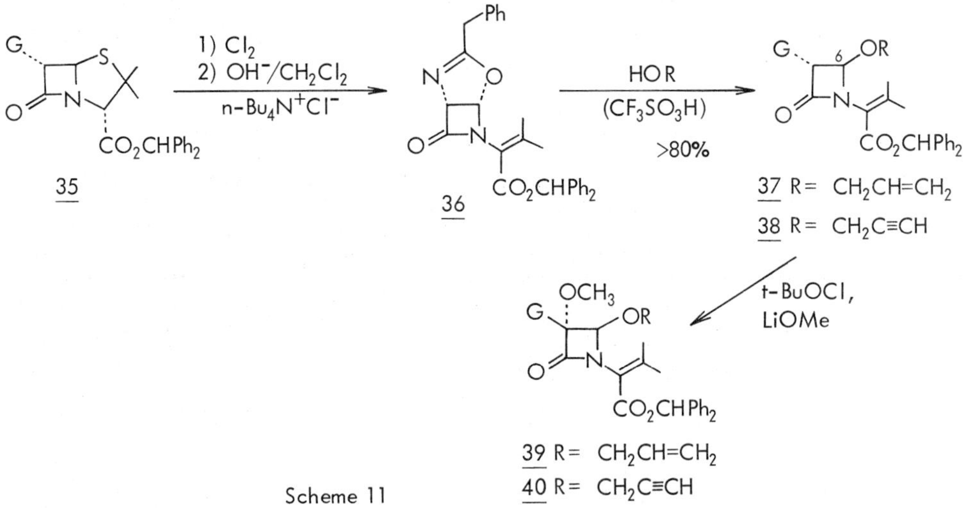

Scheme 11

enantio-oxazolinoazetidinone 36 was prepared in good yield from the 6-epi-penicillin 35 by chlorination followed by base treatment. Stereoselective cleavage of 36 with allyl alcohol or propargyl alcohol in the presence of a catalytic amount of trifluoromethanesulfonic acid proceeded smoothly as expected, producing the trans-acetal 37 or 38, respectively, in more than 80% yields with the correct configuration at C_6 (Ref. 17). The trans compounds 37 and 38 were methoxylated at C_7 by the conventional method (Ref. 18) and gave 39 and 40, respectively, in high yields. We now had the 7-acylamino group with the necessary β-configuration. In this way, these important intermediates became readily accessible from epi-penicillin under almost complete stereocontrol. Thus industrial production using this transformation seemed feasible.

Dihydrooxazine ring closure
As already discussed, the intramolecular Wittig reaction developed by Woodward and his associates (Ref. 12) can be applied to ring closure in this approach as illustrated in Scheme 12. The 6β-acetonyloxyazetidinone derivatives 43 and 44 were derived from the propargyloxyazetidinones 41 and 40 either directly by hydration followed by functionalization at the acetonyl methyl or by way of the allyloxyazetidinones 42 and 39 by successive epoxidation, cleavage with the nucleophiles (YH), and oxidation. The phosphoranes 45 and 46 were prepared in good overall yields by ozonization of 43 and 44 followed by reduction, chlorination, and triphenylphosphine treatment. These phosphoranes were finally heated and smoothly gave 1-oxacephem derivatives 47 and 48. Connection of the above ring closure sequence in the 7α-methoxy series with the efficient and practical preparation of the intermediate 39, which was discussed earlier, provided a fairly good and practical synthetic route to the 7α-methoxy-1-oxacephem nucleus. Although its feasibility for industrial production was fairly high, the route still involved too many synthetic steps and some chromatographical, though simple, purification processes.

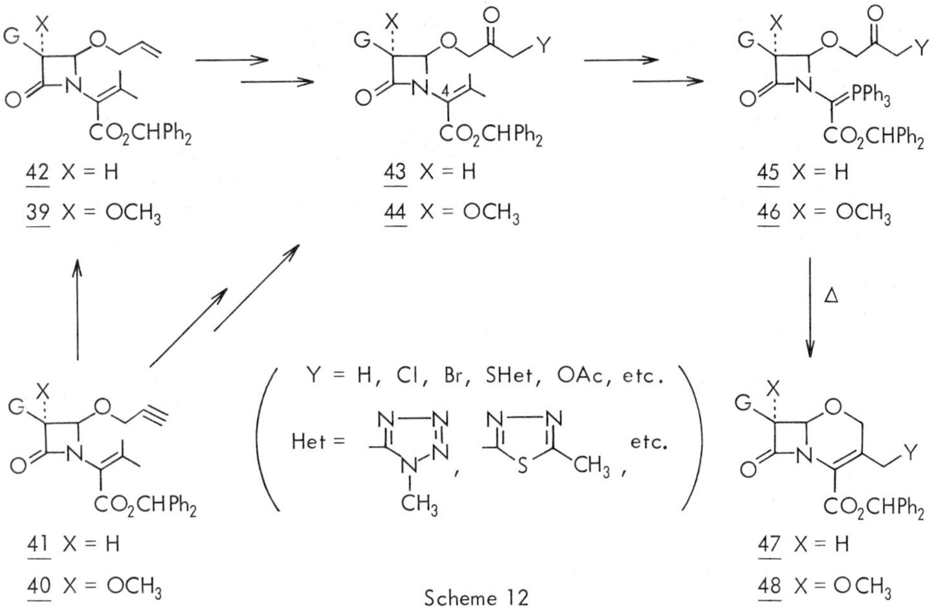

APPROACH VIA INTERMEDIATE d (A CONVERGENT SYNTHESIS)

Recently, R. W. Ratcliffe et al. (Ref. 13) reported a useful method for 5-membered ring closure in their carbapenem (thienamycin 2) synthesis. This method involves an efficient intramolecular carbene insertion to the β-lactam -NH- bond. Application of this method to construction of the 1-oxacephem nucleus provides a new approach via the intermediate d in Scheme 5, although this approach is suitable for obtaining the 3'-nor type 1-oxacephem nucleus. It is noteworthy that this approach represents a convergent synthesis and thus may be advantageous from the industrial viewpoint. The building block 50 was easily prepared from the epi-penicillin 49 by chlorination, base treatment, and ozonolytic removal of the 3-methyl-2-butenoate side chain of the resulting enantio-oxazolinoazetidinone (phenyl instead benzyl in 36; Scheme 11) in good overall yield. Reaction of the N-silyl derivative 51 with 4-hydroxy-2-diazoacetoacetate 52 prepared from 4-hydroxy-acetoacetate by diazo-transfer reaction using tosyl azide, stereoselectively gave the trans azetidinone derivatives 53 in high yield. The product after desilylation was heated in benzene with a catalytic amount of rhodium acetate affording the expected 3'-nor-1-oxacephem 55. Product 55 was identified with an authentic sample obtained from the known 56 by ozonolysis.

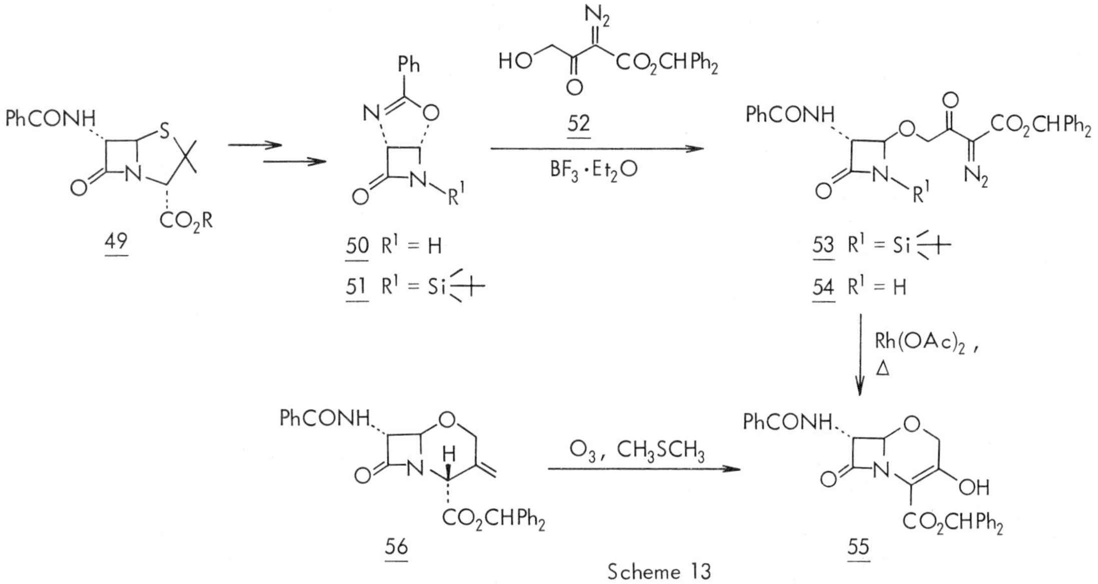

This approach thus provides a very short and efficient route to the 3-oxo-1-oxacephem nucleus.

APPROACH VIA INTERMEDIATE a (INDUSTRIALLY FEASIBLE SYNTHETIC ROUTE)

Tetrahydrooxazine ring closure and the stereochemical problem

As mentioned earlier, this approach provides the most direct and rational route to the 1-oxacephem nucleus. However, it shares the same stereochemical problem with the approach via c: the difficulty of introducing oxygen functionality from the β-side against the α-directing participation of the neighboring 7β-acylamino group. Despite this, we optimistically hoped that the severe steric interaction between the C_4-ester and the azetidinone ring would prevent the α attack via the transition state Tt and permit formation of the desired [6R]-1-oxacephem nucleus via Tc. The oxazolinoazetidinone intermediate 59 was subjected to

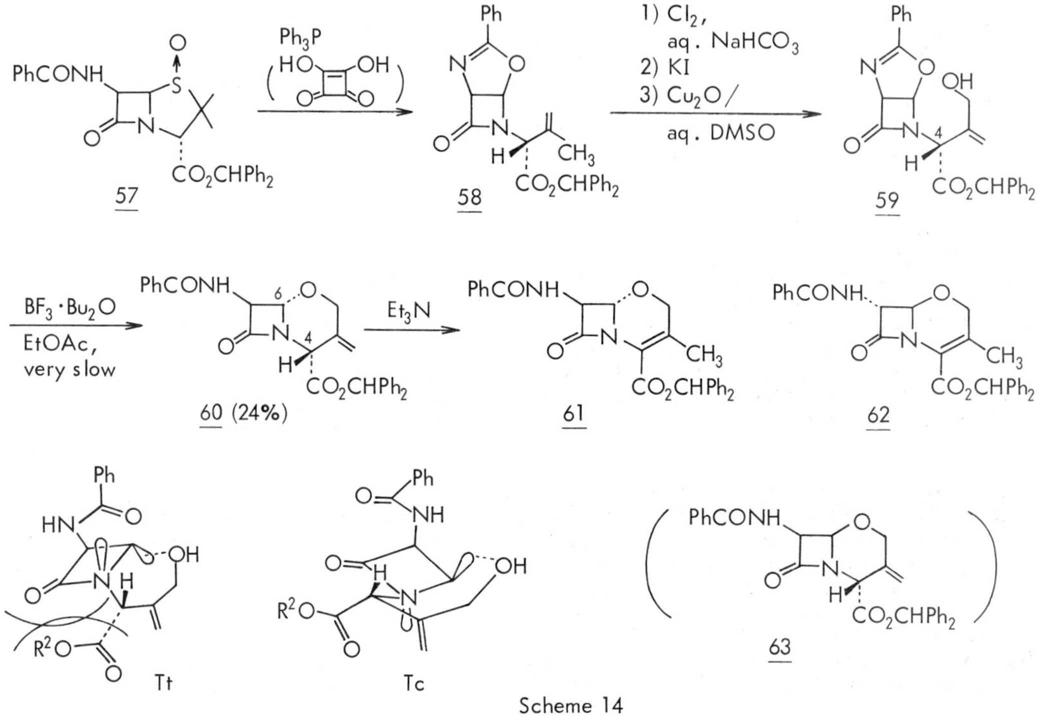

Scheme 14

boron trifluoride-catalyzed ring closure. Unfortunately, no cis product 63 was formed and the reaction proceeded sluggishly giving the isomeric 1-oxacepham 60 in only 24% yield. The structure of 60 was proved by its transformation into 61 which was identified as the enantiomer of an authentic sample of 62.

Oxazolinoazetidinone 59 was obtained by the sequence of reactions shown in Scheme 14. Penicillin S-oxide 57 was converted into oxazolinoazetidinone 58 in 85% yield by our recently developed method (Ref. 19) involving heating a toluene solution with triphenylphosphine in the presence of a catalytic amount of squaric acid. This valuable intermediate 58 having a non-conjugated 3-butenoate side chain can not be easily prepared otherwise.

The failure to obtain the desired 1-oxacepham 63 by ring closure of the normal oxazolinoazetidinone derivative 59 led us to use the corresponding enantiomeric oxazolinoazetidinone 68 in place of 59. Epi-penicillin S-oxide 64 was smoothly transformed into enantio-oxazolinoazetidinone 65 in 80% yield by heating with triphenylphosphine (Ref. 20). It was interesting to compare this result with a similar transformation to give the normal oxazoline 59 described above. Compound 65 underwent the "ene" type chlorination giving 66 which, on treatment with potassium iodide, gave allylic iodide 67. This was hydrolyzed to allylic alcohol 68 by treatment with cuprous oxide in warm aqueous dimethyl sulfoxide. In marked contrast to the case of 59, ring closure of this enantiomeric oxazoline alcohol catalyzed with boron trifluoride proceeded very smoothly giving the desired 7-epi-1-oxacepham derivative 56 in 90% yield. Despite the weak point of giving the 7α-epimeric acylamino derivative, this sequence of reactions was the most efficient and provided an important part of an industrially feasible synthetic route (Ref. 21). We therefore undertook further study to improve some of the steps.

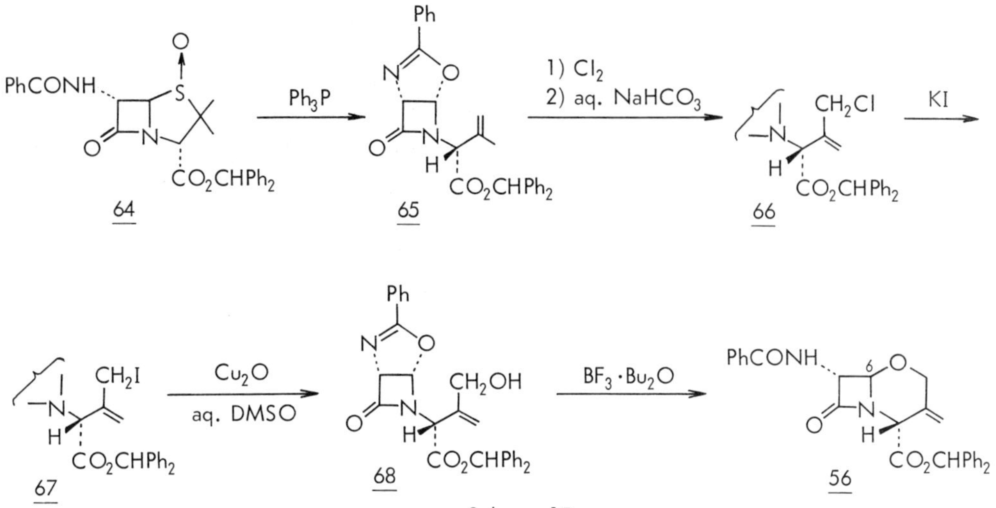

Scheme 15

Process improvements

For hydrolysis of the allylic iodide 67 to the allylic alcohol 68, various methods other than the Cu$_2$O method have been investigated. These include use of silver perchlorate in aqueous acetone, treatment with silver nitrate or a combination of sodium nitrate and methyl p-toluenesulfonate followed by reduction of one resulting allylic nitrate with zinc and acetic acid, and application of the Evans method involving sulfoxide rearrangement. As these methods have already been reported elsewhere (Ref. 21), other methods developed thereafter are described below.

Hydrogen peroxide or air oxidation of the allylic iodide under irradiation.

We observed that the allylic iodide 67 was oxidized readily with 30% hydrogen peroxide under irradiation in the presence of sodium bicarbonate to give a mixture of peroxide 70 (74%), allylic alcohol 68 (8%), and α,β-unsaturated aldehyde 71 (12%). The major peroxide 70 could be reduced to alcohol 68 in nearly quantitative yield. After intensive study in cooperation with the Process Research Division of our company to improve this two-step process, we found that allylic alcohol 68 could be obtained in 95% yield by air oxidation of allylic halide 66 or 67 under irradiation in the presence of excess amounts of dehydroacetic acid (DHA) and sodium iodide (Ref. 22). Undoubtedly, a radical mechanism is at work in these

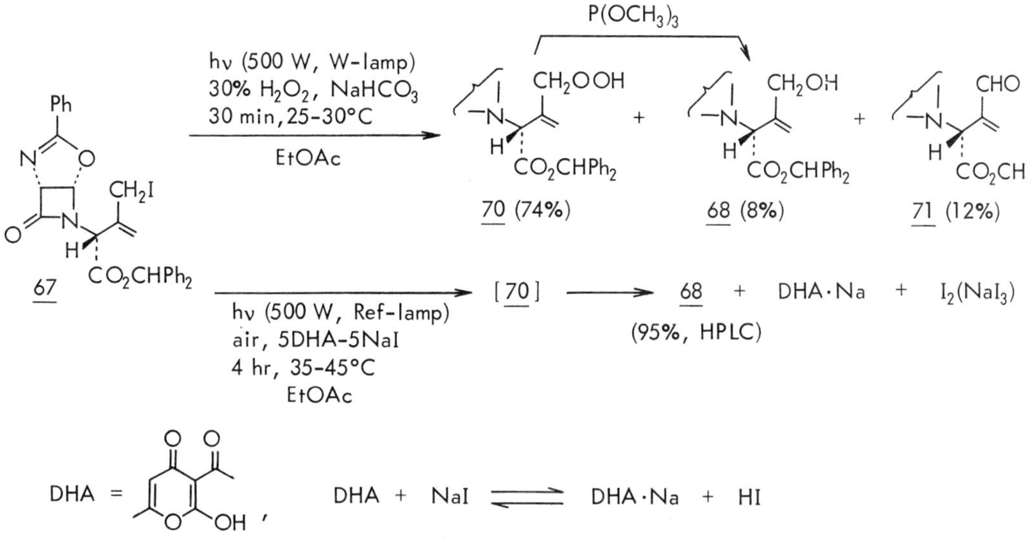

Scheme 16

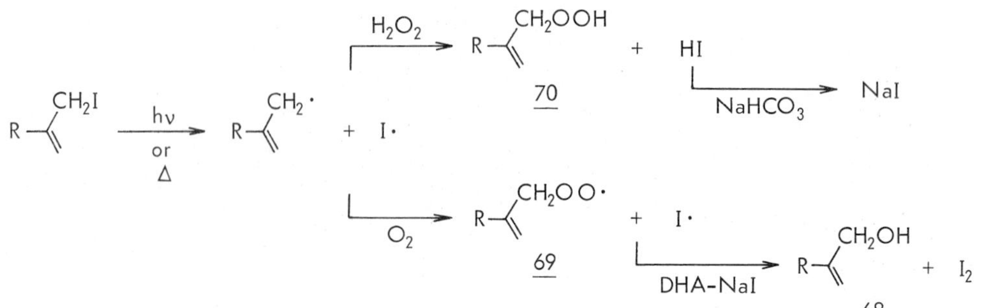

Scheme 17

procedures and the initially formed allylic radical reacts with either hydrogen peroxide or oxygen to form hydroperoxide 70 or the hydroperoxide radical 69 as shown in Scheme 17. The combination of DHA and sodium iodide seems to act as a reducing agent which converts the intermediate hydroperoxide radical into the alcohol. Of these two procedures, the latter autoxidation process is satisfactory for laboratory preparation of the allylic alcohol but not applicable to industrial production because of a low quantum yield in the photochemical step.

Peracid oxidation of the allylic iodide (a new [2,3]-sigmatropic rearrangement). Very recently, Reich and Peake (Ref. 23) and Macdonald et al. (Ref. 24) independently reported peracid oxidation of alkyl iodides to give alcohols, in some cases, accompanied by olefins, epoxides and/or ketones. According to the mechanism postulated by Macdonald et al., which is illustrated in Scheme 18, the primarily formed iodoso compound dissociates into the carbonium ion and hypoiodite ion. This carbonium ion is transformed, by either hydration or deprotonation, into the alcohol or the olefin which is further converted into the ketone or the epoxide. On the other hand, hypoiodous acid undergoes disproportionation to iodine

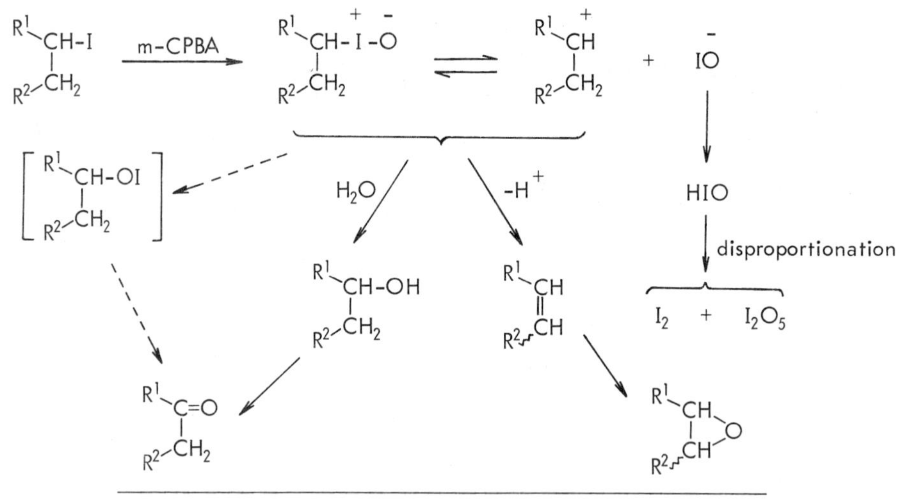

Scheme 18

and iodine pentoxide. We applied this reaction to convert the allylic iodide 67 into the alcohol 68. We expected the reaction to proceed smoothly and give a high yield of the allylic alcohol 68, since the stable allylic cation should have been produced. However, the result was disappointing. The allylic alcohol 68 was obtained but only in 40% yield and was accompanied by many other products including the α,β-unsaturated aldehyde 71. We thought that the side reactions might arise from the primarily formed iodic acid and that use of a base was essential. Accordingly, experiments were run in the presence of inorganic base such as $NaHCO_3$ or Na_2CO_3 in a two-layer system. The results were excellent, as can be seen from Table 1. In view of the high yield and the low price of peracetic acid, this new process may be applicable to industrial production, although some safety measures must be adopted.

TABLE 1. Results of peracid oxidation of allylic iodide 67 to give allylic alcohol 68

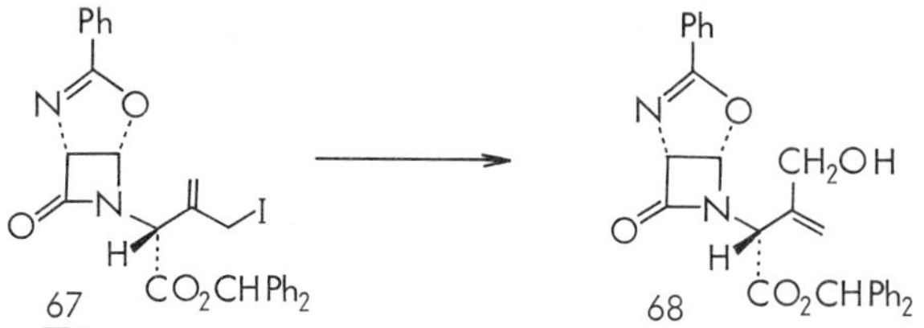

Peracid (mol)	Base (mol)	Solvents	Temp.	Time (hr)	Yield (%, HPLC)
m-CPBA (4.0)	NaHCO$_3$ (2.0)	EtOAc-H$_2$O (2:1)	r.t.	2	90
PhCO$_3$H (4.0)	NaHCO$_3$ (2.0)	EtOAc-H$_2$O (2:1)	r.t.	3	88
CH$_3$CO$_3$H (6.0)	NaHCO$_3$ (2.0)	EtOAc-H$_2$O (1:1)	r.t.	17	93
CH$_3$CO$_3$H (6.0)	NaHCO$_3$ (2.0)	CH$_2$Cl$_2$-H$_2$O (1:1)	r.t.	7	93

Apart from the industrial application, we have also been interested in the mechanism of this reaction. We observed some clear differences between our reaction and the Reich and Macdonald reaction. First, no iodine is liberated in our reaction, while Macdonald reported the formation of 1/3 mole of iodine. Second, at least 3 molar equivalents of peracid are consumed before our reaction is completed, while only 1.33 molar equivalents are necessary in their oxidation. These differences strongly suggest that our reaction follows a different mechanism. This reaction may proceed by an allylic rearrangement of the primarily formed allylic iodoso compound similar to the well-known rearrangement of the structurally related allylic selenoxide. To verify this, we checked whether allylic rearrangement takes place in this reaction. Deuterated allylic iodide 67(d), in which deuterium incorporation was 15-17% at the olefinic methylene and 25-27% at the allylic methylene, was prepared and subjected to m-chloroperbenzoic acid oxidation. The resulting allylic alcohol 68(d) was found to contain 26-28% deuterium at the olefinic methylene and 14-15% at the allylic methylene as listed in Table 2. Determination of the deuterium

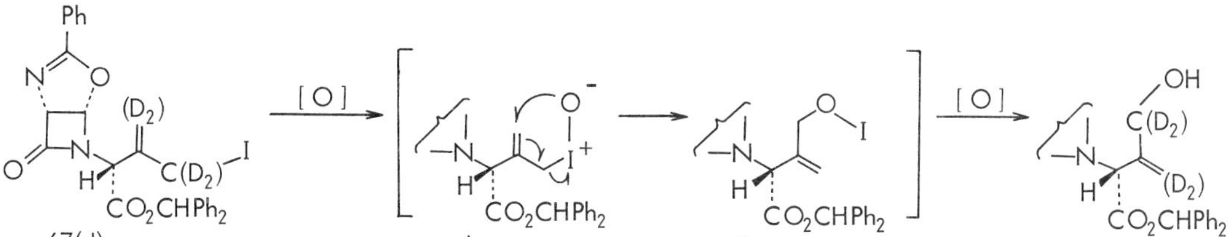

Scheme 19

TABLE 2. Deuterium incorporation (%) in 67(d) and 68(d)

		67(d)			68(d)		
Run	(1)	(2)	(3)	(1)	(2)	(3)
d (>=<H/H) %	16	17	15	28	28	26
d (—CH$_2$—) %	25	27	27	14	15	15
Total %	41	44	42	42	43	41

content and distribution was made by means of mass spectroscopic, proton n.m.r., and ^{13}C n.m.r. (deuterium isotope effect) studies (Ref. 25). This result clearly indicates that, as we assumed, this oxidation-hydrolysis reaction proceeds by [2,3]-sigmatropic rearrangement of, most probably, the allylic iodoso species A to the allylic hypoiodite B (Ref. 26). To confirm this mechanism we analyzed the products. Besides the main product 68, α,β-unsaturated aldehyde 71, α,β-unsaturated carboxylic acid 72, the epimeric allylic alcohol 73 and the lactone 74 were isolated in minor amounts. While the acid 72 proved to be formed from the aldehyde 71 by Baeyer-Villiger oxidation, this aldehyde was not formed from the allylic alcohol 68 under our reaction conditions. As for inorganic products, we isolated no iodine but did obtain sodium iodate in nearly quantitative yield.

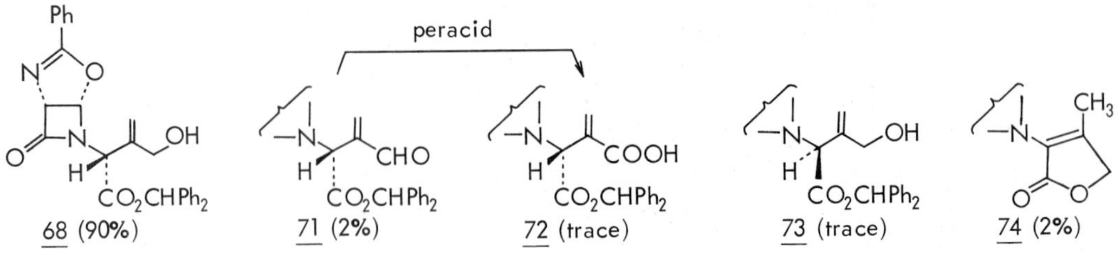
Ph
peracid
OH
CHO
COOH
OH
CH3
CO2CHPh2
68 (90%)
71 (2%)
72 (trace)
73 (trace)
74 (2%)

$NaIO_3$ (~100%) I_2 (~0%)

Scheme 20

On the basis of these facts, we propose the stoichiometry and the mechanism for our reaction as shown in Scheme 21. Significantly, 3 molar equivalents of peracid are used for oxidation of the allylic iodide and one equivalent of sodium bicarbonate for neutralization of the liberated iodic acid. The iodine atom in allylic iodide (i) is oxidized preferentially with one mole of peracid giving unstable iodoso compound (ii) which rapidly rear-

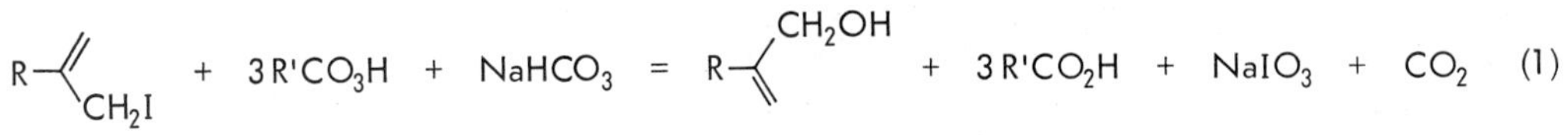
R
CH2I
+ 3R'CO3H + NaHCO3 =
CH2OH
+ 3R'CO2H + NaIO3 + CO2 (1)

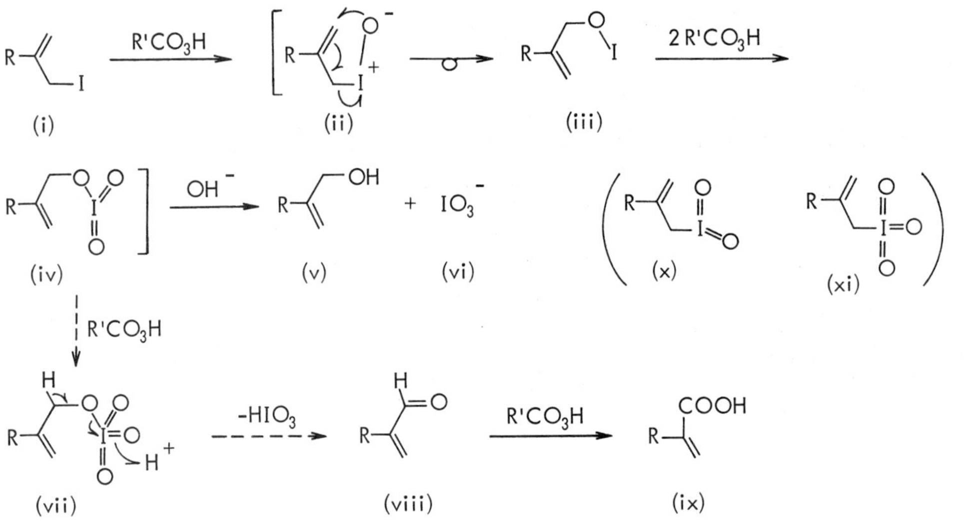
R'CO3H
2R'CO3H
OH
IO3
R'CO3H
-HIO3
R'CO3H
COOH
(i)
(ii)
(iii)
(iv)
(v)
(vi)
(x)
(xi)
(vii)
(viii)
(ix)

Scheme 21

ranges to allyl hypoiodite (iii) in a [2,3]-sigmatropic manner. The latter undergoes further oxidation with 2 moles of peracid giving allyl iodate (iv) which is then hydrolyzed with basic catalysis to give allylic alcohol (v) and iodate anion (vi). In this mechanism, we assume that, by analogy with the rearrangement of allylic sulfoxide or selenoxide, the rearranging species is allylic iodoso compound (ii) rather than other hypervalent iodine compounds such as (x) and (xi). It is plausible to assume that allylic iodate (iv) is further oxidized, though slowly, to periodate (vii) which decomposes to α,β-unsaturated aldehyde (viii) with liberation of iodic acid. This explanation agrees with the fact that a large amount of aldehyde is formed in the absence of an inorganic base. In such an acid medium, hydrolysis of allyl iodate must be slow and thus vulnerable to further oxidation. Conversion of the aldehyde (viii) into the acid (ix) was proven experimentally.

In view of the novelty in our reaction, we became interested in developing a general method for synthesizing allylic alcohols from allylic iodides. Methyl trans-γ-iodocrotonate 75 was subjected to this oxidation-hydrolysis reaction under the conditions indicated in Scheme 22. Deconjugated allylic alcohol 76 was formed in 70% yield accompanied by a trace of conjugated ketone 77. The formation of rearranged allylic alcohol 76 strongly supported the proposed mechanism of this reaction. It is worthy to note that no oxidation product (epoxide) of 76 was detected in the products. Because the product was expected to be

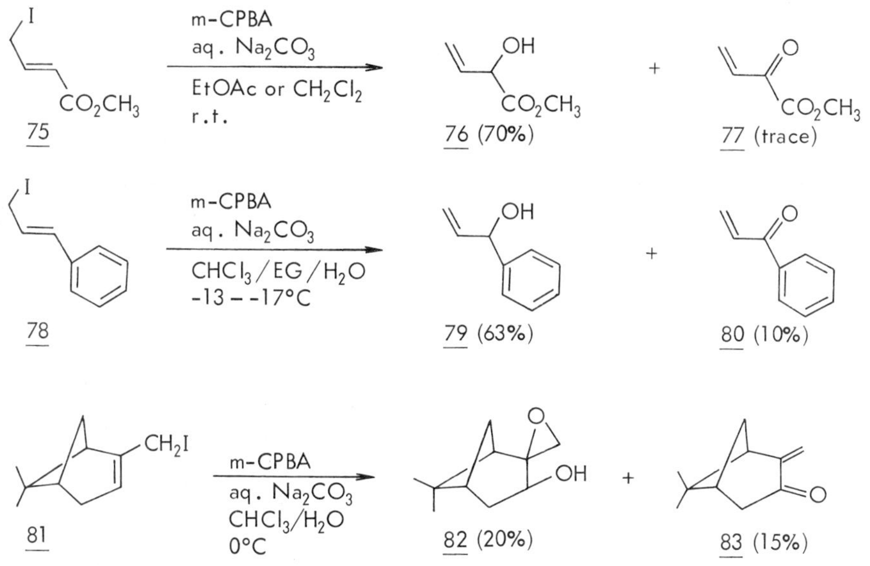

Scheme 22

unstable, oxidation of cinnamyl iodide was carried out at -13 to -17°C using ethylene glycohol as a co-solvent. The iodo group also was oxidized preferentially in this case and the rearranged allylic alcohol 79 was formed in 63% yield with 10% yield of vinyl phenyl ketone 80 as a by-product. In a preliminary experiment, m-chloroperbenzoic acid oxidation of myrtenyl iodide 81 gave 20% yield of pinocarveol β-epoxide 82 and 15% yield of pinocarvone 83. Pinocarveol β-epoxide must be derived by further epoxidation of the primarily formed pinocarveol, the expected rearranged allylic alcohol. This bicyclic olefin was expected to be sensitive to peracid. These examples demonstrate the general applicability of this reaction to syntheses of rearranged allylic alcohols.

$C_{3'}$-Functionalization and 7α-methoxylation of the 1-oxacepham nucleus

Now I would like to return to the final stage of the synthesis of 1-oxacephem antibiotics. Of concern at this stage are functionalization at $C_{3'}$ and α-methoxylation at C_7. As reported already (Ref. 21), 7-epi-3-methylene-1-oxacepham 56 was smoothly transformed into 3-chloromethyl-1-oxacephem 84 in high yield by chlorine addition under irradiation followed

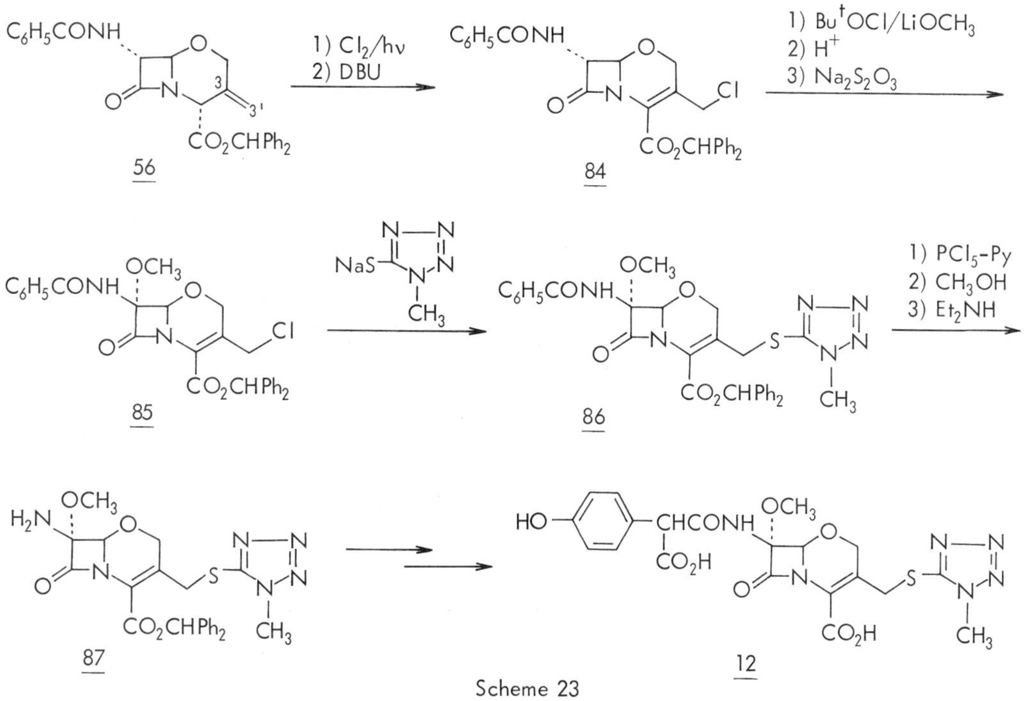

Scheme 23

by dehydrochlorination with 1,7-diazabicyclo[4.5.0]undec-6-ene (DBU). Methoxylation of 84 according to the Koppel method (Ref. 18) gave 7α-methoxy-1-oxacephem 85, which on treatment with sodium 1-methyltetrazole-5-thiolate, was led to 86 with a high overall yield. Debenzoylation by the conventional PCl_5 method gave methoxy amine 87, which was finally transformed into latamoxef (moxalactam) 12 by acylation followed by deprotection.

Most of the reactions used for the above transformation proceeded smoothly on a large scale. However, very rigorous reaction conditions were necessary in the Koppel methoxylation, because either of the reagents (t-BuOCl and $LiOCH_3$), when present in excess, cleaves the 1-oxacephem β-lactam ring. Fortunately, we have succeeded in developing the following new route which is free from this difficulty.

<u>A new, practical route to 1-oxacephems from 3-methylene-1-oxacephams.</u>   As is well known, the presence of the 3(4)-double bond, which composes the enamide system in the cephem or 1-oxacephem nucleus, is essential for the antibacterial activity. In other words, an important factor in the activity is a high acylating reactivity of the β-lactam carbonyl

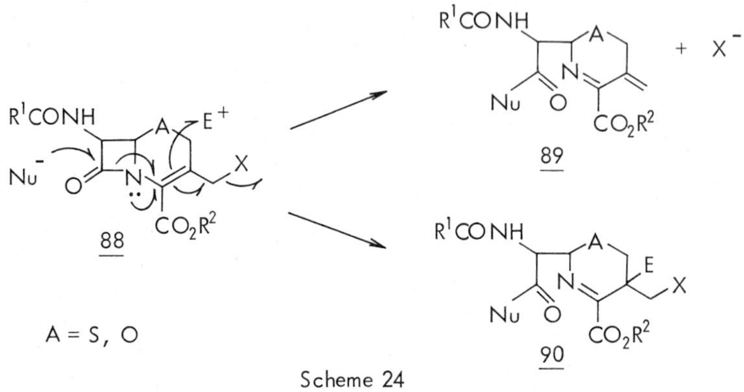

Scheme 24

enhanced by enamine resonance. Thus, the cephems 88 (A = S) and the 1-oxacephems 88 (A = O) with this enamide system are usually very susceptible to a nucleophile Nu^-, and yield 89 or 90, with simultaneous attack of an electrophile E^+, if any, as shown in Scheme 24. For example, substitution of 3'-chlorine with a nucleophile of low reactivity such as acetoxide (AcO^-) or methoxide (CH_3O^-) usually gives rise to severe decomposition, primarily producing 89. Also, in the 7α-methoxylation by the Koppel method (Ref. 18), t-butyl hypochlorite (E^+) and lithium methoxide (Nu^-) must be added at the same time in order not to cause an excess of either reagent, which would lead to decomposition to 90. Therefore, it would be advantageous if such elaborations could be made on 1-oxacephams without the 3(4)-double bond and then form this bond at the final step.

We found that anti-Markovnikov addition of methanesulfenyl chloride or benzenesulfenyl chloride to the 3-exomethylene of 7-epi-1-oxacepham 56 was effected in a highly regio- and stereoselective manner, giving 3β-methylthio- or phenylthio-3α-chloromethyl cepham, 91 or 92, respectively, as a sole product in nearly quantitative yield (Ref. 27). This adduct 91 reacted very readily with sodium 1-methyltetrazolyl-5-thiolate and even with dimethyl acetamide or methanol in the presence of silver tetrafluoroborate to give in high yields, the products substituted with N-methyltetrazolylthio 93a and acetoxy (after hydrolysis) 93b or methoxy 93c, respectively. On peracid oxidation followed by warming at 50-60°C for 10 minutes, the compounds 93a and 93b were converted, with removal of methanesulfenic acid, into the corresponding 1-oxa-3-cephem derivatives 94a and 94b in very high yields. These 1-oxacephems were deprotected by the conventional method, and gave now the free acids 95a and 95b. Alternatively, compounds 93a and 93b were first deprotected to give the free acids 96a and 96b, which on treatment with peracid or 30% hydrogen peroxide in the presence of sodium tungstate (Na_2WO_4), were converted smoothly into the 1-oxacephem-4-carboxylic acids 95a and 95b in high overall yields also. The whole sequence of reactions proved to be applicable also to 7α-methoxy-7-(2-thienyl)acetamido analog 97. Thus, this unique synthetic route has wide applicability.

It is interesting to note that the regio-selective and anti-Markovnikov addition of methane- or benzenesulfenyl chloride can be rationalized by assuming the formation of episulfonium ion 98 as an intermediate. This ion is preferentially attacked by the chloride ion at the less hindered secondary carbon ($C_{3'}$) to give 91 or 92, respectively. The exceptional ease with which substitution of 3'-chlorine with various nucleophiles occurs can also be accounted for by the initial formation of this episulfonium ion 98, as shown in

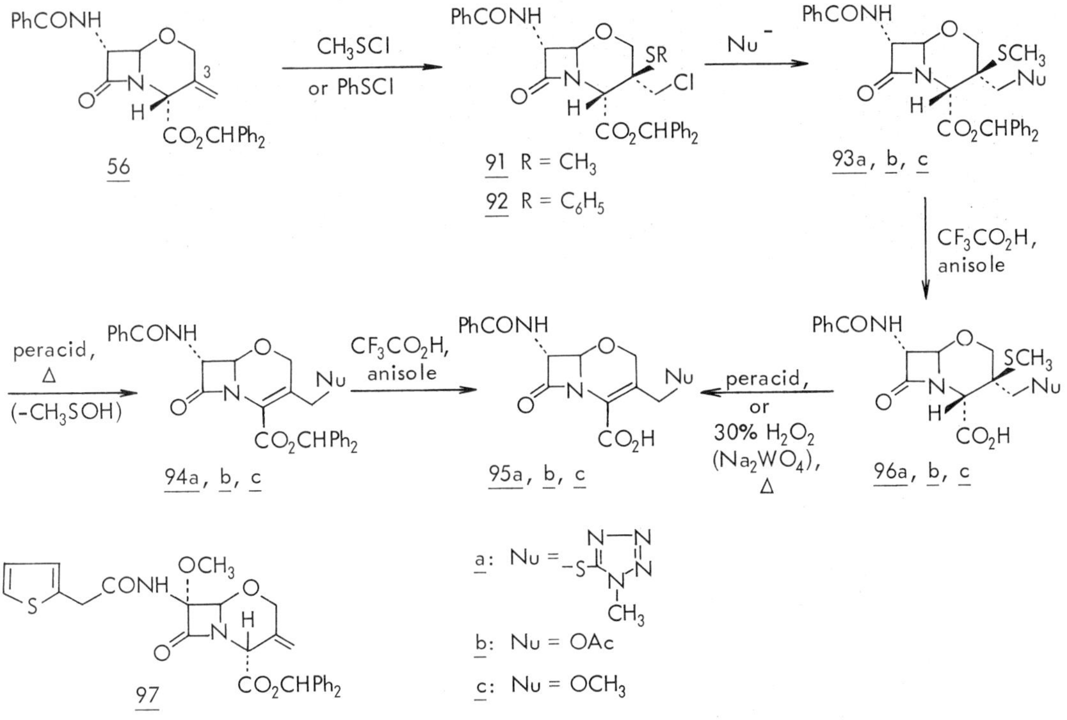

Scheme 25

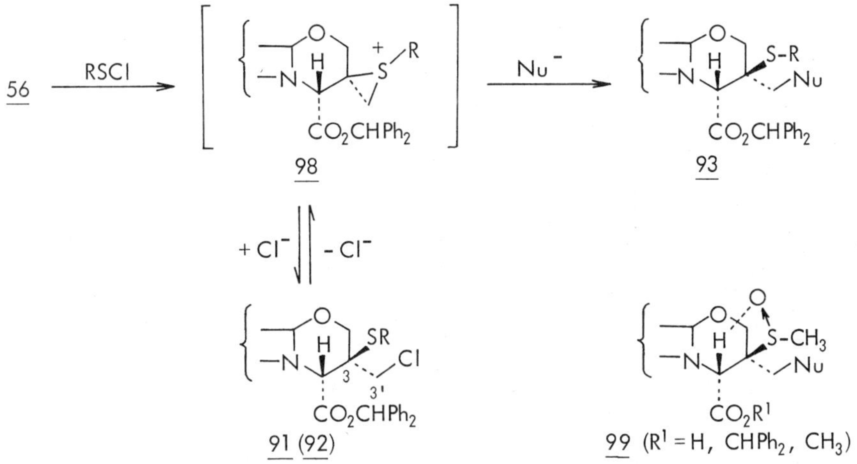

Scheme 26

Scheme 26. However, it is not clear why this sulfenyl chloride addition takes place with such a high β-directing stereoselectivity. Presumably, the 3-methylene 1-oxacepham molecule in the transition state preferably assumes a conformation in which the β-side of the 3-methylene is opened. Moreover, it should be pointed out that, only with this β-orientation of the 3-methylthio group in 99, is oxidative cis [2,3]-sigmatropic elimination with 4β-hydrogen to form the 3(4)-double bond possible.

Here is how latamoxef was synthesized by way of this new route. 7-Epi-3-methylene-1-oxacepham 56, a key intermediate in our latamoxef synthesis, was converted into 93, as shown just above. Compound 93 was debenzoylated to 7α-amino-1-oxacepham 100 in a high overall yield. This was treated with methanesulfenyl chloride in the presence of propylene oxide and molecular sieves according to the Squibb method (Ref. 28) and gave the methylthioimino derivative 101 in good yield. Treatment of this intermediate with methanol containing hydrogen chloride gave methoxy amine 102, the important nucleus, in high yield. Acylation and subsequent deprotection with aluminum chloride and anisole transformed methoxy amine

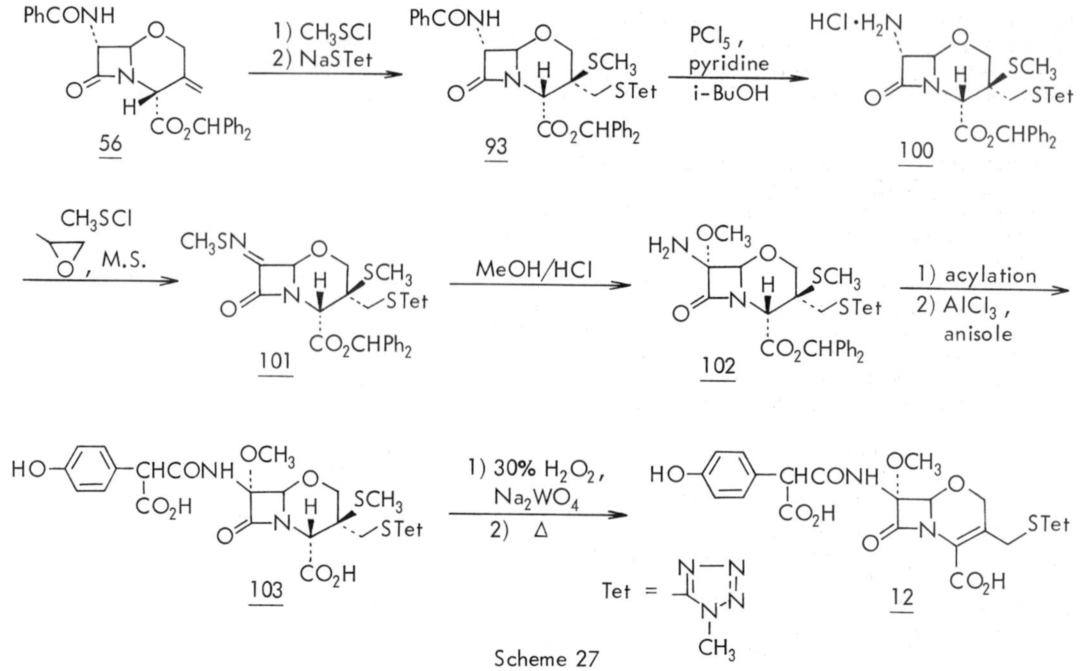

Scheme 27

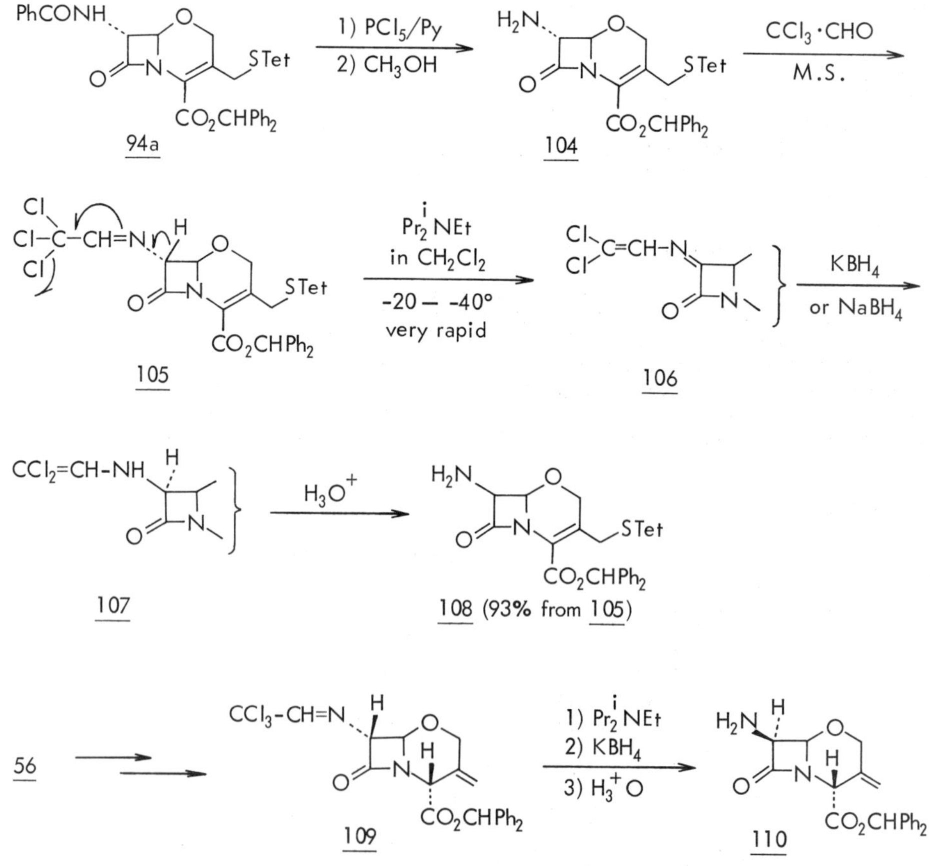

Scheme 28

102 into the cephamcarboxylic acid 103 very smoothly. Oxidation of this acid with 30% hydrogen peroxide in the presence of a catalytic amount of sodium tungstate, followed by slight warming, gave latamoxef (moxalactam) 12 in high yield. Thus, we established a useful alternative route for latamoxef synthesis.

Epimerization of the 7-amino group from α to β

The 7α-benzoylamino-3-methylene-1-oxacepham 56 is the first intermediate with the bicyclic ring system in our 1-oxacephem synthesis along the approach via a. In this intermediate, the 7-acylamino group is inevitably α-oriented as the result of stereocontrol at C_6. This 7α-orientation of the acylamino group should be changed to 7β for this compound to possess antibacterial activity. As discussed earlier, this can be done by one-step methoxylation at C_7, when the desired 1-oxacephems are of the cephamycin type like latamoxef 12. However, the epimerization does not seem so simple when the 7α-acylamino group is to be inverted to 7β without substitution at 7α. Several ingenious methods (Ref. 29) have recently been reported for this purpose in the cephalosporin field, but none has proved to be efficient and highly stereoselective.

We succeeded in developing a new method (Ref. 30) which is efficient and practical. 7α-Amino-1-oxacephem 104, obtained by debenzoylation of 94a (Scheme 25) by the conventional method, was condensed with chloral in the presence of molecular sieves to give the Schiff base 105 in good yield. When this product was treated with Hünig base (Pr^i_2NEt) at -20° to -40°C, very rapid elimination of hydrogen chloride took place to give 106 in nearly quantitative yield. This product was reduced with potassium or sodium borohydride, also in excellent yield, to 107 which was then hydrolyzed to 7β-amino-1-oxacephem 108 under acid conditions. It should be emphasized that this three-step transformation could be performed as a one-pot reaction to give 108 in an excellent overall yield of 93% from the Schiff base 105. As the dehydrochlorination step proceeds so rapidly, this reaction can be applied to preparation of 3-methylene-1-oxacepham 110 without any isomerization of the 3(3') double bond to 3(4). Thus, the Schiff base 109 obtained in two steps from 56 was subjected to this one-pot epimerization process to obtain another important intermediate 110 retaining the 3-methylene group in high overall yield.

Acknowledgement - The recent work, including the convergent synthesis along the approach via d, the peracid oxidation of the allylic iodides, development of the alternative route via the methanesulfenyl chloride adducts, and epimerization of 7α-acylamino substituents was performed with the earnest cooperation of Drs. S. Yamamoto, T. Aoki, and T. Tsuji, and Mr. H. Itani and Mr. H. Takahashi to whom I would like to extend my sincere thanks. Other earlier work, including the first synthesis of the 1-oxacephem nucleus, development of the improved stereocontrolled routes and finally establishment of the industrially feasible synthetic route was achieved in close cooperation with Drs. M. Yoshioka, M. Narisada, T. Tsuji, S. Uyeo, Y. Hamashima, S. Yamamoto, I. Kikkawa, T. Hamada, H. Onoue, M. Ohtani, T. Aoki, Y. Nishitani, S. Kamata, H. Satoh, S. Mori and their collaborators. Drs. K. Tokuyama and M. Tanaka of our Process Research Division cooperated in the photolytic autoxidation of the allylic iodide. I would like to take this opportunity to express my sincere gratitude to these chemists for their stimulating cooperation.

REFERENCES

1. J.T. Howarth and A.G. Brown, J. Chem. Soc., Chem. Comm. 266 (1976).
2. G. Albers-Schönberg, B.H. Arison, O.D. Hensen, J. Hirshfield, K. Hoogsteen, E.A. Kaczka, R.E. Rhodes, J.S. Kahan, F.M. Kahan, R.W. Ratcliff, E. Walton, L.J. Ruswinkel, R.B. Morin and B.G. Christensen, J. Am. Chem. Soc. 100, 6491 (1978).
3. A. Imada, K. Kitano, K. Kintaka, M. Muroi and M. Asai, Nature 289, 590 (1981).
4. R.B. Sykes, C.M. Cimarusti, D.P. Bonner, K. Bush, D.M. Floyd, N.H. Georgopapadokou, W.H. Koster, W.C. Liu, W.L. Parker, P.A. Principe, M.L. Rathnum, W.A. Slusarchyk, W.H. Trejo and J.S. Wells, Nature 291, 489 (1981).
5. (a) R.B. Woodward, in Recent Advances in the Chemistry of β-Lactam Antibiotics, Ed. J. Elks, Special Publication, No. 28, p. 167, The Chemical Society, London (1977). (b) R.B. Woodward, Phil. Trans. R. Soc. Lond. B289, 239 (1980).
6. R.N. Guthikonda, L.D. Cama and B.G. Christensen J. Am. Chem. Soc. 96, 7584 (1974).
7. S. Wolfe, J.-B. Ducep, K.-C. Tin and S.-L. Lee, Can. J. Chem. 52, 3996 (1974).
8. L.D. Cama and B.G. Christensen, J. Am. Chem. Soc. 96, 7582 (1974).
9. (a) E.G. Brain, C.L. Branch, A.J. Eglington, J.H.C. Nayler, N.F. Osborne, M.J. Pearson, T.C. Smale, R. Southgate and P. Tolliday, in Recent Advances in the Chemistry of β-Lactam Antibiotics, Ed. J. Elks, Special Publication, No. 28, p. 204, The Chemical Society, London (1977). (b) C. Branch and M.J. Pearson, J. Chem. Soc.,

Perkin Trans. I 2268 (1979).
10. M. Narisada, H. Onoue and W. Nagata, Heterocycles 7, 839 (1977).
11. M. Narisada, T. Yoshida, H. Onoue, M. Ohtani, T. Okada, T. Tsuji, I. Kikkawa, H. Haga, H. Satoh, H. Itani and W. Nagata, J. Med. Chem. 22, 757 (1979).
12. R. Scartazzini, H. Peter, H. Bickel, K. Heusler and R.B. Woodward, Helv. Chim. Acta 55, 408 (1972).
13. R.W. Ratcliffe, T. Salzmann and B.G. Christensen, Tetrahedron Lett. 21, 31 (1980).
14. S. Kamata, S. Yamamoto, N. Haga and W. Nagata, J. Chem. Soc., Chem. Comm. 1106 (1979).
15. D.F. Corbett and R.T. Stoodley, J. Chem. Soc., Perkin Trans. I 185 (1974).
16. M. Yoshioka, I. Kikkawa, T. Tsuji, Y. Nishitani, S. Mori, K. Ōkada, M. Murakami, F. Matsubara, M. Yamaguchi and W. Nagata, Tetrahedron Lett. 4287 (1979).
17. S. Uyeo, I. Kikkawa, Y. Hamashima, H. Ona, Y. Nishitani, K. Okada, T. Kubota, K. Ishikura, Y. Ide, K. Nakano and W. Nagata, J. Am. Chem. Soc. 101, 4403 (1979).
18. G.A. Koppel and R.E. Koeler, J. Am. Chem. Soc. 95, 2403 (1973).
19. S. Yamamoto, S. Kamata, N. Haga, Y. Hamashima and W. Nagata, Tetrahedron Lett. 22, 3089 (1981).
20. Y. Hamashima, S. Yamamoto, S. Uyeo, M. Yoshioka, M. Murakami, H. Ona, Y. Nishitani and W. Nagata, Tetrahedron Lett. 2595 (1979).
21. (a) W. Nagata, Phil. Trans. R. Soc. Lond. B289, 225 (1980). (b) M. Yoshioka, T. Tsuji, S. Uyeo, S. Yamamoto, T. Aoki, Y. Nishitani, S. Mori, H. Sato, Y. Hamada, H. Ishitobi and W. Nagata, Tetrahedron Lett. 21, 351 (1980).
22. Unpublished results from Shionogi & Co., Ltd.
23. H.J. Reich and S.L. Peake, J. Am. Chem. Soc. 100, 4888 (1978).
24. T.L. Macdonald, N. Narasimhan and L.T. Burka, J. Am. Chem. Soc. 102, 7760 (1980).
25. S. Yamamoto, H. Itani, T. Tsuji and W. Nagata, Abstract Paper, p. 89, Proceeding of the 41st Symposium on the Synthetic Organic Chemistry, Tokyo, June 3-4, (1982).
26. N.R.A. Beeley and J.K. Sutherland, J. Chem. Soc., Chem. Comm. 321 (1977). These authors observed that peracid oxidation of a norbornyl iodide derivative with m-chloroperbenzoic acid gave the corresponding olefin. They suggested that the initially formed iodoso-compound may have undergone [2,3]-sigmatropic elimination.
27. H. Yamaguchi, I. Tetrai, K. Ozawa, T. Oba and S. Ishimoto, Japanese Kokai Tokkyo Koho No. 77, 59,185; C.A. 87, 152240n (1977).
28. E.M. Gordon, H.W. Chang and C.M. Cimarusti, J. Am. Chem. Soc. 99, 5504 (1977).
29. (a) R.W. Ratcliffe and B.G. Christensen, Tetrahedron Lett. 4649 (1973). (b) T. Kobayashi, K. Iino and T. Hiraoka, J. Am. Chem. Soc. 99, 5505 (1977).
30. T. Aoki, N. Haga, Y. Sendo, T. Konoike, S. Kamata, M. Yoshioka and W. Nagata, Abstract Paper, 3S10-2, p. 445, The 101st Annual Meeting of Pharmaceutical Society of Japan, April 2-4, 1981, Kumamoto, Japan.

STUDIES RELATED TO BETA-LACTAM COMPOUNDS

Saul Wolfe

Department of Chemistry, Queen's University, Kingston, Ontario,
K7L 3N6 Canada

Abstract - Although the natural processes leading to beta-lactam compounds proceed via relatively simple amino acid and/or peptide precursors, chemical syntheses of the same compounds, or of structural analogs, have invariably required multistep processes. An example of such a process, leading to a 1-oxacephalosporin, is presented. The difficulties encountered during this work have led to the examination of a combined chemical and biological approach to the synthesis of unnatural beta-lactam-containing systems. This approach is based on the observation that certain actinomycetes will synthesize concurrently more than one type of beta-lactam system from different amino acid or peptide precursors. Validation of the new approach has been secured by the successful syntheses of cephalosporins substituted with H or Et at C3, using three enzymes isolated from Streptomyces clavuligerus, and the appropriate peptide precursors.

INTRODUCTION

The beta-lactam family of natural products (Chart 1) includes the penicillins, cephalosporins and cephamycins, in which the beta-lactam ring is fused to a five-membered or six-membered sulfur-containing ring, together with clavulanic acid, in which the beta-lactam is fused to a five-membered oxygen-containing ring, the carbapenems, in which the beta-lactam is fused to a five-membered carbon-containing ring, and the nocardicins and monobactams, which are monocyclic compounds. Although there are many individual members of this family, only two can be used directly in medicine without structural change. These are penicillin G, the penicillin in which R = benzyl; and clavulanic acid. All other clinically important beta-lactam compounds have been prepared from one or another of the natural products by some kind of structural change. For many years, these structural changes focused on the substituents around the peripheries of the various ring systems, and not on the nature of the ring systems themselves.

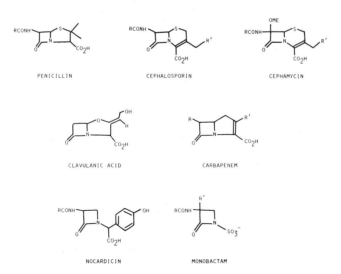

Chart 1. Some naturally occurring beta-lactam compounds

The first significant nuclear modification of a beta-lactam natural product was published in 1974 (1), with the conversion of 6-aminopenicillanic acid (<u>1</u>) to a 3-methyl-1-oxacephem (<u>2</u>). This synthesis was accomplished in three stages, involving a total of 9, 10 or 11 steps, depending on the precise route followed, and an overall yield of 14%. The first stage led to the chloroazetidinone 3, having the S-configuration at C4, which corresponds to a trans-relationship between the beta-lactam substituents. This stereochemistry was desired, so that a ring closure at C4 with inversion of configuration would afford an oxacephem possessing the necessary cis-stereochemistry. The second stage led to functionalization of the methyl group cis to the beta-lactam nitrogen (4), and the third stage consisted of the ring closure (vide infra).

During the same period, Woodward and his co-workers in Basle devised a synthesis of penems (5), also from a penicillin precursor (2). Their sequence proceeded in four stages, with a total of 13 steps, and an overall yield of 4%. The first stage led from penicillin V to the acylthioazetidinone 6, structurally analogous to the chloroazetidinone 3, but with the cis-stereochemistry of the beta-lactam substituents. The beta-lactam nitrogen substituent was detached during the second stage (6 → 7) and replaced, during the third stage (7 → 8), with a phosphoranyl-containing substituent. Cyclization to 5 was achieved by a Wittig-type ring closure.

The two sequences just summarized illustrate the two principal strategies that have now evolved for the partial syntheses of nuclear modified beta-lactam compounds (3). In the first, the isoprenoid moiety attached to the beta-lactam nitrogen of 3 is retained, and elaborated further. In the second, this moiety of 6 is removed completely, and a new ring is reconstructed. An advantage of this latter strategy is that it is easily carried out on a laboratory scale to provide products for biological evaluation; however, this approach does not lend itself to the development of an industrial process.

The 1-oxacephem syntheses of Nagata and his co-workers (4) illustrate this point. These workers used the phosphoranyl strategy to reach the C3-oxygenated azetidinone 9 from a penicillin precursor. This was then cyclized to 10, and elaborated further to oxacephems of the general structure 11. After hundreds of variations of this latter structure had been evaluated, a specific compound was selected for further development, and a process was worked out which retained the carbon skeleton attached to the beta-lactam nitrogen of the penicillin precursor. This process proceeds via the epi-oxazoline 12, which is functionalized on the methyl group (12 → 13), and cyclized to give the 7-epi-oxacephem 14. The remainder of the synthesis is straightforward.

The published synthesis of Moxalactam (Chart 2) consists of 16 steps, and proceeds in an overall yield of ca 20%. Chart 2, which is based on available data, also presents the cost of some clinically used beta-lactam antibiotics, expressed as the wholesale price to the pharmacist, in US dollars, of one day's supply of material. Moxalactam, a third generation cephalosporin, is approximately five times more expensive than cephalothin, a first generation cephalosporin; and cephalothin is, in turn, approximately fifty times more expensive than ampicillin, a semi-synthetic penicillin. Clearly, continued effort is necessary to achieve more ready access to this important class of compounds.

CHEMICAL SYNTHESES RELATED TO 1-OXACEPHEMS

Our pioneering studies with the chloroazetidinones 3 led to two general procedures for the intermolecular displacement of chlorine. One, termed the "SN_1 procedure", employs a Lewis acid in a protic solvent (5) and leads, inter alia, to alkoxy or acyloxyazetidinones 15.

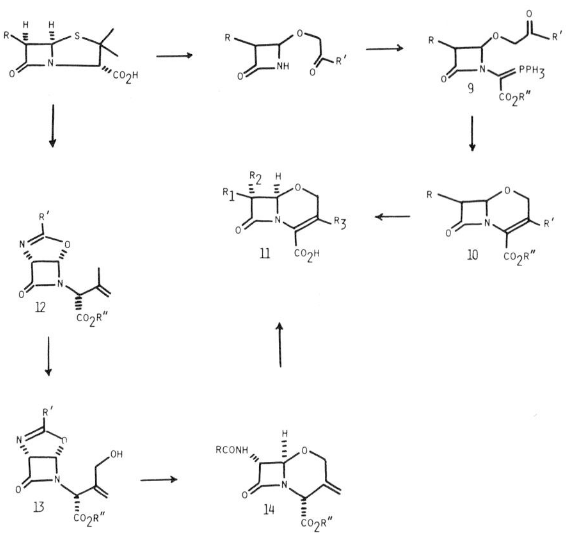

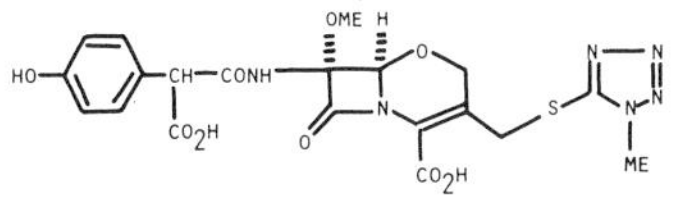

Chart 2

Cost of Some Antibiotics

Drug		Highest Recommended Dosage	Cost*
PENICILLIN V	(GENERIC)	2G/DAY	US $ 0.16
AMPICILLIN	(BRISTOL)	4G/DAY	0.60
CEPHALEXIN	(LILLY)	12G/DAY	2.16
CEPHALOTHIN	(LILLY)	12G/DAY	32.92
CEFAZOLIN	(LILLY)	6G/DAY	32.76
CEFAMANDOLE	(LILLY)	12G/DAY	58.97
CEFOXITIN	(MERCK)	12G/DAY	59.10
CEFOTAXIME	(ROUSSEL)	12G/DAY	129.66
MOXALACTAM	(SHIONOGI)	12G/DAY	146.45

*REFERS TO THE WHOLESALE COST TO THE PHARMACIST FOR ONE DAY'S TREATMENT. DATA FROM DRUG TOPICS RED BOOK 1981.

These are formed as cis/trans mixtures, the trans isomer predominating, and the same mixture of products is obtained from the 4R- and the 4S-chloroazetidinones. The second procedure, termed the "SN$_2$ procedure", employs a tetramethylguanidinium salt in a non-hydroxylic solvent (6), and results in inversion of configuration at C4 (3 → 16). With the SN$_2$ procedure, the experimental conditions and the ease of the reaction depend greatly upon the nature of the C3-azetidinyl substituent. For example, for R = H$_2$N, the displacement occurs in refluxing methylene chloride; for R = acylamino, the displacement proceeds in refluxing chloroform; and, for R = phthalimido (Ft), the reaction requires steam bath temperatures in dimethylformamide solvent.

For various reasons, the phthalimido group had been employed at C3 in our first oxacephem synthesis (1), but the allylic alcohol 17 could not be cyclized under SN$_2$ conditions. Cyclization did occur smoothly under SN$_1$ conditions, using stannous chloride in dimethoxyethane, but the undesired trans-isomer of 18 predominated. These various observations suggested that an SN$_2$-type ring closure to a 1-oxacephem would require an allylic alcohol such as 19, with an amino group at C3. It was considered that an acylamino group at C3 would not suffice; although such compounds can be manipulated (7), ring closure is achieved more readily from C3, to form an oxazolinoazetidinone (20 → 21). At the same time, groups other than methyl at C3' were now desired, to allow access to 1-oxacephems having substituents other than methyl at C3.

The reactions summarized in Chart 3 have been carried out with R = Ft, PhOCONH, Cl$_3$CCH$_2$OCONH, and ClCH$_2$CONH (8). The latter two contain amino protecting groups whose removal can be achieved under mild conditions. Bromination of 22 with two mole-equiv of NBS leads to the dibromo compounds 23, which are convertible to the diformates 24 with excess tetraethylammonium formate in chloroform at room temperature. Deformylation of 24, using hydrogen chloride in methanol-methylene chloride, leads, quantitatively, to hydroxylactones 25. Interestingly, when R = ClCH$_2$CONH, 22 cyclizes to an oxazoline during chromatography; however, once the methyl groups have been functionalized, this cyclization is no longer facile. The diformates 24 regenerate the dibromides 23 quantitatively upon treatment with 1 mole-equiv of BBr$_3$; an analogous result is obtained with BCl$_3$. However, with 0.6 mole-equiv of a boron halide, 24 are converted to E/Z mixtures of mixed halogeno-formates in which the E isomers 26 predominate. These can be isolated in pure form, and deformylated to the halogeno-hydrins 27.

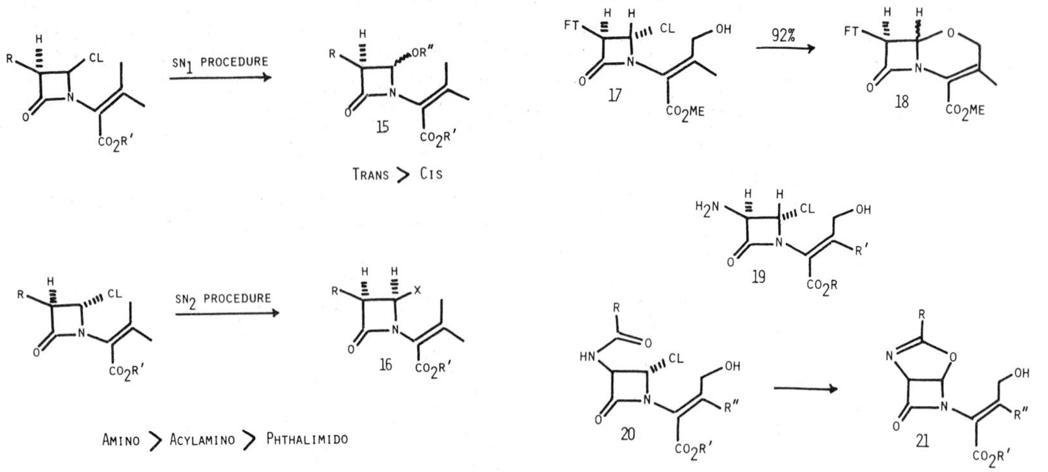

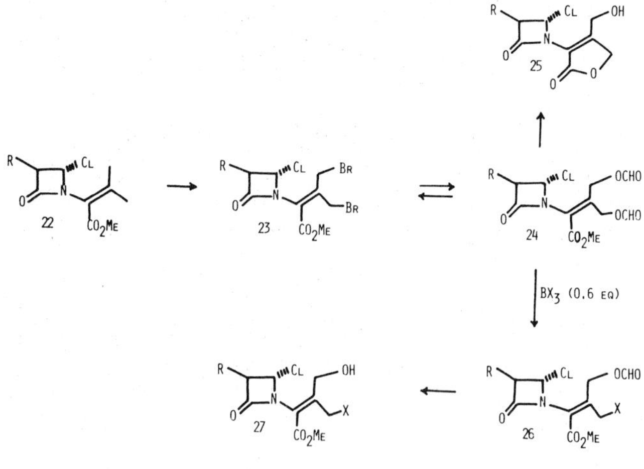

Chart 3

The dibromides 23 are converted to the deconjugated olefins 28, using zinc in acetic acid-acetonitrile (9). Under the same conditions, the E/Z bromo-formates 29 afford the deconjugated formates 30. These can be deformylated to allylic alcohols of type 31 (cf 13) or, alternatively, ozonized, methylated (CH$_2$N$_2$), and deformylated to allylic alcohols of type 32.

With a C3 amide or urethane substituent, the displacement of bromine from an E or Z bromo-formate proceeds with retention of stereochemistry about the carbon-carbon double bond (8). For example, with a tetraalkylammonium acetate in chloroform at 0°, the E-isomer is converted to the E-acetoxyformate 33, and the Z-isomer is converted to the Z-acetoxyformate 34. The configurations of these four compounds were established by deformylation experiments: thus, 33 is converted to an acetoxyhydrin, and 34 to an acetoxylactone. With mercaptomethyltetrazole in dimethylformamide, followed by deformylation, 33 affords a mercaptomethyltetrazolyl-substituted allylic alcohol.

Chart 4 summarizes the position at this stage, for sequences with R = Cl$_3$CCH$_2$OCONH. Three types of alcohols, viz., 35 (MTZ = mercaptomethyltetrazole), 36 and 37 have been secured, in 8, 8 and 10 steps, and overall yields of 7.3%, 8.2% and 5.5% respectively. Completion of these sequences would entail (i) deprotection of the C3 nitrogen; (ii) ring closure; (iii) attachment of a biologically significant acyl substituent; (iv) deprotection of the methyl ester. It is evident that the process contains too many steps, and has proceeded in an unacceptable overall yield.

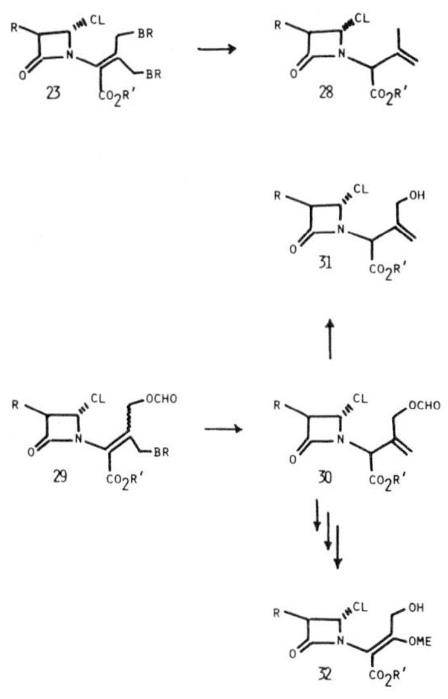

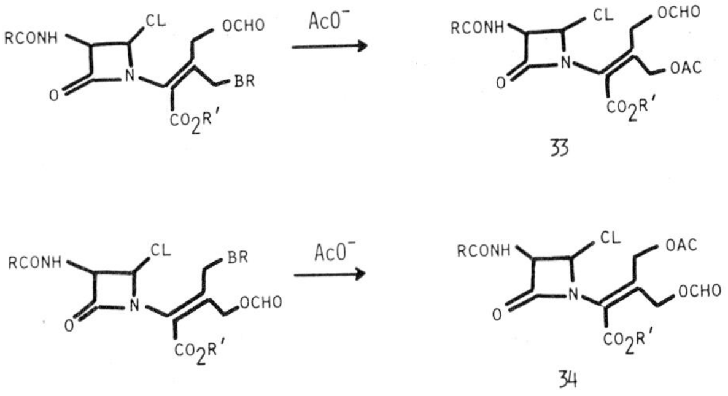

Compound 35 was carried forward. Silylation of the hydroxyl group with Hanessian's reagent (10), followed by C3-nitrogen deprotection with activated zinc, prepared (11) from zinc chloride and sodium naphthalenide, afforded the O-silylated amine. Cyclization occurred, upon treatment of the latter compound with tetrabutylammonium fluoride, to give the cis-oxacephem 38. However, 38 was formed in this cyclization step in less than 5% yield.

The chemical transformations just described have been reported here in order to illustrate: (i) the fact that a simple, economically feasible organic synthetic process to 1-oxacephems has not been achieved; (ii) the kinds of sequences that were investigated prior to a decision to abandon "traditional" organic chemical approaches to such problems in favour of a novel approach more suited to research in an academic laboratory.

A BIOLOGICAL APPROACH TO UNNATURAL BETA-LACTAM CONTAINING NUCLEI

Background
On September 18, 1944, during the Anglo-American wartime investigation of penicillin (12), researchers at Glaxo Laboratories wrote that *"there is reason to expect that (unnatural) penicillins will be amenable to biosynthesis by the use of appropriate components to intro-*

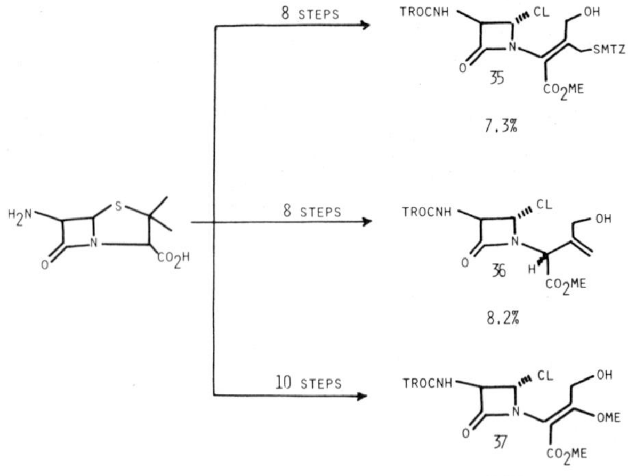

Chart 4

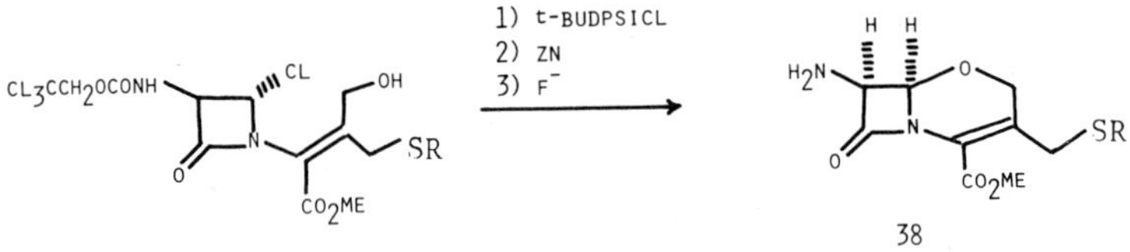

duce their side chains into the molecule". This prediction was realized at the Lilly Research Laboratories, who reported, inter alia, on May 15, 1945, that phenoxymethylpenicillin is formed upon addition of phenoxyacetic acid or its equivalent to Penicillium fermentations. Interestingly, the fact that phenoxymethylpenicillin is acid stable and orally effective was not discovered until several years later, in a laboratory which is now part of the Sandoz company (13). This compound, now named penicillin V, was introduced around 1954 as the first important unnatural beta-lactam compound. Although this kind of biological side chain incorporation is restricted to monosubstituted acetic acids, and is observed only with Penicillium spp, we regard such work as the historical precedent for the results now described, which have led to biological syntheses of unnatural beta-lactam-containing nuclei.

The naturally-occurring beta-lactam compounds are formed as secondary metabolites of both eukaryotic and prokaryotic organisms. To a naive organic chemist, a eukaryote is a higher life form, and it has a more complicated cell structure, which restricts the types of compounds that can be synthesized or metabolized. Examples of eukaryotic beta-lactam-producing organisms are the fungi Penicillium chrysogenum and Cephalosporium acremonium. A prokaryote, on the other hand, is a lower, earlier, life form, with a more primitive cell structure, which allows a greater variety of chemical transformations to take place. This suggests, again naively, that prokaryotes are more versatile at organic synthesis than are eukaryotes, provided that this versatility can be understood and controlled. Examples of prokaryotic beta-lactam-producing organisms are the actinomycetes Streptomyces clavuligerus, S. cattleya and S. lipmanii.

As an illustration of the differing capabilities of eukaryotic and prokaryotic beta-lactam-producing organisms, P. chrysogenum, a eukaryote, synthesizes a particular tripeptide (vide infra), and converts this peptide exclusively to penicillin as the only stable beta-lactam-containing end product. C. acremonium, also a eukaryote, synthesizes the same tripeptide, and converts this peptide sequentially to penicillin and cephalosporin. In contrast, the prokaryote S. clavuligerus synthesizes penicillin, cephalosporin and cephamycin from one amino acid-containing precursor and, at the same time, clavulanic acid, from a completely different precursor. The prokaryote S. cattleya synthesizes penicillin and cephalosporin from one precursor and, at the same time, the carbapenem thienamycin, from a different precursor.

In the penicillins, cephalosporins and cephamycins, all of the atoms of the bicyclic ring systems are derived from the amino acids cysteine and valine (14). The five-membered ring of thienamycin is derived from the amino acid glutamic acid, and the side chain (R' = $SCH_2CH_2NH_2$) is derived from the amino acid cysteine; the remaining four carbon atoms of thienamycin (R = CH_3CHOH) are derived from acetoacetate (15). Glutamic acid is also a precursor of clavulanic acid, but appears in this molecule in a different way from thienamycin (16); in this case, glutamic acid contributes only the nitrogen atom of the beta-lactam ring. The remaining three atoms of the beta-lactam ring of clavulanic acid are derived from glycerol or phosphoenolpyruvate. In the nocardicins, the three carbon atoms are derived from the amino acid serine, and the beta-lactam nitrogen from p-hydroxyphenyl-glycine, which is derived, in turn from tyrosine (17).

This diversity of amino acid or peptide precursors, and structural types, suggested that more than one beta-lactam-forming process exists in nature. If this hypothesis were correct, it might be possible to synthesize other types of beta-lactam compounds, or to gain more ready access to existing types of beta-lactam compounds, by an appropriate combination of amino acid or peptide precursors, and cell-free biological systems, in particular, for the reasons indicated above, systems derived from prokaryotic beta-lactam-producing organisms. Cell-free systems would be necessary for such experiments, because extracellular peptides are not accepted by the intact organisms.

Protocol
The problem was approached according to a four-stage protocol, as follows:

(i) Stage 1 comprised the preparation of cell-free systems from prokaryotic beta-lactam-producing organisms, and determination of the physical properties, cofactors and specificities associated with such systems. Ideally, purification of the enzymes of interest was desired, so that a substance could be removed from a bottle, added, together with any necessary cofactors, to a peptide, and a predictable chemical transformation achieved. Since the nature of the similarities and any differences between eukaryotic and prokaryotic systems also had to be defined, Stage 1 also required cell-free work with eukaryotic beta-lactam-producing organisms. This stage of the protocol has been realized, in collaboration with two microbiological research groups. All work with prokaryotic organisms has been performed by Professor D.W.S. Westlake and Dr. S.E. Jensen, Department of Microbiology, University of Alberta, Edmonton, Canada. Studies with the eukaryote C. acremonium have been carried out by Professor A.L. Demain and Drs. J. Kupka and Y.-Q. Shen at the Massachusetts Institute of Technology. Close contact among the chemical and microbiological research groups allowed the development of experiments which differed only in the nature or purity of the enzyme source;

(ii) Stage 2 required the development of syntheses of the amino acid or peptide derivatives to be used as substrates, together with syntheses of the beta-lactam-containing products expected from the cyclization of these substrates, so that the properties of these putative products would be known;

(iii) Stage 3 required the development of analytical procedures to monitor the cell-free experiments;

(iv) Stage 4 was needed to guide the remainder of the work. To distinguish between reasonable and unreasonable proposed biological transformations, some understanding of the detailed chemical pathways leading to the natural products was necessary. This required model studies leading to the development and testing of biogenetic theories of beta-lactam formation (18). The results obtained during Stages 1-3 are presented here. The most recent work concerning Stage 4 will be published elsewhere.

Penicillin and cephalosporin biosynthesis in whole cells
The primary penicillin produced by whole cell fermentations is isopenicillin N (39); this compound has a δ-(L-α-aminoadipyl) side chain, and is formed (19) by oxidative cyclization of the tripeptide precursor δ-(L-α-aminoadipyl)-L-cysteinyl-D-valine (ACV) (40), which is formed, in turn, from three amino acids: L-α-aminoadipic acid, L-cysteine, and L-valine. The aminoadipic acid first combines with cysteine to form δ-(L-α-aminoadipyl)-L-cysteine, to which L-valine is then added; L- to D-epimerization of the valinyl moiety occurs at this point.

The cyclization of ACV to isopenicillin N requires the loss of four hydrogen atoms and corresponds, formally, to two two-equivalent oxidative processes. During these processes, two carbon-bound hydrogens are lost, viz., the beta-hydrogen of the valinyl moiety (20), and the pro-$\underline{S}$ hydrogen of the cysteinyl moiety; and the new carbon-sulfur and carbon-nitrogen bonds are formed with retention of configuration. This stereochemical information will be utilized later.

The primary cephalosporin formed in whole cell fermentations is desacetoxycephalosporin C (41) (21). This compound, which contains a dihydrothiazine ring, is formed by oxidative ring expansion of a thiazolidine precursor, and the pro-$\underline{R}$ (beta) methyl group of penicillin becomes C2 of the cephalosporin. This stereochemical information will also be utilized later. The side chain of 41 is δ-($\underline{D}$-α-aminoadipyl), and the linking of penicillin to cephalosporin biosynthesis is accomplished by an epimerase, which converts isopenicillin N to penicillin N (42). To anticipate the results which follow later, a major difference between the cell-free systems derived from eukaryotic and prokaryotic beta-lactam-producing organisms is that the latter contain the epimerase and the former do not. Consequently, under most conditions, cell-free syntheses with the eukaryotic systems cannot be taken beyond the oxidation level of penicillin.

Syntheses of peptides and related compounds
Before work could be undertaken with peptide analogs, it was first necessary to demonstrate that the natural processes just described could be reproduced in cell-free experiments. This work required large quantities of the natural precursor ACV. This compound was synthesized by coupling of N-BoC-S-trityl-$\underline{L}$-cysteine with the benzhydryl ester of $\underline{D}$-valine, to give the fully-protected dipeptide 43. A 15 min treatment with anhydrous formic acid at room temperature led to the crystalline, partially protected peptide 44. Conversion to fully-protected ACV (45) was performed by coupling of 44 to 46 (22). Deprotection of 45 was achieved in two stages: removal of the trityl group, with iodine in methanol, led to the disulfide 47; and all other protecting groups were then removed upon overnight treatment with formic acid. This sequence leads to ACV disulfide. The compound is best stored in this form, and is converted to ACV itself, as needed, with dithiothreitol (DTT). The synthesis just described is readily adapted to systematic modification of the aminoadipyl moiety, and it has been used to examine the effects of chain length, functionality, and stereochemistry in this part of the peptide.

Penicillin N, isopenicillin N, and side chain analogs of these two compounds, were synthesized (23) by coupling of the appropriate ω-carbobenzoxy and/or benzhydryl protected dicarboxylic acids to the benzhydryl ester of 6-APA (48), followed by hydrogenolysis at atmospheric pressure in the presence of one equivalent of sodium bicarbonate. The p-toluenesulfonic acid salt of the ester 48 was prepared in 90% yield by treatment of 6-APA with two equivalents of diphenyldiazomethane in methanol-methylene chloride, followed by removal of the solvent and addition of anhydrous p-toluenesulfonic acid in acetone.

Among the compounds of interest were N-acetyl-ACV and its cyclic analog N-acetylisopenicillin N. The syntheses of these compounds required N-acetyl-$\underline{L}$-α-aminoadipic acid alpha-benzhydryl ester (49), but access to this compound by acetylation of the amino acid precursor 50 proved to be surprisingly difficult. All of the commonly-used acetylation procedures led to piperidone formation. Eventually, the acetyl analog (51) of the Fujisawa compound (24) 2-(t-butoxycarbonyloximino)-2-phenylacetonitrile (BOC-ON) solved the problem. Addition of 51, a stable crystalline compound, to an aqueous acetone solution of 50 led to 49 in 90% yield.

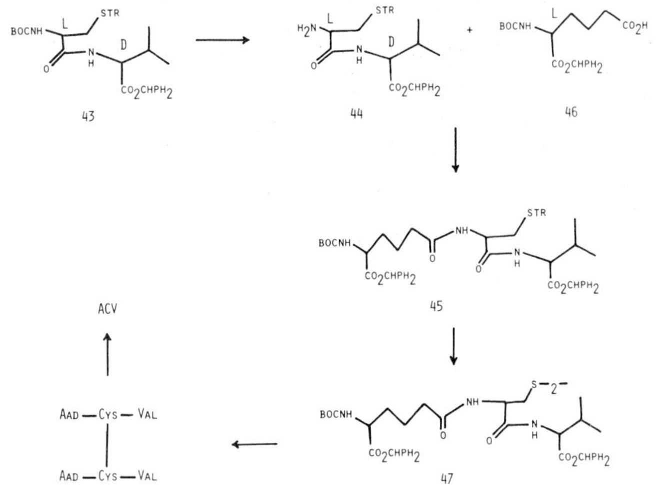

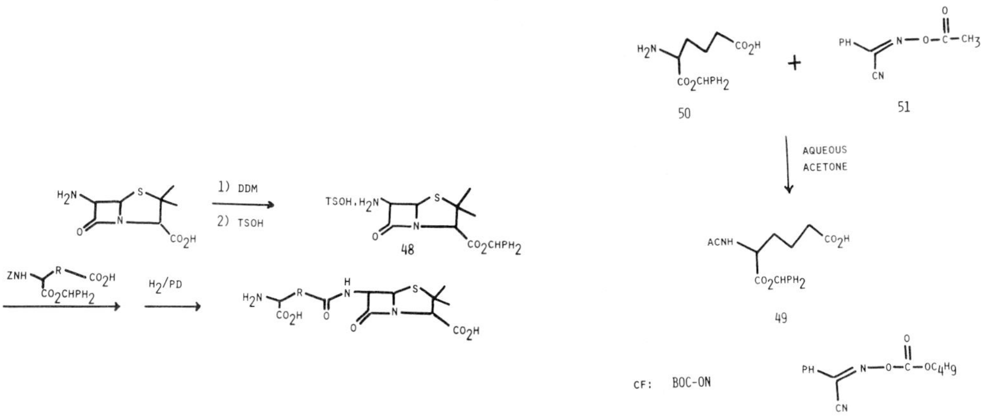

The limiting reagent in the peptide syntheses just described is L-α-aminoadipic acid, because of the present cost of this compound (US $35/g). The compound has been prepared by total synthesis (22) and by partial synthesis from L-aspartic acid, L-glutamic acid or L-lysine (25). However, none of these sequences seemed entirely satisfactory, because of the number of steps and low overall yields. Hungarian workers have recently reported a series of deamination studies, using sodium nitroprusside as a formal source of NO⁺ (26). Applied to lysine, this reagent leads to 35-40% yields of L-pipecolic acid (52), and also 35-40% yields of the L-amino acid alcohol 53. Both 52 and 53 were convertible to L-α-aminoadipic acid. The route from 53 is preferred, because the normal protecting groups for the peptide syntheses can be introduced at this stage, and the oxidation carried out, as needed, over a platinum catalyst. The route from 52 involves more steps, but is also shown here because the oxidation step 54 → 55 has been achieved, inter alia, using a new oxidizing agent.

A number of dienes have been found to undergo an abnormal oxidation by potassium permanganate, leading to anhydrotetrols (27). One of these abnormal reactions is the conversion of 1,5-hexadiene to cis-2,5-dihydroxymethyltetrahydrofuran, which is observed in aqueous acetone at pH 6. This reaction, discovered by Klein and Rojahn (28), employs carbon dioxide ebullition to maintain the pH control. However, this leads to considerable loss of the volatile diene, and the product cannot be isolated in yields higher than 20%. A buffering agent is needed in such permanganate oxidations, because the reduction of permanganate to manganese dioxide leads to hydroxyl ion formation ($MnO_4^- + 3e + 2H_2O \rightarrow MnO_2 + 4OH^-$).

It occurred to us that the problem of pH control in such reactions might be overcome by use of zinc permanganate or magnesium permanganate in place of potassium permanganate, because of the consequent precipitation of zinc hydroxide or magnesium hydroxide. Magnesium permanganate was described by Arthur Michael in 1905 (29). The reagent reacted rapidly with many

SYNTHESIS. 163 (1982)

organic compounds; however, the nature of these oxidations was not examined. Aqueous zinc permanganate converted several olefins into neutral carbonyl-containing products (30); however, these products were also not characterized. Cornforth and his co-workers (31) have employed zinc permanganate in aqueous acetone for the oxidation of a primary alcohol to a zinc carboxylate. No other references to the oxidation of organic compounds by these permanganate salts could be found.

Zinc permanganate and magnesium permanganate have now been prepared in quantity by the sequence shown in equations (1) and (2). Reaction of (water-insoluble) barium manganate (eq 1) with the stoichiometric amounts of sulfuric acid and MO (M = Zn, Mg) (eq 2) leads to zinc permanganate or magnesium permanganate, together with insoluble barium sulfate and manganese dioxide. Filtration through Celite, and evaporation, leads to the salts, as hexahydrates.

$$2KMnO_4 + Ba(NO_3)_2 + Ba(OH)_3 \rightarrow 2BaMnO_4 + \tfrac{1}{2}O_2 + 2KNO_3 + H_2O \qquad (1)$$

$$3BaMnO_4 + 3H_2SO_4 + MO \rightarrow M(MnO_4)_2 + 3BaSO_4 + MnO_2 + 3H_2O \qquad (2)$$

To our great surprise, both of these salts reacted violently with virtually every common laboratory solvent, including ethers, amines, ketones, carboxylic acids, and primary, secondary and tertiary alcohols. By comparison, potassium permanganate was innocuous. A number of procedures were examined to achieve control over these oxidations. Support on silica gel, as described by Regen for potassium permanganate (32) gave the best results. In methylene chloride solvent, tetrahydrofuran and tetrahydropyran are oxidized to lactones, benzyl alcohol to benzaldehyde, stilbene to benzaldehyde, and cyclohexene to adipoin and adipaldehyde upon shaking for 5 min with zinc permanganate on silica gel; benzoylated piperidine and pyrrolidine are oxidized to lactams, and tolan to benzil with longer reaction times. The oxidations cease after ca 50% conversion, because of the precipitation of manganese dioxide on the insoluble reagent; filtration and repetition of the treatment leads to an increase in the isolated yield. Under identical conditions, zinc permanganate and magnesium permanganate give comparable results, but the conversions achieved by potassium permanganate are less than 2%.

Analytical procedures
Transformations of the peptides were monitored by bioassay and HPLC procedures (33). The HPLC assembly consisted of a M-6000A pump, UK-6 injector, M-450 variable wavelength detector operated at 220 nm, M-420 data module, and µBondapak-C18 column (Rad Pak A in an RCM-100 radial compression module) as stationary phase, all from Waters Scientific Co. The mobile phase consisted of methanol/0.05M potassium phosphate buffer; the methanol content and pH were varied according to the particular separation. A short precolumn of Bondapak C18/Corasil protected the main column.

Figure 1 shows, in A, ACV disulfide; in B, dithiothreitol; and, in C, the conversion of the inactive substrate ACV dimer to the active substrate ACV monomer by DTT. Figure 2 shows a representative mixture of ACV analogs, all after treatment of the disulfides with DTT. Analogs of isopenicillin N are also well separated from each other, and from their putative peptide precursors.

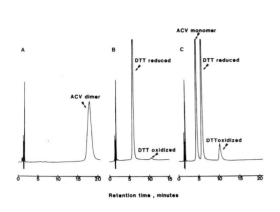

Fig. 1. HPLC data showing the conversion of ACV dimer to ACV monomer

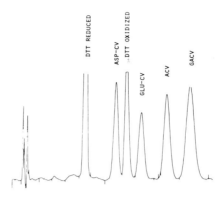

Fig. 2. HPLC data showing the separation of aspartyl, glutamyl, aminoadipyl and glycylaminoadipylcysteinylvalines from each other and from the oxidized and reduced forms of DTT

The bioassay procedures developed to complement the chromatographic experiments were based upon the following observations: (i) Micrococcus luteus is equally sensitive to isopenicillin N and penicillin N, but cephalosporins are 30 times less active against this organism; (ii) E. coli Ess is equally sensitive to penicillin N and desacetoxycephalosporin C, but isopenicillin N is 20 times less active against this organism; (iii) penicillins, but not cephalosporins, are inactivated by the enzyme penicillinase. These observations allowed the composition of mixtures of isopenicillin N, penicillin N and desacetoxycephalosporin C to be determined qualitatively by biological methods.

Cell-free systems and enzyme preparations
A number of methods of cell breakage were examined for each organism of interest, and analogous results obtained in each case. For example, with S. clavuligerus (34), relative rates of conversion of ACV to antibiotics were 1.0, 0.88 and 0.39, using extracts prepared, respectively, by sonication, French pressure cell, and Omnimixer-plastic beads. Sonication always gave the best results. This procedure consisted of a 30 sec treatment of 48 h washed cells, followed by centrifugation. The supernatant from this treatment was designated "crude cell-free extract".

The crude cell-free extract from S. clavuligerus was separated into three enzyme fractions by a three-stage treatment. The first stage consisted of addition of ammonium sulfate to 40% saturation, centrifugation, continued addition of ammonium sulfate to the supernatant to 70% saturation, and centrifugation again. The resulting pellet, resuspended in pH 7 buffer, was termed "salt-precipitated cell-free extract" (SPCFX). This SPCFX retained all enzyme activities, and showed negligible baseline contamination in HPLC assays. In the second stage, the epimerase (isopenicillin N → penicillin N) (MW 60,000) was cleanly separated from the cyclase (ACV → isopenicillin N) (MW 36,500) and ring expansion (penicillin N → desacetoxycephalosporin C) (MW 29,000) enzymes, by chromatography of the SPCFX on Sephadex G200. In the third stage, the cyclase and ring expansion enzymes were separated by ion exchange chromatography on DEAE trisacryl. A 100-fold purification of the cyclase was achieved in this manner. Analogous separations of the cyclase and ring expansion enzymes were achieved using SPCFX from C. acremonium (35). However, as noted earlier, the epimerase was absent from these systems. With these results, one of the original objectives had been reached: it was now possible to remove a substance from a bottle, and employ this substance as a reagent for peptide transformations.

Cell-free transformations of ACV
The details of a typical cyclization of ACV are as follows: to 0.4 ml of reaction mixture were added ACV dimer (0.9 mM), Tris-HCl pH 7.0 buffer (50.0 mM), and enzymes. For these quantities, the optimized amounts of the essential cofactors ferrous sulfate and ascorbic acid were found to be 45.0 μM and 2.8 mM, respectively. An absolute requirement for DTT was also seen, since the optimized amount of this compound (4.0 mM) was greater than that required to reduce ACV dimer to ACV monomer. The reaction was performed for the desired time at 20°C, and then terminated by addition of methanol (0.4 ml) to precipitate protein. When all three enzymes from S. clavuligerus were employed in this reaction mixture, the

peptide was converted to a mixture of isopenicillin N and penicillin N. Ring expansion did not occur, because this enzyme required one additional cofactor; when this cofactor, alpha-ketoglutarate (1 mM), was present ACV was converted to desacetoxycephalosporin C. Figure 3 shows HPLC data illustrating the conversion of ACV to penicillins. Figure 4 shows the effect of addition of alpha-ketoglutarate to such reaction mixtures: curve A refers to the penicillin formed in the absence of alpha-ketoglutarate; curve C refers to a synthetic mixture of penicillin N and desacetoxycephalosporin C; and B is the reaction mixture 30 min after the addition of the cofactor.

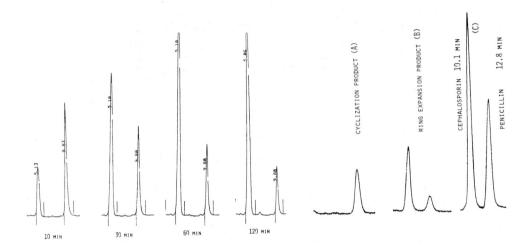

Fig. 3. Time course of the cyclization of ACV to penicillins by a cell-free system from S. clavuligerus

Fig. 4. Ring expansion of the penicillin produced in Fig. 3 following addition of alpha-ketoglutarate

Behaviour of aminoadipyl analogs
The molecular weights, cofactor requirements and enzyme kinetics exhibited by the cyclase and ring expansion enzymes of S. clavuligerus and C. acremonium were very similar. The behaviour of different substrates, modified in the aminoadipyl moiety, towards the two systems was also similar. The L-aspartyl, L-glutamyl, and D-α-aminoadipyl-containing peptides did not cyclize. Cyclization was observed with adipyl, glycyl-L-α-aminoadipyl and N-acetyl-L-α-aminoadipyl-containing peptides. The adipyl compound gave ca 20% cyclization to the corresponding penicillin, carboxybutylpenicillin, but SPCFX converted the glycyl and N-acetyl compounds to penicillin N and isopenicillin N, via an initial deacylation of these peptides to ACV. The purified cyclase from S. clavuligerus did not cyclize the glycyl-L-α-aminoadipyl-containing peptide.

Of all penicillin N and isopenicillin N analogs so far examined under ring expansion conditions, only the natural substrate penicillin N was converted to a cephalosporin. These various results revealed that the enzymatic conversion of an ACV analog to an unnatural cephalosporin nucleus would require (i) a δ-L-α-aminoadipyl side chain; (ii) an enzyme system containing the epimerase. The prokaryotic system was, therefore, employed.

Modification of the valinyl moiety. Biological synthesis of unnatural cephalosporins
On the basis of the stereochemical aspects of penicillin and cephalosporin biosynthesis reviewed earlier, if cell-free transformations of substrates modified in the valinyl moiety of ACV as in 56 were to occur, these transformations were expected to proceed in a predictable manner. Thus, carbon-sulfur bond formation with retention of configuration at the beta carbon of the valine analog, should lead to isopenicillin N analogs of type 57. Epimerization to penicillin N analogs of type 58, followed by ring expansion, should lead to cephalosporin analogs of type 59, with transfer of the beta carbon atom attached to C2 of 58 into C2 of the six-membered ring.

These expectations have been realized. Replacement of valine by alpha-aminobutyric acid in the peptide precursor ($R_1 = R_2 = H$) has led to the cephalosporin 59 ($R_1 = R_2 = H$). Replacement of valine by allo-isoleucine in the peptide precursor ($R_1 = C_2H_5$, $R_2 = H$) has led to the cephalosporin 59 ($R_1 = C_2H_5$, $R_2 = H$).

In the case of 60, the peptide in which D-valine is replaced by D-alpha-aminobutyric acid, it was not known a priore whether carbon-sulfur bond formation would proceed with loss of the pro-R methylene proton or the pro-S methylene proton of the ethyl group. Assuming retention of configuration in each case, loss of the pro-S-hydrogen would lead to 61, with a trans-relationship between the C2 and C3 thiazolidine substituents, but an endo orientation of the C2 methyl group. Loss of the pro-R hydrogen would lead to 62, with a cis-relationship between the C2 and C3 thiazolidine substituents, and an exo-orientation of the C2 methyl group. Only 61 should undergo ring expansion to a cephalosporin. Although the synthesis of the pro-R deuterated analog of 60 was contemplated, to increase any reactivity difference between the two sites, this synthesis proved unnecessary. Only one penicillin was observed by HPLC; and this penicillin, presumably 61, underwent ring expansion to 63 (36).

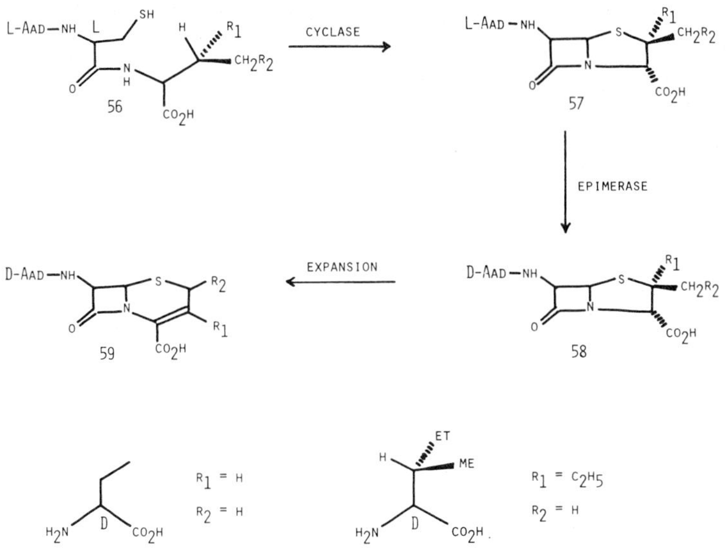

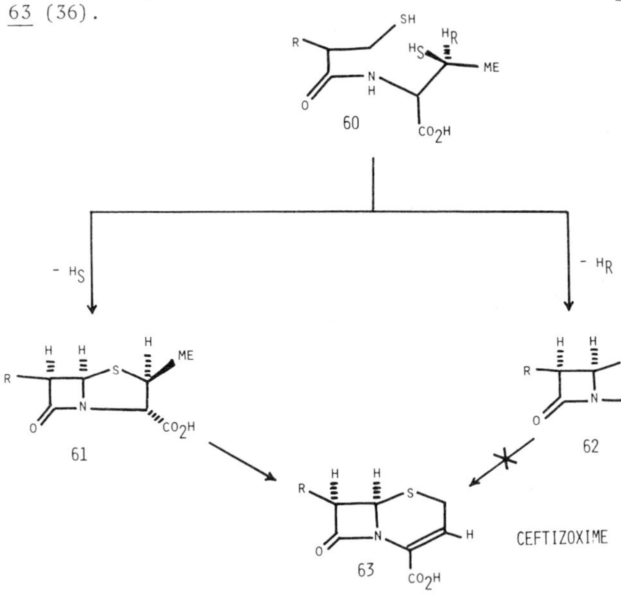

The nucleus of 63 is known (37). It was first prepared by a combined chemical and microbiological degradation of the C3 acetoxymethyl group of the natural product cephalosporin C, and is seen in the Fujisawa compound Ceftizoxime (38).

SUMMARY AND PROGNOSIS

A manipulation of the penicillin-cephalosporin biosynthetic pathway has been achieved, which provides unnatural cephalosporins modified in the substitution pattern of the six-membered ring, by replacement of the valine moiety of the natural precursor, and use of three enzymes isolated from a prokaryotic beta-lactam producing organism. It remains to be determined whether more deep-seated structural changes can be accomplished by alteration of the cysteinyl moiety of the natural precursor. At the same time, it should be clear that other beta-lactam-forming pathways remain to be elucidated and, hopefully, exploited. The strategies and techniques developed for the present work should facilitate such studies.

Acknowledgements - The microbiological studies presented here could not have been possible without the enthusiastic encouragement and collaboration of Professor D.W.S. Westlake and Dr. S.E. Jensen of the University of Alberta, and Professor A.L. Demain and Drs. J. Kupka and Y.-Q. Shen of the Massachusetts Institute of Technology, whose continuing efforts are greatly appreciated. The chemical syntheses and development of methodology were performed by Drs. H. Auksi, R.J. Bowers, M. Jouany, and C.C. Shaw, Messrs. C.F. Ingold and M.G. Jokinen, and Mrs. L. Lyubechansky. All of the work was supported by the Natural Sciences and Engineering Research Council of Canada, through its Operating, PRAI, and Strategic Grants Programmes.

REFERENCES

1. S. Wolfe, J.-B. Ducep, K.C. Tin, and S.-L. Lee, Can. J. Chem. 52, 3996-3999 (1974).
2. I. Ernest, J. Gosteli, C.W. Greengrass, W. Holick, D.E. Jackman, H.R. Pfaendler, and R.B. Woodward, J. Am. Chem. Soc. 100, 8214-8222 (1978).
3. See, e.g.: R.D.G. Cooper, in Topics in Antibiotic Chemistry. Vol. 3. Edited by P.G. Sammes. Ellis Horwood Limited, Chichester, 1979; E.T. Gunda and J. Cs. Jászberényi, in Progress in Medicinal Chemistry. Edited by G.P. Ellis and G.B. West. Elsevier/North Holland Biomedical Press. Vol. 12, Chapter 8, 1975; Vol. 14, Chapter 4, 1977.
4. W. Nagata, in New Synthetic Methodology and Biologically Active Substances. Studies in Organic Chemistry. Vol. 6. Edited by Z. Yoshida. Elsevier, 1981, pp. 95-111.
5. S. Wolfe and M.P. Goeldner, Tetrahedron Lett. 5131-5134 (1973).
6. S. Wolfe, W.S. Lee, G. Kannengiesser, and J.-B. Ducep, Can. J. Chem. 50, 2894-2897 (1972).
7. S. Wolfe, S.-L. Lee, J.-B. Ducep, G. Kannengiesser, and W.S. Lee, Can. J. Chem. 53, 497-512 (1975).
8. S. Wolfe and C.C. Shaw, Can. J. Chem. 60, 144-153 (1982).
9. S. Wolfe, U.S.P. 4,200,571 (1980).
10. S. Hanessian and P. Lavallee, Can. J. Chem. 53, 2975-2977 (1975).
11. R.T. Arnold and S.T. Kulenovic, Synth. Comm. 7, 223-232 (1977).
12. O.K. Behrens, in The Chemistry of Penicillin. Edited by H.T. Clarke, J.R. Johnson, and R. Robinson. Princeton University Press, 1949, Chapter 19.
13. H. Hamberger and P. Stütz, personal communications.
14. H.R.V. Arnstein, Ann. Rept. Progr. Chem. 54, 339-352 (1957).
15. B.G. Christensen and R.B. Morin, personal communications.
16. S.W. Elson, R.S. Oliver, B.W. Bycroft, and E.A. Faruk, J. Antibiotics. 35, 81-86 (1982).
17. C.A. Townsend and E.M. Brown, J. Am. Chem. Soc. 104, 1748-1750 (1982).
18. S. Wolfe, R.J. Bowers, S.K. Hasan, and P.M. Kazmaier, Can. J. Chem. 59, 406-421 (1981), and references cited therein.
19. D.J. Aberhart, Tetrahedron. 33, 1545-1559 (1977).
20. S. Wolfe, R.J. Bowers, D.A. Lowe, and R.B. Morin, Can. J. Chem. 60, 355-361 (1982).
21. E.P. Abraham, J. Antibiotics. Suppl. 30, 1-26 (1977).
22. S. Wolfe and M.G. Jokinen, Can. J. Chem. 57, 1388-1396 (1979).
23. S.E. Jensen, D.W.S. Westlake, R.J. Bowers, and S. Wolfe, J. Antibiotics, in press.
24. M. Itoh, D. Hagiwara, and T. Kamiya, Tetrahedron Lett. 4393-4394 (1975).
25. K. Ramsamy, R.K. Olsen, and T. Emery, Synthesis. 42-43 (1982), and references cited therein.
26. L. Kisfaludy, F. Korenczki, and A. Katho, Synthesis. 163 (1982).
27. S. Wolfe and C.F. Ingold, J. Am. Chem. Soc. 103, 940-941 (1981).
28. E. Klein and W. Rojahn, Tetrahedron. 21, 2353-2358 (1965).
29. A. Michael and W.W. Garner, Am. J. Chem. 267-271 (1905).
30. H.Z. Sable, K.A. Powell, H. Katchian, C.B. Niewoehner, and S.B. Kadlek, Tetrahedron. 26, 1509-1524 (1970); and personal communication from H.Z. Sable.
31. J.W. Cornforth, R.H. Cornforth, G. Popják, and L. Yengoyan, J. Biol. Chem. 241, 3970-3987 (1966).
32. S.L. Regen and C. Koteel, J. Am. Chem. Soc. 99, 3837-3838 (1977).
33. S.E. Jensen, D.W.S. Westlake, and S. Wolfe, J. Antibiotics, in press.
34. S.E. Jensen, D.W.S. Westlake, and S. Wolfe, J. Antibiotics, 35, 483-490 (1982).
35. J. Kupka, Y.-Q. Shen, S. Wolfe, and A.L. Demain, FEBS Lett, in press; J. Kupka, Y.-Q. Shen, S. Wolfe, and A.L. Demain, J. Antibiotics, submitted for publication. A labile epimerase appears to be present in such systems, but can be detected only in crude cell-free preparations. See: T. Konomi, S. Herchen, J.E. Baldwin, M. Yoshida, N.H. Hunt, and A.L. Demain, Biochem. J. 184, 427-430 (1979); G.S. Jayatilake, J.A. Huddleston, and E.P. Abraham, Biochem. J. 194, 645-647 (1981).
36. Analogous cyclizations have been reported by G.A. Bahadur, J.E. Baldwin, J.J. Usher, E.P. Abraham, G.S. Jayatilake, and R.L. White, J. Am. Chem. Soc. 103, 7650-7651 (1981), using C. acremonium.
37. H. Peter and H. Bickel, Helv. Chim. Acta. 57, 2044-2054 (1974).
38. H. Nakano, Med. Res. Rev. 1, 127-157 (1981).

STEREOCHEMISTRY OF PALYTOXIN

Yoshito Kishi

Department of Chemistry, Harvard University, 12 Oxford Street, Cambridge, Mass. 02138, USA

Abstract - The complete assignment of the stereochemistry of palytoxin, based primarily on organic synthesis, is described.

Palytoxin, the toxic principle isolated from marine soft corals of the genus Palythoa, is the most poisonous substance known to date, except for a few polypeptides and proteins found in bacteria and plants. Pioneering investigations by the Nagoya group[1] and by the Hawaii Group[2] have recently led them independently to suggest structure 1 for palytoxin. Thus, it is evident that palytoxin is uniquely distinct from molecules which organic chemists have routinely dealt with in terms of magnitude of molecular size, of structural complexity, and so on. Shortly after the structure became available, we undertook studies toward the total synthesis of palytoxin. The very first step of our investigation was to establish the stereochemistry, in a relative as well as absolute sense, of palytoxin. In addition to 1 exo, 3 cis and 4 trans double bonds, 64 asymmetric centers exist on the carbon backbone of palytoxin; however, the relative and absolute stereochemistry of all except 13 had not been established unambiguously.

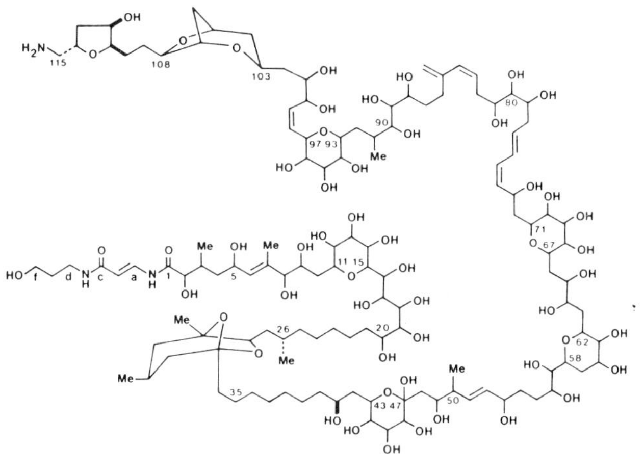

1 : palytoxin

Our overall plan of research is summarized as follows:

Step 1. Assignment of both relative and absolute stereochemistry.

    a. By using only well established organic reactions, synthesize all the stereoisomers possible for a given degradation product of palytoxin from an optically active substance with the known absolute configuration.
    b. Confirm that all the stereoisomers can be distinguished by spectroscopic methods.
    c. Find which stereoisomer is identical with the degradation product.
    d. Repeat the same procedure for other degradation products.
    e. Establish the complete structure of palytoxin.

Step 2. Total Synthesis.

Once the stereochemistry is completely established, synthetic efforts follow the usual course of organic synthesis.

Needless to point out, the most crucial and essential requirement for Step 1 was the availability of palytoxin and/or its degradation products. From their end, Professors Hirata and Uemura have kindly agreed to carry out collaborative investigations on the stereochemistry assignment. In this article, we would like to review our collaborative efforts which have successfully resulted in the complete assignment of the stereochemistry of palytoxin.[3]

We chose to study degradation product 2a[1d,2a] as our first target. The structure, including the absolute configuration, of acetal 3b, which is a more advanced degradation product of 2b, was determined by Hirata, Uemura and their co-workers using X-ray analysis.[1a,c] A stereospecific, practical synthesis of 3a via 4 and 5 was recently achieved in our laboratory.[4] Thus, we were ready to turn our attention to the degradation product 2a, which had two unassigned asymmetric centers.

2a : $R^1 = C_6H_5$,

2b : $R^1 = $ p-BrC_6H_4, $R^2 = $

3a : $R^1 = C_6H_5$,

3b : $R^2 = $ p-BrC_6H_4, $R^2 = $

Some comments on the stereochemistry assignment of these types of compounds by spectroscopic methods such as ^{1}H-NMR spectroscopy would appear to be appreciated here. Discussing compounds with a -CHX-CHY- group in acyclic systems, it is widely recognized that the spin-spin coupling constant for the erythro isomer (J ≅ 6-10 Hz) is larger than that for the corresponding threo isomer (J ≅ 1-4 Hz). These observations are usually explained on the basis of preferred conformations for erythro or threo isomers.[5] However, when this method is applied to the stereochemistry assignment of a -CHX-CHY- system of acyclic compounds, the preferred conformation for a given system must be assumed, which is not trivial particularly for polyfunctional molecules like palytoxin. Primarily for this reason, we had serious reservations about accepting stereochemistry assignments made in this manner. During the course of this study, we have synthesized a large number of compounds with a $[-CH(OAc)-]_n$ group and analyzed the ^{1}H-NMR spectra. The more compounds analyzed, the more serious were our doubts about this approach. The examples given in Table 1 demonstrate this clearly. Of course, there is little doubt about applying ^{1}H-NMR methods for the stereochemistry assignment of cyclic compounds. Unfortunately, palytoxin is available in very limited amounts, which makes it practically impossible to attempt preparation of a cyclic derivative of 2a.

Table 1

Spin-Spin Coupling Constants

	J_1	J_2
	5.5 Hz (threo)	5.5 Hz (threo)
	4.0 or 6.0 Hz (threo)	6.0 or 4.0 Hz (erythro)
	9.0 Hz (erythro)	2.1 Hz (threo)
	6.8 Hz (erythro)	3.0 Hz (erythro)

Under these circumstances, we decided to synthesize all the possible stereoisomers of 2a or its equivalents, to establish unambiguously the configuration of unassigned asymmetric centers. For this and other reasons, the carbohydrate chain-extension method had been developed in our laboratory (Scheme 1).[6] Using this method, we synthesized two erythro isomers 7 and

Scheme 1

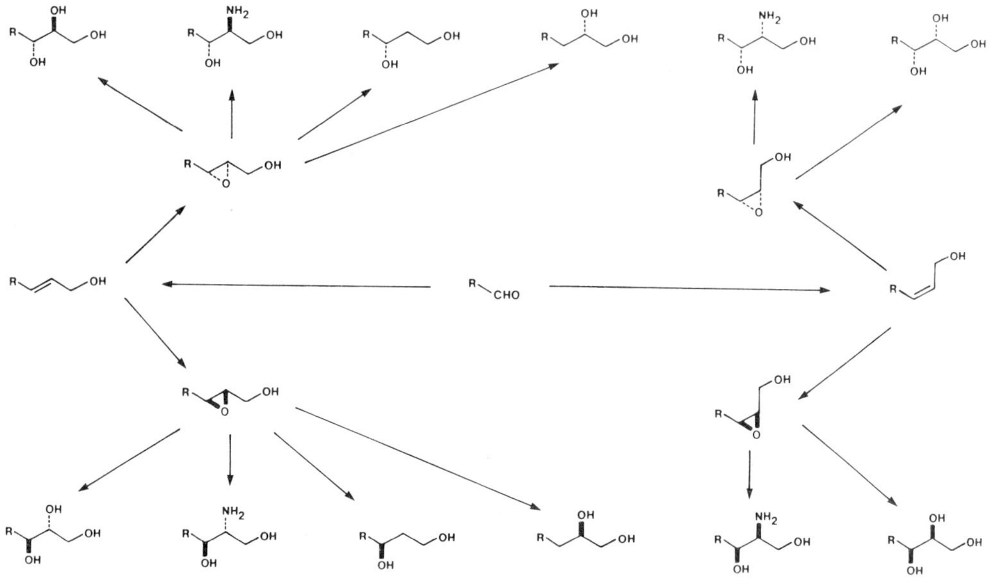

8 from trans allylic alcohol 6. The assignment of the absolute configuration of these compounds depended upon Sharpless asymmetric epoxidation,[7] which is reliable for trans allylic alcohols.[8] Since the stereochemical outcome of Sharpless asymmetric epoxidation of cis allylic alcohols is not always predictable,[8] two threo alcohols 9 and 10 were prepared from the intermediates used in the transformation of 6 into 7 and 8.[9] The chemical shifts and spin-spin coupling constants for the C.99, C.100 and C.101 protons of the acetate of only isomer 10 were found to closely resemble those of 2a, suggesting that 10 possesses the natural configuration at these asymmetric centers. Indeed, the tetraacetate 2a, synthesized from 4 and acetonide alcohol 11, was found to be identical with degradation product 2a[10] on comparison of spectroscopic data, establishing the stereochemistry of palytoxin at C.100 and C.101 as shown in 2a.

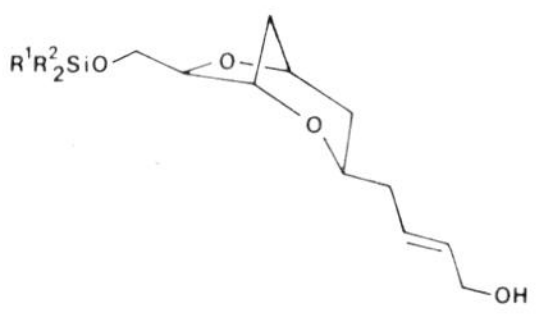

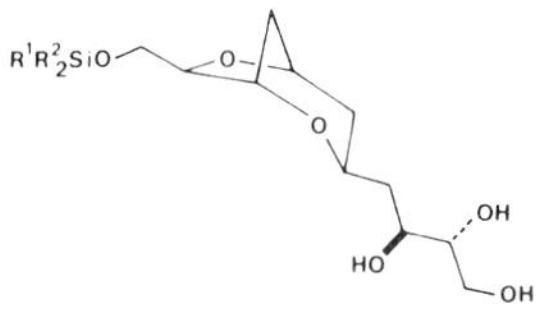

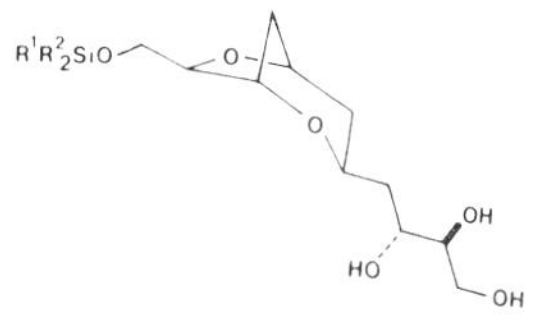

R^1 = t-Bu, R^2 = Ph

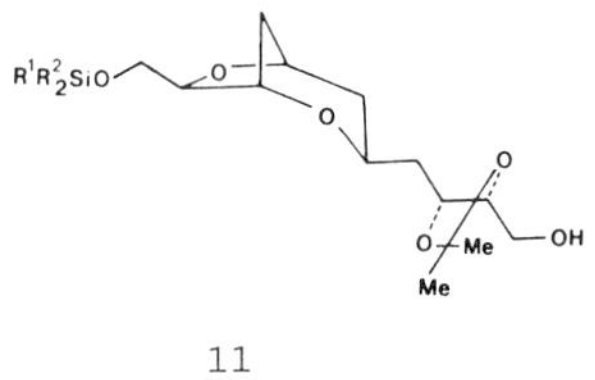

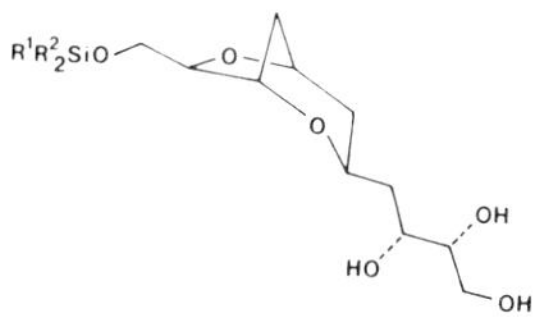

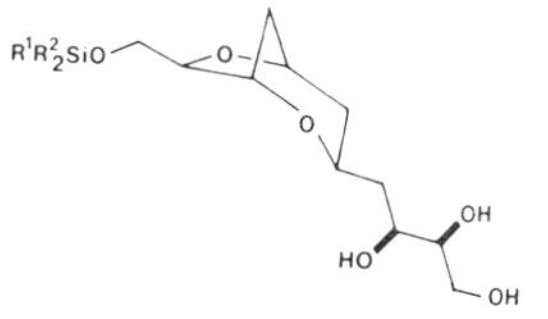

We next turned our attention to nonaacetate 12,[11] a degradation product known to contain the C.84 through C.98 carbon backbone.[1d,2a] The ^{1}H-NMR spectrum of 12 suggested that the relative stereochemistry of the tetrahydropyran portion was as indicated in the structure.[2a,1d] However, the relative stereochemistry of the acyclic portion remained unknown. To determine the relative and absolute stereochemistry of 12 efficiently, we initially studied the more advanced degradation product 13.[12]

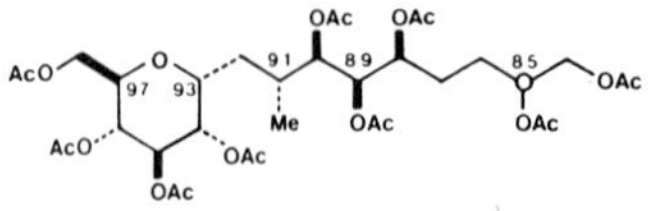

12

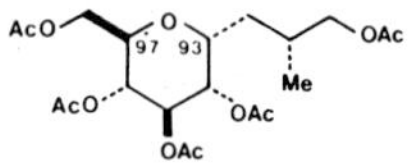

13

It was obvious that degradation product 13 should be synthesized from a naturally occurring, common carbohydrate. We chose to use 3,4,6-tribenzyl-D-mannose 1,2-epoxide (14)[13] to introduce the critical axial C-glycoside bond of 13. Thus, 14 was reacted with the Grignard reagent prepared from (S)-(+)-3-tert-butoxy-2-methyl-bromopropane[14] in the presence of Li_2CuCl_4 to yield stereoselectively alcohol 15. Swern oxidation of 15, followed by diborane reduction, furnished an 8:1 mixture of alcohols 16 and 15. Debutylation of 16, followed by debenzylation and acetylation, gave pentaacetate 13. Starting with 14 and (R)-(-)-3-tert-butoxy-2-methyl-1-bromo-propane, the same sequence of reactions produced the C.91 diastereomer of 13. On comparison of the spectroscopic data and optical rotations, synthetic pentaacetate 13 was found to be identical with degradation product 13, establishing the stereochemistry at C.91, C.93, C.94, C.95, C.96 and C.97.

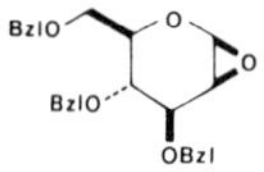

14

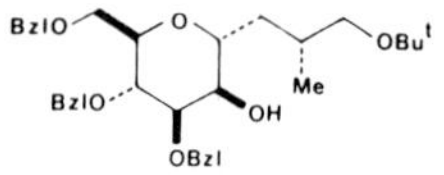

15

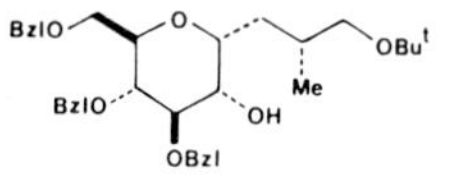

16

In order to study the stereochemistry at C.88, C.89 and C.90, 16 was converted to trans allylic alcohol 17 and then subjected to the carbohydrate chain-extension method[6] in a manner identical with that used for allylic alcohol 6. The triols 18, 19, 20 and 21 resulted. The threo relationship of triols 20 and 21, and consequently the erythro relationship of triols 18 and 19, was further confirmed by the fact that OsO_4 oxidation of the t-butyldiphenylsilyl ether of trans allylic alcohol 17, followed by (n-Bu)$_4$NF treatment, furnished a 1:1 mixture of triols 20 and 21.

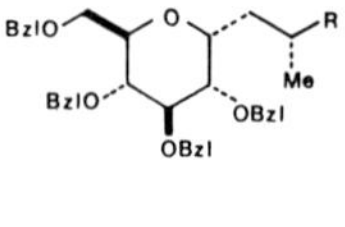

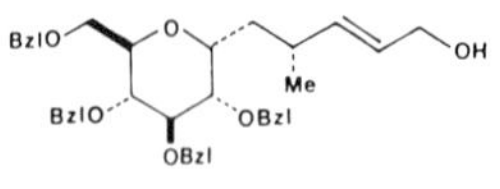

17

18 : R =

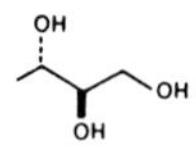

19 : R =

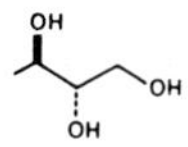

20 : R =

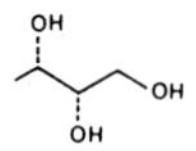

21 : R =

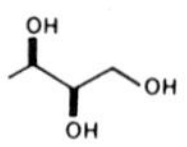

The acetonide alcohol 22, prepared from threo triol 21, was subjected to Swern oxidation, followed by addition of 3-butenylmagnesium bromide to the resulting aldehyde, to yield approximately a 1:2 mixture of alcohols 23 and 24. Osmium tetroxide oxidation of the minor alcohol 23, followed by aqueous acid hydrolysis, debenzylation and acetylation, furnished nonaacetate 12. The same sequence of reactions on 24 gave the corresponding nonaacetate. On comparison of the spectroscopic data, synthetic nonaacetate 12 was found to be identical with degradation product 12, while the nonaacetate derived from 24 was different. The corresponding nonaacetates were likewise synthesized from derivatives of the remaining triols 18, 19 and 20, and none was found to be identical with degradation product 12.

22

23

24

26

25

The C.88 stereochemistry of alcohols 23 and 24 was established by the following experiments. Benzylation of the major alcohol 24, followed by aqueous acid hydrolysis, periodate oxidation and sodium borohydride reduction, afforded 2-benzyloxyhex-5-en-1-ol (25). The α_D value of 25 was found to be -9.4°. The specific rotations of R- and S-2-benzyloxyhex-5-ene-1-ols, prepared from D-and L-glyceraldehyde ketals,[15] were found to be -11.7° and +10.4°, respectively. Thus, C.88 has the S configuration in the minor alcohol 23. This conclusion was further confirmed by comparison of the ^{1}H-NMR spectra of the MTPA[16] esters of 25 obtained from the above-mentioned sources.

The experiments summarized above allowed us to assign the stereochemistry of degradation product 12 as indicated. Since this assignment of the relative stereochemistry between C.90 and C.91 was based solely on the results of Sharpless asymmetric epoxidation[7] of 17, we felt it was desirable to have additional evidence. For this reason, the stereochemistry assignment at C.88, C.89 and C.90 by an alternative method was performed. Thus, cis allylic alcohols 26 were prepared from 16 and subjected to OsO$_4$ oxidation, aqueous acid hydrolysis, debenzylation and acetylation to yield a mixture of nonaacetates corresponding to 12. However, none of these nonaacetates was identical with degradation product 12, thus establishing the relative stereochemistry at C.88 and C.89 as threo. This information, along with knowledge of the relative stereochemistry between C.89 and C.90 and of the absolute stereochemistry at C.88 and C.91 (vide supra), excluded all possible structures for degradation product 12 except the one shown.

The pentaacetate 27, which contains C.77 through C.83, is a degradation product of palytoxin. The relative stereochemistry of 27 was found by NMR experiments to be as indicated, but the absolute stereochemistry remained unknown.[1b,2a] Synthesis of pentaacetate 27 from 2-deoxy-D-glucose via nitrile 28 was performed straightforwardly. Upon comparison of α_D values, synthetic pentatacetate 27 was found to be identical with degradation product 27. To determine the connectivity of 27 with the remaining portions of the palytoxin structure, a stepwise degradation reaction was performed. Palytoxin was partially degraded to 29,[1b] which was then ozonized and treated with acetic anhydride/pyridine to yield trans-α,β-unsaturated aldehyde 30. Synthesis of the two possible diastereomers, i.e. 30 and its C.80 diastereomer, was achieved from 28. Upon comparison of the ^{1}H-NMR spectra, synthetic 30 was found to be identical with degradation product 30, establishing the stereochemistry of C.79, C.80 and C.81 as shown in 29.

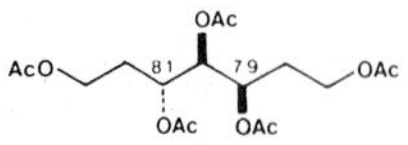

27

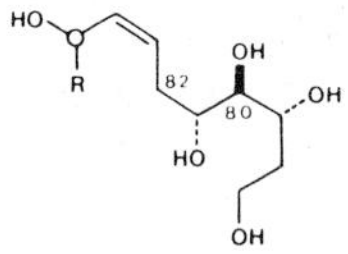

29 : R = C.86-C.115 segment

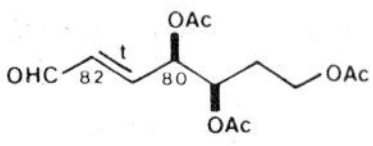

30

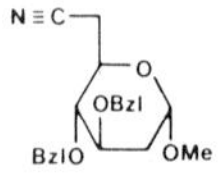

28

We then turned our attention to degradation product 31, containing the C.52 through C.74 carbon backbone of palytoxin.[1b,2a] The ^{1}H-NMR spectrum of 31 provided valuable information on the relative stereochemistry of the cyclic portions, indicating it to be as shown in 31. In addition, two more advanced degradation products 32[1b,2a] and 33[17] were useful in this study.

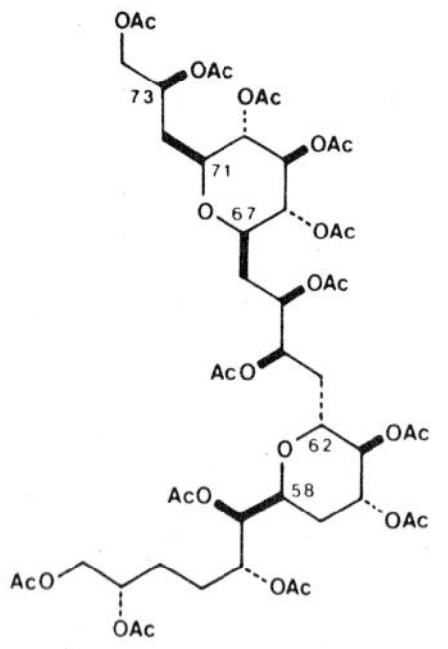

31

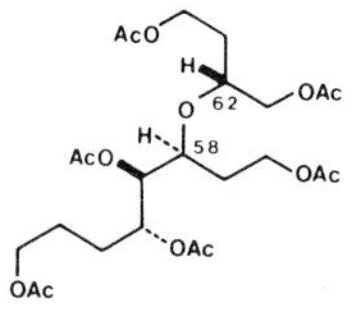

32

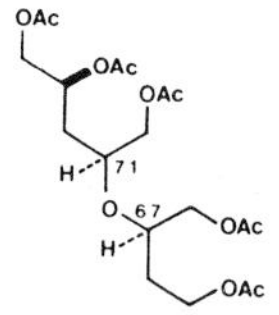

33

The synthesis of 37 (the antipode of 33) was next accomplished. The trans allylic alcohol 35, synthesized from D-glucose (34), was subjected to Sharpless asymmetric epoxidation[7] using L-(+)-diethyl tartrate to yield the expected epoxide 36. Regioselective reductive epoxide opening of 36,[6] followed by acetylation, debenzylation, periodate oxidation, borohydride reduction and acetylation, furnished pentaacetate 37. The C.73 epimer of 37 was also synthesized from 35 by treatment with the same sequence of reagents except for D-(-)-diethyl tartrate in the asymmetric epoxidation. Upon comparison of spectroscopic data and optical rotations, 37 was found to be the antipode of degradation product 33, establishing the stereochemistry at C.67, C.71 and C.73. The stereochemistry C.68, C.69 and C.70 was not determined directly by this synthesis, but was concluded with reasonable confidence from the aforementioned ^{1}H-NMR data and was later confirmed by synthesis (vide infra).

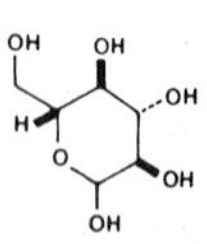

34

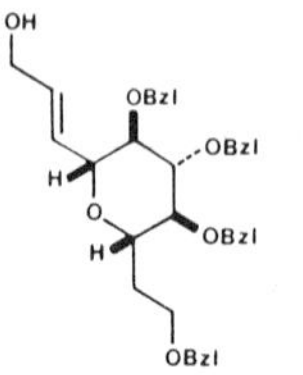

35

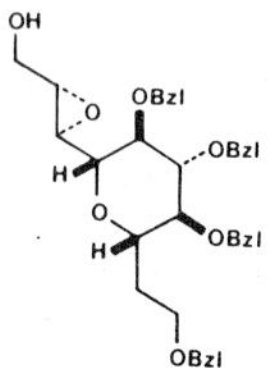

36

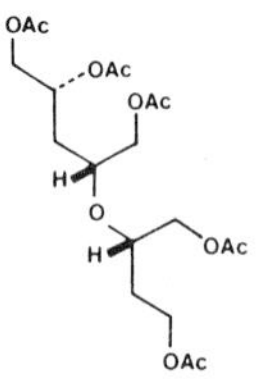

37

The stereochemistry of degradation product 32 was established by the following synthesis. Trans allylic alcohol 39, prepared from 2-deoxy-D-glucose (38), was subjected to asymmetric epoxidation[7] using D-(−)-diethyl tartrate to yield the expected epoxide 40 as the major product. Camphorsulfonic acid treatment of 40 in acetone at room temperature gave acetonide 41 as the sole product. Exclusive formation of the axial C-glycoside bond in this cyclization was dictated by the stereochemistry of 40. Grignard reaction of aldehyde 42, prepared from 41, with 3-butenylmagnesium bromide in ether gave a 1:6 mixture of alcohols 43 and 44. Upon comparison of spectroscopic data and optical rotations, hexaacetate 32, obtained from the minor alcohol 43 in 7 steps, was found to be identical with degradation product 32 while the corresponding hexaacetate derived from the major alcohol 44 was different, establishing the stereochemistry at C.57, C.58 and C.62. The relative stereochemistry between C.56 and C.57, which was not confirmed by this synthesis, was found to be erythro as follows. Wittig reaction of aldehyde 45, prepared from 41, and phosphonium salt 46 (for the configuration at C.53, see below), prepared from D-glyceraldehyde acetonide, yielded cis olefin 47 ($J_{56,57}$ = 11.0 Hz). After modification of the upper side chain, 47 was subjected to osmium tetroxide oxidation to furnish a 7:1 mixture of the two possible erythro diols. Upon comparison of spectroscopic data, hexaacetate 32, derived from the major erythro diol, was found to be identical with degradation product 32.

Information about the stereochemistry of C.53 was obtained from the following study. The tetraacetate 48, containing C.47 through C.56, was isolated as a degradation product of palytoxin.[1f,1d] By using the carbohydrate chain-extension method,[6] triacetate 49 and its C.49 diastereomer were synthesized from (S)-(+)-3-hydroxy-2-methylpropionic acid.[14] On comparison of spectroscopic data and optical rotations, triacetate 49 was found to be identical with an advanced degradation product 49.[18] Wittig reaction of aldehyde 51, prepared from D-glyceraldehyde acetonide in 8 steps, with phosphonium salt 50, prepared from an intermediate of the synthesis of 49, followed by debenzylation and acetylation gave trans olefin 48. Starting with 50 and the antipode of 51, the C.53 diastereomer of 48 was also prepared. Upon comparison of ^{1}H-NMR data, synthetic tetraacetate 48 was found to be identical with degradation product 48, establishing the stereochemistry at C.49, C.50 and C.53.

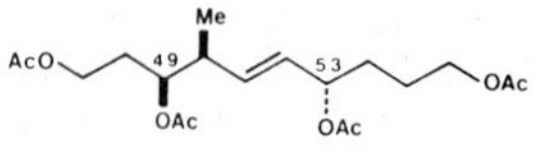

The relative stereochemistry between C.64 and C.65 was determined by the following experiments. Wittig reaction of phosphonium salt 53 with aldehyde 52 gave cis olefin 54. The stereochemistry of the olefinic bond was established by the spin-spin coupling constant ($J_{64,65}$ = 11.3 Hz) of the olefinic protons. Osmium tetroxide oxidation of 54, followed by acetylation, aqueous acid hydrolysis, debenzylation and acetylation, yielded a 1:1 mixture of two erythro acetates. Neither was found to be identical with degradation product 31 upon comparison of spectroscopic data. Photochemically-induced double bond isomerization of 54 gave a mixture of trans olefin 55 (J = 15.5 Hz) and cis olefin 54. Osmium tetroxide oxidation of 55, followed by conversion to the polyacetates as before, yielded a 1:1 mixture of threo acetate 31 and its C.64-C.65 diastereomer. Upon comparison of spectroscopic data and optical rotations, one of the threo isomers was found to be identical with degradation product 31.

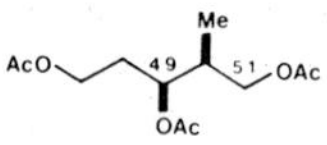

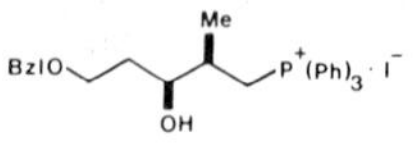

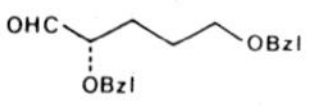

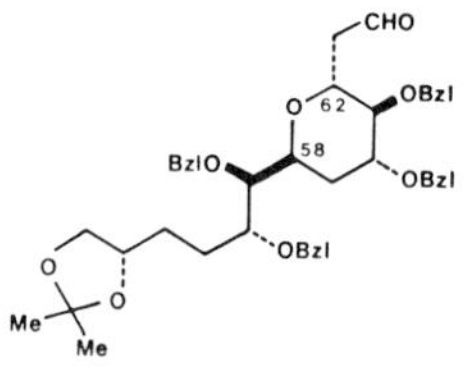

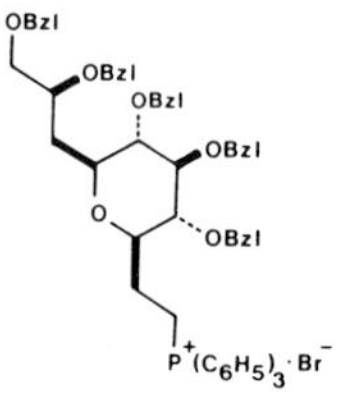

The absolute configuration at C.65 was established by the following experiments. Wittig reaction of phosphonium salt 56 with aldehyde 57 produced cis olefin 58, which was sequentially subjected to hydroboration, acetylation, aqueous acid hydrolysis, debenzylation and acetylation to yield a mixture of acetates. Upon comparison of spectroscopic data, one of these acetates was found to be identical with degradation product 31. Assignment of the absolute configurations at C.65 and C.73 was based on Sharpless asymmetric epoxidation[7] used in the course of the synthesis of 57. This assignment was further supported by the fact that the heptaacetate 59, prepared from 57, was a meso compound -- note the symmetry of this substance. Since the relative stereochemistry between C.64 and C.65 was shown to be threo (vide supra), the absolute stereochemistry of C.64 was now known. Thus, the stereochemistry of the degradation product 31 is as shown in the structure.

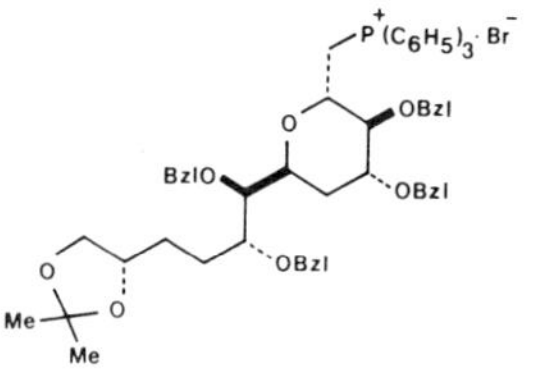

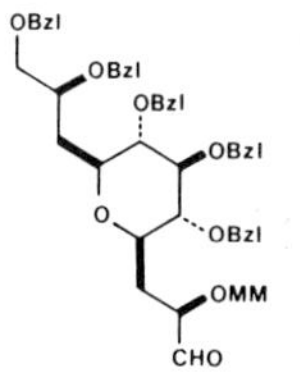

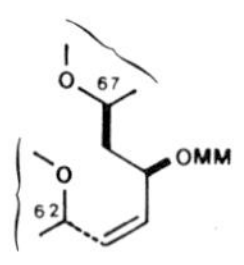

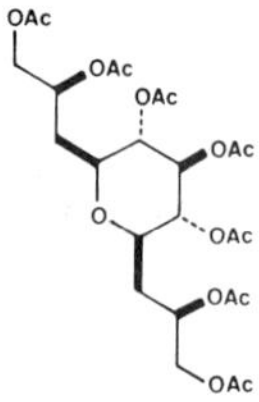

For investigation of the configuration of C.7 through C.51 of palytoxin, degradation products 60, 61, 62, 63 and 64 were available.[19] Of these, 64 deserves special comment. Hirata, Uemura and their co-workers established its structure, including the absolute configuration, by X-ray analysis.[1c] We have recently developed a practical, stereoselective synthetic route from (S)-(-)-citronellal to the optically active bicyclic acetal alcohol 65,[20,21] which provided a solid foundation to study the stereochemistry of C.18 through C.51. First, we worked with degradation product 63. The ^{1}H-NMR data suggested that the relative stereochemistry between C.43 and C.44 was as shown in 63.[22] Routine synthetic operations allowed the transformation of 65 into aldehyde acetate 66. Wittig reaction of 66 with phosphonium salt 67, followed by hydrogenation-hydrogenolysis and acetylation, gave pentaacetate 63. Upon comparison of the spectroscopic data and optical rotations, the synthetic substance was found to be identical with degradation product 63, establishing the stereochemistry at C.43 and C.44.

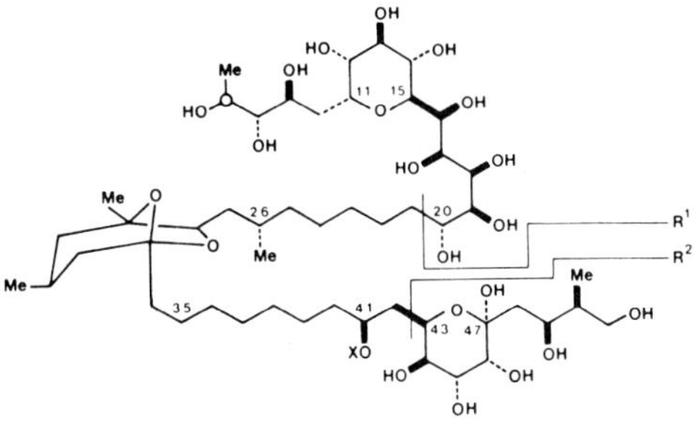

60 : X = H

61 : R^1 = same as acetylated form of 60,

X = Ac,

R^2 = (structure shown)

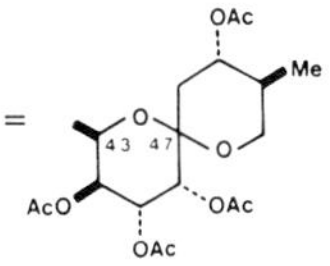

62 : R^1 = same as acetylated form of 60,

X = Ac,

R^2 = (structure shown)

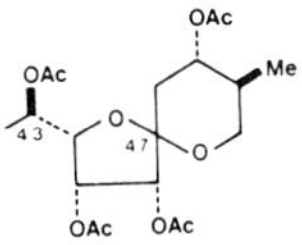

63 : R^1 = (structure), X = Ac, R^2 = (structure shown)

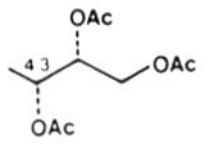

64 : R^1 = (structure with OCOC$_6$H$_4$Br-p), X = H, R^2 = (structure shown)

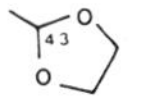

Having already determined the stereochemistry at C.49 and C.50 (vide supra), we next studied the configuration at C.45 and C.46. NMR studies on degradation products 61 and 62 suggested that the stereochemistry at these centers was most likely as shown in 60.[22] This assignment was confirmed by the following experiments. Aqueous acetic acid treatment of cis-α,β-unsaturated ketone 70, synthetized from 68 and 69, resulted in the formation of a 3:2 mixture of two unsaturated spiro-6,6-ketals 71a and 71b whose structures differed only in their configurations at the spiro center. Acetylation of the major isomer 71a, followed by OsO$_4$ oxidation and acetylation, yielded a single tetraacetate. Since one face of the olefinic bond of acetylated 71a was more sterically hindered, structure 72 was tentatively assigned to this product. This assignment was confirmed by further experiments utilizing cis-α,β-unsaturated ketone 70. Osmium tetroxide oxidation of 70 prior to spiroketalization led to a 2:1 mixture of two erythro diols 73a and 73b. Acetic acid treatment of the major diol 73a, followed by acetylation, yielded a 1:2 mixture of spiro-6,6-ketal tetraacetates. The minor product was identical with 72, while the major product was identical with the corresponding spiro-6,6-ketal tetraacetate derived from 71b, establishing that 73a had the same relative stereochemistry between C.44 and C.45 as did 72, and consequently 73b had the opposite. The stereochemistry at C.44 and C.45 of 73b was, in turn, proven by its successful transformation to 74 in 7 steps and correlation with 2-deoxy-D-glucose.

65

66

67

70

71a,b

72

68

73a : X = OH, Y = H

73b : X = H, Y = OH

74

69

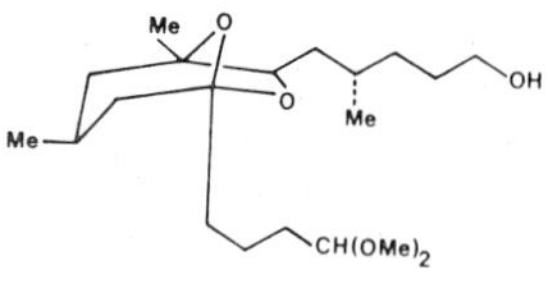

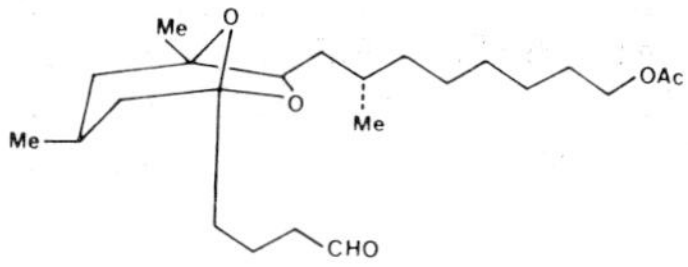

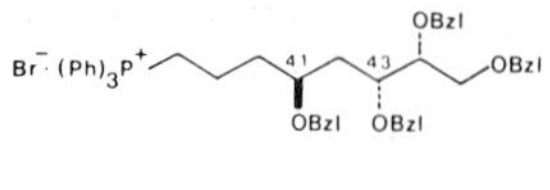

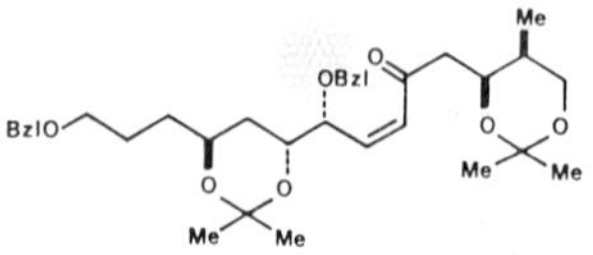

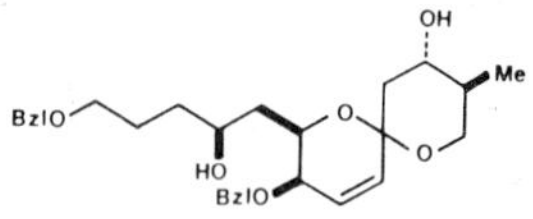

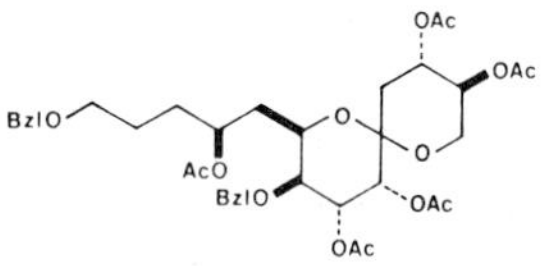

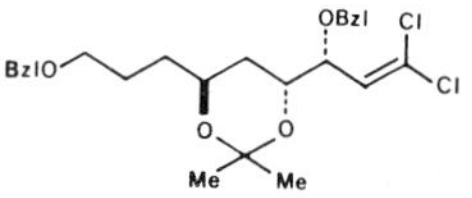

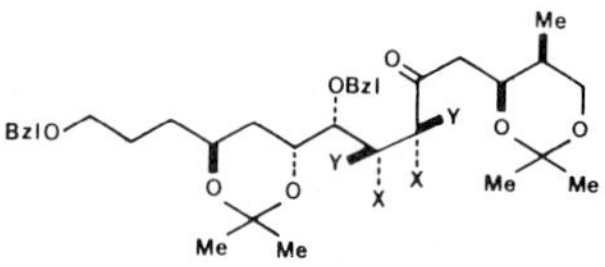

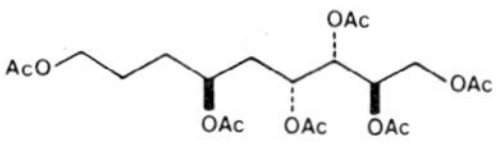

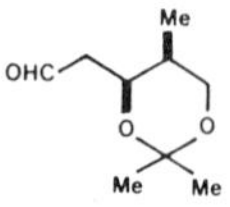

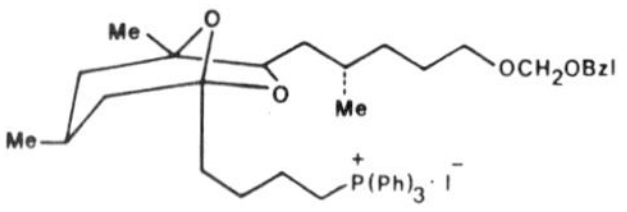

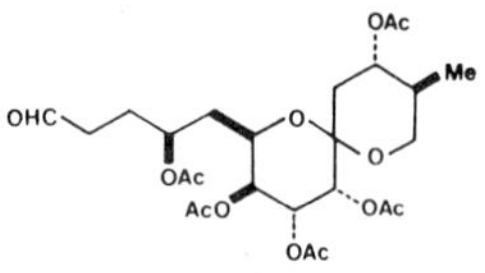

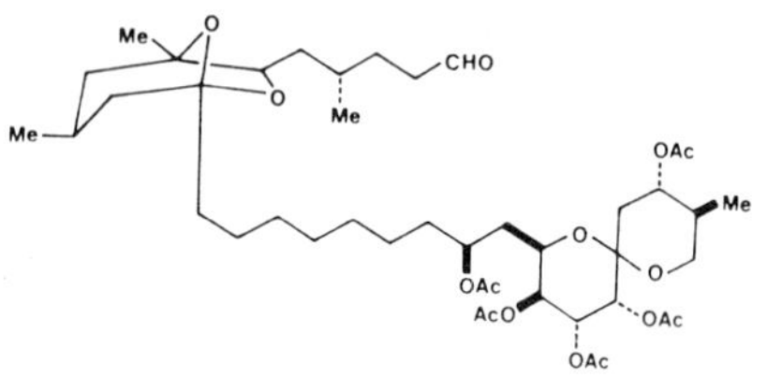

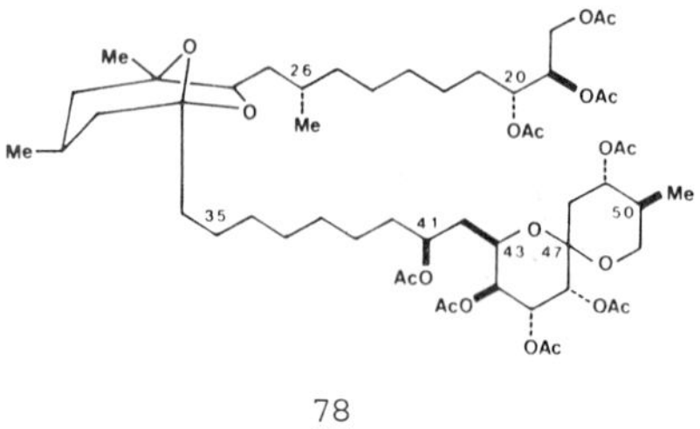

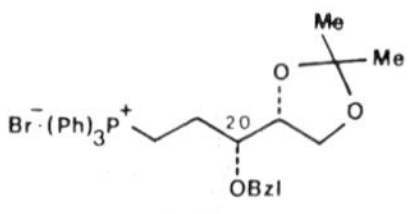

Having determined the stereochemistry of 72, the coupling of the spiroketal segment with the bicyclic segment was studied. The bicyclic segment 65 was converted to phosphonium salt 75 and reacted with aldehyde 76, derived from 72, under standard Wittig reaction conditions. Subsequent hydrogenation-hydrogenolysis and Swern oxidation yielded aldehyde 77. In order to establish the stereochemistry at C.19 and C.20, we had originally planned to couple 77 with a suitable segment to synthesize degradation product 61. However, further efforts by Professors Hirata and Uemura fortunately led to the isolation of a new degradation product, the gross

75

76

77

78

79

structure of which was shown to be 78.²³ Wittig reaction of aldehyde 77 with phosphonium salt 79, prepared from D-xylose, followed by hydrogenation-hydrogenolysis and acetylation, furnished octaacetate 78. The same sequence of reactions was performed using the antipode of 79, prepared from L-xylose, to yield the C.19 and C.20 diastereomer of 78.²⁴ Upon comparison of the spectroscopic data, synthetic octaacetate 78 was found to be identical with degradation product 78, establishing the stereochemistry at C.19, C.20, C.45 and C.46.

The three acetates 80, 81 and 82 were known to be more advanced degradation products of 60.²⁵ The ¹H-NMR spectrum of 80 suggested that the relative stereochemistry of the tetrahydropyran ring was as indicated in the structure.²⁵ The stereochemistry of the acyclic portions

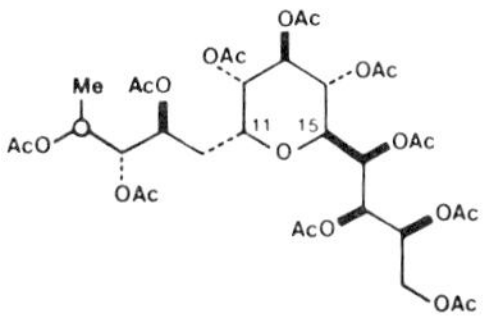

remained unknown. We planned to synthesize 80, 81 and 82 from a naturally occurring, common carbohydrate. For this and other reasons, we urgently needed to develop a selective method to form an axial C-glycoside bond at the anomeric position. In short, this could effectively be achieved by nucleophilic addition to the pyran oxonium ion derived from readily available tetrabenzylpyranose derivatives. By using a nucleophile equivalent with a hydride anion, it was also possible to form an equatorial C-glycoside bond at the anomeric position (Scheme 2).²⁶ This method provided a highly selective and practical method to synthesize precursors required for construction of rings A, F and G in large quantity. Synthesis of 82 was straightforwardly achieved from benzyl ether 85. Upon comparison of spectroscopic data and optical rotations, synthetic triacetate 82 was found to be identical with degradation product 82, establishing the absolute configuration at C.11 and C.15.

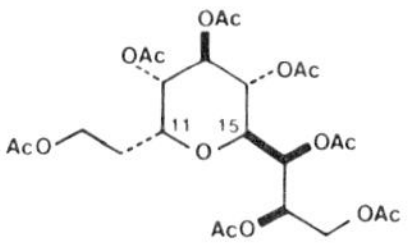

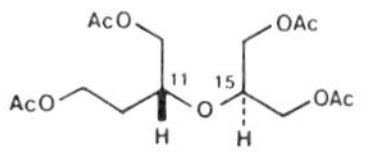

Scheme 2

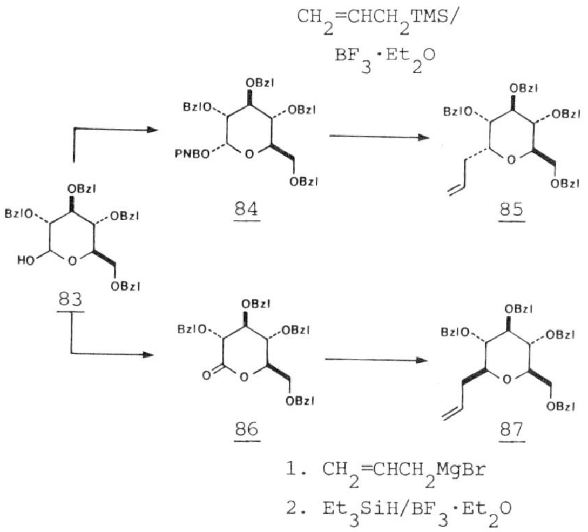

Since the stereochemistry of the tetrahydropyran ring was known from the ¹H-NMR data, only four diastereomers remained as structural possibilities for degradation product 81. By using the carbohydrate chain-extension method,⁶ all four diastereomeric heptaacetates, 81, 89, 90 and 91, were synthesized from alcohol 88. Upon comparison of ¹H-NMR spectra, synthetic threo heptaacetate 81 was found to be identical with degradation product 81, establishing the stereochemistry at C.12, C.13, C.14, C.16 and C.17. Using similar methods, alcohol 92 and its C.18 diastereomer were synthesized. The C.18 stereochemistry of 92 was unambiguously established by synthesis of one of the intermediates from L-glyceraldehyde.

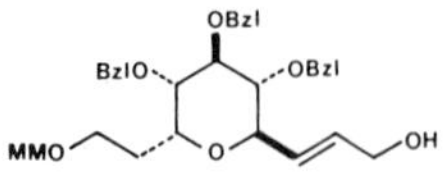

In order to study the stereochemistry at C.8, C.9 and C.18, 92 was transformed into cis- and trans-α,β-unsaturated ketones 93 and 94 via routine synthetic operations. Osmium tetroxide oxidation of cis-α,β-unsaturated ketone 93, followed by separation of isomers, borohydride reduction, deacetonization, debenzylation and acetylation, furnished two pairs of decaacetates with an erythro relationship between C.8 and C.9. Likewise, two pairs of decaacetates with a threo relationship between C.8 and C.9 were obtained from trans-α,β-unsaturated ketone 94. Upon comparison of ¹H-NMR spectra, one pair of the erythro decaacetates was found to be identical with degradation product 80, establishing the relative stereochemistry between C.8 and C.9 and the absolute stereochemistry at C.18.

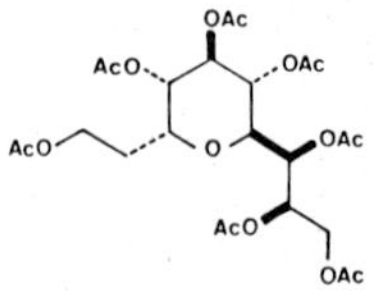

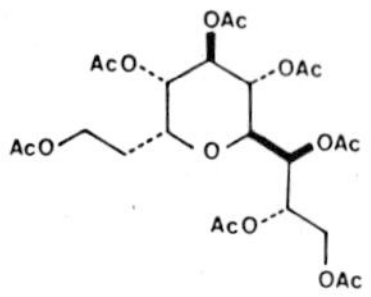

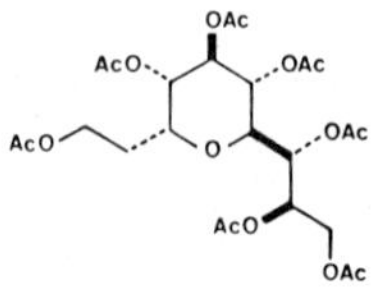

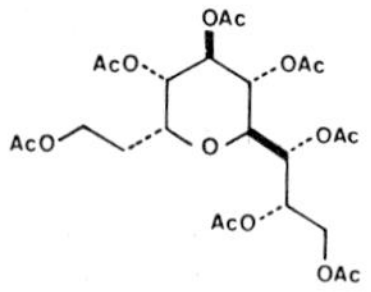

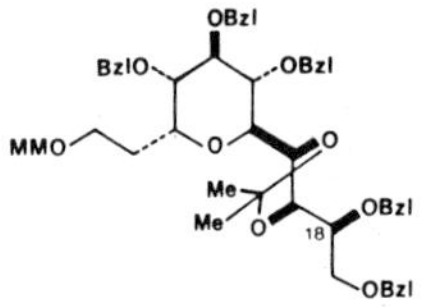

The absolute configuration at C.8 was concluded by the following experiments. The erythro diol with the unnatural configuration at C.8 and C.9, obtained by OsO₄ oxidation of 93 (vide supra), was transformed into heptaacetate 95. The absolute configuration of 95 was determined as follows. Trans allylic alcohol 96 was subjected to Sharpless asymmetric epoxidation using D-(-)-diethyl tartrate to yield the expected epoxide 97, which was then converted to heptaacetate 95. Heptaacetate 95 thus prepared was found to be identical with heptaacetate 95 derived from 93, establishing the absolute stereochemistry at C.8 and consequently at C.9.

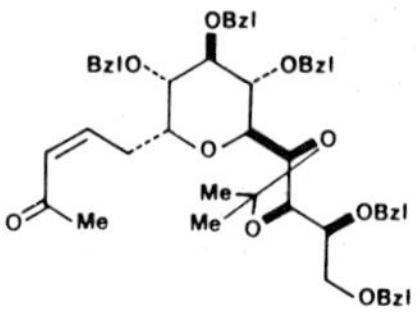

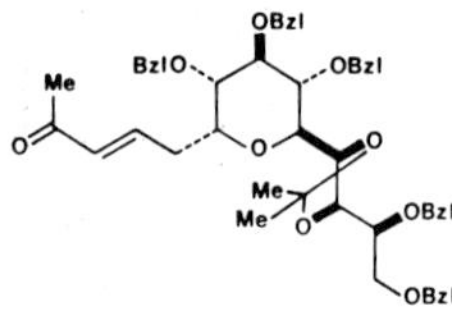

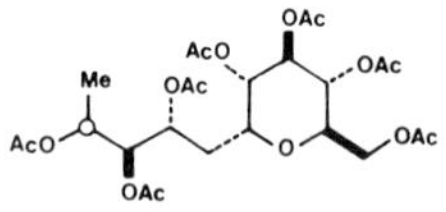

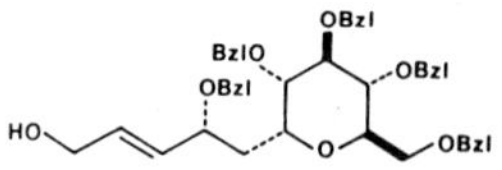

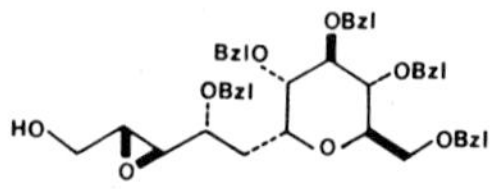

Successful assignment of the stereochemistry of degradation products 78 and 80 allowed us to define the stereochemistry of degradation product 60 as shown in the structure (for the stereochemistry at C.47, see below).

The lactone diacetate 98, containing C.1 through C.6, is a known degradation product of palytoxin. The ¹H-NMR spectrum of 98 suggested that the relative stereochemistry at C.2, C.3 and C.5 was as indicated in 98, but the absolute stereochemistry was unknown.[27] By using the carbohydrate chain-extension method, tetraacetate 99 was synthesized from (S)-(+)-3-hydroxy-2-methylpropionic acid.[14] Upon comparison of spectroscopic data and optical rotations, tetraacetate 99 was found to be identical with the tetraacetate prepared from the degradation product,[28] establishing the absolute stereochemistry of C.2, C.3 and C.5.

98

99

In the preceding paragraphs of this article, we have disclosed the stereochemistry of key degradation products of palytoxin. It is important to note that all of the asymmetric centers existing in palytoxin are found intact[29] in these degradation products so that we are now in a position to describe the complete structure of the toxin. Owing to extensive investigations by the Nagoya group[1] and by the Hawaii group[2], no ambiguity remains concerning either the connectivity of these degradation products or the stereochemistry of all of the olefinic bonds. The spin-spin coupling constants in the ^{1}H-NMR spectrum of palytoxin itself[30] provided confirmation of the stereochemistry assignment of the double bonds.[31]

The palytoxin used in this study was extracted from Okinawan Palythoa tuberculosa,[1] whereas the palytoxin used by the Hawaii group was extracted from Hawaiian Palythoa toxica or a Tahitian Palythoa species.[2] Although direct comparison of Okinawan palytoxin with Hawaiian or Tahitian palytoxin has not been made, it seems from ^{13}C-NMR spectra of palytoxin and ^{1}H-NMR spectra of degradation products[32] that they are identical. There is, however, one discrepancy between the gross structure assigned to Tahitian palytoxin by the Hawaii group[2a] and that assigned to Okinawan palytoxin by the Nagoya group.[1a] The latter suggested a hemiketal structure at the C.47 position, while the Hawaii group suggested an anhydro structure at the C.44 and C.47 positions. The ^{13}C-NMR spectrum of Tahitian palytoxin in D_2O at 25°C has been reported to show a signal at 100.2 ppm for the C.47 ketal carbon, while the ^{13}C-NMR spectrum of Hawaiian palytoxin has been reported not to show hemiketal carbon signals. The ^{13}C-NMR spectrum of Okinawan palytoxin in D_2O at room temperature shows an intense signal at 100.3 ppm. The FAB mass spectrum of this sample shows an ion at 2678.9 ± 0.2, which corresponds well with the molecular ion of the hemiketal structure (cal'd. for $C_{129}H_{223}N_3O_{54}$ = 2678.5).[33] The ^{13}C chemical shifts of C.2 in tagatose, in which the stereochemistry is similar to that at C.44, C.45 and C.46 of palytoxin, and 2,5-anhydrotagatose[34] indicate that it is very unlikely that a hemiketal carbon and a ketal carbon would show identical chemical shifts.[35] The chemical reactivity observed for palytoxin (this includes: 1. smooth hydrogen-deuterium exchange at the C.48 position in D_2O at room temperature, 2. facile bond cleavage between C.46 and C.47 upon periodate oxidation at 0°C and 3. formation of an α,β-unsaturated ketone at C.47 through C.49) is easily explained by the hemiketal structure, but not by the ketal structure unless hydrolysis of the anhydro form to the corresponding hemiketal is unusually facile. It should be noted that 2,5-anhydrotagatose remained unchanged for 24 hours at 55°C in D_2O.

As is known for ketoses,[36] the six-membered hemiketals 1A must exist in equilibrium with the five-membered hemiketals 1B. However, the equilibrium of palytoxin is weighted heavily in favor of 1A for steric reasons -- note that four substituents must remain on one face of the tetrahydrofuran ring of 1B. With regard to the stereochemistry at C.47, 1A-a seems to be preferred over 1A-b primarily for two reasons: (1) the two bulky side chains can take an equatorial orientation in 1A-a and (2) 1A-a has anomeric stabilization.

1A-a

1A-b

1B-a

1B-b

Palytoxin can now be defined as structure 100! Thus, a solid foundation has been laid for the next steps in the chemical investigation of the toxin. One final comment on the stereochemistry assignment of palytoxin recently proposed[37] seems warranted here. The misassignment of a large number of stereocenters through primary dependence on NMR methods alone points out the limitation of these types of experiments and the continuing importance of organic synthesis in structure elucidation.

100 : palytoxin

Acknowledgment: The work reviewed in this article resulted from the fruitful collaborative investigations between the Nagoya and Harvard groups. I would like to express my sincere appreciation to Professors Hirata and Uemura for their part in the stimulating, exciting and challenging venture in which we have been engaged for the past two years. Equally, I would like to express sincere appreciation to my very able and dedicated co-workers at Harvard, Dr. Cha, Dr. Christ, Mr. Finan, Dr. Fujioka, Dr. Klein, Dr. Ko, Mr. Leder, Mr. McWhorter, and Dr. Pfaff. Without their great determination and efforts, we couldn't even begin to dream of the conclusion which I have presented here. Financial assistance from the National Institutes of Health (NS-12108) and the National Science Foundation (CHE 78-06296) is gratefully acknowledged.

REFERENCES AND FOOTNOTES

1. a. D. Uemura, K. Ueda, Y. Hirata, H. Naoki, T. Iwashita, Tetrahedron Lett. 2781 (1981).
   b. D. Uemura, K. Ueda, Y. Hirata, H. Naoki, T. Iwashita, Tetrahedron Lett. 1909 (1981).
   c. D. Uemura, K. Ueda, Y. Hirata, C. Katayama, J. Tanaka, Tetrahedron Lett. 4861 (1980).
   d. D. Uemura, K. Ueda, Y. Hirata, C. Katayama, J. Tanaka, Tetrahedron Lett. 4857 (1980).
   e. R. D. Macfarlane, D. Uemura, K. Ueda, Y. Hirata, J. Am. Chem. Soc. 102, 875 (1980).
   f. Y. Hirata, D. Uemura, K. Ueda, S. Takano, Pure Appl. Chem. 51, 1875 (1979).
2. a. R. E. Moore, G. Bartolini, J. Am. Chem. Soc. 103, 2491 (1981). b. R. E. Moore, F. X. Woolard, G. Bartolini, J. Am. Chem. Soc. 102, 7370 (1980). c. R. E. Moore, F. X. Woolard, M. Y. Sheikh, P. J. Scheuer, J. Am. Chem. Soc. 100, 7758 (1978). d. R. E. Moore, R. E. Dietrich, B. Hatton, T. Higa, P. J. Scheuer, J. Org. Chem. 40, 540 (1975).
   e. R. E. Moore, P. J. Scheuer, Science 172, 495 (1971).
3. a. L. L. Klein, W. W. McWhorter, Jr., S. S. Ko, K.-P. Pfaff, Y. Kishi, D. Uemura, H. Hirata, manuscript submitted for publication. b. S. S. Ko, J. M. Finan, Y. Kishi, D. Uemura, Y. Hirata, manuscript submitted for publication. c. H. Fujioka, W. J. Christ, J. K. Cha, J. Leder, Y. Kishi, D. Uemura, Y. Hirata, manuscript submitted for publication. d. J. K. Cha, W. J. Christ, J. M. Finan, H. Fujioka, Y. Kishi, L. L. Klein, S. S. Ko, J. Leder, W. W. McWhorter, Jr., K.-P. Pfaff, D. Uemura, Y. Hirata, manuscript submitted for publication.
4. S. S. Ko, L. L. Klein, K.-P. Pfaff, Y. Kishi, manuscript submitted for publication.
5. For example, see H. O. House, D. S. Crumrine, A. Y. Teranishi, H. D. Olmstead, J. Am. Chem. Soc. 95, 3310 (1973).

6. N. Minami, S. S. Ko, Y. Kishi, J. Am. Chem. Soc. 104, 1109 (1982) and J. M. Finan, Y. Kishi, Tetrahedron Lett. 2719 (1982). Similar methods were also reported by Corey (E. J. Corey, P. B. Hopkins, A. M. Munroe, A. Marfat, S. Hashimoto, J. Am. Chem. Soc. 102, 7986 (1980)), Roush (W. R. Roush, R. J. Brown, J. Org. Chem. 47, 1371 (1982)), and Masamune and Sharpless (T. Katsuki, A. W. M. Lee, P. Ma, V. S. Martin, S. Masamune, K. B. Sharpless, D. Tuddenham, F. J. Walker, J. Org. Chem. 47, 1373 (1982), P. Ma, V. S. Martin, S. Masamune, K. B. Sharpless, S. M. Viti, J. Org. Chem. 47, 1378 (1982), and A. W. M. Lee, V. S. Martin, S. Masamune, K. B. Sharpless, F. Walker, J. Am. Chem. Soc. 104, 3515 (1982)).
7. T. Katsuki, K. B. Sharpless, J. Am. Chem. Soc. 102, 5974 (1980). The latest publication on this subject from Sharpless is J. Am. Chem. Soc. 103, 6237 (1981).
8. For examples, see H. Nagaoka, Y. Kishi, Tetrahedron 37, 3873 (1981), and the references cited in footnote 6.
9. This was performed by the following steps: 1. Swern oxidation of five-membered carbonate alcohols (cf. compound 18 in our J. Am. Chem. Soc. publication cited in reference 6), 2. $NaBH_4$ or $Zn(BH_4)_2$ reduction, 3. TLC separation, 4. aqueous base hydrolysis. For an alternative solution, see the last Masamune-Sharpless publication cited in reference 6.
10. Degradation product 2a was prepared by hydrogenolysis (Pd-C/AcOH-MeOH) of $2b^{1d,2a}$. We are indebted to Professors Hirata and Uemura for a sample of degradation product 2b.
11. This substance was a diastereomeric mixture at C.85. In this paper a small circle such as the one in structure 12 indicates that a degradation product (or its related compound) is a diastereomeric mixture due to the asymmetric center introduced during degradation reactions.
12. This substance was prepared from 12. We are indebted to Professors Hirata and Uemura for a sample of degradation product 12.
13. S. J. Sondheimer, H. Yamaguchi, C. Schuerch, Carbohydr. Res. 74, 327 (1979).
14. This substance was prepared from (S)-(+)-3-hydroxy-2-methyl-propionic acid (N. Cohen, W. F. Michel, R. J. Lopresti, C. Neukom, G. Saucy, J. Org. Chem. 41, 3505 (1976)). We are indebted to Dr. Cohen for a generous gift of the acid.
15. D-Glyceraldehyde acetonide was prepared according to the method reported by H. O. L. Fischer and E. Baer (Helv. Chim. Acta 17, 622 (1934)). Following the procedure reported for the D-series (H. Zinner, J. Milbradt, Carbohydr. Res. 2, 470 (1966)), L-glyceraldehyde cyclohexanone ketal was prepared from L-arabinose.
16. J. A. Dale, D. L. Dull, H. S. Mosher, J. Org. Chem. 34, 2543 (1969), G. R. Sullivan, J. A. Dale, H. S. Mosher, J. Org. Chem. 38, 2143 (1973).
17. This substance was prepared from the degradation product reported as compound 12 in reference 1b in 2 steps. We are indebted to Professors Hirata and Uemura for a sample of degradation products 32 and 33.
18. This substance was prepared from the degradation product reported as compound 3 in reference 1a in 3 steps. We are indebted to Professors Hirata and Uemura for a sample of degradation products 48 and 49.
19. For degradation products 60, 61, 63 and 64, see references 1a, 1c and 2a. Hydrochloric acid treatment (3.5% HCl/RT/5 min) of 60, followed by acetylation, yielded an approximately 1:1 mixture of two major products, 61 and 62, along with small amounts of their stereoisomers with respect to the spiroketal center. We are indebted to Professors Hirata and Uemura for a sample of degradation products 60, 61, 62, 63 and 64.
20. J. Leder, H. Fujioka, Y. Kishi, manuscript in preparation.
21. For synthetic work related to this segment, see W. C. Still, I. Galynker, J. Am. Chem. Soc. 104, 1774 (1982).
22. The approximately 5% NOE observed between the C.44 and C.45 protons and also between the C.45 and C.46 protons of 62 suggested that these three protons were cis-oriented on the 5-membered ring: private communication from Professors Hirata and Uemura. The spin-spin coupling constants $J_{43,44} = 2.0$ Hz, $J_{44,45} = 3.6$ Hz and $J_{45,46} = 4.0$ Hz observed for 61 are consistent with this assignment.
23. Aqueous base hydrolysis [10% aq. NaOH-MeOH (1-1)/RT] of 61 gave the corresponding polyalcohol, which was subjected to $NaIO_4$ oxidation (10 wt.% $NaIO_4/H_2O/0°C/2$ min.), followed by $NaBH_4$ workup, acetylation and TLC separation, to yield 78. We are indebted to Professors Hirata and Uemura for a sample of degradation product 78.
24. ^{1}H-NMR signals due to the C.18 through C.21 portion of 78 were found to correspond exceptionally well to those of threo nonan-1,2,3-triol triacetate but not to those of erythro nonan-1,2,3-triol triacetate.
25. For acetates 80 and 82, see references 1a and 2a, respectively. Acetate 81 was isolated as a minor product of periodate oxidation of 60. We are indebted to Professors Hirata and Uemura for a sample of degradation products 80 and 82, and for a ^{1}H-NMR spectrum of 81.
26. M. D. Lewis, J. K. Cha, Y. Kishi, J. Am. Chem. Soc. 104, (1982), in press.
27. This assignment was made based on the spin-spin coupling constants $J_{2,3} = 11.6$ Hz and $J_{4,5} = 12.0$ and 3.5, given in the supplementary material for reference 2a.
28. This substance was prepared in 3 steps from the degradation product reported as compound 1 in reference 1f. We are indebted to Professors Hirata and Uemura for a sample of this degradation product.

29. Some degradation products which we have dealt with in the preceding paragraphs were prepared via intermediates with asymmetric centers adjacent to carbonyl groups, e.g. preparation of 78 from 60. The possibility that epimerization of these asymmetric centers occurred during preparation and/or isolation of these degradation products was excluded by careful control experiments for each case.
30. The olefinic region of the ^{1}H-NMR spectrum of palytoxin was virtually first order. The following spin-spin coupling constants were observed: $J_{a,b}$ = 14.4 Hz, $J_{51,52}$ = 15.6, $J_{74,75}$ = 10.9, $J_{76,77}$ = 14.8, $J_{83,84}$ = 11.7 and $J_{98,99}$ = 10.9: private communication from Professors Hirata and Uemura.
31. The stereochemistry of the trisubstituted olefin at C.6 to C.7 had been determined by an NOE experiment on the aldehyde, which was obtained by periodate cleavage between C.8 and C.9 of palytoxin[1f]: private communication from Professors Hirata and Uemura. The possibility that double bond isomerization occurred during preparation and/or isolation of this aldehyde was excluded by the fact that no deuterium was incorporated when the oxidation was carried out in deuterated solvents.
32. For ^{13}C-NMR spectra of palytoxin, see references 2d and 1f. For ^{1}H-NMR spectra of degradation products 2b, 12, 27, 31, 63 and the acetylated dihydro derivative of 60, see references 2a and 2b.
33. Private communication from Profesors Hirata and Uemura.
34. P. Kozz, S. Deyhim, K. Heynes, Chem. Ber. 111, 2909 (1978).
35. The chemical shift of the ketal carbon of 2,5-anhydrotagatose in D_2O was found to be 109.6 ppm, whereas that of the 6-membered hemiketal carbon was 99.9 ppm and that of the 5-membered hemiketal carbon was 103.6 ppm. Similar results were observed for fructose (6-membered hemiketal carbon: 98.9 ppm and 5-membered hemiketal carbon: 102.3 ppm) and 2,5-anhydrofructose (108.6 ppm), and for sorbose (6-membered hemiketal carbon: 98.8 ppm and 5-membered hemiketal carbon: 102.8 ppm) and 2,5-anhydrosorbose (111.2 ppm). For ^{13}C-NMR spectra of ketoses, see reference 36.
36. L. Que, Jr., G. R. Gray, Biochemistry 13, 146 (1974).
37. R. E. Moore, G. Bartolini, J. Barchi, A. A. Bothner-By, J. Dodok, J. Ford, J. Am. Chem. Soc. 104, 3776 (1982).

DIASTEREO- AND ENANTIO-SELECTIVE CYCLOADDITION AND ENE REACTIONS IN ORGANIC SYNTHESIS

Wolfgang Oppolzer

Département de Chimie Organique, Université de Genève, CH - 1211 Genève 4, Switzerland

Abstract

1. Highly enantioselective C,C-bond formation has been achieved in a predictable manner

1.1 by Lewis-acid-mediated processes such as intramolecular ene-type reactions (synthesis of (+)-α-allokainic acid), bimolecular Diels-Alder- and 1,4-additions using enoates derived from various chiral alcohols followed by regeneration of the auxiliary group;

1.2 in the "chiral pool"-synthesis of (-)-α-kainic acid from (S)-(-)-glutamic acid *via* a thermal intramolecular ene reaction, proving the absolute configuration of the neurophysiologically interesting natural products (-)-α-kainic acid, (-)-α-allokainic acid and domoic acid.

2. Intramolecular "magnesium-ene" reactions of 2-alkenyl-magnesium chlorides to olefinic bonds have been systematically studied and applied, as illustrated by

2.1 the development of a reliable and efficient preparation of 2-alkenyl Grignard reagents using slurries of evaporated magnesium;

2.2 a synthesis of (±)Δ9,12-capnellene using an iterative cyclopentane-annulation based on type-I-ene reactions;

2.3 a synthesis of the unusual sesquiterpene (±)sinularene *via* a type-I-reaction;

2.4 regio- and stereochemical studies of unprecedented type-II-reactions;

2.5 a direct and highly selective synthesis of the olfactively interesting norsesquiterpene (±)-khusimone exploiting a type-II-reaction.

INTRODUCTION

During the last decade synthesis has advanced considerably toward the objective of constructing complex organic molecules in a minimum number of steps with high overall yields. There is no doubt that this progress is due largely to the development and refinement of key reactions which provide most of the required structural complexity efficiently with predictable as well as high regio- and stereo-selectivity. To this end we started many years ago a general program focused on studies and applications of intramolecular cycloaddition-[1] and ene-reactions[2]. Today's presentation shall cover two major topics: 1) asymmetrically induced C,C-bond closures using either a removable chiral auxiliary group or a readily available chiral building block and 2) regio- and stereo-selective intramolecular "magnesium-ene" reactions. Crucial applications of these methods to the syntheses of natural products will exemplify their potential. All this work had been initiated by our interest in intramolecular ene-reactions, classified according to the mode by which the enophilic chain is attached, either at the olefinic terminal (type I), at the central atom (type II), or at the allylic terminal (type III) of the ene unit (Scheme 1).[2]

Scheme 1 POSSIBLE MODES OF <u>INTRAMOLECULAR</u> ENE REACTIONS.

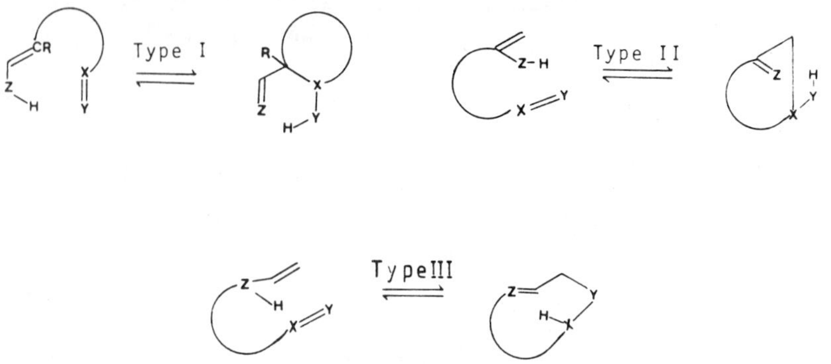

1. **HIGHLY ENANTIOSELECTIVE C,C-BOND FORMATION**

   1.1 ASYMMETRIC INDUCTION IN LEWIS-ACID-MEDIATED PROCESSES

   1.1.1 Synthesis of (+)-α-Allokainic Acid

Recently we have reported a 100% diastereo- and 90% enantio-selective $AlMe_2Cl$-mediated ene-type I-reaction $\underline{1} \rightarrow \underline{2}$ (Scheme 2). In the subsequent conversion

Scheme 2

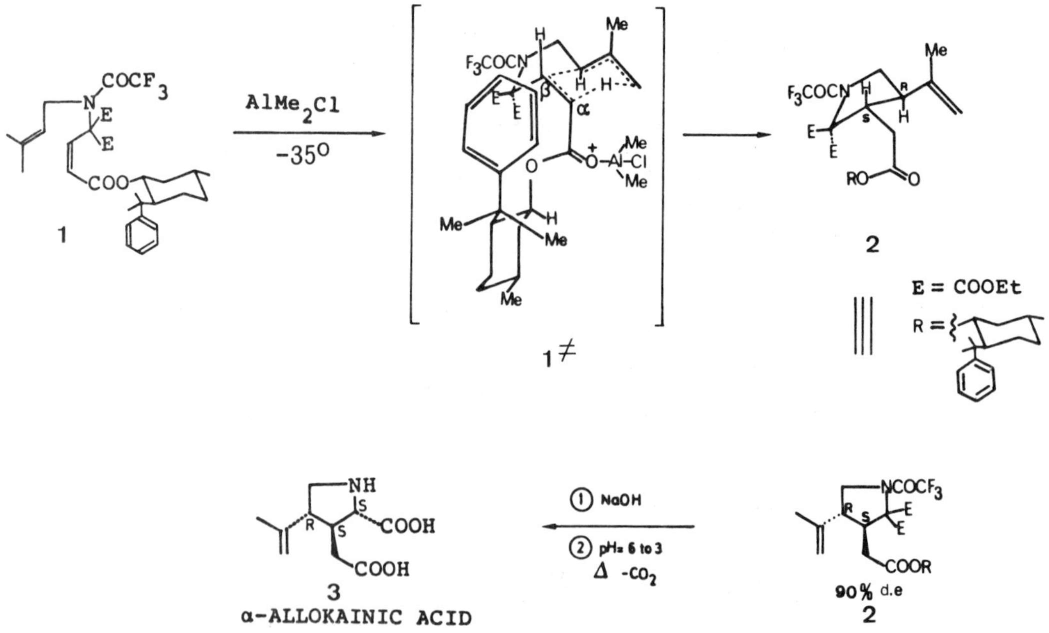

of $\underline{2}$ to the pure natural (+)-α-allokainic acid $\underline{3}$ the auxiliary (−)-8-phenyl-menthol was easily recovered. The remarkable asymmetric induction during the ene process $\underline{1} \rightarrow \underline{2}$ may be rationalised by a transition state geometry $\underline{1}^{\neq}$ where the ester-carbonyl group is antiplanar with the $C_\alpha=C_\beta$- and synplanar with the alkoxy-C,H-bonds. Accordingly, the phenyl ring shields the C_β-*si* face of the enophile.[3]

1.1.2 Asymmetric Diels-Alder Reactions

Our choice to use (-)-8-phenylmenthol as a chiral auxiliary has been prompted by previous asymmetric Diels-Alder addition of acrylates derived therefrom. We attribute these very encouraging results of *E.J. Corey*[4] to an analogous phenyl-shielding of the acrylate C_α-*re* face as depicted in 4 (Scheme 3).

Scheme 3

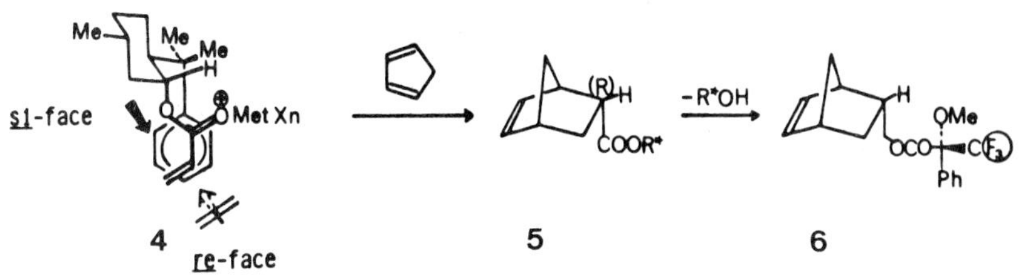

Lewis Acid	Solvent	Temp.(0°)	Yield %	%e.e. of (R)<u>endo</u>-Adduct
1.5 eq. SnCl$_4$	toluene	0	80	89
1.5 eq. Me$_2$AlCl	CH$_2$Cl$_2$	0	85	64
0.7 eq. AlCl$_3$	CH$_2$Cl$_2$	-20	89	65
1.5 eq. TiCl$_4$	CH$_2$Cl$_2$	-20	74	90

Reinvestigation of the addition 4 → 5 using ^{19}F-NMR analysis of the *Mosher* derivatives 6 indicated an asymmetric induction of 64 to 90% whose extent depended mostly on the Lewis acid used (Scheme 3)[5]. Moreover, the need to purify the oily (-)-8-phenylmenthol by careful chromatography and the difficult access to its more interesting (+)-enantiomer stimulated a search for more versatile and effective chiral auxiliary alcohols.

As an ultimate goal we aimed at the preparation and use of such alcohols which 1) lead to quantitative asymmetric induction in addition reactions to their acrylates, 2) are easily accessible from inexpensive precursors in both antipodal forms or as an equivalent pair of *si*- and *re*-face directing isomers, 3) may be purified by crystallisation and 4) are recovered easily after introduction of chirality has been achieved. Apparent options include cyclohexanols which are locked in a rigid chair conformation with both an equatorial hydroxyl group and an α-*trans* positioned aryl-substituted chain, as well as norbornanols which carry these crucial functionalities either in a *cis/endo* or *cis/exo* relationship. Accordingly, the *si*-face directing acrylates A, B and C and their *re*-face directing counterparts D, E and F may be schematically envisaged (Scheme 4). Our first step toward this target was

Scheme 4

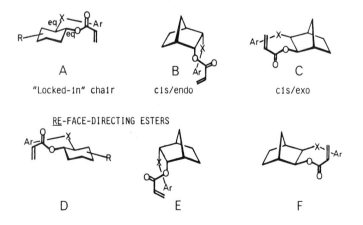

a study of TiCl$_4$-mediated additions of cyclopentadiene to acrylates derived from conformationally rigid cyclohexanols (Scheme 5)[6]. Several examples show

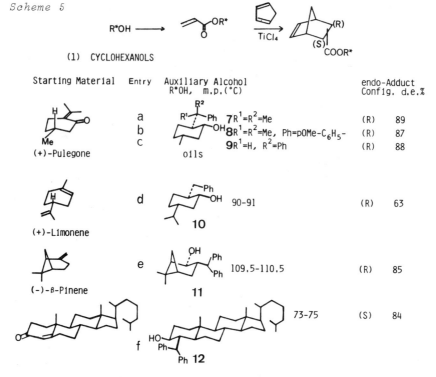

Scheme 5

equal.but not higher induction (≈ 90% d.e.) than the reference entry a. Noteworthy is entry f since the alcohol 12, accessible in two steps from cholestenone, constitutes the first readily available "re-face directing" auxiliary. For the preparation of the envisaged norbornanols camphor is a particularly suitable chiral source, because both enantiomers are easily available, and its derivatives are frequently crystalline. Indeed, entries h and j (Scheme 6)

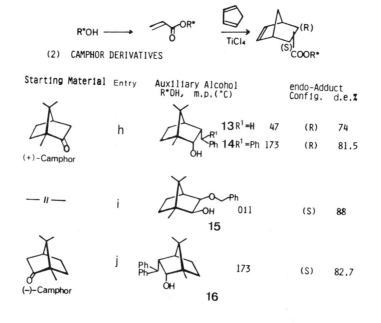

Scheme 6

exhibit the use of antipodal cis-diphenylmethyl borneols and entry i the
selectively prepared re-face directing cis-benzyloxyisoborneol 15 giving the
(S)-cycloadduct with 88% d.e. Pursuing the latter lead (Scheme 7), diphenyl-

Scheme 7

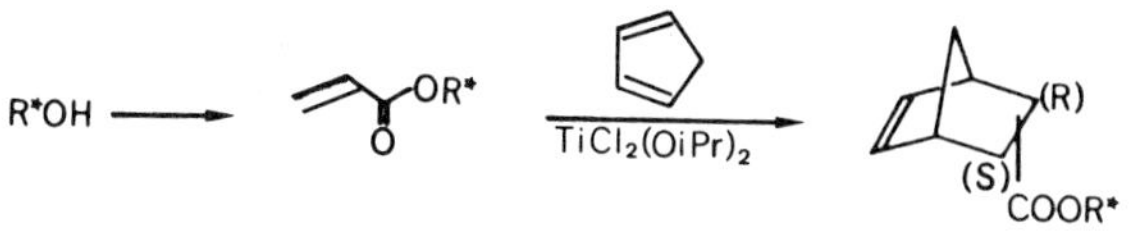

Entry	Auxiliary Alcohol R*OH	R¹	R²	m.p. °C	Temp. °C	yield %	endo/exo	Endo-Adduct Config.	d.e.%
a	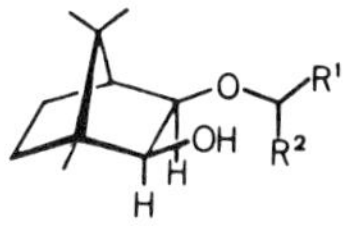 17	Ph	H	oil	0	91	86/14	S	46
b		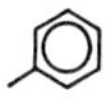 Ph	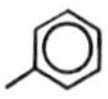 Ph	74	0	94	86/14	S	64
c		//	//	//	-20	94	90/10	S	72
d		 β-naphthyl	H	73-74	-20	98	90/10	S	69
e		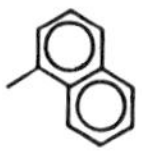 α-naphthyl	//	74-75	0	97	85/15	S	54

methoxy- (entry b), α-naphthylmethoxy- (entry e) and β-naphthylmethoxy-(entry
d) isoborneols 17 were prepared from cis-3-hydroxyisoborneol. Although these
alcohols are nicely crystalline, their acrylates suffered from rapid ether-
cleavage on exposure to TiCl₄. Furthermore the use of TiCl₄ generally
required painstaking care to avoid the presence of traces of water which may
decrease the chemical yield of the acrylate cycloaddition due to polymerisation. On the other hand, high chemical yields of cycloadducts were safely
obtained at -20°C using TiCl₂(OiPr)₂, a mild Lewis acid previously developed
by Mukaiyama⁷. However, under these reaction conditions the optical yield of
the benzyl ether addition (entry a) dropped to 46% d.e. Nevertheless, according to our expectations an increase of the aromatic surface improved the
shielding and therefore the induction up to 72% d.e. in favor of the (S)-
adduct. Again, using TiCl₂(OiPr)₂ the acrylates of the cis-3-hydroxyisobornyl-
arylmethyl ethers 18 (Scheme 8) gave higher optical yields (≤92% d.e.)

Scheme 8

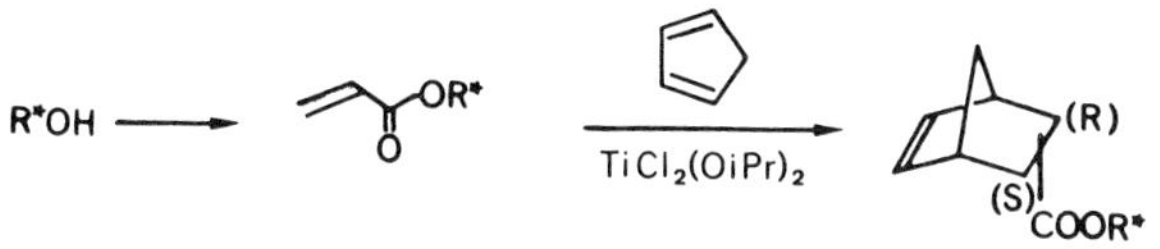

Entry	Auxiliary Alcohol R*OH	R¹	R²	m.p. °C	Temp. °C	yield %	endo/exo	Endo-Adduct Config.	d.e.%
a	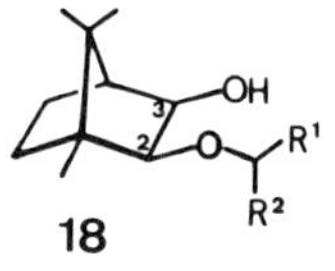 18	 Ph	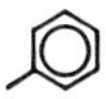 Ph	57	-20	74	95/5	R	91
b		//	//	//	-40	36	97/3	R	92
c		 β-naphthyl	H	70-71	-20	98	95/5	R	92
d		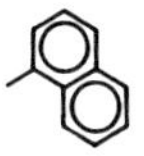 α-naphthyl	H	69-70	0	97	93/7	R	88

in favor of the (R)-adduct).[8] We then wondered to what extent the observed shielding was due to π-π-aryl/acrylate-overlap, or to mere steric crowding. In fact, we were pleased to find that the neopentyl ethers 20 and 22 (Scheme 9) displayed a chirality directing ability dramatically superior to the

Scheme 9

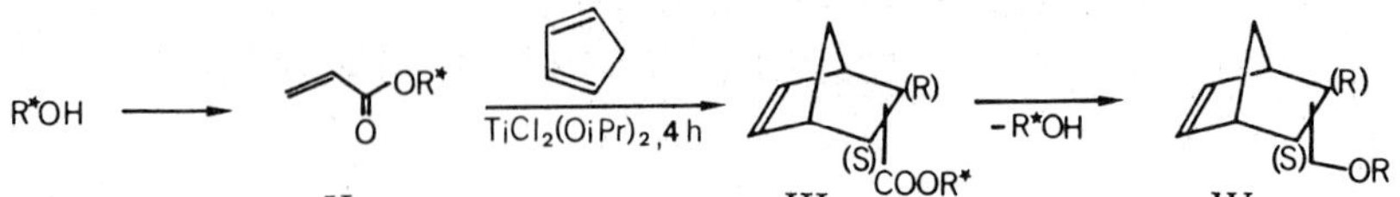

Starting Material	Auxiliary Alcohol R*OH I	m.p. °C	Reaction Temp.	yield %	endo/exo	Endo-Adduct Config.	d.e.%
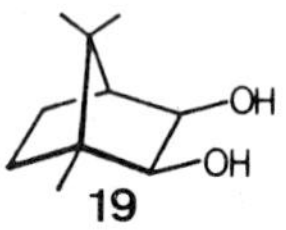 19	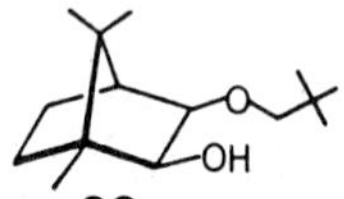 20	oil	-20°	95	96/4	S	97
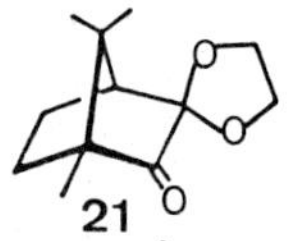 21	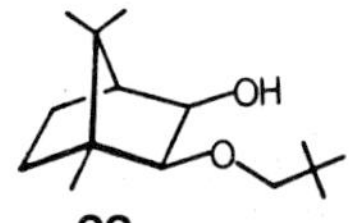 22	4-5	-20°	94	96/4	R	99.4[a] 99.2[b]
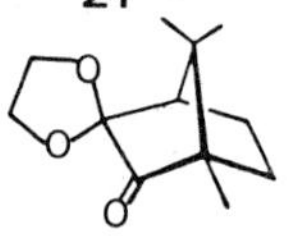 23	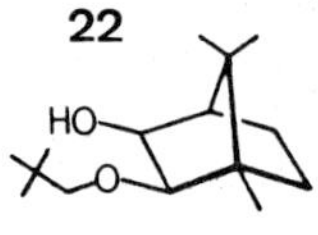 24	4-5	-20°	98	95/5	S	99.4[a] 99.1[b]

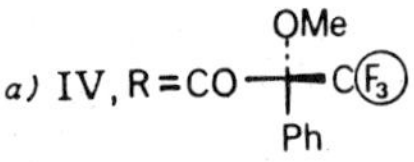

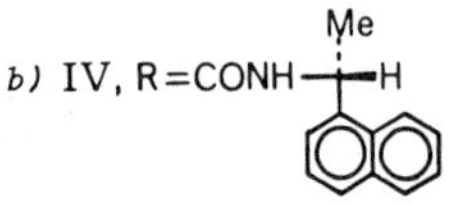

a) IV, R=CO–C(OMe)(CF$_3$)(Ph)   b) IV, R=CONH–CH(Me)(naphthyl) (HPLC)

examples listed previously. Notably 22, selectively prepared from 21[8] (65% total yield) gave after esterification and acrylate cycloaddition the (R)-adduct III in 99.2 to 99.4% d.e. as determined by ^{19}F-NMR- and HPLC-analyses of IVa and IVb, respectively. Crystallisation of 22 (pentane -30°C) permits its facile purification and purity control despite the low melting point. Its antipode 24, readily accessible from (-)-camphor furnished the (S)-adduct IV in equally high chemical and optical yield. It thus follows that the alcohols 22 and 24 constitute the first reported chiral auxiliaries which on acrylate/cyclopentadiene-cycloaddition furnish efficiently and predictably either (R)- or (S)-adducts in virtually quantitative enantiomeric excess.[8] Examination of models assuming a staggered conformation of the neopentyloxy-chain indicates indeed strong steric blocking of the 22-acrylate C_α-re-face by the t-butyl group. The superiority of 22 over 20 may be attributed to a buttressing by

Scheme 10

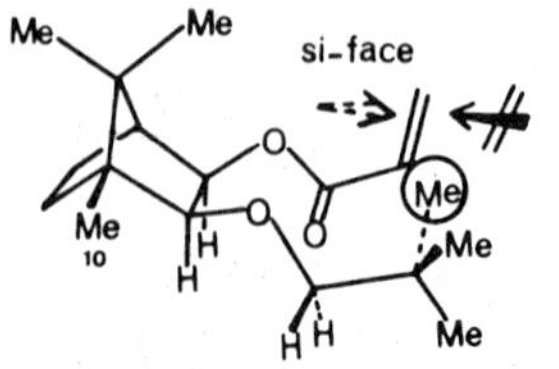

the (C,10)-methyl group which pushes the ether chain closer to the acrylate. (The latter argument explains also the higher induction of the acrylates derived from 18 (Scheme 8) as compared to those derived from 17 (Scheme 7)). Presently we are applying our findings to the synthesis of natural products and continue to develop even more practical chiral alcohols, generally applicable to a range of enantioselective reactions.

An indication that the scope of this concept extends ene-type and Diels-Alder additions will be illustrated briefly as follows.

1.1.3 Asymmetric 1,4-Additions

Recently smooth BF_3-mediated 1,4-additions of organocopper reagents to α,β-unsaturated carbonyl compounds have been reported by *Yamamoto*.[9] (Scheme 11)

Scheme 11

Scheme 12

	R	yield %	e.e.%
a	C_6H_5	76	> 99 [1)] 99.3 [2)]
b	nBu	75.4	> 99 [1)]

	R	yield %	e.e.%
c	C_6H_5	36(72)	24
d	nBu	76	70 [1)]

1) HPLC and ^{1}H-NMR of A

2) $[\alpha]_D$ of acid

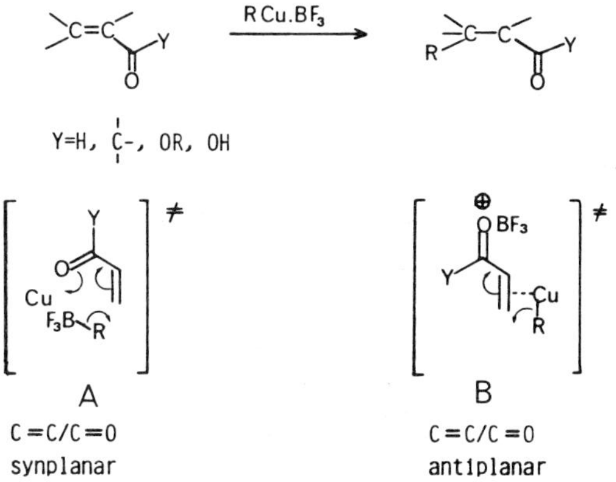

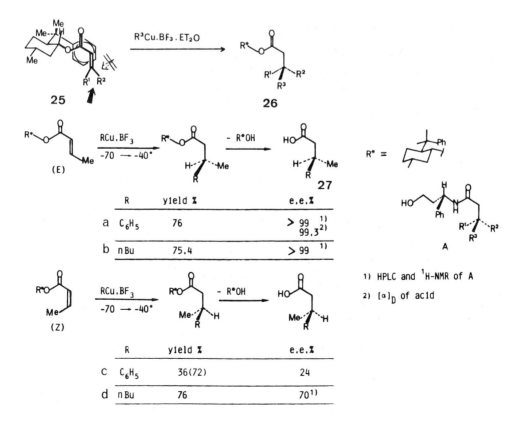

In contrast to the proposed cyclic transition state A we anticipated again an antiplanar C=C/C=O-disposition (B). Accordingly, owing to phenyl-shielding, 8-phenylmenthyl enoates 25 would react with a reagent $R^3Cu \cdot BF_3$ at the olefinic face opposite to the phenyl ring (Scheme 12). The experimental results in support of this hypothesis are summarised in Scheme 12.[10] Indeed, starting from *trans*-crotonate 25 β-substituted carboxylic acids 27 may be obtained in virtually quantitative optical yield by successive addition and saponification as indicated by entries a and b. Although the sign of asymmetric induction in the analogous treatment of *cis*-crotonate agrees with our stereochemical model it is definitely lower (entries c,d). Nevertheless, we have evidence that by reversing the order of group introduction either enantiomer of a β-substituted carboxylic acid 27 may be accessible in high chemical and optical yield *via* the corresponding *trans*-esters 25.[11] Further options include the use of new auxiliary alcohols mentioned above. Experiments are under way to explore the scope of this asymmetric 1,4-addition as well as its applicability to the synthesis of natural products.

1.2 SYNTHESIS OF (−)-α-KAINIC ACID

The above mentioned seaweed constituent (+)-α-allokainic acid (3) co-occurs with the structurally related but much more powerful neuroexcitants (−)-α-kainic acid (28) and (−)-domoic acid (29)[12b] (Scheme 13).

Scheme 13

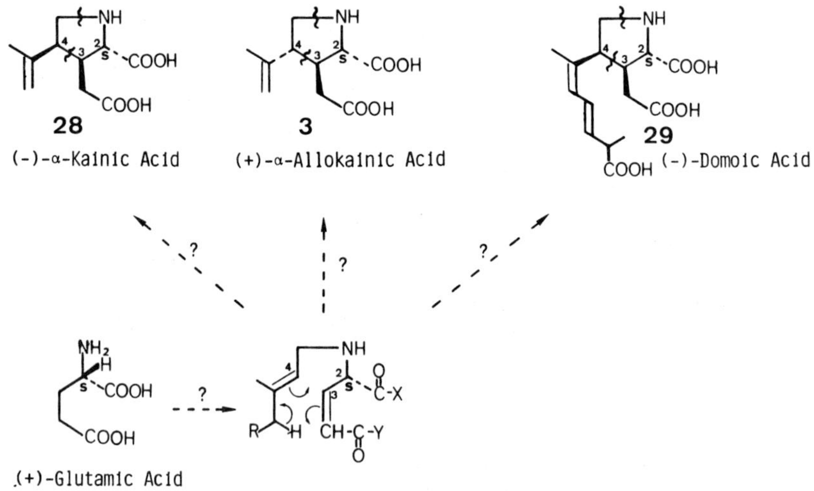

28 (−)-α-Kainic Acid   3 (+)-α-Allokainic Acid   29 (−)-Domoic Acid

(+)-Glutamic Acid

Correlation of 28 with 3 and 29 indicated that they possess the same absolute configuration of C(2), which has not been properly assigned[12a] before accomplishment of this work[12b]. Advancing the hypothesis that all those acids are biosynthesised from (S)-glutamic acid and an isoprenoid unit *via* an ene-type cyclisation, we aimed at an analogous enantioselective synthesis of (−)-α-kainic acid (28) which establishes unambiguously its absolute configuration. To this end we expected the chiral center C(2) arising from glutamic acid to control sterically the formation of the C(3)/C(4)-bond in the strategic type-I-ene reaction. To avoid loss of the stereochemical integrity by a thermal olefin migration as previously observed in the process 30 → 31[13] (Scheme 14) a

Scheme 14

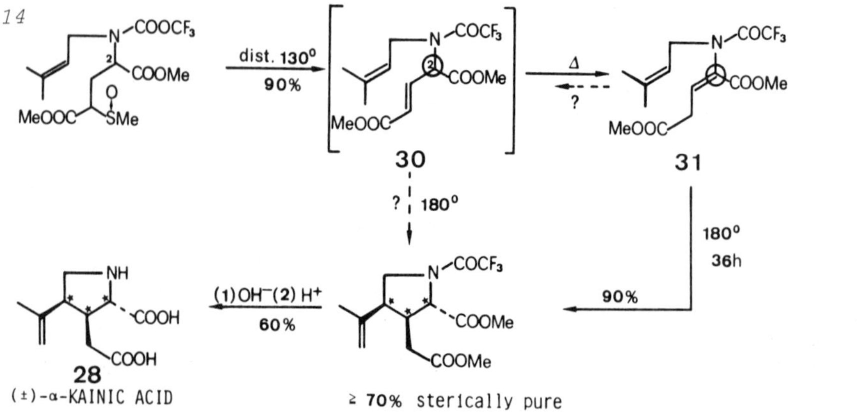

28 (±)-α-KAINIC ACID   ≥ 70% sterically pure

suitable "protection" of the α-carboxyl group by a reduction-oxidation sequence was envisaged (Scheme 15). As a control experiment N-benzoyl prolinol was

Scheme 15

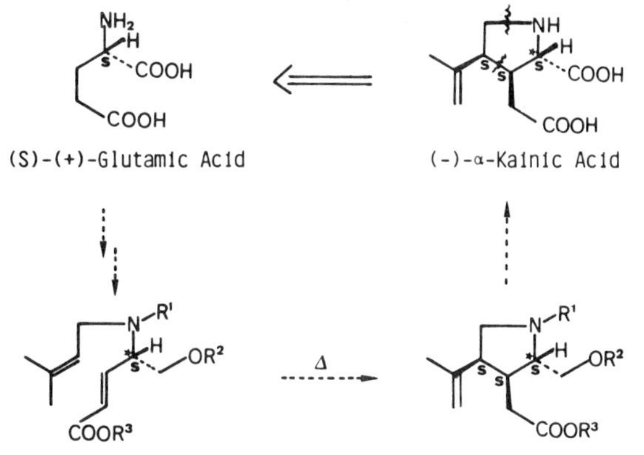

oxidised with Jones' reagent to give N-benzoyl-proline with virtually quantitative retention of configuration. Starting from commercially available (+)-5-ethyl glutamate successive N-protection, selective reduction of the free carboxyl group with diborane at -15°C, silylation of the resulting alcohol and N-alkylation (HMPA) gave the mono-olefin 33 (Scheme 16). Introduction of

Scheme 16

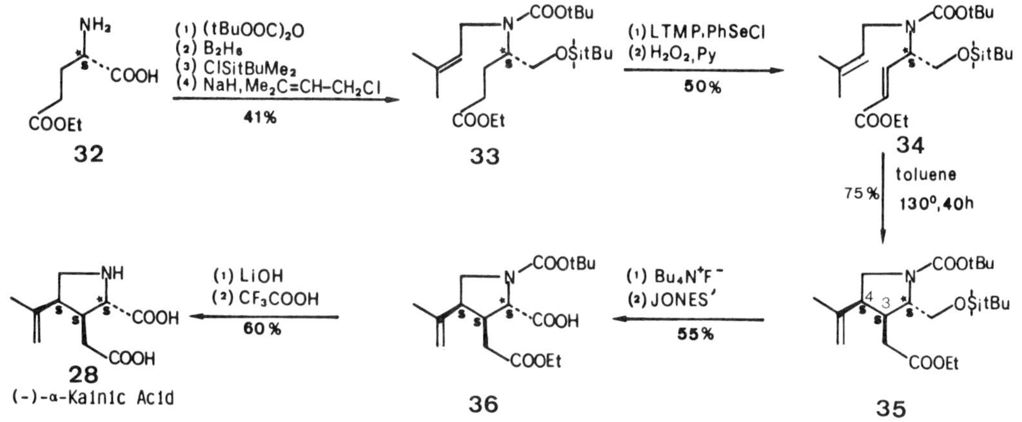

the conjugated double bond was achieved by deprotonation of 33 with 2 equivalents of lithium 2,2,6,6-tetramethyl piperidide, successive selenation, oxidation and selenoxide elimination producing the α,β-unsaturated ester 34. The stage was now set for the crucial closure of the five-membered ring. Heating a 5% solution of the 1,6-diene 34 in toluene at 130° for 40 h using a sealed Pyrex tube gave the desired pyrrolidine 35 in 75% yield. As we had anticipated the configurations of the newly formed centers C(3) and C(4) in 35 were nicely controlled in the ene process. Proof of the depicted stereochemistry was obtained by the conversion of the ene product 35 to (-)-α-kainic acid as follows. Cleavage of the silylether and subsequent oxidation of the resulting primary alcohol with Jones' reagent furnished the carboxylic acid 36. Saponification of 36, followed by removal of the tert.-butoxycarbonyl group with trifluoroacetic acid and subsequent treatment of an aqueous solution of the evaporated reaction mixture with ion exchange resins furnished enantiomerically pure (-)-α-kainic acid (28). ^{1}H-NMR-analysis of crude 28 and of the mother liquor after its crystallisation showed not even a trace of other stereoisomers thus confirming the high stereoselectivity of the key step 34 → 35. The synthetic (-)-α-kainic acid was shown to be identical with natural 28 by mixed m.p., IR, chiroptic and ^{1}H-NMR (360 MHz) evidence, as well as by comparison of the corresponding dimethylester in the presence of a chiral shift reagent.

In summary, this direct approach affords (−)-α-kainic acid from (S)-(+)-5-ethylglutamate in 5% overall yield[14] and establishes the absolute configuration of the natural products α-kainic acid (28), α-allokainic acid (3) and domoic acid (29). It furthermore illustrates the potential to achieve steric control in thermal intramolecular ene reactions.

2. INTRAMOLECULAR "MAGNESIUM-ENE" REACTIONS

2.1 RELIABLE PREPARATION OF 2-ALKENYLMAGNESIUM CHLORIDES

Allylic Grignard reagents II constitute a versatile class of synthetic intermediates. However their preparation by direct metalation of an allyl halide I suffers frequently from the propensity of II to couple with unused halide I forming 1,5-hexadienes III (Scheme 17).

Scheme 17

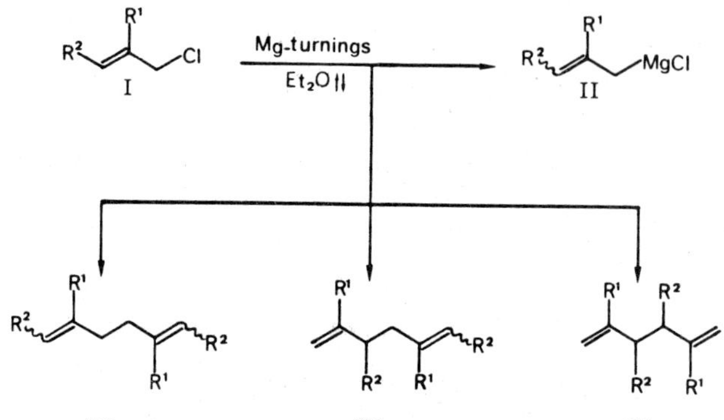

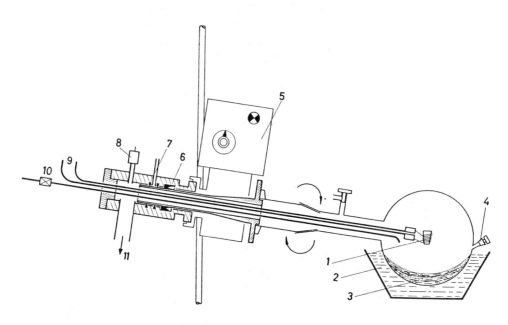

Among various procedures to minimise coupling the use of highly activated magnesium obtained by reduction of magnesium(II)halide with potassium seems to be rather encouraging.[15] Our own attempts to submit the thus obtained slurries containing II to thermal intramolecular additions to olefinic bonds (60 to 130°C) led only to extensive polymerisation. Therefore we turned our attention to the preparation of clean, alkali-metal and -halide free solutions of 2-alkenyl Grignard reagents employing extremely reactive magnesium suspensions obtained by evaporation of the metal into cooled THF using a practical and simple rotating solution reactor[16] (Scheme 18). To test this

Scheme 18

idea a variety of allylic chlorides was added slowly to a slurry of precondensed magnesium in THF at -65°C. Subsequent carbonation of the resulting alkenylmagnesium chloride solutions with CO_2 at low temperature furnished β,γ-unsaturated carboxylic acids in reliably high yields as depicted in Scheme 19.

Scheme 19

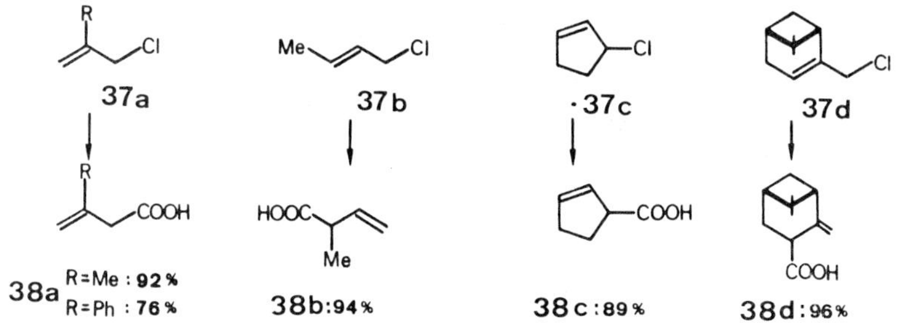

The neutral fractions contained none or only little (1.2% from 37c, 13% from 37a, R=Ph) 1,5-hexadienes III. It follows that this method permits to suppress the otherwise detrimental coupling I + II → III working at a convenient scale of 0.001 to 0.1 mole which signifies its potential for the synthesis of more complex 2-alkenylmagnesium halides.[17]

2.2 SYNTHESIS OF (±)-Δ9,12-CAPNELLENE

Despite the extensive pioneering work of Lehmkuhl[18] the formal ene reaction of allylic Grignard reagents to olefins (Scheme 20) has not yet been applied

Scheme 20

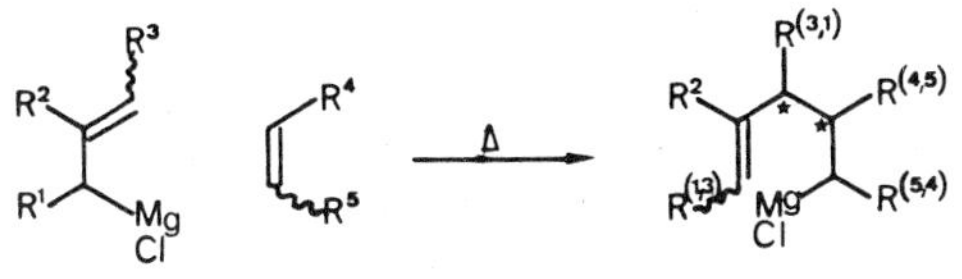

to strategically devised organic synthesis. Apart from the difficulty to prepare the 2-alkenylmagnesium halides by conventional procedures problems of efficiency and regiochemistry may limit the applicability of the bimolecular reaction as exemplified in Scheme 21.[19] More encouraging in terms of regio-

Scheme 21

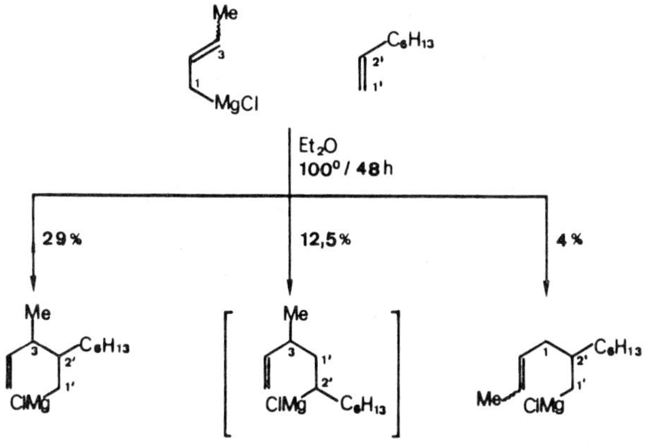

and stereoselection is the intramolecular type I-"magnesium-ene" reaction depicted in Scheme 22.[20] Nonetheless, this field remains virtually unexplored.

Scheme 22

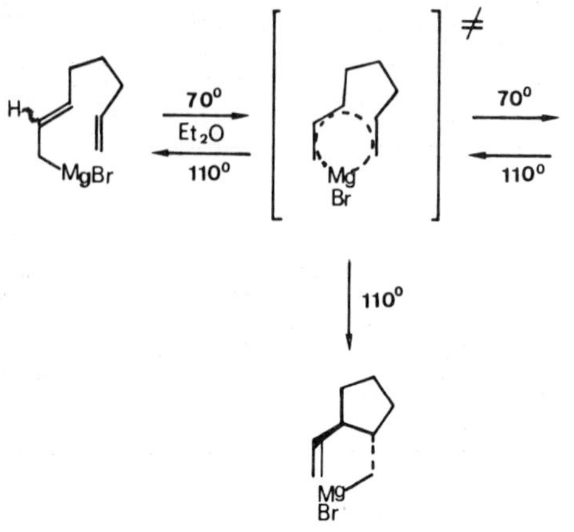

This is surprising in view of the numerous imaginable possibilities offered by subsequent functionalisation of the cyclised Grignard products. We were particularly intrigued by the idea of rendering this process iterative in order to assemble rapidly polycyclopentanoid systems (Scheme 23). Thus, it

Scheme 23

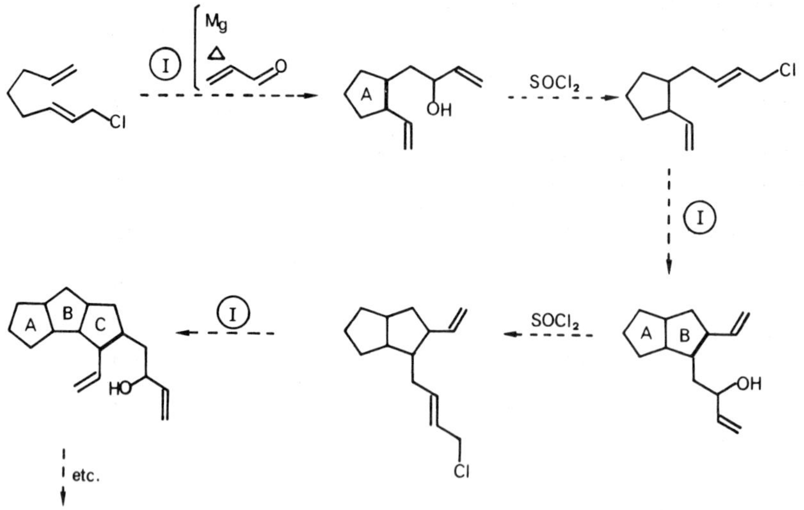

seemed conceivable to annelate three carbons by two successive synthetic operations which involve Grignard trapping with acrolein and treatment with $SOCl_2$.

Scheme 24 describes the first application of this approach to a synthesis of (±)-Δ9,12-capnellene (47) which has been recently synthesised by *R.D. Little et al.*[21] and by *L.A. Paquette et al.*[22]

Scheme 24

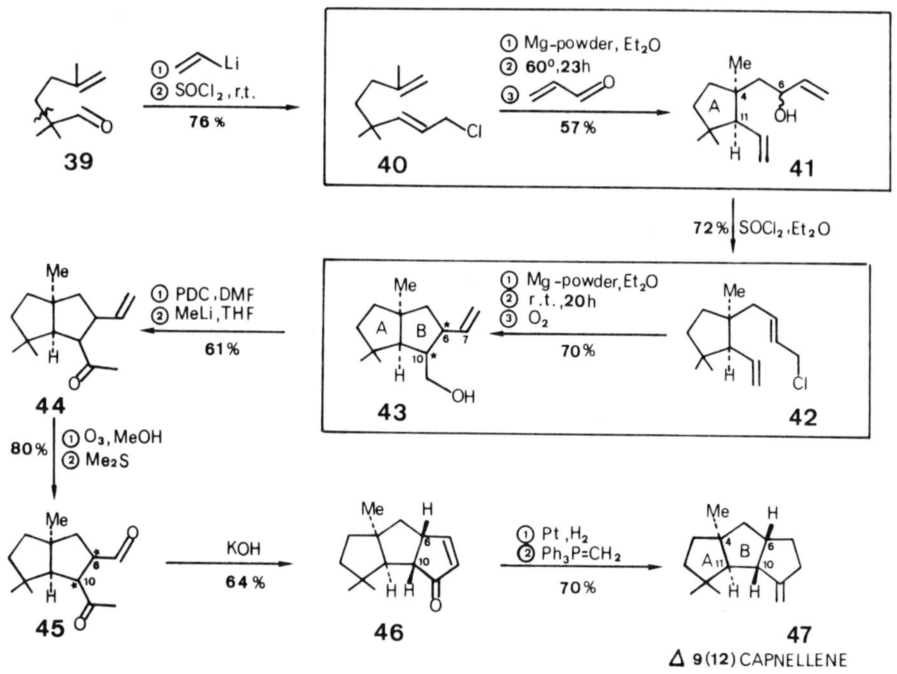

Successive treatment of the known aldehyde 39 [21] with vinyllithium and thionyl chloride gave the allylic chloride 40. Metalation of 40 using commercially available magnesium powder (Merck), heating of the thus obtained 2-alkenyl-magnesium chloride solution at 60° for 23 h and subsequent addition of acrolein furnished the cyclopentane 41 in 57% overall yield. It is worth noting that this key step 40 → 41 involves the first known successful "magnesium-ene" addition to a non-activated 1,1-disubstituted olefin which closes the congested bond C(4)/C(11) with high stereochemical control. Consequently, the allylic chloride 42 was obtained as a single stereoisomer on reaction of 41 with $SOCl_2$ (2 equiv, ether r.t., 2h). The prerequisites are now secured for the second key step. In fact, the Grignard reagent prepared from 42 (Mg-powder, ether) cyclised smoothly at r.t. (20 h) to give after trapping with O_2 a 3:2-mixture of the C(6)/C(10)-*cis/trans*-stereoisomeric alcohols 43 (70% yield). The kinetic nature of this surprisingly poor stereoselection follows from metalation of 42 (precondensed Mg) and cyclisation in ether at 0° (18 h) which furnished again a 3:2-ratio of the isomers 43 together with 75% of non-cyclised material. Having assembled rings A and B by a sequence of two "magnesium-ene" reactions we aimed at the remaining closure of ring C with concomitant thermodynamic control over the configuration at C(6) and/or at C(10). Oxidation of the primary alcohols 43 to the corresponding carboxylic acids followed by addition of methyllithium (2 equiv) afforded the methylketones 44 (61% overall yield) identified by comparison (^{1}H-NMR, IR) with an authentic, sterically pure sample[22]. The final transformation 44 → 47 relies on the protocol previously employed by *Paquette et al.*[22] More specifically, during the aldolisation 45 → 46 (5% aq. KOH, EtOH 1:1, 50°, 2.5 h) both isomers 45 were channeled into the same *cis-anti-cis*-tricyclo[6.3.0.0^{2,6}]-undecenone 46 *via* base-induced epimerisation at C(6) and/or at C(10). The enone 46 as well as (±)-Δ9,12-capnellene (47) display spectra (^{1}H-NMR, ^{13}C-NMR, IR) identical to those of authentic (±)-46[22] and natural (-)-47.[23]

2.3 SYNTHESIS OF (±)-SINULARENE

Another structurally attractive target molecule which we decided to synthesise is the unusual sesquiterpene sinularene 59 recently synthesised in a non-stereoselective manner by *D. Wege et al.*[24] Again our basic strategy was centered on an intramolecular type-I "Mg-ene" reaction. Thus, starting from

the known norbornene 48[25] we envisaged to form in the critical step the bonds
C(7),C(15) and C(6),C(5) thereby closing a six-membered ring (Scheme 25).

Scheme 25

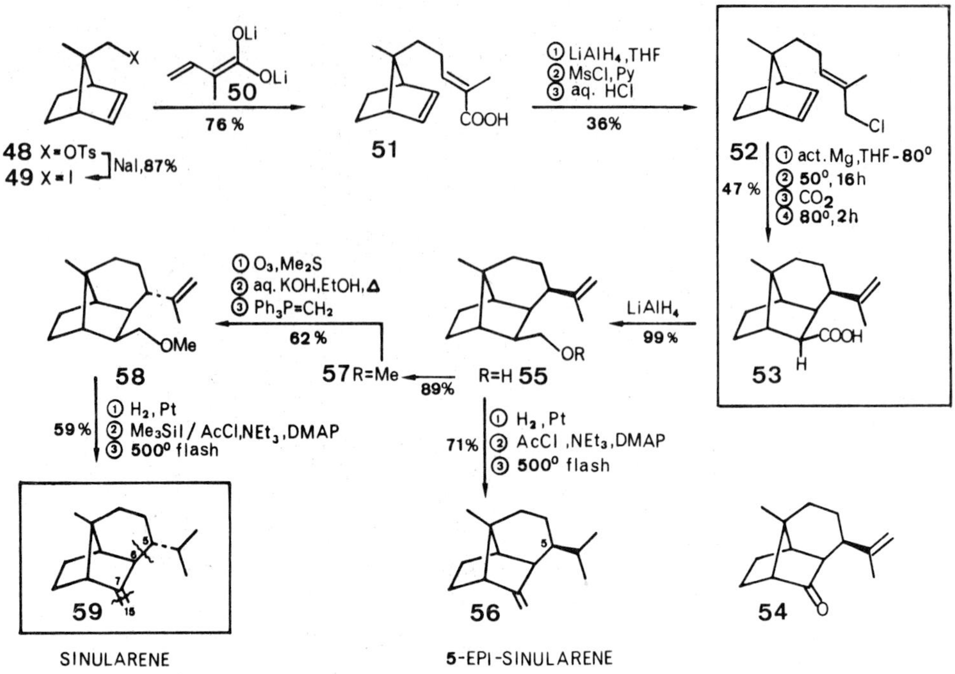

Founded on previous experience[20] and model examinations we stood a good chance
to establish the desired relative configuration of the centers C(5)/C(6) by
thermodynamic control. Selective γ-alkylation of the dienolate dianion 50
(Note a) by the iodide 49 gave the (E)-carboxylic acid 51 (76%). 51 was
converted to the key-precursor 52 by reduction with LiAlH₄ and successive
treatment of the resulting allylic alcohol with MsCl, pyridine and aq. HCl[27].
Slow addition of the allylic chloride 52 to precondensed magnesium in THF at
-80°, heating of the 2-alkenylmagnesium chloride solution at 50° for 16 h in
a closed Carius tube followed by trapping of the cyclised Grignard product
with CO_2 (80°, closed tube) furnished stereoselectively the crystalline carboxylic acid 53. The configuration of 53 follows from its smooth iodo-
lactonisation as well as from trapping of the cyclised alkenylmagnesium
chloride with O_2 followed by Jones' oxidation which furnished the ketone 54
identified by comparison (^{1}H-NMR, IR) with authentic 54[24]. Accordingly reduction of 53 with LiAlH₄, catalytic hydrogenation of 55, acetylation and ester
pyrolysis gave pure 5-epi-sinularene (56). Our initial hope to achieve thermodynamic control over C(5) by carrying out the cyclisation of the Grignard
reagent derived from 52 at higher temperature did not materialise due to
destruction. Consequently, we had recourse to a less direct thermodynamic
stereoselection. O-Methylation (i) NaH, DMF ii) MeI, HMPA) of 55 gave the
ether 57 (89%). Ozonolysis of 57, base-induced epimerisation of the crude
methylketone and Wittig reaction afforded sterically pure 58 (62% overall
yield). Having arrived at the desired configuration of C(5) there remained
only hydrogenation of 58, ether cleavage (Me_3SiI), acetylation and acetate
pyrolysis to obtain (±)-sinularene[28] showing IR, ^{1}H-NMR- and MS-spectra iden-
ticyl to those of naturally occurring (-)-59.

2.4 STUDIES OF TYPE-II "MAGNESIUM-ENE" REACTIONS

We turned next to examine the unprecedented type-II "metallo-ene" reaction
(Scheme 20, R^2, R^4= $(CH_2)_n$). The first regiochemical question (Scheme 26)
was whether on variation of the distance between ene and enophile, A would
cyclise to give B or C implying Mg-transfer either to the terminal enophilic
site C(1') or, alternatively, to the closer site C(2').

Note a: Usually bislithium salts of dienolate dianions are α-alkylated.[26]

Scheme 26

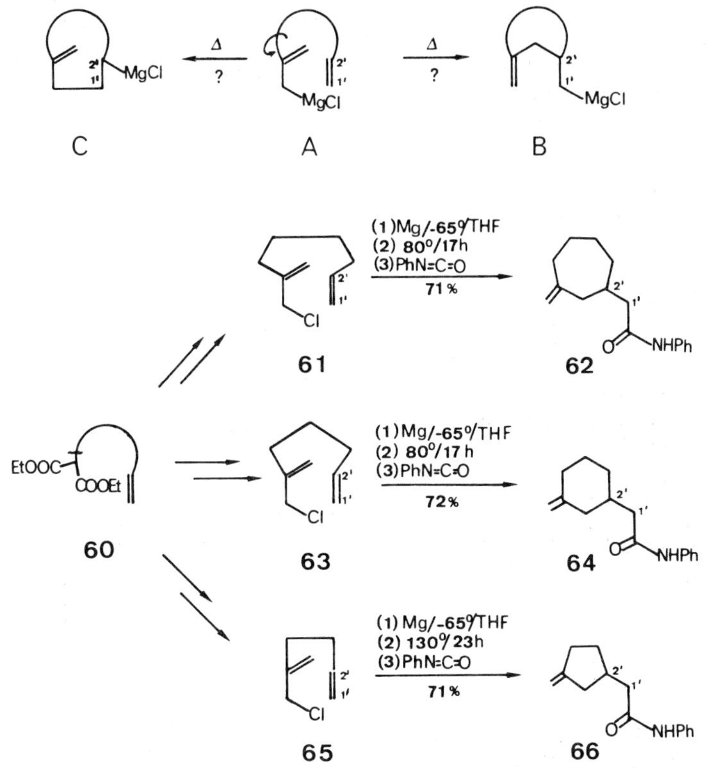

In fact, heating the 2-alkenylmagnesium chlorides, prepared from 61 and 63 using precondensed magnesium, in THF at 80° for 17 h followed by quenching of the cyclised Grignard products with phenylisocyanate gave in each case a single product 62, and 64, respectively which arise from exclusive metal-transfer to C(1') (A → B). GC- and ^{1}H-NMR-analysis showed no evidence for a product relating to the illusive isomer C. Surprisingly, also the lower homologue prepared from 65 was efficiently cyclised at a higher reaction temperature to give the cyclopentane 66; again the same regiochemistry was observed irrespective of the presumed angle strain in the transition state. Scheme 27 outlines another regiochemical uncertainty of type II-"magnesium-ene" reac-

Scheme 27

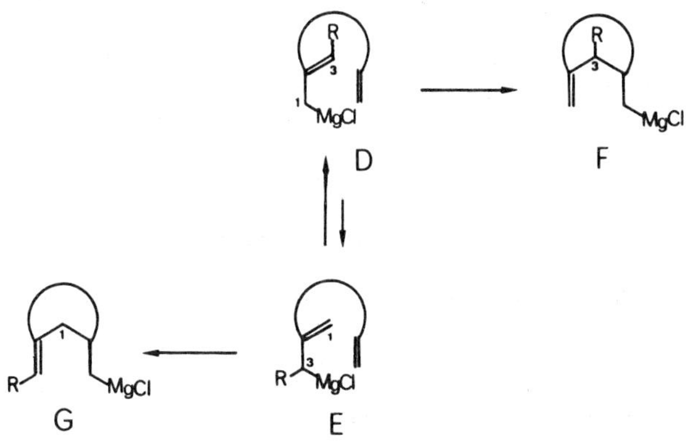

tions which remains to be clarified. On thermal cyclisation of unsymmetrically substituted allylmagnesium halides rapid 1,3-metal migration[29] $\underline{D} \rightleftarrows \underline{E}$ leaves two possibilities: either C,C-bond formation with the more or less substituted ene terminal C(3) ($\underline{D} \rightarrow \underline{F}$) or C(1) ($\underline{E} \rightarrow \underline{G}$), respectively. Accordingly, 2-alkyl-2-alkenylallylmagnesium chlorides were prepared from $\underline{67}$, $\underline{69}$ and $\underline{71}$ using precondensed Mg; subsequent heating and quenching gave in either case a single product $\underline{68}$, $\underline{70}$ and $\underline{72}$, respectively (Scheme 28). Remarkably

Scheme 28

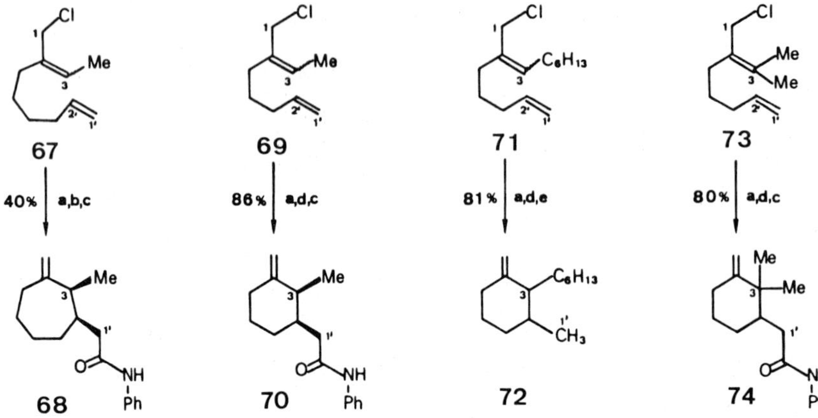

(a): Mg/-65°/THF   (b): 90°/60h   (c): PhN=C=O   (d): 80°/17h   (e): H$_2$O

this holds also for the 3,3-dimethyl-2-pentenylallylmagnesium chloride derived from $\underline{73}$ which cyclised solely by joining a quaternary with a tertiary carbon to yield $\underline{74}$. ^{1}H-NMR, ^{13}C-NMR and IR evidence unambiguously confirms the presence of an *exo*-methylene group in the products; this proves clearly that in all cases C,C-bonding occurs with the more substituted C(3) of the ene unit ($\underline{D} \rightarrow \underline{F}$). Moreover, the ^{1}H-NMR spectra of $\underline{68}$ and $\underline{70}$ exhibit the signal of a *sec.* methyl group which on irradiation decouples with H$_A$. The vicinal coupling constants J_{AB} = 2Hz ($\underline{68}$) and J_{AB} = 4 Hz ($\underline{70}$) indicate the *cis*-disposition of the CH$_3$ and the CH$_2$CONHPh groups in $\underline{68}$ and $\underline{70}$. Partial cyclisation (6 to 10%) of the Grignard reagents derived from $\underline{65}$, $\underline{67}$ and $\underline{71}$ at lower reaction temperature gave no isomeric cyclisation products (GC) consistent with a kinetic regio- and stereo-selection. From a mechanistic standpoint it is interesting to note that the striking stereoselectivity of the transformations $\underline{67} \rightarrow \underline{68}$ and $69 \rightarrow 70$ agrees with a concerted reaction involving a (Z)-ene unit (Note b) as depicted in Scheme 29.

Scheme 29

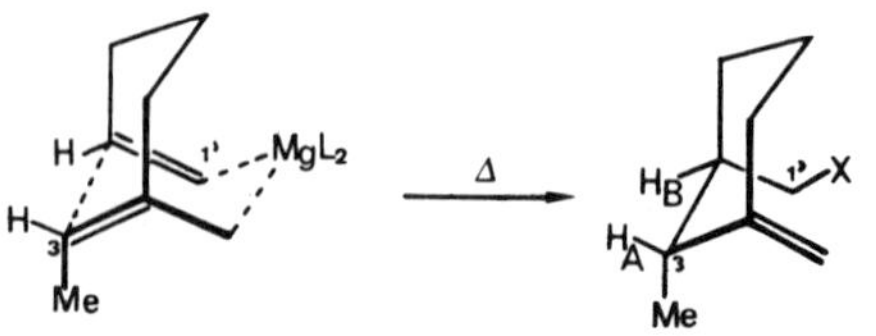

Note b: In crotylmagnesium halides the (Z)-configuration is already favored over the (E)-configuration[30]. This preference should be even more pronounced in 2,3-dialkyl-2-alkenylmagnesium halides.

We assume that the observed closure of a 5-membered ring (65 → 66) and the clear preference of a sterically more crowded C,C-bonding process (73 → 74) may be explained by a dominating coordination of the migrating magnesium with the least substituted enophile site in the transition state. Applications of this work[31] to the synthesis of natural products are presently being explored in our laboratory as illustrated below by the synthesis of (±)-khusimone.

2.5 SYNTHESIS OF (±)-KHUSIMONE

The norsesquiterpene (-)-khusimone (75, Scheme 30), a minor but olfactively interesting constituent of vetiver oil[32] is a fascinating challenge to organic synthesis. Apart from degradations of natural zizanoic acid to (-)-75[33] two imaginative but non-stereoselective total syntheses have been accomplished by *Büchi et al.*[34] and *Chan et al.*[35]. Particular difficulties thereby encountered concerned the relative configuration C(5)/C(8) as well as the positional control over the sterically encumbered *exo*-methylene group. The cornerstone of our synthetic strategy (Scheme 30) is the closure of the bond C(7),C(8)

Scheme 30

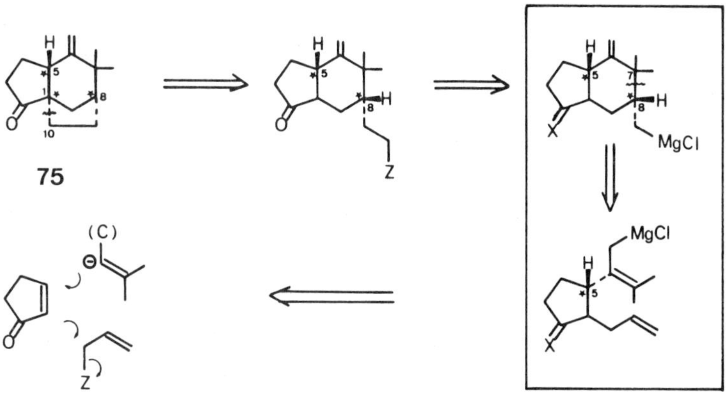

with concomitant generation of the methylene group by an intramolecular type-II "magnesium-ene" reaction.[31] Starting from cyclopentenone (76) (Scheme 31)

Scheme 31

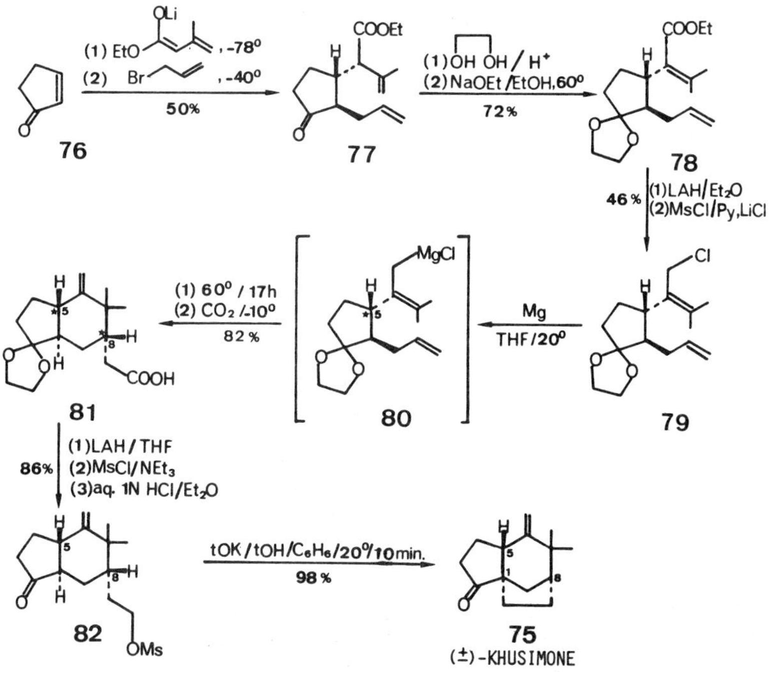

conjugate addition of the dienolate derived from 3,3-dimethylacrylate coupled with enolate-trapping by alkylation with allyl bromide furnished directly the 2,3-disubstituted cyclopentanone 77 in 50% yield. Accordingly, all but one of the carbon atoms of 75 have been aligned in a single synthetic operation. 77 was converted to the key precursor 79 by successive protection of the carbonyl group as an ethyleneacetal, EtONa-induced olefin migration, reduction of the conjugated ester with LiAlH₄ and treatment of the allylic alcohol with MsCl, pyridine, LiCl. The unstable allylic chloride 79, purified by rapid filtration through silica gel, gave smoothly the Grignard reagent 80 on slow addition to a stirred suspension of commercially available magnesium powder (Merck) in THF. Heating the resulting 0.6$\underline{N}$ solution of 80 at 60° for 17 h in a closed Carius tube followed by trapping of the cyclised organomagnesium chloride with CO_2 at -10° furnished, after crystallisation (ether/pentane), the carboxylic acid 81 in high overall yield (82% from 79). No isomer of 81 could be found in the mother liquor (^{1}H-NMR, GC). Whereas the unidirectional nature of the process 80 → 81 agrees with our previous results[31] its virtually quantitative stereoselectivity is particularly noteworthy. Assuming kinetic stereoselection the alternative transition states A and B have been examined (Scheme 32). Indeed, B shows a boat conformation of the developing

Scheme 32

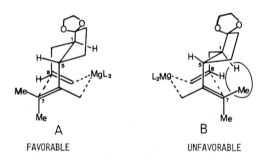

A
FAVORABLE

B
UNFAVORABLE

cyclohexane causing severe flagpole repulsion of one C(7)-methyl and the C(1)-hydrogen, whereas the evolving chair in A is perfectly attainable. We thus predicted A to be favored over B which entails the desired *cis*-disposition of the H-C(5) and H-C(8) in 81.

Unambiguous evidence for this stereochemical assignment was provided by the transformation of 81 into (±)-khusimone as follows. Reduction of the carboxylic acid 81 with LiAlH₄, mesylation of the primary alcohol (MsCl, NEt₃) and subsequent acetal cleavage (aq. HCl, ether) furnished after crystallisation the ketomesylate 82 (86% yield from 81). Finally, intramolecular alkylation of 82 by brief exposure to *t*-BuOK, *t*-BuOH, C_6H_6 at reflux gave after sublimation pure crystalline (±)-khusimone (75, 98% yield), identified by comparison with authentic (-)-75 (GC, IR, ^{1}H-NMR, ^{13}C-NMR and MS). In summary (±)-khusimone was obtained from cyclopentenone by a sequence of 9 synthetic operations in 11% overall yield[36]. This strategic application of an outstandingly regio- and stereoselective intramolecular "magnesium-ene" reaction exemplifies the potential value of this method in synthesis.

Acknowledgement - It is a pleasure to acknowledge the contributions of my very able collaborators Kurt Bättig, Christian Chapuis, Mao Dao Guo, Rita Pitteloud, Dana P. Simmons, Heinrich F. Strauss and Klaus Thirring. Their names are cited in the appropriate references. We thank the Swiss National Science Foundation, Sandoz Ltd, Basel and Givaudan S.A., Vernier for generous financial support of our work. We are indebted to Prof. H. Morimoto, Prof. L.A. Paquette, Prof. D. Wege and Dr. B. Maurer for kindly providing reference spectra and an authentic sample.

REFERENCES

1. A. Padwa, Angew. Chem. 88, 131-144 (1976); Angew. Chem. Int. Ed. Engl. 15, 123-136 (1976); W. Oppolzer, Angew. Chem. 89, 10-24 (1977); Angew. Chem. Int. Ed. Engl. 16, 10-23 (1977); W. Oppolzer, Synthesis 793-802 (1978); C. Brieger and J.N. Bennett, Chem. Rev. 80, 63-97 (1980); W. Oppolzer, Pure Appl. Chem. 53, 1181-1201 (1981); W. Oppolzer, Acc. Chem. Res. 15, 135-141 (1982).
2. W. Oppolzer and V. Snieckus, Angew. Chem. 90, 506-516 (1978); Angew. Chem. Int. Ed. Engl. 17, 476-486 (1978); W. Oppolzer and K. Bättig, Helv. Chim. Acta, 64, 2488-2491 (1981).
3. W. Oppolzer, C. Robbiani and K. Bättig, Helv. Chim. Acta 63, 2015-2018 (1980).
4. E.J. Corey and H.E. Ensley, J. Am. Chem. Soc. 97, 6908-6909 (1975).
5. W. Oppolzer, M. Kurth, D. Reichlin and F. Moffatt, Tetrahedron Lett. 2545-2548 (1981).
6. W. Oppolzer, M. Kurth, D. Reichlin, C. Chapuis, M. Mohnhaupt and F. Moffatt, Helv. Chim. Acta 64, 2802-2807 (1981).
7. T. Mukaiyama, Angew. Chem. 89, 858-866 (1977); Angew. Chem. Int. Ed. Engl. 16, 817-826 (1977); M.T. Reetz, Angew. Chem. 94, 97-109 (1982); Angew. Chem. Int. Ed. Engl. 21, 96-108 (1982).
8. W. Oppolzer, C. Chapuis and M.D. Guo, unpublished work.
9. Y. Yamamoto and K. Maruyama, J. Am. Chem. Soc. 100, 3240-3241 (1978); Y. Yamamoto, S. Yamamoto, H. Yatagai, Y. Ishihara and K. Maruyama, J. Org. Chem. 47, 119-126 (1982).
10. W. Oppolzer and H.J. Löher, Helv. Chim. Acta 64, 2808-2811 (1981).
11. W. Oppolzer, H.J. Löher and A. Meunier, unpublished work.
12. a) H. Morimoto, J. Pharm. Soc. Japan 75, 941-943 (1955); H. Morimoto and R. Nakamori, Ibid. 76, 26-30 (1956); R. Nakamori, Ibid. 76, 291-294 (1956); b) Y. Ohfune and M. Tomita, J. Am. Chem. Soc. 104, 3511-3513 (1982).
13. W. Oppolzer and H. Andres, Helv. Chim. Acta 62, 2282-2284 (1979).
14. W. Oppolzer and K. Thirring, J. Am. Chem. Soc. in press.
15. Y.H. Lai, Synthesis, 585-604 (1981).
16. E.P. Kündig and C. Perret, Helv. Chim. Acta 64, 2606-2613 (1981).
17. W. Oppolzer, E.P. Kündig, P.H. Bishop and C. Perret, Tetrahedron Lett. in press.
18. H. Lehmkuhl, Bull. Soc. Chim. Fr. Part II, 87-95 (1981).
19. H. Lehmkuhl, D. Reinehr, D. Hennerberg, G. Schomburg and G. Schroth, Liebigs Ann. Chem. 119-144 (1975).
20. H. Felkin, L.D. Kwart, G. Swierczewski and J.D. Umpleby, J. Chem. Soc. Chem. Commun. 242-243 (1975).
21. R.D. Little and G.L. Carroll, Tetrahedron Lett. 4389-4392 (1981); R.D. Little, G.W. Muller, G. Venegas, G.L. Carroll, A. Bukhari, L. Patton and K. Stone, Tetrahedron 37, 4371-4383 (1981).
22. K.E. Stevens and L.A. Paquette, Tetrahedron Lett. 4393-4396 (1981).
23. W. Oppolzer and K. Bättig, unpublished work.
24. P.A. Collins and D. Wege, Aust. J. Chem. 32, 1819-1826 (1979).
25. J.A. Berson, D.S. Donald and W.J. Libbey, J. Am. Chem. Soc. 91, 5580-5593 (1969).
26. P.E. Pfeffer, L.S. Silbert and E. Kinsel, Tetrahedron Lett. 1163-1166 (1973); J.A. Katzenellenbogen and A.L. Crumrine, J. Am. Chem. Soc. 98, 4925-4935 (1976).
27. C.A. Bunton, D.L. Hachey, J.-P. Leresche, J. Org. Chem. 37, 4036-4039 (1972).
28. W. Oppolzer, H.F. Strauss and D.P. Simmons, unpublished work.
29. J.E. Nordlander, W.G. Young and J.D. Roberts, J. Am. Chem. Soc. 83, 494-495 (1961).
30. O.A. Hutchinson, K.R. Beck, R.A. Benkeser and J.B. Grutzner, J. Am. Chem. Soc. 95, 7075-7082 (1973).
31. W. Oppolzer, R. Pitteloud and H.F. Strauss, unpublished work.
32. D.C. Umrani, R. Seshadri, K.G. Gore and K.K. Chakravarti, Flavour Ind. 1, 623-624 (1970); B. Maurer, M. Fracheboud, A. Grieder and G. Ohloff, Helv. Chim. Acta 55, 2371-2382 (1972).
33. B. Maurer, Swiss Patent 575 362 (1972); Ibid. 583 162 (1974).
34. G. Büchi, A. Hauser and J. Limacher, J. Org. Chem. 42, 3323-3324 (1977).
35. H.-J. Liu and W.H. Chan, Can. J. Chem. 57, 708-709 (1979).
36. W. Oppolzer and R. Pitteloud, unpublished work.

REGIO-, DIASTEREO-, AND ENANTIOSELECTIVE C-C COUPLING REACTIONS USING METALATED HYDRAZONES, FORMAMIDES, ALLYLAMINES, AND AMINONITRILES

D. Enders

Institut für Organische Chemie und Biochemie, Rheinische Friedrich-Wilhelms-Universität, Gerhard-Domagk-Str. 1, D-5300 Bonn 1,
Federal Republic of Germany

Abstract - Highly enantioselective α-alkylations of ketones and aldehydes using metalated chiral hydrazones of (S)- or (R)-1-amino-2-methoxymethyl-pyrrolidine ("SAMP/RAMP-hydrazones") are described. Furthermore, reduction/N-N bond cleavage of SAMP-hydrazones leads to primary amines of (R)- or (S)-configuration with selectivities of up to 94% ee depending on the reducing agent. Proline derived chiral hydrazines can also be used efficiently as agents for chromatographic (HPLC) resolution of aldehydes. Lithiated chiral (thio)formamides and allylamines are used in asymmetric nucleophilic acylation processes and as chiral homoenolate equivalents, respectively. In addition, new simple and generally applicable syntheses of spiroacetals and α-aminoketones are presented.

The development of efficient and highly selective methods for carbon-carbon bond formation has been and continues to be a challenging and exciting endeavour in organic chemistry. Especially, in the field of asymmetric synthesis the chemical community has seen a real breakthrough in the past 7 years and several processes routinely allowing asymmetric inductions of greater than 90% enantiomeric excess are now at our disposal (Ref. 1). Our contribution to this progress in recent years was the design of a general method for asymmetric C-C bond formation α to the carbonyl group of aldehydes and ketones via metalated chiral hydrazones of (S)- or (R)-1-amino-2-methoxymethyl-pyrrolidine (SAMP/RAMP-hydrazones, Ref. 2).

In a continuation of our efforts to utilize chiral organometallics in asymmetric synthesis, this lecture will first focus on further applications and valuable extensions of the SAMP/RAMP-hydrazone method. In addition, a novel application of lithiated dimethylhydrazones (DMH's, Ref. 3), the achiral representatives of these highly reactive enolate equivalents or d^2-reagents $\underline{1}$ (Ref.4), will be described. (S)-Proline derived chiral hydrazines, such as $\underline{2}$, can also serve as efficient agents for chromatographic resolution of aldehydes, using high pressure liquid chromatography (HPLC). In the second part, the use of chiral d^1-reagents, e.g. lithiated (thio)formamides $\underline{3}$ and aminonitriles $\underline{4}$, in asymmetric nucleophilic acylation reactions will be discussed, as well as a new and simple synthesis of α-dialkylamino-ketones via metalated aminonitriles $\underline{4}$. Finally, it will be demonstrated that metalated allylamines or enamines $\underline{5}$ constitute d^3-reagents and thus represent the first chiral homoenolate equivalents (Fig.1).

SYNTHESIS OF SPIROACETALS

Spiroacetal structures of the general type $\underline{6}$ are characteristic and crucial structural features of many natural products ranging from simple insect pheromones to complex polyether antibiotics. By breaking the bonds shown in bold print, the retrosynthetic analysis leads to ketodiols $\underline{7}$ and then to simple building blocks, like ketones and hydroxyalkylating reagents $\underline{8}$. Suitable examples of the latter are oxiranes (n,m= 1), oxetanes (n,m= 2), and protected ω-haloalcohols (n,m= 1,2,3...). Thus, in the synthetic direction a regioselective α,α'-double hydroxyalkylation of simple ketones should be a general entry to $\underline{6}$. Since enantiomerically pure building blocks $\underline{8}$ are readily available and in addition the SAMP/RAMP-hydrazone method may be applied, control of absolute stereochemistry should be possible (Fig.2).

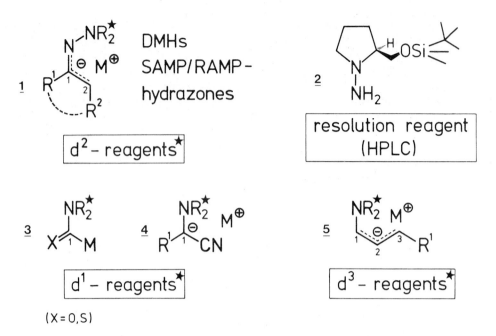

Fig.1. Reagents for regio-, diastereo-, and enantioselective electrophilic substitutions and for chromatographic resolution.

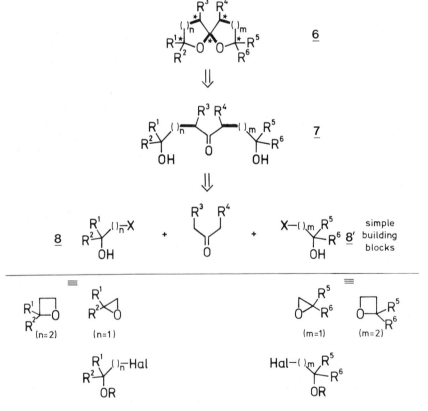

Fig.2. General synthesis of spiroacetals from ketones and hydroxyalkylating agents - retrosynthetic analysis (new formed bonds are shown in bold print throughout this article).

Indeed, we recently had success in applying this general route. The basic procedure is demonstrated with the synthesis of racemic and optically active chalcogran. This substance is the principal aggregation pheromone of *Pityogenes chalcographus* (L.), a bark-beetle named "Kupferstecher", which recently created a serious pest problem in European spruce forests (Ref. 5).

Acetone-dimethylhydrazone is metalated with n-butyllithium and then treated with 1,2-epoxy-butane. In the same flask, the other methyl group can be regiospecifically metalated and then trapped with ethylene oxide. Quenching with acetic acid yields a hydrazonodiol. By refluxing this intermediate over an acidic cation exchange resin, hydrazone cleavage and ketalization occurs. In this manner, racemic and optically active chalcogran was prepared on a 10 g scale, using racemic or (S)-1,2-epoxy-butane, in an overall yield of 70%. Steps one to five are performed conveniently in an one pot procedure (Fig.3, Ref. 6).

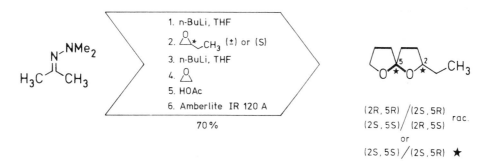

steps 1-5 in one pot, 10g scale!
rac. chalcogran biologically active in field tests   [CELAMERCK, Ingelheim]

Fig.3. Synthesis of racemic and optically active chalcogran from acetone and oxiranes - aggregation pheromone of *Pityogenes chalcographus* (L.).

Initial field tests already revealed that racemic chalcogran has a good attractive activity. Further tests with the optically active pheromone are now in progress (Celamerck, Ingelheim). Table 1 summarizes a few spiroacetals, prepared from a combination of the oxirane/oxetane building blocks and acetone including some wasp and fly pheromones. Since relatively low yields were observed with oxetane ring openings, protected ω-haloalcohols are the synthons of choice for building blocks of type 8. Enantioselective syntheses of more complex spiroacetals are now in progress.

TABLE 1. Spiroacetals from acetone and oxiranes/oxetanes.

spiroacetal	prepared from	yield [%]	species (function)
	a + a	50-70	UNNATURAL
	a + b	60	UNNATURAL
	b + b	65	UNNATURAL
	a + c	70	PITYOGENES CHALCOGRAPHUS L. (AGGREGATION PHEROMONE)

TABLE 1. continued.

spiroacetal	prepared from	yield [%]	species (function)
H₃C⟨⟩O⟨⟩CH₃	c + c	65	ANDRENA WILKELLA (UNKNOWN)
H₃C⟨⟩O⟨⟩	b + d	38	DOLICHOVESPULA SAXONICA (UNKNOWN) PARAVESPULA GERMANICA (UNKNOWN) PARAVESPULA VULGARIS (REPELLENT)
⟨⟩O⟨⟩	d + d	16	DACUS OLEAE (SEXUAL PHEROMONE)

a: △   b: △   c: △*   d: □

SAMP/RAMP-HYDRAZONE METHOD

C-C bond formations α to the carbonyl group, the "backbone of organic chemistry", belong to the most important synthetic operations. In recent years, we have reported an efficient method for asymmetric C-C coupling reactions of this type based on metalated chiral hydrazones of (S)- or (R)-1-amino-2-methoxymethyl-pyrrolidine (SAMP or RAMP, Ref. 2). SAMP is prepared from (S)-proline in four steps and 50% overall yield. RAMP is obtained from (R)-glutamic acid in six steps and yields 35%.

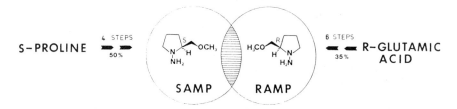

To further demonstrate the synthetic utility of our method, I would like to present new applications and extensions. Most difficult to achieve are asymmetric electrophilic substitutions of conformationally flexible acyclic ketones and aldehydes. (E)-4,6-Dimethyl-6-octene-3-one 9 is a defense substance of "daddy longlegs" *Leiobunum vittatum* and *L.calcar*, which belong to the class of spiders (Ref. 7). When attacked, the animals visibly wet themselves with secretion and release a powerful odor. Since only a few microliters of this secretion are available from nature and the absolute configuration of 9 is unknown, the asymmetric synthesis of both enantiomers was undertaken.

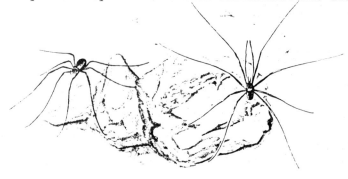

The retrosynthetic analysis leads to diethylketone and bromide 10. In fact, transformation of diethylketone to the corresponding SAMP- [(S)-11] or RAMP-hydrazone [(R)-11] followed by metalation with lithium-diisopropylamide (LDA) in ether, trapping with 10 at -110°C and acidic hydrolysis in a two phase system - first salt formation, then 3N hydrochloric acid/n-pentane - gave the (S)- or (R)-enantiomer of 9 in an overall chemical yield of 70% and in very high selectivities of ≥95% e e , respectively. This means that almost complete stereoselection in the deprotonation and alkylation steps have been achieved and virtually no racemization occurred during hydrazone cleavage (Fig.4, Ref.8). In this way, g-amounts of both optical antipodes, as well as racemic 9 (via the corresponding dimethylhydrazone), were prepared. In collaboration with Prof. Meinwald and Prof. Eisner (Cornell) appropriate bioassays are now planned.

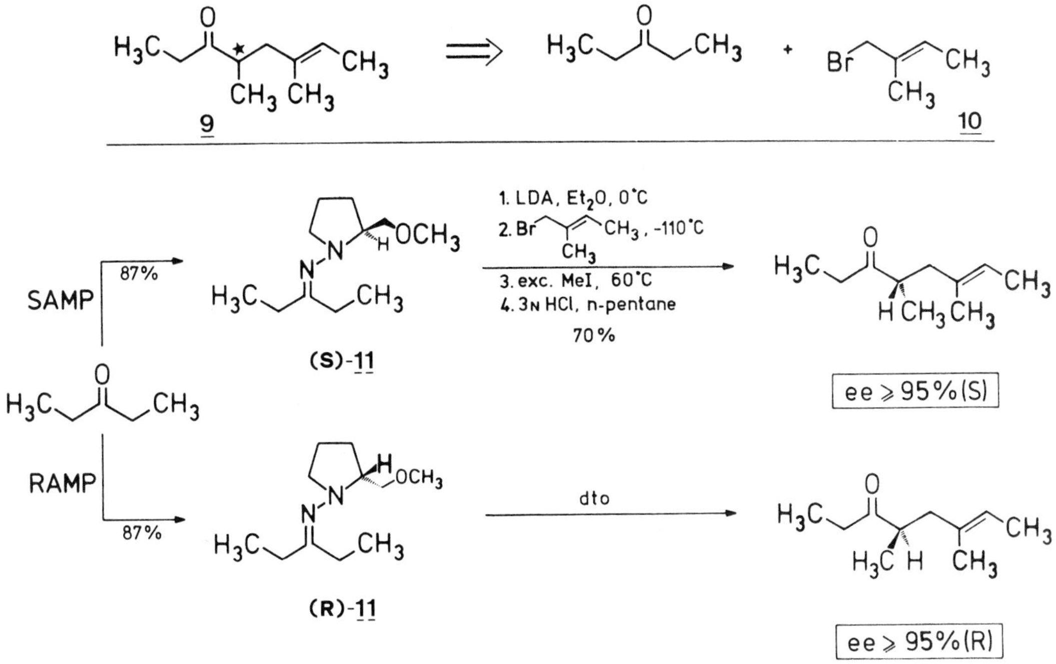

Fig.4. Asymmetric synthesis of both enantiomers of (E)-4,6-dimethyl-6-octane-3-one - defense substance of "daddy longlegs" *Leiobunum vittatum* and *L.calcar* (opilionids).

Metalated SAMP/RAMP-hydrazones are very reactive nucleophiles and a great variety of activated and unactivated electrophiles can be introduced. Of special interest are those leading to polyfunctional molecules. A few examples are illustrated in Fig.5. For instance, diethylketone was alkylated with bromoacetic acid-t-butylester to give the optically active γ-ketoester 12; in this case the hydrazone was oxidatively cleaved with ozone in n-pentane at -78°C. Similarly, bromide 13 furnished the polyfunctional ketone 14. It is interesting that the ozonolysis occurred chemoselectively at the CN-double bond of the hydrazone, leaving the enolether function untouched. In both cases the effected enantioselection was again at the 95% level. In asymmetric hydroxyethylations with oxiranes, however, relatively low inductions were observed, as can be seen from the reaction of cyclohexanone and ethylene oxide to form 15 (ee= 30%) (Ref. 8).

In summary, the SAMP/RAMP-hydrazone method, as a modern version of the classical enolate chemistry, allows α-alkylations of ketones and aldehydes in excellent enantioselectivities and good overall chemical yields. To date, when ether was employed as the solvent during C-C bond formation, the enantiomeric excess was greater than 90% for all the acyclic cases tested so far. The ant alarm pheromone 16 (ee= 99.5%), the defense substance 9 (ee≥ 95%) and the simple aldehydes 20 and 21 (both ee= 95%) are typical examples. In α-alkylations of cyclic 5 to 8-membered ring ketones, the asymmetric inductions range from 60-99% ee. In addition, the reactions are completely regioselective, e.g. in the case of menthone 17 (Ref. 9) and the cyclohexenone 18 (Ref.10). Qua-

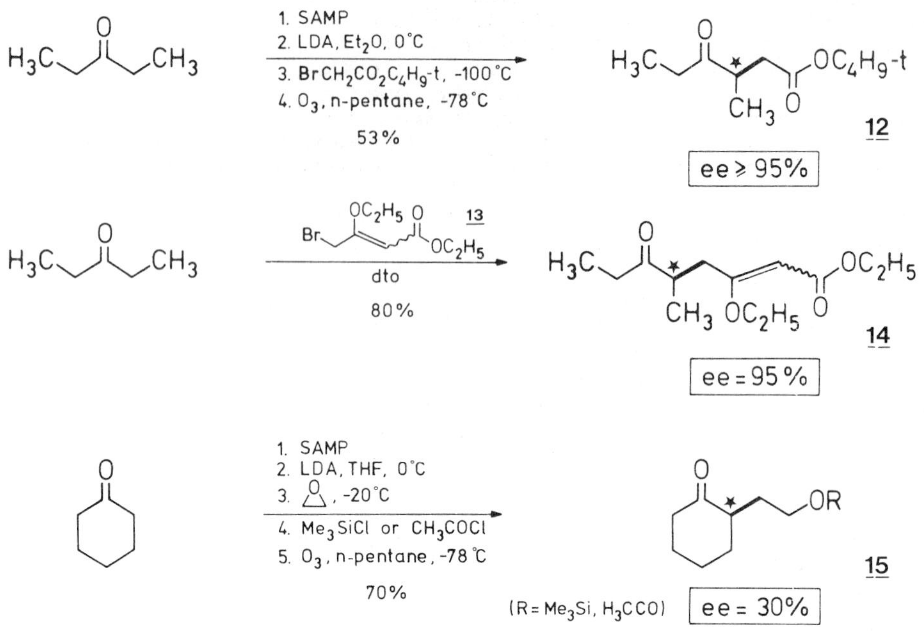

Fig.5. Formation of polyfunctionalized molecules by asymmetric C-C bond coupling reactions.

ternary chiral carbon centers, such as in 19, may also be generated with high enantioselectivity (Fig. 6).

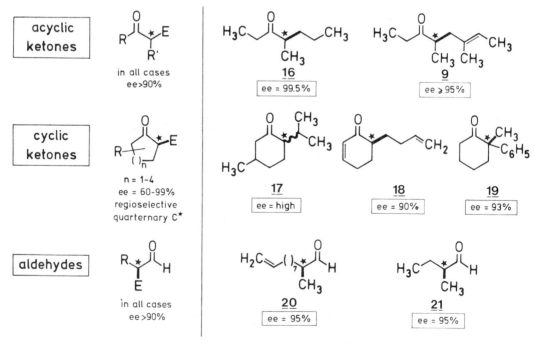

Fig.6. Asymmetric α-alkylations of ketones and aldehydes via SAMP/RAMP-hydrazones.

From our early results, first published in 1978 (Fig.7, Ref. 2c, 2f), it is obvious that asymmetric aldol reactions via SAMP/RAMP-hydrazones need to be optimized. In principle, high asymmetric inductions at the position β to the carbonyl group, using simple, α-unsubstituted enolates as chiral nucleophiles, are difficult to achieve in aldol reactions; this is also true for other approaches (Ref. 10,11). Nevertheless, simple β-ketols like [6]-gingerol, the major pungent principle of ginger, 22 or ketol 23 were prepared with enantioselectivities of 36-62%; after two recrystallizations 23 was obtained as a pure enantiomer. In the meantime, excellent stereoselectivities for α-substituted systems were obtained in the exciting results of Evans et al. (11), Heathcock et al. (12), Masamune et al. (13), Meyers et al. (14) and many other groups. Recently, we were able to start further optimizations of our methodology mentioned above. Diastereo- and enantioselectivities observed by changing the gegenion of metalated SAMP/RAMP-hydrazones are very promising (Ref. 15). Medium enantioselectivities were obtained in asymmetric alkylations of β-ketoesters, e.g. 24 (ee= 60%) (Fig.7).

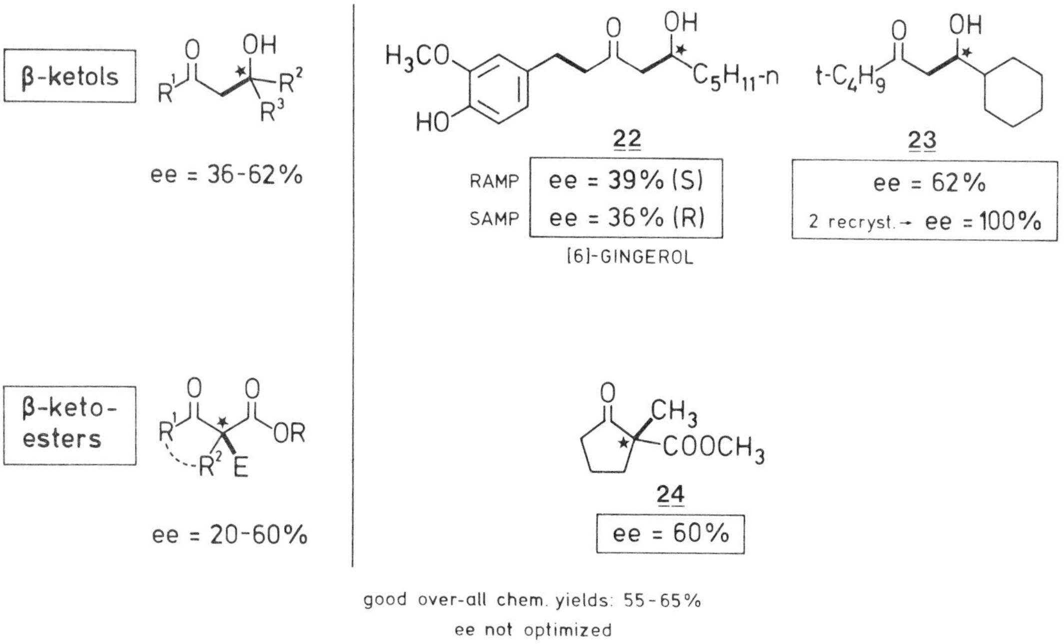

Fig.7. Asymmetric aldol reaction to form β-ketols and α-alkylations of β-ketoesters via SAMP/RAMP-hydrazones.

A new synthetic method "lives or dies" with reproducable and successful applications by others. Thus, we were very pleased to see highly successful and sophisticated applications of the SAMP/RAMP-hydrazone method in enantioselective syntheses of various natural products, such as pheromones 25 and 28 [Bestmann et al. (16) and Mori et al. (17)], an ionophore antibiotic 26 [Nicolaou et al. (18)] and a sesquiterpene 27 [Pennanen (19)]. In all cases, a key step of the syntheses was the enantioselective α-alkylation of a ketone or aldehyde, indicated by bold print and an arrow (Fig.8).

Asymmetric reductive amination of ketones

Because of the many reports in the literature describing asymmetric reductions of the CN-double bond of imino derivatives (Ref.1), including a few chiral hydrazones (Ref. 20), it seemed possible to further expand the synthetic utility of our method. Besides asymmetric syntheses of α-substituted carbonyl compounds, regio-, diastereo- and enantioselective syntheses of amines should be possible, too. In principal, a selective reduction of the CN-double bond followed by N-N-cleavage of SAMP/RAMP-hydrazones should lead to α-chiral amines A. In combination with our previously described procedure for α-alkylation α,β-chiral amines B or , starting from aldehydes, β-chiral amines C should be obtained (Fig. 9).

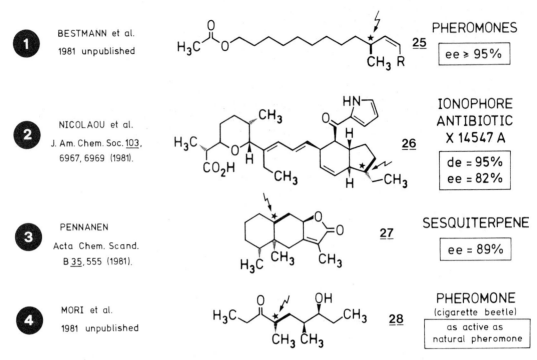

Fig.8. Highly successful synthetic applications of the SAMP/RAMP-hydrazone method in natural product synthesis.

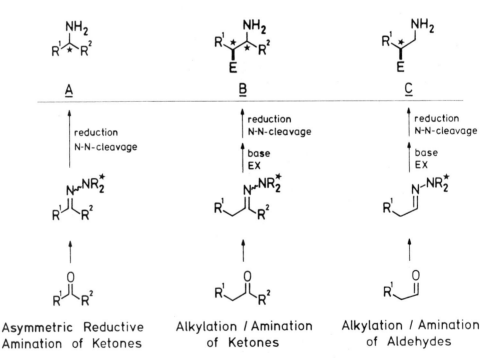

Fig.9. General asymmetric syntheses of α- and/or β-chiral primary amines via SAMP/RAMP-hydrazones.

We decided to first work out the asymmetric reductive amination of ketones (29 → 30). The corresponding SAMP-hydrazones (E/Z)-31 were reduced with lithium aluminum hydride (LAH) or catechol borane (CB) to give hydrazines 32, which upon N-N-cleavage with Raney nickel/hydrogen led to the desired primary amines 30. In the first test cases the overall yields were 42-60% and selectivities between 50 and 94% ee were reached. A recycling of the chiral information via

33 is also possible.

To our surprise, in all experiments, run with lithium aluminium hydride, primary amines with the (S)-configuration in excess were obtained, whereas reductions with catechol borane led predominantly to (R)-amines. This indicates an unique reaction mechanism in both cases, but with opposite diastereoface differentiation. Thus, the (E/Z)-ratio of geometrical isomers of the unsymmetrical hydrazones 31 should be a limiting factor for the final enantioseletivity reached in 30. One of the advantages of metalated hydrazones is the fact that these anionic species are usually stereochemically uniform, independent of the (E/Z)-ratio of the starting material. Under mild work up conditions only one stereoisomer can be generated (Ref. 21). This should allow us, therefore, to encrease the percent ee by a deprotonation/reprotonation trick (Fig. 10).

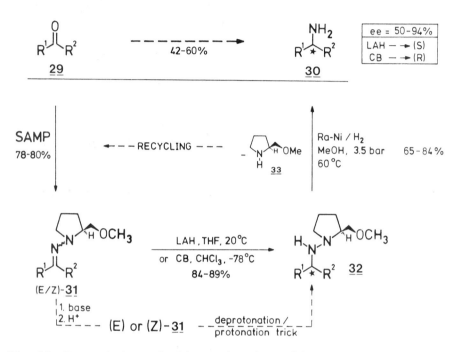

Fig.10. Asymmetric reductive amination of ketones via SAMP-hydrazones.

In fact, the preliminary data illustrated in Fig.11 show a correlation between the (E/Z)-ratio of 31 and the enantiomeric excess of the product amines 30. Indeed, in the asymmetric synthesis of (R)-phenethylamine the asymmetric induction was enhanced from 60 to 81% ee, using the deprotonation/reprotonation trick mentioned above. To explain the synthetically useful result that both enantiomers of a primary amine can be prepared in excess via SAMP-hydrazones by a simple change of the reducing agent, we assume the following: The predominant N-N conformation of SAMP/RAMP-ketohydrazones in solution is the one with the pyrrolidine ring plane perpendicular to the CN-double bond (Ref. 22). Lithium aluminium hydride and other alkali metal containing reducing agents like sodium borohydride and sodium cyanoborohydride, are chelated as shown in 34. Thus, a hydride is transferred predominantly from below the plane leading to (S)-amines in excess. With catechol borane, where this type of chelation is not possible, the sterically less hindered topside is attacked predominantly, and (R)-amines are formed in excess (35).

R*-NH$_2$	(E/Z) ratio	ee [%] LAH	ee [%] CB
C$_6$H$_5$-*C(NH$_2$)-CH$_3$	≈ 8:1	70 (S)	60 (R)
	(E)		81$^+$(R)
t-C$_4$H$_9$-*C(NH$_2$)-CH$_3$	(E) only	90 (S)	88 (R)
C$_6$H$_5$-CH$_2$-*C(NH$_2$)-CH$_3$	≈ 3:1	50 (S)	35 (R)

$^+$deprotonation/protonation-trick was used

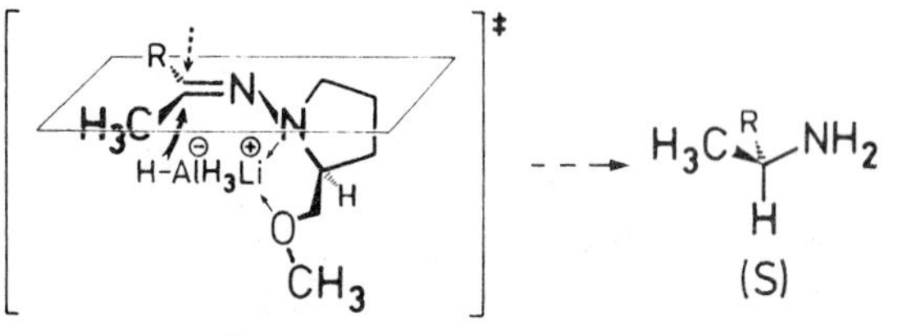

Fig.11. Asymmetric synthesis of both enantiomers of primary amines by simple change of reducing agent - enhancement of % ee using a deprotonation/protonation-trick.

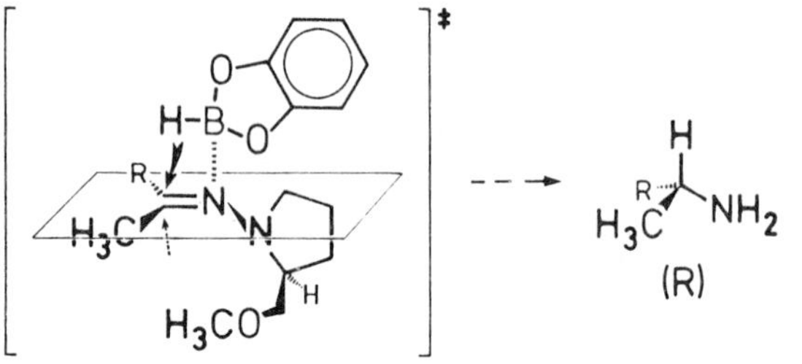

Structure of lithiated SAMP-hydrazones

Whereas alkali metal enolates are known to exist as tetrameric aggregates in solution (Ref. 24) and in the crystalline form (Ref. 25), definitive structural information about alkali metal azaenolates is still lacking. In general, lithiated hydrazones, as ambident anionic species, may be drawn as an α-lithiohydrazone, a π-1-azaallylanion-lithiumcation-contact species, or a lithio-enehydrazinate. In this context, MNDO-calculations with lithium-parametrization were carried out (Ref. 26). Interestingly, the calculated relative energy (RE) of the classical α-lithiohydrazone structure 36 with intramolecular chelation was lowest. However, the π-species 37, preferred by us so far, came out close (RE= 2.5 kcal/mol). To our surprise, the lithio-enehydrazinate 38 was calculated to be about 25 kcal/mol less favourable. The intramolecular chelation energy, shown in all three structures, was estimated to about 6 kcal/mol. In addition, the dimerization energy was estimated to be smaller than the solvation energy, suggesting a monomeric structure for lithiated hydrazones in solution (Fig.12). To learn more experimentally, the determination of the aggregation in solution and an X-ray analysis of a metalated hydrazone are now planned.

CHROMATOGRAPHIC RESOLUTION OF ALDEHYDES

Despite the great importance of aldehydes and ketones in organic chemistry, simple and generally applicable methods for their resolution are rather rare (Ref. 27). Using modern chromatographic techniques like high pressure liquid chromatography (HPLC), it should be possible to develop such a general procedure. We first focused on the synthetically useful, but highly sensitive α-chiral aldehydes 39. Hydrazones 40 should again be ideal candidates as diastereomeric derivatives; they can be cleaved after separation as a 1:1 mixture of α-epimers by several methods without any racemization of the carbonyl products (Ref. 2). Of the enantiomerically pure hydrazines tested so far, best results were obtained with silylated derivatives of type 41 (Fig.13).

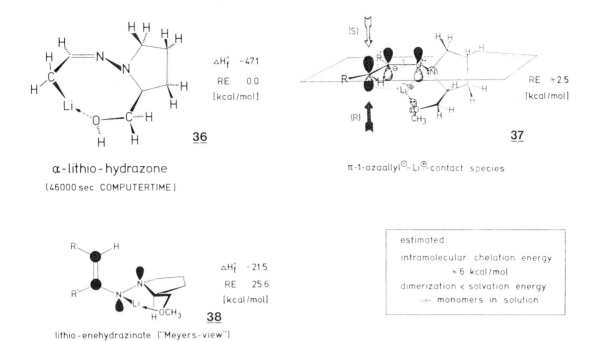

Fig.12. MNDO-calculations of the structure of lithiated SAMP-hydrazones (with Li-parametrization).

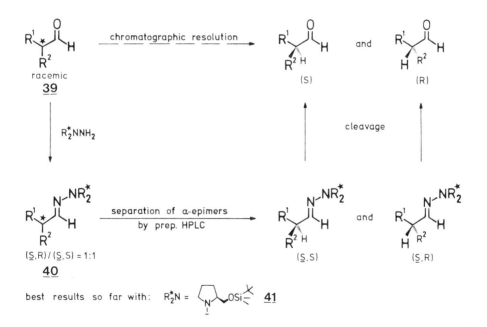

Fig.13. Chromatographic resolution (HPLC) of α-substituted aldehydes via chiral hydrazones.

The procedure and two typical examples are illustrated in Fig.14. Racemic aldehydes 39 are transformed into 1:1 mixtures of epimers 42 by treatment with (S)-1-amino-2-hydroxymethyl-pyrrolidine, followed by a silylation. As is demonstrated in the case of 2-methyl-pentanal 43 (X=O) and 2-methyl-undecenal 44 (X=O) under the analytical HPLC-conditions shown, separation factors of α= 1.25-1.4 are reached and the (S,R)- and (S,S)-epimers are separated by minutes. To carry out a resolution on a preparative scale, simply a longer column and higher pressure is used. In this way, for instance, 0.6 g of 44 were separated per injection and thus, g-amounts of pure epimers can be obtained in one working day (Ref. 28).

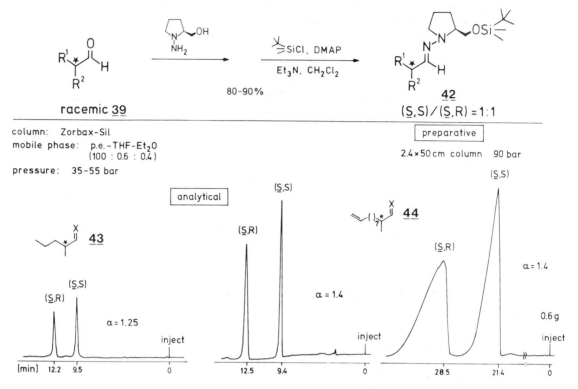

Fig.14. Chromatographic resolution of α-substituted aldehydes by preparative HPLC.

CHIRAL d^1-REAGENTS

In view of the great synthetic importance of nucleophilic acylation reactions using d^1-reagents (donor reactivity at position one, Umpolung of classical a^1-carbonyl reactivity), the time is now ripe to develop asymmetric versions of these processes. Metalated chiral (thio)formamides 45 and aminonitriles 46 should be suitable chiral d^1-reagents, since the "chiral information" can easily be introduced via the amino function.

We recently reported the synthesis of enantiomerically pure (R)- and (S)-α-hydroxyketones 48 and vicinal diols 49 by trapping carbamoyl-lithium reagents 45 (X= O, M= Li) with ketones, followed by a simple chromatographic separation of the hydroxyamide epimers 47 and subsequent reaction with methyllithium (Ref. 29). Hydroxyamides 47 can also be reduced to α-hydroxyaldehydes 50 or hydrolysed under somewhat drastic conditions to α-hydroxy acids 51. In principle therefore, nucleophilic asymmetric carboxylations, formylations, acylations, or α-hydroxyalkylations are possible (Fig.15). The final problem to be solved was to increase the diastereoselectivity of the nucleophilic asymmetric carbamoylation.

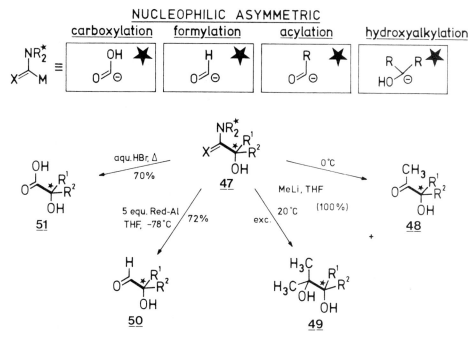

Fig.15. Nucleophilic asymmetric carboxylation, formylation, acylation, and α-hydroxyalkylation via metalated chiral (thio)formamides.

In order to optimize the diastereomeric excess (de), we first varied the chiral auxiliary in a test reaction 52 → 53 with t-butylmethyl-ketone as the electrophile. Percent de was determined by ¹H-nmr-shift experiments. As can be seen from Fig.16, many chiral formamides were tested. Probably due to the extraordinarily high reactivity of the intermediate lithium compounds, notoriously low selectivities of only 5-20% de were reached in the cases A-H. When a methoxyethoxymethyl (MEM) protecting group (I), a thioformamide J and a combination of both as in K was used, de could be increased from 30 to 40 to 60% (Ref. 30). We are now convinced that the 95% de level can be reached by further variation of the chiral auxiliary and/or the metal.

In our first attempts to use chiral α-aminonitriles as d^1-reagents in asymmetric nucleophilic acylation reactions, we metalated 54 with lithium diisopropylamide (LDA) or potassium diisopropylamide (KDA) and trapped the "anion" with benzaldehyde. After addition of ammonium chloride or methyliodide the adducts 55 were obtained in good yield. Unfortunately, after aminonitrile cleavage, the enantiomeric excess of the hydroxy- or methoxy-ketone 56 was only 10% - so, optimization by variation of all parameters is again necessary. However, we found that during distillation (R= H) hydrogen cyanide was eliminated and after tautomerization the α-aminoketone 57 was formed (Fig.17). Thus, it seemed possible to develop a new general route to α-aminoketones, which are biologically significant substances and important compounds in synthetic organic chemistry.

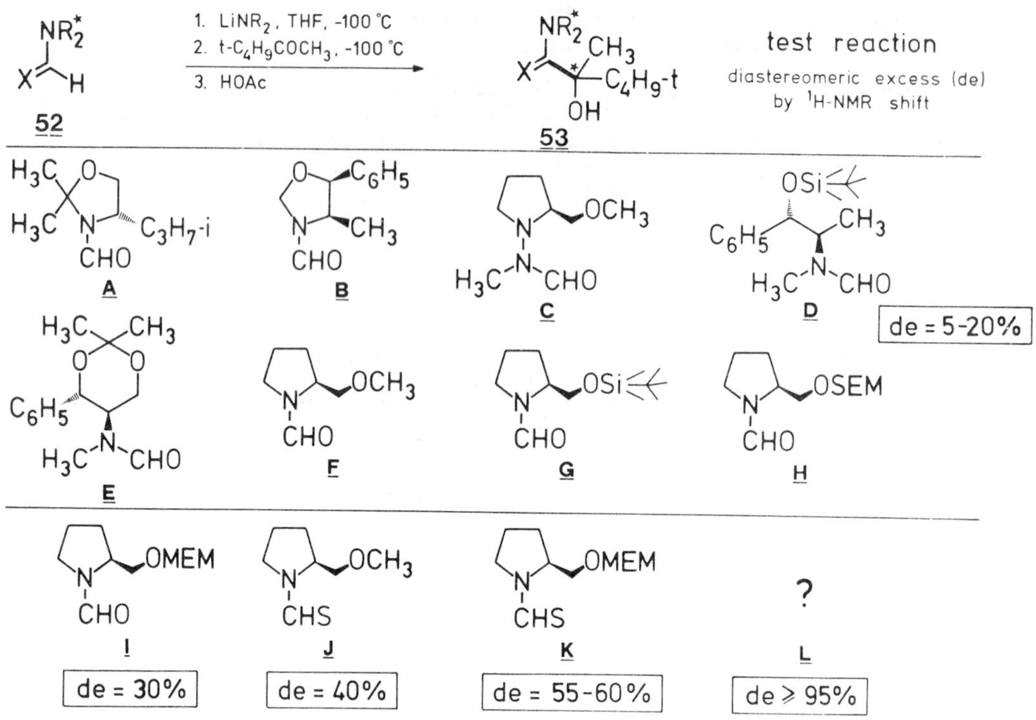

Fig. 16. Optimization of the diastereomeric excess in a test reaction by variation of the chiral auxiliary.

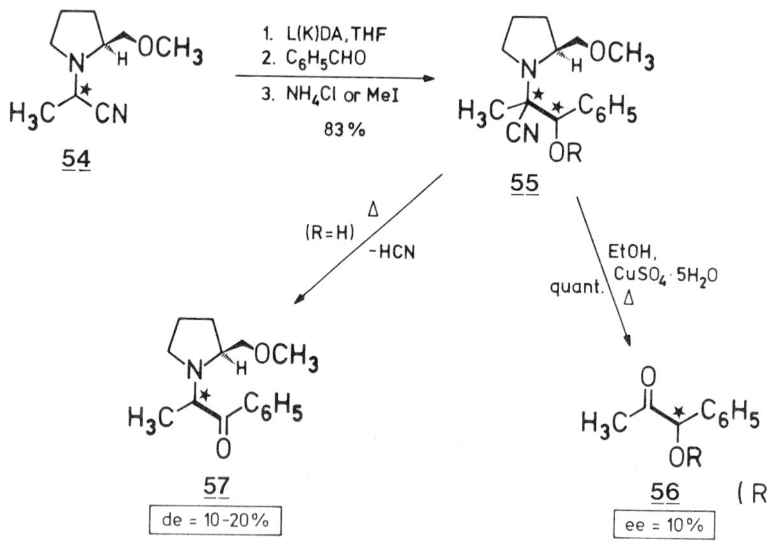

Fig. 17. First test of a metalated α-aminonitrile as chiral d¹-reagent.

C-C-Connective synthesis of α-aminoketones

As it turned out, such a C-C connective synthesis of α-dialkylamino-ketones, starting from easily accessible aldehydes and secondary amines, could indeed be worked out. As shown in Fig.18, the procedure involves the transformation of an aldehyde and a secondary amine into the corresponding α-aminonitrile 58, followed by metalation and subsequent addition of a second aldehyde. During distillation of the adducts 59 hydrogen cyanide is eliminated and the resulting aminoenol 60' immediately tautomerizes to the desired α-aminoketones 60 in overall yields of 55-81%.

Thus, the deprotonated α-aminonitriles function as equivalents of α-aminocarbanion synthons B, previously demonstrated by Stork et al., and the aldehydes R^2CHO function as equivalents of acylcations C (Fig.18). Since the reaction of metalated α-aminonitriles with aldehydes can alternatively be used to prepare aminoalcohols (Ref. 31) or α-hydroxyketones (Ref.32), this new route again enlarges the already great synthetic potential of α-aminonitriles (Ref. 33).

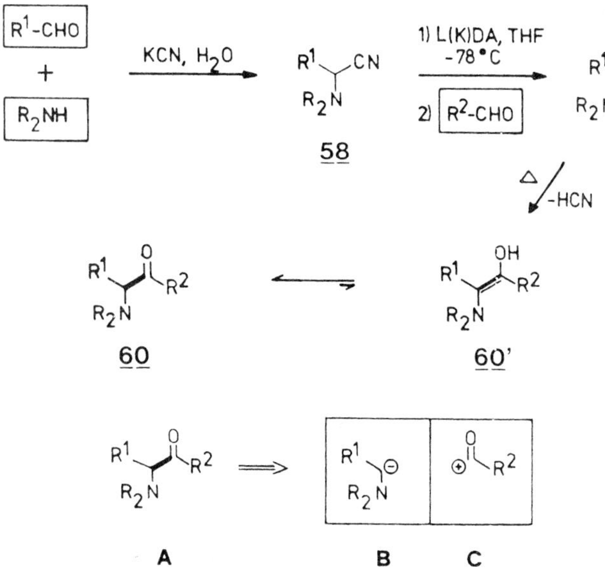

Fig.18. C-C-Connective synthesis of α-dialkylamino-ketones from aldehydes and secondary amines.

One major advantage of the method described above is the possibility of preparing α-dialkylamino derivatives of unsymmetrical ketones as pure regioisomers by synthon control. This may be demonstrated by the synthesis of the two regioisomers 3-dimethylamino- (61) and 5-dimethylamino-octan-4-one (62), starting from dimethylamine and either n-propanal/n-pentanal or two equivalents of n-butanal as building blocks, in yields of 67% and 60% respectively. With cinnamonaldehyde 1.2-addition occured exclusively leading to 65 in an overall yield of 79% (Fig.19).

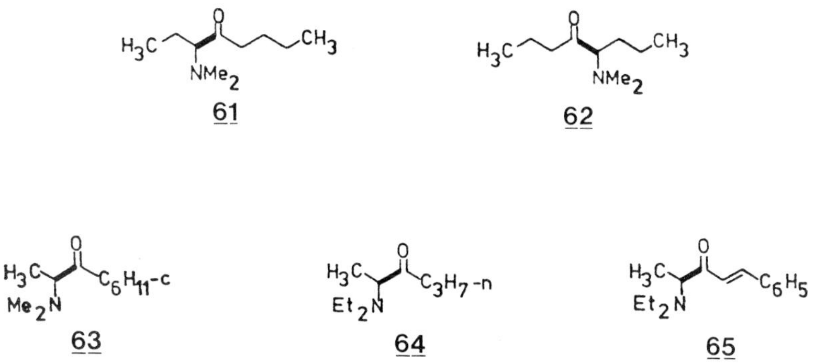

Fig.19. Regioisomerically pure unsymmetric α-aminoketones by synthon control.

CHIRAL HOMOENOLATE EQUIVALENTS (d^3-REAGENTS)

Let us finally focus on metalated allylamines or enamines of the type 66, which can serve as chiral homoenolate equivalents with d^3-reactivity, again an Umpolung.

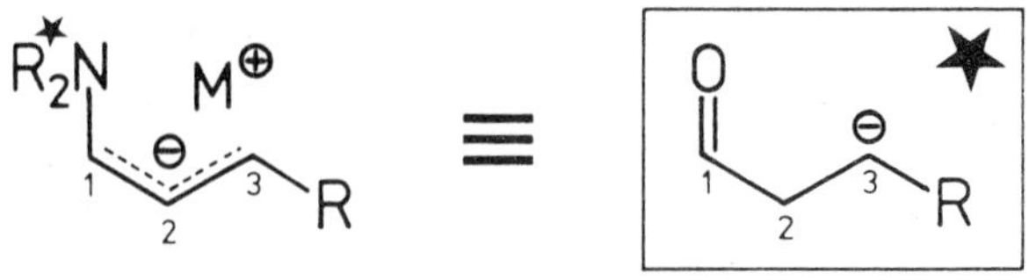

The method involves a deprotonation of allylamines 67 or enamines 68 with a strong base like tert-butyllithium/potassium-tert-butoxide to "allylic anions" 66. As shown in Fig.20, 66 is formed in a "sickle"-configuration. Alkylation with alkylhalides followed by non-aqueous work up affords the (Z)-enamines 69, which are hydrolyzed in a two phase system to optically active β-substituted aldehydes 70. Our recently published results using the chiral auxiliary 71, were limited to R^1= phenyl and enantiomeric excesses of 50-67% (Ref. 34). Since then, we learned that R^1 can also be an alkyl group and selectivities of up to 80% are reached. Although the chemistry works fine with chiral phosphoramidates of type 72, the asymmetric inductions were too low. Unfortunately, the more rigid, heterocyclic oxazaphospholidines of type 73 are too unstable under the strongly basic metalation conditions (Ref.35). The development of other, more efficient, chiral homoenolate equivalents is now in progress.

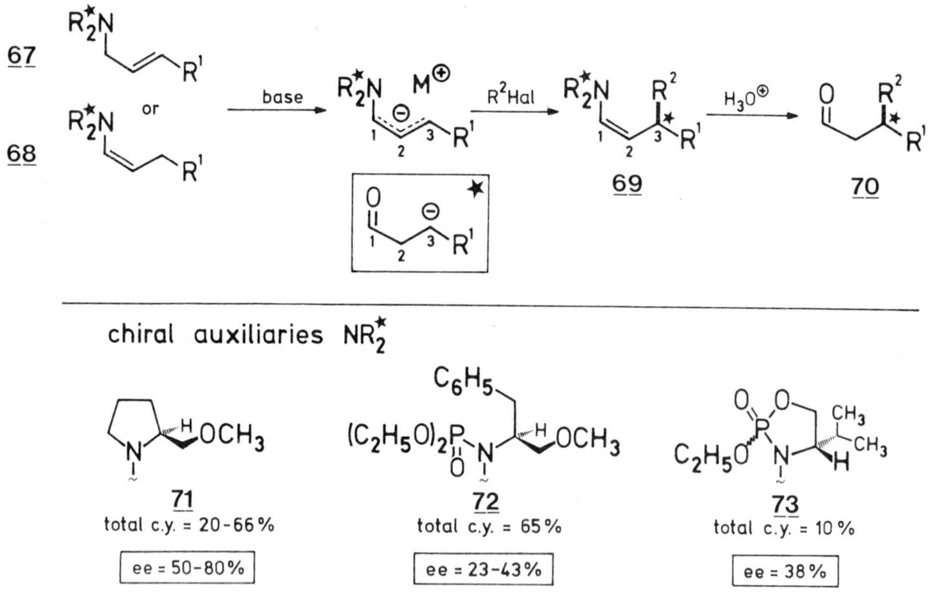

Fig.20. Asymmetric synthesis of β-substituted aldehydes via chiral homoenolate equivalents - preliminary results.

I hope I have convinced you with our more or less successful attempts to develop new methods for regio- and stereoselective carbon-carbon bond formations, that our field is exciting and full of promise for the future.

Acknowledgements - This work was supported by the Deutsche Forschungsgemeinschaft, the Fonds der Chemischen Industrie and the Minister für Wissenschaft und Forschung des Landes Nordrhein-Westfalen. I thank the German chemical industry - Degussa, BASF, Hoechst AG, Bayer AG and Celamerck for the support with chemicals and Dr. Mengel (Celamerck) for valuable discussions and his collaboration in the chalcogran-project. The skill and hard work of my coworkers Dr. H. Eichenauer, Dr. P. Weuster, H. Lotter, U. Baus, Dr. V. N. Pathak, W. Dahmen, H. Schubert, W. Mies, R. Pieter, E. Dederichs and the assistance of R. Pieter, Dr. K. A. M. Kremer and Mrs B. Jendrny in the preparation of this manuscript are gratefully acknowledged.

REFERENCES

1. (a) D.Valentine, Jr. and J.W.Scott, Synthesis 329 (1978). (b) H.B.Kagan and J.C.Fiaud, Top.Stereochem. 10, 175 (1978). (c) J.W.ApSimon and R.P.Seguin, Tetrahedron 35, 2797 (1979). (d) E.L.Eliel and S.Otsuka, Asymmetric Reactions and Processes in Chemistry, ACS Symposium Series 185, Am.Chem. Soc., Washington, D.C. (1982).
2. (a) D.Enders and H.Eichenauer, Angew.Chem., Int.Ed. 15, 549 (1976). (b) D. Enders and H.Eichenauer, Tetrahedron Lett. 191 (1977). (c) H.Eichenauer, E. Friedrich, W.Lutz and D.Enders, Angew.Chem., Int.Ed. 17, 206 (1978). (d) D.Enders and H.Eichenauer, ibid. 18, 397 (1979). (e) D.Enders and H.Eichenauer, Chem.Ber. 112, 2933 (1979). (f) D.Enders, H.Eichenauer and R.Pieter, ibid. 112, 3703 (1979). (g) K.G.Davenport, H.Eichenauer, D.Enders, M.Newcomb and D.E.Bergbreiter, J.Am.Chem.Soc. 101, 5654 (1979). (h) D.Enders, Chemtech. 11, 504 (1981), published without corrections.
3. (a) E.J.Corey and D.Enders, Chem.Ber. 111, 1337 (1978). (b) E.J.Corey and D.Enders, ibid. 111, 1362 (1978).
4. D.Seebach, Angew.Chem., Int.Ed. 18, 239 (1979).
5. W.Francke, V.Heemann, B.Gerken, J.A.A.Renwick and J.P.Vité, Naturwiss. 64, 590 (1977).
6. D.Enders, P.Weuster, W.Dahmen and E.Dederichs, unpublished results.
7. (a) J.Meinwald, A.F.Kluge, J.E.Carell and T.Eisner, Proc.natu.Acad.Sci., USA 68, 1467 (1971). (b) M.S.Blum and A.L.Edgar, Insect Biochem. 1, 181 (1971). (c) T.H.Jones, W.E.Conner, A.F.Kluge, T.Eisner and J.Meinwald, Experientia 32, 1234 (1976).
8. D.Enders and U.Baus, unpublished results.
9. D.Enders and R.Pieter, unpublished results.
10. A.I.Meyers and G.Knaus, Tetrahedron Lett. 1333 (1974).
11. (a) D.A.Evans, J.M.Takacs, L.R.McGee, M.D.Ennis, D.J.Mathre and J.Bartroli, Pure Appl.Chem. 53, 1109 (1981). (b) D.A.Evans, J.V.Nelson and T.R.Taber, Top.Stereochem. 13, 1 (1982).
12. (a) C.H.Heathcock, Ref. 1d, p. 55. (b) C.H.Heathcock, Science 214, 395 (1981). (c) C.H.Heathcock, Comprehensive Carbanion Chemistry Vol. 2, Elsevier, Amsterdam, in press.
13. S.Masamune, Organic Synthesis, Today and Tomorrow, p.197, Pergamon Press, Oxford (1981).
14. A.I.Meyers, Ref. 1d, p. 83.
15. D.Enders, U.Baus and V.N.Pathak, unpublished results; boron azaenolates and titanated SAMP/RAMP-hydrazones are most promising.
16. H.J.Bestmann et al., private communication (1981).
17. K.Mori et al., private communication (1981).
18. (a) K.C.Nicolaou, D.P.Papahatjis, D.A.Claremon and R.E.Dolle, III, J.Am. Chem.Soc. 103, 6967 (1981). (b) K.C.Nicolaou, D.A.Claremon, D.P.Papahatjis and R.L.Magolda, ibid. 103, 6969 (1981).
19. S.I.Pennanen, Acta Chem.Scand. B35, 555 (1981).
20. (a) A.N.Kost, R.S.Sagitullin and M.A.Yurovskaja, Chem.Ind. (London) 1496 (1966). (b) E.J.Corey, R.J.McCaully and H.S.Sachdev, J.Am.Chem.Soc. 92, 2476 (1970); E.J.Corey and H.S.Sachdev, ibid. 92, 2488 (1970). (c) S.Kiyooka, K.Takeshima, H.Yamamoto and K.Suzuki, Bull.Chem.Soc.Jpn. 49, 1897 (1976).
21. H.Ahlbrecht, E.O.Düber, D.Enders, H.Eichenauer and P.Weuster, Tetrahedron Lett. 3691 (1978), and literature cited therein.
22. P.Rademacher, H.-U.Pfeffer; D.Enders, H.Eichenauer and P.Weuster, J.Chem. Research (S) 222 (1979), (M) 2501 (1979).
23. D.Enders and H.Schubert, unpublished results.
24. (a) L.M.Jackman and N.M.Szeverenyi, J.Am.Chem.Soc. 99, 4954 (1977). (b) L. M.Jackman and B.C.Lange, Tetrahedron 33, 2737 (1977). (c) L.M.Jackman and B.C.Lange, J.Am.Chem.Soc. 103, 4494 (1981).
25. (a) R.Amstutz, W.B.Schweizer, D.Seebach and J.D.Dunitz, Helv.Chim.Acta 64, 2617 (1981). (b) D.Seebach, R.Amstutz and J.D.Dunitz, ibid. 64, 2622 (1981).
26. J.G.Andrade and P.v.R.Schleyer, unpublished results.
27. J.Jacques, A.Collet and S.H.Wilen, Enantiomers, Racemates and Resolutions, Wiley, New York (1981).
28. D.Enders and W.Mies, unpublished results.
29. D.Enders and H.Lotter, Angew.Chem., Int.Ed. 20, 795 (1981).

30. D.Enders and H.Lotter, unpublished results.
31. G.Stork, R.M.Jacobson and R.Levitz, Tetrahedron Lett. 771 (1979).
32. V.Reutrakul, P.Ratananukul and S.Nimgirawath, Chem.Lett. 71 (1980).
33. D.Enders and H.Lotter, Tetrahedron Lett. 23, 639 (1982) and literature cited therein.
34. H.Ahlbrecht, G.Bonnet, D.Enders and G.Zimmermann, Tetrahedron Lett. 21, 3175 (1980).
35. D.Enders, H.Lotter and R.Pieter; H.Ahlbrecht and G.Zimmermann, unpublished results.

A NEW METHODOLOGY FOR THE GENERATION OF o-QUINODIMETHANES AND RELATED INTERMEDIATES — AN APPROACH TO ASYMMETRIC SYNTHESIS OF POLYCYCLES

Yoshihiko Ito

Department of Synthetic Chemistry, Faculty of Engineering, Kyoto University, Yoshida, Kyoto 606, Japan

Abstract - Stereoselective synthesis of polycycles on the basis of the new generation of α-alkyl-o-quinodimethanes and o-quinone methide N-alkylimines by the fluoride anion induced 1,4-elimination of o-[α-(trimethylsilyl)-alkyl]benzyltrimethylammonium halides and o-[N-(trimethylsilyl)-N-alkyl-amino]benzyltrimethylammonium halides is described. Cycloaddition reactions of α-alkyl-α'-[2-(N,N-dimethylamino)alkoxy]-o-quinodimethanes having an asymmetric center in the α'-alkoxy substituent, which are generated from the 1,4-elimination of 2-[o-[1-(trimethylsilyl)alkyl]phenyl]-3,3-dimethyloxazolidinium salts, may provide a new approach to asymmetric synthesis of polycycles.

INTRODUCTION

Intramolecular Diels-Alder reaction has been widely used as a key step in the stereocontrolled construction of complex frameworks. o-Quinodimethane as an enophile in the Diels-Alder reaction is very reactive, because of the restoration of aromaticity on the cycloaddition, and has been utilized for the synthesis of polycyclic ring systems which are otherwise difficult to prepare. Especially, the successful applications of o-quinodimethane intermediate for the syntheses of steroid and alkaloid structures having aromatic A ring have profoundly aroused the interest of synthetic organic chemists. Some useful methodologies for the generation of o-quinodimethanes have been so far developed (Ref. 1). The generation of o-quinodimethanes by thermal electrocyclic ring-opening of benzocyclobutene precursors has found wide applications in construction of various complex molecules including steroids and alkaloids (Ref. 2 & 3). Elegant preparations of the benzocyclobutene precursors by cobalt catalyzed cotrimerization of acetylene derivatives have made the benzocyclobutene methodology more attractive (Ref. 4). But the generation of o-quinodimethanes by the ring opening of benzocyclobutenes as well as cheletropic desulfurization from 1,3-dihydrobenzo[c]thiophene-2,2-dioxides (Ref. 5) needs relatively higher temperature around 200°C, although it is not a serious drawback.
We found a mild and efficient generation of o-quinodimethane intermediates (3), in which 1,4-conjugative elimination of o-[α-(trimethylsilyl)alkyl]benzyltrimethylammonium halide was triggered by fluoride anion at ambient temperature (Ref. 6). Moreover, the fluoride induced 1,4-elimination was also extended to 2-[N-(trimethylsilyl)-N-alkylamino]benzyltrimethylammonium halides to generate o-quinone methide N-alkylimine intermediates (21) (Ref. 7). Described herein is the synthesis of polycyclic ring systems including steroidal structures on the basis of the new methodology for the generation of 3 and 21.

GENERATION AND CYCLOADDITION OF o-QUINODIMETHANES

1,4-Elimination of o-[α-(trimethylsilyl)alkyl]benzyltrimethylammonium salts
Intermolecular cycloaddition. On treatment of o-(trimethylsilylmethyl)benzyltrimethylammonium halide (2a) in acetonitrile or in methylene chloride with tetrabutylammonium fluoride (TBAF) at room temperature, 1,4-elimination took place instantaneously to generate o-quinodimethane (3a) almost quantitatively with the concomitant formations of trimethylfluorosilane and trimethylamine. A suspension of cesium fluoride in acetonitrile can also be used instead of an acetonitrile solution of TBAF in the generation of o-quinodimethane. o-Quinodimethane (3a) thus generated in situ was dimerized in the absence of dienophile to produce a spiro-dimer (4) in a fairly good yield. As a leaving group in the 1,4-elimination of 2, the trimethylamine leaving group is not essential. o-(Trimethylsilylmethyl)benzyl halide also underwent the fluoride anion-induced 1,4-elimination to generate o-quinodimethane. However, the trimethylamine leaving group is the best choice in the present o-quinodimethane generation, because of the ready synthetic availability of o-(trimethylsilylmethyl)benzyldimethylamine (1a) and its feasible elaboration through the silicon stabilized benzylic carbanion (1, R=Li) to lead a requisite precursor for the generation of α-substituted o-quinodimethane (3b).

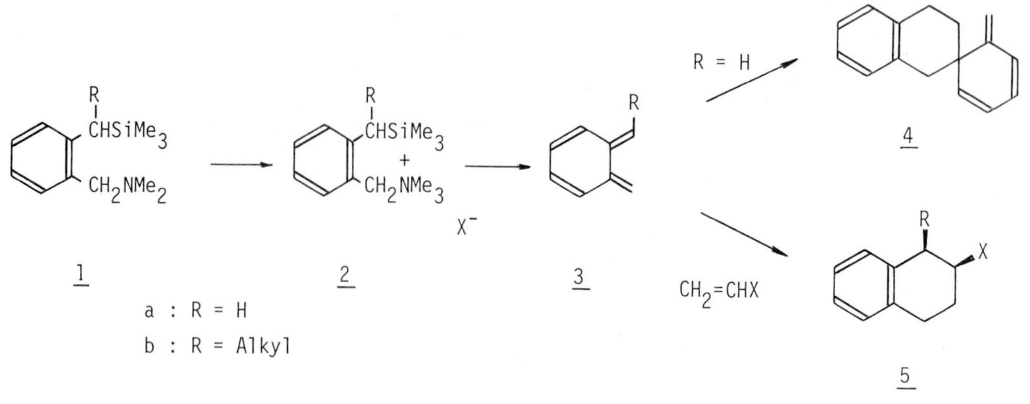

The generation of o-quinodimethane (3) in the presence of electron deficient olefins and acetylenes such as acrylate, acrylonitrile, fumarate and acetylenedicarboxylate afforded Diels-Alder cycloadducts (5) in high yields at ambient temperature. In the reaction of α-alkyl-o-quinodimethane (3b) with dienophiles was produced a mixture of regio- and stereo-isomeric cycloadducts with cis-1,2-disubstituted tetrahydronaphthalene (5b) predominating. The formation of cis-1,2-disubstituted tetrahydronaphthalene (5b) as a major product may be predicted from selective formation of (E)-α-alkyl-o-quinodimethane and its cycloaddition according to the Diels-Alder endo rule, taking into consideration the electronic effect of α-alkyl substituent in 3b.

α,α'-Disubstituted o-quinodimethanes (3c) were also quantitatively generated by the fluoride anion induced 1,4-elimination of the corresponding 1-[o-[1-(trimethylsilyl)alkyl]phenyl]alkyl-trimethylammonium halides (2c). Noteworthy was that (E,E)-α,α'-disubstituted o-quinodimethane (3c) was selectively generated regardless of threo or erythro stereochemistry of the starting materials (2c), as demonstrated in the cycloaddition between α,α'-dimethyl-o-quinodimethane (3c-i) and dimethyl fumarate. On treatment of a 3 : 2 diastereoisomeric mixture of 1-[o-[1-(trimethylsilyl)ethyl]phenyl]ethyltrimethylammonium iodide (2c-i) with cesium fluoride in the presence of dimethyl fumarate, a single stereoisomeric cycloadduct (6) was produced in a quantitative yield, which can be derived only from (E,E)-α,α'-dimethyl-o-quinodimethane (3c-i)

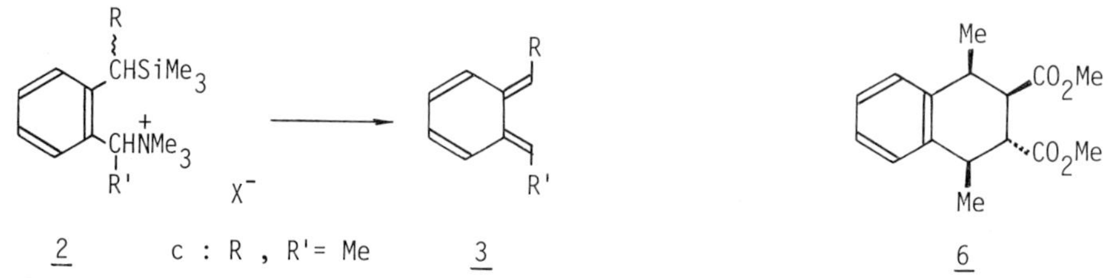

However, the fluoride induced 1,4-elimination of 4-trimethylsilyl-1,2,3,4-tetrahydronaphth-1-yltrimethylammonium iodide (2d), which is only permitted to generate (Z,Z)-α,α'-disubstituted o-quinodimethane (3d), proceeded in the presence of dienophile as well, producing a large amount of dimeric product (7) and a mixture of the expected bicyclic endo- and exo-cycloadducts (5d-endo) and (5d-exo) in low to moderate yields with endo-cycloadduct (5d-endo) predominating.

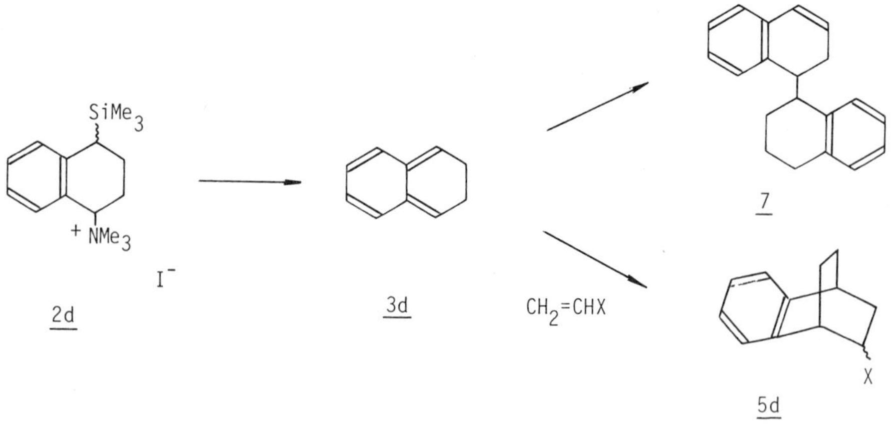

A pyridine analogue of o-quinodimethane (3e) was similarly generated by treating 2-(trimethylsilylmethyl)picol-3-yltrimethylammonium bromide (2e) with cesium fluoride. The cycloaddition reactions of 3e with acrylate and acrylonitrile gave rise to two isomeric cycloadducts (8) with no regioselectivity.

Intramolecular cycloaddition. α-Alkenyl-o-quinodimethanes generated from the corresponding ammonium halides underwent the intramolecular Diels-Alder reaction to give polycycles. When o-(1-trimethylsilyl-6-heptenyl)benzyltrimethylammonium iodide (2f) was added in an acetonitrile solution of TBAF at room temperature, the intramolecular cyclization of the α-(hex-5-enyl)-o-quinodimethane (3f) generated took place to give 22% yield of octahydrophenanthrene along with a dimeric product of α-hexenyl-o-quinodimethane. But the generation and cyclization of 3f in a refluxing acetonitrile produced a 14 : 1 mixture of trans- and cis-octahydrophenanthrene in 77% yield with the dimeric by-product (8%). The same degree of trans stereoselectivity has also been observed in the intramolecular cyclizations via the corresponding benzocyclobutene and 1,3-dihydrobenzo[c]thiophene-2,2-dioxide precursors and explained by assuming an exo transition state for the cyclization (Ref. 4 & 5).
Next, stereoselective syntheses of steroidal strucutures such as estrone methyl ether (12a) and 6β-methylestra-1,3,5(10)-trien-17-one (12b) were constructed on the basis of the selective generation of (E)-α-alkyl-o-quinodimethane 3b and (E,E)-α,α'-dialkyl-o-quinodimethane 3c, which has been described above. The key in the construction of the steroidal structures was a selective generation of the silicon-stabilized carbanions from 2-(trimethylsilyl)methyl-5-methoxybenzyldimethylamine (9a) and α-[o-[(trimethylsilyl)methyl]phenyl]ethyldimethylamine (9b) followed by alkylations with the cyclopentanone moiety to lead to the requisite precursors 10a and 10b. Unlike the formation of the silicon stabilized carbanion from 9b (n-BuLi in THF), the selective lithiation at the benzylic carbon α to the silicon of 9a was achieved by the treatment with n-BuLi in THF in the presence of cosolvent HMPA. The ammonium iodides 11a and 11b obtained by quaternization of 10a and 10b with methyl iodide were reacted with cesium fluoride in a refluxing acetonitrile to give estrone methyl ether (12a) and 6β-methylestra-1,3,5(10)-trien-17-one (12b) in 80-90% overall yield based upon the starting 9.

1,4-Elimination of 2-[o-[1-(trimethylsilyl)alkyl]phenyl]oxazolidinium salts

A generation of o-quinodimethanes with electron donating heteroatom substituents at the α-position, which may be expected to exert higher reactivity and regioselectivity in the Diels-Alder cycloaddition, has been scarcely known (Ref. 8). It was now found that the fluoride induced 1,4-elimination technique was extended to 2-[o-[1-(trimethylsilyl)alkyl]phenyl]oxazolidinium salts (14) to generate α-alkyl-α'-alkoxy-o-quinodimethane intermediates (15) (Ref. 9).

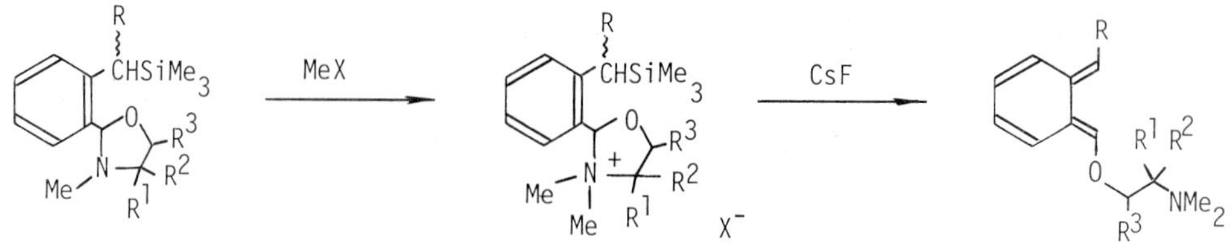

A regio- and stereo-selective intermolecular cycloaddition of 15 is demonstrated by the trapping of 15 with methyl acrylate. When 2-[o-(trimethylsilylmethyl)phenyl]-3,3,5-trimethyl-oxazolidinium iodide (14a) was reacted with CsF in acetonitrile in the presence of methyl acrylate at room temperature, a 1 : 1 diastereoisomeric mixture of cis-1-[2-(N,N-dimethylamino)-1-methylethoxy]-2-(methoxycarbonyl)-1,2,3,4-tetrahydronaphthalene (16a) was produced in ca. 90% yield, uncontaminated with its regioisomeric cycloadducts.

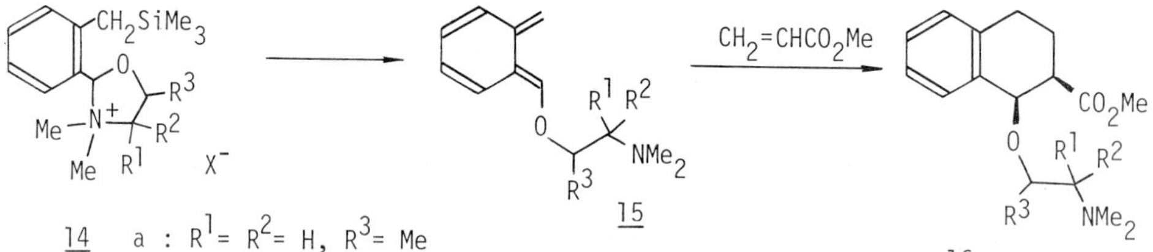

14   a : $R^1 = R^2 = H$, $R^3 = Me$
     b : $R^1, R^2 = H$, (S)-Me, $R^3 =$ (S)-Ph

Intramolecular cyclization of 2-[o-[1-(trimethylsilyl)alkenyl]phenyl]oxazolidinium salts (14c)-(14h) also proceeded even at room temperature to give polycycles 17 in satisfactory yields. It is interesting that the generation of α-(hex-5-enyl)-o-quinodimethane from o-[1-(trimethylsilyl)hept-6-enyl]benzyltrimethylammonium salt at the same temperature afforded octahydrophenanthrene in only low yield with the spiro-dimer predominating (Ref. 6).

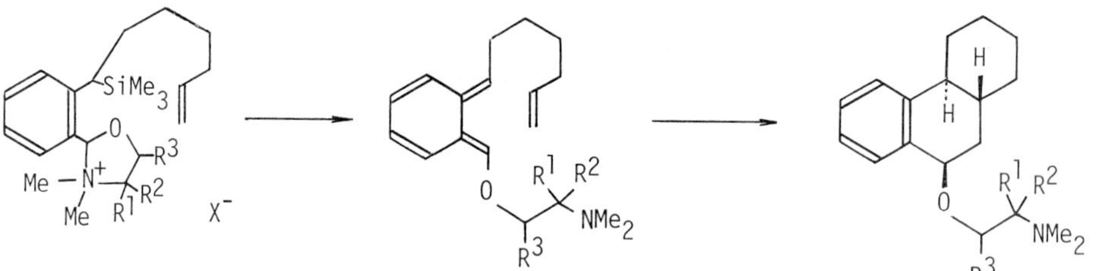

For instance, when 2-[o-[1-(trimethylsilyl)hept-6-enyl]phenyl]-3,3,5-trimethyloxazolidinium iodide (14c) was stirred with a suspension of CsF (2-3 fold excess) in acetonitrile at room temperature overnight, 8 : 9 trans-6-[2-(N,N-dimethylamino)-1-methylethoxy]octahydrophenanthrene (17c) was produced as a 1 : 1 diastereoisomeric mixture (80%). Removal of the 2-(N,N-dimethylamino)-1-methylethoxy substituent in 17c was achieved by hydrogenolysis with 10% Pd/C in acetic acid containing 2% of aqueous 70% $HClO_4$ to give trans-octahydrophenanthrene (18) in 75% overall yield from 14c, which was contaminated by a few percent of the C(9) epimer.

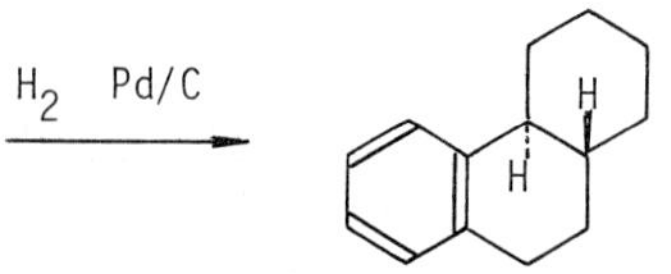

Asymmetric synthesis of polycycles. The formation of 1 : 1 mixtures of two diastereoisomers of cis-cycloadduct 16a and of trans-cycloadduct 17c mentioned above may be rationalized by assuming inter- and intra-molecular Diels-Alder reactions with o-quinodimethane intermediates 15a and 15c through the endo transition state and the exo transition state, respectively, in which dienophiles approach equally either of the two enantiotopic faces of the reacting diene moiety of 15. If the assumption is correct, the proper choice of 2-(N,N-dimethylamino)-alkoxy substituents on 15 might lead to the asymmetric formation of polycycles. It was actually found that some 2-[o-[1-(trimethylsilyl)hept-6-enyl]phenyl]-3,3-dimethyl-5-phenyl-oxazolidinium salts (14d)-(14f) were cyclized enantioselectively via the corresponding 15 to give polycycles with asymmetric induction (Table 1).
The intramolecular cyclization of 2-[o-[1-(trimethylsilyl)hept-6-enyl]-3,3-dimethyl-4(R)-methyl-5(R)-phenyloxazolidinium triflate (14f) at 0°C produced 6-[2(R)-(N,N-dimethylamino)-1(R)-phenylpropoxy]-trans-octahydrophenanthrene (17f) as a mixture of two diastereoisomers. The NMR singlets at $\delta 2.12$ and 2.25 ppm, which are ascribed to methyl protons on nitrogen of 17f, in a 1 : 3 ratio indicate the degree of asymmetric induction. The removal of the 2-(N,N-dimethylamino)alkoxy substituent from 17f by hydrogenolysis on Pd/C gave trans-octahydrophenanthrene (18) in 73% overall yield, which showed $[\alpha]_D^{19}$ +46.6° (c 1.11, C_6H_{12}). On the other hand, two diastereoisomers of 17f were separated by preparative tlc on silica gel, and the major diastereoisomer was treated with H_2 on Pd/C to give and optically pure trans-octahydrophenanthrene (18) with $[\alpha]_D^{19}$ +92.9° (c, 1.06, C_6H_{12}). The rotation also indicated that the cycloadduct 17f consisted of a 75 : 25 ratio of diastereoisomers.

TABLE 1. Intramolecular cyclizations of oxazolidinium salts (14c)-(14h)

| Oxazolidinium salt | | | | Reaction | Octahydrophenanthrene(18) | |
R^1,	R^2 (C4)	R^3 (C5)	X	Temp(°C)	$[\alpha]_D^{19}$ (C_6H_{12})	%ee[a]
H,	H	Me	I (2c)	20		
H,	H	(R)-Ph	I (2d)	20	+25.8	28
H,	H	(R)-Ph	I (2d)	0	+31.6	34
H,	(S)-Me	(R)-Ph	I (2e)	20	+36.3	39
H,	(S)-Me	(R)-Ph	OTf (2e)	20	+33.4	36
H,	(R)-Me	(R)-Ph	OTf (2f)	20	+40.5	44
H,	(R)-Me	(R)-Ph	OTf (2f)	0	+46.6	50
H,	(R)-Ph	H	I (2g)	20	-0.8	
H,	(S)-PhCH$_2$	H	I (2h)	20	-1.7	

a) %ee was calculated on the basis of the maximum rotation $[\alpha]_D^{19}$ 92.9° (c 1.06, C_6H_{12})

As seen in Table 1, a phenyl substituent at the C-5 on the oxazolidinium ring of 14 (14d, 14e and 14f) remarkably increased the asymmetric induction in the intramolecular Diels-Alder cycloaddition via 15. Phenyl and benzyl substituents at the C-4 on the oxazolidinium ring of 14 (14g and 14h) have no significant effect on the enantioselection in the intramolecular cycloaddition. Similarly, the intermolecular cycloaddition of 2-[o-(trimethylsilylmethyl)-phenyl]-3,3-dimethyl-4(S)-methyl-5(R)-phenyloxazolidinium triflate (14b) with methyl acrylate afforded a 2 : 1 diastereoisomeric mixture of cis-1-[2(S)-(N,N-dimethylamino)-1(R)-phenyl-propoxy]-2-(methoxycarbonyl)-1,2,3,4-tetrahydronaphthalene (16b) in 92% yield. The major diastereoisomer, which was separated by preparative tlc on silica gel, was treated with H_2 on Pd/C followed by hydrolysis to furnish 1,2,3,4-tetrahydronaphthalene-2-carboxylic acid (19) with the R configuration as determined from the known rotation and configuration of 19 (Ref. 10). Consequently, the major diastereoisomer of 16b was assigned to the 1(R), 2(R) configuration depicted in 16b-i. Although the absolute configuration of 17 or/and 18 produced has not been determined, the enantioselection in the Diels-Alder cycloaddition via o-quino-dimethane intermediates may be accounted for in accordance with the Trost's observation that π-stacking interaction may serve as steric steering group to direct the incoming dienophile to one of the two enantiotopic faces of the diene (Ref. 11). Two conformations A and B depicted below (Fig. 1) may be envisioned for the π-stacking interaction in the present Diels-Alder cycloaddition. Inspection of molecular models reveals that the former encounters a severe nonbonded interaction between the aromatic hydrogen and the benzylic hydrogen adjacent to the ether oxygen as indicated in A and the latter would be favored.

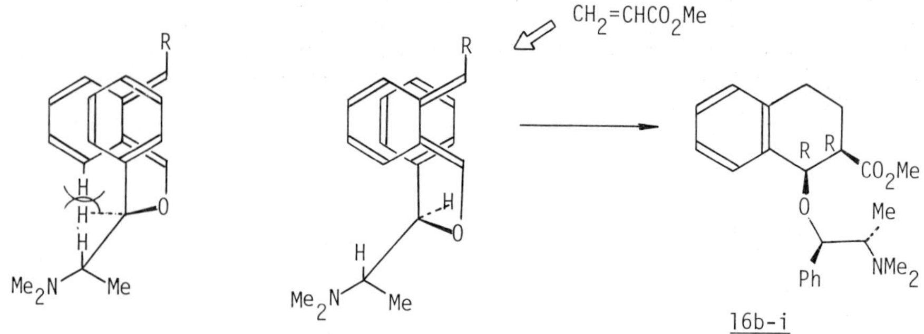

Fig. 1   A   B   16b-i

GENERATION AND CYCLOADDITION OF o-QUINONE METHIDE IMINES

Utilization of o-quinone methide imine intermediates (21) in organic synthesis has been limited, because of the lack of an efficient methodology for the generation of 21. The extension of the fluoride anion induced 1,4-elimination to o-[N-(trimethylsilyl)-N-alkylamino]-benzyltrimethylammonium halide (20) also provided a new and versatile generation of o-quinone methide N-alkylimine intermediate (21) (Ref. 7).
When o-[N-(trimethylsilyl)-N-methylamino]benzyltrimethylammonium iodide (20a) was treated with CsF or tetrabutylammonium fluoride in acetonitrile at room temperature, spiro-tetrahydroquinoline derivative (22a) was produced, which may be derived from [4+2]cyclodimerization of the o-quinone methide imine (21a).

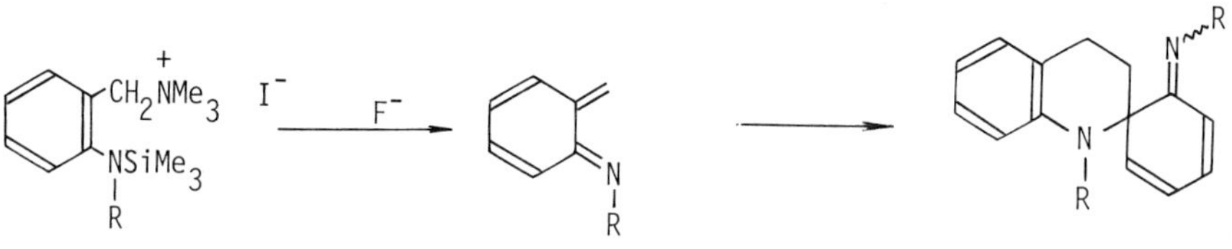

20   a : R = Me   21   22

However, attempts to trap the o-quinone methide imine (21a) with dienophiles such as acrylate, fumarate, acetylenedicarboxylate and N-phenylmaleimide all failed, resulting in the formation of 22a as the sole isolable product. However, intramolecular Diels-Alder reaction of o-quinone methide N-alkenylimine intermediates (24) provided a useful synthetic method for construction of nitrogen-containing polycycles. Nitrogen-containing heterocycles, benzo[e]indolizidine (25a) and benzo[c]quinolizidine (25b) were prepared in moderate yields by the fluoride anion induced 1,4-eliminations of o-[N-(trimethylsilyl)-N-pent-4-enylamino]-benzyltrimethylammonium bromide (23a) and o-[N-(trimethylsilyl)-N-hex-5-enylamino]benzyl-trimethylammonium bromide (23b).

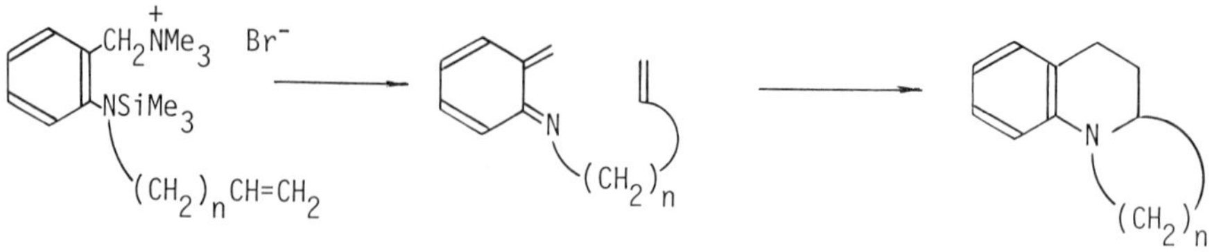

23   a : n = 3   24   25
     b : n = 4

Finally, the synthesis of nitrogen-containing polycycles based on the new o-quinone methide imine generation has been extended to the stereoselective synthesis of 9-aza-estra-1,3,5(10)-trien-17-one (29). The requisite precursor (28) for construction of 29 was prepared by the reaction of o-aminobenzyldimethylamine (26) with the cyclopentanone moiety followed by N-silylation and quaternization. On treatment of 28 with CsF in a refluxing acetonitrile, 8,14-anti-13,14-trans-9-aza-estra-1,3,5(10)-trien-17-one (29) was selectively produced in 60% yield based on 27.

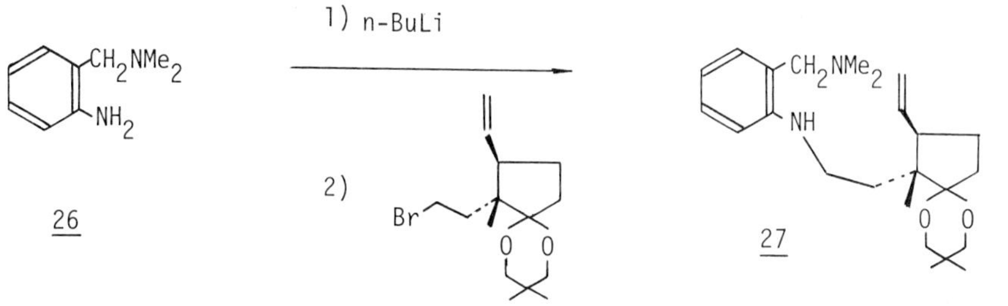

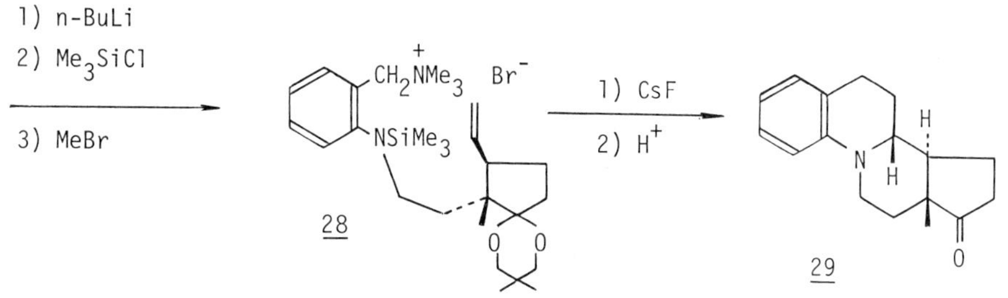

REFERENCES

1. W. Oppolzer, Synthesis 793 (1978).
2. (a) W. Oppolzer, K. Bättig and M. Petrzilka, Helv. Chim. Acta. 61, 1945 (1978).
   (b) T. Kametani, H. Nemoto, H. Ishikawa, K. Shiroyama, H. Matsumoto and K. Fukumoto, J. Am. Chem. Soc. 99, 3461 (1977).
3. (a) W. Oppolzer, J. Am. Chem. Soc. 93, 3833 (1971). (b) T. Kametani, M. Takemura, M. Ogasawara and K. Fukumoto, J. Heterocycl. Chem. 11, 179 (1974).
4. R.L. Funk and K.P.C. Vollhardt, J. Am. Chem. Soc. 102, 5245 and 5253 (1980).
5. K.C. Nicolaou, W.E. Barnette and P. Ma, J. Org. Chem. 45, 1463 (1980).
6. (a) Y. Ito, M. Nakatsuka and T. Saegusa, J. Am. Chem. Soc. 102, 863 (1980). (b) Y. Ito, M. Nakatsuka and T. Saegusa, J. Am. Chem. Soc. 103, 476 (1981).
7. Y. Ito, S. Miyata, M. Nakatsuka and T. Saegusa, J. Am. Chem. Soc. 103, 5250 (1981).
8. P.G. Sammes, Tetrahedron, 32, 405 (1976).
9. Y. Ito, Y. Amino, M. Nakatsuka and T. Saegusa, Unpublished results.
10. A. Schoofs, J.P. Guette and A. Horeau, Bull. Soc. Chim. Fr. 1215 (1976).
11. B.N. Trost, D. O'krongly and J.L. Belletire, J. Am. Chem. Soc. 102, 7595 (1980).

ASYMMETRIC SYNTHESIS USING ENANTIOMERICALLY PURE SULFOXIDES

Gary H. Posner*, Kyo Miura, John P. Mallamo, Martin Hulce and Timothy P. Kogan

Department of Chemistry, The Johns Hopkins University, Baltimore, Maryland 21218, USA

Abstract - High asymmetric induction has been observed during conjugate addition of various organometallic reagents to some enantiomerically pure 2-arylsulfinyl-2-cycloalkenones. These stereocontrolled carbon-carbon bond forming reactions are discussed in terms of their mechanistic and synthetic aspects, with emphasis on (1) using the same enantiomerically pure sulfoxide to produce either an (R)- or an (S)-3-substituted cycloalkanone, and (2) rationalization of the results in terms of either chelated or non-chelated ketosulfoxide reactants.

INTRODUCTION

Within the past two years, we have demonstrated that enantiomerically pure α-carbonyl α,β-ethylenic sulfoxides can be prepared efficiently and can undergo organometallic conjugate addition leading to 3-substituted carbonyl compounds in high enantiomeric purity (Ref. 1). The major significance of this work is its application to asymmetric synthesis of valuable 3-substituted cycloalkanones (Ref. 2,3). We have now established that these stereocontrolled carbon-carbon bond forming reactions are of general scope and that the same enantiomerically pure α-sulfinylcycloalkenone can be used to prepare either an (R)- or an (S)-3-substituted cycloalkanone. These results are rationalized in terms of chelated or non-chelated reactants.

RESULTS AND DISCUSSION

We have prepared several enantiomerically pure 2-tolylsulfinyl-2-cycloalkenones on 10-gram scale as shown in eq. 1 (Ref. 4). Each of these sulfinyl cyclopentenones and cyclohexenones is a crystalline solid which is stable for at least several months, and (—)-menthol, the original source of chirality for the entire sequence, is recovered and can be re-used if very large (e.g. kilogram) quantities of these sulfinyl cycloalkenones are needed.

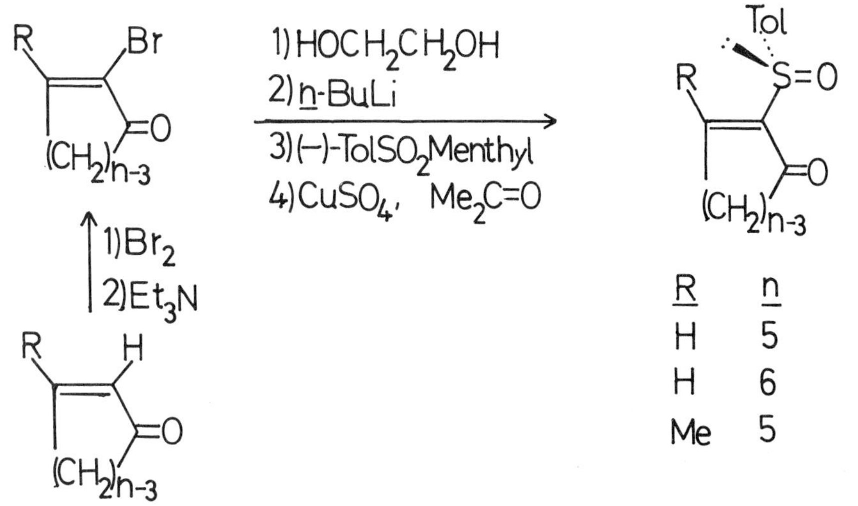

R	n
H	5
H	6
Me	5

As expected based on chelate model 1 in which a metal ion (M) complexes with the bidentate α-sulfinylcycloalkenone and therefore locks the chiral sulfinyl group into the conformation shown, organometallic conjugate addition occurred from that diastereotopic face of the enone opposite to the bulky p-tolyl group; various (R)-3-substituted cycloalkanones were formed with very high enantiomeric excess.

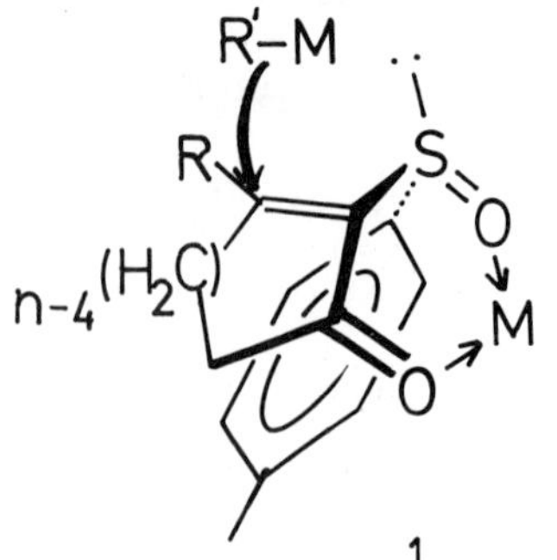

For example, enantiomerically pure α-sulfinylcyclopentenone 2 underwent complex formation with zinc dibromide and then asymmetric conjugate addition with various Grignard reagents; after reductive cleavage of the chiral sulfinyl group, diverse (R)-(+)-3-substituted cyclopentanones were produced in excellent enantiomeric purity (eq. 2). Virtually complete asymmetric induction was observed in the absence of zinc dibromide during ethyl triisopropoxytitanium conjugate addition and during methylmagnesium chloride conjugate addition; methylmagnesium chloride is known to form relatively strong complexes with Lewis bases.

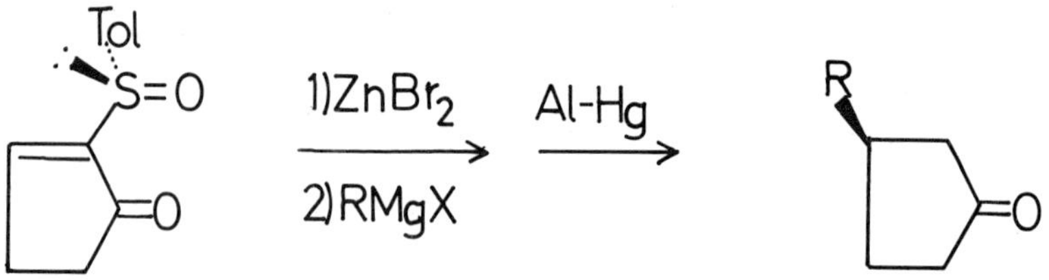

X	R	% Yield	% e.e.
I	Me	89	87
Cl	Et	84	80
Br	$CH_2=CH$	75	>98
Cl	Ph	70	92
Br	1-Naph	90	>98

Likewise, enantiomerically pure α-sulfinylcyclohexenone 3 underwent zinc-mediated asymmetric Grignard conjugate addition to afford (R)-(+)-3-substituted cyclohexanones in good enantiomeric purity (eq. 3).

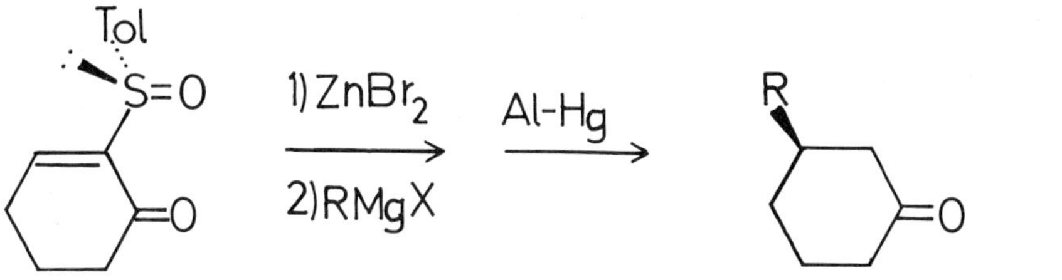

Tol
S=O
O
1) ZnBr2
2) RMgX
Al-Hg
R
O

(3)

3

X	R	% Yield	% e.e.
Br	Me	95	62
Cl	i-Pr	62	55

Even enantiomerically pure α-sulfinyl-β-methylcyclopentone 4 underwent asymmetric conjugate addition of ditolylcopperlithium to afford (R)-(+)-3-methyl-3-tolylcyclopentanone in high enantiomeric purity with absolute stereochemistry in accord with that predicted by chelate model 1 (eq. 4). This highly stereocontrolled synthesis of a quaternary carbon center is particularly noteworthy because so many complex organic synthons and natural products contain such a structural element (Ref. 5).

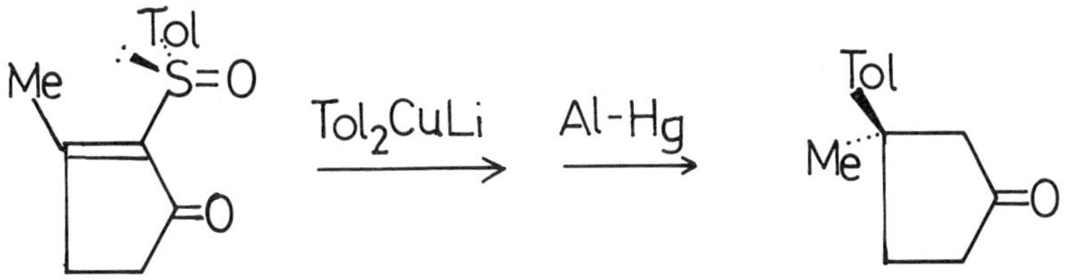

Tol
Me
S=O
O
Tol2CuLi
Al-Hg
Tol
Me
O

(4)

4

51% Yield
87% e.e.

NMR spectroscopic evidence, including NOE results and ^{1}H and ^{13}C chemical shifts which are changed upon complexation of these α-sulfinylcyclopentenones with zinc dibromide, support the intermediacy of chelate 1. In contrast, the preferred conformation of such α-sulfinylcycloalkenones, in the absence of complexing metal ions, appears to be such that the dipoles of the sulfinyl and carbonyl groups are oriented in opposite directions, as in non-chelate model 1' (Ref. 6).

O=S
Tol
O
(CH2)n-3

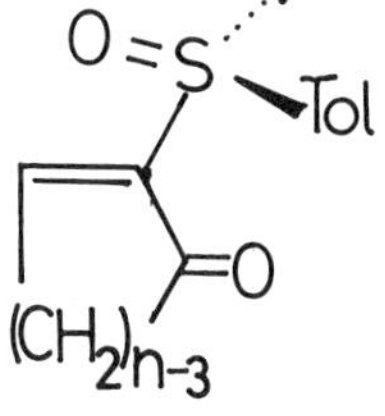

1'

If nucleophilic conjugate addition were to occur to non-chelate conformer 1' with its diastereotopic β-face blocked by the tolyl group, then an (S)-(−)-3-substituted cycloalkanone would be expected. Indeed, non-complexing diorganomagnesium reagents added predominantly in a conjugate and stereocontrolled fashion as shown in eqs. 5 and 6.

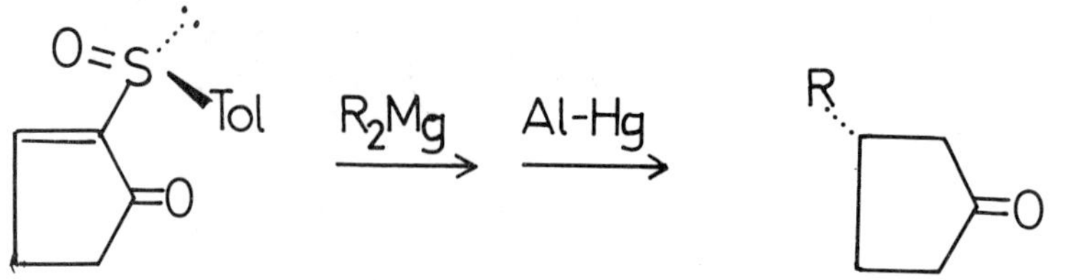

(5)

R	%Yield	% e.e.
Me	60	91
Et	88	81
Ph	72	>98

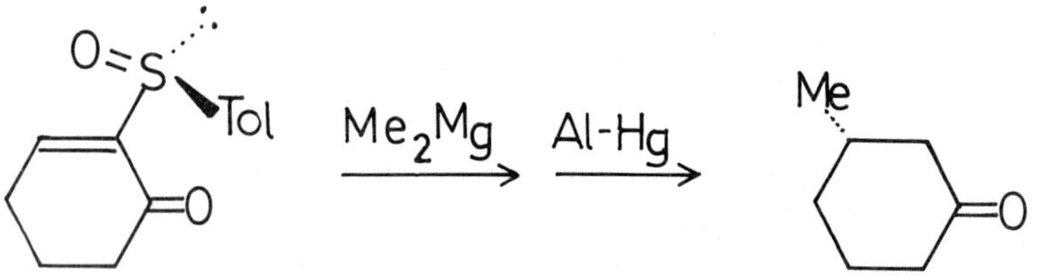

(6)

67% Yield
79% e.e.

Equations 2-6 clearly indicate that the same enantiomerically pure α-sulfinylcycloalkenone can be used in a rational way to prepare either enantiomer of an (R)-(+)- or an (S)-(−)-3-substituted cycloalkanone; simple variation of reaction conditions allows a chelate or a non-chelate conformer to react in a highly enantiocontrolled fashion. This new asymmetric synthesis of 3-substituted cycloalkanones of high enantiomeric purity should find many applications in synthetic organic chemistry.

Acknowledgement - We gratefully acknowledge financial support from the National Science Foundation (CHE 79-15161), from the Donors of the Petroleum Research Fund, administered by the American Chemical Society, from G. D. Searle and Co., from Merck, Sharp and Dohme, and in the form of a SERC (NATO) Science Fellowship to Timothy P. Kogan.

REFERENCES

1. G.H. Posner, J.P. Mallamo and K. Miura, J. Amer. Chem. Soc., 103, (1981).
2. G.H. Posner, M. Hulce, J.P. Mallamo, S.A. Drexler and J. Clardy, J. Org. Chem., 46, 5244 (1981).
3. G.H. Posner, J.P. Mallamo, M. Hulce and L.L. Frye, J. Amer. Chem. Soc., 104, 000 (1982).
4. M. Hulce, L.L. Frye, T.P. Kogan and G.H. Posner, Org. Syntheses, manuscript in preparation.
5. S.F. Martin, Tetrahedron, 36, 419 (1980).
6. B.M. Trost, T.N. Salzmann and K. Hiroi, J. Amer. Chem. Soc., 98, 4887 (1976).

STEREOCONTROL VIA CYCLIZATION REACTIONS

Paul A. Bartlett

Department of Chemistry, University of California, Berkeley, California 94720, USA

Abstract - Electrophilic addition to chiral alkenes generates two new stereorelationships: one arises from the *anti* specificity of the addition reaction itself, and the other is the relationship between the new chiral centers and the ones present in the alkene substrate. A direct way to control this latter relationship is to couple the electrophilic addition reaction with a cyclization process. Such a strategy is useful for controlling both acyclic and cyclic stereorelationships in synthesis. A number of specific cyclization reactions will be discussed, as well as their application to the construction of precursors and subunits of a variety of macrolide and polyether natural products.

INTRODUCTION

An important element in research directed toward the stereocontrolled synthesis of acyclic (1) and saturated heterocyclic compounds has been the development of reactions which proceed with relative asymmetric induction. Such processes allow the introduction of remote chiral centers without the need for resolution of chiral intermediates or for the discovery of reagents for absolute asymmetric induction. For some time we have studied reactions of the general form depicted below, for the purposes of both acyclic and heterocyclic stereocontrol.

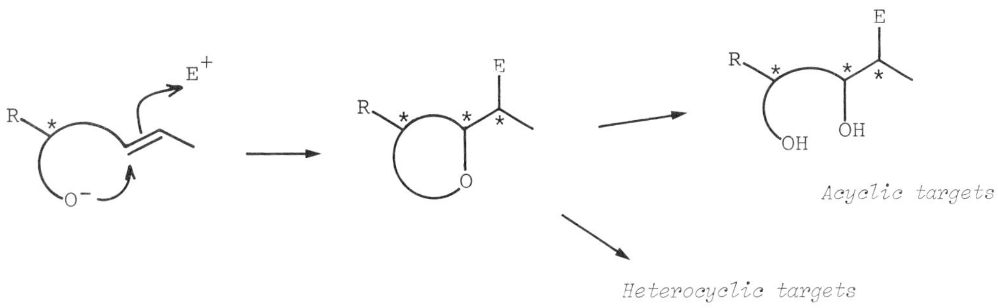

Acyclic targets

Heterocyclic targets

Essentially any olefin electrophilic addition reaction can proceed with intramolecular participation of an appropriately positioned nucleophile (2, 3). The important question, both from the point of view of mechanistic understanding and synthetic development, is whether the reaction can be performed in such a way as to allow chiral centers in the substrate to control the relative configuration of new chiral centers generated during the course of the reaction.

IODOLACTONIZATION

The iodolactonization reaction has been applied to a wide variety of cyclic and acyclic unsaturated carboxylic acids in the eighty or so years since it was first discovered (4). The generally accepted mechanism of the reaction involves initial formation of a cyclic iodonium intermediate which undergoes intramolecular attack by the carboxylate group. The stereochemical features of iodolactonization have long been known to involve an *anti* addition process. Although attack of iodine is felt to precede attack of the carboxyl group, in many cyclic systems it is clear that the carboxyl group actually directs the stereochemical outcome (4). At the time we began our work, there had been no reports of stereocontrol in the iodolactonization of acyclic unsaturated acids.

Under the conditions traditionally employed for the iodolactonization reaction (KI_3, aq. $NaHCO_3$, ether), we observed products of kinetic control with the simple substrates 1a and 1b (5). Even under relatively vigorous conditions (refluxing KI/acetonitrile/18-crown-6, for example), we could see no significant isomerization of the *cis*-products to the more stable *trans*-lactones. With the disubstituted olefinic acid 4 on the other hand, iodolactonization under

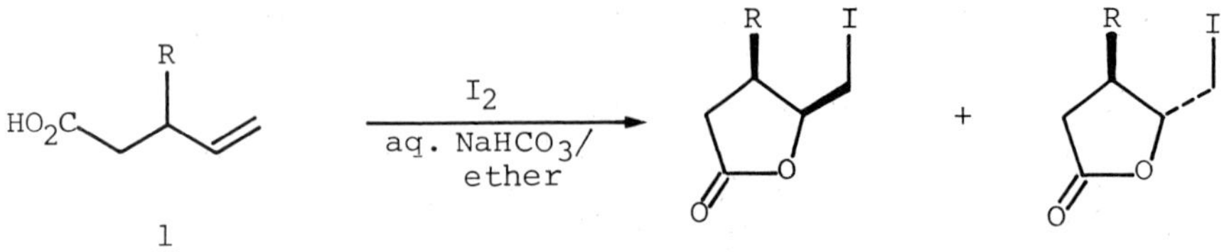

a: R = Me   Ratio 2:3 = 3:1
b: R = Ph   = 2:1

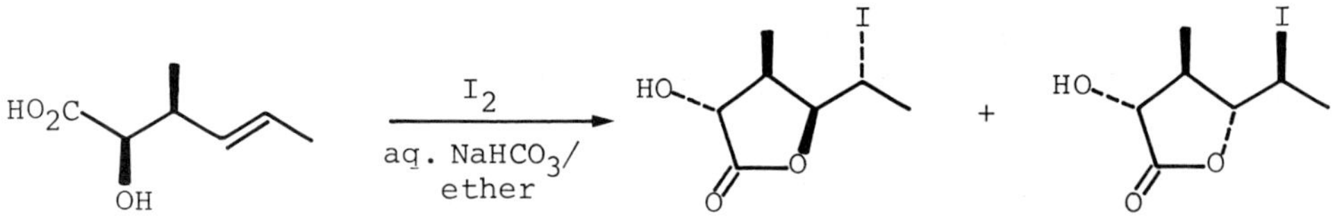

Ratio *cis*:*trans* = 1:3

the traditional alkaline conditions led initially to a mixture of all four possible regio- and stereoisomers, which equilibrated in the reaction mixture over a period of hours to a 3:1 mixture of 3,4-*trans* to 3,4-*cis* γ-lactones (6).

The first indication that the iodolactonization reaction could be controlled in a predictable fashion was obtained by cyclizing the methyl esters. In this case, the intermediate oxocarbonium ion equilibrates with acyclic starting material before demethylation takes place, and the thermodynamic product is favored (see below).

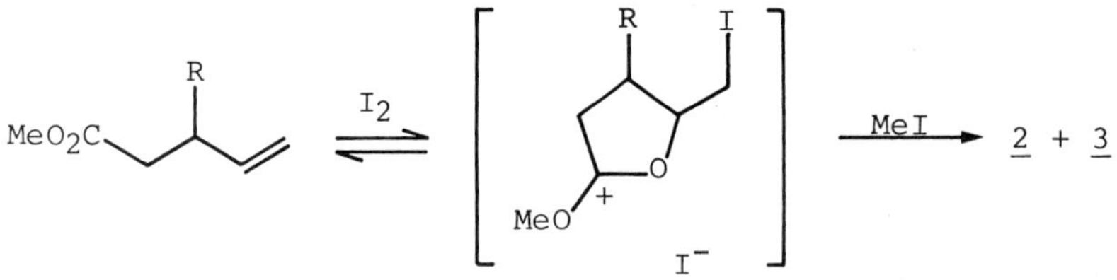

Ratio 2:3
R = Me   1:10
R = Ph   < 1:10

The most generally useful method for obtaining high stereocontrol is to perform the iodolactonization reaction on the free carboxylic acid with iodine in acetonitrile in the absence of base. The *cis* and *trans* products can be observed to equilibrate under these conditions, via the protonated form, and very high stereoselectivity can be obtained. The examples outlined in Table 1 (following page) illustrate that stereocontrol can be achieved predictably in both five-membered ring (*trans*-1,2) and six-membered ring lactones (*trans*-1,2, and *cis*-1,3)(6).

In a recent application, the iodolactonization reaction was used to introduce the γ-hydroxy group stereoselectively in a synthesis of the lactone derived from the unusual amino acid γ,δ-dihydroxyisoleucine found as a constituent of the Amanita mushroom toxins (7).

TABLE 1. Stereochemistry of iodolactonization

Entry	Reaction	Conditions	Ratio	Yield
1	HO$_2$C-CH$_2$-CHR-CH=CH$_2$ $\xrightarrow{\text{I}_2, \text{ acetonitrile}, 0°C}$ **2** + **3**	R = Me R = Ph	2:3 <1:10 2:3 1:20	84% 98%
2	HO$_2$C-(CH$_2$)$_2$-CH(Me)-CH=CH$_2$ → cis-lactone + trans-lactone	kinetic[a] thermodynamic[b]	1:2.3 10:1	83% 77%
3	HO$_2$C-CH$_2$-CH(Me)-CH$_2$-CH=CH$_2$ → cis + trans pyranones	kinetic[a] thermodynamic[b]	3:1 6:1	97% 81%
4	HO$_2$C-CH(Me)-(CH$_2$)$_2$-CH=CH$_2$ → two diastereomeric lactones	kinetic[a] thermodynamic[b]	1.8:1 1.1:1	78% 68%
5	HO$_2$C-CH(Me)-CH$_2$-CH(Me)-CH=CH$_2$ → two diastereomeric lactones	kinetic[c] thermodynamic[b]	3.5:1 20:1	89%

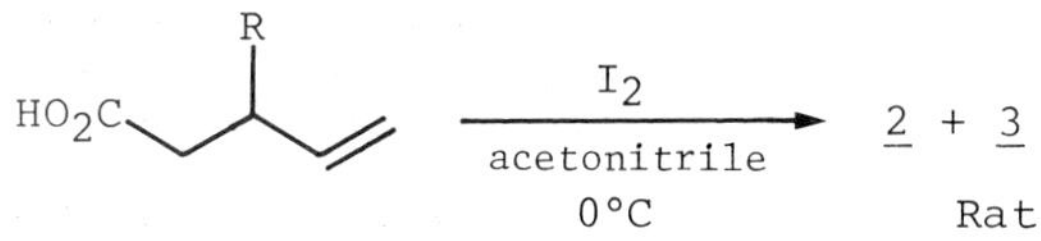

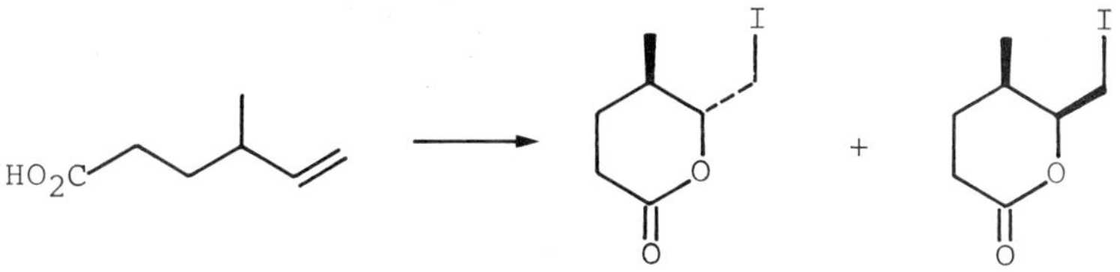

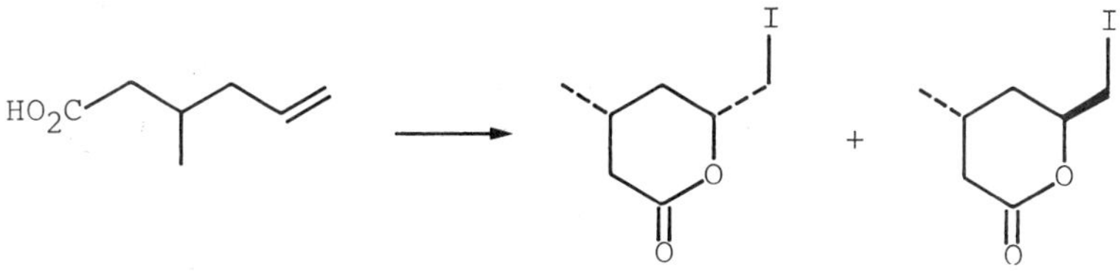

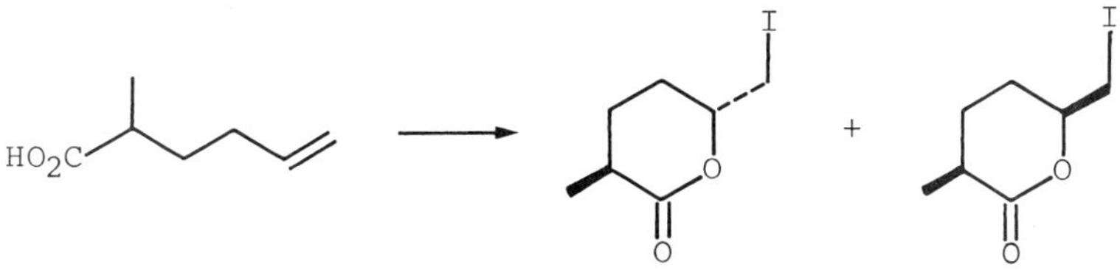

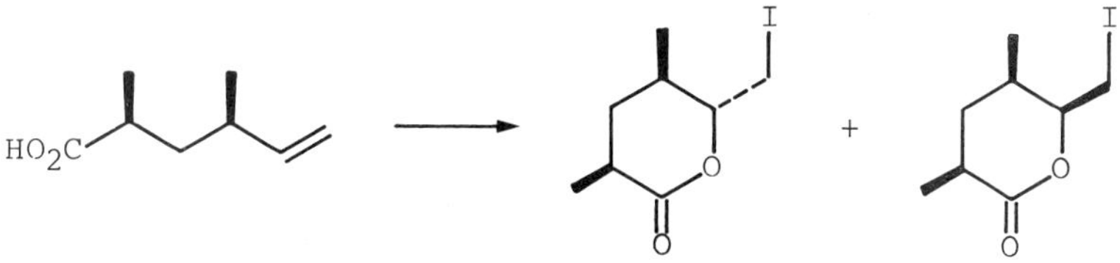

[a] I$_2$, NaHCO$_3$, MeCN, 0 °C
[b] I$_2$, MeCN, 0 °C
[c] N-iodosuccinimide, chloroform, 21 °C

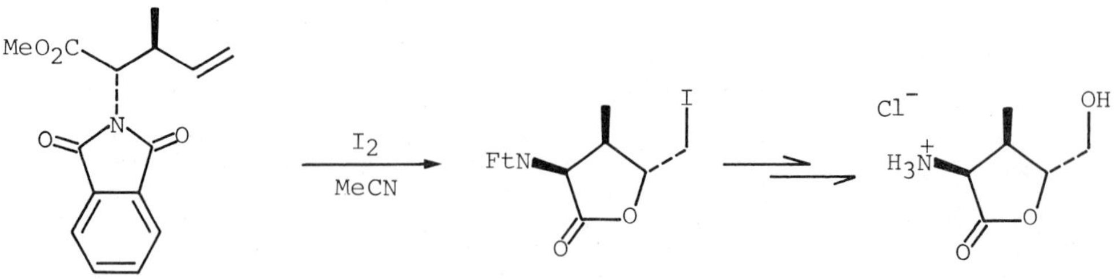

Iodolactones are readily transformed into epoxy esters, completing a two-step sequence (depicted below) for the epoxidation of acyclic olefinic acids in a stereoselective manner.

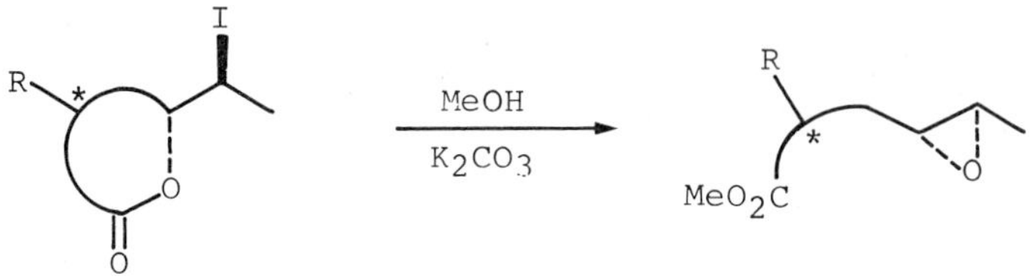

These epoxy esters in turn are useful as building blocks in the synthesis of complex acyclic systems having alternating methyl and oxygen substituents, as illustrated in Schemes 1 and 2 which outline approaches to the stereochemically complex portions of tirandamycin and rifamycin.

SCHEME 1.

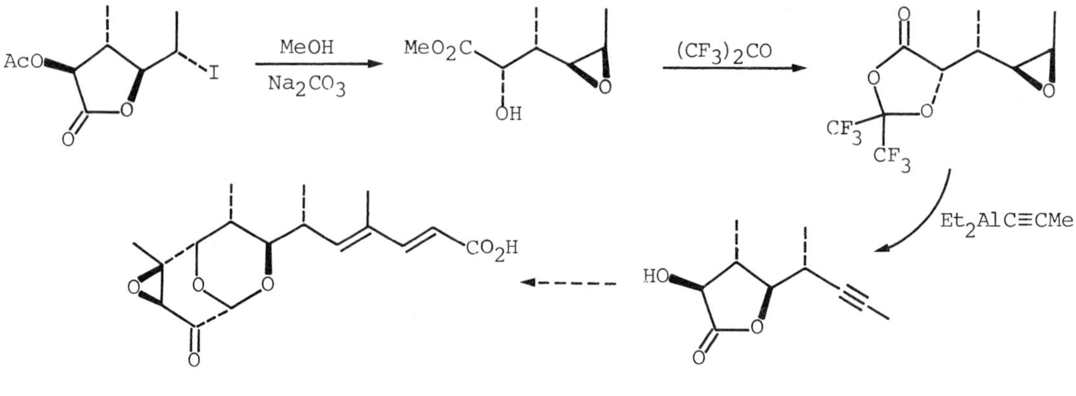

SCHEME 2.

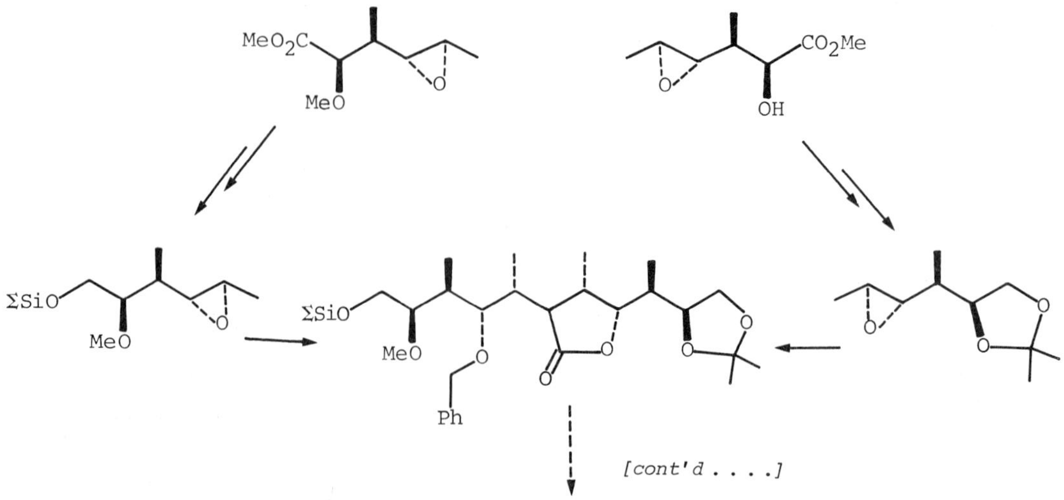

[cont'd . . . .]

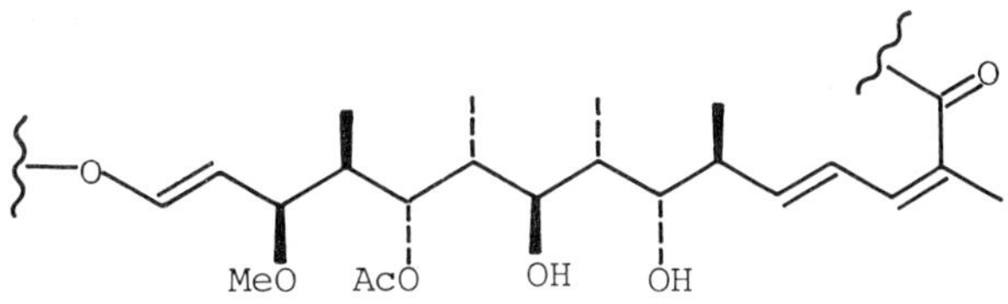

We have since employed the iodolactone from R*S*-2,4-dimethyl-5-hexenoic acid (Table 1, entry 5) in a highly stereoselective synthesis of α-multistriatin, one of the components of the aggregation pheromone of the European elm bark beetle (8).

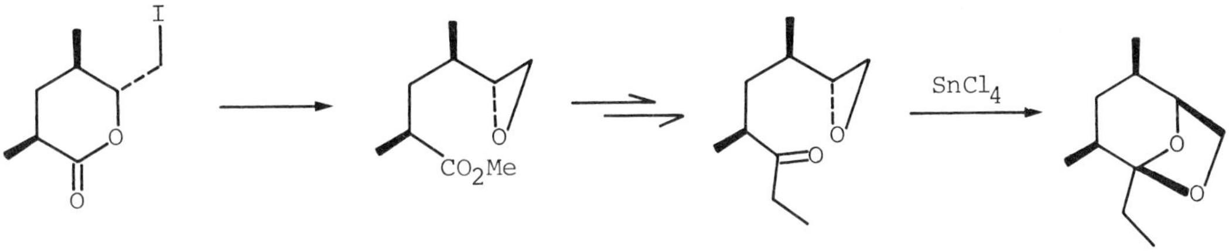

A related unsaturated acid is being used to evaluate other electrophilic cyclization reactions, and to complete the stereochemical assignment of serricornin, the sex pheromone of the cigarette beetle (9,10).

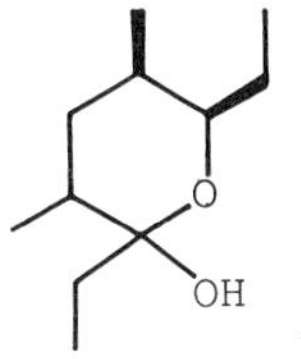

α-Multistriatin

Serricornin

MERCURICYCLIZATION

Mercurilactonization of acyclic unsaturated acids is not as stereoselective as iodolactonization (10,11). For example, 3-phenyl-4-pentenoic acid gives only a 6:1 ratio of *trans/cis* isomers on treatment with mercuric acetate in methanol, whereas iodolactonization of this material gives up to a 20:1 ratio (Table 1). Furthermore, under a variety of conditions for reduction of the mercuri substituent, a significant amount of reductive elimination to give back starting material is encountered.

Nonetheless, mercuricyclization of the unsaturated diacid 5 appeared to be an attractive route to the Prelog-Djerassi lactone (12). Such a process clearly would provide the all-equatorial δ-lactone, and possibly allow the side chain stereochemistry to be controlled as well. As it turned out, the diacid 5 could not be induced to cyclize in the presence of mercuric ion. However, the

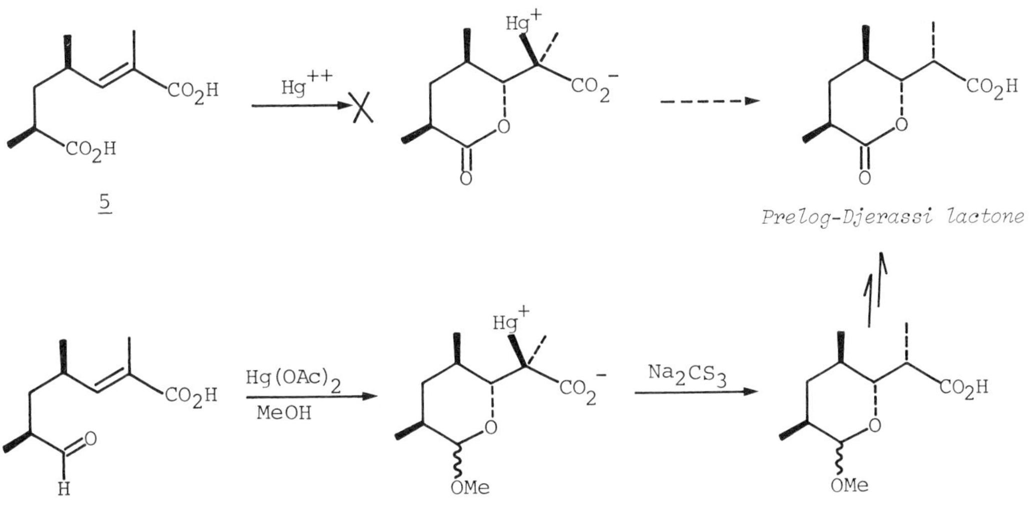

Prelog-Djerassi lactone

(7:2 mixture at side chain)

corresponding aldehyde 6, as its methyl hemiacetal, afforded the cyclic mercuriacetal 7 stereospecifically. Reductive cleavage of the mercuri substituent was accomplished predominantly with retention of configuration using sodium trithiocarbonate in cold methanol, opening the way to stereoselective synthesis of the target.

Recently, we have investigated the related mercuricyclization of unsaturated hydroperoxides, as a route to substituted tetrahydrofurans of the type found in the ionophore antibiotics (13). Under the appropriate conditions, dienoate 8 is converted to the *trans*-disubstituted 1,2-dioxane 9. Unfortunately, in

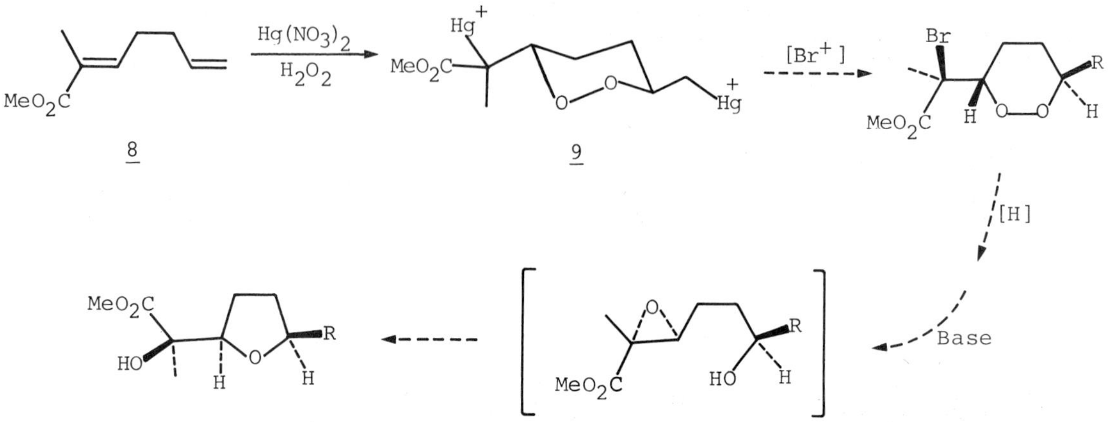

contrast to the behavior of secondary alkylmercury compounds (14), the tertiary mercuri group of 9 cannot be brominated stereoselectively. In addition, the cyclic peroxide moiety has proved to be unexpectedly resistant to reductive cleavage.

SYNTHESIS OF 1,3-DIOLS; NONACTIC ACID

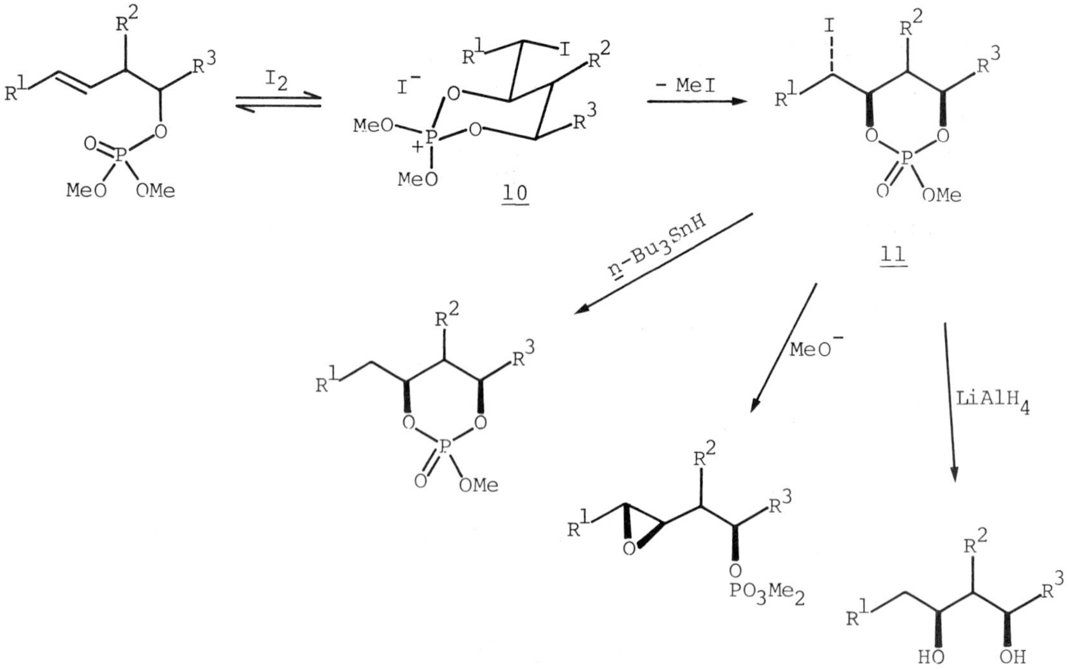

In order to allow a carbinol chiral center to control the stereochemistry of a cyclization reaction, an "extending" functional group is required. One example of such a group is the hydroperoxy moiety illustrated above. The first such extending group we investigated, however, was the phosphate ester moiety, for the stereocontrolled functionalization of homoallylic alcohols (15). Iodine-induced cyclization of homoallylic phosphates proceeds via the cyclic tetraalkoxyphosphonium intermediate (10) to give the cyclic phosphates (11) after loss of methyl or ethyl iodide. The reaction is highly stereoselective, affording products in which the carbinol centers are generated in an *erythro* relationship. The high stereoselectivity arises from equilibration between the starting material and the first cyclic intermediate, phosphonium ion 10, before dealkylation takes place. This mechanism is analogous to the iodolactonization of carboxylic esters, in that thermodynamic control over the stereochemistry is obtained in a formally irreversible reaction. The iodophosphate products undergo a number of transformations, including deiodination, conversion to the acyclic epoxyphosphates, and reduction to the *erythro* diols.

We have applied this reaction to a synthesis of nonactic acid, the subunit of the ionophoric macrolide nonactin (16). This route utilizes the phosphate cyclization to prepare the 1,3-diol 12 stereoselectively. Further elaboration of this material involves hydrogenation of an enol ether intermediate (13) to

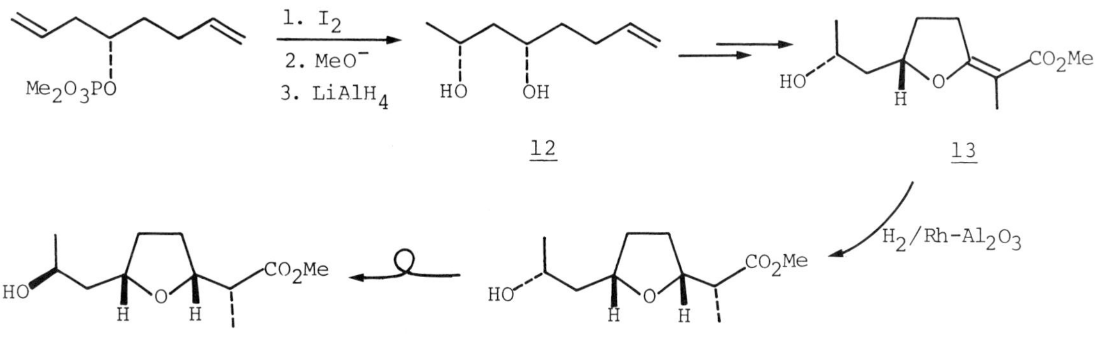

Methyl nonactate

introduce the chiral centers at C-2 and C-3 with the correct stereochemistry relative to that at C-6. Because of the *erythro* specificity of the phosphate cyclization, the carbinol center at C-8 is generated with the wrong configuration, and must be inverted before the correct stereoisomer of nonactic acid is obtained.

As an alternative to the phosphate cyclization, we investigated the cyclization of analogous carbonate esters (17). To our surprise, the methyl carbonates proved to be completely unreactive toward iodine in acetonitrile, in contrast to the behavior of the corresponding esters (see above). The t-butyl and benzyl carbonates, on the other hand, cyclize readily and in high yield to

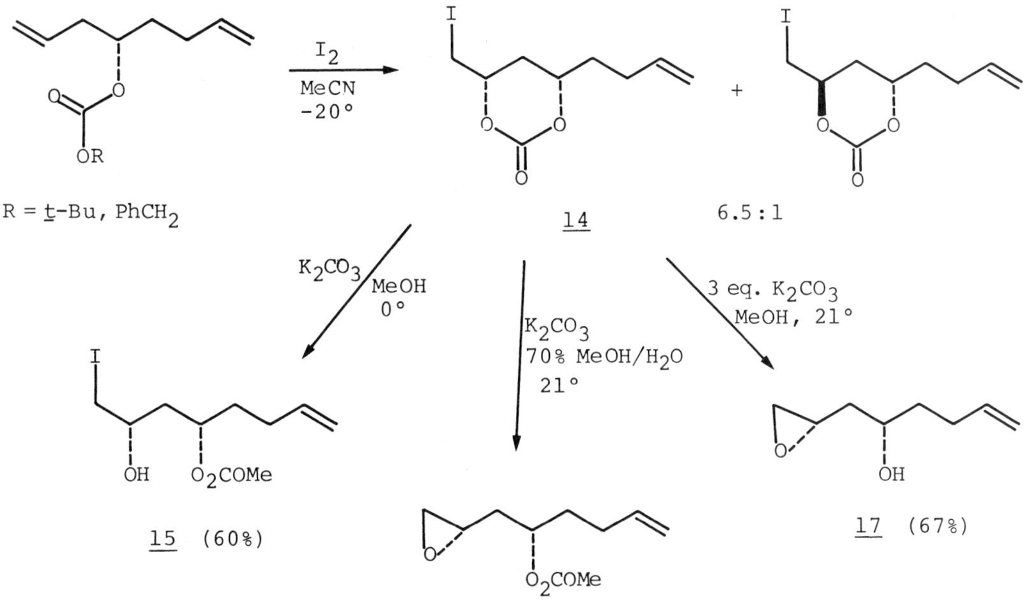

R = t-Bu, PhCH$_2$

15 (60%)

16 (92%)

17 (67%)

give the expected cyclic iodocarbonates. This process is not as stereoselective as the phosphate cyclization because the intermediate cation 18 has a trigonal center in place of the tetrahedral phosphorus of 10. Cyclizations of methoxy-substituted benzyl carbonates show a rate (but not a stereochemical) dependence on the stability of the corresponding benzyl cation. This behavior indicates that loss of the R-group from the cationic intermediate 18 is rate-limiting, and that thermodynamic control over the formation of this intermediate is likely to be occurring in this cyclization as well.

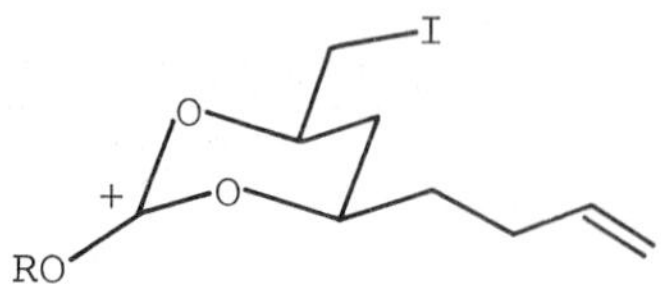

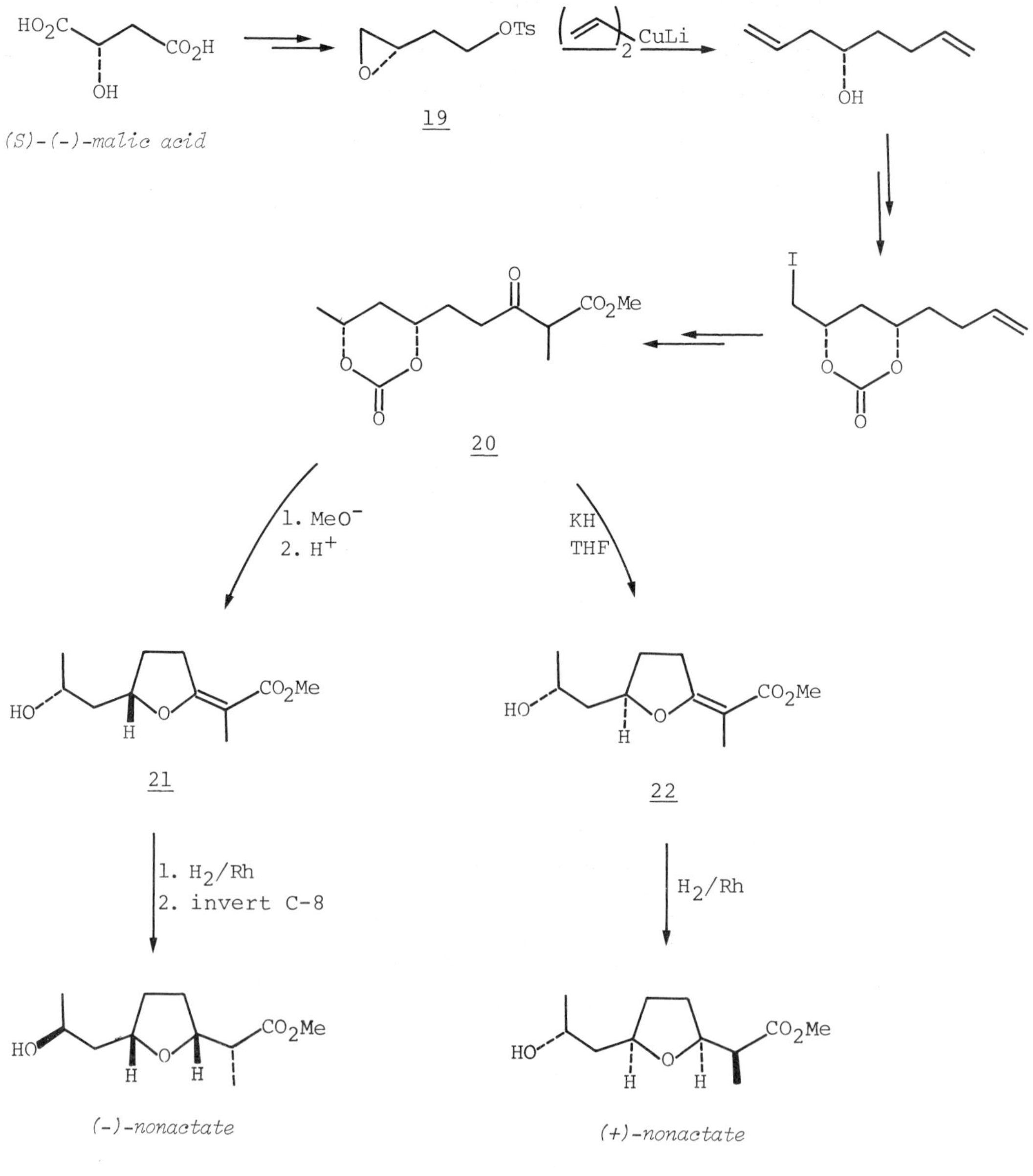

The cyclic carbonate diastereomers can be readily separated, and have the advantage of undergoing a number of selective reactions which are not possible for the cyclic phosphates. As illustrated on the previous page, iodocarbonate 14 can be sequentially opened to the iodohydrin 15, converted to the epoxy carbonate 16, or cleaved to the epoxy alcohol 17, depending on how vigorously it is treated with methanolic carbonate. We have incorporated the carbonate cyclization in an improved synthesis of nonactic acid which furnishes both of the enantiomers from a single optically active precursor (18). In contrast to most natural product syntheses, both enantiomers of nonactic acid are required, since the natural product nonactin is comprised of alternating enantiomers of this subunit.

Starting with S-(-)-malic acid, the S-enantiomer of the octadienol precursor is prepared in high yield by the reaction of lithium divinylcuprate with the epoxy tosylate 19. After carbonate cyclization, tin hydride deiodination, and conversion to the β-keto ester (20), two modes of ring closure are possible. Methanolysis of the carbonate moiety followed by acid-catalyzed dehydration provide enol ether 21, with the 6S,8S stereochemistry. After hydrogenation and inversion at C-8, this material gives the (-)-enantiomer of methyl nonactate.

In contrast, formation of the enolate of β-keto ester 20 results in ring closure by an intramolecular O-alkylation reaction with concomitant inversion at C-6, and affords the 6R,8S enol ether 22 in 92% yield. Hydrogenation of this material leads directly to the (+)-enantiomer of methyl nonactate. The ability of the carbonate moiety to act as such an effective leaving group must result from the unfavorable *syn* conformation that both of the ester linkages are forced to adopt. For comparison, the enolate of the corresponding diacetate is completely inert under far more vigorous conditions.

SYNTHESIS OF 2,5-SUBSTITUTED TETRAHYDROFURANS

The widespread occurrence of 2,5-di- and 2,2,5-trisubstituted tetrahydrofuran units in polyether natural products prompted us to investigate methods for their stereoselective construction. Although cyclic ethers are readily synthesized by the electrophilic cyclization of unsaturated alcohols, in the tetrahydrofuran ring system the 1,3-stereorelationship between the two α-positions is difficult to control because of the flexible nature of the five-membered ring. Moreover, the product that does predominate in the general case is the *trans* isomer, while one often requires the *cis*. We reasoned that improved selectivity, as well as a preference for formation of the *cis* product, might be obtained by cyclizing an ether instead of an alcohol, and introducing the ill-defined 1,3-relationship as two 1,2-relationships, as depicted below (19).

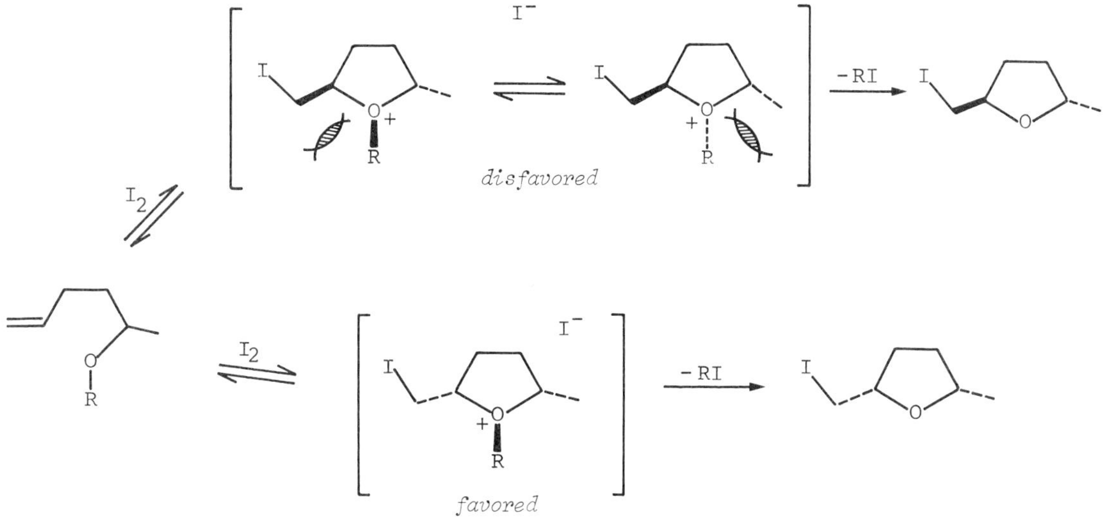

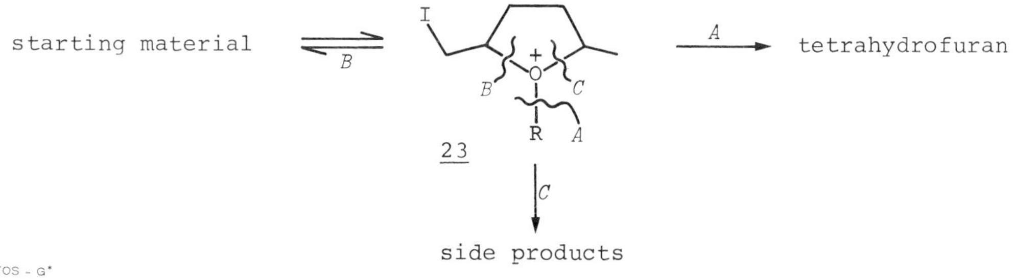

The success of this strategy depends critically on the correct balance between the rates at which the oxonium intermediates break down to give products (cleavage A), to give back starting material (cleavage B), or to give side products (cleavage C); see structure 23. For example, when R = methyl, the rate of A-cleavage is too slow relative to that of C-cleavage, and only a low yield of tetrahydrofuran is obtained. On the other hand, if A-cleavage is too fast relative to B-cleavage, as is the case for R = benzyl, the intermediate oxonium ions do not have sufficient opportunity to equilibrate, and poor stereoselectivity is observed. For a variety of olefinic alcohols, we found that substituted benzyl ethers exhibit the correct balance of reactivity so that both high *cis* selectivity and good yields can be obtained (results are summarized in Table 2)(19).

TABLE 2. Synthesis of *cis*-Tetrahydrofurans

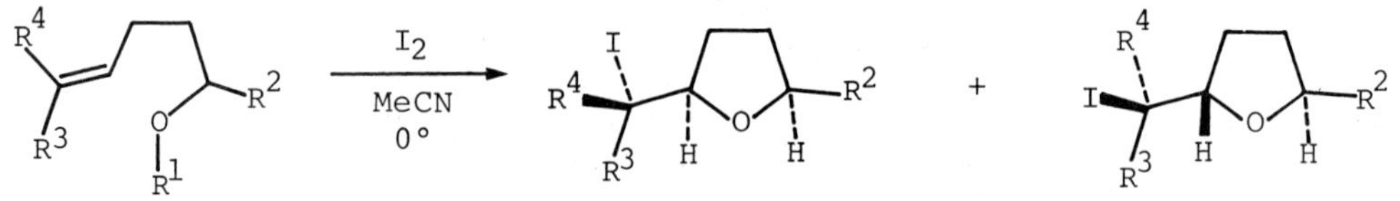

Entry	R^1	R^2	R^3	R^4	Ratio, cis/trans	Yield
1	PhCH	Me	H	H	2	60%
2	2,6-Cl$_2$PhCH$_2$	Me	H	H	21	63%
3	2,6-Cl$_2$PhCH$_2$	Me	Me	H	25	75%
4	2,6-Cl$_2$PhCH$_2$	Me	H	Me	12	47%
5	2,6-Cl$_2$PhCH$_2$	Me	CO$_2$Me	H	50	60%
6	4-BrPhCH$_2$	Me	CO$_2$Me	Me	10	44%

This strategy is applicable to the synthesis of 2,2,5-trisubstituted tetrahydrofurans as well, as demonstrated by the results which are outlined below (19, 20). In these compounds, the side chains are sufficiently bulky that cyclization of the alcohols themselves gives the *trans* products with good selectivity. Cyclization of the various benzyl ethers, on the other hand, affords the *cis* isomers with similar selectivity. These iodoethers are useful for the synthesis of the linalyl oxides and the sesquiterpene davanone.

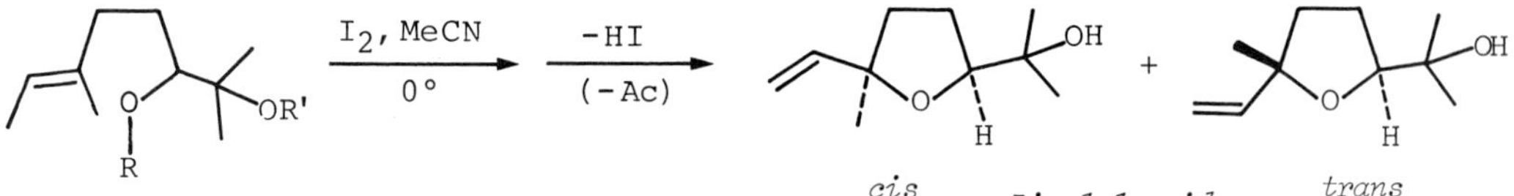

		Linalyl oxide	
R = R' = H		1:20	(70% yield)
R = PhCH$_2$, R' = Ac		13:1	(70% yield)

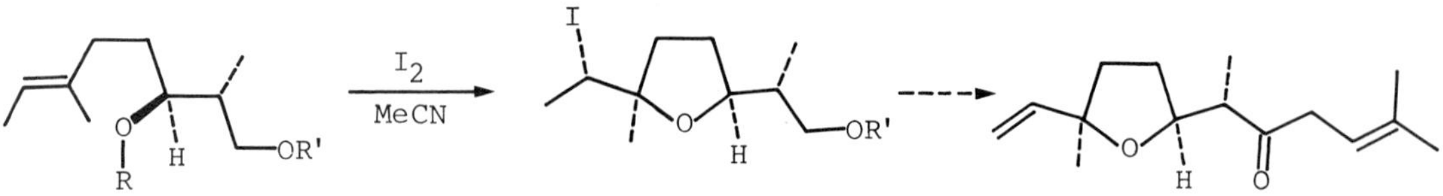

R	R'	Ratio, cis/trans
H	H or CPh$_3$	1:5
PhCH$_2$	CPh$_3$	5:1
PhCH$_2$	PhCH$_2$	6:1
4-BrPhCH$_2$	4-BrPhCH$_2$	10:1

Acknowledgment. Support for this research was provided by a grant from the National Institutes of Health (CA-16616).

REFERENCES

1. P.A. Bartlett, *Tetrahedron*, **36**, 2-72 (1980).
2. See for example, B. Capon and S.P. McManus, *Neighboring Group Participation*, Plenum, New York (1976).
3. V.I. Staninets and E.A. Shilov, *Russ. Chem. Rev.*, **40**, 272-283 (1971).
4. M.D. Doyle and D.I. Davies, *Chem. Soc. Rev.*, 171-197 (1979).
5. P.A. Bartlett and J. Myerson, *J. Am. Chem. Soc.*, **100**, 3950-3952 (1978).
6. J. Myerson, Ph.D. Thesis; University of California, Berkeley (1980).
7. P.A. Bartlett, D.J. Tanzella, and J.F. Barstow, *Tetrahedron Letters*, **23**, 619-622 (1982).
8. P.A. Bartlett and J. Myerson, *J. Org. Chem.*, **44**, 1625-1627 (1979).
9. K. Mori, H. Nomi, T. Chuman, M. Kohno, K. Kato, and M. Noguchi, *Tetrahedron Letters*, **22**, 1127-1130 (1981).
10. Unpublished work of D.M. Richardson, University of California, Berkeley.
11. C.F. Pizzo, Ph.D. Thesis; University of California, Berkeley (1981).
12. P.A. Bartlett and J.L. Adams, *J. Am. Chem. Soc.*, **102**, 337-342 (1980).
13. Unpublished work of P.C. Ting, University of California, Berkeley.
14. T.R. Hoye and M.J. Kurth, *J. Am. Chem. Soc.*, **101**, 5065-5067 (1979).
15. P.A. Bartlett and K.K. Jernstedt, *J. Am. Chem. Soc.*, **99**, 4829-4830 (1977).
16. P.A. Bartlett and K.K. Jernstedt, *Tetrahedron Letters*, 1607-1610 (1980).
17. P.A. Bartlett, J.D. Meadows, E.G. Brown, A. Morimoto, and K.K. Jernstedt, *J. Org. Chem.*, **47**, 0000 (1982).
18. Unpublished work of J.D. Meadows, University of California, Berkeley.
19. S.D. Rychnovsky and P.A. Bartlett, *J. Am. Chem. Soc.*, **103**, 3963-3964 (1981).
20. Unpublished work of C.P. Holmes, University of California, Berkeley.

STEREOCONTROLLED SYNTHESIS OF 25S,26-DIHYDROXY- AND 1α,25S,26-TRIHYDROXYCHOLECALCIFEROL SIDE CHAINS

Andrew D. Batcho, John Sereno, Naresh K. Chadha,
John J. Partridge, Enrico G. Baggiolini and Milan R. Uskoković

*Chemical Research Department, Hoffmann-La Roche Inc., Nutley,
New Jersey 07110, USA*

The 25S,26-dihydroxy (1) and 1α,25S,26-trihydroxy (2) metabolites of vitamin D_3 exhibit an interesting separation of activities. They stimulate intestinal calcium absorption but do not elicit any influence on bone resorption or mineralization.

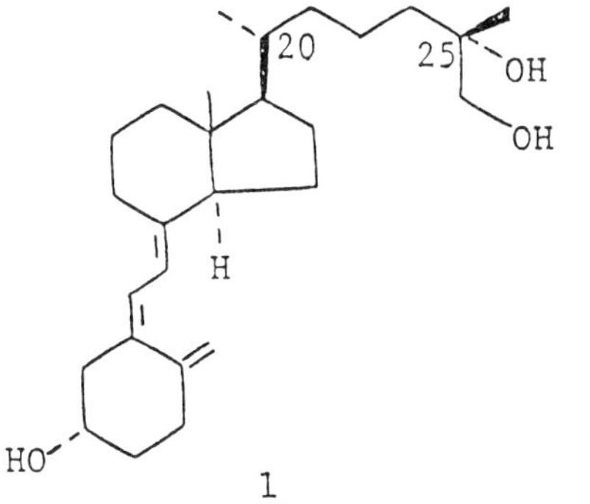

Two different synthetic approaches to these metabolites have been investigated in our laboratory. The side chain containing C-20 and C-25 chiral centers, separated from each other by a three carbon chain, was first formed by a chiral convergent synthesis, which has been described previously.[1,2] More recently, we have studied a chiral linear synthetic approach in which both the vicinal C-20 and the distant C-25 chiral centers are formed by asymmetric induction under the influence of the steroid-ring chirality. In this case, the C-20 chiral center is first formed by a stereospecific *ene* addition of formaldehyde to the corresponding 17Z-ethylidene steroid moiety as described in a recent paper.[3] The C-25 chiral center is then introduced by a thermodynamically stereocontrolled 1,3-dipolar cycloaddition of the C-23 nitrone to methyl methacrylate.

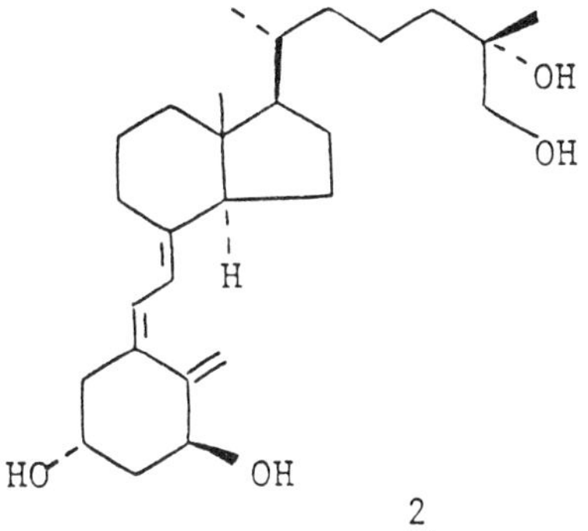

This remote asymmetric induction was first invetigated in the case of one-carbon shorter C-22' nitrone 3 as shown in Scheme 1. This reaction proceeded regiospecifically to give in high yield four stereoisomeric isoxazolidines with the 22R,24S-isomer 5 being the major product (60%).

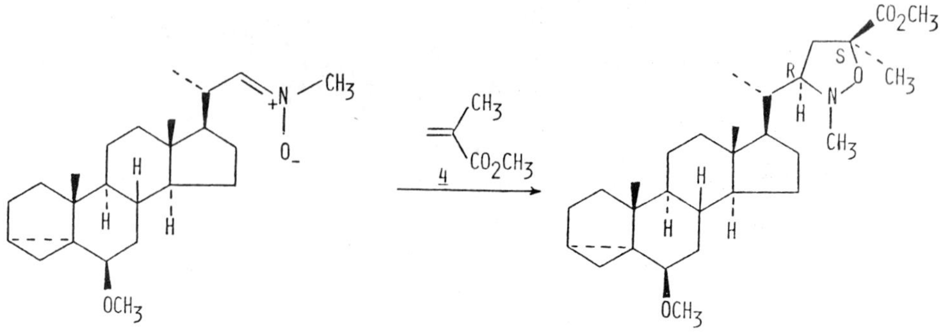

Scheme 1

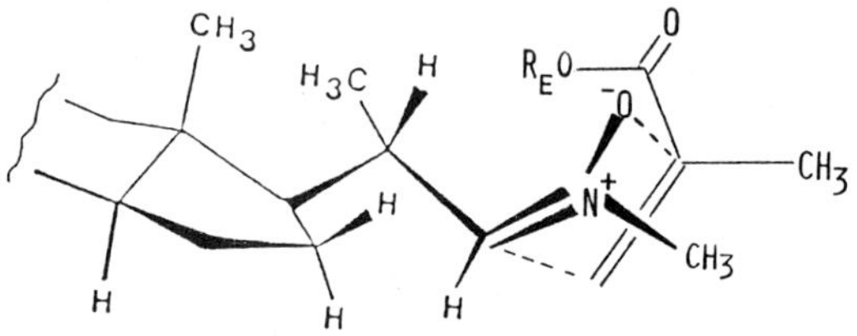

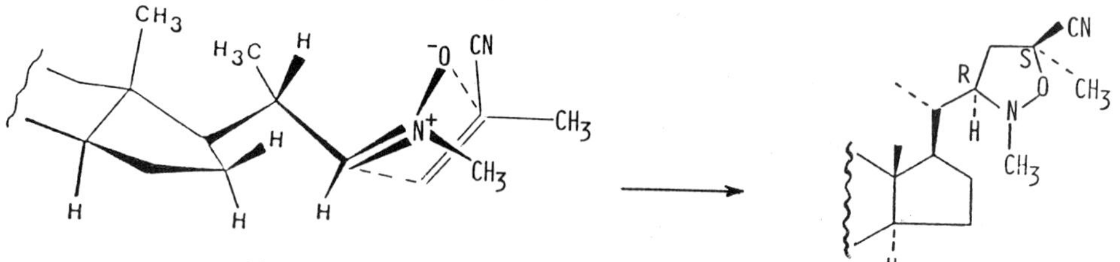

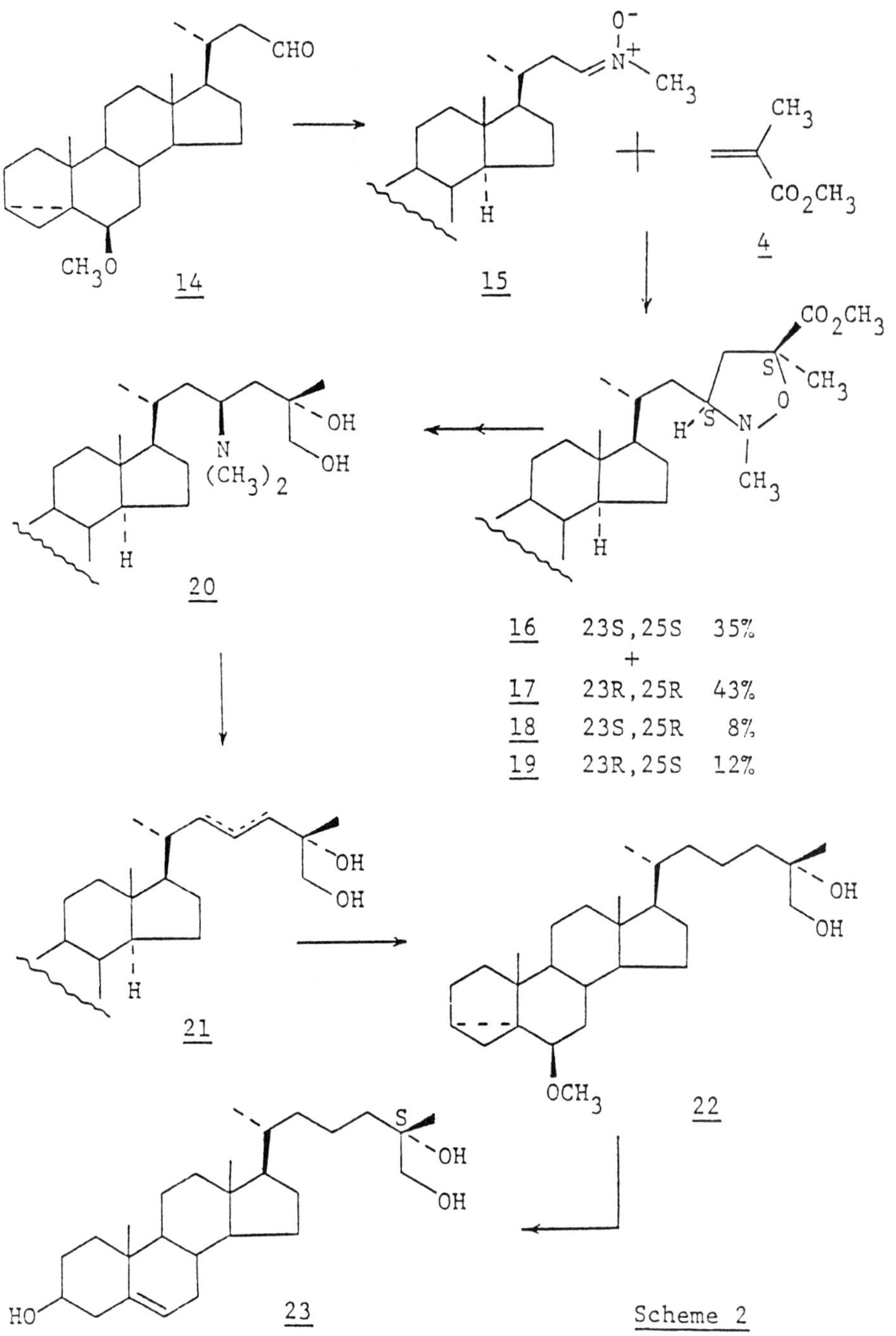
CHO
CH_3O
14
O^-
N^+
CH_3
15
CH_3
CO_2CH_3
4
CO_2CH_3
S
CH_3
O
H
S
N
CH_3
OH
N
$(CH_3)_2$
OH
20
16 23S,25S 35%
+
17 23R,25R 43%
18 23S,25R 8%
19 23R,25S 12%
OH
OH
21
OH
OH
OCH_3
22
S
OH
OH
HO
23

Scheme 2

This result can be rationalized by the preference for the Z-exo transition state 9, giving 22R,24S (5) + 22S,24R (6) isomers, over the Z-endo, giving 22R,24R (7) + 22S,24S (8) isomers, and by preferred steric approach of methyl methacrylate from the back-side of the nitrone producing more of the 22R than the 22S isomers. The use of the bulkier ester R_E = $CH(CH_3)_2$ reduced the yield of the Z-exo in favor of the Z-endo products. The use of the less bulky methacrylonitrile (see transition state 10) increased the yield of the back-side Z-exo product 11 significantly at the expense of the Z-endo product 13.

In the case of homologous C-23 nitrone 15, the stereocontrol on the approach of methyl methacrylate exerted by the C-20 chiral center is completely lost (Scheme 2). This addition, which proceeds smoothly at room temperature, gives preferentially the 23S,25S (16) and 23R,25R (17), products of Z-exo transition state, but in almost equal 35% and 43% yields, respectively. However, the stereocontrol favoring the desired 23S,25S-isomer (16) is regained by equilibration of the undesired isomers 17 - 19. As the least polar, the 23S,25S-isomer (16) is easily separated from the mixture by preparative HPLC. The remaining three isomers are heated with an excess of methyl methacrylate at 140° in xylene to give almost quantitatively the equilibrated mixture containing 41% of the desired 23S,25S-isomer (16).

23R,25R (17)		140°C	23S,25S (16)	41%
23S,25R (18) +	CH₃ / CO₂CH₃	xylene →	23R,25R (17)	25%
23R,25S (19)			23S,25R (18)	13%
			23R,25S (19)	21%
ratio 43:8:12	4			

The completion of the (20R,25S)-25,26-dihydroxy side chain required reduction of the ester group to the alcohol, opening of the isoxazolidine ring and elimination of the amino group, 16 → 22. The tandem methylation with methyl iodide and reduction with lithium aluminum hydride smoothly gave the dimethyl-amino diol 20. Hoffmann degradation produced 21 mainly as the Δ^{22}-olefin, while Cope elimination gave the Δ^{23}-olefin preferentially. Olefins 21 were not separated but directly hydrogenated to the desired 25S,26-dihydroxy side chain (22). Compound 22 was used in the synthesis of 25S,26-dihydroxycholecalciferol (1) as previously described[1]. The same methodology has been applied more recently in the synthesis of 1α,25S,26-trihydroxycholecalciferol (2).

REFERENCES

1. J.J. Partridge, S.-J. Shiuey, N.K. Chadha, E.G. Baggiolini, J.F. Blount, and M.R. Uskokovic, J. Amer. Chem. Soc. 103, 1253-1255 (1981).
2. J.J. Partridge, S.-J. Shiuey, N.K. Chadha, E.G. Baggiolini, B.M. Hennessy, M.R. Uskokovic, J.L. Napoli, T.A. Reinhardt, and R.L. Horst, Helv. Chim. Acta 64, 2138-2141 (1981).
3. A.D. Batcho, D.E. Berger, S.G. Davoust, P.M. Wovkulich, and M.R. Uskokovic, Helv. Chim. Acta 64, 1682-1687 (1981).

CARBOHYDRATE DERIVATIVES IN THE ASYMMETRIC SYNTHESIS OF NATURAL PRODUCTS

Bert Fraser-Reid*, Leon Magdzinski and Bruce Molino

Chemistry Department, University of Maryland, College Park, Maryland 20742, USA

Judging from the current literature,[1] carbohydrate-based approaches to the synthesis of the multichiral arrays found in macrolides and ionophores has now gained wide acceptance among organic chemists. The idea for such a synthetic approach was first expressed in published form in 1974 by Miljovic[2];however, the most impressive experimental advances were first achieved in Hanessian's laboratory, culminating with the synthesis of a potential intermediate for erythronolide 1, containing all ten chiral centers.[3] The target seemed ideal to these pioneering workers,[2,3] since it could be disconnected between C-8 and C-9 to give two segments, both of which could be prefabricated from hexopyranosides.

Generally, the attractiveness of sugars for synthesis arises from the fact that asymmetric centers can be generated upon a sugar heterocycle with great stereocontrol; and an important bonus is the ease with which configuration at a given site can be established through mechanistic and/or NMR considerations.[4,5] However, centers remote from the ring, i.e. "off-template stereocenters", are less reliably created and/or authenticated. For example, in the aforementioned approach to erythronolide, 1, the C-8 stereocenter formed upon connecting the two sugar precursors, was obtained as a mixture of epimers.[3] However C-8 is alpha to a carbonyl and can be epimerized into the desired orientation. It is needless to say that this fortuitous circumstance does not always exist,[6] and therefore the low control with "off-template stereocenters" constitutes a serious limitation of the applicability of carbohydrates to the synthesis of polychiral arrays.

Scheme I.

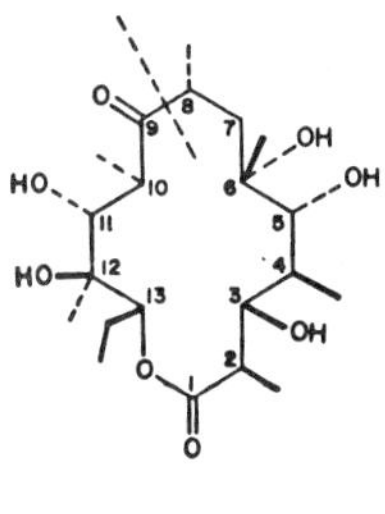

1

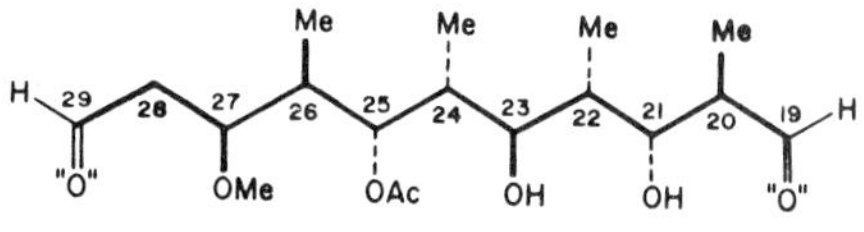

2

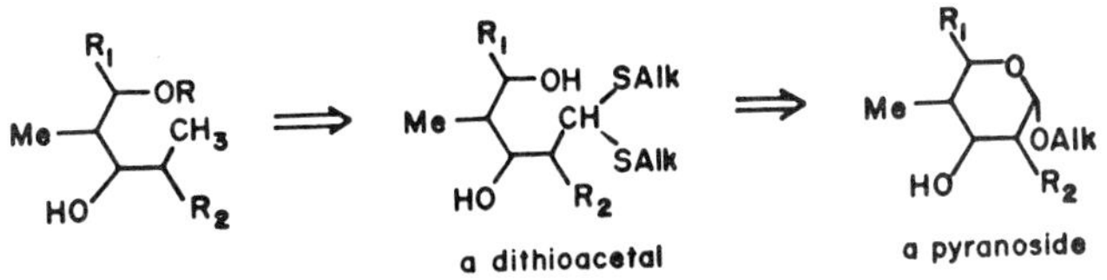

3 4 5

O27 → C24

R_1 = C28–C29

R_2 = C23–C19

6

O23 → C20

R_1 = C24–C28

R_2 = C19

7

This limitation is fully appreciated when the ansa chain (2) of rifamycin[7] is examined. In contrast to erythronolide, 1, the concatenation of eight contiguous chiral centers does not present any logical point(s) for disconnection and the concept of using prefabricated sub-structures carries with it an inevitable threat of providing a mixture of four diastereomers at each point of union. The processes of separation and identification which would ensue might not be trivial.

Basically, the problem is that the most appropriate sugars substrates, the hexopyranoses, offer only three contiguous chiral centers (C2, C3, C4) for manipulation; the fourth (C5-OH) is inaccessible being sequestered in the ring, and is indeed responsible for maintaining the integrity of the template. The problem, therefore, is how to extend the range of the sugar to encompass the eight contiguous chiral centers required by 2.

It is appropriate to take note of the spate of excellent processes for acyclic diastereoselection now available,[8] and it is conceivable that these could be wedded to traditional carbohydrate synthetic practices in order to provide an answer to this predicament.[9] However, given our early proselytism on the merits of sugars as stereochemical templates,[10] we preferred to determine whether a solution to this vexing problem could not be found within the domain of carbohydrate chemistry.

In addressing this problem, we were mindful of the fact that the most reliable carbohydrate stereochemical templates are the α-glycopyranosides wherein the anomeric effect is operative.[11] This circumstance was therefore germane to our deliberations.

Scheme II.

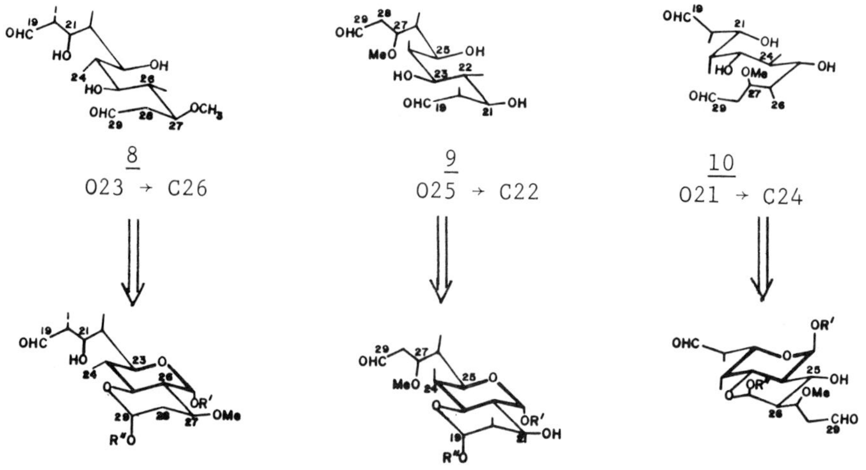

8
O23 → C26

9
O25 → C22

10
O21 → C24

11

12

13

Being biosynthesized from propionate,[12] the ansa chain 2 has several sets of OH and CH$_3$ groups in 1-4 relationships, as summarized in structure 3. The retrosynthesis leads to the dithioacetal, 4, a type of compound readily obtainable from a pyranoside 5.[13] In this concept, therefore, each set of 1-4 related CH$_3$ and OH groups constitutes a "hidden pyranoside". Of the five possible options, 6-10, the two "end" options 6 and 7 (Scheme I) must be rejected since each has four "off-template stereocenters", which from our discussion above, would present problems.

The three remaining options, 8, 9 and 10 (Scheme II) are drawn conformationally, and it is seen that they generate "backbone" pyranoside templates, shown in heavy type as the α anomers 11, 12 and 13, respectively, having the configurations indicated. Structure 13 is immediately rejected, since it would require the use of an L-sugar for the synthesis, a highly uneconomical business (usually!).

In deciding how to proceed, there were two significant observations. Firstly, in both 11 and 12 the R$_2$ appendage (c.f. 3 Scheme I) allows for a "satelite" pyranoside to be fused at the front" of the "backbone" template. Secondly, in both the R$_2$ and the vicinal OH substituents are trans-diequatorial. For creating this structural feature, the opening of an epoxide by a carbon nucleophile would seem to be a logical choice; however, since epoxides open preferentially in a trans diaxial sense, a $^1C_4$ system 14 would be more appropriate than the $^4C_1$ counterpart 16, since the desired chiral arrangement 15 is obtained directly.[14] The desired $^1C_4$ conformation exists in the 1,6-anhydro pyranoses, one of the oldest families of sugar derivatives. Thus, 1,6-anhydro-β-D-glucopyranose (levoglucosan), 17, may be

Scheme III.

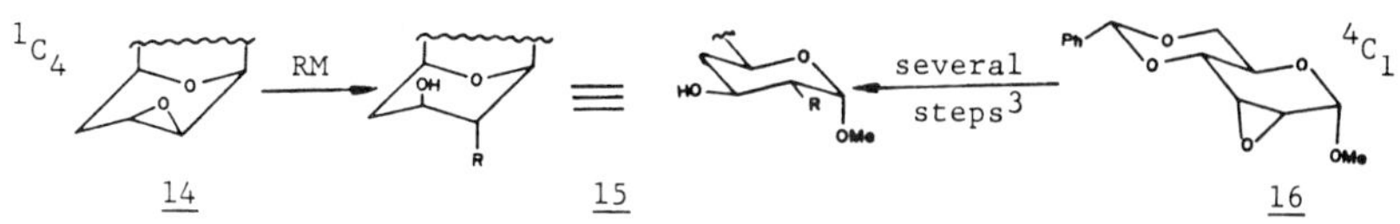

obtained directly by pyrolysis of starch,[15] or in three easy steps from D-glucose[16] and there is an extensive body of knowledge about its chemistry.[17] For example, Czechoslovak workers have developed efficient syntheses of the dianhydro sugar 18.[18]

The epoxide 18 was processed as indicated to give the Z-olefin 20a. Acid catalyzed solvolysis with benzyl alcohol cleaved the 1,6-anhydro ring, giving the 2-deoxy glycoside 21 in 58% yield. Controlled oxidation with manganese dioxide then gave the hemiacetal 23 directly.

Scheme IV.

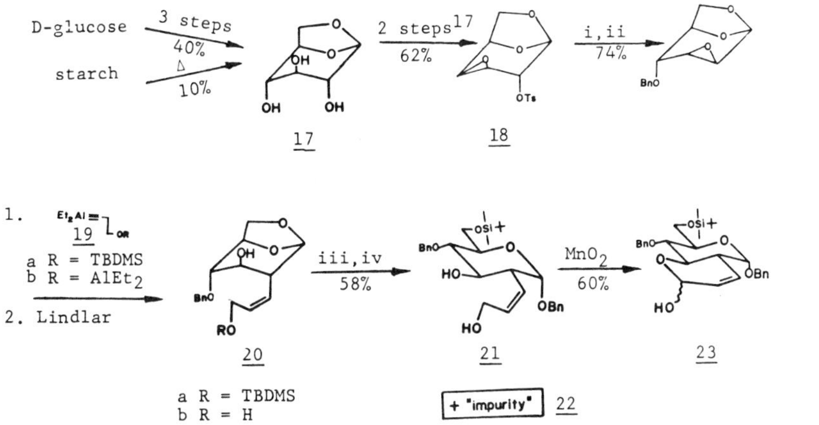

(i) PhCH$_2$OH/TsOH. (ii) NaOMe. (iii) PhCH$_2$OH/BF$_3$·OEt. (iv) TBDMSiCl.

From the summary in Scheme V our options for the olefin 23 would be to introduce H and OCH$_3$, if the glucoside 11 was our target, or alternatively CH$_3$ and OH group if the galactoside 12 was our choice. The merits of 23 as a synthon for these tasks takes on special significance by noting that the "satelite" ring is in fact a hex-2-enopyranose akin to compound 27. These unsaturated sugars had played a central role in our early researches,[10] and the wealth of knowledge available about them[19] meant that most of the transformations conceivable for 23 had probably already been explored.[20-23] If not, compounds 28 and 29 constitute readily available models.[19]

Therefore, in the bis-pyranose system 23, there are six (potential) contiguous chiral centers, indicated by asterisks, all of which can be manipulated confidently and predictably. This bis-pyranose system has therefore extended the capability of the simple hexopyranose 17; and a tris-pyranose would extend it even further! Hence the manifold 24 possesses eight (potential) contiguous chiral centers b-i, exactly the number that exists in the ansa chain (2) of rifamycin!!

With this realization, a clear choice can be now made between the two

Scheme V.

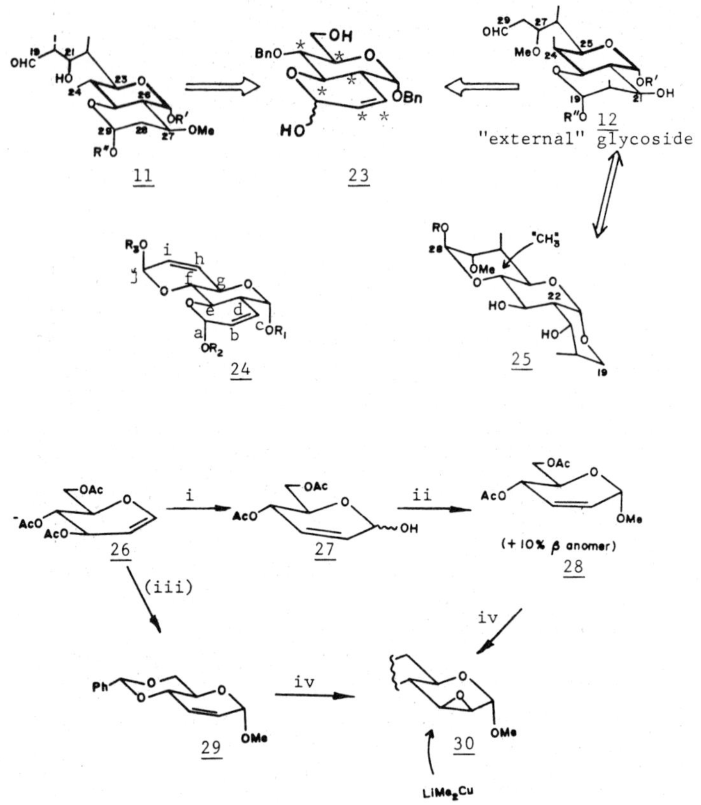

(i) H₂O/dioxan/reflux[21] (ii) HC(OMe)₃[20]
(iii) PhCH(OMe)₂[22] (iv) MCPBA[23]

coils, 11 and 12. Thus, the eight contiguous chiral chiral centers of 2 must be made to coincide with (or be derivable from) the eight "potential" sites "b" to "i" of 24. None can be "wasted" on achiral centers. Hence, 11 had to be rejected since it would have placed the C28-CH₂ at position "b" of 24. In the case of 12, the C28-CH₂ occurs at position "j", an aldehydic center which is achiral anyway.

With our choice of 12 therefore clearly defined, we returned to the solvolysis of 20b Scheme IV. We had been disappointed with the relatively low yield of 21 and were therefore prompted to investigate the structure of a major "impurity". Not surprisingly, the substance was found to 22, Scheme VI, resulting from competitive glycosylation by the internal hydroxyl group.[24] The yield of 22 could undoubtedly be improved by removing the external competition from benzyl alcohol. Accordingly, epoxide 18 (Scheme IV) was cleaved with the dianion 19b, and when the anhydro sugar 20b (Scheme VI) was treated with neat trifluoroacetic acid for two minutes, 22 was isolated in 78% yield.

In light of this development, the ansa chain would have to be coiled "forward", the corresponding tricyclic synthon being 25 (Scheme V). In order to provide for the axial C20-CH₃, it was necessary to achieve epoxidation from the convex surface of 22. As indicated, the stereoselectivity with m-chloroperbenzoic acid was dependent on the protecting group chosen for the C23-OH, the acetyl group giving 31 exclusively. On the other hand, Sharpless epoxidation[25] of the free homoallylic alcohol (22, R=H) afforded

Scheme VI.

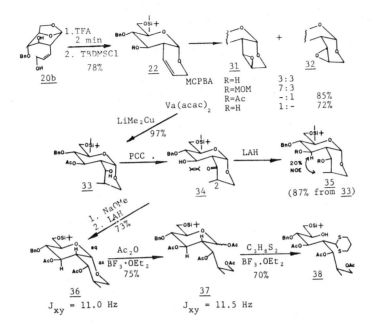

the endo epoxide 32 only.

Cleavage of 31 went smoothly giving 33 and the configuration of the hydroxyl group was inverted by oxidation to ketone 34, following with immediate reduction so as to forestall epimerization at C20. Success with the latter objective was educed by observing a 20% nOe indicated for 35. Furthermore, treatment of 34 with sodium methoxide for two minutes followed by lithium aluminium hydride reduction now led to the diasteromer 36.

This was a convenient stage at which to develop a strategy for converting the anomeric carbon into a dithioacetal as required by the plan in Scheme I. Treatment of the unwanted diastereomer 36 with propane 1,3-dithiol and boron trifluoride etherate caused lysis of the benzyl ether. On the other hand, acetolysis cleaved the acetal ring at the axial rather than the equatorial site giving 37. This conclusion follows from the fact that the coupling constant (see Scheme VI) remained virtually unchanged, indicating the preservation of the "central" pyranose ring. The dithioacetal 38 was then readily obtained.

Concurrently with the foregoing investigations into the "frontside" ring, we have been undertaking studies towards the "backside" ring. The readily prepared aldehyde 39 was converted into the enal 40 which was irradiated at 350nm in a medium containing methanol and pyridinium p-toluenesulfonate (PPTS)[26] in the expectation of obtaining (i) isomerization of the double bond, (ii) cleavage of the ethoxyethyl group with concomitant cyclization to the pyranose, and (iii) acid catalyzed conversion to the methyl glycoside 42. Isolation of the latter indicated that our expectations were rewarded. However, the low yield and the lengthy reaction time, caused us to examine stronger acids than PPTS; but when acetic acid, HCl or TsOH were tried, no 42 was obtained!

The specific effect of Grieco's acid (PPTS)[26] caused us to suspect that the double bond isomerisation was not the photochemical event that we had planned, but rather the result of conjugate addition of pyridine 40 to give 43 and 42. Indeed removal of the assembly from the photoreactor brought about better yields of 42.

Nevertheless, a better yielding sequence was desired, and so the Bestmann-Trippett reagent 44 was examined.[27] Our early experiments were carried out with the diethyl acetal, but the yields proved variable and

Scheme VII.

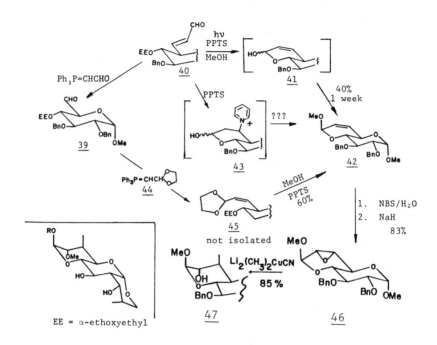

EE = α-ethoxyethyl

the cause of this was soon traced to the great difficulty of purifying the diethyl phosphonium salt precursor. Accordingly, we turned our attention to the ethylene acetal 44, the salt for which is readily purified. Gratifyingly, the Z-olefin 45 was obtained in excellent condition. However, it was best to carry out the subsequent acid-catalyzed methanolysis without purification, and in this way the diglycoside was obtained, exclusively as the axial anomer shown in 42.

Comparison of 42 and the target 25 indicated that epoxide 46 would allow for entry of the C26-CH3 syn to the OCH3. Previously Hicks, in our laboratory, had shown that 30 (Scheme V) could be opened with lithium dimethyl cuprate in the desired sense.[28] Accordingly 42 was epoxidized to 46 and this has been opened to 47. However the higher order cuprate reagent of Lipshutz,[29] Li2Me2CuCN, proved to be the reagent of choice.

Scheme VIII

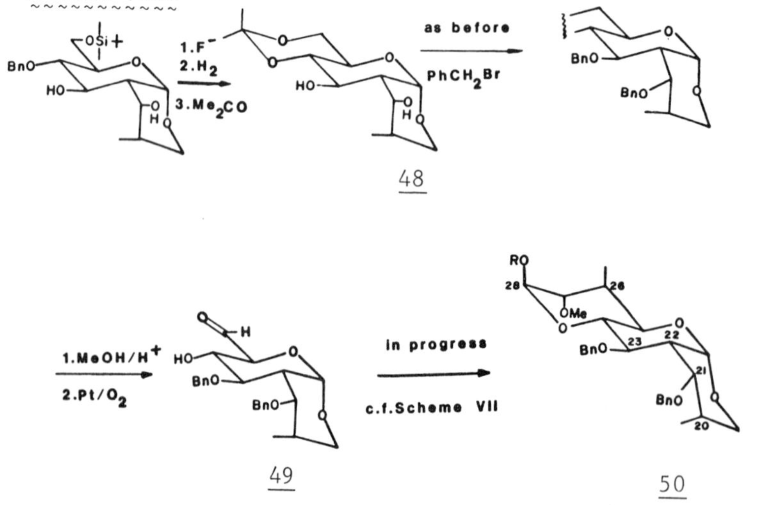

The best method which has been found for utilizing the model studies explored in Scheme VII is shown in Scheme VIII. Thus the epimerization at C-21 is delayed until the acetonide 48 has been formed, and methanolysis and selective oxidation[30] has afforded hydroxy aldehyde 49 to which we are now applying the upper ring pyranosidic homologation (c.f. 39 → → 47), which would afford the target 50.

Scheme IX

Kishi intermediate

51

52

53

template and lower glycoside

anomeric centre

upper glycoside

● created with complete stereoselectivity
■ from glucose template

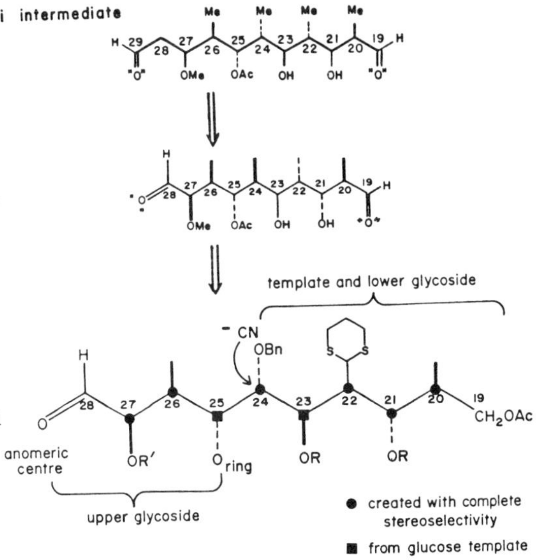

Scheme IX summarizes the work which has been carried out on the model systems. The tricyclic system 50 is equivalent to 53. As indicated, six stereocenters have been introduced with complete stereocontrol, without any need to fractionate epimeric mixtures. Of the remaining two, C-25 is the original C-5 of D-glucose, and it is responsible for the stereochemical integrity of the entire assembly; there has been no tampering with C-23.

With respect to the carbon substituent required at C-24, we have developed a procedure for efficient displacement of an equatorial C-4 triflate using a tetraalkylammonium cyanide. (As a point of interest the corresponding stereocenter in the streptovaricins is carboxyl group.) Compound 53 is therefore a synthon of dialdehyde 52 which differs from the Kishi intermediate, 51, by an aldehyde homologation.

Acknowledgement

We are grateful to the University of Maryland, Merck Sharp & Dohme, and the National Science Foundation for financial support. A generous gift of levoglucosan from Professor Hans Paulsen of the University of Hamburg, West Germany enabled us to initiate this project. The assistance of our undergraduate collaborator, Bruce Cwiber, has proved invaluable for the progress of this work.

References

1. See for example: (a) R.E. Ireland, P.G.M. Wuts, and B. Ernst, J. Am. Chem. Soc., 103, 3205 (1981); (b) K.C. Nicolaou, M.R. Pavia, and S.P. Seitz, ibid, 104, 2027, 2030 (1982); (c) K.C. Nicolaou, M.R. Pavia, and S.P. Seitz, ibid, 103, 1222, 1224 (1981); (d) F.E. Ziegler, and P.J. Gilligan, J. Org. Chem., 46, 3874 (1981); (e) K. Tatsuta, Y. Amemiya, S. Maniwa and M. Kinoshita, Tetrahedron Lett., 21, 2837 (1980); (f) K. Tatsuta, Y. Amemiya, Y. Kanemura, and M. Kinoshita, ibid, 22, 3997 (1981); (g) S. Hanessian, P.C. Tyler, and Y. Chapleur, ibid, 22, 4583 (1981); (h) D.H.R. Barton, M. Benechie, F. Khuong-Huu, P. Potier, and V. Reyna-Pinedo, ibid, 23, 651 (1982); (i) R. Bonjouklian and B. Ganem, Carbohydrate Res., 76, 245 (1979); (j) E.J. Corey L.O. Weigel, A.R. Chamberlin, and B. Lipschutz, J. Am. Chem. Soc., 102, 1439 (1980); (k) P.-T. Ho, Can. J. Chem., 58, 858 (1980).
2. (a) M. Miljkovic, M. Gligorijevic, T. Satoh, and D. Miljkovic, J. Org. Chem., 39, 1830 (1974); (b) M. Miljkovic and D. Glisin, ibid, 40, 3357 (1975); (c) Bull Soc. Chim. Beograd., 42, 659 (1977).
3. (a) S. Hanessian, and G. Rancourt, ibid, 55, 111 (1977); (b) S. Hanessian, G. Rancourt and Y. Guindon, Can. J. Chem., 56, 1843 (1978).
4. B. Fraser-Reid, and R.C. Anderson, Progr. Chem. Org. Natural Prods., 39, 1, 1980.
5. S. Hanessian, Acc. Chem. Res., 12, 159 (1979).
6. See for example: S. Hanessian, P.C. Tyler, G. Demailly, and Y. Chapleur, J. Am. Chem. Soc., 103, 6243 (1981).
7. H. Nagaoka, W. Rutsch, G. Schmid, H. Iio, M.R. Johnson, and Y. Kishi, ibid, 102, 7962 (1980); H. Iio, H. Nagaoka, and Y. Kishi, ibid, 102, 7965 (1980).
8. See for example: C.H. Heathcock, C.T. White, J.J. Morrison, and D. VanDerveer, J. Org. Chem., 46, 1296 (1981); D.A. Evans, J. Bartroli, and T.L. Shih, J. Am. Chem. Soc., 103, 2127 (1981); S. Masamune, W. Choy, F.A.J. Kerdesky, and B. Bimperiali, ibid, 103, 1566 (1981); Y. Yamamoto, H. Yatagai, and K. Maruyama, ibid, 103, 3229 (1981).
9. See for example, M. Nakata, H. Takao, Y. Ikeyama, T. Sakai, K. Tatsuta, and M. Kinoshita, Bull. Chem. Soc. Japan, 54, 1749 (1981).
10. B. Fraser-Reid, Accts. Chem. Res., 8, 192 (1975).
11. The Anomeric Effect, Origin and Consequences, Ed., W.A. Szarek and D. Horton, A.C.S. publication, 1979.
12. R.B. Woodward, Agnew. Chem., 69, 50 (1959).
13. M.L. Wolfrom and A. Thompson, Methods Carbohyd. Chem., 2, 427 (1963); P.E. Sum and L. Weiler, Can. J. Chem., 56, 2700 (1978).
14. N.K. Kotchetkov, A.F. Sviridov, and M.S. Ermolenko, Tetrahedron Lett., 22, 4315 (1981).
15. R.B. Ward, Methods Carbohyd. Chem., 2, 394 (1963).
16. G.H. Coleman, ibid, 2, 398 (1963).
17. M. Cerny, and J. Stanet, Jr., Advan. Carbohyd. Chem., 32, 24 (1977).
18. T. Traha, and M. Cerny, Coll. Czech. Chem. Commun., 36, 2216 (1971).
19. R.J. Ferrier, Advan. Carbohyd. Chem., 24, 199 (1969); 20, 67 (1965).
20. See for example: E. Fischer, M. Bergmann, and H. Schotte, ibid, 53, 509 (1920); E.L. Albano, D. Horton, and J.H. Lauterbach, Carbohydrate Res., 9, 149 (1969); S. McNally, and W.G. Overend, J. Chem. Soc., 1978 (1966); C.L. Stevens, J.B. Filippi, and K.G. Taylor, J. Org. Chem.,31, 1292 (1966); O. Achmatowicz in "Organic Synthesis Today and Tomorrow", Ed., B.M. Trost and C.R. Hutchinson (IUPAC), Pergamon Press, 1981, p. 307.
21. B. Fraser-Reid, and B. Radatus, J. Am. Chem. Soc., 92, 5288 (1970); M. Bergmann, Ann., 443, 223 (1925).
22. B. Fraser-Reid, D.L. Walker, S.Y.-K. Tam, and N.L. Holder, Can. J. Chem., 51, 3950 (1973).
23. R.J. Ferrier, and N. Prasad, J. Chem. Soc., C, 570, 575 (1969).
24. G. Descotes, M. Lissac, J. Delmau and T. Duplan, C.R. Acad. Sc. Paris. Ser C., 267, 1240 (1968).
25. K.B. Sharpless and R. Michaelson, J. Am. Chem. Soc., 95, 6136 (1973).
26. M. Miyashita, A. Yoshikoshi, and P.A. Grieco, J. Org. Chem., 42, 3772 (1977).
27. S. Trippett, and D.M. Walker, J. Chem. Soc., 1266 (1961); H.J. Bestmann, K. Rogh, and M. Ettzinger, Agnew. Chem., 91, 748 (1979); H.J. Bestmann, K. Roth, and M. Ettlinger, Chem. Ber., 115, 161 (1982).
28. D.R. Hicks and B. Fraser-Reid, Can. J. Chem., 53, 2017 (1975).
29. B.H. Lipshutz, J. Kozlowski and R.S. Wilhelm, J. Am. Chem. Soc., 104, 2305 (1982).
30. K. Heyns and P. Koll, Methods Carbohyd. Chem. 6, 342 (1972).

CARBOHYDRATES AS "CHIRAL TEMPLATES" IN ORGANIC SYNTHESIS — TARGET: BOROMYCIN

Stephen Hanessian, Daniel Delorme, Peter C. Tyler*, Gilles Demailly* and Yves Chapleur*

Department of Chemistry, University of Montreal, Montreal, Quebec, H3C 3V1 Canada

Abstract - Progress in our studies directed toward the total synthesis of the antibiotic boromycin is discussed.

SYNTHETIC DESIGN WITH "CHIRAL TEMPLATES"

Over the years, the synthetic chemist has relied on a variety of conceptually intriguing approaches to the construction of the carbon frameworks present in natural products (1). Antithetic or retrosynthetic analysis, pioneered by Corey (2), has become an integral part of synthetic design, and it forms the basis for the generation of synthons (retrons). An operationally different strategy which we have been using (3) relies on the recognition in the target structure of certain symmetry elements, functional group interrelations and carbon framework patterns. Appropriate bond disconnections generate a "chiral template" which can be derived from a readily available chiral precursor by systematic chemical manipulation. Thus, a "chiral template" corresponds to an actual synthetic intermediate (not necessarily a synthon) which represents a replica of a segment of the target and carries its stereochemical code. Figure 1 illustrates in schematic form,

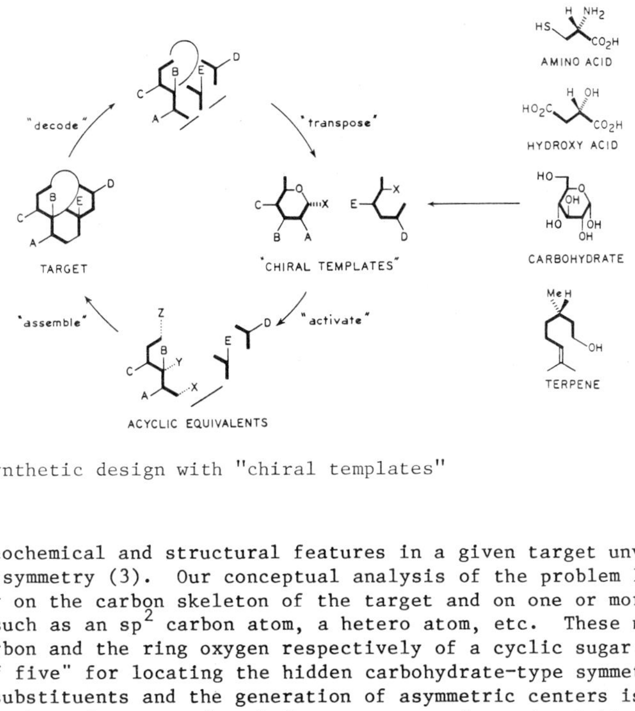

Fig. 1 Synthetic design with "chiral templates"

how decoding stereochemical and structural features in a given target unveils hidden carbohydrate-type symmetry (3). Our conceptual analysis of the problem leads us to place emphasis primarily on the carbon skeleton of the target and on one or more internal reference points such as an sp^2 carbon atom, a hetero atom, etc. These may coincide with the hemiacetal carbon and the ring oxygen respectively of a cyclic sugar. We can thus advance a "rule of five" for locating the hidden carbohydrate-type symmetry. The incorporation of substituents and the generation of asymmetric centers is an operational consequence which takes advantage of such features as conformational bias, topography and stereoelectronic properties inherent to the carbohydrate structure.

---
* Post-doctoral associate

We have previously classified targets according to a "visual dialogue" leading to the emergence of carbohydrate-type symmetry which may be apparent, partially hidden, or hidden (Fig. 2). The latter category has provided us with much challenge and stimulation.

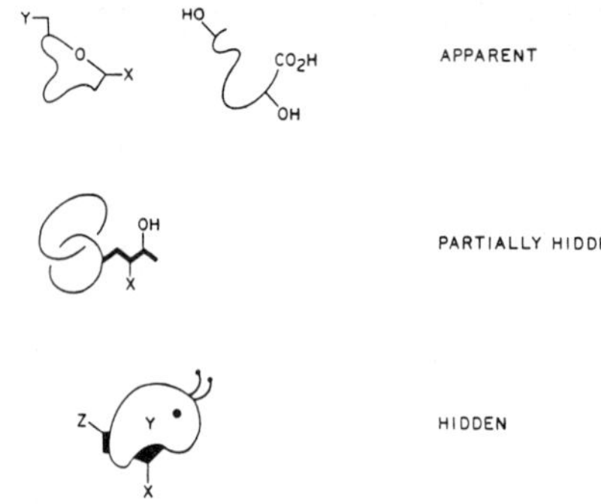

Fig. 2 Carbohydrate-type symmetry

Compared to other chiral precursors such as amino acids, hydroxy acids and terpenes (4), carbohydrates offer some advantages in natural product synthesis and have been extensively used in recent years (5).

TARGET: BOROMYCIN

In 1967, Prelog and coworkers (6) isolated a boron-containing metabolite of Streptomyces antibioticus, which they named boromycin. Structural investigations (7) involving limited chemical transformations and more demanding X-ray crystallographic studies led to the establishment of the constitutional structure of boromycin 1 to be as shown in Fig. 3. In addition to being the first boron-containing natural product, it proved to be a

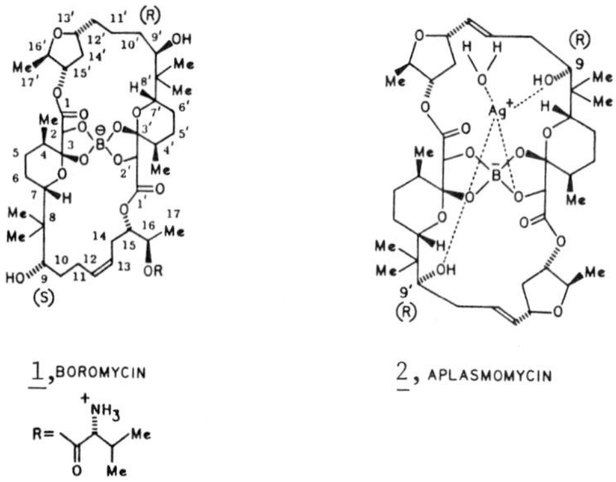

Fig. 3 Structures of boromycin and aplasmomycin

novel macrolide structure. More recently another related class of ionophoric antibiotics, aplasmomycin 2, was discovered (8). Treatment of boromycin with aqueous base led to cleavage of the ester linkage with D-valine and subsequent acid treatment removed the boron atom to give des-boron des-valine boromycin, 3 (7) (Fig. 4). Remarkably,

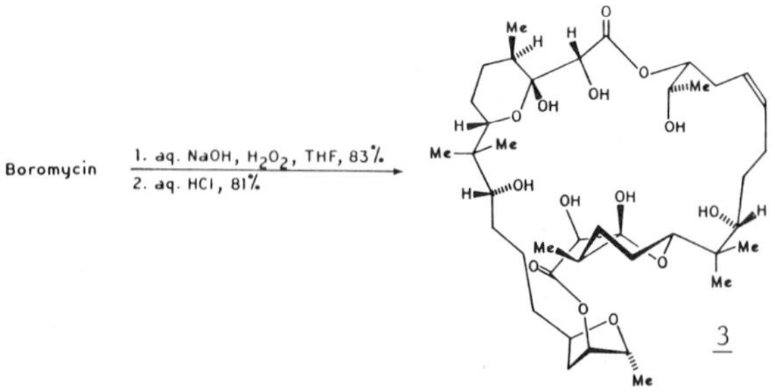

Fig. 4 Formation of des-boron des-valine boromycin

X-ray analysis of this compound showed that the three dimensional structural integrity of the molecule was still maintained (7). Recent efforts in our laboratory (P. Tyler) have shown that cleavage of the D-valine ester linkage can be done more efficiently with the highly nucleophilic (and less basic) hydroperoxide ion (Fig. 4). As previously mentioned, some degradative studies on boromycin (7) produced a neutral $C_{18}H_{32}O_5$ lactone 4 and an acidic $C_{13}H_{24}O_4$ fragment 5 as shown in Fig. 5. The lactone structure deduced mainly from

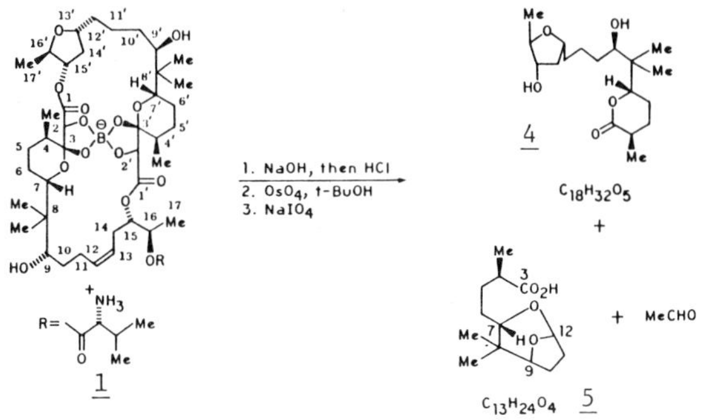

Fig. 5 Chemical degradation of boromycin and the formation of a C_{18} lactone corresponding to the "upper" segment of the target (C-3'−C-17')

its mode of formation and later confirmed from X-ray data, constitutes the "upper" segment of the target and can be a useful relay substance in a projected synthesis of boromycin (9). The acid fragment 5, on the other hand is the result of a more intricate transformation in its oxidative pathway of formation and does not offer much in the way of synthetic utility as an intermediate.

STRUCTURAL ANALYSIS

In spite of its somewhat awesome structure, appropriate bond disconnections of the target reveal elements of apparent and partially hidden carbohydrate-type symmetry. Thus, in Fig. 6, we show two lactone-type structures corresponding to the "upper" 6 and

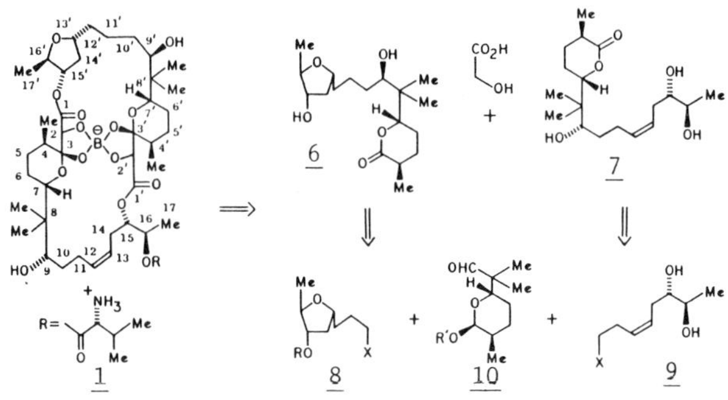

Fig. 6 Retrosynthetic analysis of boromycin

"lower" 7 segments of the target, which can be further simplified to produce a tetrahydrofuran 8, an acyclic 9, and a tetrahydropyran unit 10. The latter represents a common moiety in the parent lactones and in combination with nucleophilic equivalents of the chiral tetrahydrofuran and acyclic units could be envisaged to provide access to the "upper" and "lower" segments of boromycin. Examination of the structure of boromycin reveals apparent carbohydrate-type symmetry located in the tetrahydrofuran and tetrahydropyran units, and partially hidden symmetry in the acyclic segment (C-12−C-17, note application of the "rule of four"). Since the relationship of the cyclic units to furanose and pyranose type sugar precursors is evident, the main challenge in preparing intermediates such as 8 and 10 resides in devising highly stereocontrolled and efficient processes for their production. Segment C-10−C-17 (or C-12−C-17) on the other hand contains two vicinal asymmetric centers having the 15 S, 16 R configuration, which could be related to the tartaric acid with the same sense of chirality. As we shall see later, this segment can be readily prepared from a carbohydrate precursor. In comparing the structures of boromycin 1 and aplasmomycin 2, (Fig. 3), one can note obvious convergence of the cyclic units but different senses of chirality at 9 and 9' as well as the presence of an E olefinic linkage in 2 which has C2 symmetry. It is of interest to point out that a second tetrahydrofuran ring (C-13-0-16) in boromycin would have produced a C2 symmetrical structure also except for the opposed configurations at C-9,9'. Nature, it seems, traced a different course for boromycin.

A SYNTHETIC BLUEPRINT FOR THE UPPER SEGMENT OF BOROMYCIN - THE $C_{18}H_{32}O_5$ LACTONE 4.

In a program directed at the total synthesis of the target boromycin, a reasonable first objective appears to be the $C_{18}H_{32}O_5$ lactone 4 which constitutes the upper segment of the antibiotic (9). Two modes of bond disconnection, namely at A and B (Fig. 7) produce

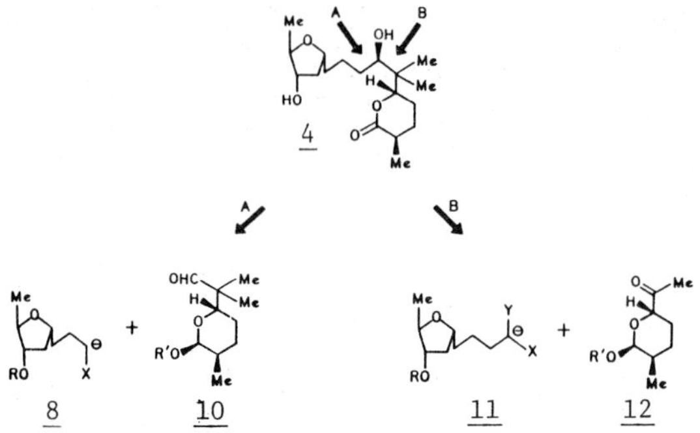

Fig. 7 Bond forming strategies for the $C_{18}H_{32}O_5$ lactone A

the reactive partners 8/10, and 11/12 respectively, in which the tetrahydrofuran units are nucleophilic components. If for the time being access to 8 and 10 is conceded, then their union is expected to produce 4 and/or its epimer. In the second instance however, (site B) the problem of generating the gem dimethyl group must be addressed since access to structures with functionalized quaternary carbon atoms is not a trivial process (10). In Fig. 8, we illustrate potential routes to 4 based on two such types of bond forming

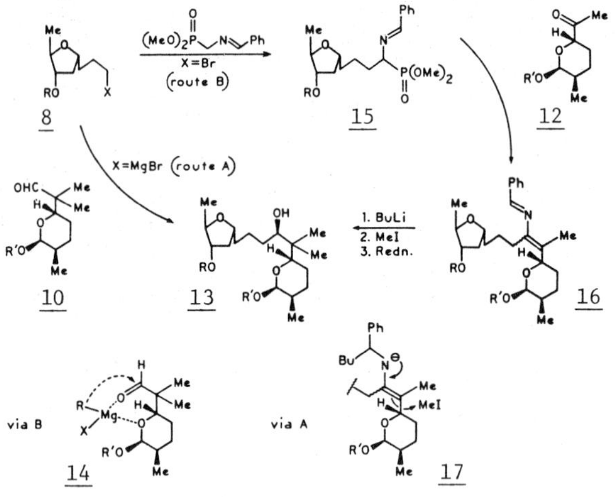

Fig. 8 Grignard and azadiene strategies

strategies. Thus, in route A, a Grignard reagent generated from 8 could be allowed to react with aldehyde 10, possibly via a coordinated species such as 14, to produce the desired 13. Controlling the stereochemical outcome of such Grignard reactions in related systems has been possible (11), hence the prospect of generating a preponderance of one epimer, as in 13. Union via route B presents the intriguing possibility of exploring the merits of azadiene chemistry (10). Thus, treatment of 8 (X=Br) with dimethyl benzylideneaminomethylphosphonate should produce the phosphonate 15 (Fig. 8), which in turn should lead to the azadiene intermediate 16. Alkylation of the latter with methyl iodide as described for other systems (10) should occur via species 17 to produce initially a carbonyl analog of 13, hence the desired alcohol after reduction. In actual practice, we were unable to reproducibly generate Grignard reagents from halides of 8 in spite of many efforts (P. Tyler) and route A was therefore abandoned. As it will be discussed later, we also had

difficulties in executing some of the chemistry proposed in route B and we sought
alternative strategies for the construction of structures corresponding to 13.

SYNTHESIS OF THE FUNCTIONALIZED TETRAHYDROFURAN UNIT

The C-10'-C-17' segment of boromycin comprises a 2,4,5-trisubstituted tetrahydrofuran
incorporating a three-carbon appendage as depicted in expression 8, (Fig. 9). Among several

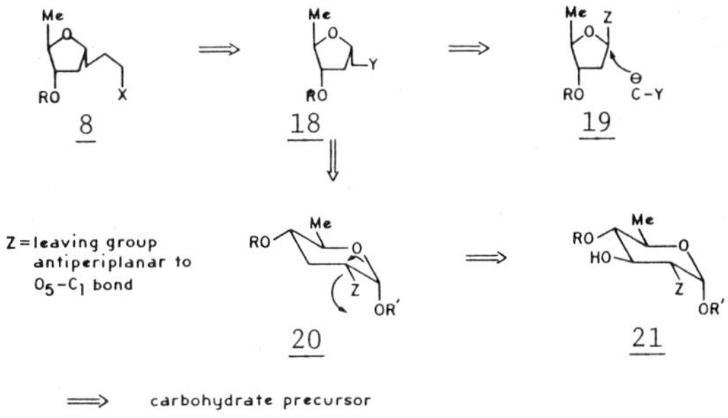

Fig. 9 Retrosynthetic analysis of the tetrahydrofuran unit 8

approaches to such chiral tetrahydrofurans one can envisage a C-glycoside approach (12) in
which an appropriately activated precursor (Z=leaving group) can be made to react with a
carbon nucleophile in stereoselective manner to give an α-substituted chiral tetrahydrofuran
18. Because of the lack of precedents in systems related to 19, it was decided to explore
another approach. The ring contraction of 2-aminocyclohexanols is a well known process (13)
which has been applied to 2-amino sugar derivatives (14). Thus, as shown in Fig. 9, if it
were possible to generate an entity such as 20, in which the orientation of the leaving
group Z is antiperiplanar to the O-5-C-1 bond, the conditions for ring contraction could be
present to give the tetrahydrofuran structure 18. If one further assumes that Z is an amino
group, then the analysis shown in Fig. 9 leads to the readily available
2-amino-2,6-dideoxy-D-glucose derivative 21 as a suitable starting material from which one
would have to generate 20 by an efficient deoxygenation reaction.

Development of a new leaving group - a planned diversion

Although there are a number of efficient procedures for the conversion of a hydroxyl
function into a deoxy group (15), factors concerned with neighboring group reactions, steric
hindrance and propensity for elimination may interfere at times. In a program concerned
with the design and reactivity of organic functional groups, we found that imidazolyl-
sulfonates (imidazylates), (Fig. 10), readily prepared from alcohols are excellent leaving

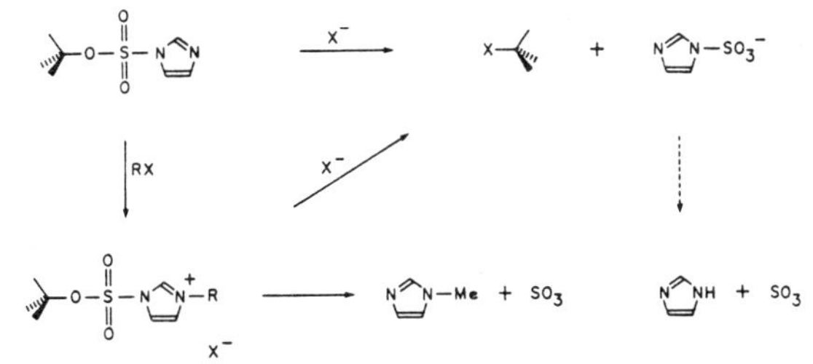

Fig. 10 Design and reactivity of imidazolylsulfonate - a new leaving group

groups in S_N2-type displacement reactions (16). The design of this entity was
predicated upon its anticipated nucleofugal reactivity based on kinetic and thermodynamic
criteria. While much attention has been paid to the nature of the nucleophile and the
solvent in such reactions, the need to accentuate the nucleofugal properties of the exiting
entity (sulfonate, etc) has not been fully exploited except for enhancement due to pKa
considerations and charge delocalization (triflates, p-nitrobenezenesulfonates, etc).
Thus, in addition to the inherent sulfonate-type character of imidazylates, consider the
possibility of reactivity enhancement by remote activation. Treatment of primary and some
secondary imidazylates with methyl iodide at room temperature led to the corresponding
iodides in high yield (16). Imidazylates have been found to be highly efficient in

displacement reactions of "difficult" cases where steric, and stereoelectronic effects on the one hand, and elimination on the other have played havoc with tosylates for example. Triflates (17) have been immensely useful, but unlike most imidazylates, their shelf-life is limited. The imidazylate methodology was found to be extremely useful in generating structures related to 20.

The readily available 2-amino-2-deoxy-D-glucopyranose derivative 22 (Fig. 11), was a

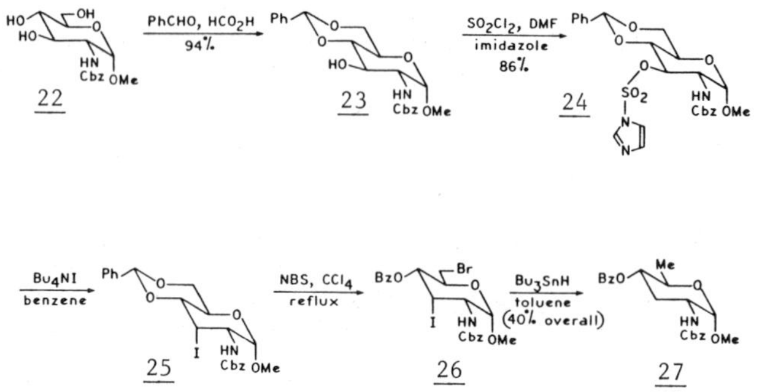

Fig. 11 Systematic manipulation of 2-amino-2-deoxy-D-glucopyranose

convenient starting material for the elaboration of the tetrahydrofuran unit 8 according to the plan shown in Fig. 9 (9). A series of routine transformations led to the crystalline imidazylate 24 which was efficiently displaced by iodide ion to give 25. It is noteworthy that unlike related cases of displacements vicinal to a neighboring amide or urethane, no participation took place and no oxazolidone derivatives were formed (18). Ring opening of the benzylidene acetal with NBS (19) and tri-n-butyltin hydride mediated reduction of both halogen-containing sites led to the desired intermediate 27 in good overall yield, which was further transformed into the amino sugar derivative 30, (Fig. 12). We now had to confront the

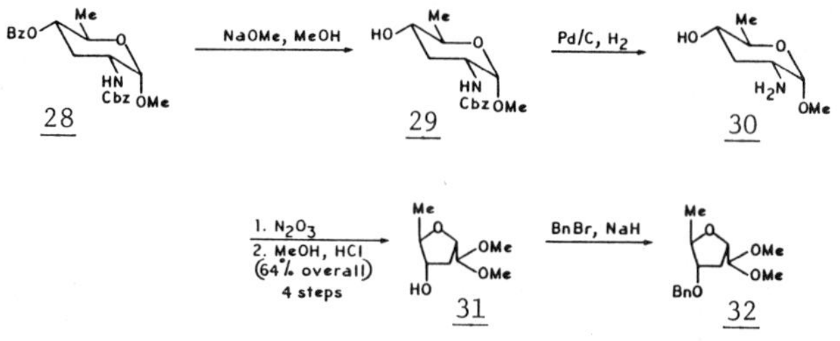

Fig. 12 Stereocontrolled ring contraction and generation of a chiral tetrahydrofuran unit.

critical issue of stereocontrolled ring contraction. This process took place under mild conditions particularly by using dinitrogen trioxide (20) and led to the trisubstituted tetrahydrofuran derivative 31, and 32 after benzylation. There now remained to extend the α-chain with appropriate end-group functionality that would enable us to effect the necessary bond forming reactions later in the synthesis. Since a three carbon appendage was required for one of our bond forming strategies (route A, Fig. 7), the Wittig procedure was clearly the preferred option. Figure 13 illustrates how 32 was chain-extended by standard methodology to produce the allylic alcohol 35 and the saturated bromide 36 via the aldehyde 32 and ester 34. Note that DIBAH reduction of 34 led to the allylic alcohol, while lithium aluminum hydride caused complete reduction. In either case bromination with triphenylphosphine and N-bromosuccinimide (21) proceeded in excellent yield. Many attempts to generate Grignard reagents from bromides 35 and 36 were unsuccessful, even by application of a number of recommended procedures (22). Having access to the alcohols 37 and 40 (Fig. 14), it was decided to explore the prospects of generating carbanionic species from the corresponding sulfoxides. Treatment of 37 and 40 individually with diphenyldisulfide and tributylphosphine (23) produced the corresponding thioethers in excellent yield and the latter were oxidized to isomeric sulfoxides 39 and 41. Since 41 could be a synthetic precursor to the "upper" segment of aplasmomycin (see Fig. 3), its reactivity vis-a-vis a model aldehyde such as trimethylacetaldehyde was studied. Unfortunately the sulfoxide anion underwent reversible ring opening presumably via an intermediate 43, leading to a mixture of open-chain and cyclic epimerized products. In an effort to stabilize the carbanion by internal coordination of the cation, we prepared the 2-pyridylsulfoxide

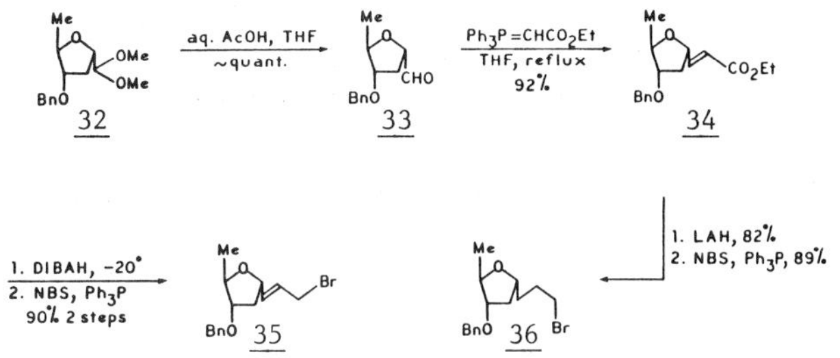

Fig. 13 Functionalized tetrahydrofuran units for coupling

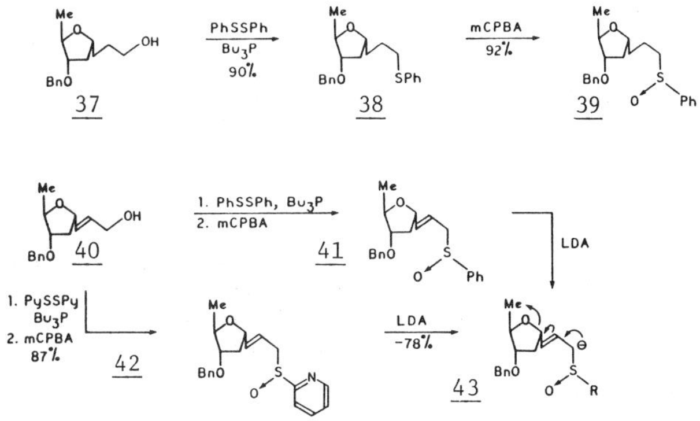

Fig. 14 Sulfoxide intermediates derived from the tetrahydrofuran unit

derivative 42 in excellent overall yield. Initial studies showed that such a derivative was also subject to ring-opening via 43 and we focussed our attention on the reactions of the boromycin-related intermediate 39.

SYNTHESIS OF THE FUNCTIONALIZED TETRAHYDROPYRAN UNIT

Having achieved a viable synthesis of a chiral and appropriately functionalized tetrahydrofuran unit for subsequent elaboration of the "upper" segment of boromycin, we now turned to the tetrahydropyran portion, which, as previously shown (Fig. 4), is a common unit to the "upper" and "lower" segments of our target. Examination of the structure in question 10, reveals two main challenges, namely regioselective incorporation of carbon-branching with control of stereochemistry, and elaboration of the quaternary carbon center. While other approaches can be envisaged, retrosynthetic analysis related the structure in question to a D-hexopyranose (Fig. 15). If one chooses to start with D-glucose, then the task ahead

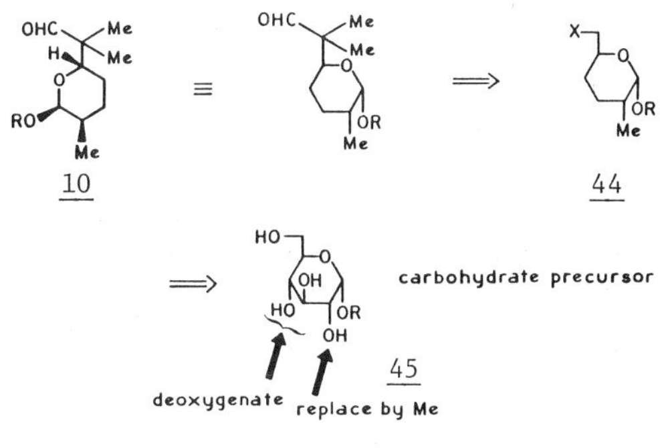

Fig. 15 Retrosynthetic analysis of the tetrahydropyran unit of boromycin

would be to deoxygenate at C-3/C-4, replace the C-2 hydroxyl by methyl and elaborate C-6 into the quaternary aldehyde unit. Provided that such operations can be performed efficiently, the D-glucose route could be very attractive, since the sense of chirality at C-5 in the sugar (ex. 44, 45) and C-4 (C-4') in the target 10 (boromycin numbering) is coincident. In Fig. 16 we illustrate how D-glucose was transformed into a variety of alkyl

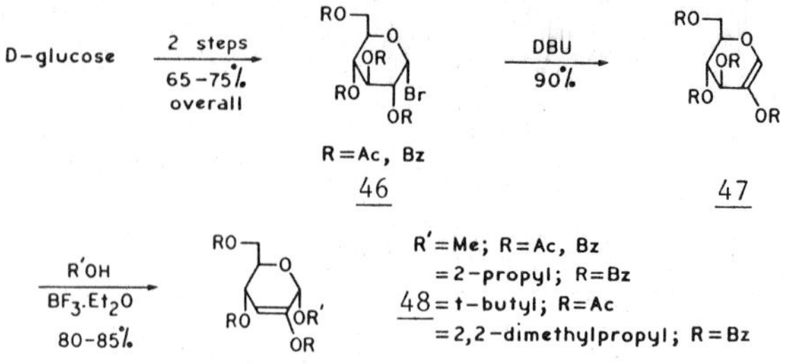

Fig. 16 Synthesis of alkyl tri-O-acyl-3-deoxy-α-D-hex-3-enopyranosides by the Ferrier procedure

tri-O-acetyl and O-benzoyl-3-deoxy-α-D-hex-3-enopyranosides such as 48 (R=Ac, R'=t-Bu). The necessity to prepare and use glycosides having bulky aglycones will be discussed in coming paragraphs, but suffice it to mention that all intermediates in the sequence were highly crystalline and the Ferrier procedure (24) to produce such alkyl 3-deoxy-α-D-hex-3-enopyranosides was adaptable to sterically demanding alcohols as well. Based on previous experience in our laboratory (25), a remarkable reaction was discovered (Y. Chapleur), in which a β-acetoxy enol ester such as 48 was transformed into the diene 49 upon treatment with methylenetriphenylphosphorane (Fig. 17). The enol ester carbonyl in 48 appears to

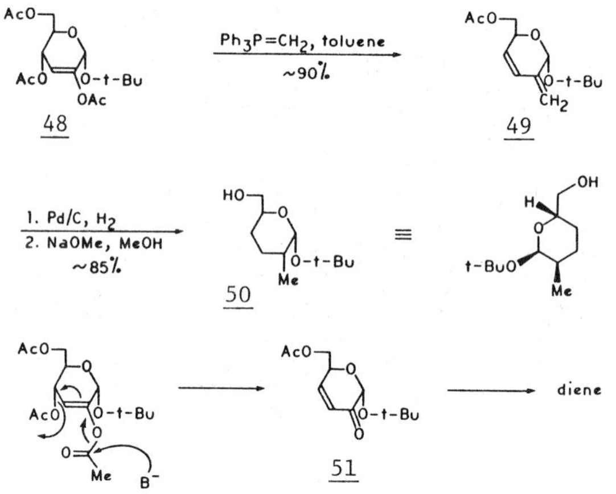

Fig. 17 Synthesis of the chiral tetrahydropyran unit and establishment of stereochemistry.

behave more like a ketone function and we envisage a process whereby a basic entity generates the enone 51, which undergoes the anticipated Wittig reaction to give the diene. As expected, treatment of the enone 51, prepared by a different route, under the same conditions, led to 49. Having achieved efficient deoxygenation at C-3/C-4 from D-glucose in a minimum number of steps, the challenge was now to effect stereocontrolled reduction of 49 so as to produce a chiral tetrahydropyran unit containing a C-methyl substituent with the desired sense of chirality. It is in anticipation of such a process that we have elected to work with a glycoside carrying a bulky aglycone such as 48. Catalytic reduction of the diene 49 followed by deacetylation produced the tetrahydropyran derivative 50 in which the desired isomer was present in greater than 9:1 ratio.

With the chiral tetrahydropyran unit 50 in hand, we were now faced with the challenge of elaborating the side-chain. Our initial plans called for the application of azadiene chemistry(10) as shown in Fig. 18. Standard methodology gave the methyl ketone 52 which was found to react with the required phosphonate reagent to give the azadiene 53. The anticipated course of events upon reacting 53 with n-butyllithium and addition of methyl iodide to create the quaternary center, are depicted in expression 54. Unfortunately after limited experimentation it was found that the anion in 54 chose to follow a different course of events as shown in 56, leading to ring opening (G. Demailly). It is possible that with a more careful study of reaction parameters such a temperature and time, this reaction could

Fig. 18   The azadiene approach for the elaboration of the side-chain in 10

be turned into a successful interprise leading to the desired 56 and on to 10 by mild hydrolysis of the Schiff's base. In the interest of pursueing our synthetic objectives however, we chose to forgoe the elegance of the Martin methodology (10) in favor of practicality and a more classical approach. Thus, ketone 51 was transformed into the enone 57 which was treated with lithium dimethylcuprate and the resulting enolate trapped to give the silyl enol ether 58 having the double-bond in the desired position. Ozonolysis of 57 produced the desired 59 in very good overall yield (Fig. 19).

Fig. 19   Synthesis of the chiral tetrahydropyran 59

ASSEMBLY OF THE "UPPER" SEGMENT OF BOROMYCIN

Having access to the chiral and functionalized units 39 and 59, we were now ready to address the critical issue of assembling the "upper" segment of the target. In Fig. 20 we

Fig. 20   Coupling methodology for the "upper" segment of boromycin illustrate the synthesis of the β-hydroxysulfoxide 60, its desulfurization and the formation of a 1.5:1 mixture of alcohols 61 and 62 respectively. By subsequent transformations, it was established that the slightly preponderant product 61 was related to the natural product. In the interest of operational efficiency, it was found that the unwanted isomer 62 could be recycled by an oxidation-reduction process, to give additional quantities of 61. There remained to transform 61 into the target lactone 65 (Fig. 21). In view of the

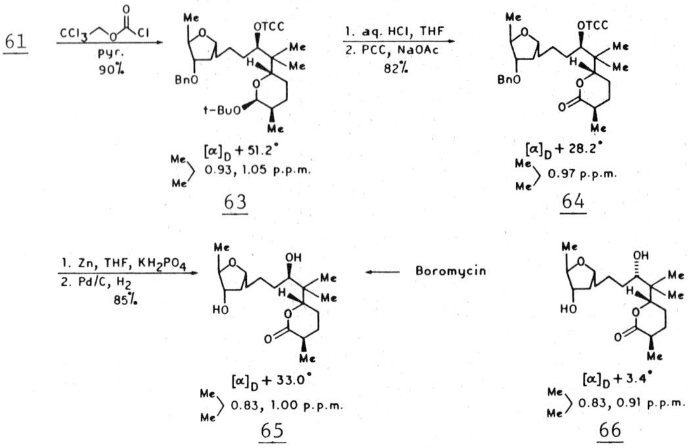

Fig. 21 Synthesis of the $C_{18}H_{32}O_5$ degradation product 65 from boromycin and its 9' epimer

type of transformations envisaged, a convenient O-protecting group was sought and the trichloroethoxycarbonyl group was selected. In spite of its hindered nature, the 9' hydroxyl group was easily trichloroethoxycarbonylated to give 63. Mild acid treatment produced the corresponding lactol, which after oxidation with pyridinium chlorochromate (26), gave lactone 64. Removal of the carbonate and benzyl protecting groups afforded lactone 65 which was found to be identical (400 MHz n.m.r., mass spec. etc) with material obtained from the degradation of boromycin (7). In the interest of comparison, the 9' epimeric lactone 66 was also prepared via the same sequence. Note the surprisingly different magnitudes of optical rotation and the different chemical shift for one of the two methyl groups in the gem-dimethyl system, which can be attributed to different conformers (rotamers) of the side-chain. With a viable synthesis of the $C_{18}H_{32}O_5$ lactone completed and its stereochemical and constitutional identity established, we turned our attention to the completion of the entire upper portion of the target. The next synthetic hurdle was the introduction of the glycolic acid chain at the lactone carbonyl by a Claisen-type or related condensation. The choice of an O-protected glycolic acid derivative was evident provided that the corresponding anion could be made and its reactivity exploited. Coincident with these plans was a report by Meinwald and coworkers (27) who demonstrated the addition of glycolate anions to lactones and the isolation of the corresponding adducts (Fig. 22).

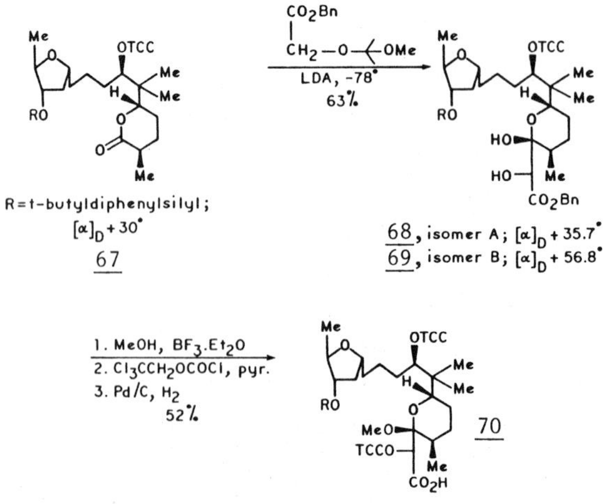

Fig. 22 Elaboration of the C-1'-C-2' glycolic acid chain - Assembly of the "upper" segment of boromycin

After extensive experimentation, we found that benzyl-O-(methoxy-2-propyl) glycolate was a convenient source for the anion. In addition the O-trichloroethoxycarbonyl and t-butyldiphenylsilyl groups appeared to be compatible with the reaction conditions. Condensation with 67 gave two compounds in a 1:1 ratio corresponding to the 2'(R) and 2'(S) isomers which could be separated by chromatography. Treatment of the mixture with methanol in the presence of boron trifluoride etherate gave the corresponding methyl glycosides. It should be pointed out that the stereochemical depiction for the anomeric center is based on the reasonable assumption that attack of the glycolate anion would take place from the β-(least hindered)side of the carbonyl function in 67. Glycoside formation is also expected to proceed under the thermodynamic control to provide the α-D-glycoside. Protection of the glycolic acid hydroxyl group and catalytic debenzylation provided the entire "upper" segment of boromycin in protected form and in good overall yield. Although compound 70 exists as a mixture of epimers, it is possible to separate them and to recycle the unwanted 2'(S)-isomer. During the course of this study we prepared a number of glycolate esters in order to study the possibility of stereocontrolled Claisen condensation and mild cleavage of the ester function to the corresponding acid. Thus the O-benzyl pyridylthio 71, methylthiomethyl 72, methoxyethoxymethyl 73, and phenylthiomethyl 74 glycolates were prepared from O-benzyl glycolic acid 76 by standard procedures (Fig. 23). Esters 72 and 73 were found not to be stable during conditions of methanolysis in the presence of boron trifluoride etherate. The glycolate anions produced from 72 and its phenyl analog were found to be unsuitable as a result of side-reactions. In view of our interest in mild conditions of ester cleavage, we prepared the sulfone ester 75 and found that it was transformed into the parent acid 76 in the presence of aqueous sodium bicarbonate.

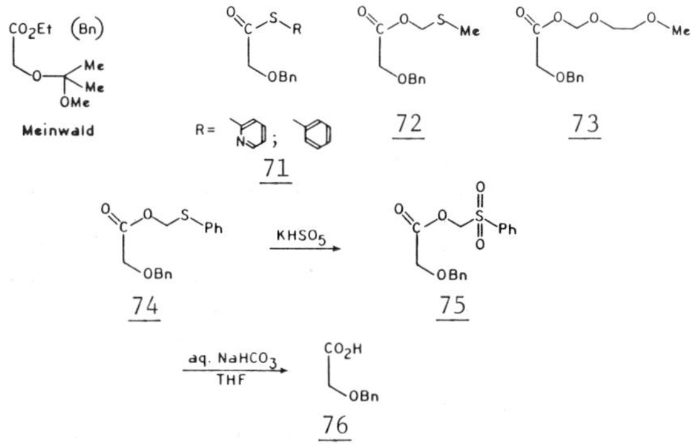

Fig. 23 Glycolate esters

Unfortunately anion formation from 75 was not regioselective as judged from condensations with model lactones. In an effort to study the stereochemistry of glycolate anion additions, we looked at the reaction of O-tetrahydropyranyl menthyl and bornyl esters 78 with a model lactone 77. While condensation was very effective, there did not seem to be an overwhelming difference in isomer ratio (Fig. 24).

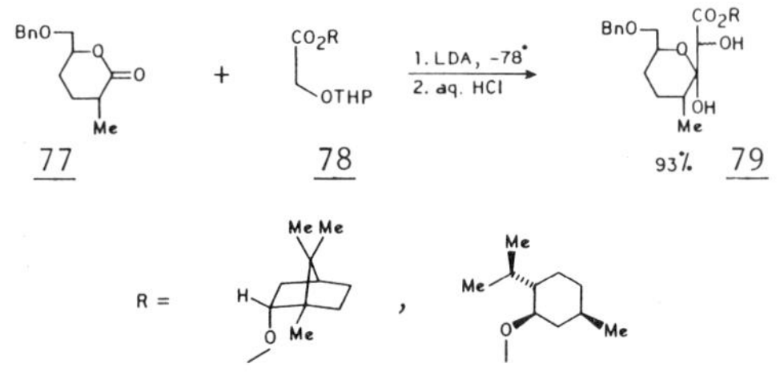

Fig. 24 Glycolate anions with chiral ester groups

A STRATEGY FOR THE "LOWER" SEGMENT OF BOROMYCIN

The C-1-C-17 segment of boromycin can be subdivided into three units - the glycolic acid chain (C-1-C-2); the tetrahydropyran unit and the acyclic unit (C-10-C-17) (Fig. 25). The latter can be related to a chiral starting material containing two contiguous asymmetric

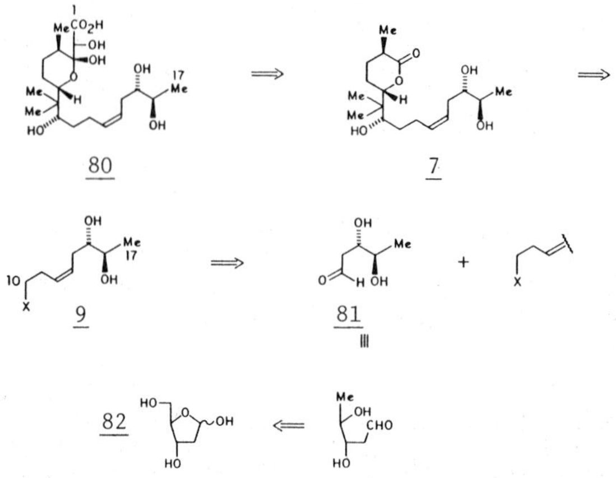

Fig. 25 Retrosynthetic analysis of the C-1-C-17 "lower" segment of boromycin

centers. Considering the type of functionality leads one to propose (R,S)tartaric acid as a convenient starting material. In spite of this, it should be appreciated that such a choice is not without its manipulative problems, since both ends have to be differentiated and elaborated individually. Another possibility could be based on the recognition of partially hidden carbohydrate-type symmetry. Thus bond disconnection of 9 at C-12/C-13 produces a dideoxypentose 81 which can be derived from 2-deoxy-D-erythro-pentose 82, in turn readily available from D-arabinose by the Gray procedure (28). Figure 26 illustrates how this process leads to the ketene dithioacetal 84 which can be efficiently reduced to 85

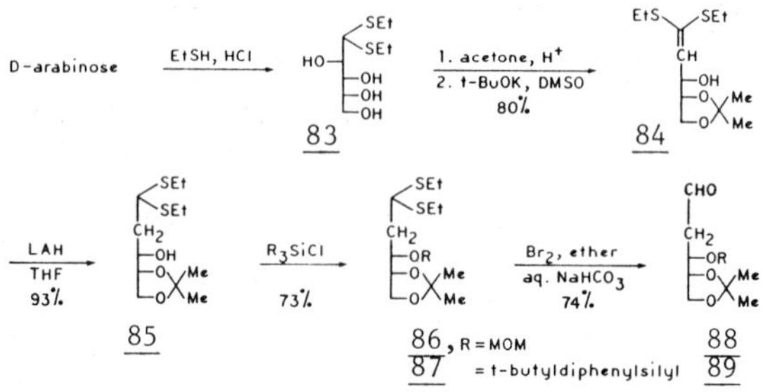

Fig. 26 Synthesis of protected 2-deoxy-D-erythro-pentose diethyl dithioacetals

and protected as methoxymethyl 86, or t-butyldiphenylsilyl 87, ethers (D. Delorme). Deprotection of the dithioacetal derivatives gave the corresponding aldehydo sugars 88 and 89. These were useful models for the incorporation of a cis-double bond as required in the target. It was found that Wittig reaction in a mixture of THF and HMPA (3:1 ratio) led to the olefin 90 in good yield (Fig. 27). Next, we looked at methods for the deoxygenation at

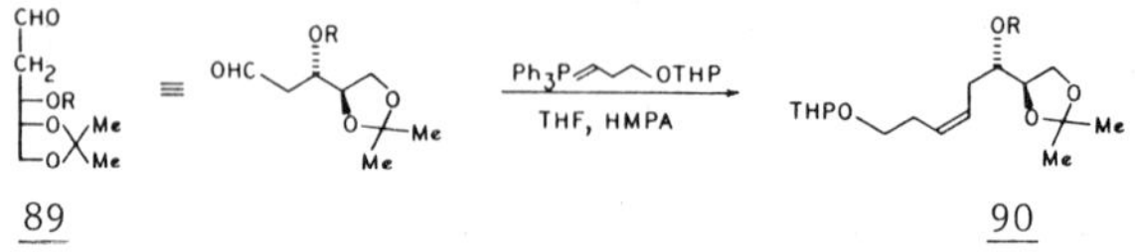

Fig. 27 Formation of a cis-olefin in a model

C-5 to produce the desired five-carbon chiral unit (C-13-C-17). As shown in Fig. 28, one such route was based on the preparation of the 5-tosylate 91 which could be reductively detosylated with lithium triethylborohydride (29). However, migration of the silyl group had also occurred in the process to give 92. Furthermore, regeneration of the aldehyde led

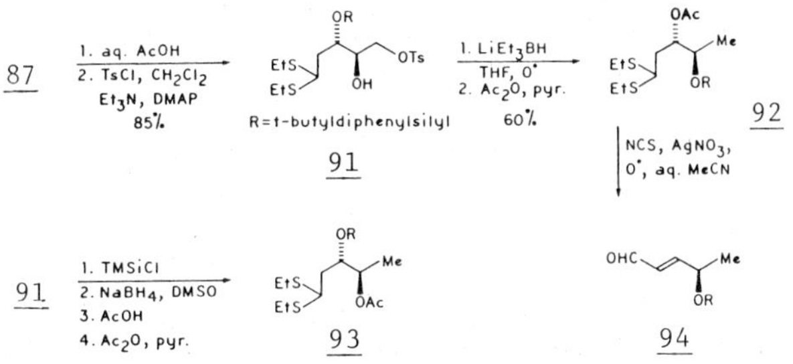

Fig. 28 Attempts to deoxygenate at C-5

to the elimination product 94. Protection of the hydroxyl group in 91 as the trimethylsilyl ether, followed by reduction with sodium borohydride in dimethylsulfoxide (30) and functional group manipulation gave the desired intermediate 93. Further chemistry proved problematic and other avenues were explored with the objective of preparing the C-13-C-17 aldehyde in a suitably protected form.

The literature reports examples of intramolecular participation by an alkylthio group in 5-O-p-tolylsulfonyl pentose dithioacetals (31) which results in the migration of the alkylthio group to the terminal position. This process is most prevalent when the acyclic conformational features of the substrate is conducive to such intramolecular attack. We therefore reasoned that tosylation of the dithioacetal derivative 95 (Fig. 29) would proceed

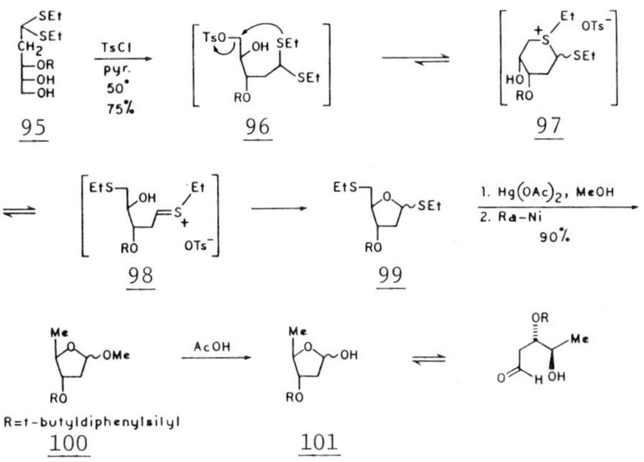

Fig. 29 Synthesis of 2,5-dideoxy-D-erythro-pentose via sulfonium and thionium ion intermediates

through 96 which could rearrange through the intermediacy of sulfonium 97 and thionium 98 ions, leading ultimately to the thioglycoside 99. Indeed treatment of 95 under normal tosylating conditions, but at 50° led to the transposed derivative 99 which has a potential deoxy function at C-5. Mercuric ion catalyzed methanolysis followed by desulfurization gave the desired methyl 2,5-dideoxy-D-erythro-pentofuranoside 100 protected as the 3-O-t-butyl-diphenylsilyl ether. Mild acid hydrolysis produced the lactol 101.

By a combination of the methods described above, we were able to construct the C-1-C-17 "lower" segment of boromycin 104 via intermediates 102 and 103 (Fig. 30).

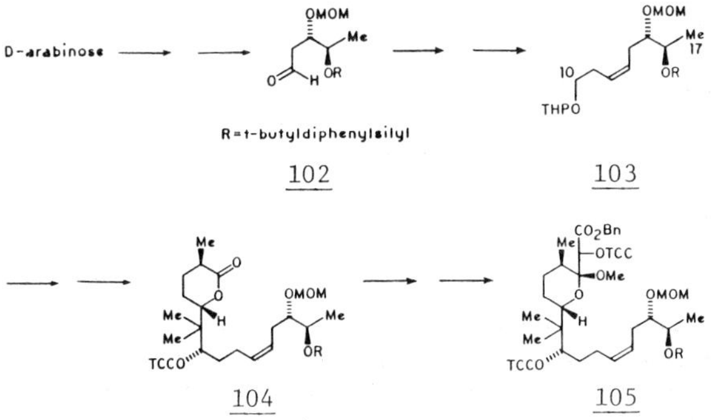

Fig. 30 Assembly of the "lower" segment (C-1-C-17) of boromycin

Glycolate condensation using the protected lactone 104 gave the adduct as a mixture of epimers which could be easily separated. Methanolysis and protection then led to 105 which represents the "lower" segment of boromycin as a differentially protected derivative. The stereochemical and constitutional identity of lactone 104 was confirmed by comparison with material obtained from the degradation of boromycin (Fig. 31).

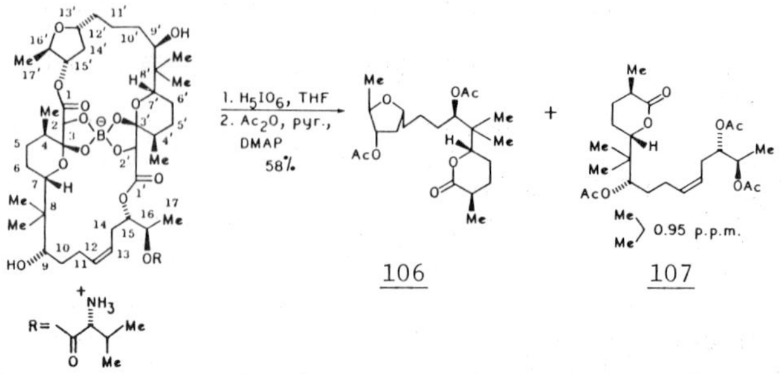

Fig. 31 Degradation of boromycin and the isolation of "upper" and "lower" lactones

SEMISYNTHESIS OF DES-VALINE BOROMYCIN

As a parallel activity, we have been successful in synthesizing des-valine boromycin from des-valine-des-boron boromycin, thus achieving a semi-synthesis of the des-valino antibiotic (P. Tyler) (Fig. 32). A related study was recently reported by White and coworkers (32) as part of their elegant synthetic studies toward the same objective (33).

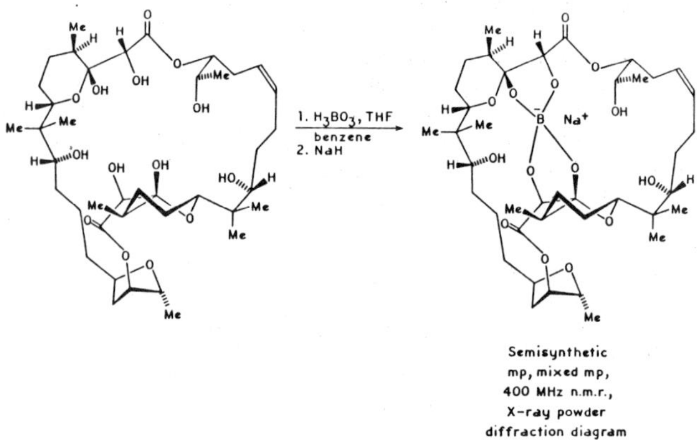

Fig. 32 Semi-synthesis of des-valine boromycin from boromycin

Acknowledgment - We gratefully acknowledge financial assistance from the National Science and Engineering Council of Canada, from the Ministry of Education of Quebec and fellowships to D. Delorme (NSERC), G. Demailly, Y. Chapleur (CNRS, France). We also thank Drs. H. Beierbeck and P.M. Tan for registering 400 MHz n.m.r. spectra, Prof. V. Prelog and Dr. H. Gschwend for samples of boromycin.

REFERENCES

1. See for example, R.B. Woodward in, Perspectives in Organic Chemistry, A.R. Todd, ed., Interscience, 1956, pp. 155-184; R.E. Ireland, Organic Synthesis, Prentice Hall Inc., Englewood Cliffs, New Jersey, 1969; S. Turner, The Design of Organic Synthesis, Elsevier Scientific Publishing Co., Amsterdam, 1976; S. Warren, Designing Organic Syntheses - A Programmed Introduction to the Synthon Approach, J. Wiley & Sons Inc., New York, 1978.
2. E.J. Corey, Pure Appl. Chem., 14, 30 (1967).
3. S. Hanessian, Acc. Chem. Res., 12, 159-165 (1979); Plenary lecture, North American Medicinal Chemistry Symposium, Toronto, Canada, June 20-24, 1982.
4. See for example, W.A. Szabo and H.T. Lee, Aldrichimica Acta, 13, 13-20 (1980); D. Seebach, in Modern Synthetic Methods, p. 91-171, R. Scheffold, ed., Otto Salle Verlag, Frankfurt am Main, Germany (1980).
5. See for example, B. Fraser-Reid and R.C. Anderson, Progress Chem. Org. Nat. Prod., 39, 1-61 (1980); A. Vasella, in Modern Synthetic Methods, pp. 173-267, R. Scheffold, ed., Otto Salle Verlag, Frankfurt am Main, Germany (1980).
6. R. Hütter, W. Keller-Schierlein, F. Krüsel, V. Prelog, G.C. Rodgers, Jr., P. Suter, G. Vogel, W. Voser and H. Zähner, Helv. Chim. Acta, 50, 1533-1539 (1967).

7. J.D. Dunitz, D.M. Hawley, D. Miklos, D.N.J. White, Y. Berlin, R. Marusic and V. Prelog, Helv. Chim. Acta, 54, 1790-1713 (1971); W. Marsh, J.D. Dunitz and D.N.J. White, Helv. Chim. Acta, 57, 10-17 (1974).
8. H. Nakamura, Y. Iitaka, T. Kitahara, T. Okazaki and Y. Okami, J. Antibiot., 30, 714-719 (1977).
9. S. Hanessian, P.C. Tyler, G. Demailly and Y. Chapleur, J. Am. Chem. Soc., 103, 6243-6246 (1981); see also Abstr. Papers 27th Natl. Org. Symp., Nashville, Tenn., June 21-25, 1981, pp. 67-69.
10. S.F. Martin and G.W. Phillips, J. Org. Chem., 43, 3792-3794 (1978); S.F. Martin, Tetrahedron, 36, 419-460 (1980).
11. S. Hanessian, G. Rancourt and Y. Guindon, Can. J. Chem., 56, 1843-1846 (1978); M.L. Wolfrom and S. Hanessian, J. Org. Chem., 27, 1800-1804 (1962); D.J. Cram and K.R. Kopecky, J. Am. Chem. Soc., 81, 2748-2755 (1959).
12. S. Hanessian and A.G. Pernet, Advan. Carbohydr. Chem. Biochem., 33, 111-188 (1976).
13. M. Chérest, H. Felkin, J. Sicher, F. Sipos and M. Tichy, J. Chem. Soc., 2513-2520 (1965) and references cited therein.
14. See for example, J. Defaye, Advan. Carbohydr. Chem. Biochem., 25, 181-228 (1970); J.M. Williams, Advan. Carbohydr. Chem. Biochem., 31, 9-79 (1975) and references cited therein.
15. See for example, N.R. Williams and J. Wander in The Carbohydrates, pp. 761-798; W. Pigman and D. Horton, eds., Academic Press, N.Y. (1980); S. Hanessian, Advan. Carbohydr. Chem., 21, 143-208 (1966).
16. S. Hanessian and J.-M. Vatèle, Tetrahedron Lett., 22, 3579-3582 (1981).
17. See for example, R.W. Binkley, M.G. Ambrose and D. G. Hehemann, J. Org. Chem., 45, 4387-4391 (1980); T.H. Haskell, P.W.K. Woo and D.R. Watson, J. Org. Chem., 42, 1302-1305 (1977).
18. P.H. Gross, K. Brendel and H.K. Zimmerman, Angew. Chem., Internat. Edit., 3, 379-380 (1964).
19. S. Hanessian and N.R. Plessas, J. Org. Chem., 34, 1035-1044 (1969).
20. P. Angibeaud, J. Defaye and H. Franconie, Carbohydr. Res., 78, 195-204 (1980).
21. S. Hanessian, M.M. Ponpipom and P. Lavallée, Carbohydr. Res., 24, 45-56 (1972).
22. See for example, R.M. Coates and M.W. Johnson, J. Org. Chem., 45, 2685-2697 (1980).
23. I. Nakagawa and T. Hata, Tetrahedron Lett., 1409-1412 (1975).
24. R.J. Ferrier, N. Prasad and G.H. Sankey, J. Chem. Soc., C., 581-591 (1969); see also R.J. Ferrier, Methods Carbohydr. Chem., 6, 307-311 (1972).
25. S. Hanessian, P.C. Tyler and Y. Chapleur, Tetrahedron Lett., 22, 4583-4586 (1981); S. Hanessian, G. Demailly, Y. Chapleur and S. Léger, J.C.S. Chem. Commun., 1125-1126 (1981).
26. E.J. Corey and J.W. Suggs, Tetrahedron Lett., 2647-2650 (1975).
27. A.J. Duggan, M.A. Adams, P.J. Byrnes and J. Meinwald, Tetrahedron Lett., 4323-4326 (1978).
28. M.Y.H. Wong and G.R. Gray, J. Am. Chem. Soc., 100, 3548-3553 (1978).
29. S. Krishnamurthy and H.C. Brown, J. Org. Chem., 41, 3065-3066 (1976).
30. H. Weidmann, N. Wolf and W. Timp, Carbohydr. Res., 24, 184-187 (1972).
31. N.A. Hughes and R. Robson, J. Chem. Soc., 6, 2366-2368 (1966).
32. M.A. Avery, J.D. White and B.H. Arison, Tetrahedron Lett., 22, 3132-3126 (1981).
33. J.D. White, Abstr. Papers 179th National Meeting of the American Chemical Society, Houston, Texas, March 24-28, 1980; ORG 48.

STEREOCHEMICAL CONTROL IN MACROCYCLE SYNTHESIS

E. Vedejs, J. M. Dolphin, D. M. Gapinski and H. Mastalerz

S. M. McElvain Laboratory of Organic Chemistry, University of Wisconsin,
Madison, Wisconsin 53705, USA

Abstract - Stereocontrolled synthesis of the C_1-C_9 segment of erythronolide is described. Introduction of stereochemistry at C_5 and C_6 depends on a prediction of transition state conformation for osmylation of a 9- or 10-membered ring alkene. Local conformational effects are believed to account for selectivity, and their importance is probed in several other medium ring alkenes.

Crystal structures of natural products having rings of 8- or more members are now available for numerous carbocycles, lactones, and lactams. We were interested to find that among these structurally diverse molecules, certain conformational features can be attributed to the substitution pattern along a short ring segment. For example, olefins having an alkyl group in the allylic position prefer similar <u>local conformations</u> in spite of large variations in ring size and substitution.[1] In the case of α-branched Z-olefins, the favored solid state geometry can be represented by the structure 1-A where the alkyl branch point occupies a pseudoequatorial orientation close to the olefin plane.

In the case of α-branched E-olefins, a striking preference for a crownlike local geometry similar to 1-C is observed. The E-double bond is turned perpendicular to the approximate plane of the ring to minimize transannular interactions, and the alkyl branch point is in the pseudoequatorial arrangement.[1]

If similar local conformational preferences are dominant in the transition state for olefin addition reactions, it would be possible to predict the stereochemistry of newly formed bonds with respect to the alkyl branch point. Ideally, predictions might be feasible using simple transition state models based on local conformation, and without resorting to a more detailed approximation of the actual conformational equilibria by computational methods.[2]

Best results using this approach are expected for addition reactions having a reactant-like transition state. Thus, attack on a Z-olefin conformer 1-A should occur from the olefin face opposite to the ring carbons ("peripheral attack").[2] Conformations based on the alternative local geometry 1-B should be less reactive (as well as less stable) because the direction of peripheral attack is now hindered by the pseudoaxial R substituent.

Similar arguments can be advanced for E olefins, but caution must be used in making predictions. Although individual conformers such as 1-C should have a large bias for peripheral attack, it is not safe to assume that the favored local conformer will also be the most reactive. The pseudoequatorial alkyl branch point extends forward with respect to the olefin plane, and may interfere with approach by external reagents. Alternative local conformers with C-R eclipsed with the olefin plane may be more reactive. Also, the unique factors which favor "perpendicular" olefin rotamers similar to 1-C should be very sensitive to the degree of rehybridization in the transition state. Relatively late transition states would certainly encounter important transannular interactions.

Figure 1

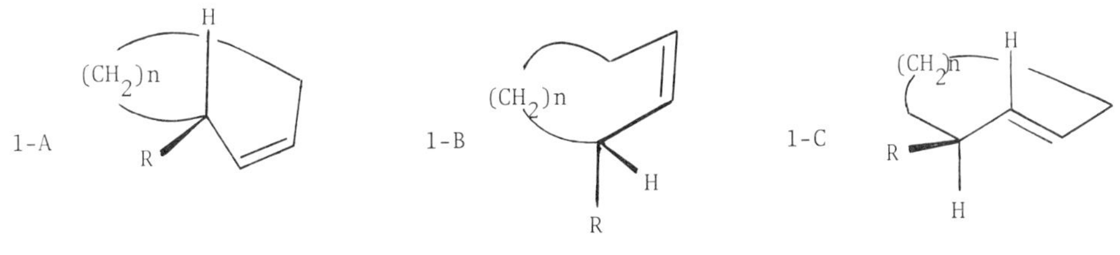

Our goal is to learn how large and how general such local conformational factors might be. To this end, we have looked at several olefin addition reactions using simple medium ring substrates.

Isomeric 1,3-dimethyl cyclododecenes are available from 2,12-dimethylcyclododecanone as shown in Fig. 2. The meso "syn" dimethylcyclododecanol 2-A affords a mesylate which elminates to the Z olefin upon reaction with KOtBu. The isomeric mesylate obtained from d,l alcohol 2-B is quite resistant to base-induced elimination, but the corresponding xanthate ester eliminates thermally to a 2:1 E:Z mixture.

Figure 2

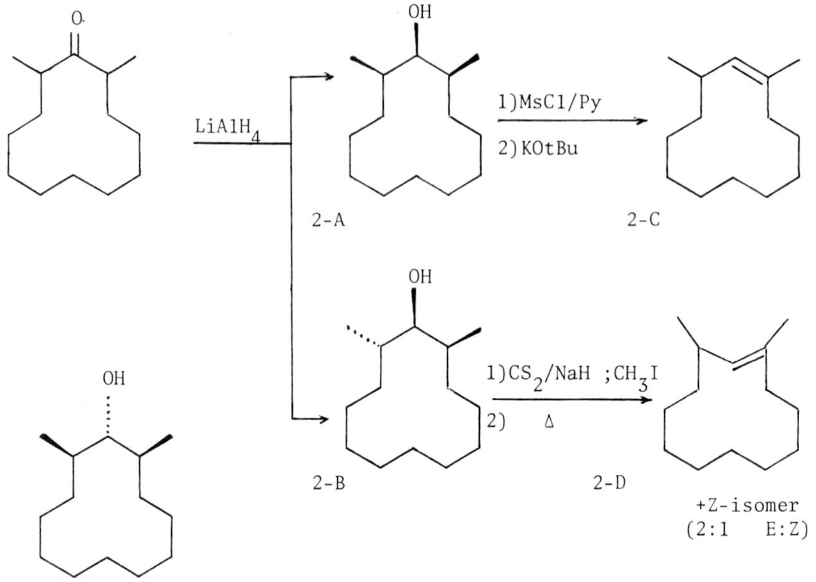

For our first attempt to deduce stereochemical preferences in this series, we chose the hydroboration of Z-olefin 2-C. Depending on which olefin face is attacked, one would expect either the d,l alcohol 2-B, or a new meso "anti" isomer (Fig. 3). Hydroboration of the Z alcohol gives meso "anti" dimethylcyclododecanol 3-B as the major product, as predicted by peripheral attack on local conformer 3-A. A surprisingly large amount (20% of products) of a tertiary alcohol 3-C of unknown stereochemistry is also formed in this experiment.

Two other cis addition reactions (epoxidation; OsO_4 hydroxylation) have been studied. Thus, epoxidation of Z-dimethylcyclododecene affords a single epoxide 4-A within limits of NMR detection (≥95% one isomer). Elimination occurs to the methylene alcohol 4-B using the diethylaluminum-TMP reagent[3] and hydrogenation gives a mixture of meso and d,l alcohols 4-C and 4-D, ca. 1:1. Meso "anti" alcohol 4-C can only arise from epoxide stereochemistry 4-A; the isomeric epoxide would have given meso "syn" product, none of which can be detected. Thus, epoxidation occurs from the same olefin face as hydroboration, and with even greater selectivity.

Figure 3

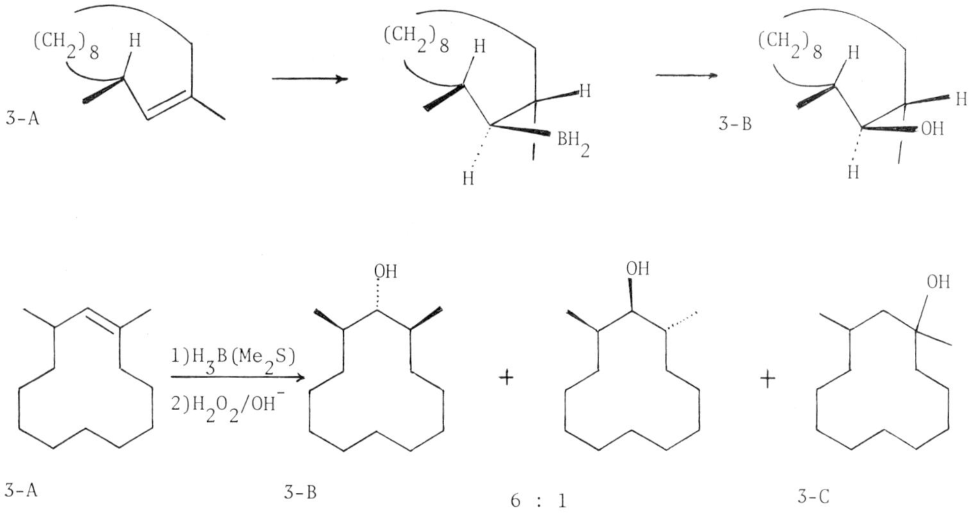

Figure 4

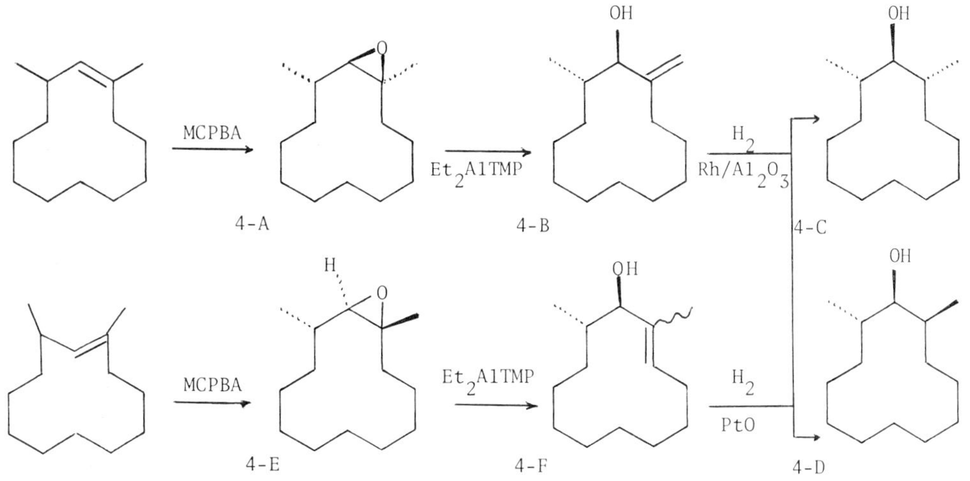

A similar sequence of reactions starting from E-1,3-dimethylcyclododecene is also illustrated in Fig. 4. In this case, epoxide 4-E eliminates with Et_2Al-TMP to a 1:1 E,Z mixture of endocyclic olefins 4-F, hydrogenation of which affords a mixture of the same two alcohols 4-C and 4-D as obtained from the Z olefin series. Again, the stereochemistry of meso alcohol 4-C proves the stereochemistry assigned to epoxide 4-E. In Fig. 5, we show that this stereochemistry is expected from peripheral attack on a local crownlike geometry with pseudoequatorial methyl.

Figure 5

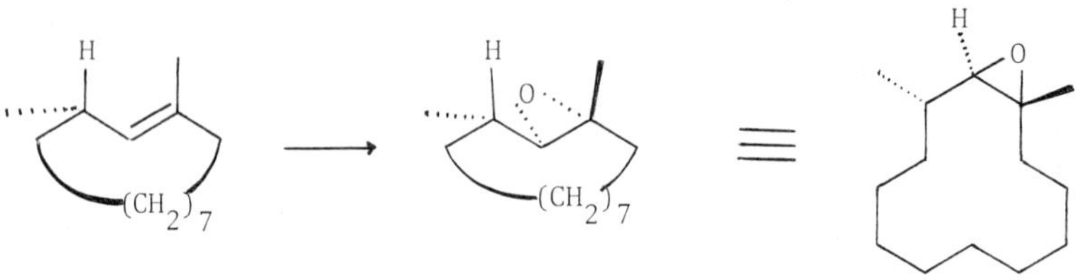

Osmylation of the E and Z olefins is also highly selective (Fig. 6). The stereochemistry of diol 6-A is proved by correlation with known products from earlier experiments. Thus, allylic alcohol 6-B (identical to 4-B) affords a 1:1 mixture of epoxides 6-C and 6-D upon treatment with MCPBA. Reduction of 6-C + 6-D with $LiAlH_4$ gives two diols, one of which is identical with 6-A. The other diol 6-E is formed upon osmylation of E-1,3-dimethylcyclododecene.

Figure 6

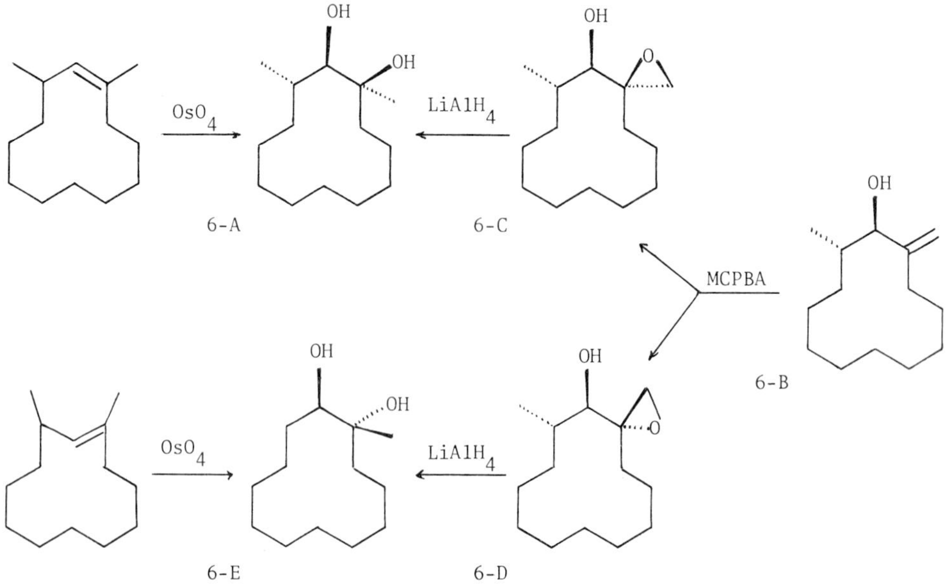

In summary, all of these cis-additions of E or Z 1,3-dimethyldodecene occur with high
selectivity as predicted using a simple model of local conformation having pseudoequatorial
alkyl substituents. One consequence of this geometry of addition is that the relative
stereochemistry of attack at the olefinic C_2 (secondary carbon) is the same for both E and
Z olefins.

The selectivity of addition is decreased in the analogous disubstituted E-olefin isomer
(Fig. 7). Thus, MCPBA epoxidation of 7-A gives a 6:1 ratio of epoxides in favor of the
predicted 7-B while osmylation is nonselective (1:1 diol mixture). In contrast, the
Z olefin 7-C (obtained using an olefin inversion sequence from 7-A) reacts with good selectivity. As discussed earlier, the favored local conformer of E olefins is not necessarily
the most reactive. Apparently, the disubstituted E-olefin is quite capable of reacting
via geometries differing from crownlike 7-D. The absence of a C_1-methyl group in this
system raises the possibility that alternative eclipsed geometries such as 7-E might become
more appealing to the bulky reagent OsO_4L_2. In the case of Z olefin 7-C, the absence of
a C_1 methyl group causes no great perturbation relative to the Z-trisubstituted olefin
studied earlier. This result can be understood by inspection of the pseudoequatorial local
conformer 7-F. There is little reason to believe that changing R from methyl to hydrogen
would significantly alter the local geometry.

Figure 7

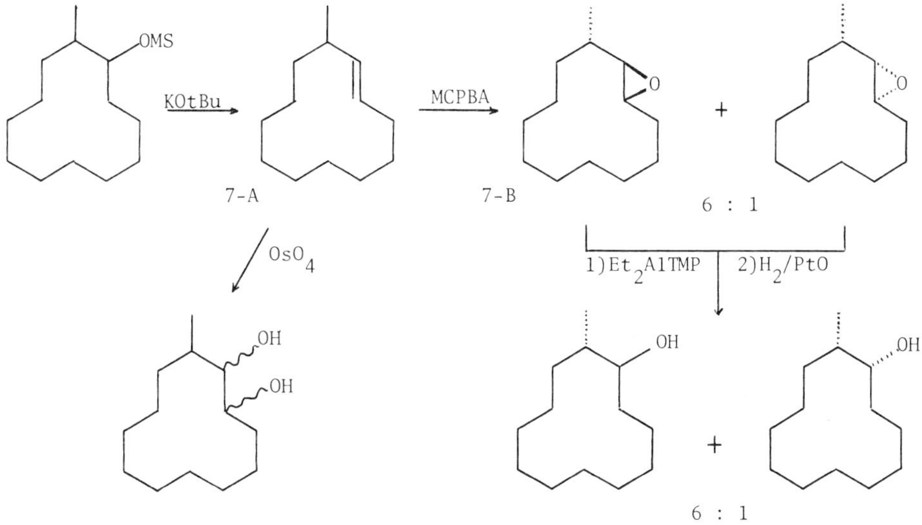

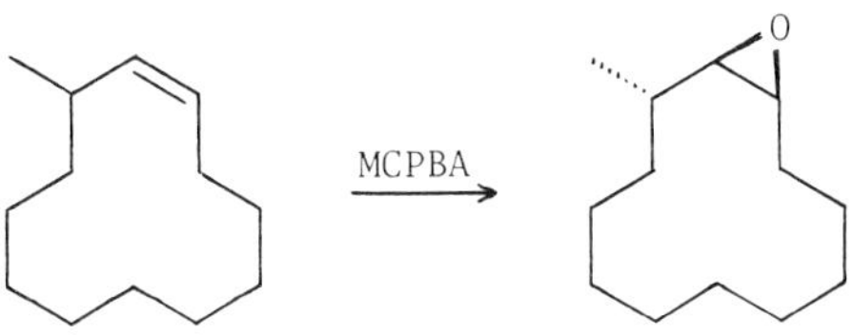

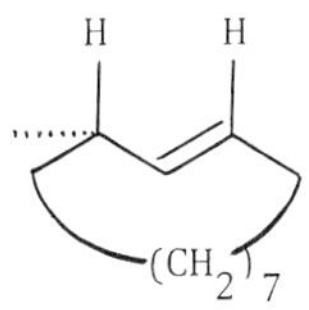

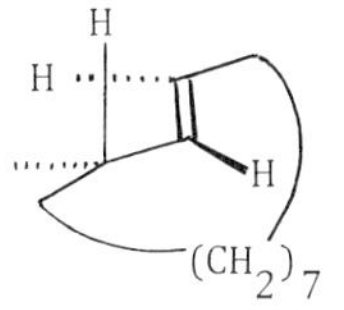

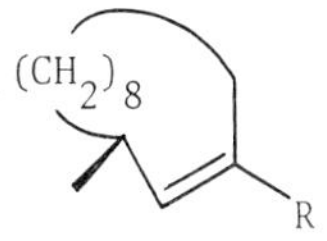

We have applied principles of local conformational control of stereochemistry to the synthesis of a C_1-C_9 fragment of erythronolide (Fig. 8). Inspection of the erythronolide substitution pattern (8-A) reveals several options for introducing stereochemistry by cis addition to E or Z olefins. For strategic reasons, we have chosen an approach based on the osmylation of a C_5-C_6 double bond in some intermediate equivalent to 8-B.

Figure 8

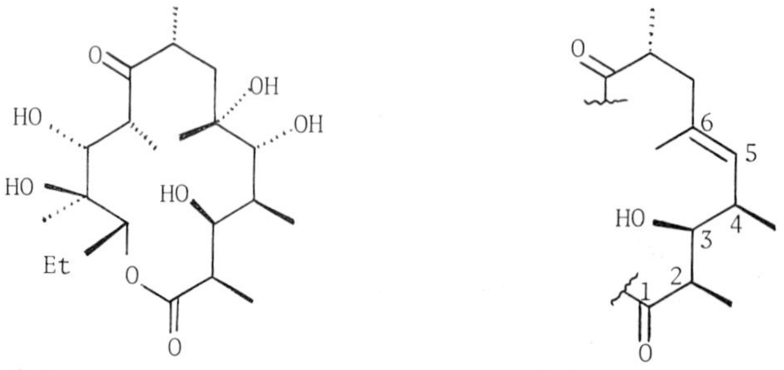

Our approach follows a sulfur-mediated ring expansion sequence.[4] A stereospecific synthesis of the required ring expansion substrate 9-D is described in Fig. 9. Key steps include a regiospecific Diels-Alder addition of α-oxodithio ester 9-B, which is available by ylide cycloreversion from 9-A.[5] Least hindered hydroboration of 9-C using thexylborane controls three of the asymmetric centers of erythronolide, and a series of routine steps gives 9-D in 42% overall yield.

Figure 9

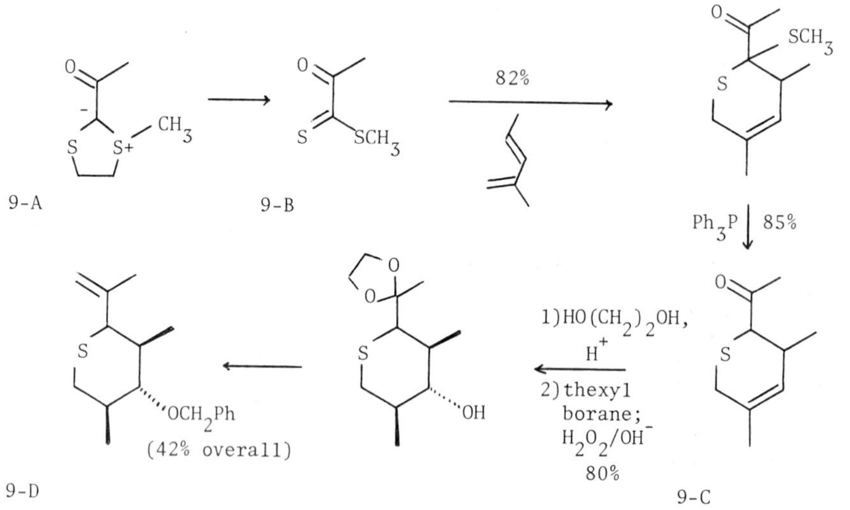

Ring expansion by 2,3-shift occurs smoothly upon treatment of the α-propenyl thiane 9-D with the trifluorosulfonate ester of ethyl lactate followed by DBU (Fig. 10). The only significant product 10-B (86%) has E olefin geometry according to NOE evidence, a stereochemical result which is expected from rearrangement of an equatorial ylide 10-A.[4,6] In order to convert 10-B into the erythronolide substitution pattern as in 10-C, it is now necessary to hydroxylate the alkene. The erythronolide stereochemistry is predicted by the usual arguments based on local conformational control.

Figure 10

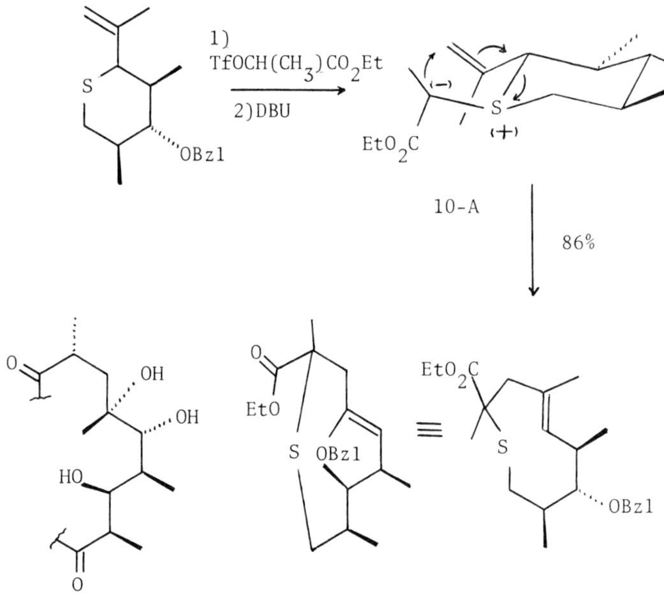

In Fig. 11, we illustrate the local geometry predicted for the E-olefin 11-A. NMR evidence (NOE effects; methine-methine J-values) suggests that crownlike geometry extends from C_2 to C_7. Reaction of 11-A with $OsO_4 \cdot Py_2$ gives a single diol 11-B, together with a small amount of a lactone 11-C in a ratio of about 30:1. The lactone does not arise from 11-B, and is probably derived from the other diol diastereomer. The stereochemistry of 11-B is proved by a sequence of desulfurization and acetonide formation, resulting in 11-D. This substance is closely related to erythronolide segments prepared by Heathcock and by Masamune, and the stereochemistry follows from comparison of acetonide methine coupling constants.[7] Thus, 11-A undergoes osmylation with high selectivity from the predicted olefin face.

Figure 11

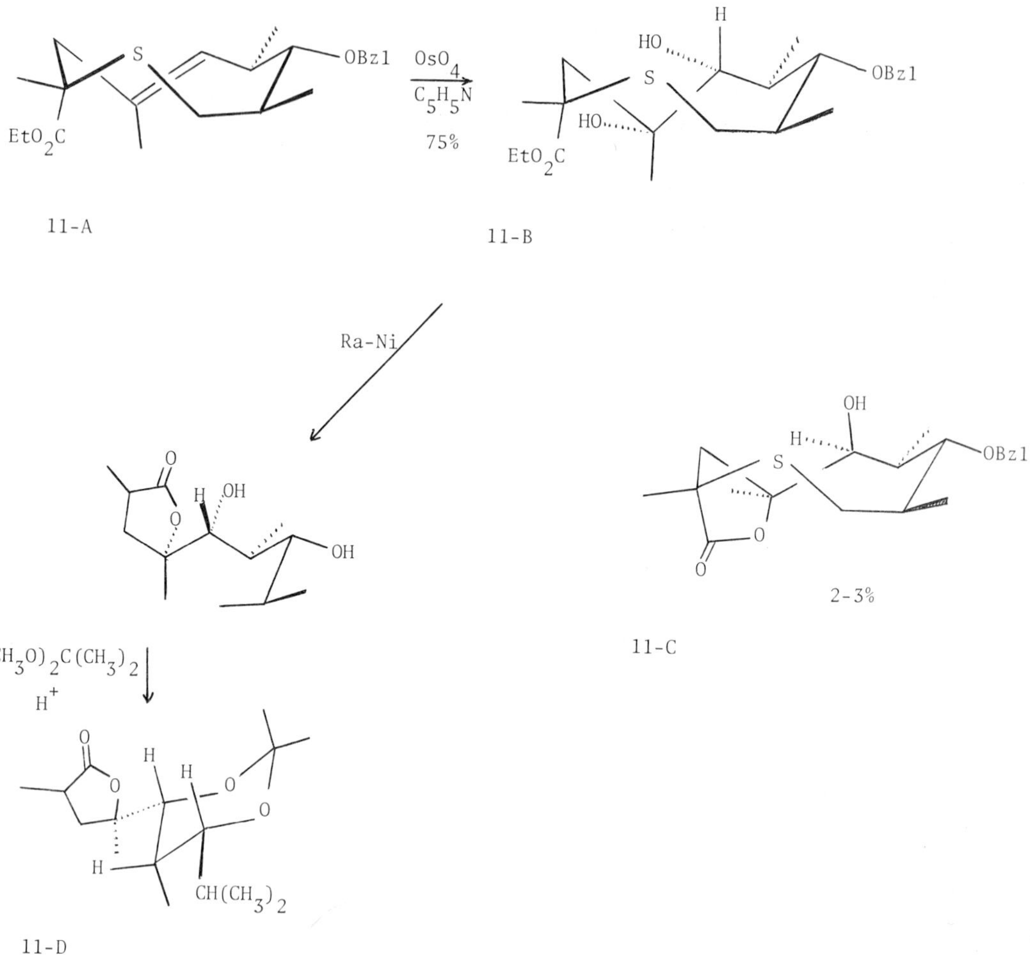

With the natural stereochemistry established, we turned to the problem of removing sulfur and introducing the C_1 oxygen function (Fig. 12). This can be done easily by known methods based on sulfide chlorination (NCS), solvolysis of the α-chlorosulfide, and reduction of a labile aldehyde intermediate.

Figure 12

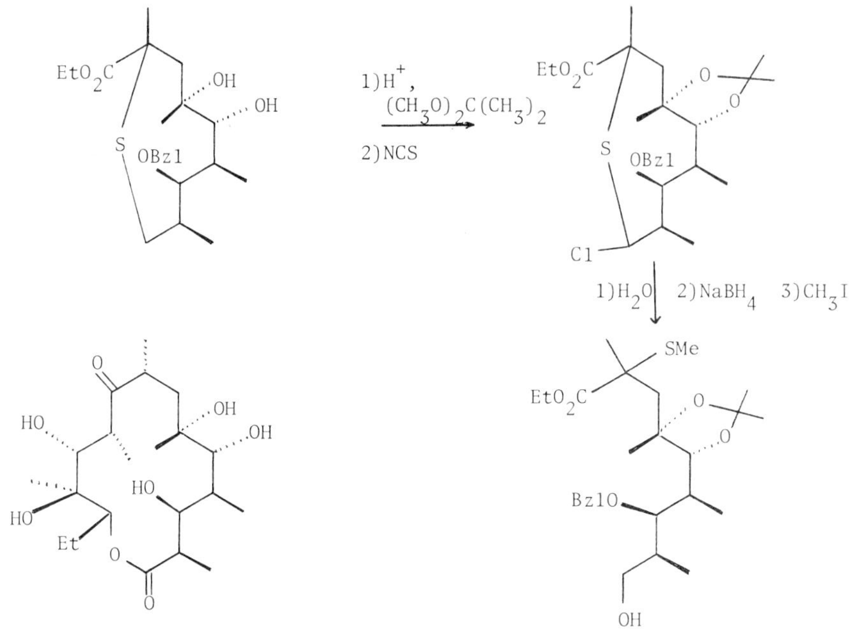

Work continues in our laboratory to introduce the remaining carbons of erythronolide, but for now I would like to return to the issue of stereochemical control to insert a cautionary note. We have explored one alternative ring expansion strategy to the C_1-C_9 erythronolide fragment as described in Fig. 13. The key ring expansion step (reaction of 13-A with dichloroketene, 3,3-shift)[8] occurs in 85% yield to give thiol lactone 13-B.

When the dechlorination product (Zn/HOAc) 13-C is treated with osmium tetroxide, the intermediate diol undergoes spontaneous acyl transfer to afford a γ-lactone 13-F. Stereochemical correlation as before (Ra-Ni; acetonide formation) reveals that 13-F has the unnatural stereochemistry at both new asymmetric centers! The initially formed diol must have been 13-E; we cannot find the "correct" diastereomer 13-D or derived products. For reasons not yet fully understood, the 10-membered ring thiol lactone osmylates with opposite olefin face selectivity compared to the 9-membered cyclic sulfide which was described earlier.[9]

Our first response to this result was to doubt the assignment of E-olefin geometry to either the 9- or the 10-membered ring compounds. However, there is no error in assignments. Thus, osmylation of the acyclic ester sulfide 13-G gives a 1.3:1 ratio of 13-F and 13-H. The latter has been converted into 11-D, a derivative of the 9-membered sulfide series, by treatment with Ra-Ni, dimethoxypropane, and LDA/CH_3I. Therefore, both ring expansion methods must produce E-olefins.

Figure 13

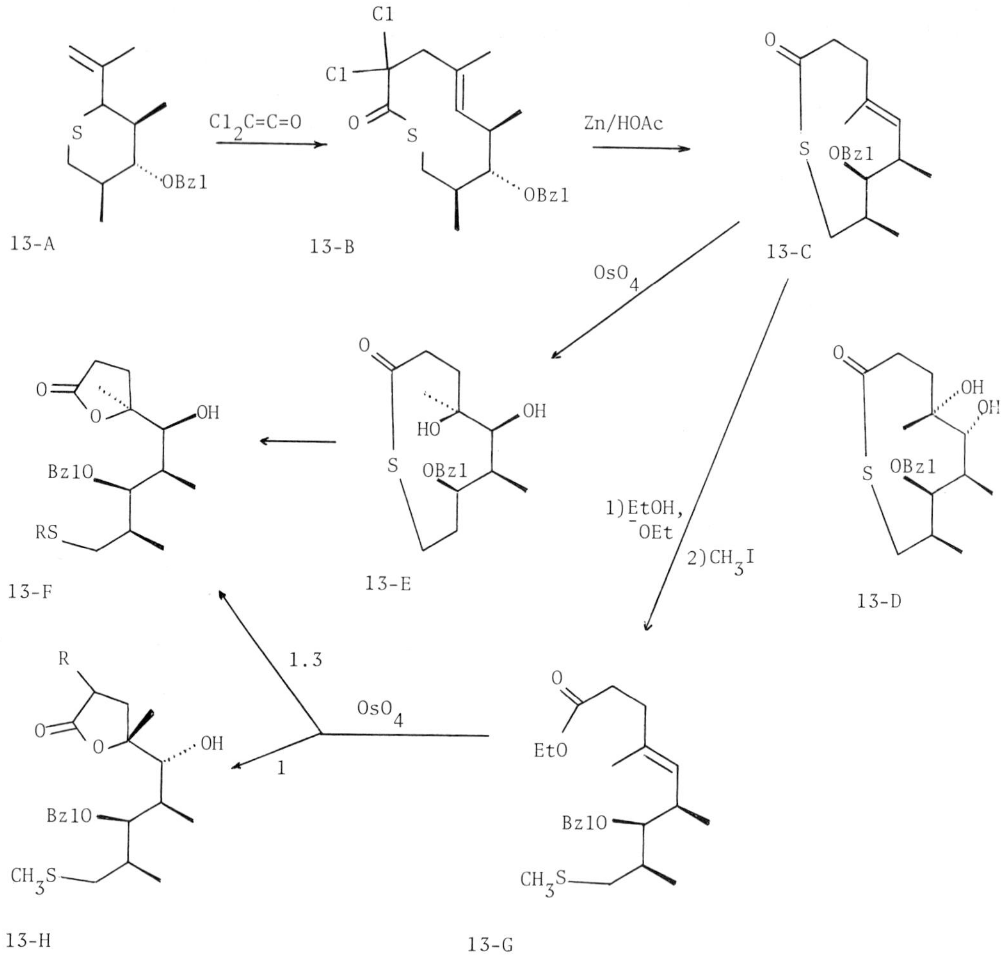

One final series of experiments was performed to complete our exploration into the conformational aspects of osmylations. The Z olefin 14-A was prepared from 13-C by photosensitized isomerization. Osmylation occurs from the predicted olefin face to give an intermediate diol 14-B which rearranges to the lactone 14-C during chromatography. The stereochemistry is proved as usual via Ra-Ni desulfurization and acetonide formation. Apparently, the influence of a transannular thiol lactone group does not interfere with normal stereoselectivity in the Z olefin series.

Figure 14

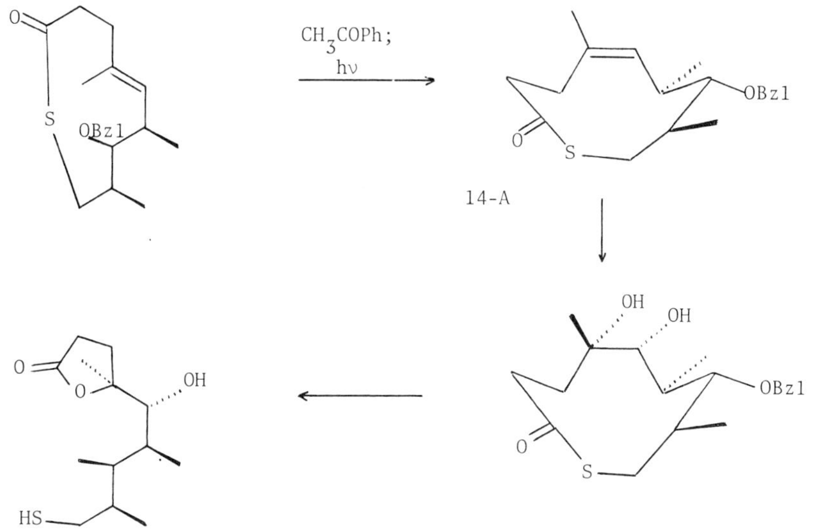

In conclusion, stereochemical predictions based on local geometry have proved reliable in all cases involving α-branched Z-olefins. The analogous E-olefins pose a more subtle problem. Predictable stereocontrol has been obtained with simple trisubstituted E-alkenes, and with the erythronolide series based on 9-membered cyclic sulfides. The contrasting behavior of 10-membered thiol lactone 13-C and the lower selectivity of addition to disubstituted E-alkene 7-A provide evidence for more than one accessible transition state geometry. Much remains to be learned regarding origins of these effects.

REFERENCES

1. Examples including endocyclic E or Z alkene, epoxide, lactone: A. J. Weinheimer, J. A. Matson, D. van der Helm, and M. Poling, Tetrahedron Lett. 1295 (1977); H. A. Whaley, C. G. Chidester, S. A. Mizsak, and R. J. Wnuk, ibid. 3659 (1980).

2. W. C. Still and I. Galynker, Tetrahedron 37, 3981 (1981). This paper also includes many interesting examples of kinetic stereochemical control by remote substituents, as well as useful representations of computer generated medium ring conformations.

3. A. Yasuda, H. Yamamoto, and H. Nozaki, Bull. Chem. Soc. Jpn. 52, 1705 (1979).

4. E. Vedejs, J. P. Hagen, B. L. Roach, and K. L. Spear, J. Org. Chem., 43, 1185 (1978); E. Vedejs, M. J. Arco, D. W. Powell, J. M. Renga, and S. P. Singer, ibid. 43, 4831 (1978).

5. E. Vedejs, M. J. Arnost, J. M. Dolphin, and J. Eustache, J. Org. Chem. 45, 2601 (1980).

6. V. Ceré, C. Paolucci, S. Pollicino, E. Sandri, and A. Fava, J. Org. Chem. 46, 3315 (1981).

7. S. Masamune, M. Hirama, S. Mori, Sk. Asrof Ali, and D. S. Garvey, J. Am. Chem. Soc. 103, 1568 (1981); C. H. Heathcock, J. P. Hagen, E. T. Jarvi, M. C. Pirrung, and S. D. Young, ibid. 103, 4972 (1981). We are grateful to Prof. Masamune and Prof. Heathcock for comparison spectra.

8. For analogous reactions, see G. Rossini, G. G. Spineti, E. Foresti, and G. Pradella, J. Org. Chem. 46, 2228 (1981); R. Malherbe and D. Bellus, Helv. Chim. Acta 61, 3096 (1978).

9. The transannular proximity of thiol lactone and olefin π systems is the most obvious perturbation in 13-C compared to 11-A. An endocyclic ester is in effect a second E-double bond, and conformational changes due either to strain or electronic effects are quite conceivable.

Acknowledgement. This work was supported by the National Science Foundation.

STEREOCHEMICAL CONTROL IN MACROCYCLIC COMPOUNDS

W. Clark Still

Department of Chemistry, Columbia University, New York, NY 10027, USA

ABSTRACT: Macrocyclic compounds have conformational properties which are quite useful for stereochemical control in the synthesis of natural products. This paper describes recent work using macrocyclic stereocontrol which is directed toward the synthesis of the germacrane eucannabinolide and the 16-membered macrolide antibiotic rosaramicin.

Since completion of my work some years ago on the cyclodecanoid pheromone periplanone B, I have been convinced that macrocycles offer some extremely interesting opportunities for stereocontrol in organic synthesis. In order to obtain preliminary information on the potential of such macrocyclically controlled asymmetric induction, Igor Galynker examined a variety of reactions on monomethylated macrocyclic ketones and lactones. His study demonstrated that high remote asymmetric induction is a common feature of macrocyclic addition and substitution reactions and that both the extent and direction of stereoselection are related to the strain energy of the macrocycles involved. Some of his results are shown below and are for kinetic enolate methylation, dimethyl cuprate addition or catalytic hydrogenation. With the exception of the final hydrogenation which proceeded with a 10:1 trans:cis ratio, all reactions below gave the product shown with at least 20:1 stereoselection.

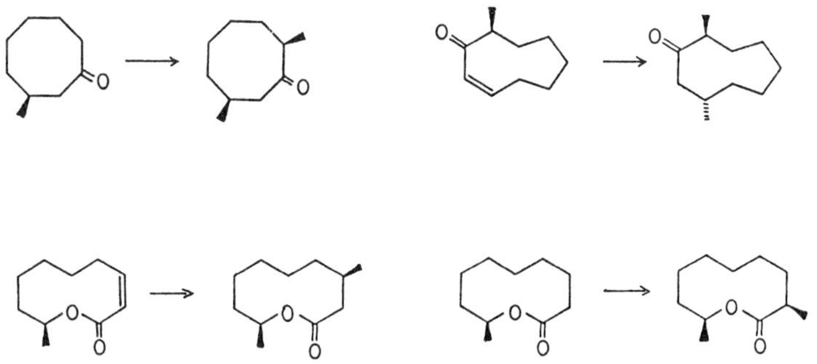

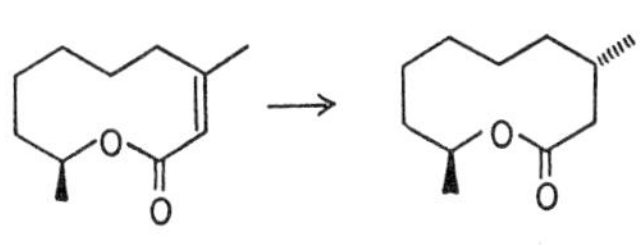

Igor Galynker
Tetrahedron, 37, 3981 (1981)

Study of the low energy conformations of the macrocycles involved in these reactions provides insight into the stereoselectivity observed. With the 10-membered lactone, for example, the products of enolate alkylation and cuprate addition appear to derive from the conformations shown below. The darkened positions show the less strained conformations of 7-, 8- and 9-methylated nonanolides and the major products were found to correspond to peripheral reagent addition to those predicted conformations.

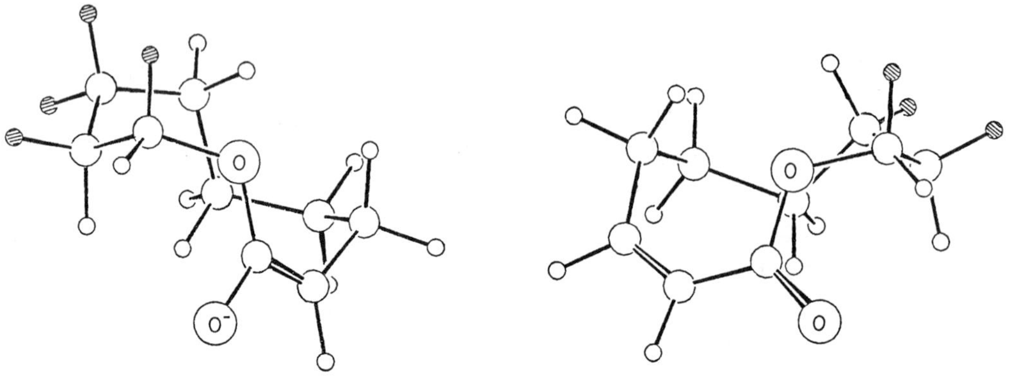

Since simple macrocycles seem to provide predictable and useful asymmetric induction, we have begun syntheses of a number of natural products using macrocyclic stereocontrol as a central strategy. Two of these target molecules are eucannabinolide, a cytotoxic germacranolide, and rosaramicin, a 16-membered macrolide antibiotic. In this paper, I will describe our synthetic work on these compounds.

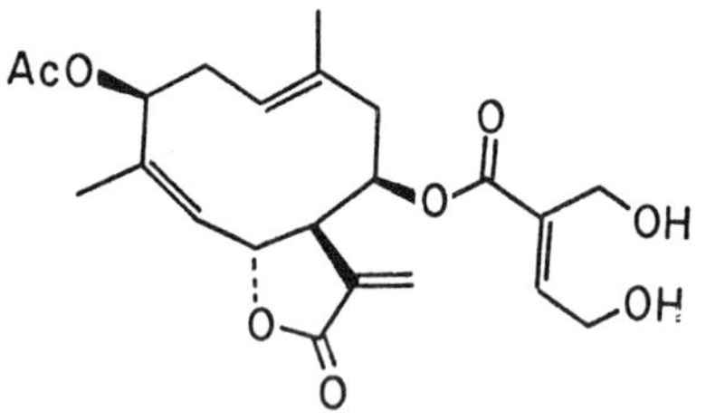

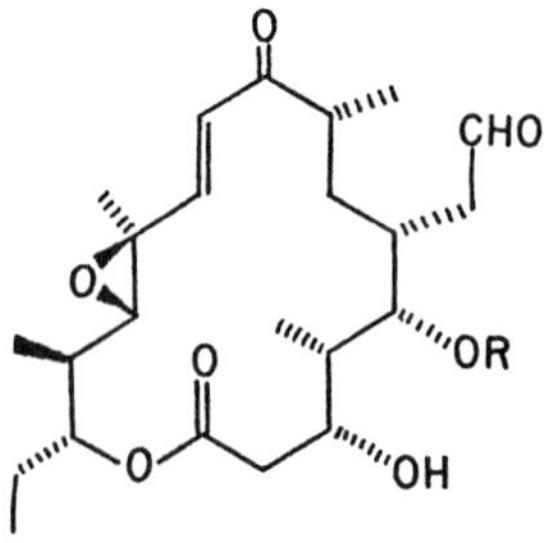

Eucannabinolide   Rosaramicin

SYNTHESIS OF EUCANNABINOLIDE

Previous results from our laboratory have demonstrated that the oxy-Cope rearrangement provides a smooth pathway from monoterpenoid starting materials to appropriately functionalized germacrane-like intermediates as shown below. Application of the route to germacranolide synthesis, however, is problematic because of the characteristic oxygenation at C8 which threatens beta-elimination after the ring expansion takes place.

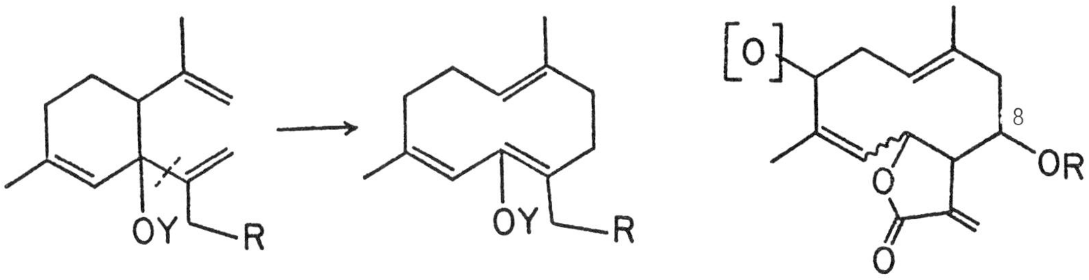

Stereochemical and regiochemical uncertainties are also present since the configurations at C6 and C7 (at least) and the direction of lactonization of the acrylic acid appendage would need to be set while on a conformationally flexible macrocyclic framework. Effective solutions to these problems have been found which allow rational construction of the germacranolide eucannabinolide (1). An account of these solutions follows.

Our synthesis began with (+)-carvone. Reduction (LiAlH$_4$, Et$_2$O, 0 deg), epoxidation (MCPBA, CH$_2$Cl$_2$, 25 deg) and protection (PhCH$_2$OCH$_2$Cl, i-Pr$_2$NEt, 25 deg) gave an epoxyether in >70% yield. The epoxide was eliminated via the selenoxide (1. PhSeK-LiBr, THF, 25 deg; 2. 30% H$_2$O$_2$, NaHCO$_3$, NaOAc, THF, 60 deg, 16 hrs) to a tertiary allylic alcohol which was oxidized (Jones' reagent, 0 deg, 1.5 hrs) to the required enone 3.

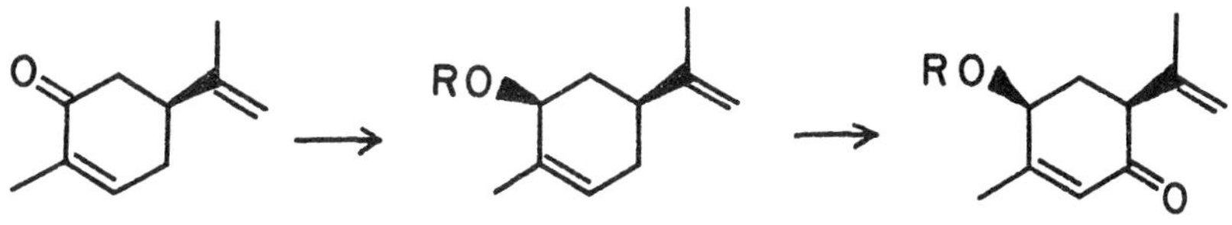

2

3

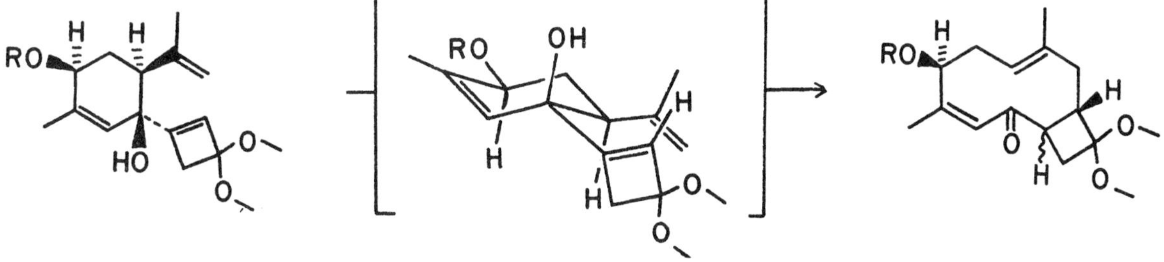

7

8a CIS
8b TRANS

An appropriate equivalent of the required alkoxyvinylacrylic acid appendage was found to be a cyclobutenone acetal which was prepared from the known acetal of the ketene/ethoxyacetylene cycloadduct. Low temperature addition of $Bu_3SnMgCl$ (Bu_3SnLi, $MgCl_2$, THF, -70 deg, 5 min) followed by direct in situ mesylation (1.2 equiv. MsCl) led to a 1,1-mesyloxystannane. Elimination was accomplished with excess powdered K_2CO_3 in DMSO (100 deg, 1 hr) and led to the desired cyclobutenyl tin reagent (96% yield).

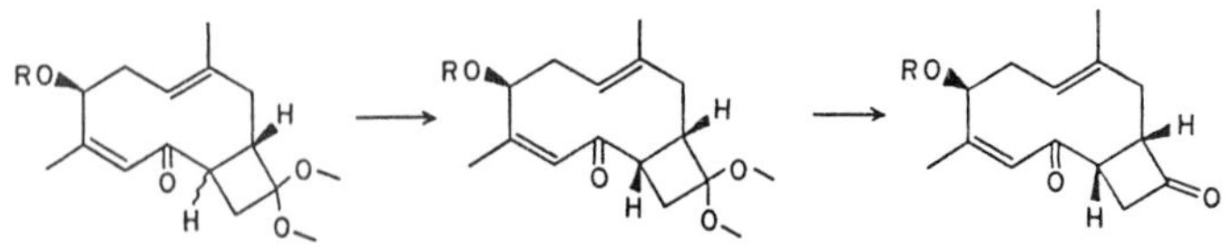

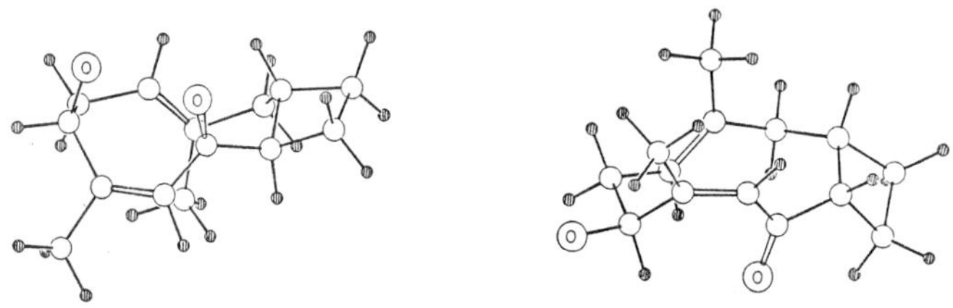

Coupling of enone 3 and the cyclobutenyl tin reagent described above proceeded via lithiation (n-BuLi, THF, -70 deg, 30 min) and addition of the enone 3. The adduct 7 which formed was a single diastereomer to the extent of at least 6:1 and was isolated by flash chromatography in 82% yield (85% conversion).

Oxy-Cope ring expansion was effected using 5 equivalents of $KN(TMS)_2$ in dimethoxyethane at 85 deg (14 hrs) and led to formation of 8 in high yield. Assuming that the cyclobutenyllithium adds trans to the bulky isopropenyl substitutent and that the oxy-Cope rearrangement proceeds via a chairlike transition state, then the stereochemistry of the C3 and the C8 substitutents must be trans as shown above. The C7 stereochemistry is the result of a kinetic protonation step and turned out to be nearly a 1:1 mixture of diastereomers 8a and 8b under a variety of protonation conditions. It was found however that when the mixture was stirred in dry MeOH containing powdered K_2CO_3 (25 deg, 24 hrs), a 15:1 ratio (270 MHz ^{1}H NMR and HPLC) of isomers was produced. The major isomer was tentatively assigned as cis (8a) and its isolated yield based on 7 was 90% at 80% conversion. At this point, our cis assignment rested largely on an MM2 molecular mechanics evaluation of the most stable conformations of 8a and 8b (above). The MM2 force field places (cis) 8a approximately 3 kcal more stable than (trans) 8b and furthermore shows in 8a a nicely aligned array of atoms suitably arranged for a long range W-type coupling between the C7-hydrogen and the equatorial C9-hydrogen. Such coupling was displayed in the 250 MHz ^{1}H NMR as a 4.5 Hz doublet. The structures shown above are the minimum energy ones for each compound and were generated by a ring-making computer program which will be briefly described later in this paper.

With homogeneous 8a in hand, the butyrolactone moiety was demasked by gentle acid hydrolysis (aq HOOCCOOH/silica gel, CH_2Cl_2, 35 deg, 2 hrs) and Baeyer-Villiger oxidation (anh H_2O_2, $Ti(Oi-Pr)_4$, i-Pr_2NEt, Et_2O, -30 deg, 15 min). The resulting ketolactone 9 was not purified but was immediately reduced with sodium borohydride (MeOH, 0 deg, 30 min) to yield 10 (55% yield based on 8a) as the only isolable hydroxylactone.

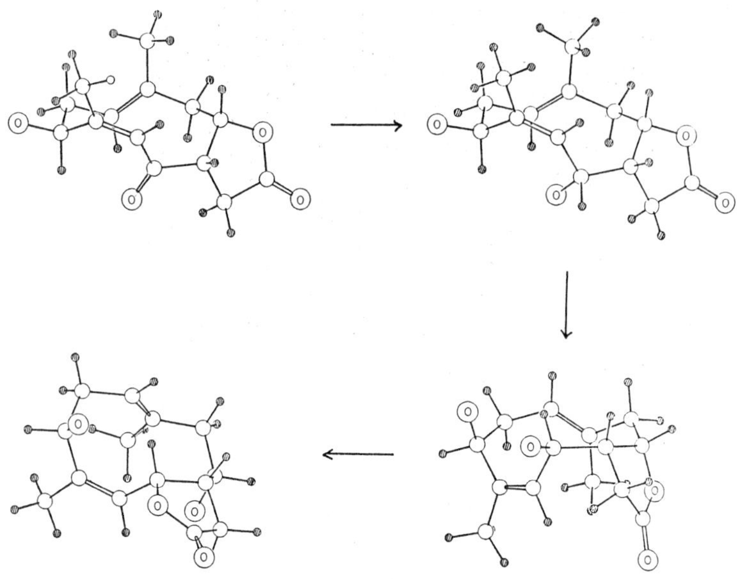

This reduction is an interesting one with respect to the lowest energy conformations of the species involved. The MM2 ground state structure of 9 is shown above and is compatible with observed ^{1}H NMR coupling constants for an axial C3-H (dd, J = 12.3, 4.5 Hz) and the C7-H/C8-H splitting (J = 5.8 Hz). A plausible mechanism for the formation of 10 involves peripheral addition of hydride to a low energy conformation of 9. This addition would lead kinetically to a relatively strained conformation (10k) having the hydroxyl pushed over the center of the ring. Conformational equilibration finally leads to the ground state structure (10t) which is predicted by the MM2 force field to have exchanged the transannular C6-OH interaction for an energetically less demanding axial C3-OH. Again, the ^{1}H NMR is consistent with the 10t geometry as indicated by the coupling constants shown above.

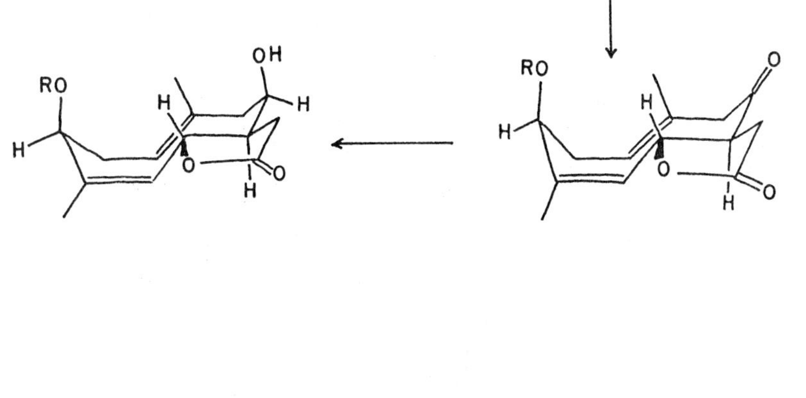

In order to convert 10 to the natural ring substitution and stereochemistry, both the direction of lactonization and the configurations at C7 and C8 had to be changed. Mechanisms for these adjustments were suggested by MM2 calculations which predicted the C6-lactone 11 to be >1 kcal more stable than 10 and which also predicted the trans keto lactone 13 to be of similar energy to 12. While these energy differences are only marginally significant, they do suggest that equilibrations to the desired relative stereochemistry should be feasible. Guided by this information, we treated 10 with catalytic K_2CO_3 in MeOH (25 deg, 5 hrs) to yield the sensitive isomeric lactone 11 in 61% yield. Equilibration in CD_3OD gave an equilibrium ratio of 9:1 as judged by ^1H NMR. Collins oxidation (CH_2Cl_2, 25 deg, 5 min) then gave 12 (71%) which was equilibrated with DBU in THF (25 deg, 3 hrs). The equilibrium ratio was found to be 7:1 by ^1H NMR and the pure trans isomer 13 was isolated by flash chromatography in 70% yield. Peripheral reduction ($NaBH_4$, MeOH) gave 14 as the only product (93%) which should possess the natural regiochemistry and stereochemistry. The C3 benzyloxymethyl protecting group was removed catalytically (H_2-20% $Pd(OH)_2$/C, 97% EtOH, 25 deg, 22 psi) to give a crystalline diol (15, 78%, mp = 146-147 deg) which was subjected to X-ray crystallographic analysis. The expected structure was confirmed and is shown below.

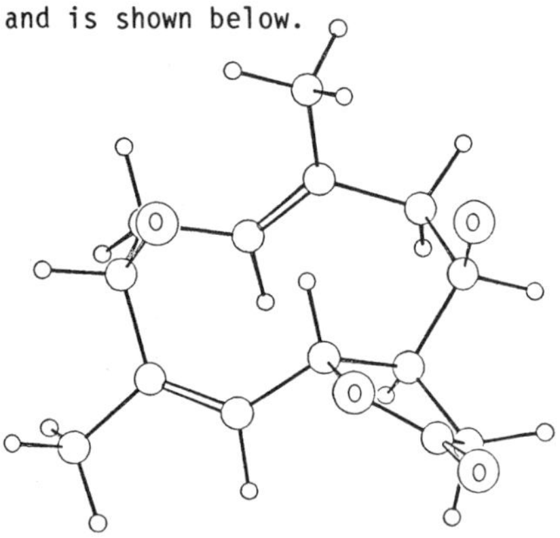

The synthesis was completed by means of straightforward conversions. Thus silylation (TMS-imidazole, C_5H_5N, CH_2Cl_2, 25 deg) and hydroxymethylation (a. LDA, THF; b. HCHO(g), -70 deg) gave the aldol adduct (75% at 77% conversion). Mesylation (MsCl, Et_3N, 4-dimethylaminopyridine, CH_2Cl_2, 25 deg) followed by elimination (DBU, dioxane, 70 deg, 30 min) formed the desired methylene lactone 16 (R,R'=TMS, 82% yield). Desilylation (Bu_4NF, THF, 25 deg) and acetylation of the more reactive C3 hydroxyl (AcOH, DCC, 4-pyrrolidinopyridine) gave 16 (R=Ac, R'=H; mp = 122-123 deg, 82% yield) which was finally esterified (DCC, 4-pyrrolidinopyridine; 52%) at C8 with the acetonide of dihydroxytiglic acid and deprotected ($C_5H_5NH+OTs-$, MeOH, $HOCH_2CH_2OH$; 54%) to yield 1. Synthetic eucannabinolide thus prepared was found by IR, ^1H NMR (270 MHz), TLC and CD to be identical with a sample of authentic 1 kindly supplied by Professor Koji Nakanishi at Columbia University.

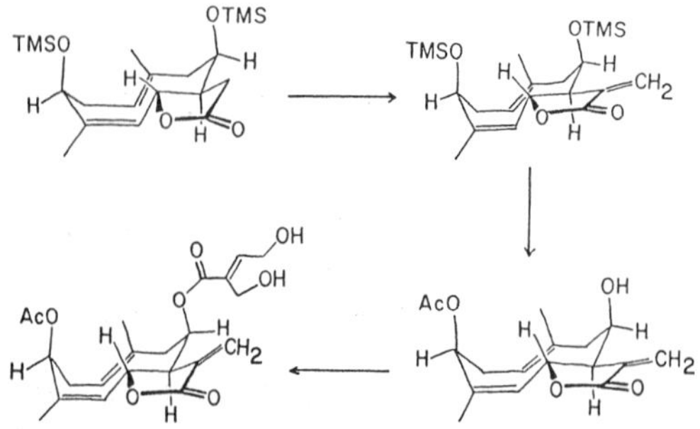

COMPUTATIONAL GENERATION OF MACROCYCLE GEOMETRIES

Many of the strategies and structure determinations described above made use of the low energy conformations of the molecules involved. Although mechanical molecular models serve quite well for the study of small and normal ring structures, macrocyclic molecules are best modeled computationally using the molecular mechanics calculations pioneered by Westheimer and refined by Allinger, Schleyer and others (review: <u>J. Chem. Ed.</u>, 59, 269 (1982)). All of these computations operate by starting with some initial crude molecular geometry which the chemist must supply and then by moving the atoms repeatedly toward positions of lower energy to ultimately produce a stable conformation. This stable conformation is of course related to the initial geometry and to find all the low energy conformations of a compound, a number of initial geometries must be supplied and energy minimized. With medium and large rings, it is quite difficult to produce all possible starting geometries manually since such molecules are so flexible. For this reason we have written a computer program called RINGMAKER which automatically produces all starting geometries of a macrocycle having any specified dihedral angle resolution and giving rings which have any specified closure bond distance and angles. The resulting initial geometries are then automatically read by a modified version of the Allinger MM2 program. That program minimizes each initial geometry, eliminates any duplicate conformations and stores the results for examination.

Shown below is a summary of the ringmaking operation for cyclononane. The object is to determine three-dimensional x, y and z coordinates for each atom in the ring for all possible starting geometries and the calculation goes something like this:

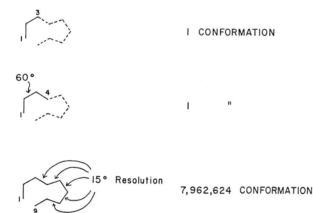

| CONFORMATION

| "

7,962,624 CONFORMATIONS

Constraints:

1) 1.0-2.0 Å   Closure Distance
2) 100-120°   Closure Angles
3) 2.1 Å   Transannular Contact

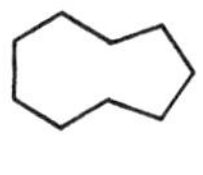

2,895   CONFORMATIONS

Eliminate Near Duplicates

1,050   CONFORMATIONS

The first three atoms have their relative positions defined by the two bond distances and the associated bond angle. The fourth atom may be defined knowing the next bond distance, angle and a dihedral angle. Although for complex structures, this first dihedral angle may have a variety of possible values, we may assume it to be 60 degrees for cyclononane since all conformations must have at least 1 gauche linkage somewhere in the ring. We have thus defined the positions of the first 4 atoms of cyclononane. The remaining atoms are defined by the bond lengths and angles and by allowing all the remaining independent (of closure) dihedral angles to vary around full circle with say 15 degree resolution. All the remaining atoms are thus defined and approximately 8 million conformations are produced. Of course only a small percentage of these conformations have the two ends of the chain within a reasonable bonding distance. If we apply the constraints of a 1-2 angstrom closure distance (atoms 1-9) and 100-120 degree closure bond angles (atoms 2-1-9 and 1-9-8), then only 2895 of the original 8000000 conformations form rings. It is possible to eliminate 2/3 of these since many are near duplicates (all the dihedral angles are the same except for one which differs only by 15 degrees). The resulting 1000 conformations are then submitted to a modified version of the MM2 program which minimizes each of the initial geometries and then stores all of the unique ones within some specified number of kilocalories (say 5 kcal/mole) of the ground state. When the calculation is complete (approximately 2 hours of computing), the five structures shown below are produced. These structures include the four previously known conformations of cyclononane and one other one related to the lowest energy [333] form.

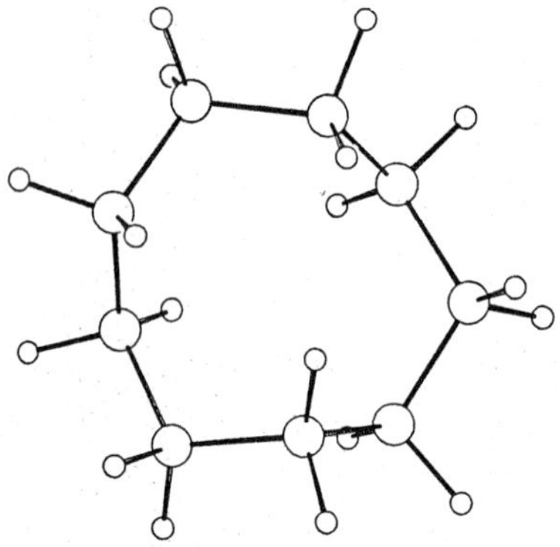

0.8

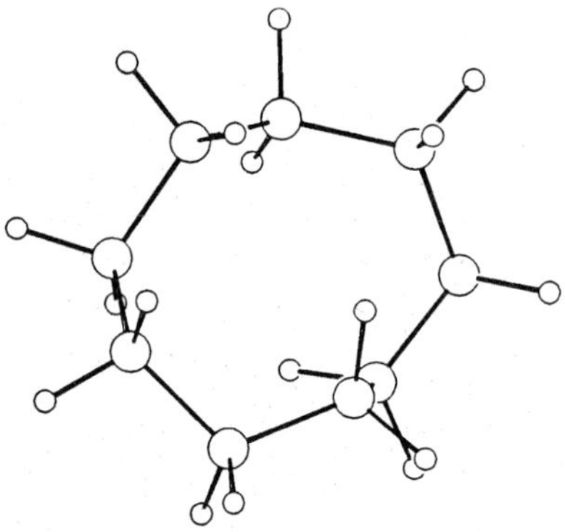

2.2

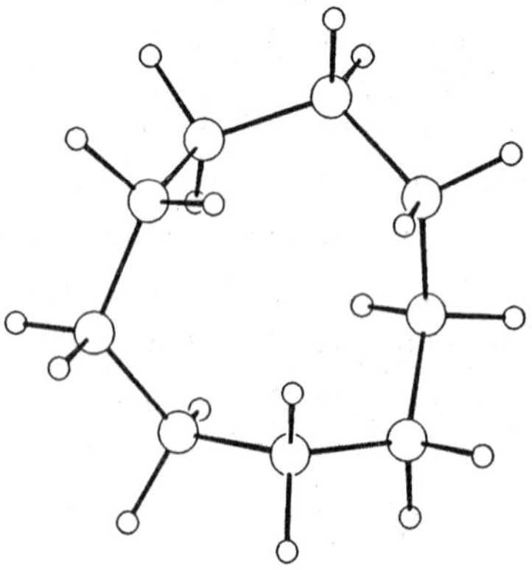

0.0 (333)

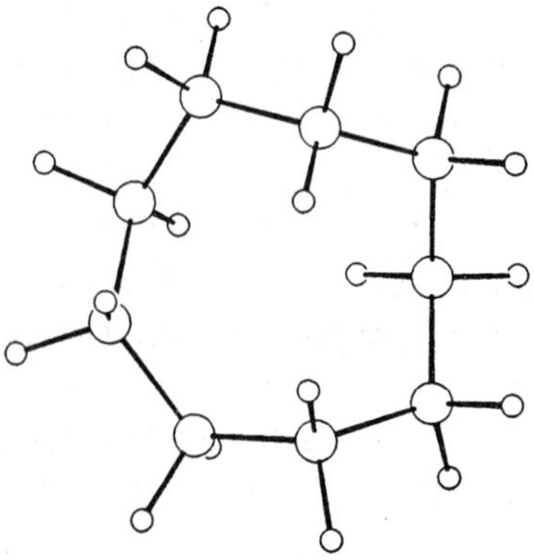

0.7 (225)

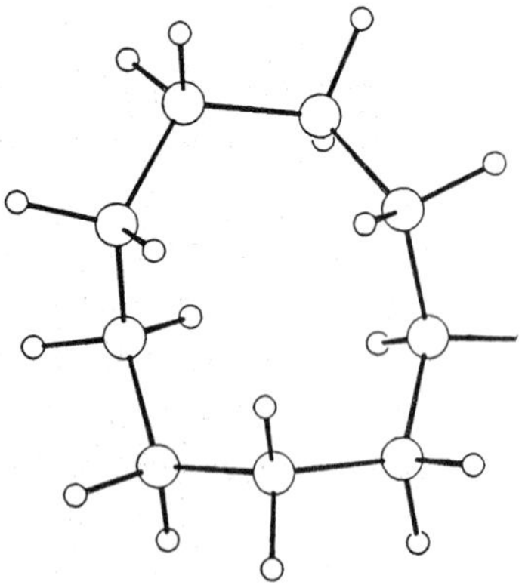

3.1 (234)

This approach to macrocycle conformation is quite a powerful one. We have examined a number of macrocycles having ring sizes 8-14 and in every case the calculation described above found all the known conformations as well as some new ones. While it is not difficult to model simple cycloalkanes in this way, one might worry that the method might break down with complex molecules due to inadequacies in the force field. In order to test the validity of the approach to molecules like those we have been examining in the laboratory, we have applied the calculation to a variety of macrocycles whose structures are known by x-ray crystallography. These molecules are shown below.

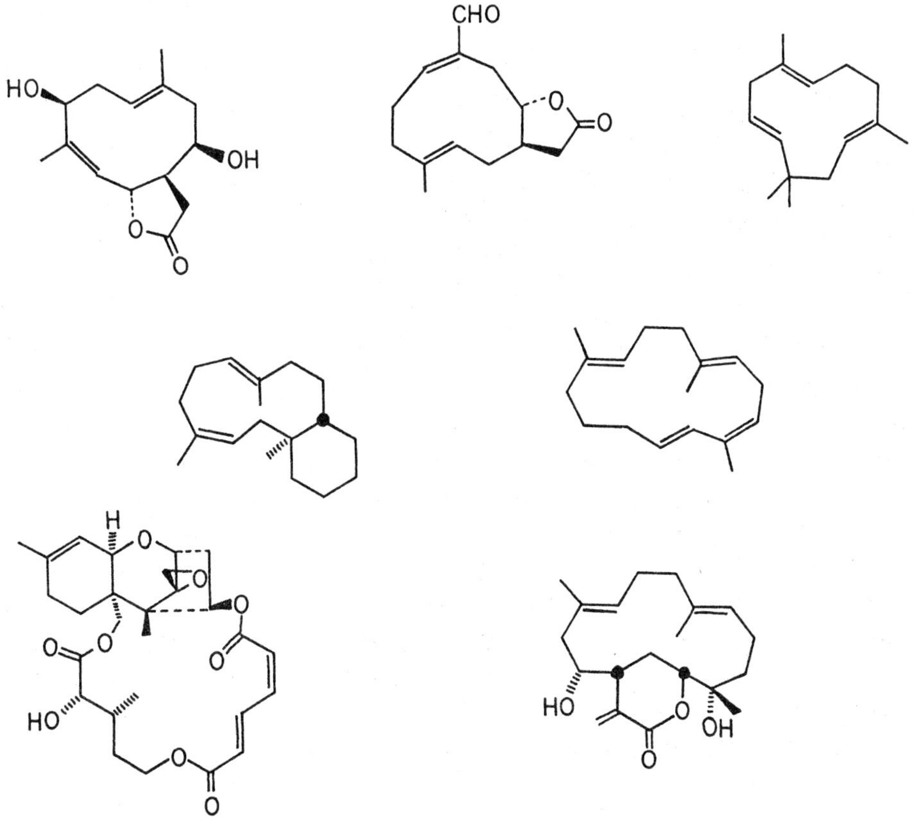

In every case except for humulene (upper right), the minimum energy conformation predicted automatically was the same as that found in the crystal state. In the case of humulene (whose x-ray structure was that of a disilver complex), the x-ray conformation was also found but was calculated to be approximately 1 kcal/mole above the ground state in energy. Only in the case large rings having more than 14 members are significant problems found. In such systems, the MM2 program often does not give good convergence with the result that a number of very similar structures are produced having different energies. My final topic involves just such a structure, the 16-membered lactone ring of rosaramicin.

SYNTHESIS OF ROSARAMICIN

Our approach to the 16-membered macrolide rosaramicin is somewhat different than earlier strategies for construction of such molecules. Previous work on related molecules like tylosin involved the independent synthesis of several optically pure fragments followed by coupling of the fragments and closure of the ring. Although this is an effective approach, we thought that the 16-membered lactone ring of rosaramicin might have conformational properties which would allow the creation of much of the substitution and relative stereochemistry directly. In particular, we felt that the two adjacent stereocenters at C14 and C15 would provide enough bias to add all the remaining substituents with remote asymmetric control by these two primordial asymmetric centers.

The simple macrocyclic template we chose is shown below as structure 1. It was constructed by standard chemistry as shown below from simple precursors and was cyclized using conditions recently reported by K.C. Nicolaou (K_2CO_3, 18-crown-6, DME) in better than 75% yield.

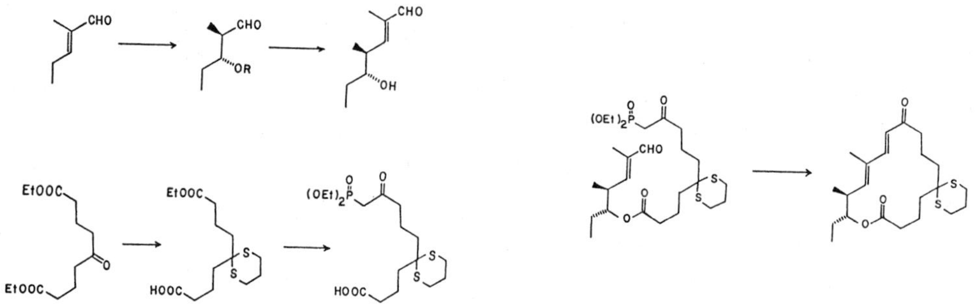

The conformational structures of this molecule and rosaramicin are of special interest. In the rosaramicin crystal structure (below), it may be seen that all the substituents are in relatively stable pseudoequatorial orientations and that the ring itself is composed of long stretches of anti arrangements of atoms with only a few gauche turns. In 1, the s-trans lactone linkage, the gauche turn at the primordial asymmetric centers C14 and C15, and the planarity of the dienone system favor conformations which resemble rosaramicin itself. Furthermore, the introduction of the (gem-substituted) thioketal at C5 favors a pair of gauche turns at precisely the location of a similar pair of turns in rosaramicin. Thus we expected that the natural rosaramicin conformation might also be found in 1. Shown below are x-ray structures of rosaramicin (the attached sugar has been deleted for clarity) and our lactone 1. It can be readily seen that the gross conformations of the two molecules are the same in spite of extensive polar and protic functionality found in rosaramicin.

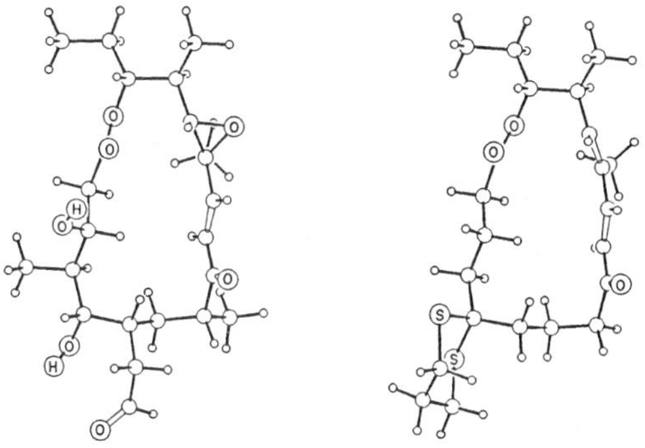

Our synthetic plan at this point was quite a simple one. We wished to methylate the enolate of the ketone at C8, then deprotect the C5 ketone and add the methyl and the acetaldehyde substitutents adjacent to it via enolate alkylations. Finally, the C5 ketone would be reduced and the C3 hydroxyl would be added via an epoxide. Stereochemical control is obviously the crucial concern and we hoped that the conformation of the 16-membered ring would serve as an effective medium for transmission of stereochemical information at C14 and C15.

The first alkylation was accomplished with $KN(TMS)_2$ and methyl iodide in THF. The product was isolated in high yield and turned out to be a single diastereomer to the extent of at least 20:1! This compound could be equilibrated with base to yield the epimer (approximately 2:1 epimer:original diastereomer). We have some evidence that the kinetic alkylation product has the natural stereochemistry shown below.

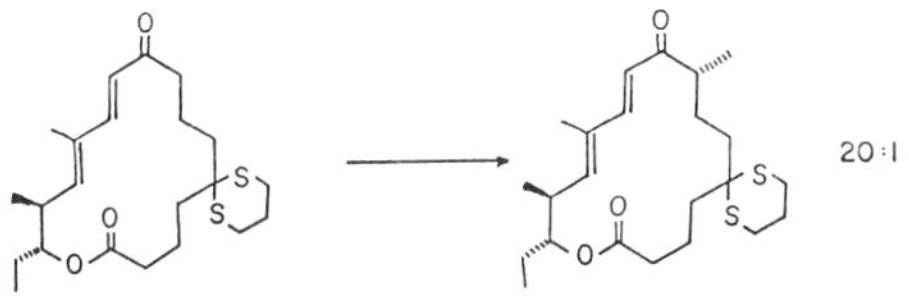

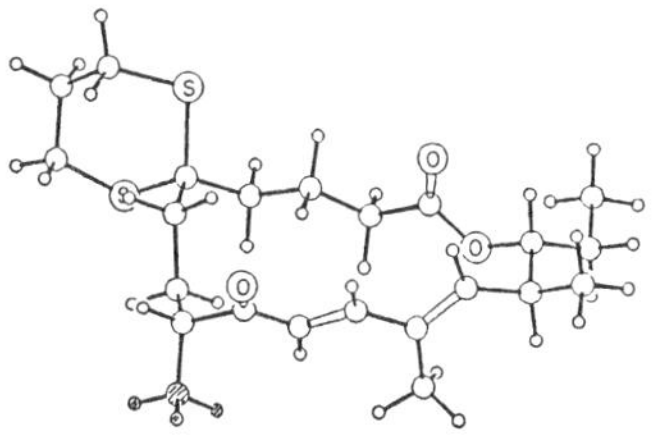

The tentative structural assignment is based on the NMR observation of a large (anti) and a small (gauche) coupling constant for the C9 methine hydrogen and on a NOE of the C11 vinyl hydrogenation on irradiation of the newly added C9 methyl. Although this assignment is not yet firm since we are cannot rule out other conformations of the methylated macrocycle, it is clear that the ring provides an excellent medium for operation of 1,7-asymmetric induction.

In order to add the remaining substituents, the rigidifying thioketal must be removed. When this is done (HgO, HBF_4), a diketone is produced which probably exists as a mixture of conformations. Although we were not able to produce a unique set of low energy conformers due to the size and flexibility of the ring, molecular mechanics did suggest a number of low energy forms with a kcal or so of each other. Some of these structures are shown below.

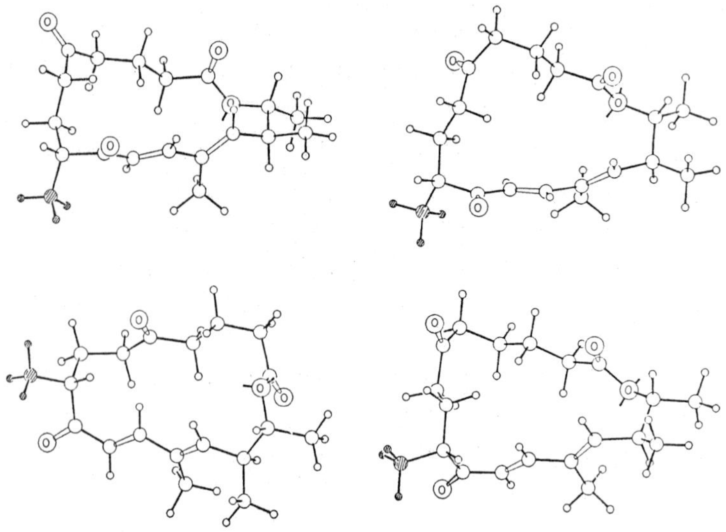

Without adequate molecular modeling, it is difficult to predict the outcome of subsequent alkylation reactions although examination of the low energy forms above provided some expectation that the most stable form of the enolate of the C5 ketone would be in the direction of C6. When the alkylation was attempted using LiN(TMS)$_2$ and CH$_3$I, a single regioisomer was produced with a 6:1 diastereomeric ratio. There was no doubt that the second alkylation had taken place at C6 since the complete splitting pattern for C6-C8 could be easily seen in the 270 MHz ^{1}H NMR. It was also clear that the two compounds were diastereomers since they could be interconverted with acid or base.

Finally to test the feasibility of introduction of the third methyl group, the major dialkylated product was isolated chromatographically and methylated once more using kinetic enolate methodology. A single isomer was formed in high yield! An analogous alkylation of the minor dialkylated product also produced a single diastereomer! It is thus clear that the 16-membered macrolide can provide substantial stereochemical control to reactions which produce new remote asymmetric centers.

Our synthetic problem now became the determination of the complete stereochemistry of the trialkylated macrolide. Although we originally planned to use X-ray crystallography at this point, we have be unable to produce satisfactory crystals and so an alternative scheme was taken up. This new scheme involved direct correlation with a rosaramicin degradation product but had the obvious deficiency that our stereochemistry would be determined only if it turned out to be the same as that of rosaramicin.

In order to prepare a molecule having rosaramicin-like substitution, our monomethylated diketone was alkylated as before except that tert.-butyl iodoacetate was used as the alkylating agent. The same 6:1 ratio of diastereomers was produced in high yield. The third alkylation was then effected as before to give a single diastereomer starting from the major iodoacetate alkylation product. This compound was converted to the free acid with trifluoroacetic acid. The stereochemistry shown below is that of the natural material but is not necessarily that of our trialkylated product.

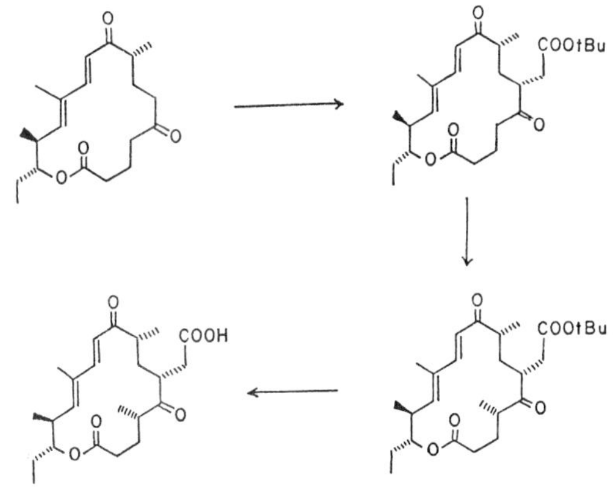

The mixed anhydride was then prepared with ethyl chloroformate and reduced with sodium borohydride to give a triol. Silylation (TMS-imidazole) and epoxidation (MCPBA) gave a single epoxide which was treated with fluoride and then MnO_2 to produce the dihydroxy enone shown below. Finally, silylation ($tBuPh_2SiCl$) and oxidation (Collins' reagent) gave a diketone which could be compared to the rosaramicin degradation product prepared as outlined in the following paragraph.

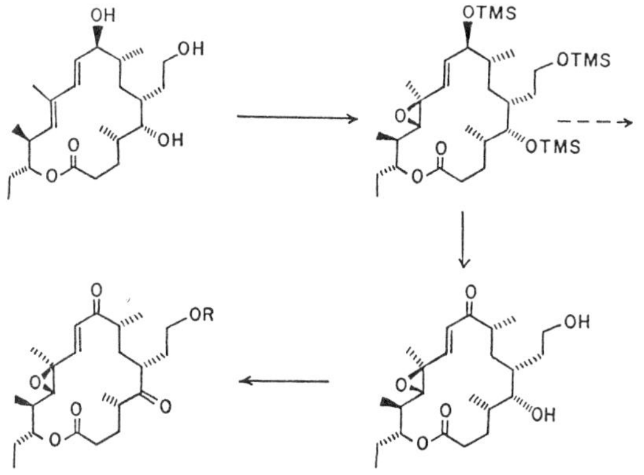

Removal of the sugar moity from rosaramicin was accomplished by a published procedure and was followed by reduction ($NaBH_4$, MeOH) and silyation (TMS-Imidazole). Treatment of this material with $KN(TMS)_2$ in THF at -40 degrees resulted in smooth elimination of the C3 acetate to yield an unsaturated lactone. This alpha,beta unsaturation could be removed by conjugate addition of hydride using $LiBH(s-Bu)_3$ in THF. Finally desilylation, MnO_2 oxidation, primary alcohol silylation and oxidation essentially as before gave a product which should be identical with our synthetic material provided that the stereochemistry is the same. Careful 270 MHz ^1H NMR comparision showed the two materials to be quite similar but distinguishable. Thus our trialkylated material does not have the natural stereochemistry. If our first methylation does indeed give the natural configuration, then there is no doubt that the departure from the natural series occurred in the last (C4) methylation since we have also transformed the minor second alkylation product to the correlation point. While our major trialkylated product is very similar by nmr (except in the region of C4), the minor trialkylated product appears quite different.

Although our synthesis at this point has not yielded the natural stereoisomer of rosaramicin, we believe that we will be able to use macrocyclic stereocontrol to produce all of the other diastereomers in a rational and efficient way. Not only is it possible in systems like these to change stereoselection by varying the order of atom substitution but it is also possible to gain access to different conformations of the macrocycle by reordering the sequence of steps or by making small changes in substitution or functionality. I believe that this approach to remote asymmetric induction is a particularly powerful one whose major problem at this point is one of product prediction. While contemporary molecular mechanics can be quite helpful with ring sizes up to 12 or 14, larger rings will probably require refinements in both the force fields and the way in which minimization is accomplished. Even with smaller rings, more work is needed to confidently relate strain to kinetic diastereoselection. Significant advances in this area are expected in the near future.

ACKNOWLEDGEMENT: I cannot close this paper without expressing my sincere gratitude to a number of my colleagues whose work is described above. The eucannabinolide project has been an ongoing concern for several years now and it is due to the efforts of Kazuo Yoshihara of the Suntory Institute for Bioorganic Research, of Gilbert Revial from Professor Ficini's labortory in Paris and most recently of Shizuaki Murata from Professor Noyori's group here in Japan that I am able to describe the completion of our work in the germacrane area to you. Finally, I wish to make special note the diligent and careful work of Vance Novack, a third year graduate student at Columbia, who in one year has proven the validity of our approach to rosaramicin.

NEW HYDROBORATING AGENTS

Herbert C. Brown

Department of Chemistry, Purdue University, West Lafayette, Indiana 47907, USA

Abstract - Hydroboration with diborane or borane generating reagents is a fast reaction leading to organoboranes, R_3B, in essentially quantitative yields. Yet it suffers from certain disadvantages. Many of these difficulties can be overcome by use of substituted boranes with but one or two active B–H bonds. Among the reagents explored are dicyclohexylborane, disiamylborane, diisopinocampheylborane, catecholborane, dichloroborane-etherate, dichloroborane-methyl sulfide, dibromoborane-methyl sulfide, thexylchloroborane-methyl sulfide, thexylborane, isopinocampheylborane, monochloroborane etherate, monochloroborane-methyl sulfide, and monobromoborane-methyl sulfide. These studies have revealed remarkable differences in selectivities. For example, it is now possible to hydroborate $RC \equiv CR$ in the presence of $RC \equiv CH$, or *vice versa*. Similarly, it is possible to hydroborate $RCH=CH_2$ in the presence of $R(CH_3)C=CH_2$, or *vice versa*. Indeed, one of the reagents reveals selectivities of cis alkenes over trans of ~ 100. These differences can often be utilized to advantage in complex syntheses.

INTRODUCTION

The discovery in 1956 and 1957 that unsaturated organic compounds are rapidly converted into organoboranes by treatment with diborane or diborane precursors in ethereal solvents (Ref. 1) provided a convenient new route to these valuable derivatives (Refs. 1-10). The reaction is practically instantaneous and usually essentially quantitative. It is a remarkably clean reaction (Ref. 1).

DIFFICULTIES WITH DIBORANE HYDROBORATION

With investigation, however, certain problems became apparent. For example, the hydroboration of terminal olefins gave 94% of the primary borane derivative and 6% of the secondary (Eqn. 1) (Ref. 11).

$$\text{CH}_2=\text{CHCH}_2\text{CH}_2\text{CH}_3 \xrightarrow{BH_3} \text{6\%} + \text{94\%} \quad (1)$$

Oxidation would then produce a mixture of 94% of the primary alcohol and 6% of the secondary. It can involve major efforts to remove 6% of an isomeric impurity.

Similarly, hydroboration is insensitive to steric effects. It exhibits little discrimination between the two carbon atoms of the carbon-carbon double bond in *cis*-4-methyl-2-pentene (Eqn. 2) (Ref. 11).

$$\xrightarrow{BH_3} \quad 43\% \quad + \quad 57\% \quad (2)$$

In certain cases, even more serious problems are encountered with directive effects, as indicated by the distributions realized in the hydroborating of styrene and allyl chloride (Eqns. 3, 4) (Ref. 12).

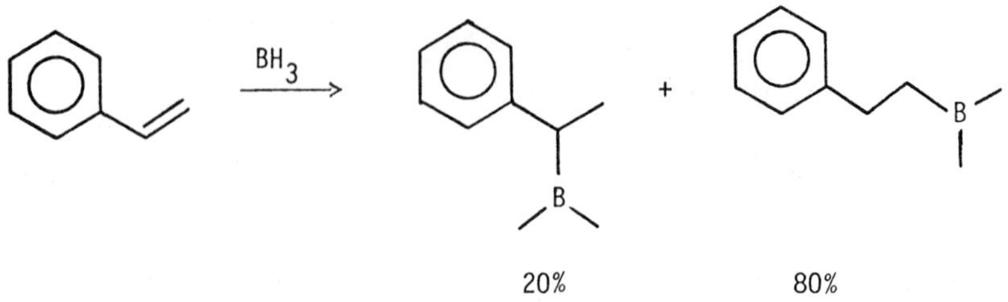

$$\text{styrene} \xrightarrow{BH_3} \text{PhCH(B)CH}_3 \ (20\%) + \text{PhCH}_2\text{CH}_2\text{B} \ (80\%) \quad (3)$$

$$\text{ClCH}_2\text{CH=CH}_2 \xrightarrow{BH_3} \text{ClCH}_2\text{CH(B)CH}_3 \ (40\%) + \text{ClCH}_2\text{CH}_2\text{CH}_2\text{B} \ (60\%) \quad (4)$$

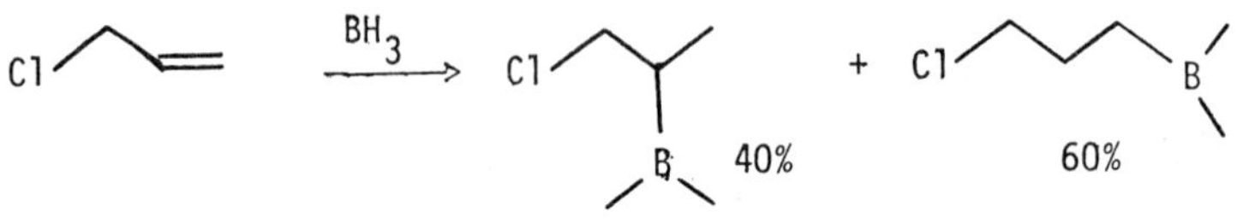

The hydroboration of terminal acetylenes yields only minor amounts of the desired vinylic boranes (Eqn. 5) (Ref. 13).

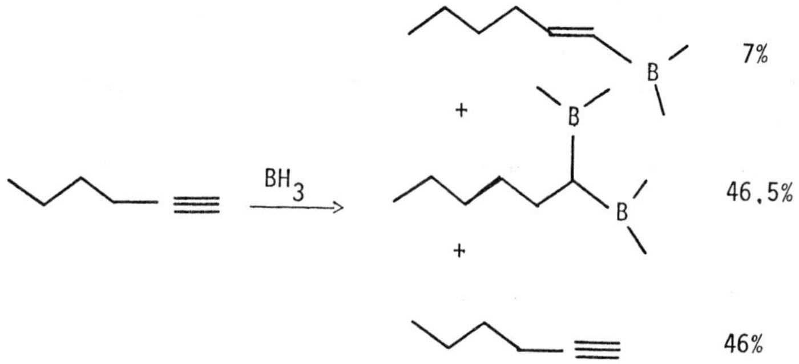

$$\text{1-hexyne} \xrightarrow{BH_3} \text{vinylic borane (7\%)} + \text{internal dihydroboration product (46.5\%)} + \text{recovered alkyne (46\%)} \quad (5)$$

Additional problems are encountered with conjugated dienes, such as 1,3-butadiene. Here, there is 24% placement of boron on the 2-position. Moreover, the monoolefin produced is more reactive than the diene. Consequently, hydroboration tends to proceed almost completely to the dihydroboration stage (Eqn. 6) (Ref. 14).

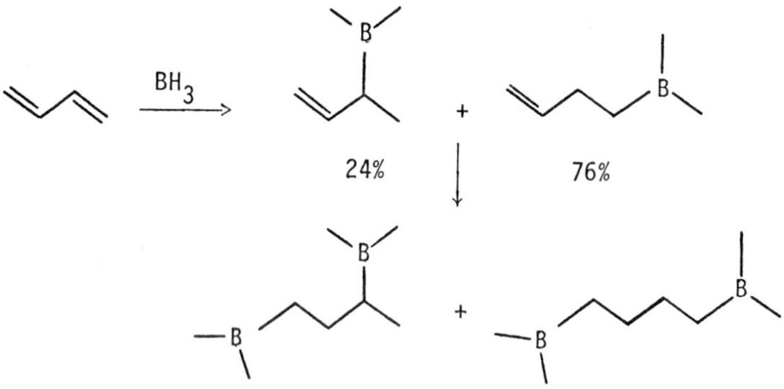

$$\text{1,3-butadiene} \xrightarrow{BH_3} \text{2-B (24\%)} + \text{1-B (76\%)} \rightarrow \text{dihydroboration products} \quad (6)$$

In the hydroboration of other dienes, the polyfunctional nature of borane can lead to the preferred formation of cyclic derivatives or resistant polymers (Eqn. 7) (Ref. 15).

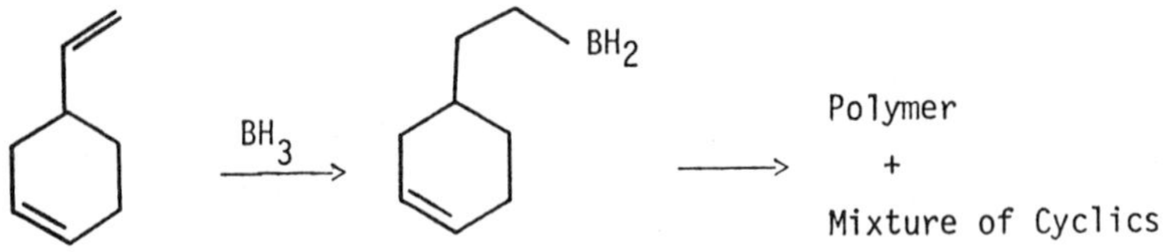

$$\text{4-vinylcyclohexene} \xrightarrow{BH_3} \text{cyclohexenyl-CH}_2\text{CH}_2\text{BH}_2 \longrightarrow \text{Polymer + Mixture of Cyclics} \quad (7)$$

Another problem is the small range of reactivities realized with diborane (Chart I) (Ref. 16).

Chart I

Alkene	Relative Reactivity
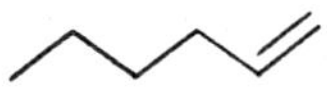 1-butene (CH₂=CHCH₂CH₃)	100
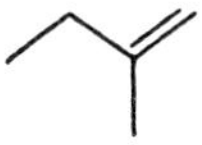 2-methyl-1-butene	121
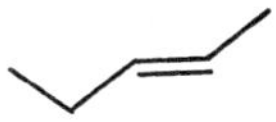 2-pentene	34
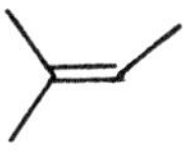 2-methyl-2-butene	50
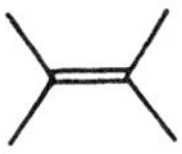 2,3-dimethyl-2-butene	7

This makes it difficult to achieve the selective hydroboration of one type of double bond in the presence of a second type.

Another problem is provided by the reactions of the organoborane intermediate, R_3B. In some reactions, such as oxidation, all three groups are utilized (Eqn. 8) (Ref. 17). In other

$$R_3B + 3\ HOOH \xrightarrow{NaOH} 3\ ROH + NaB(OH)_4 \quad (8)$$

reactions, such as iodination, only two of the three groups are readily utilized (Eqn. 9) (Ref. 18).

$$R_3B + I_2 \xrightarrow{NaOH} 2\ RI + NaRB(OH)_3 \quad (9)$$

In still other reactions, only one of the three groups on boron is utilized (Eqn. 10) (Ref. 19).

$$R_3B + N_2CHCOOEt \longrightarrow RCH_2COOEt + R_2BOH \quad (10)$$

A problem of another type is indicated by the elegant Zweifel syntheses of *cis*-alkenes (Eqn. 11) (Ref. 20). The reaction requires a dialkylborane as the intermediate, but only a

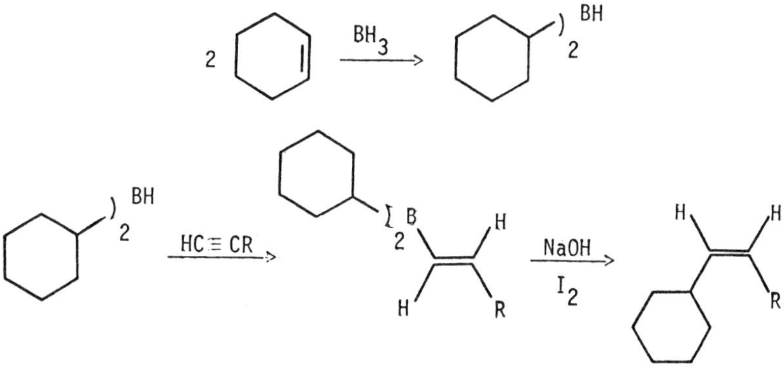

$$2\ \text{cyclohexene} \xrightarrow{BH_3} (\text{Cy})_2BH$$

$$(\text{Cy})_2BH \xrightarrow{HC\equiv CR} \text{Cy}_2B\text{-CH=CHR} \xrightarrow{NaOH,\ I_2} \text{Cy-CH=CHR (cis)} \quad (11)$$

relatively few hindered alkenes undergo hydroboration to these dialkylboranes.

The same problem is encountered in the important Zweifel synthesis of trans alkenes (Eqn. 12) (Ref. 21).

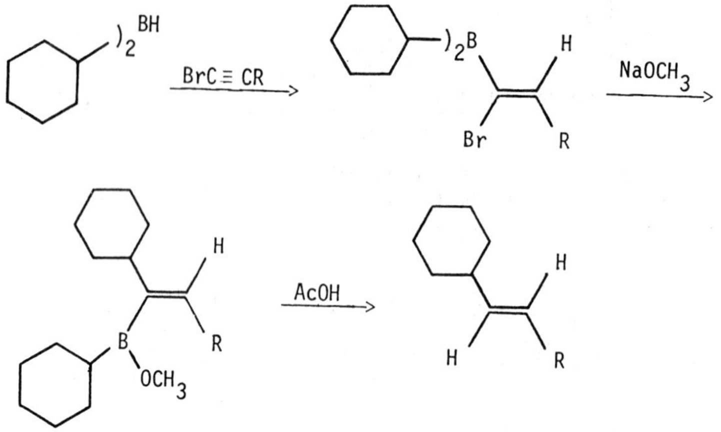

(12)

MONO- AND DIALKYLBORANES

These experiences stimulated us to undertake a search for substituted boranes which would still be good hydroborating agents, but might circumvent certain of these problems. This search was begun early, in 1960, but it is still continuing. It is the purpose of this lecture to summarize our accomplishments in this area.

One of the early reagents we explored was dicyclohexylborane (Eqn. 13) (Ref. 22).

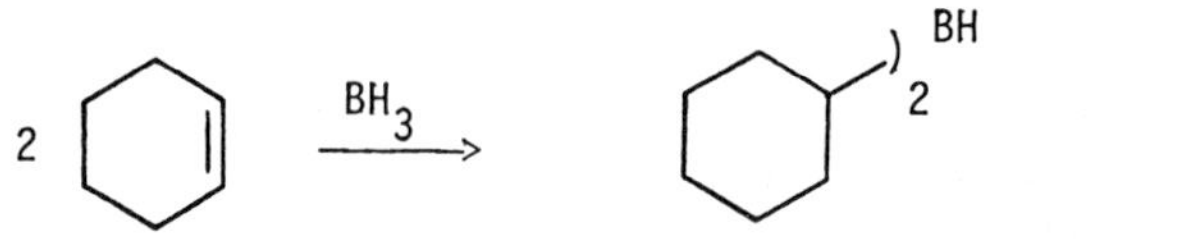

dicyclohexylborane (Chx_2BH)

(13)

An even more hindered hydroborating agent was disiamylborane (Eqn. 14) (Ref. 23).

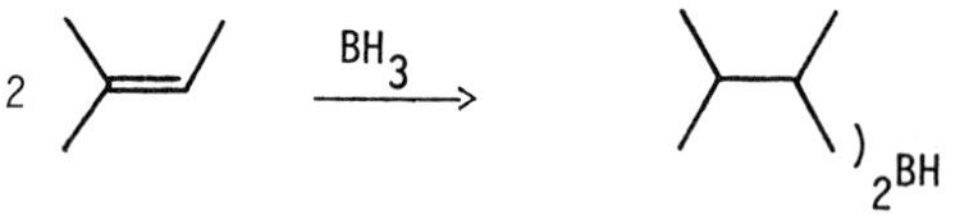

disiamylborane (Sia_2BH)

(14)

Transannular hydroboration of 1,5-cyclooctadiene produces a bicyclic borane of unusual stability. This reagent, 9-borabicyclo[3.3.1]nonane (9-BBN), has proven exceptionally valuable, both for hydroboration and as a blocking group (Eqn. 15) (Refs. 24, 25).

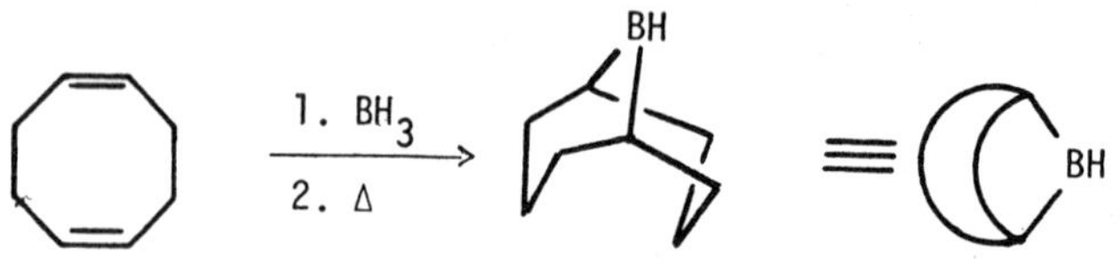

9-borabicyclo[3.3.1]nonane
(9-BBN)

(15)

The hydroboration of 2,3-dimethyl-2-butene can be controlled to give a valuable monoalkylborane, thexylborane (Eqn. 16) (Refs. 26-28).

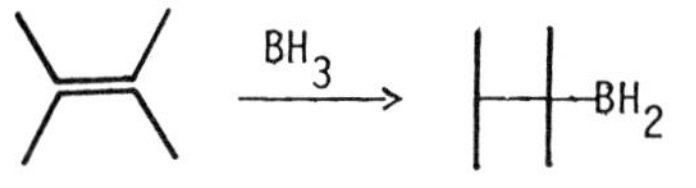

$$\underset{}{\diagup\!\!\!=\!\!\!\diagdown} \xrightarrow{BH_3} \text{H}{-}BH_2 \qquad (16)$$

thexylborane (ThxBH$_2$)

The hydroboration of optically active α-pinene provides diisopinocampheylborane, an optically active hydroborating agent. This reagent can be converted into monoisopinocampheylborane, also a valuable optically active hydroborating agent (Eqn. 17) (Refs. 29-31).

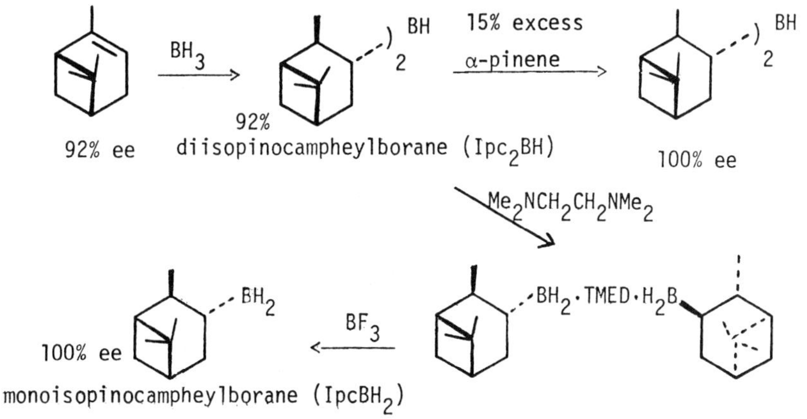

$$\text{α-pinene} \xrightarrow{BH_3} \underset{92\%}{\text{Ipc}_2BH} \xrightarrow[\text{α-pinene}]{15\% \text{ excess}} \text{Ipc}_2BH \text{ (100\% ee)} \qquad (17)$$

92% ee   diisopinocampheylborane (Ipc$_2$BH)   100% ee

$\xrightarrow{Me_2NCH_2CH_2NMe_2}$ IpcBH$_2$·TMED·H$_2$B·Ipc $\xrightarrow{BF_3}$ monoisopinocampheylborane (IpcBH$_2$) 100% ee

HETEROSUBSTITUTED BORANES

We also undertook to explore potential heterosubstituted boranes. Unfortunately, dimethoxyborane (Eqn. 18) is not an active hydroborating agent. Apparently electron-donation from the methoxy groups so satisfy the electron deficiency of the boron atom that it no longer attacks carbon-carbon double bonds. On the other hand, catecholborane is effective (Eqn. 19) (Ref. 32).

$$2\ CH_3OH \xrightarrow{BH_3} (H_3CO)_2BH \qquad (18)$$

$$\text{catechol} \longrightarrow \text{catecholborane (CtO}_2\text{BH)} \qquad (19)$$

Mono- and dichloroborane etherates are readily prepared by reaction of controlled amounts of boron trichloride with lithium borohydride in ethyl ether (Eqn. 20) (Ref. 33).

$$3\ BCl_3 + LiBH_4 \xrightarrow{Et_2O} 4\ HBCl_2 \cdot OEt_2$$
dichloroborane etherate

$$BCl_3 + LiBH_4 \xrightarrow{Et_2O} 2\ H_2BCl \cdot OEt_2$$
monochloroborane etherate

(20)

These proved to be valuable hydroborating agents. However, they possessed one serious disadvantage. They were unstable on storage (the ether linkage evidently underwent a slow cleavage with the boron-chloride bond). The carbon-sulfur bond is quite stable toward boron trihalides and halosubstituted boranes. The desired chloro- and bromoboranes were readily synthesized by a simple redistribution (Eqns. 21, 22) (Ref. 34).

$$2\ Cl_3B \cdot SMe_2 + H_3B \cdot SMe_2 \rightarrow 3\ HBCl_2 \cdot SMe_2$$
dichloroborane methyl sulfide

$$Cl_3B \cdot SMe_2 + 2\ H_3B \cdot SMe_2 \rightarrow 3\ H_2BCl \cdot SMe_2$$
monochloroborane methyl sulfide

(21)

$$2\ Br_3B \cdot SMe_2 + H_3B \cdot SMe_2 \rightarrow 3\ HBBr_2 \cdot SMe_2$$
dibromoborane methyl sulfide

$$Br_3B \cdot SMe_2 + 2\ H_3B \cdot SMe_2 \rightarrow 3\ H_2BBr \cdot SMe_2$$
monobromoborane methyl sulfide

(22)

On the other hand, the most convenient route to mono- and diiodoborane methyl sulfide is the reaction of iodine with borane methyl sulfide (Eqn. 23) (Ref. 35).

$$H_3B \cdot SMe_2 + I_2 \rightarrow HBI_2 \cdot SMe_2$$
diiodoborane methyl sulfide

$$2\ H_3B \cdot SMe_2 + I_2 \rightarrow 2\ H_2BI \cdot SMe_2$$
monoiodoborane methyl sulfide

(23)

MIXED DISUBSTITUTED BORANES

Thexylchloroborane is a new disubstituted borane containing one alkyl group and one hetero-substituent. It is the first member of a new class and has already proven to possess exceptional utility. It is readily prepared by two different routes (Eqns. 24, 25) (Refs. 36, 37).

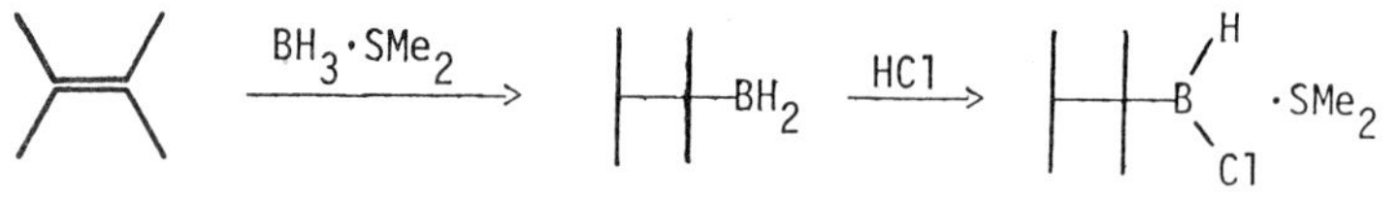

$$\text{alkene} \xrightarrow{BH_3 \cdot SMe_2} \text{ThexBH}_2 \xrightarrow{HCl} \text{ThexBHCl} \cdot SMe_2 \quad (24)$$

$$\text{alkene} \xrightarrow{H_2BCl \cdot SMe_2} \text{ThexBHCl} \cdot SMe_2 \quad (25)$$

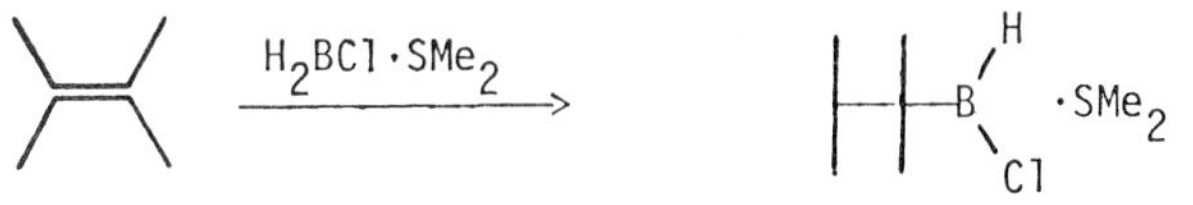

thexylmonochloroborane methyl sulfide

DIRECTIVE EFFECTS

These new hydroborating agents solve many of the problems encountered with simple hydroboration based on diborane. For example, the hydroboration of terminal alkenes with disiamylborane gives 99% of the terminal derivative, only 1% of the secondary (Eqn. 26) (Ref. 38).

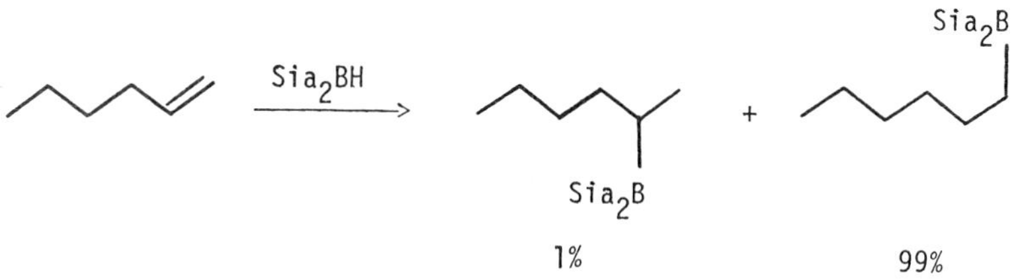

$$\text{1-alkene} \xrightarrow{Sia_2BH} \text{secondary (Sia}_2\text{B, 1%)} + \text{terminal (Sia}_2\text{B, 99%)} \quad (26)$$

The large steric requirements of the reagent results in a desirable regioselectivity between the two carbon atoms of the carbon-carbon double bond in *cis*-4-methyl-2-pentene (Eqn. 27) (Ref. 38).

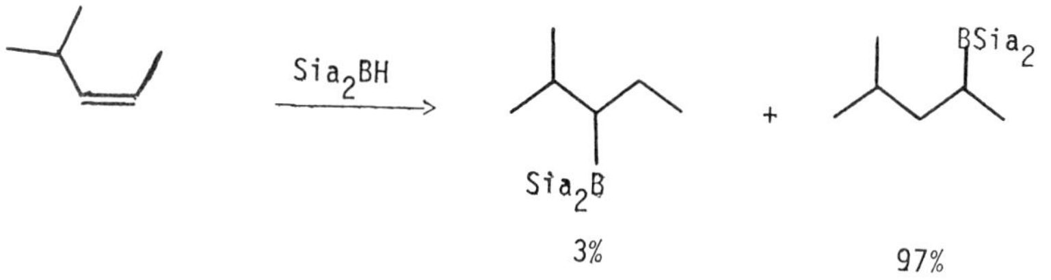

$$\text{cis-4-methyl-2-pentene} \xrightarrow{Sia_2BH} \text{(Sia}_2\text{B, 3%)} + \text{(BSia}_2\text{, 97%)} \quad (27)$$

9-BBN exhibits a truly astounding regioselectivity. For example, 1-alkenes, such as 1-hexene, give 99.9% of the primary derivative (Eqn. 28) (Ref. 39).

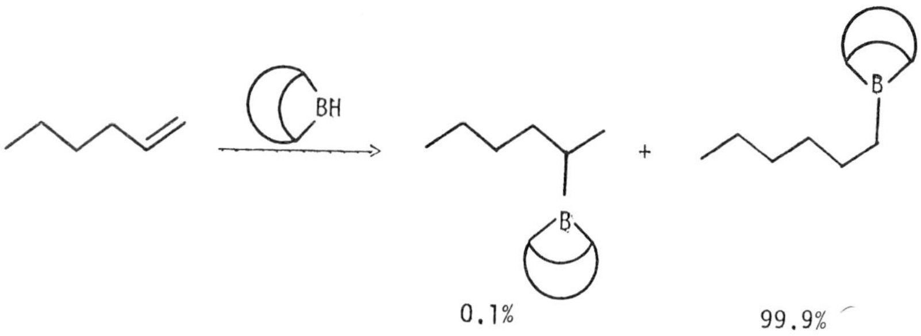

$$\text{1-hexene} \xrightarrow{9\text{-BBN}} \text{(B, 0.1%)} + \text{(B, 99.9%)} \quad (28)$$

It achieves a regioselectivity of 99.8% in the hydroboration of *cis*-4-methyl-2-pentene (Eqn. 29) (Ref. 39).

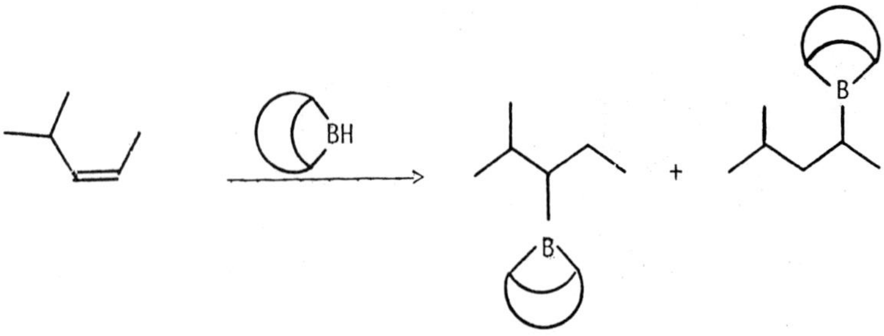

(29)

0.2%   99.8%

Monochloroborane ethyl etherate also exhibits a high regioselectivity for the terminal position of 1-alkenes (Eqn. 30) (Ref. 40).

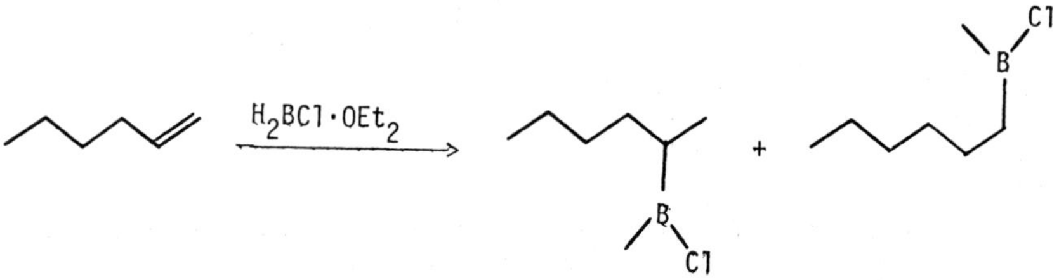

(30)

0.5%   99.5%

However, its steric requirements are similar to those of borane. It does not distinguish significantly between the carbon atoms of the double bond in *cis*-4-methyl-2-pentene (Eqn. 31) (Ref. 40).

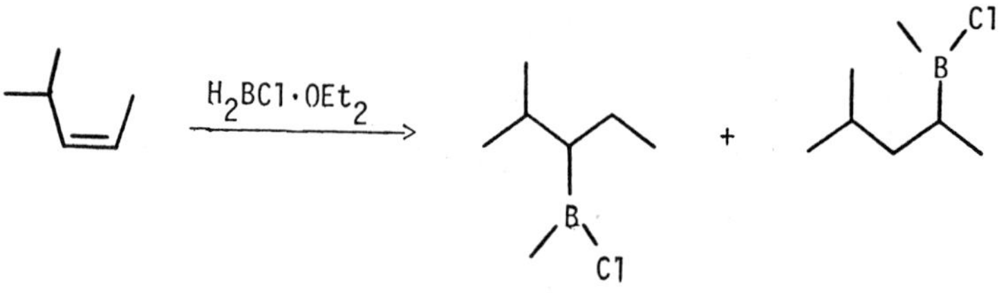

(31)

40%   60%

Dibromoborane methyl sulfide exhibits an unusual characteristic. It reveals an enhanced tendency to place the boron atom at the tertiary position (Chart II) (Refs. 41, 42).

Chart II

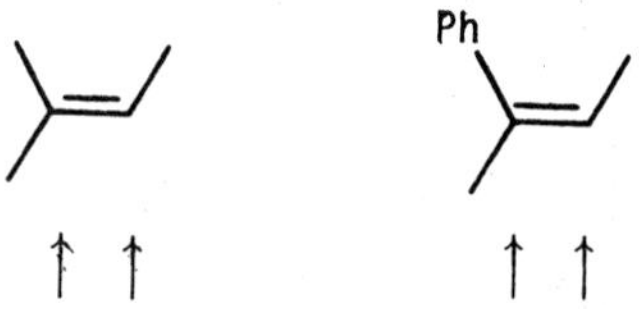

	↑ ↑		↑ ↑	
BH_3	2	98	2.8	97.2
$Br_2BH \cdot SMe_2$	7	93	20	80

The powerful regioselectivity of 9-BBN also overcomes the problems referred to earlier in the hydroboration of styrene (Eqn. 32) and allyl chloride (Eqn. 33) (Refs. 43, 44).

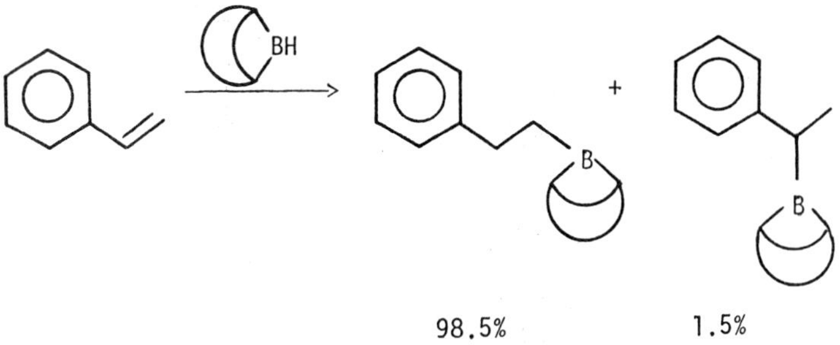

$$PhCH=CH_2 \xrightarrow{\text{9-BBN-H}} PhCH_2CH_2-B(9\text{-BBN}) \; (98.5\%) \; + \; PhCH(CH_3)-B(9\text{-BBN}) \; (1.5\%) \tag{32}$$

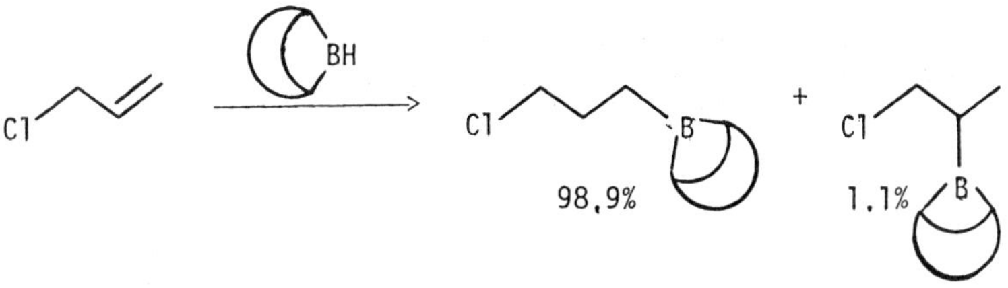

$$ClCH_2CH=CH_2 \xrightarrow{\text{9-BBN-H}} ClCH_2CH_2CH_2-B(9\text{-BBN}) \; (98.9\%) \; + \; ClCH_2CH(CH_3)-B(9\text{-BBN}) \; (1.1\%) \tag{33}$$

STOICHIOMETRIES

These new hydroborating agents also overcome many of the problems in controlling stoichiometry in hydroborating with borane, a trifunctional reagent. For example, the monohydroboration of 1,3-pentadiene is easily achieved with disiamylborane (Eqn. 34) (Ref. 45).

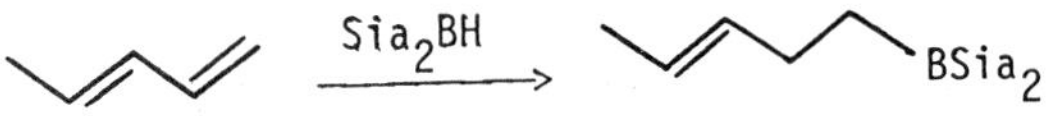

$$CH_3CH=CHCH=CH_2 \xrightarrow{Sia_2BH} CH_3CH=CHCH_2CH_2-BSia_2 \tag{34}$$

Similarly, 4-vinylcyclohexene is readily monohydroborated by 9-BBN (Eqn. 35) (Ref. 46), as well as other disubstituted boranes.

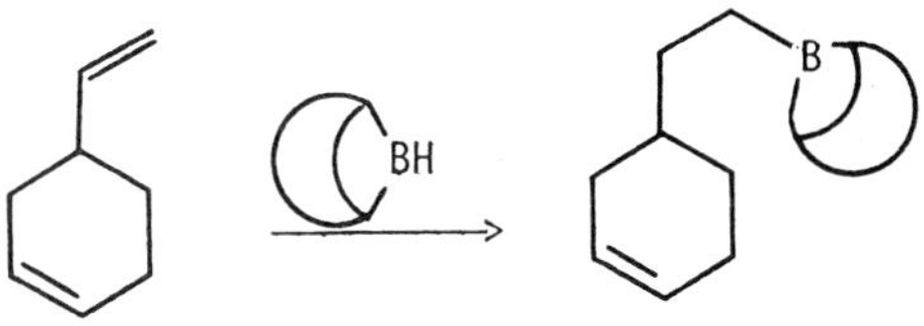

$$\text{4-vinylcyclohexene} \xrightarrow{\text{9-BBN-H}} \text{4-(2-(9-BBN)ethyl)cyclohexene} \tag{35}$$

These reagents overcome the problem of achieving the clean monohydroboration of acetylenes, making the vinylboranes available in excellent yields (Eqns. 36, 37) (Refs. 20, 13).

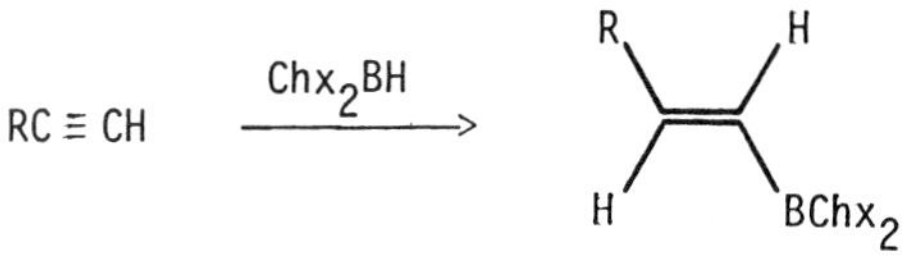

$$RC\equiv CH \xrightarrow{Chx_2BH} \underset{H}{\overset{R}{}}C=C\underset{BChx_2}{\overset{H}{}} \tag{36}$$

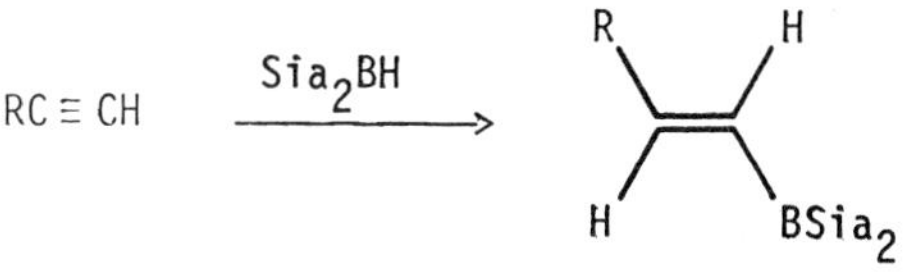

$$RC\equiv CH \xrightarrow{Sia_2BH} \underset{H}{\overset{R}{}}C=C\underset{BSia_2}{\overset{H}{}} \tag{37}$$

9-BBN tends to react further. However, by using a two-fold excess of acetylene, it is possible to prepare the B-vinyl-9-BBN derivative in satisfactory yield (Eqn. 38) (Ref. 47).

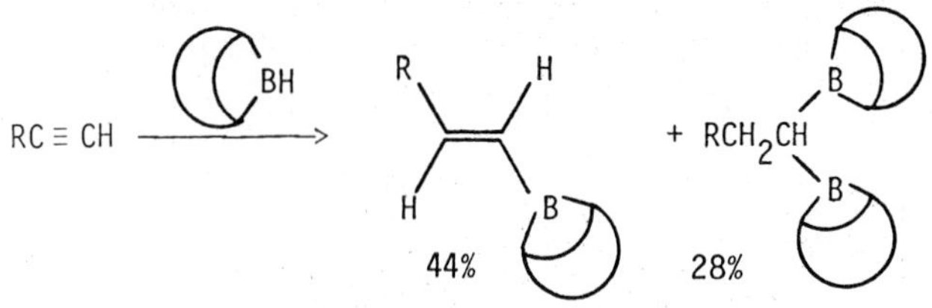

$$RC \equiv CH \xrightarrow{\text{( )}_2 BH} \underset{44\%}{\overset{R \quad H}{\underset{H \quad B( )_2}{C=C}}} + RCH_2CH(B( )_2)_2 \quad (38)$$
$$ 28\%$$

The monohydroboration of acetylenes also proceeds cleanly with catecholborane and dibromoborane methyl sulfide (Eqns. 39, 40) (Refs. 48, 49). These derivatives have many valuable applications for synthesis.

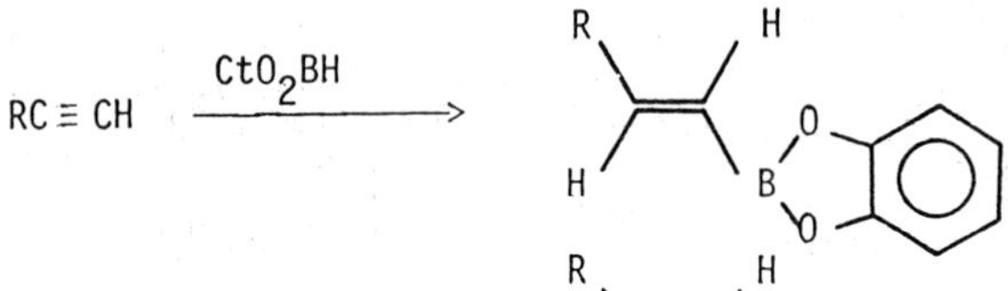

$$RC \equiv CH \xrightarrow{CtO_2BH} \underset{H}{\overset{R}{C}}=\underset{B(O_2C_6H_4)}{\overset{H}{C}} \quad (39)$$

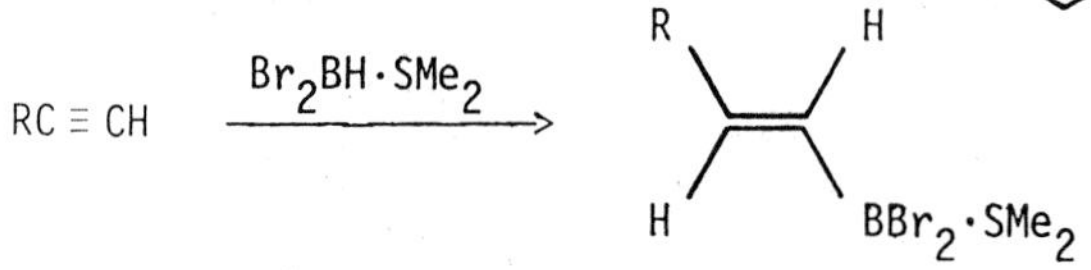

$$RC \equiv CH \xrightarrow{Br_2BH \cdot SMe_2} \underset{H}{\overset{R}{C}}=\underset{BBr_2 \cdot SMe_2}{\overset{H}{C}} \quad (40)$$

Thexylmonochloroborane can also be utilized for the monohydroboration of acetylenes and the intermediates find valuable application (Eqn. 41) (Ref. 50).

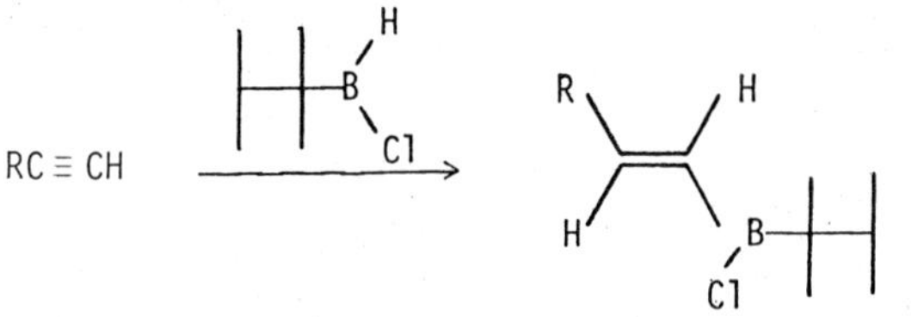

$$RC \equiv CH \xrightarrow{\text{ThxBHCl}} \underset{H}{\overset{R}{C}}=\underset{B(Cl)(Thx)}{\overset{H}{C}} \quad (41)$$

Thexylborane, as a bifunctional hydroborating agent, is especially valuable to achieve cyclic hydroboration (Eqn. 42) (Ref. 51).

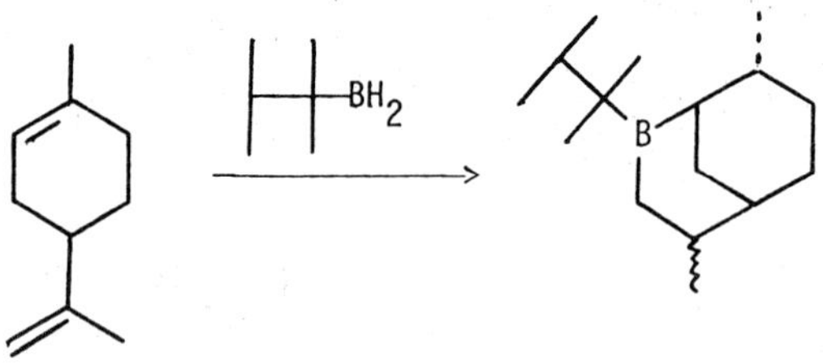

$$\text{limonene} \xrightarrow{\text{ThxBH}_2} \text{bicyclic borane product} \quad (42)$$

In the same system, limonene, thexylmonochloroborane achieves monohydroboration (Eqn. 43) (Ref. 52).

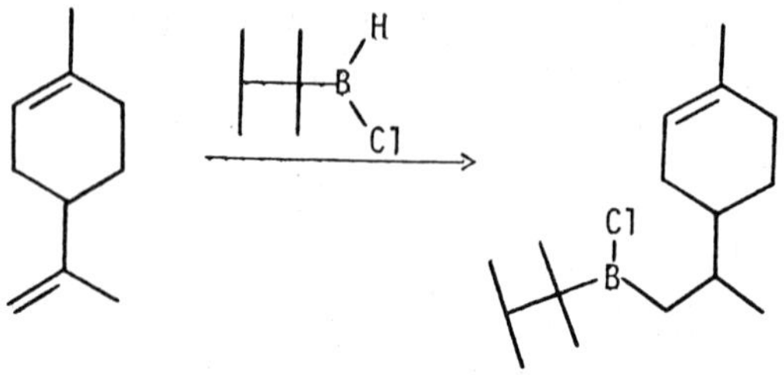

$$\text{limonene} \xrightarrow{\text{ThxBHCl}} \text{monohydroboration product} \quad (43)$$

Most alkenes other than the unhindered 1-alkenes can react with thexylborane to achieve the synthesis of the desired thexylmonoalkylboranes (Ref. 54). Unfortunately, simple terminal alkenes react to give a mixture (Eqn. 44) (Ref. 27).

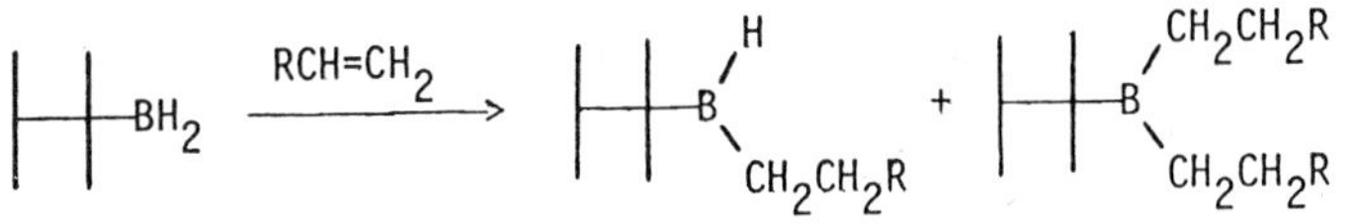

$$\text{Thx-BH}_2 \xrightarrow{RCH=CH_2} \text{Thx-B}(H)(CH_2CH_2R) + \text{Thx-B}(CH_2CH_2R)_2 \quad (44)$$

Consequently, it was never possible to synthesize thexyldialkylboranes with the two alkyl groups representing two straight-chain moieties. However, this is now feasible with thexylmonochloroborane (Eqn. 45) (Ref. 37).

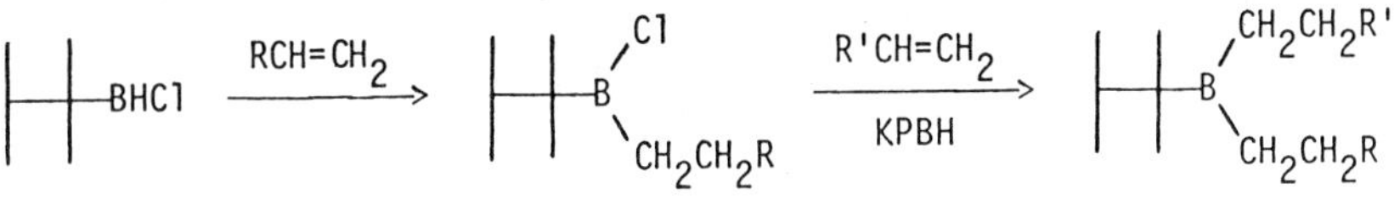

$$\text{Thx-BHCl} \xrightarrow{RCH=CH_2} \text{Thx-B}(Cl)(CH_2CH_2R) \xrightarrow[KPBH]{R'CH=CH_2} \text{Thx-B}(CH_2CH_2R')(CH_2CH_2R) \quad (45)$$

RELATIVE REACTIVITIES

These new hydroborating agents exhibit major differences in their sensitivity to different structures. This makes possible certain highly selective hydroborations. For example, disiamylborane is far more reactive to a terminal double-bond than to a 2-methyl-1-alkene (Refs. 38, 45). The reverse is true for dibromoborane methyl sulfide (Chart III) (Ref. 53).

Chart III

	Sia_2BH	$Br_2BH \cdot SMe_2$
$RCH=CH_2$	100	100
$\underset{\|}{CH_3}$ $RC=CH_2$	4.9	2040

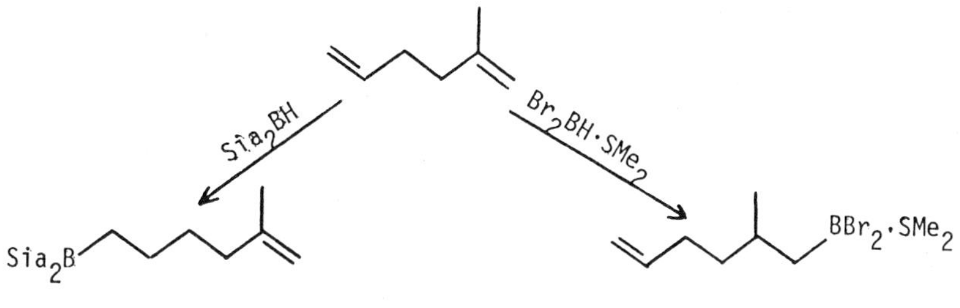

Thus, by careful choice of reagent, it becomes possible to hydroborate selectively either of the two double bonds in 2-methyl-1,5-hexadiene. 9-BBN is surprisingly sluggish toward acetylenes (Ref. 54). On the other hand, dibromoborane methyl sulfide is exceptionally reactive towards acetylenes, especially disubstituted derivatives, $RC \equiv CR$ (Ref. 53). This makes possible some remarkably selective hydroborations (Chart IV).

Chart IV

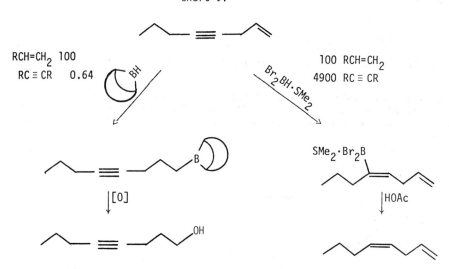

It was considered exceptional when disiamylborane revealed a preference for cis alkenes over the corresponding trans by factors of approximately 10 (Ref. 55). However, thexylchloroborane now reveals a preference of approximately 100 (Chart V) (Ref. 56).

Chart V

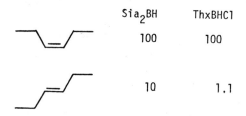

	Sia_2BH	ThxBHCl
(cis)	100	100
(trans)	10	1.1

It is now possible to hydroborate selectively a terminal alkyne in the presence of an internal alkyne, or *vice versa* (Chart VI) (Refs. 49, 57).

Chart VI

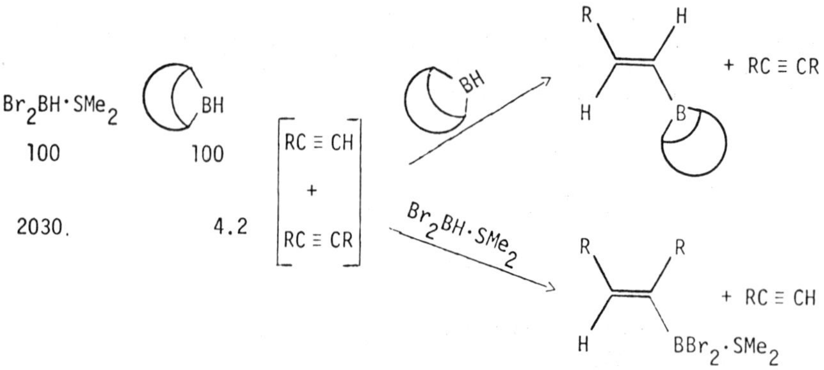

Clearly these new reagents expand greatly the versatility of the hydroboration reaction.

APPLICATIONS

It is appropriate to point out some representative applications of these new hydroborating agents.

For many years the synthesis of 1,2-diphenyl-*exo*-norbornanol escaped us. Every synthesis we tried failed. However, the observation that dibromoborane methyl sulfide gives a significant amount of hydroboration at the tertiary position solved the problem (Eqn. 46) (Ref. 58).

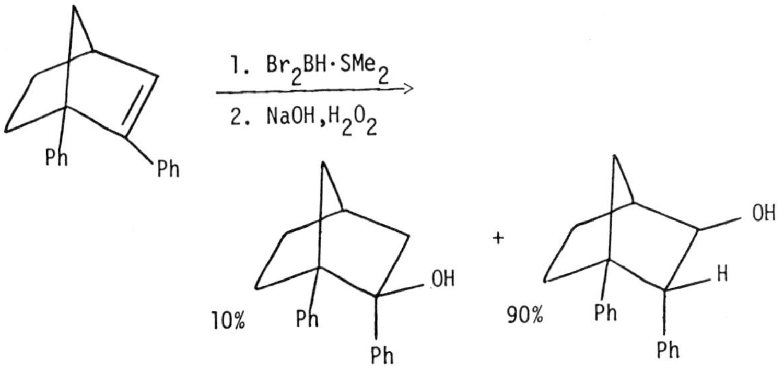

(46)

There was little difficulty in separating the desired tertiary alcohol from the more plentiful secondary.

Thexylborane makes possible an elegant synthesis of ketones that is almost general (Eqn. 47) (Ref. 59).

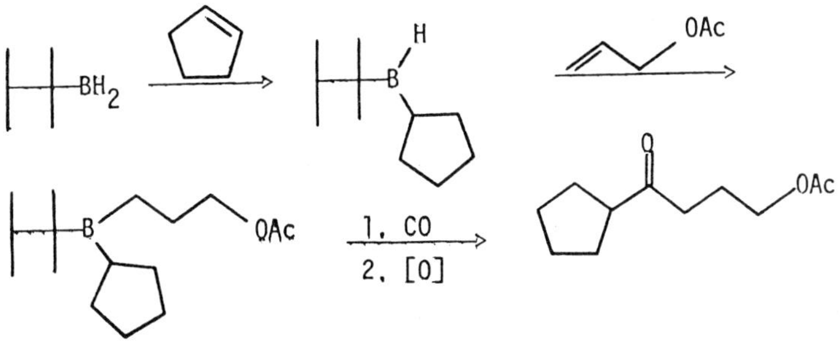

(47)

Only the synthesis of ketones of the type $RCH_2CH_2\overset{O}{\overset{\|}{C}}CH_2CH_2R'$ was not possible. This was especially unfortunate because it excluded the synthesis of straight-chain ketones. However, thexylchloroborane has solved this problem (Eqn. 48) (Ref. 60).

(48)

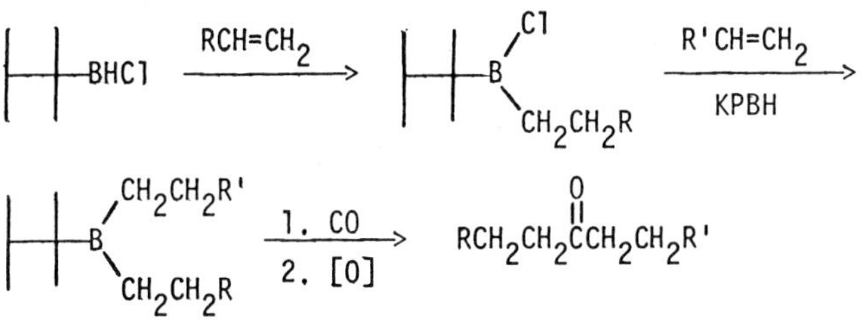

Diisopinocampheylborane has made possible a simple synthesis of chiral boronic acids (Eqns. 49, 50) (Ref. 61).

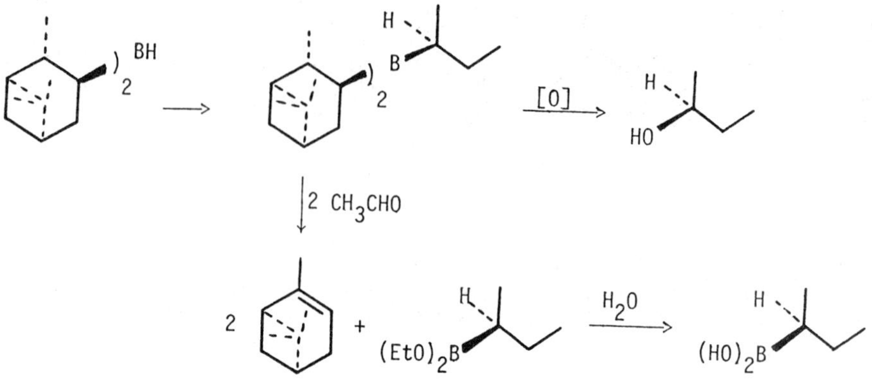

(S)-(+)-boronic ester, 97% ee (49)

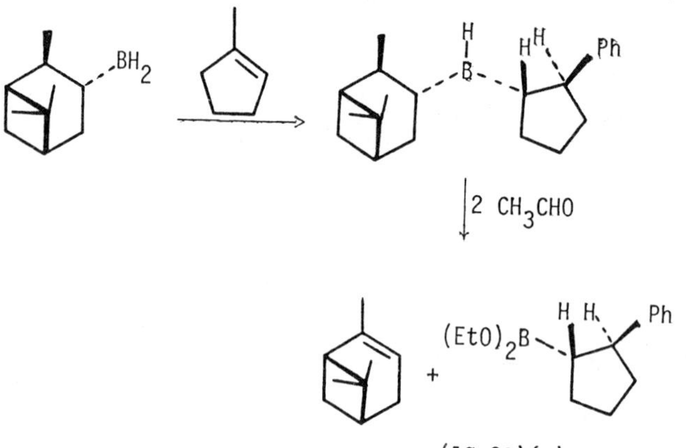

(1S,2S)(+)-boronic ester, 98% ee (50)

It has been established by M. M. Midland and coworkers that B-Ipc-9-BBN possesses remarkable properties for the asymmetric reduction of aldehydes (Eqn. 51) and acetylenic ketones (Eqn. 52) (Ref. 62).

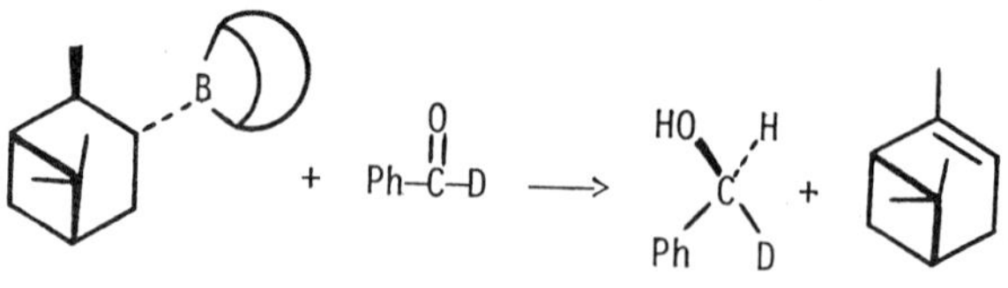

100% ee (51)

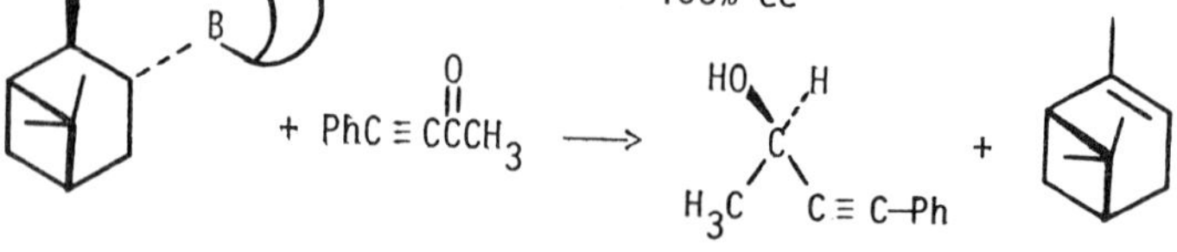

100% ee (52)

Unfortunately, the reagent could not be used for the reduction of simple ketones, such as acetophenone (Eqn. 53) (Ref. 63).

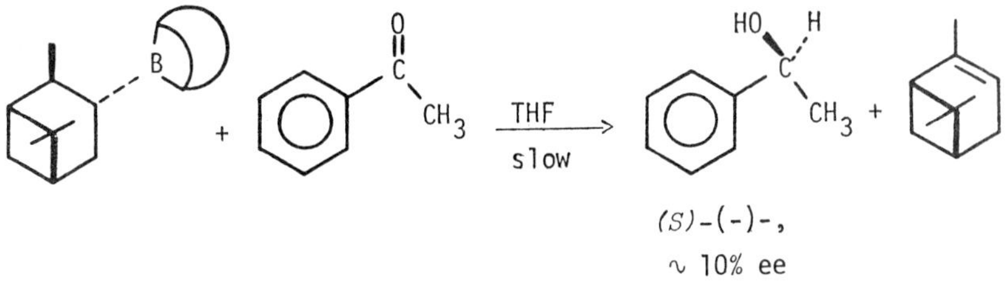

$$\text{Ipc}_2\text{BH} + \text{PhCOCH}_3 \xrightarrow[\text{slow}]{\text{THF}} \text{Ph-CH(OH)-CH}_3 \, (S)\text{-}(-)\text{-}, \sim 10\% \text{ ee} + \alpha\text{-pinene} \quad (53)$$

Fortunately, by shifting from dilute solutions of the reagent (0.5 M) in THF to the neat reagent, or concentrated solutions (2 M), the reaction can be extended to acetophenone ketones (Eqn. 54) (Ref. 64).

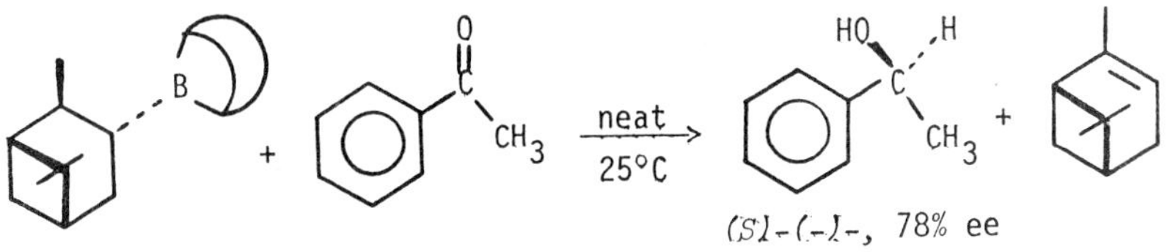

$$\text{Ipc}_2\text{BH} + \text{PhCOCH}_3 \xrightarrow[25°C]{\text{neat}} \text{Ph-CH(OH)-CH}_3 \, (S)\text{-}(-)\text{-}, 78\% \text{ ee} + \alpha\text{-pinene} \quad (54)$$

The utility of the reagent with this modified procedure is indicated by the results realized with phenacyl bromide (and other α-haloketones) (Eqn. 55) (Ref. 65).

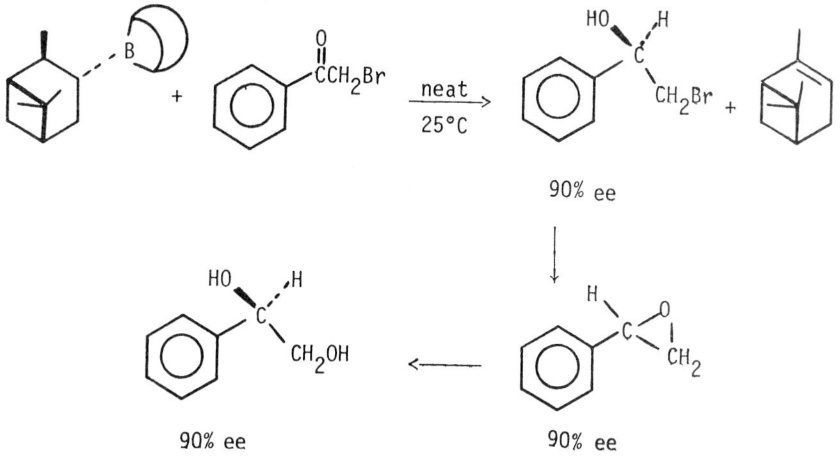

$$\text{Ipc}_2\text{BH} + \text{PhCOCH}_2\text{Br} \xrightarrow[25°C]{\text{neat}} \text{Ph-CH(OH)-CH}_2\text{Br} \, (90\% \text{ ee}) + \alpha\text{-pinene}$$

$$\text{Ph-CH(OH)-CH}_2\text{OH} \, (90\% \text{ ee}) \longleftarrow \text{Ph-CH(-O-)CH}_2 \, (90\% \text{ ee}) \quad (55)$$

The facile synthesis of cyclopropanes, including cyclopropylcarbinyl chloride, illustrates the remarkable versatility of these derivatives (Eqn. 56) (Ref. 66).

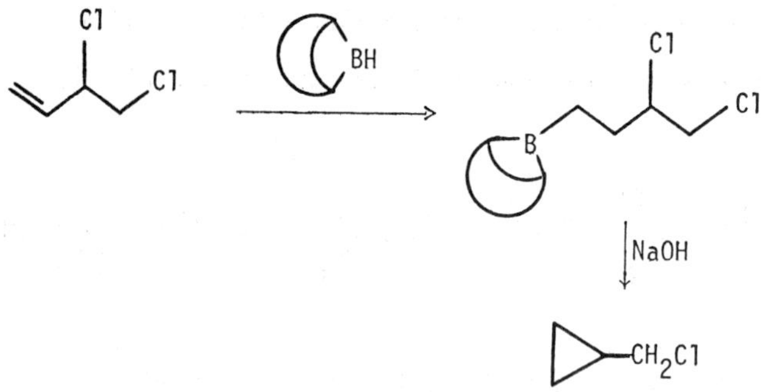

(56)

GENERALIZATION OF THE ZWEIFEL SYNTHESES OF CIS AND TRANS ALKENES

The elegant Zweifel syntheses of cis and trans alkenes were pointed out earlier (Eqns. 11, 12) (Refs. 20, 21). Unfortunately, these syntheses suffer from two serious disadvantages:

1. The reactions require R_2BH. Consequently, they were limited to those hindered alkenes which undergo hydroboration to R_2BH.

2. Only one of the R groups is utilized. This can be a serious difficulty in cases where R is a valuable intermediate.

Accordingly, we undertook a systematic program to utilize these new hydroborating agents to overcome these difficulties.

Hydroboration of alkenes with monohaloborane-methyl sulfide leads easily to the dialkyl-haloboranes. These intermediates are readily hydrided by lithium aluminum hydride to give R_2BH (Eqn. 57) (Ref. 67).

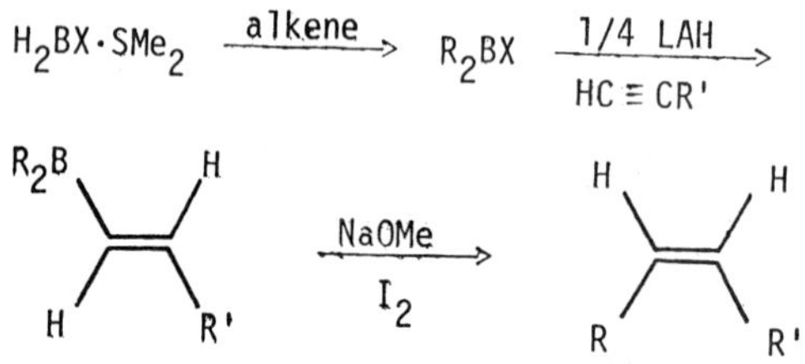

(57)

This has the advantage that R is now general. However, it still possesses the disadvantage that one R group is lost. Moreover, in cases where R is relatively unhindered, excess alkyne must be used to minimize dihydroboration.

The same approach makes the Zweifel synthesis of trans alkenes quite general (Eqn. 58) (Ref. 68).

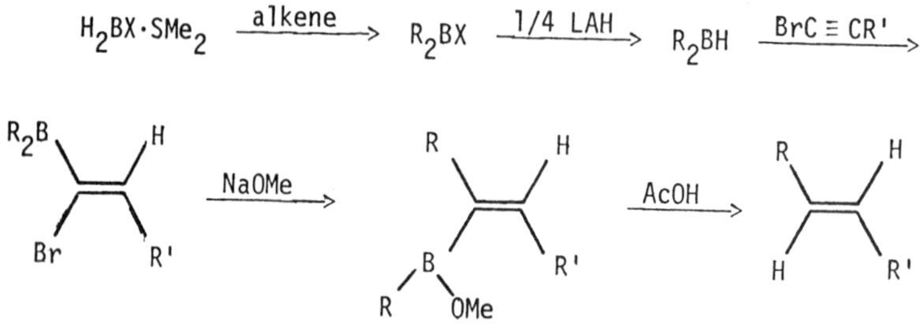

(58)

In this case, dihydroboration is no problem, but there is still the loss of one R group. As discussed earlier, the synthesis of thexylmonoalkylborane is now general (Eqn. 59).

$$\text{ThxBHCl} \xrightarrow{\text{alkene}} \text{ThxB(Cl)R} \xrightarrow{\text{KPBH}} \text{ThxB(H)R} \xleftarrow{\text{alkene}} \text{ThxBH}_2 \quad (59)$$

We explored the applicability of these thexylmonoalkylboranes for the Zweifel synthesis of cis and trans alkenes (Eqn. 60) (Refs. 69, 70).

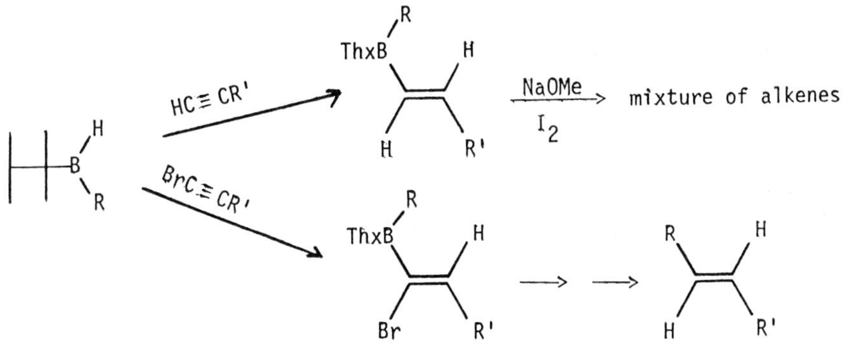

$$(60)$$

ThxBHR + HC≡CR' → ThxB(R)(H)C=C(H)(R') $\xrightarrow{\text{NaOMe}, I_2}$ mixture of alkenes

ThxBHR + BrC≡CR' → ThxB(R)(H)C=C(Br)(R') → → R(H)C=C(H)R'

The synthesis of trans alkenes is satisfactory. However, in the cis synthesis, the thexyl group migrates competitively with the alkyl group. The loss of a thexyl group is also undesirable, if it can be avoided.

Dibromoborane methyl sulfide provided the final solution for the cis alkenes (Eqn. 61), (Ref. 71).

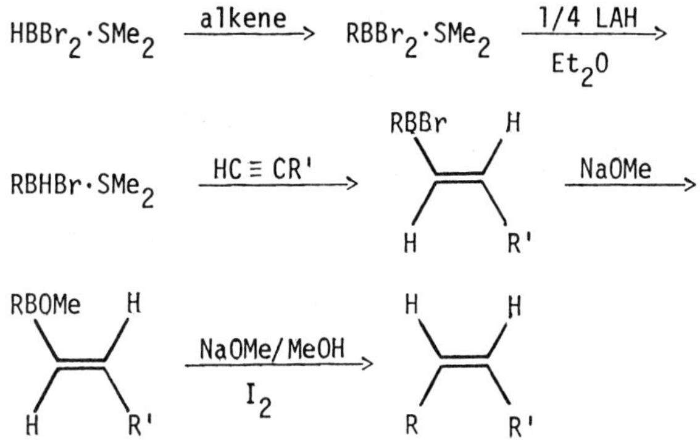

$$\text{HBBr}_2 \cdot \text{SMe}_2 \xrightarrow{\text{alkene}} \text{RBBr}_2 \cdot \text{SMe}_2 \xrightarrow[\text{Et}_2\text{O}]{1/4 \text{ LAH}} \text{RBHBr} \cdot \text{SMe}_2 \xrightarrow{\text{HC} \equiv \text{CR}'} \text{RBBr(H)C=C(H)R'} \xrightarrow{\text{NaOMe}} \text{RBOMe(H)C=C(H)R'} \xrightarrow[I_2]{\text{NaOMe/MeOH}} \text{(H)(R)C=C(H)(R')} \quad (61)$$

This solution also proved satisfactory for the synthesis of trans alkenes (Eqn. 62) (Ref. 72).

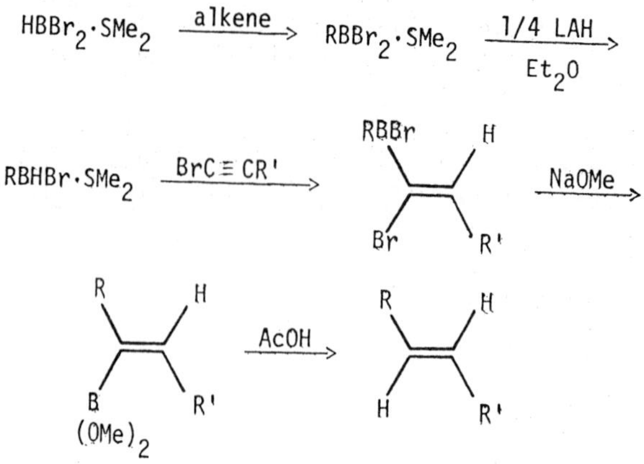

$$\text{HBBr}_2 \cdot \text{SMe}_2 \xrightarrow{\text{alkene}} \text{RBBr}_2 \cdot \text{SMe}_2 \xrightarrow[\text{Et}_2\text{O}]{1/4 \text{ LAH}}$$

$$\text{RBHBr} \cdot \text{SMe}_2 \xrightarrow{\text{BrC}\equiv\text{CR'}} \underset{\text{Br}\quad\text{R'}}{\overset{\text{RBBr}\quad\text{H}}{\text{C=C}}} \xrightarrow{\text{NaOMe}} \quad (62)$$

$$\underset{\text{B(OMe)}_2\quad\text{R'}}{\overset{\text{R}\quad\text{H}}{\text{C=C}}} \xrightarrow{\text{AcOH}} \underset{\text{H}\quad\text{R'}}{\overset{\text{R}\quad\text{H}}{\text{C=C}}}$$

These procedures appear to be general for essentially all R groups. There is no loss of R group. At this time we are not aware of any disadvantages.

The utility of these new procedures was tested by applying the synthesis of cis alkenes to the synthesis of muscalure, the sex pheromone of the house fly *(Musca domestica)*. It operated entirely satisfactorily (Eqn. 63) (Ref. 71).

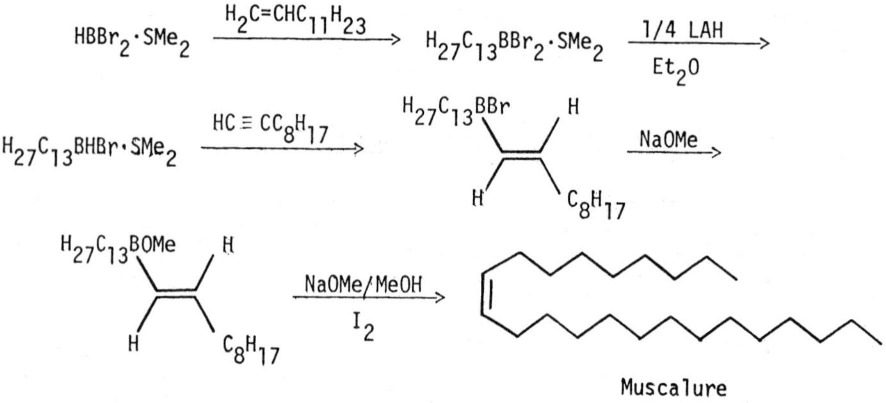

$$\text{HBBr}_2 \cdot \text{SMe}_2 \xrightarrow{\text{H}_2\text{C=CHC}_{11}\text{H}_{23}} \text{H}_{27}\text{C}_{13}\text{BBr}_2 \cdot \text{SMe}_2 \xrightarrow[\text{Et}_2\text{O}]{1/4 \text{ LAH}}$$

$$\text{H}_{27}\text{C}_{13}\text{BHBr} \cdot \text{SMe}_2 \xrightarrow{\text{HC}\equiv\text{CC}_8\text{H}_{17}} \underset{\text{H}\quad\text{C}_8\text{H}_{17}}{\overset{\text{H}_{27}\text{C}_{13}\text{BBr}\quad\text{H}}{\text{C=C}}} \xrightarrow{\text{NaOMe}} \quad (63)$$

$$\underset{\text{H}\quad\text{C}_8\text{H}_{17}}{\overset{\text{H}_{27}\text{C}_{13}\text{BOMe}\quad\text{H}}{\text{C=C}}} \xrightarrow[\text{I}_2]{\text{NaOMe/MeOH}} \text{Muscalure}$$

STEREOSPECIFIC SYNTHESIS OF TRISUBSTITUTED OLEFINS

Zweifel also introduced a stereospecific synthesis of trisubstituted alkenes (Eqn. 64) (Ref. 20).

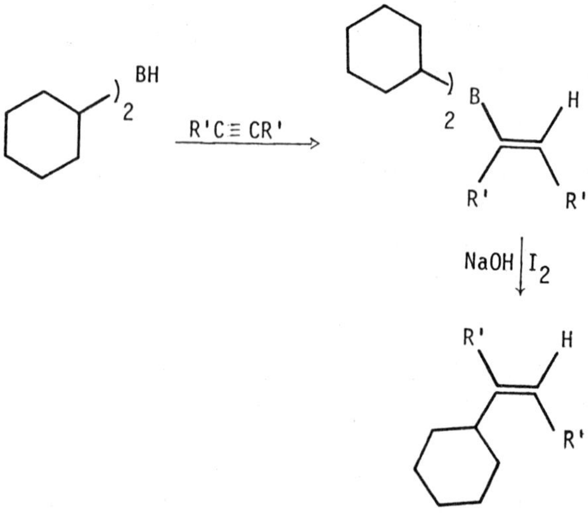

$$(\text{Cy})_2\text{BH} \xrightarrow{\text{R'C}\equiv\text{CR'}} (\text{Cy})_2\text{B}\underset{\text{R'}\quad\text{R'}}{\overset{\quad\text{H}}{\text{C=C}}} \xrightarrow{\text{NaOH} | \text{I}_2} \underset{\text{Cy}\quad\text{R'}}{\overset{\text{R'}\quad\text{H}}{\text{C=C}}} \quad (64)$$

Unfortunately, this procedure suffers from the same disadvantage as his syntheses of cis and trans alkenes. It requires R_2BH as reagent. Consequently, it was limited to those few hindered alkenes that could be converted to R_2BH by direct hydroboration with borane. In addition, only one of the two R groups was utilized.

Application of our new synthesis of R_2BH *via* hydridation of R_2BX made the procedure general for essentially all R groups available through hydroboration (Eqn. 65) (Ref. 73).

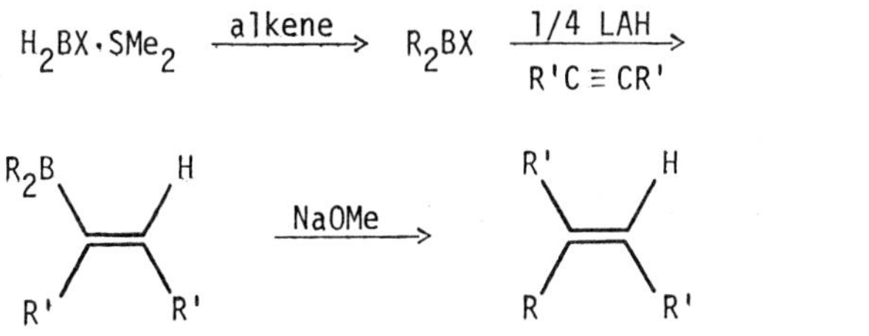

$$H_2BX \cdot SMe_2 \xrightarrow{\text{alkene}} R_2BX \xrightarrow[R'C \equiv CR']{1/4 \text{ LAH}} \tag{65}$$

$$\underset{R'}{\overset{R_2B}{>}}=\underset{R'}{\overset{H}{<}} \xrightarrow{\text{NaOMe}} \underset{R}{\overset{R'}{>}}=\underset{R'}{\overset{H}{<}}$$

Stereochemistry established

The procedure still suffered from the disadvantage that only one R group was utilized. Fortunately, our new development, proceeding through dibromoborane methyl sulfide, worked ideally (Eqn. 66) (Ref. 74).

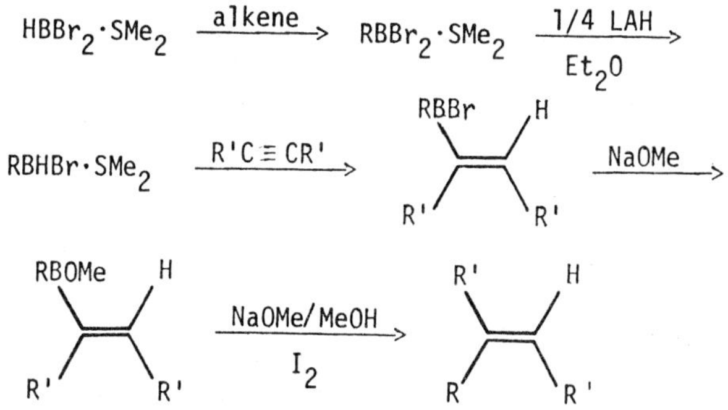

$$HBBr_2 \cdot SMe_2 \xrightarrow{\text{alkene}} RBBr_2 \cdot SMe_2 \xrightarrow[\text{Et}_2O]{1/4 \text{ LAH}}$$

$$RBHBr \cdot SMe_2 \xrightarrow{R'C \equiv CR'} \underset{R'}{\overset{RBBr}{>}}=\underset{R'}{\overset{H}{<}} \xrightarrow{\text{NaOMe}} \tag{66}$$

$$\underset{R'}{\overset{RBOMe}{>}}=\underset{R'}{\overset{H}{<}} \xrightarrow[I_2]{\text{NaOMe/MeOH}} \underset{R}{\overset{R'}{>}}=\underset{R'}{\overset{H}{<}}$$

With this new procedure, the reaction appears to be general for all R groups and all R groups are utilized. There are no difficulties evident at this time.

GENERAL PROCEDURES FOR PHEROMONE SYNTHESIS

A current program is the development of one-pot synthetic procedures for the synthesis of pheromones utilizing such borane-based chemistry. Space will not permit a detailed discussion of this program. However, the following two examples will illustrate the versatility of these methods (Eqns. 67, 68) (Refs. 75, 76).

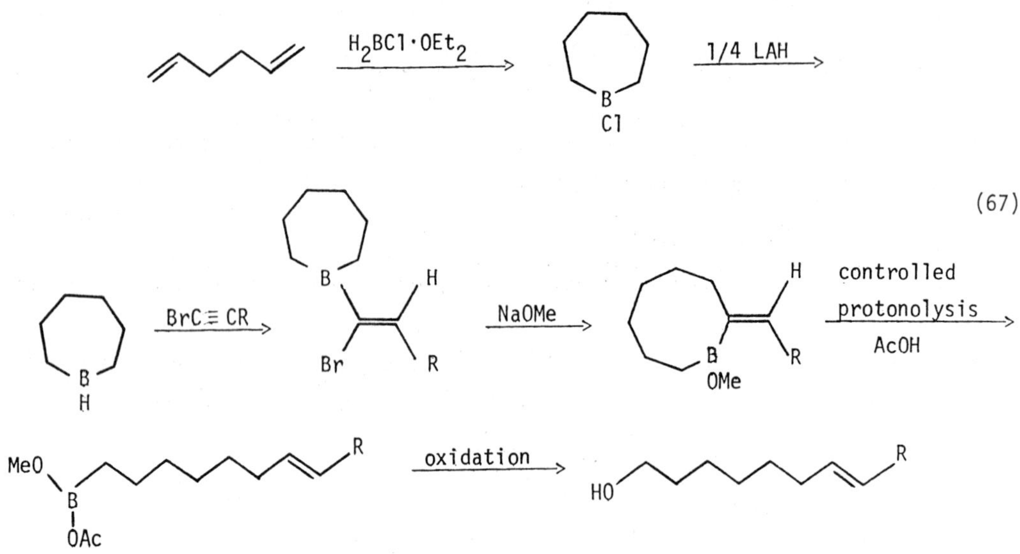

CONCLUSION

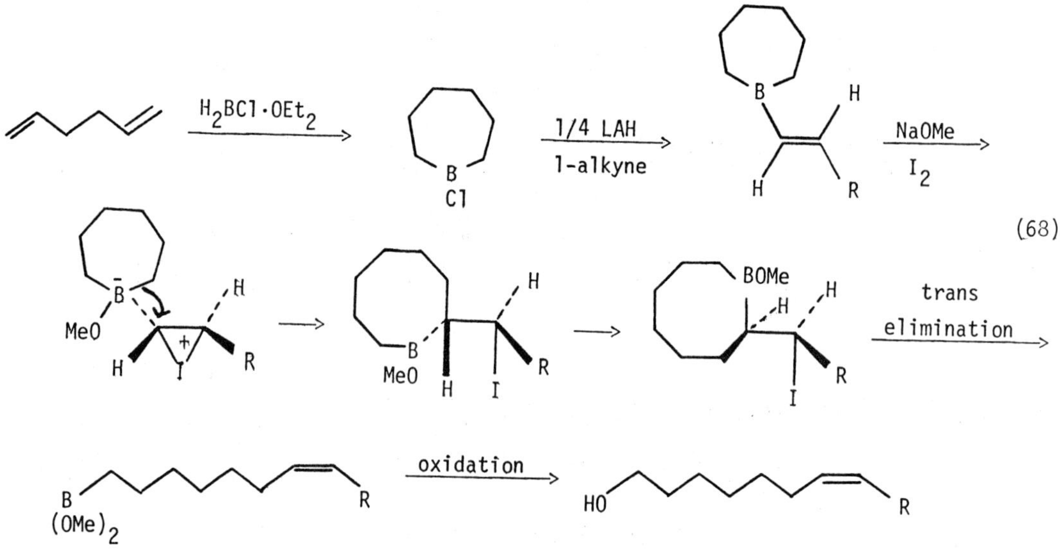

It is evident that hydroboration and organoborane chemistry continue to expand. The new hydroboration reagents overcome many of the difficulties encountered in hydroboration with diborane itself. Their use makes it possible to control hydroboration to achieve remarkable selectivities and essentially quantitative yields of the desired intermediates. The rapidly expanding chemistry of organoboranes greatly simplifies the task of the synthetic chemist in converting these intermediates into desired products. There appears to be no slowing of the pace—we can look forward with confidence to the continued rapid development of this powerful synthetic tool.

REFERENCES

1. H. C. Brown, Hydroboration, Benjamin, New York (1963); second printing (with Nobel Lecture), Benjamin/Cummings, Reading, Mass. (1980).
2. E. L. Mutterties, ed., Chemistry of Boron and Its Compounds, Wiley, New York (1967).
3. N. Nesmeyanov and R. A. Sokolik, Methods of Elemento-Organic Chemistry, North-Holland, Amsterdam (1967), vol. 1.
4. M. Grassberger, Organische Borverbindungen, Verlag Chemie, Wien (1971).
5. H. C. Brown, Boranes in Organic Chemistry, Cornell University Press, Ithaca, New York (1972).
6. G. M. L. Cragg, Organoboranes in Organic Synthesis, Dekker, New York (1973).
7. H. C. Brown, et. al., Organic Syntheses via Boranes, Wiley-Interscience, New York (1975).
8. T. Onak, Organoborane Chemistry, Academic Press, New York (1975).
9. A. Pelter and K. Smith, Comprehensive Organic Chemistry, ed., D. H. Barton and W. D. Ollis, Pergamon, Oxford (1979), vol. 3, pp. 689-940.
10. E. Negishi, Organometallics in Organic Synthesis, Wiley-Interscience, New York (1980), vol. 1.
11. H. C. Brown and G. Zweifel, *J. Am. Chem. Soc.* **82**, 4708 (1960).
12. H. C. Brown and K. A. Keblys, *J. Am. Chem. Soc.* **86**, 1791 (1964).
13. H. C. Brown and G. Zweifel, *J. Am. Chem. Soc.* **83**, 3834 (1961).
14. G. Zweifel, K. Nagase and H. C. Brown, *J. Am. Chem. Soc.* **84**, 183 (1962).
15. P. L. Burke, Ph.D. Thesis, Purdue University (1973).
16. H. C. Brown and A. W. Moerikofer, *J. Am. Chem. Soc.* **85**, 2063 (1963).
17. G. Zweifel and H. C. Brown, *Org. React.* **13**, 1 (1963).
18. H. C. Brown, M. W. Rathke and M. M. Rogić, *J. Am. Chem. Soc.* **90**, 5038 (1968).
19. J. Hooz and S. Linke, *J. Am. Chem. Soc.* **90**, 6891 (1968).
20. G. Zweifel, H. Arzoumanian and C. C. Whitney, *J. Am. Chem. Soc.* **89**, 3652 (1967).
21. G. Zweifel and H. Arzoumanian, *J. Am. Chem. Soc.* **89**, 5086 (1967).
22. H. C. Brown and G. J. Klender, *Inorg. Chem.* **1**, 204 (1962).
23. H. C. Brown and G. Zweifel, *J. Am. Chem. Soc.* **82**, 3222 (1960).
24. E. F. Knights and H. C. Brown, *J. Am. Chem. Soc.* **90**, 5280 (1968).
25. J. A. Soderquist and H. C. Brown, *J. Org. Chem.* **46**, 4599 (1981).
26. G. Zweifel and H. C. Brown, *J. Am. Chem. Soc.* **85**, 2066 (1963).
27. H. C. Brown, E. Negishi and J.-J. Katz, *J. Am. Chem. Soc.* **97**, 2791 (1975).
28. H. C. Brown, J.-J. Katz, C. F. Lane and E. Negishi, *J. Am. Chem. Soc.* **97**, 2799 (1975).
29. G. Zweifel and H. C. Brown, *J. Am. Chem. Soc.* **86**, 393 (1964).
30. H. C. Brown and N. M. Yoon, *Israel J. Chem.* **15**, 12 (1977).
31. H. C. Brown, J. R. Schwier and B. Singaram, *J. Org. Chem.* **43**, 4395 (1978).
32. H. C. Brown and S. K. Gupta, *J. Am. Chem. Soc.* **97**, 5249 (1975).
33. H. C. Brown and N. Ravindran, *J. Am. Chem. Soc.* **98**, 1785, 1798 (1976).
34. H. C. Brown and N. Ravindran, *Inorg. Chem.* **16**, 2938 (1977).
35. K. Kinberger and W. Siebert, *Z. Naturforsch.* **30B**, 55 (1975).
36. G. Zweifel and N. R. Pearson, *J. Am. Chem. Soc.* **102**, 5919 (1980).
37. H. C. Brown, J. A. Sikorski, S. U. Kulkarni and H. D. Lee, *J. Org. Chem.* **47**, 863 (1982).
38. H. C. Brown and G. Zweifel, *J. Am. Chem. Soc.* **83**, 1241 (1961).
39. H. C. Brown, E. F. Knights and C. G. Scouten, *J. Am. Chem. Soc.* **96**, 7765 (1974).
40. H. C. Brown and N. Ravindran, *J. Am. Chem. Soc.* **98**, 1785 (1976).
41. H. C. Brown, N. Ravindran and S. U. Kulkarni, *J. Org. Chem.* **45**, 384 (1980).
42. H. C. Brown and J. B. Campbell, Jr., unpublished results.
43. C. G. Scouten and H. C. Brown, *J. Org. Chem.* **38**, 4092 (1973).
44. H. C. Brown and J. C. Chen, *J. Org. Chem.* **46**, 3978 (1981).
45. G. Zweifel, K. Nagase and H. C. Brown, *J. Am. Chem. Soc.* **84**, 190 (1962).
46. H. C. Brown, E. F. Knights and R. A. Coleman, *J. Am. Chem. Soc.* **91**, 2144 (1969).
47. H. C. Brown, C. G. Scouten and R. Liotta, *J. Am. Chem. Soc.* **101**, 96 (1979).
48. H. C. Brown and S. K. Gupta, *J. Am. Chem. Soc.* **94**, 4370 (1972).
49. H. C. Brown and J. B. Campbell, Jr., *J. Org. Chem.* **45**, 389 (1980).
50. H. C. Brown and J. A. Sikorski, unpublished results.
51. H. C. Brown and C. D. Pfaffenberger, *Tetrahedron* **31**, 925 (1975).
52. J. A. Sikorski, Ph.D. Thesis, Purdue University (1981).
53. H. C. Brown and J. Chandrasekharan, manuscript in preparation.
54. C. A. Brown and R. A. Coleman, *J. Org. Chem.* **44**, 2329 (1979).
55. H. C. Brown and A. W. Moerikofer, *J. Am. Chem. Soc.* **83**, 3417 (1961).
56. H. A. Sikorski and H. C. Brown, *J. Org. Chem.* **47**, 872 (1982).
57. K. K. Wang, C. G. Scouten and H. C. Brown, *J. Am. Chem. Soc.* **104**, 531 (1982).
58. H. C. Brown, M. Ravindranathan, C. Gundu Rao, F. J. Chloupek and M.-H. Rei, *J. Org. Chem.* **43**, 3667 (1978).
59. E. Negishi and H. C. Brown, *Synthesis* 196 (1972).
60. S. U. Kulkarni, H. D. Lee and H. C. Brown, *J. Org. Chem.* **45**, 4542 (1980).
61. H. C. Brown, P. K. Jadhav and M. C. Desai, *J. Am. Chem. Soc.* **104**, 0000 (1982).

62. M. M. Midland, S. Grur, A. Tramontano and S. A. Zderić, *J. Am. Chem. Soc.* 101, 2352 (1979).
63. Private communication from M. M. Midland.
64. H. C. Brown and G. G. Pai, *J. Org. Chem.* 47, 1606 (1982).
65. H. C. Brown and G. G. Pai, manuscript in preparation.
66. H. C. Brown and S. P. Rhodes, *J. Am. Chem. Soc.* 91, 2149 (1969).
67. S. U. Kulkarni, D. Basavaiah and H. C. Brown, *J. Organomet. Chem.* 225, C1 (1982).
68. H. C. Brown and D. Basavaiah, *J. Org. Chem.* 47, 754 (1982).
69. E. Negishi, J.-J. Katz and H. C. Brown, *Synthesis* 555 (1972).
70. H. C. Brown, H. D. Lee and S. U. Kulkarni, *Synthesis* 195 (1982).
71. H. C. Brown and D. Basavaiah, *J. Org. Chem.* 47, 0000 (1982).
72. H. C. Brown, D. Basavaiah and S. U. Kulkarni, *J. Org. Chem.* 47, 0000 (1982).
73. H. C. Brown, D. Basavaiah and S. U. Kulkarni, *J. Org. Chem.* 47, 171 (1982).
74. H. C. Brown and D. Basavaiah, manuscript in preparation.
75. D. Basavaiah, *Heterocycles* 18, 153 (1982).
76. D. Basavaiah and H. C. Brown, *J. Org. Chem.* 47, 1792 (1982).

PALLADIUM- OR NICKEL-CATALYZED CROSS COUPLING INVOLVING PROXIMALLY HETEROFUNCTIONAL REAGENTS

Ei-ichi Negishi

Department of Chemistry, Purdue University, W. Lafayette, Indiana 47907, USA

Abstract - A wide variety of proximally heterofunctional organometallics and organic electrophiles participate in the Pd- or Ni-catalyzed cross coupling. Those reagents that contain α-heterofunctional groups include metal cyanides, acyl halides, α-metallovinyl ethers, α-heteroarylmetals, and α-silylorganometals. Since α-hetero-organometallics containing alkali metals or Mg do not readily react with alkenyl or aryl electrophiles, the Pd or Ni catalysis provides an attractive solution to this general problem. Metal enolates and α-halocarbonyl compounds represent a synthetically interesting class of β-heterofunctional reagents. Ketone and aldehyde enolates containing B and Zn undergo a remarkably facile and selective allylation catalyzed by Pd-phosphine complexes. The reaction promises to significantly expand the scope of the enolate alkylation. 1,4- and 1,5-Dicarbonyl compounds suitable for annulation are readily and selectively obtained by using 2,3-dichloropropene and 1,3-dichloro-2-butene, respectively. The Pd-catalyzed reaction of chloroiodo-ethene with alkynylzincs provides the corresponding chloroenynes readily convertible into terminal diynes which, in turn, are convertible into unsymmetrical diynes. The Pd-catalyzed cross-coupling reaction of γ-heterosubstituted allylic electrophiles provides a synthetic equivalent to conjugate addition via allylation, while that of β-halo-α,β-unsaturated carbonyl compounds permits conjugate substitution.

INTRODUCTION

The discovery in 1972 by Kumada (1) and Corriu (2) that the reaction of the Grignard reagents with organic halides can be markedly catalyzed by Ni-phosphine complexes opened up a new chapter for the carbon-carbon bond formation via cross coupling (3) and led to the discovery and development by us and others of the highly stereo-, regio- and chemoselective Pd-catalyzed cross coupling (4) over the past several years. These developments, together with the Cu-based methodology (5), have provided a general solution to a long-standing problem associated with the use of unsaturated organic halides and related electrophiles in cross coupling. Interestingly, we have found that, whereas organometallics containing highly electropositive metals, e.g., Li, show an unexpectedly low reactivity in these catalytic reactions, those containing metals of intermediate electronegativity, e.g., Zn (6), Al (7), and Zr (8), display a generally high reactivity. Since these metals readily participate in the hydrometallation (9) and/or carbometallation (10) of acetylenes to form stereo- and regio-defined alkenylmetals, the tandem use of these addition reactions and the Pd- or Ni-catalyzed cross coupling permits expeditious and selective syntheses of various olefins. We and others have also shown that even relatively electronegative metals, e.g., B (11) and Sn (12), may participate in the Pd- or Ni-catalyzed cross coupling.

Favorable results observed in the Pd- or Ni-catalyzed cross coupling prompted us to investigate the use of proximally heterofunctional reagents. Despite extensive developments of such reagents (13), especially acyl anion equivalents, over the past few decades, α- or β-hetero-substituted organometals cannot be readily cross-coupled with alkenyl, aryl, or alkynyl halides and related electrophiles.

Heteroatoms (Z) of our interest include halogens (F, Cl, Br, I) and other electronegative elements, such as O, S, Se, N, and P, as well as some metals, such as B, Al, Zn, Si and Sn. These heteroatoms may be incorporated in positions that are α, β and/or γ to either the metal or the leaving group which participates in the cross-coupling reaction. Thus, for example, α-heterosubstituted reagents may be represented by the following structures. β- and γ-Heterosubstituted reagents may also be classified in a similar manner.

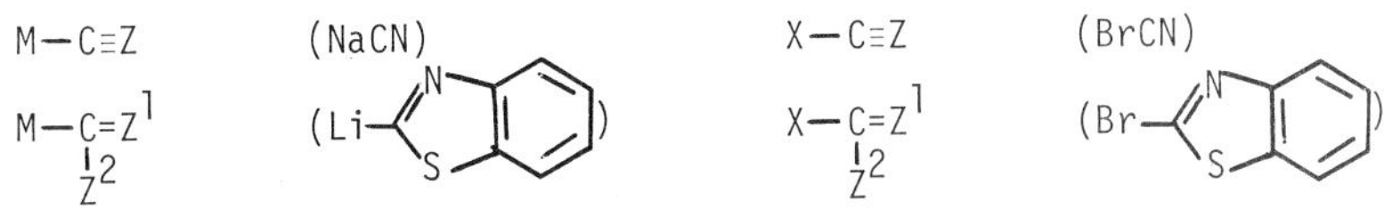

$$M-C\equiv Z \quad (NaCN) \quad \text{(Li-benzothiazole)} \quad X-C\equiv Z \quad (BrCN) \quad \text{(Br-benzothiazole)}$$

$$M-\underset{Z^2}{\overset{Z^1}{C}}= \quad \quad X-\underset{Z^2}{\overset{Z^1}{C}}=$$

$$M-\underset{R}{\underset{|}{C}}=Z \quad (Me_3SiCOCH_3) \quad X-\underset{R}{\underset{|}{C}}=Z \quad (ClCOPh)$$

$$M-\underset{Z}{\underset{|}{C}}=CR^1R^2 \quad (Li\underset{OEt}{\underset{|}{C}}=CH_2) \quad X-\underset{Z}{\underset{|}{C}}=CR^1R^2 \quad (I\underset{SiMe_3}{\underset{|}{C}}=CHMe)$$

$$M-\overset{R^1}{\underset{Z}{\underset{|}{\overset{|}{C}}}}-R^2 \quad (LiCH_2SMe) \quad X-\overset{R^1}{\underset{Z}{\underset{|}{\overset{|}{C}}}}-R^2 \quad (ClCH_2OMe)$$

$$M-\overset{R}{\underset{Z^2}{\underset{|}{\overset{|}{C}}}}-Z^1 \quad (LiCHCl_2) \quad X-\overset{R}{\underset{Z^2}{\underset{|}{\overset{|}{C}}}}-Z^1 \quad (BrHC\langle\overset{S}{\underset{S}{}}\rangle)$$

$$M-\overset{Z^1}{\underset{Z^3}{\underset{|}{\overset{|}{C}}}}-Z^2 \quad (LiCCl_3) \quad X-\overset{Z^1}{\underset{Z^3}{\underset{|}{\overset{|}{C}}}}-Z^2 \quad (BrCCl_3)$$

α-HETEROSUBSTITUTED REAGENTS

Acylation of organometallics with acyl halides and related derivatives is one of the synthetically useful transformations of α-heterosubstituted reagents (13a). Indeed, a wide variety of organometallics containing Mg (14), Zn (15), Cd (15), Hg (16), B (17), Al (18), Si (19), Mn (20), Fe (21), Rh (22), and Cu (23) are known to participate in this reaction. Thus, it appears that this methodology collectively is highly satisfactory. Nontheless, a few papers describe the use of Pd-phosphine complexes as effective catalysts in the reaction of acyl halides with organometallics containing Zn (24), Hg (25), and Sn (26). While the effectiveness of Pd-phosphine catalysts in these reactions seems indisputable, their merits and demerits relative to the previously developed methods mentioned above are not very clear.

More challenging from the synthetic viewpoint is to develop procedures for coupling various carbonylanion equivalents. Of particular interest to us are α-heterosubstituted alkenyl-metals represented by 1.

$$-\underset{Z}{\underset{|}{C}}=\underset{}{\underset{|}{C}}-M \quad (Z = Cl, Br, OR, SR, SOR, SO_2R, SeR, BR_2, AlR_2, SiR_3, SnR_3, etc.)$$

(1)

We have recently found that α-ethoxyvinylzinc chloride (1a), readily obtainable by reacting ethyl vinyl ether with t-BuLi·TMEDA followed by treatment with dry $ZnCl_2$, reacts with aryl and alkenyl halides in the presence of a catalytic amount of Pd-phosphine complexes (27, 28) (e.g., eq 1). The products can readily be hydrolyzed to give acetylated arenes.

$$H_2C=\underset{OEt}{\underset{|}{CH}} \xrightarrow[\text{2. dry } ZnCl_2]{\text{1. } t\text{-BuLi·TMEDA}} \left[H_2C=\underset{OEt}{\underset{|}{CZnCl}} \right] \xrightarrow[\text{5% Pd(PPh}_3)_4]{ArI} H_2C=\underset{OEt}{\underset{|}{CAr}} \xrightarrow{H_3O^+} H_3CCOAr \quad (1)$$

(1a)

$Ar = C_6H_5$ (90%), p-ClC_6H_4 (70%)

Likewise, α-silylalkenylalanes (1b) readily undergo the Pd-catalyzed cross coupling with unsaturated organic halides (27) (e.g., eq 2). Our preliminary results also indicate that the Pd-catalyzed cross-coupling reactions of α-thiosubstituted alkenylzinc chlorides appear to be highly promising (27).

$$n\text{-}C_6H_{13}C\equiv CSiMe_3 \xrightarrow{i\text{-}Bu_2AlH} \underset{H}{\overset{n\text{-}C_6H_{13}}{{>}}}C=C\underset{Al(Bu\text{-}i)_2}{\overset{SiMe_3}{{<}}} \xrightarrow[\text{5% Pd(PPh}_3)_4]{BrCH=CH_2} \underset{H}{\overset{n\text{-}C_6H_{13}}{{>}}}C=C\underset{CH=CH_2}{\overset{SiMe_3}{{<}}} \quad (2)$$

(80%, ≥98% Z)

The Pd-catalyzed reaction of α-phenylthioalkenyl phosphates (2) with various organoalanes give substituted alkenyl sulfides (3), which can be readily hydrolyzed with $TiCl_4$- water to give the corresponding ketones (29) (Scheme 1).

Scheme 1

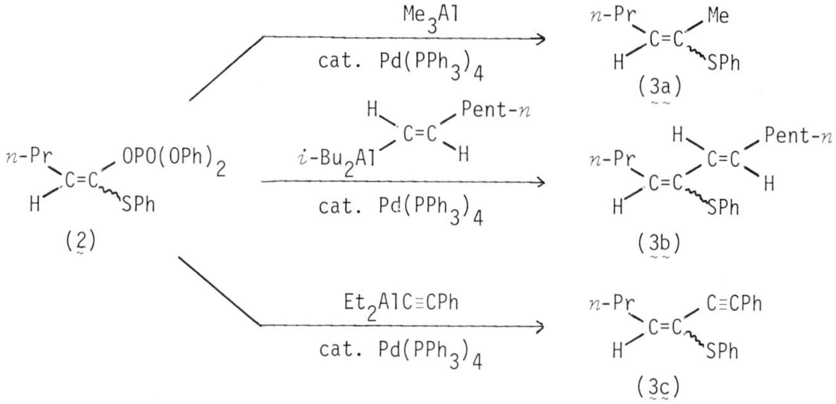

$$n\text{-Pr}, H\text{-C=C-OPO(OPh)}_2, SPh \quad (2)$$

$$\xrightarrow{Me_3Al, \text{cat. Pd(PPh}_3)_4} \quad n\text{-Pr}, H\text{-C=C-Me, SPh} \quad (3a)$$

$$\xrightarrow{i\text{-Bu}_2Al\text{-C(H)=C(H)-Pent-}n, \text{cat. Pd(PPh}_3)_4} \quad n\text{-Pr}, H\text{-C=C-C(H)=C(H)-Pent-}n, SPh \quad (3b)$$

$$\xrightarrow{Et_2AlC\equiv CPh, \text{cat. Pd(PPh}_3)_4} \quad n\text{-Pr}, H\text{-C=C-C}\equiv CPh, SPh \quad (3c)$$

Turning our attention to α-heteroaromatic reagents, a relatively large number of α-heteroarylmetals and α-heteroaryl halides have been shown to participate in the Ni-catalyzed cross coupling. Thus, for example, 2,6-dichloropyridine can be converted to muscopyridine in one step in 20% yield (30) (eq 3).

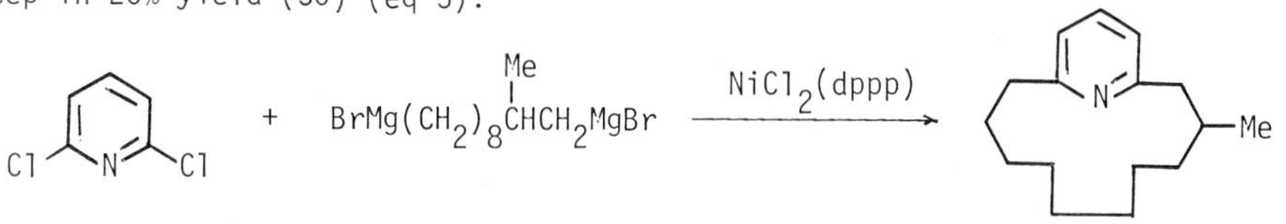

$$\text{2,6-Cl}_2\text{-pyridine} + BrMg(CH_2)_8\overset{Me}{\underset{|}{CH}}CH_2MgBr \xrightarrow{NiCl_2(dppp)} \text{muscopyridine-Me} \qquad (3)$$

dppp = $Ph_2P(CH_2)_3PPh_2$ 20%

The results of the Ni-catalyzed cross coupling involving α-heteroaromatic reagents were reviewed recently (3). Much less had been known until recently about the Pd-catalyzed cross-coupling reaction of heteroaryl reagents. We earlier found that α-bromo- or α-iodothiophene reacts with alkynylzinc chlorides in the presence of Pd-phosphine complexes to give α-alkynylthiophenes (31) (eq 4).

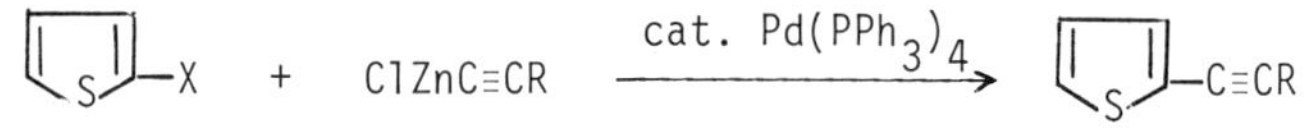

$$\text{(2-thienyl)-X} + ClZnC\equiv CR \xrightarrow{\text{cat. Pd(PPh}_3)_4} \text{(2-thienyl)-C}\equiv CR \qquad (4)$$

X = Br, I; R = Me, n-Pr, etc.   70-92%

Our recent systematic study (32) indicates that a wide variety of α- or β-heteroaromatic reagents may readily participate in the Pd-catalyzed cross coupling. Alkenyl-, aryl-, alkynyl-, and alkylmetals have been successfully employed. As in many other Pd-catalyzed reactions (4), those organometals that contain Zn generally exhibit the highest reactivity. Alkenylalanes and alkenylzirconium derivatives, readily obtainable via hydrometallation (9) or carbometallation (10) of alkynes, also react satisfactorily. The superiority of Zn over Mg is clearly shown in Scheme 2. Some representative α- and β-heterosubstituted aromatic derivatives that have been prepared by this method include 4 - 8.

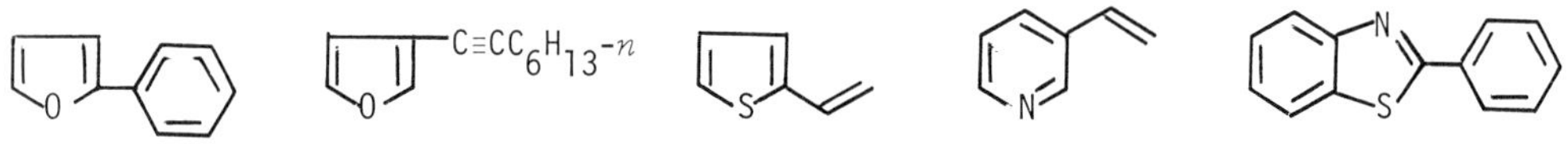

(4) 2-phenylfuran   (5) 2-(C≡CC_6H_13-n)furan   (6) 2-vinylthiophene   (7) 3-vinylpyridine   (8) 2-phenylbenzothiazole

In most cases either heteroarylmetals or the corresponding heteroaryl halides may be used to produce the desired heteroaryl derivatives, as in eq 5.

In some cases in which β-heteroaryl reagents are used, however, the Pd-catalyzed reaction of β-heteroarylzinc chlorides usually give favorable results, while that of the corresponding β-heteroaryl halides tends not to produce the desired cross-coupled products, as in eq 6.

Scheme 2

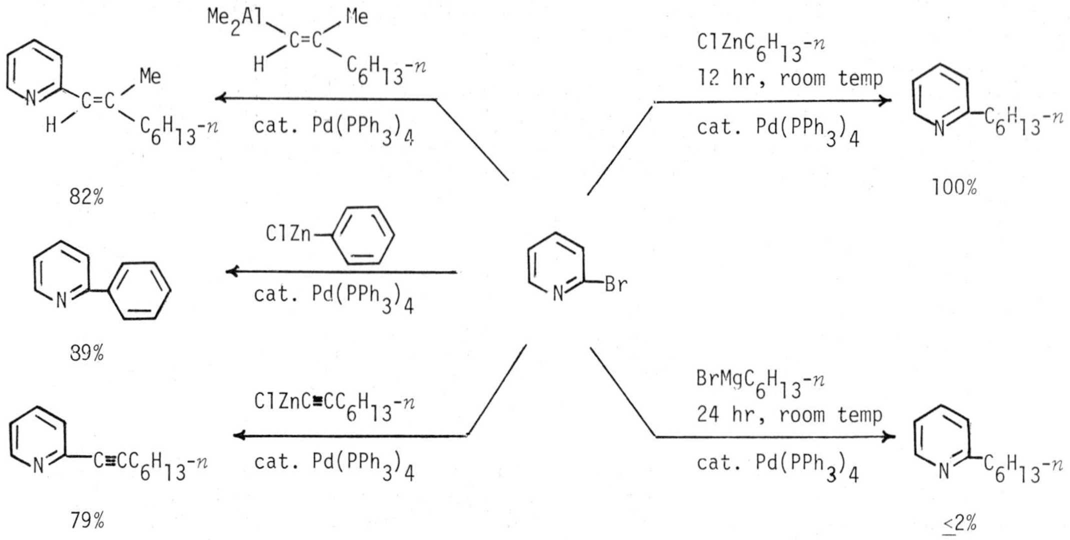

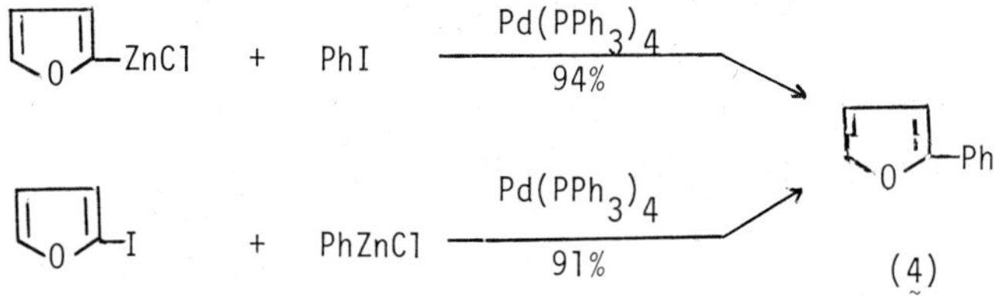

[Scheme 2 showing cross-coupling reactions of 2-bromopyridine:
- With Me₂Al-C(Me)=CH-C₆H₁₃-n / cat. Pd(PPh₃)₄ → 2-(1-methyl-1-octenyl)pyridine, 82%
- With ClZnC₆H₁₃-n, 12 hr, room temp / cat. Pd(PPh₃)₄ → 2-hexylpyridine, 100%
- With ClZn-Ph / cat. Pd(PPh₃)₄ → 2-phenylpyridine, 39%
- With ClZnC≡CC₆H₁₃-n / cat. Pd(PPh₃)₄ → 2-(1-octynyl)pyridine, 79%
- With BrMgC₆H₁₃-n, 24 hr, room temp / cat. Pd(PPh₃)₄ → 2-hexylpyridine, ≤2%]

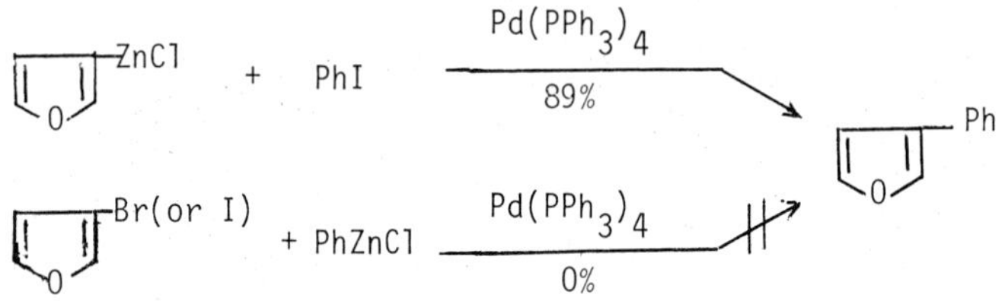

$$\text{furyl-ZnCl} + \text{PhI} \xrightarrow[94\%]{\text{Pd(PPh}_3)_4} \text{2-phenylfuran} \quad (5)$$

$$\text{furyl-I} + \text{PhZnCl} \xrightarrow[91\%]{\text{Pd(PPh}_3)_4} \text{2-phenylfuran} \quad (4)$$

$$\text{3-furyl-ZnCl} + \text{PhI} \xrightarrow[89\%]{\text{Pd(PPh}_3)_4} \text{3-phenylfuran} \quad (6)$$

$$\text{3-furyl-Br(or I)} + \text{PhZnCl} \xrightarrow[0\%]{\text{Pd(PPh}_3)_4} \cancel{\rightarrow}$$

One particularly attractive application of the Pd-catalyzed cross coupling to the synthesis of heteroaromatics involves the reaction of a 5-iodouracil derivative (9) with alkenyl-zirconium derivatives to produce (E)-5-alkenyluracil derivatives (10) in 30-96% yields (33) (eq 7).

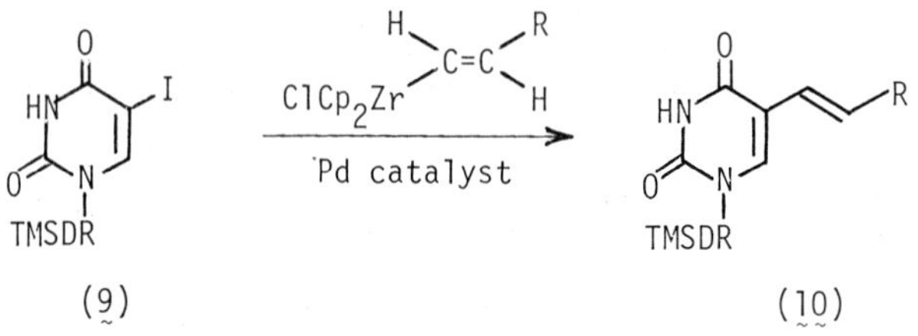

$$(9) \xrightarrow[\text{Pd catalyst}]{\text{ClCp}_2\text{Zr-CH=CHR}} (10) \quad (7)$$

TMSDR = O-3',5'-bis(trimethylsilyl)deoxyribose

Both 2-metallo- and 2-iodobenzothiazoles readily undergo the Pd-catalyzed cross coupling (27) (eq 8). A similar Ni-catalyzed reaction is also known (34). Since benzothiazole can act as a carboxyl synthon (35), the above reactions now provide a novel route to α,β-unsaturated carboxylic acids and their derivatives. In this connection, however, it should be pointed out that both the Ni-catalyzed (36) and the Pd-catalyzed (37) reactions of sodium or potassium cyanide with aryl halides provide aryl cyanides in good yields, although the corresponding reactions of alkenyl halides do not appear to have been reported.

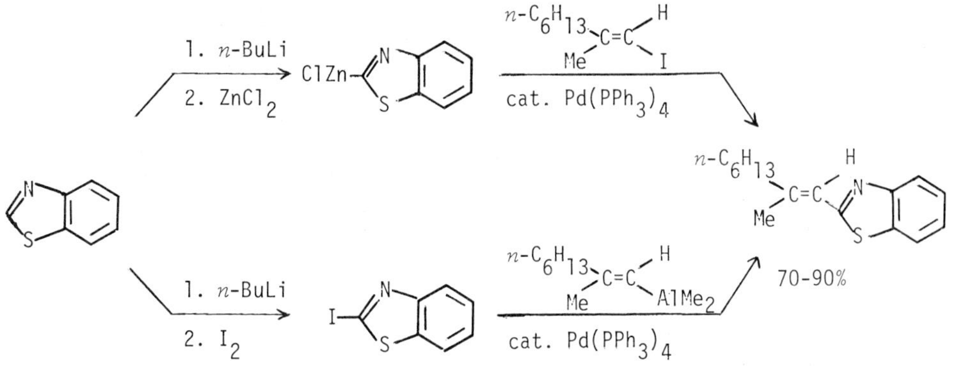

$$\text{(8)}$$

In contrast to the α-heterosubstituted reagents discussed above, all of which contain one or more M-C$_{sp2}$, M-C$_{sp}$, X-C$_{sp2}$ and X-C$_{sp}$ bonds, those which contain only M-C$_{sp3}$ and/or X-C$_{sp3}$ bonds have so far tended to fail to undergo the Pd- or Ni-catalyzed cross coupling, with a notable exception of α-silylalkylmetals. Trimethylsilylmethylmagnesium chloride reacts readily with various aryl (38) and alkenyl halides (39, 40) in the presence of a Ni- or Pd-phosphine complex to provide benzyl- and allylsilanes, respectively. The reaction of alkenyl halides provides an essentially 100% stereospecific route to allylsilanes (40) (eq 9). In this particular reaction, the use of the corresponding organozinc chloride does not seem to offer any advantage over the Grignard reagent.

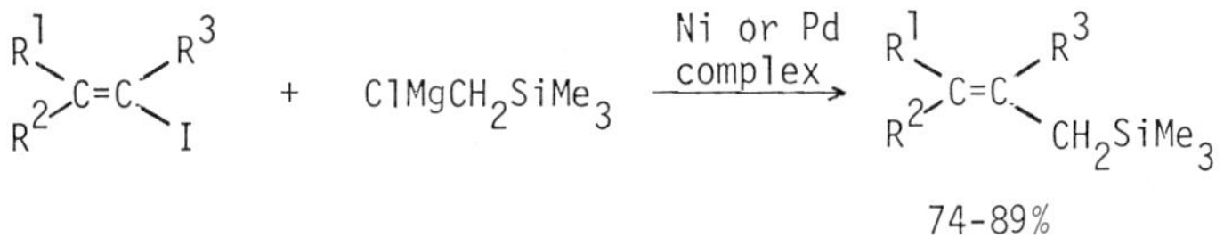

$$\text{(9)}$$

74-89%

Although Pd-catalysts do not appear to offer advantages over Ni-catalysts in simple cases, they seem to offer a significant advantage over their Ni analogues in the following asymmetric synthesis of chiral allylsilanes (41) (eq 10). As a few representative results indicate, some remarkably high % ee figures have been observed.

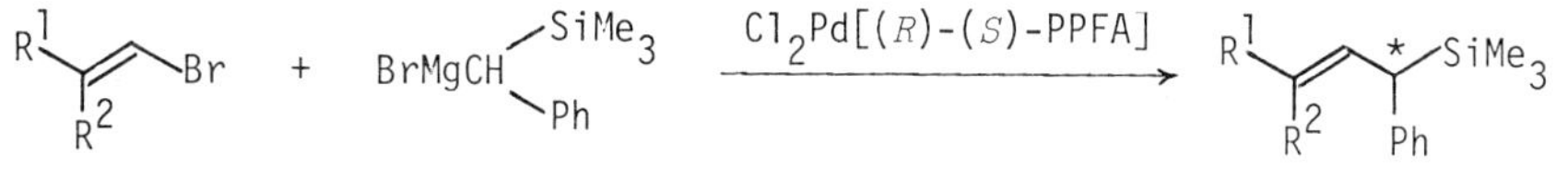

$$\text{(10)}$$

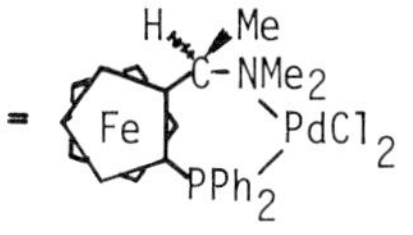

R^1	R^2	Yield(%)	% ee
H	H	42	95
Ph	H	90	95
Me	H	77	85
H	Ph	95	13

The chiral allylsilanes thus formed can be further transformed into a variety of chiral alkenes in a highly stereospecific manner (41).

Although attempts are being made to promote the cross-coupling reaction of α-heteroalkylmetals or α-heteroalkyl halides with alkenyl, aryl or alkynyl reagents using Ni- or Pd-catalysts, the results have been largely disappointing. The reagents which we have so far tested include those shown below. In these reactions organic halides used remain largely unreacted under those reaction conditions that are satisfactory for the favorable cases discussed above. Further investigation is necessary to clarify this matter.

O_2NCH_2M $MeOCH_2M$

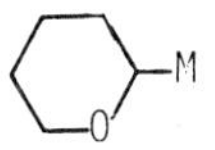

$MeSCH_2M$ $MeSO_nCH_2M$
                    (n = 1 or 2) 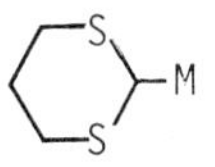

M = Li, ZnCl, etc.

Despite many frustrating results it may be said that various α-heterosubstituted reagents, such as α-heteroalkenylmetals that can serve as acylanion equivalents and α-heteroaryl derivatives, readily participate in the Pd- or Ni-catalyzed cross coupling.

β-HETEROSUBSTITUTED REAGENTS

Metal enolates and various types of enolonium ion equivalents represent two of the most interesting classes of β-heterosubstituted reagents. Strictly speaking, metal enolates may not be classified as organometallics. In organic synthesis via metal enolates, however, one normally starts with the corresponding carbonyl compounds and achieve α-substitution via metal enolates. It is therefore useful to view them as β-oxoalkylmetals.

Allylation of enolates (42) is an important synthetic methodology, since it not only serves as an obvious route to γ,δ-unsaturated carbonyl compounds, but also provides an attractive route to α-alkylated carbonyl compounds as well as 1,4- and 1,5-dicarbonyl compounds. Extensive studies by Trost and others (43) have established that the Pd-catalyzed allylation of "stabilized" enolates, such as those derived from acetoacetic and malonic esters, is a generally satisfactory and useful reaction, whereas "unstabilized" alkali metal enolates derived from ketones tend not to readily undergo this allylation. In light of our previous finding that alkali metals are far less effective than Zn, Al, Zr, etc. in the Pd-catalyzed cross coupling (3), we have investigated the effect of metals in the Pd-catalyzed allylation of enolates with geranyl acetate in THF. Although the metal enolate containing Li, Si, or Zr does not produce any significant amount of the expected allylation product (11), that containing MgCl, ZnCl, BEt$_3$K (44), or AlEt$_3$Li gives 11 in 15, 70, 81, or 65% yield, respectively (45) (eq 11).

Although the mechanism of the reaction is not very clear, the reaction is genuinely catalyzed by Pd-phosphine complexes. Thus, the reaction of 12 with a 1:1 mixture of isoprenyl chloride and 2,3-dichloropropene in the absence of Pd(PPh$_3$)$_4$ gives only the isoprenylated cyclohexanone in 51% yield with no sign of the formation of 2-chloroallyl substituted cyclohexanone (13). On the other hand, when the reaction is carried out in the presence of 5 mol % of Pd(PPh$_3$)$_4$, the major product is 13 formed in 92% yield with only 6% of the isoprenylated product (45) (eq 12).

The dramatic reversal of the product ratio appears to be consistent only with the involvement of 2-chloroallylpalladium derivative, e.g., 14, as an intermediate. The corresponding reaction of 1,3-dichloro-2-butene readily gives 15 in 80-90% yield.

During the course of our study we became aware of a paper by Fiaud and Malleron (46) reporting allylation of lithium ketone enolates with allylic acetates in the presence of 1 mol % each of bis(dibenzylideneacetonato)palladium and 1,2-bis(diphenylphosphino)ethane at room temperature. This suggests that the difficulties associated with the low reactivity of lithium enolates may be overcome by choosing a proper Pd catalyst. Unfortunately, the regio- and stereo-selectivity of the Pd-catalyzed allylation of lithium enolates has not yet been clarified.

The Pd-catalyzed allylation of either potassium enoxyborates, readily obtainable by treating potassium enolates with BEt_3, or zinc enolates, readily obtainable by treating lithium or potassium enolates with dry $ZnCl_2$, exhibits a high stereo- and regiospecificity with respect to the allylic electrophiles and a high regiospecificity with respect to the enolates, as indicated by the results summarized in Scheme 3.

Scheme 3

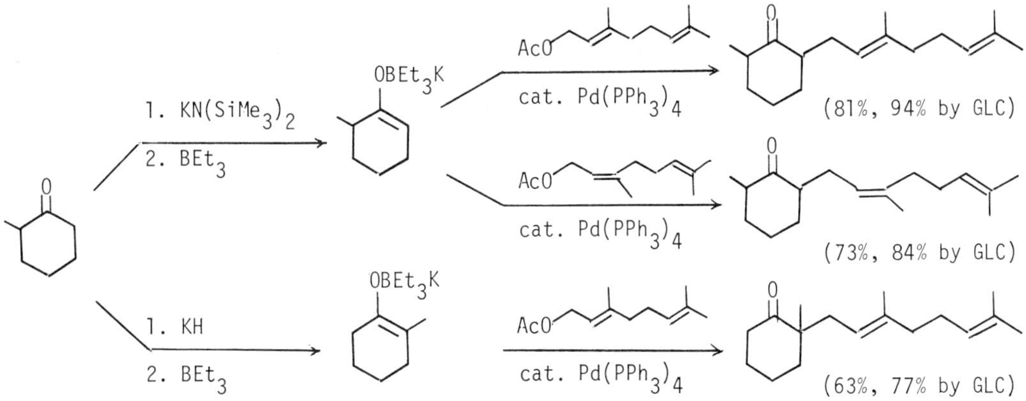

In all these reactions both the stereospecificity and regiospecificity with respect to the C_{10} side chain are $\geq 98\%$. The remarkably high stereospecificity observed in these reactions is in sharp contrast with that reported recently for a related Pd-catalyzed allylation of enoxystannanes (47) in which an essentially complete isomerization of *cis*-allylic groups into the *trans*-allylic groups occurs.

Since α-allylated ketones can readily be reduced to give α-alkylated ketones, the Pd-catalyzed allylation of ketone enolates containing Zn, B, or Al provides a selective method for α-alkylation of ketones. 2-Chloroallylated ketones, e.g., 13, and 3-chloro-2-butenylated ketones, e.g., 15, are readily convertible to the corresponding 1,4- and 1,5-diketones, respectively, by treating them with $Hg(OAc)_2$ and formic acid (48).

Although aldehydes having an α-tertiary carbon center can also be readily allylated via the Pd-catalyzed allylation of their potassium enoxyborates (49), e.g., eq 13, allylation of α-unbranched aldehydes by this method is not yet very satisfactory. Aldol condensation appears to be competitive with the desired allylation in such cases.

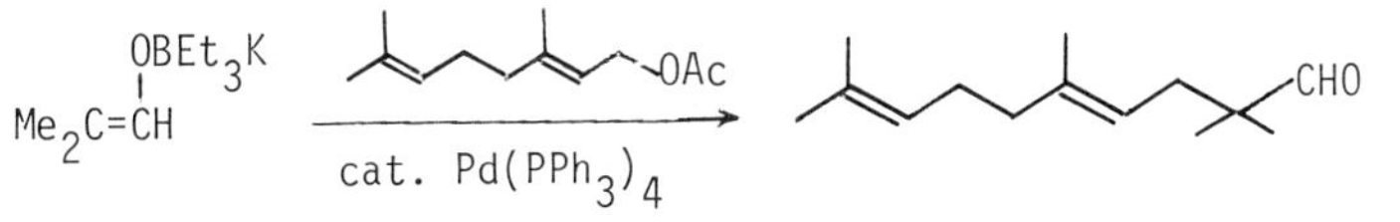

$$Me_2C=CH\text{-}OBEt_3K \xrightarrow[\text{cat. }Pd(PPh_3)_4]{\text{-OAc}} \text{CHO} \quad (13)$$

60%

At present, it is not clear whether the Pd catalysis is effective in promoting allylation of other types of enolates, such as those derived from esters, amides, and other related compounds.

β-Heterosubstituted alkenylmetals (16) and alkenyl electrophiles (17) can serve as enolate anion and enolonium ion synthons, respectively.

$$\begin{array}{cc} -C=C-M \\ | \quad | \\ Z \end{array} \cong \begin{array}{c} -C=CH \\ | \\ MO \end{array} \qquad \begin{array}{cc} -C=C-X \\ | \quad | \\ Z \end{array} \cong \begin{array}{c} -C-CH^+ \\ \| \\ O \end{array}$$

(16) (17)

Relatively little is known about the Pd- or Ni-catalyzed cross coupling involving β-heteroalkenylmetals other than that of β-heteroarylmetals discussed earlier. Hydrozirconation of ethoxyethyne produces 18, which reacts with iodobenzene in the presence of $Ni(PPh_3)_4$ to give 19 in 99% yield (8a) (eq 14).

On the other hand, a fair number of β-heterosubstituted alkenyl electrophiles have been cross-coupled with various organometallics in the presence of a Ni or Pd catalyst. Thus, β-ethoxyvinyl bromide reacts with arylmagnesium bromides in the presence of $Cl_2Ni(dppp)$ to give the corresponding cross-coupled products in good yields (50) (e.g., eq 15).

$$EtOC\equiv CH \xrightarrow{Cl(H)ZrCp_2} \underset{(18)}{\underset{H}{\overset{EtO}{C=C}}\overset{H}{\underset{ZrCp_2Cl}{}}} \xrightarrow[cat.\ Ni(PPh_3)_4]{PhI} \underset{(19)}{\underset{H}{\overset{EtO}{C=C}}\overset{H}{\underset{Ph}{}}} \quad (14)$$

$$\text{(MeO)}_2C_6H_3\text{-MgBr} + BrCH=CHOEt \xrightarrow{cat.\ Cl_2Ni(dppp)} \text{(MeO)}_2C_6H_3\text{-CH=CHOEt} \quad (15)$$
68%

The use of β-heteroalkenyl halides as enolonium ion synthons was demonstrated by the Ni-catalyzed reaction of bromotrimethylsilyloxyalkenes with various types of Grignard reagents to form 20, which can be readily hydrolyzed to form ketones and aldehydes (51) (eq 16).

$$\underset{Br}{\overset{R^1}{C=C}}\overset{R^2}{\underset{OSiMe_3}{}} \xrightarrow[cat.\ Cl_2Ni(dppp)]{R^3MgX} \underset{R^3}{\overset{R^1}{C=C}}\overset{R^2}{\underset{OSiMe_3}{}} \xrightarrow{H_3O^+} \underset{R^3}{\overset{R^1}{CHCOR^2}} \quad (16)$$
(20)

Related to this reaction is the following conversion of 4-t-butylcyclohexanone into 2-methyl-5-t-butylcyclohexanone, which provides a novel carbonyl transposition reaction with simultaneous α-alkylation (29) (eq 17).

$$\text{4-t-Bu-cyclohexanone} \xrightarrow[2.\ NaH,\ ClPO(OPh)_2]{1.\ LDA,\ PhSSPh} \text{enol phosphate-SPh} \xrightarrow[cat.\ Pd(PPh_3)_4]{Me_3Al,\ 80°C,\ 4\ hr} \text{Me-SPh-cyclohexene} \xrightarrow[CH_2Cl_2]{TiCl_4-H_2O} \text{2-Me-5-t-Bu-cyclohexanone} \quad (17)$$
78%

Bromoboration of alkynes has recently been shown to produce 21, which undergoes the Pd- or Ni-catalyzed cross coupling with various organometallics to form 22 (52) (eq 18). Since alkenylboranes can be readily oxidized to give aldehydes or ketones, 21 can serve as enolonium ion equivalents.

$$R^1C\equiv CH \xrightarrow{BrBR_2} \underset{Br}{\overset{R^1}{C=C}}\overset{H}{\underset{BR_2}{}} \xrightarrow[Pd\ catalyst]{R^2M} \underset{R^2}{\overset{R^1}{C=C}}\overset{H}{\underset{BR_2}{}} \quad (18)$$
(21)   (22)

1,2-Dihaloalkenes also offer interesting synthetic possibilities. (E)-1-Chloro-2-iodoethene, readily obtainable by treating acetylene with ICl, can undergo the Pd-catalyzed cross coupling exclusively at the iodine-bearing carbon atom. Thus, alkynylzinc chlorides can readily be converted into 1-chloroenynes (23) (53) (eq 19).

$$RC\equiv CZnCl + \underset{H}{\overset{I}{C=C}}\overset{H}{\underset{Cl}{}} \xrightarrow{cat.\ Pd(PPh_3)_4} RC\equiv C\underset{H}{\overset{H}{C=C}}\overset{H}{\underset{Cl}{}} \xrightarrow[2.\ H^+]{1.\ NaNH_2} RC\equiv CC\equiv CH \longrightarrow RC\equiv CC\equiv CR^1 \quad (19)$$
(23)

Treatment of 23 with $NaNH_2$ provides terminal diynes, which can be alkylated to form various unsymmetrical diynes by known methods. Since the Pd-catalyzed reaction of alkynylmetals with alkynyl halides tends to give nearly statistical mixtures of cross- and homo-coupled diynes, this indirect route provides a reasonable solution to this problem.

Although β-heteroalkenyl reagents containing various other elements, such as Si, N, P, and Se, appear to offer attractive synthetic possibilities, their use in the Pd- or Ni-catalyzed cross coupling does not appear to have been investigated.

γ-HETEROSUBSTITUTED REAGENTS

Conjugate addition and conjugate substitution are two of the important synthetic operations that involve carbon-carbon bond formation at a position γ to the oxo group.

$$RM + C=C-C=O \longrightarrow R-C-CH-C=O \quad (20)$$

$$RM + X-C=C-C=O \longrightarrow R-C=C-C=O \quad (21)$$

Although there are a number of useful conjugate addition reactions involving organometallics, most notably organocoppers (54), the use of β-unsubstituted α,β-unsaturated carbonyl compounds often presents difficulties due to competitive polymerization of such carbonyl compounds. To cope with these difficulties various indirect methods have also been developed. Allylation with γ-heterosubstituted allylic reagents is one such method (42).

Our approach to the development of a Michael addition equivalent suitable for annulation was discussed in the preceding section. Our preliminary results indicate that 1,2-dichloro-2-butene also undergoes the Pd-catalyzed cross coupling with phenylzinc chloride and (E)-2-(methyl-1-octenyl)dimethylalane to give the desired cross-coupled products in excellent yields (55) (eq 22). Since these products are readily convertible into the corresponding ketones, the overall transformation amounts to the conjugate addition of the organometallic reagents to methyl vinyl ketone.

$$(22)$$

Similarly, 3-trimethylsilyl-2-propenyl acetate (24) undergoes the Pd-catalyzed cross coupling (55) (e.g., eq 23).

$$(23)$$

To develop a synthetically more attractive conjugate addition equivalent it is desirable to be able to start with α,β-unsaturated carbonyl compounds and end up with the corresponding conjugate addition products. Since a few reactions that convert enones into γ-oxyallylic electrophiles including the one shown in eq 24 (56) are now known, we are currently investigating the Pd-catalyzed allylation with such allylic electrophiles.

$$C=C-C=O \xrightarrow{ISiMe_3} I-C-C=C-OSiMe_3 \quad (24)$$

Conjugate substitution as defined by eq 21 is another useful carbon-carbon bond-forming process, the significance of which does not appear to have been fully recognized. In addition to providing a route to β-substituted α,β-unsaturated carbonyl compounds, it can also provide an indirect method for conjugate addition, when coupled with conjugate reduction. We have found that β-bromosubstituted α,β-unsaturated carbonyl compounds, such as 25 - 27, react readily with aryl-, alkenyl-, alkynyl-, and homoallylmetals containing Zn (6), Al (7), and Zr (8) in the presence of a Pd catalyst. A few terpenoids, such as dendrolasin (28) and mokupalide (29) have been synthesized by this method (6f).

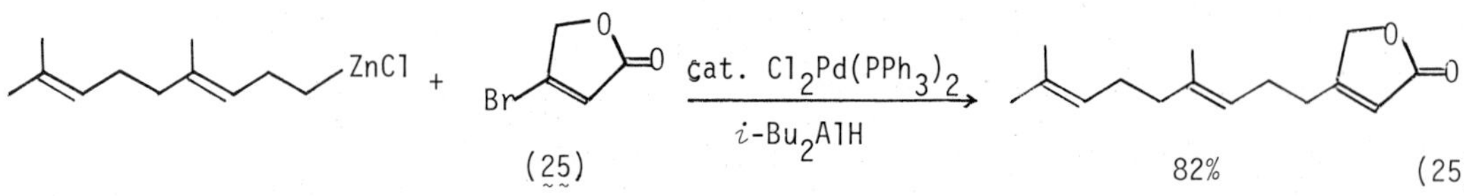

More intricate applications of the conjugate substitution to the synthesis of natural products are being investigated in our laboratories.

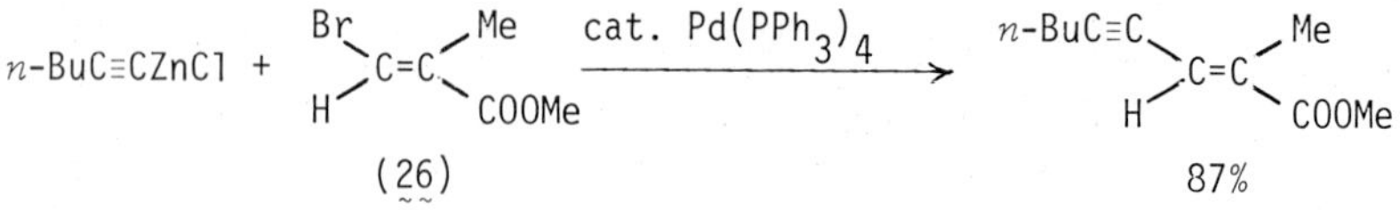

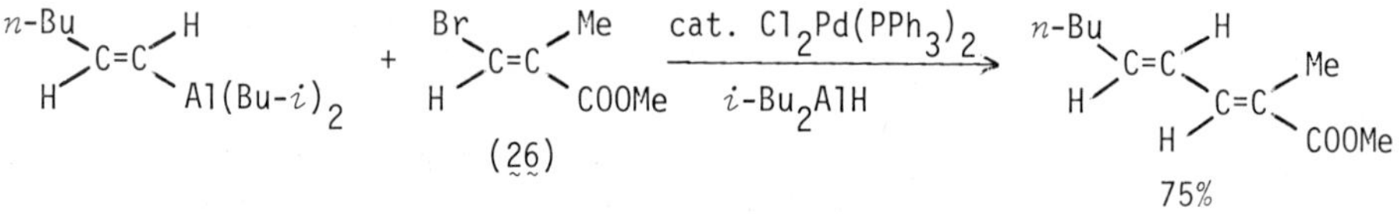

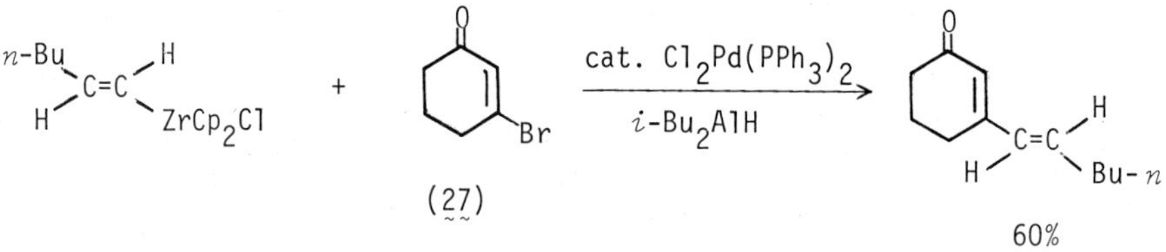

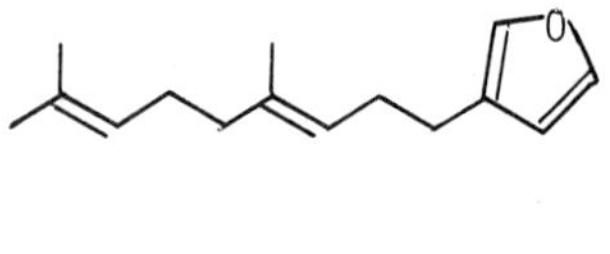

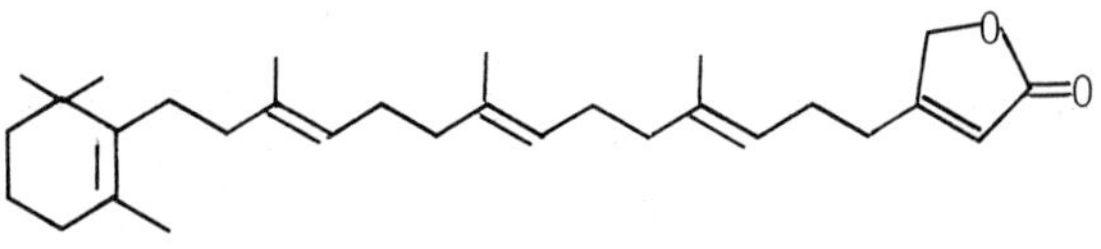

CONCLUSIONS

Although many of the results discussed in this review are preliminary and fragmentary, it has become increasing clear that the Pd- or Ni-catalyzed cross coupling is widely applicable even to those cases in which proximally heterosubstituted reagents are involved. Many of these cross-coupling reactions promise to provide viable alternatives to classical reactions involving carbonyl compounds, such as acylation, α-substitution of enolates, and conjugate addition. Additional efforts are being made to further delineate their full scope and synthetic potential.

ACKNOWLEDGMENTS

I am deeply indebted to my coworkers, whose names appear in our papers cited in this review, especially Drs. S. Baba, A.O. King, D.E. Van Horn, N. Okukado, M. Kobayashi, and H. Matsushita. My current coworkers active in this area are Dr. R.A. John, C.L. Rand, S. Chatterjee, F.T. Luo, L.D. Boardman, A.T. Stoll, R. Frisbee, J.A. Tour, and V. Bagheri. I also wish to acknowledge a fruitful collaboration with Professor A. Silveira, Jr. of SUNY College at Oswego, N.Y. Our research has been mainly supported by the National Science Foundation, the National Institutes of Health, and the Petroleum Research Fund, administered by the American Chemical Society.

REFERENCES

1. K. Tamao, K. Sumitani, and M. Kumada, *J. Am. Chem. Soc. 94*, 4374 (1972).
2. R.J.P. Corriu and J.P. Masse, *J.C.S. Chem. Comm.*, 144 (1972).
3. K. Tamao and M. Kumada, *Organometal. React.* in press.
4. (a) E. Negishi in *Aspects of Mechanism and Organometallic Chemistry*, J.H. Brewster, ed., Plenum, New York, 1978, p. 285. (b) E. Negishi, *Accounts Chem. Res.* in press.
5. G.H. Posner, *Org. React. 22*, 253 (1975).
6. (a) E. Negishi, A.O. King, and N. Okukado, *J. Org. Chem. 42*, 1821 (1977). (b) A.O. King, N. Okukado, and E. Negishi, *J.C.S. Chem. Comm.*, 683 (1977). (c) A.O. King, E. Negishi, F.J. Villani, Jr., and A. Silveira, Jr., *J. Org. Chem. 43*, 358 (1978). (d) E. Negishi, N. Okukado, A.O. King, D.E. Van Horn, and B.I. Spiegel, *J. Am. Chem. Soc. 100*, 2254 (1978). (e) E. Negishi, L.F. Valente, and M. Kobayashi, *J. Am. Chem. Soc. 102*, 3298 (1980). (f) M. Kobayashi and E. Negishi, *J. Org. Chem. 45*, 5223 (1980). (g) E. Negishi, H. Matsushita, and N. Okukado, *Tetrahedron Lett. 22*, 2715 (1981).
7. (a) E. Negishi and S. Baba, *J.C.S. Chem. Comm.*, 596 (1976). (b) S. Baba and E. Negishi, *J. Am. Chem. Soc. 98*, 6729 (1976). (c) H. Matsushita and E. Negishi, *J. Am. Chem. Soc. 103*, 2882 (1981). (d) E. Negishi, S. Chatterjee, and H. Matsushita, *Tetrahedron Lett. 22*, 3737 (1981).
8. (a) E. Negishi and D.E. Van Horn, *J. Am. Chem. Soc. 99*, 3168 (1977). (b) N. Okukado, D.E. Van Horn, W.L. Klima, and E. Negishi, *Tetrahedron Lett.*, 1027 (1978).
9. (a) For a review of hydroalumination, see T. Mole and E.A. Jeffery, *Organoaluminum Compounds*, Elsevier, Amsterdam, 1972. (b) For a review of hydrozirconation, see J. Schwartz, *J. Organometal. Chem. Library 1*, 461 (1976).
10. For a review of carboalumination, see E. Negishi, *Pure Appl. Chem. 53*, 2333 (1981).
11. (a) For the use of alkynylborates in cross coupling, see Ref. 4a. (b) N. Miyaura, K. Yamada, and A. Suzuki, *Tetrahedron Lett.*, 3437 (1979). (c) N. Miyaura and A. Suzuki, *J.C.S. Chem. Comm.*, 866 (1979). (d) N. Miyaura, T. Yano, and A. Suzuki, *Tetrahedron Lett. 21*, 2865 (1980). (e) N. Miyaura, H. Suginome, and A. Suzuki, *Tetrahedron Lett. 22*, 127 (1981). (f) N. Miyaura and A. Suzuki, *J. Organometal. Chem. 213*, C53 (1981).
12. (a) For the use of alkynyltins in cross coupling, see Ref. 4a. (b) M. Kosugi, K. Sasazawa, Y. Shimizu, and T. Migita, *Chem. Lett.*, 301 (1977). (c) B.M. Trost and E. Keinan, *Tetrahedron Lett. 21*, 2595 (1980). (d) J.K. Stille and J. Goldschalx, *Tetrahedron Lett. 21*, 2599 (1980).
13. (a) E. Negishi, *Organometallics in Organic Synthesis*, Vol. I, Wiley, New York, 1980, p. 151. (b) B.T. Gröbel and D. Seebach, *Synthesis*, 357 (1977). (c) O.W. Lever, Jr., *Tetrahedron 32*, 1943 (1976).
14. F. Huet, G. Emptoz, and A. Jubier, *Tetrahedron 29*, 479 (1973).
15. For a review, see D.A. Shirley, *Org. React. 8*, 28 (1954).
16. For a review, see R.C. Larock, *J. Organometal. Chem. Library 1*, 257 (1976).
17. E. Negishi, K.W. Chiu, and T. Yoshida, *J. Org. Chem. 40*, 1676 (1975).
18. For a review, see H. Reinheckel, K. Haage, and D. Jahnke, *Organometal. Chem. Res. A 4*, 47 (1969).
19. For a review, see P.F. Hudrlik, *J. Organometal. Chem. Library 1*, 127 (1976).
20. For a review, see J.F. Normant, *Synthesis*, 130 (1977).
21. For a review, see J.P. Collman, *Accounts Chem. Res. 8*, 342 (1975).
22. L.S. Hegedus, P.M. Kendall, S.M. Lo, and J.R. Sheats, *J. Am. Chem. Soc 97*, 5448 (1975).
23. G.H. Posner, C.E. Whitten, and J.J. Sterling, *J. Am. Chem. Soc. 95*, 7788 (1973).
24. F. Sato, K. Naruse, M. Enokiya, and T. Fujisawa, *Chem. Lett.*, 1135 (1981).
25. K. Takagi, T. Okamoto, Y. Sakakibara, A. Ohno, S. Oka, and N. Hayama, *Chem. Lett.*, 951 (1975).
26. D. Milstein and J.K. Stille, *J. Am. Chem. Soc. 100*, 3636 (1978).
27. E. Negishi, F.T. Luo, and V. Bagheri, unpublished results.
28. We have been informed by Professor L.S. Hegedus of Colorado State University that he and his coworkers have also made an extensive study of this reaction.
29. M. Sato, K. Takai, K. Oshima, and H. Nozaki, *Tetrahedron Lett. 22*, 1609 (1981).
30. K. Tamao, S. Kodama, K. Nakatsuka, Y. Kiso, and M. Kumada, *J. Am. Chem. Soc. 97*, 4405 (1975).
31. A.O. King, E. Negishi, F.J. Villani, Jr., and A. Silveira, Jr., *J. Org. Chem. 43*, 358 (1978).
32. E. Negishi, F.T. Luo, R. Frisbee, and H. Matsushita, *Heterocycles 18*, 117 (1982).
33. P. Vincent, J.P. Beaucourt, and L. Pichat, *Tetrahedron Lett. 23*, 63 (1982).
34. H. Takei, M. Miura, H. Sugimura, and H. Okamura, *Chem. Lett.*, 1447 (1979).
35. E. J. Corey and D.L. Boger, *Tetrahedron Lett.*, 5, 9 and 13 (1978).
36. L. Cassar, *J. Organometal. Chem. 54*, C57 (1973).
37. A. Sekiya and N. Ishikawa, *Chem. Lett.*, 951 (1975).
38. K. Tamao, K. Sumitani, Y. Kiso, M. Zembayashi, A. Fujioka, S. Kodama, I. Nakajima, A. Minato, and M. Kumada, *Bull. Chem. Soc. Japan 49*, 1958 (1976).
39. K. Tamao, M. Zembayashi, and M. Kumada, *Chem. Lett.*, 1239 (1976).
40. E. Negishi, F.T. Luo, and C.L. Rand, *Tetrahedron Lett. 23*, 27 (1982).
41. T. Hayashi, M. Konishi, and M. Kumada, private communication.
42. For a review, see M.E. Jung, *Tetrahedron 32*, 3 (1976).
43. For a review, see B.M. Trost, *Accounts Chem. Res. 13*, 385 (1980).

44. (a) E. Negishi, M.J. Idacavage, F. DiPasquale, and A. Silveira, Jr., *Tetrahedron Lett.*, 1225 (1978). (b) M.J. Idacavage, E. Negishi, and C.A. Brown, *J. Organometal. Chem. 186*, C55 (1980).
45. E. Negishi, H. Matsushita, S. Chatterjee, and R.A. John, *J. Org. Chem.*, in press.
46. M.C. Fiaud and J.L. Malleron, *J. C. S. Chem. Comm.*, 1159 (1981).
47. B.M. Trost and E. Keinan, *Tetrahedron Lett. 21*, 259 (1980).
48. (a) E. Negishi, F.T. Luo, A. Pecora, and A. Silveira, Jr., unpublished results. (b) For a paper describing the use of $Hg(OAc)_2$ and formic acid for this transformation, see H. Yoshioka, K. Takasaki, M. Kobayashi, and T. Matsumoto, *Tetrahedron Lett.*, 3489 (1979).
49. E. Negishi and R.A. John, unpublished results.
50. K. Tamao, M. Zembayashi, and M. Kumada, *Chem. Lett.*, 1237 (1976).
51. K. Tamao, M. Zembayashi, and M. Kumada, *Chem. Lett.*, 1239 (1976).
52. A. Suzuki, private communication.
53. E. Negishi, N. Okukado, S. Lovich, unpublished results.
54. G.H. Posner, *Org. React. 19*, 1 (1972).
55. E. Negishi and H. Matsushita, unpublished results.
56. R.B. Miller and D.R. McKean, *Tetrahedron Lett.*, 2305 (1979).

SELECTIVE REACTIONS USING ORGANOALUMINUM REAGENTS

Hisashi Yamamoto and Keiji Maruoka

Department of Applied Chemistry, Nagoya University, Chikusa, Nagoya 464, Japan

Abstract- New aspects of Beckmann rearrangement are described: (1) A new approach has been demonstrated for the regioselective synthesis of α-alkylated amines by treatment of a wide variety of oxime sulfonates with alkylaluminum reagents, followed by excess diisobutylaluminum hydride. Analogous reactions provide a new route to imino thioethers, imino selenoethers, and imino nitriles. These new processes provide a simple route to dl-pumiliotoxin C and other alkaloids in a highly stereoselective fashion. (2) Intramolecular cyclization initiated by Beckmann rearrangement provides new entry to a wide variety of heterocyclic derivatives. Utility of new methodologies is illustrated by the straightforward synthesis of muscone and muscopyridine.

INTRODUCTION

The acid-catalyzed conversion of ketoximes to amides is known as the "Beckmann rearrangement" after its discoverer. The reaction and its widely accepted mechanism are shown below.

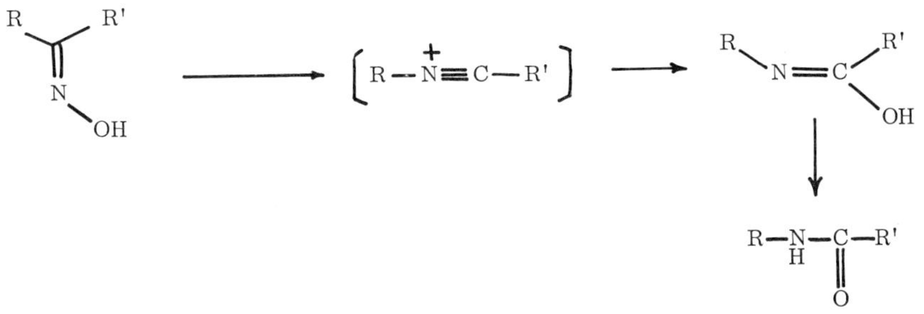

The mechanism of the Beckmann rearrangement consists essentially of the formation of an electron-deficient nitrogen atom by the partial ionization of the oxygen-nitrogen bond of the oxime with a simultaneous intramolecular migration of the group anti to the departing hydroxy group, producing the iminocarbocation, which then reacts with water to give the corresponding amide.
The direct capture of the intermediary iminocarbocation by carbon-nucleophiles is not possible using classical synthetic reactions. Since the amphoteric character of organoaluminum reagents was established (ref. 1), the iminocarbocation might undergo carbon-carbon bond formation with organo-aluminum reagents given the proper circumstances. Verification of this hypothesis has been obtained as follows.

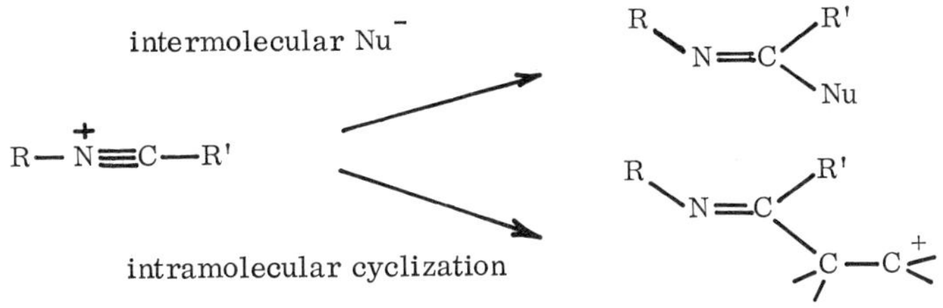

INTERMOLECULAR REACTIONS

Treatment of a wide variety of oxime sulfonates with several equivalents of alkylaluminum reagents in methylene chloride resulted in formation of the imines, which were directly reduced with excess diisobutylaluminum hydride (DIBAH) to give the corresponding amines (ref. 2). This new synthetic approach provides a simple route to many substances hitherto accessible only by lengthy or complicated syntheses. The examples cited in Table 1 illustrate the reaction.

Table 1. Alkylative Beckmann Rearrangement

Entry	Starting Material (mp °C)	Method	Product	R	Yield (%)
1	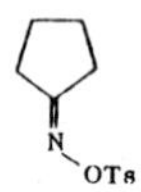 (cyclopentanone oxime OTs, 75–77)	A	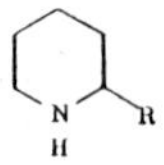 (2-substituted piperidine)	n-Pr	55
2		B		n-Pr	58
3		C	 (N-substituted cyclopentylamine)	n-Pr	62
4		D		n-Pr	23
5	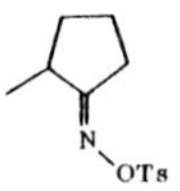 (2-methylcyclopentanone oxime OTs)	B	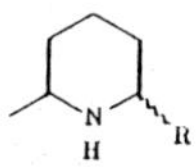 (2,6-disubstituted piperidine)	n-Pr	70
6		C		n-Pr	67
7	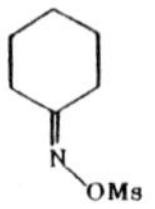 (cyclohexanone oxime OMs, 43–45)	D	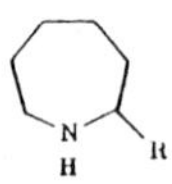 (2-substituted azepane)	Me	70
8		D		Et	47
9		D		n-Pr	64
10		D		i-Bu	52
11		E		C≡C-Bu	67
12	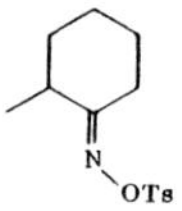 (2-methylcyclohexanone oxime OTs)	D	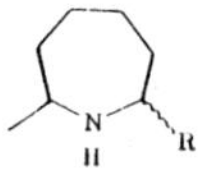 (2,7-disubstituted azepane)	n-Pr	48
13	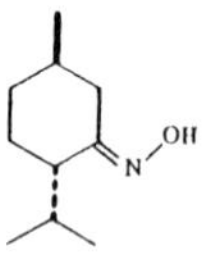 (menthone oxime OH)	F	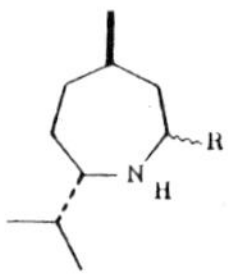	Me	57
14		G		H	82
15	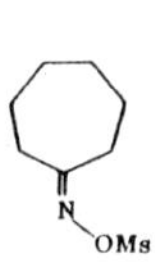 (cycloheptanone oxime OMs, 38–40)	D	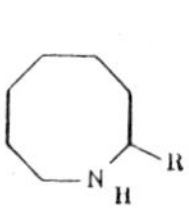	n-Pr	68
16	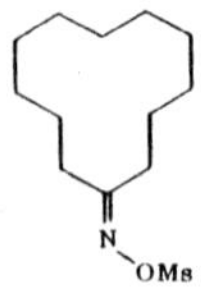 (cyclododecanone oxime OMs)	D	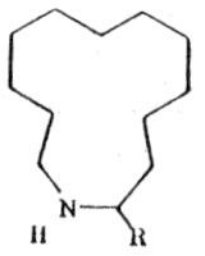	Me	60
17		H		H	73

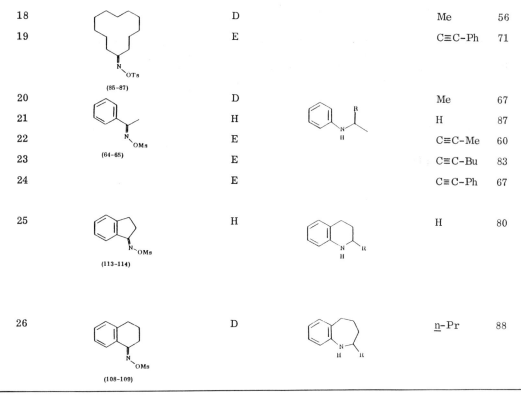

Method A: Treatment with n-Pr₃Al (2 equiv) in CH₂ClCH₂Cl at 80°C for 15 min, or n-Pr₃Al (3 equiv) in CH₂ClCH₂Cl at 25°C for 30 min, then reduction with DIBAH (1.5 equiv) at 0°C for 1 h. Method B: Treatment with n-Pr₃Al (2 equiv) in CH₂Cl₂ at 40°C for 15 min followed by DIBAH (1.5 equiv) at 0°C for 1 h. Method C: Treatment with n-Pr₃Al (3 equiv) in hexane at -78°C for 5 min and at 0°C for 1 h, then reduction with DIBAH (1.5 equiv) at 0°C for 1 h. Method D: Treatment with n-Pr₃Al (2 equiv) in CH₂Cl₂ at -78°C for 5 min and at 0°C for 1 h, then reduction with DIBAH (1.5 equiv) at 0°C for 1 h. Method E: Addition of oxime sulfonate to Et₂AlC≡CR' (2-3 equiv) at -78°C for 5 min and at 0-25°C for 1-3 h, then reduction with DIBAH (1.5 equiv) at 0°C for 1 h. Method F: Treatment of menthone oxime with n-BuLi (1.05 equiv) at 0°C followed by MsCl (1.05 equiv) in toluene at 0-25°C for 1 h, then Me₃Al (3 equiv) at -20°C for 1-3 h and at 0°C for 1h, and reduction with DIBAH at 0°C for 1 h. Method G: Treatment of menthone oxime as method F but using DIBAH (3 equiv) at 0°C for 1 h at the alkylation and reduction steps. Method H: Reaction with DIBAH (3.5 equiv) in CH₂Cl₂ at -78°C for 5 min and at 0°C for 1 h.

Thus, the aluminum reagents appear to function not only as Lewis acids to start the rearrangements but also as effective nucleophiles for their terminations. This concept suggested the possibility of finding other aluminum reagents of type R₂AlX which would function in a similar way as trialkyl-aluminums. Scheme 1 illustrates the scope of this possibility (ref. 3).

Imino thioether is an important class of compounds in view of its high synthetic utility serving as activated forms of an amide. The preparation of imino thioethers has been effected usually by two steps from amides: conversion to thioamides with phosphorus pentasulfide, followed by transformation to imino thioether by Meerwein reagent. The present method would provide a more direct approach which is operationally simple and requires only very mild conditions. Furthermore, synthetically more important feature of the new process is that an introduction of various alkylthio- as well as arylthio-groups could be easily attained.

Treatment of oxime sulfonates with organoaluminum selenolates produced imino selenoethers. Scope of the synthetic potential of this highly reactive functionality awaits further investigation.

The combination of diethylaluminum chloride with commercially available trimethylsilyl cyanide (ref. 4) was highly effective for the synthesis of imino nitriles.

Scheme 1. Beckmann rearrangement by organoaluminum reagents

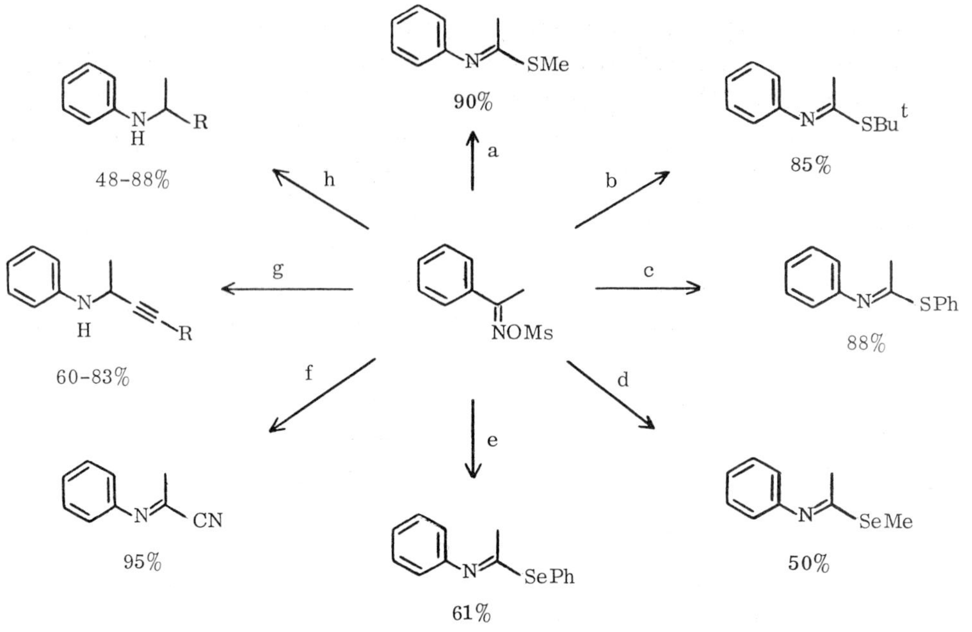

a) i-Bu$_2$AlSMe ; b) Me$_2$AlSBu^t ; c) Me$_2$AlSPh ; d) Me$_2$AlSeMe ; e) i-Bu$_2$AlSePh
f) Me$_3$SiCN-Et$_2$AlCl ; g) Et$_2$AlC≡CR, then DIBAH ; h) R$_3$Al, then DIBAH

Despite the generally superior results obtained in the reaction of organoaluminum reagents, we have sought to extend the investigation to other organometallics with the object of finding the ultimate with regard to efficiency and flexibility of this transformation. A survey was therefore made of a variety of organometallics which would have the latent possibilities of a similar function to organoaluminum reagents. Although most of these were either totally ineffective or less satisfactory as reagents, a promising result was obtained with simple Grignard reagents in nonpolar solvents (ref. 5). Thus, treatment of oxime sulfonates with Grignard reagents furnished the imines which were converted to α-alkylamines with diisobutylaluminum hydride, or further alkylated using allylic or propargylic Grignard reagents to give α,α-dialkylamines.

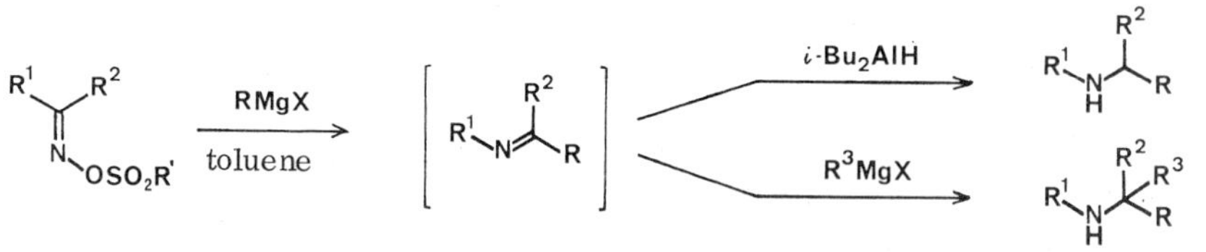

As listed in Table 2, this method has a wide utility for a large range of oximes and Grignard reagents. In pure ether or tetrahydrofuran, no or very little α-alkylation products were produced. Aromatic hydrocarbons, e.g., benzene or toluene were preferable as solvent. Notably, the propargyl group can be introduced in a regiospecific manner producing an important precursor for the synthesis of γ-ketoamines.

The efficiency of our new alkylation-Beckmann rearrangement process is highlighted by the short synthesis of dl-pumiliotoxin C, one of a variety of alkaloids isolated from toxic skin secretions of neotropical frogs of Dendrobates pumilio and D. auratur (ref. 6) (ref. 7). The synthesis was summarized in Scheme 2. Thus, 4-methyltetrahydroindenone was envisaged as an ideal starting material for ease of large scale preparation via the Stobbe condensation of 2-methylcyclohexanone. The unusually selective hydrogenation of this enone was realized using palladium black as a catalyst in dioxane in the presence of propionic acid (12 mol %) at 20°C for 12 h and 1 atm of H$_2$.

Table 2. Synthesis of α-alkyl- and α,α'-dialkylamines using Grignard reagents

Entry	Oxime Sulfonate	RMgX (equiv)	Conditions (°C, h)	R'MgX (equiv)[a]	Product	Yield (%)
1	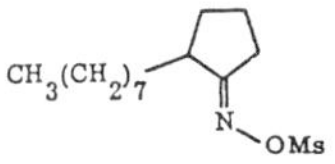 CH₃(CH₂)₇-cyclopentanone N-OMs	CH₃MgI (4)	0, 1; 25, 1		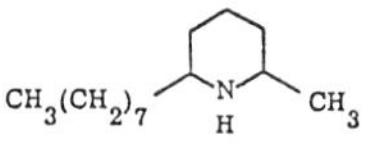 CH₃(CH₂)₇-piperidine-CH₃	36
2	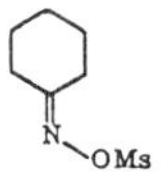 cyclohexanone N-OMs	BuMgBr (1.5)	0, 1		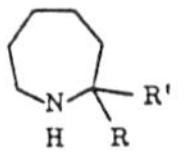 azepane R,R'	63
3		CH₃MgI (1.5)	0, 1	CH₂=CHCH₂MgBr (2)		72
4		CH₃MgI (1.5)	0, 1	HC≡CCH₂MgBr (4)		66
5	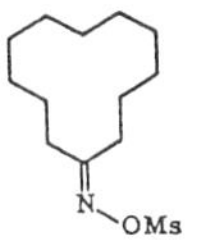 cyclododecanone N-OMs	CH₃MgI (1.5)	0, 1		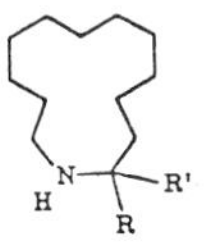 macrocyclic amine R,R'	66
6		BuMgI (1.5)	0, 1			63
7		BuMgBr (1.5)	0, 1			67
8		BuMgCl (1.5)	0, 1			40
9		OctylMgBr (1.5)	0, 1			68
10		CH₃MgI (1.5)	0, 1	CH₂=CHCH₂MgBr (2)		76
11		CH₃MgI (1.5)	0, 1	HC≡CCH₂MgBr (4)		79
12	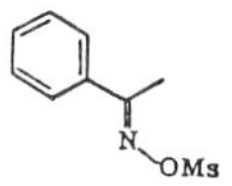 PhC(=NOMs)CH₃	CH₃MgI (1.5)	-78, 1		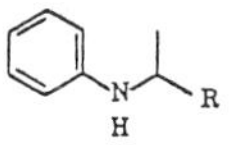 PhNH-CHR-CH₃	52
13		BuMgBr (4)	-78, 1			55
14		BuC≡CMgBr (3)	0, 1			47

[a] Unless otherwise stated, the resulting ketimine was reduced with DIBAH (4 equiv) at 25°C for 1 h.

Scheme 2. Short synthesis of dl-pumiliotoxin C.

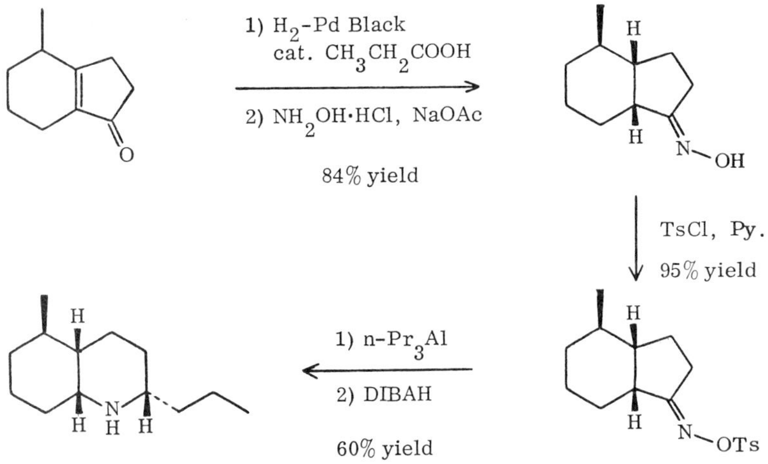

1) H₂-Pd Black, cat. CH₃CH₂COOH
2) NH₂OH·HCl, NaOAc
84% yield

TsCl, Py.
95% yield

1) n-Pr₃Al
2) DIBAH
60% yield

PUMILIOTOXIN C

The crude ketone was reacted with hydroxylamine in methanol to furnish after one recrystallization the desired oxime in 84% yield. Treatment of the oxime with TsCl-pyridine produced the oxime tosylate in 95% yield, which then transformed to dl-pumiliotoxin C with tri-n-propylaluminum-DIBAH (>99% pure and 60% yield).

Solenopsin A and B, naturally occuring piperidine alkaloids isolated from the venom of the fire ant, Solenopsis savissima, are members of 2,6-dialkylated piperidines which exhibit pronounced hemolytic, insecticidal, and antibiotic activity (ref. 8). Scheme 3 shows a simple route to Solenopsin A, which heavily depends upon the use of the Beckmann rearrangement-alkylation sequence by alkylaluminum reagents (ref. 9).

Scheme 3 Synthesis of solenopsin A.

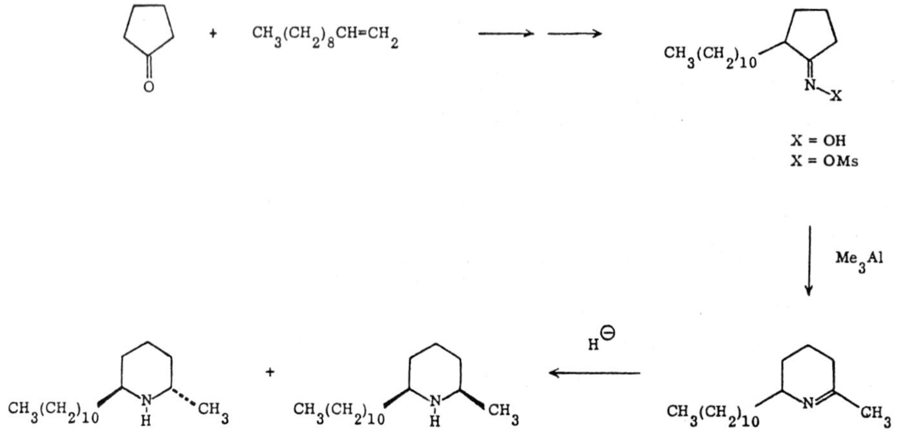

Completion of the synthesis requires reduction of the C=N double bond with correct stereochemistry. Unfortunately, however, it was soon apparent from examination of the literature that existing methodology was totally inadequate for the selective reduction of the imine into trans-2,6-disubstituted piperidine structure. Excellent stereoselective reduction was finally attained by using $LiAlH_4$-Me_3Al system as shown below:

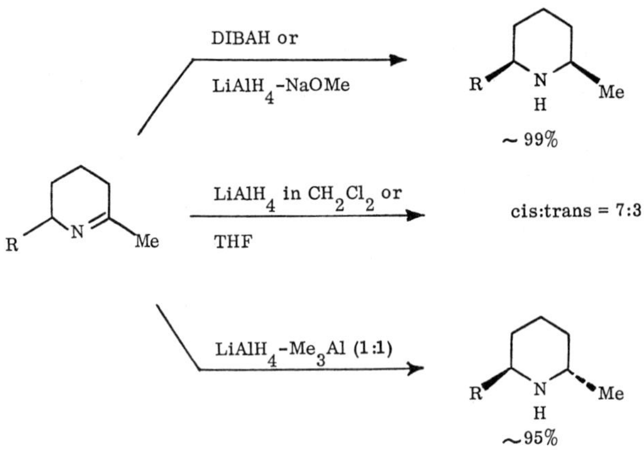

Possible explanation of the above results may be as follows:

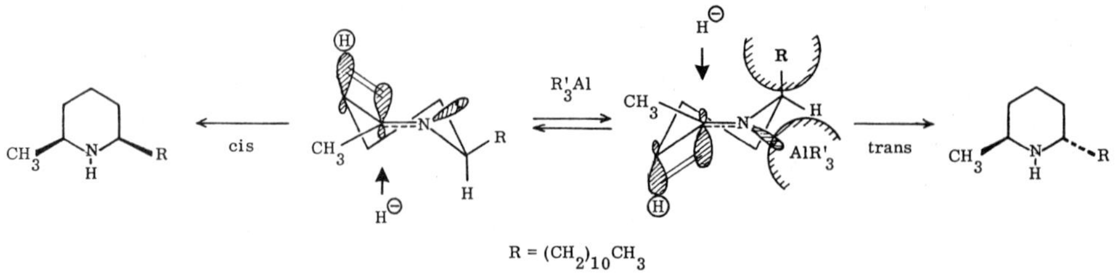

INTRAMOLECULAR REACTIONS

Olefinic ketoxime sulfonates can be rearranged to the iminocarbocation which, in principle, may be intramolecularly alkylated to the ring structures. Since these alkylations proceed by migrating the group anti to the departing sulfonate group in either exocyclic or endocyclic modes, we refer to them as Exo(B) and Endo(B) rearrangements, respectively. Further, when the breaking double bond is exocyclic to the so formed ring, we refer to them as suffix exo and endo correspondingly as in Scheme 4 (ref. 10).

Scheme 4

Exo(B)-exo

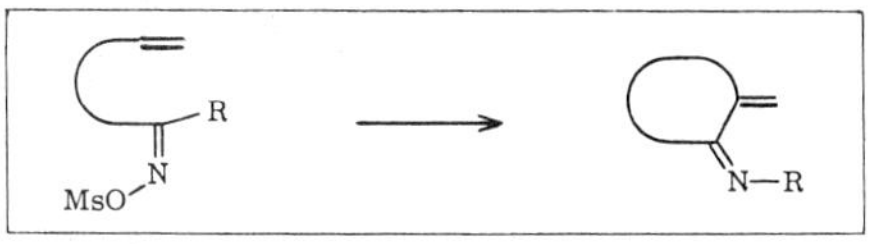

Exo(B)-endo

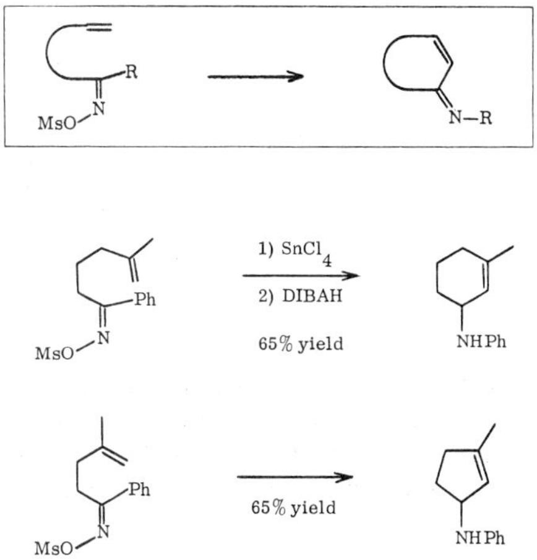

Endo(B)-exo

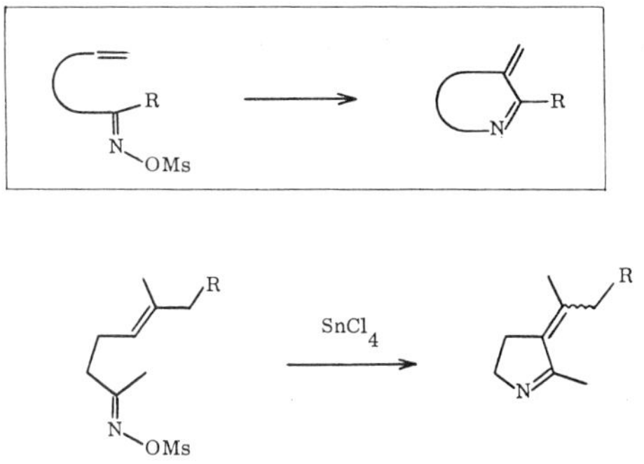

69% yield (R = H)
70% yield (R = Prenyl)

Endo(B)-endo

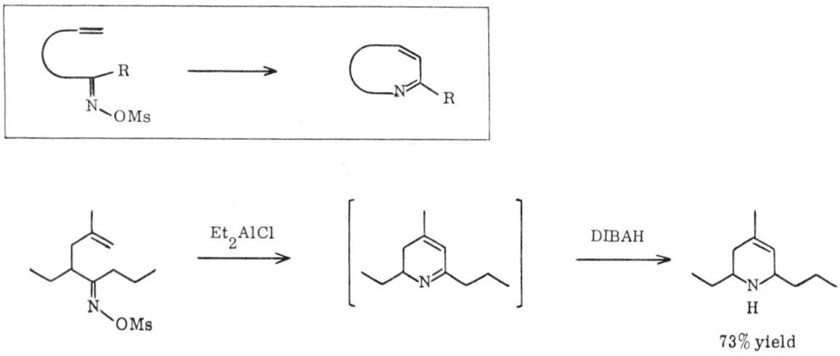

73% yield

With the above new versatile cyclizations in hand, highly efficient and unique syntheses of muscone and muscopyridine are accessible. Efficient Endo(B)-endo cyclization of 2-allyl- or 2-methallyl-cyclododecanone oxime mesylates have been developed. Scheme 5 illustrates the synthesis of muscone by two different routes.

Scheme 5

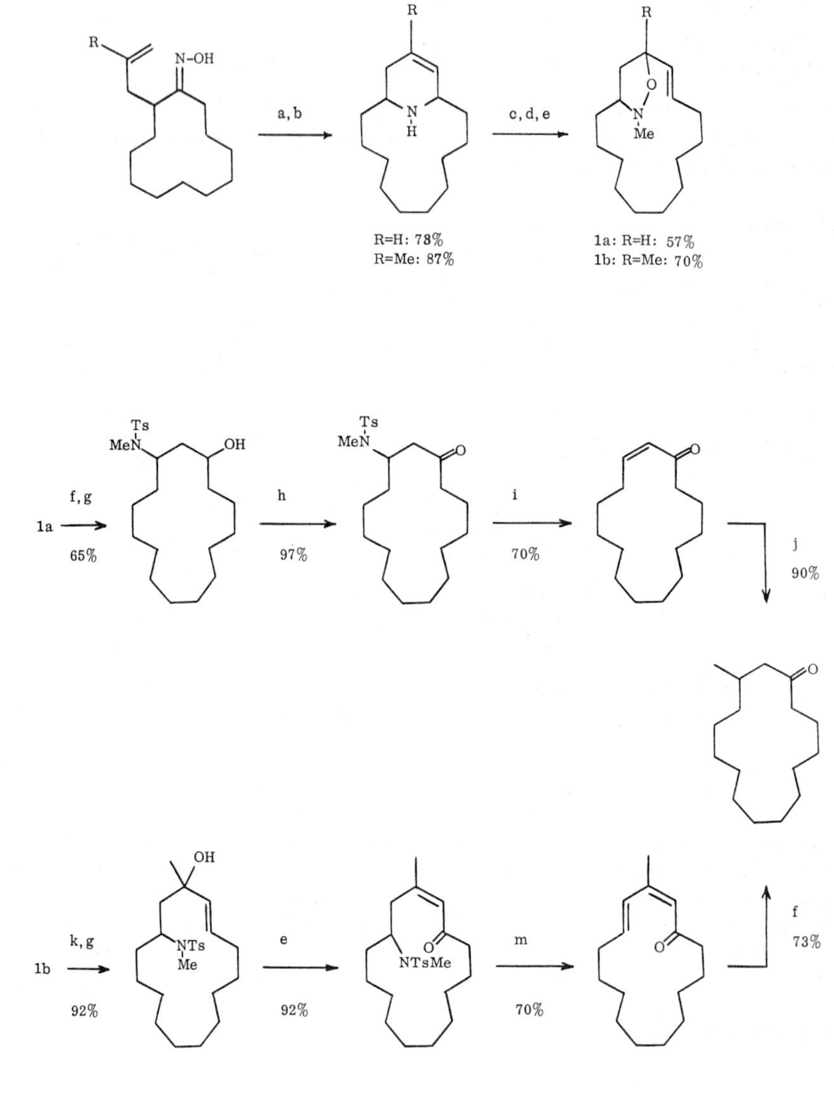

a: MsCl, NEt$_3$; b: Et$_2$AlCl or Me$_3$SiOTf, then DIBAH; c: aq. CH$_2$O, NaBH$_3$CN; d: MCPBA; e: heat; f: H$_2$, Pd/C
g: p-TsCl, NEt$_3$; h: CrO$_3$-py$_2$; i: K$_2$CO$_3$, i-PrOH; j: Me$_2$CuLi; k: Na, EtOH; l: Na$_2$Cr$_2$O$_7$; m: K$_2$CO$_3$, s-BuOH

The other and probably more direct application of Endo(B)-endo 6-membered-ring cyclization is the direct oxidation of unsaturated imine to pyridine ring as illustrated in the synthesis of [10](2.6)-pyridinophane. The cyclization conditions using several different Lewis acids are given in Table 3.

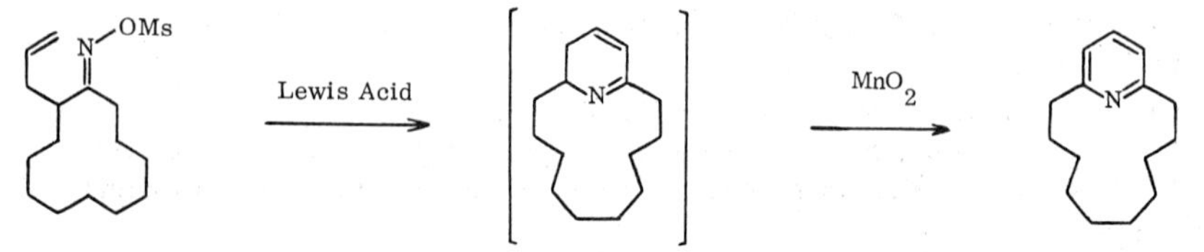

Table 3. Synthesis of [10](2,6)Pyridinophane.

Lewis Acid (equiv)	Cyclization Conditions		Yield (%)
	Temp (°C)	Time (min)	
Et_2AlCl (1.1)	20	60	39
$SnCl_4$ (1.1)	0	40	65
Me_3SiI (1.1)	20	60	68
Me_3SiOTf (0.3)	20	150	55
Me_3SiOTf (1.1)	20	50	80

An especially crucial and significant application of the new pyridine synthesis is the facile preparation of muscopyridine as illustrated in Scheme 6.

Scheme 6

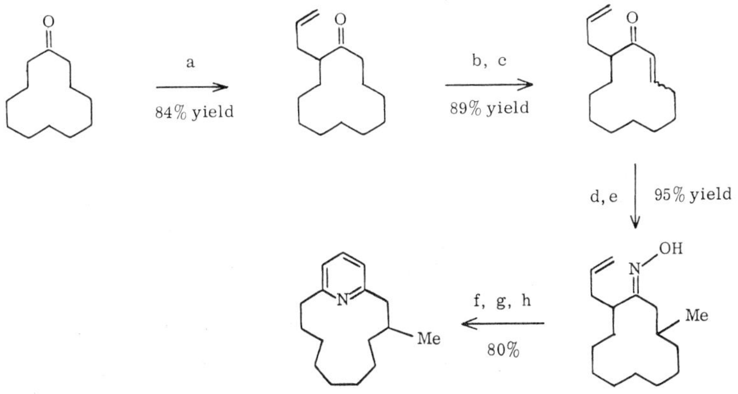

a) $i\text{-}Pr_2NLi$, $CH_2=CHCH_2Br$; b) $i\text{-}Pr_2NLi$, PhSeBr ; c) 30% H_2O_2
d) Me_2CuLi ; e) $NH_2OH \cdot HCl$, NaOH ; f) MsCl, NEt_3 ; g) Me_3SiOTf
h) MnO_2

Acknowledgment - It is a great pleasure to acknowledge the valuable contributions of our students, Miss Y. Ishida, Messrs. M. Ando, T. Miyazaki, Y. Yamamura, Y. Matsumura, S. Sakane, and K. Hattori. Financial support of this work by the Ministry of Education, Japanese Government (Grant 209010, 102008, 118006), Yamada Science Foundation, Naito Science Foundation, and Asahi Glass Foundation are gratefully appreciated. We also wish to thank Nippon Aluminum Alkyls, Ltd., for generous gift samples of aluminum reagents.

REFERENCES

1. Yamamoto, H.; Nozaki, H. Angew. Chem. Int. Ed. Engl., 1978, 17, 169.
2. Hattori, K.; Matsumura, Y.; Miyazaki, T.; Maruoka, K.; Yamamoto, H. J. Am. Chem. Soc., 1981, 103, 7368.
3. Maruoka, K.; Miyazaki, T. Ando, M.; Matsumura, Y.; Sakane, S.; Hattori, K.; Yamamoto, H. in preparation.
4. Utimoto, K.; Obayashi, M.; Shishiyama, Y.; Inoue, M.; Nozaki, H. Tetrahedron Lett., 1980, 21, 3389.
5. Hattori, K.; Maruoka, K.; Yamamoto, H. Tetrahedron Lett., in press.
6. Daly, J. W.; Tokuyama, T.; Habewehl, G.; Karle, I. L.; Witkop. B. Liebigs Ann. Chem., 1969, 729, 198.
7. Daly, J. W.; Witkop, B.; Tokuyama, T.; Nishikawa, T. Karle, I. L. Helv. Chim. Acta, 1977, 60, 1128.
8. MacConnell, J. G.; Blum, M. S.; Fales, H. M. Tetrahedron, 1971, 27, 1129.
9. Matsumura, Y.; Maruoka, K.; Yamamoto, H. Tetrahedron Lett., 1982, 23, 1929.
10. Ishida, Y.; Yamamura, Y.; Matsumura, Y.; Sakane, S.; Maruoka, K.; Yamamoto, H. in preparation.

NEW ACCESS TO CONJUGATED DIENES VIA CARBOCUPRATION OF ALKYNES

Jean F. Normant and Alexandre Alexakis

*Laboratoire de Chimie des Organo Eléments, Université P. et M. Curie,
4 Place Jussieu, 75230 Paris Cédex 05, France*

Abstract - A short review of the available methods, for constructing dienic conjugated systems shows that access to the E,Z or Z,Z isomers is still a challenging process. Carbocupration of alkynes by organo-copper - or cuprate reagents leads to vinyl-copper synthons which may be used to alkylate vinylic halides under precise conditions, depending on the presence of various salts (lithium-, magnesium-, zinc-, halides) and catalytic amounts of palladium. On the other hand carbocupration of acetylenes by vinylic copper reagents offers a new strategy for such syntheses.

INTRODUCTION

Several well known methods are commonly used for the preparation of conjugated dienes. The important development of Diels-Alder condensation reactions, these last few years, has required new methodologies for an access to such substrates in a high state of purity. However, for this particular purpose, the E,E isomers are the most useful, since they may adopt the s-cis configuration. They are also the more thermodynamically stable isomers, as compared to their E-Z or Z-Z conterparts. These later substrates are, accordingly, available only by a limited number of methods. This talk is mainly concerned by the approach to structures like the followings :

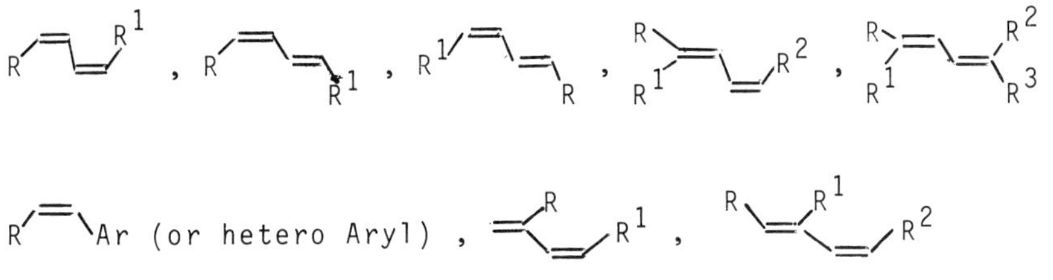

with the severe limitation that they should be obtained as a single isomer, and need not tedious G.C.or H.P.L.C. separations (although E,E isomers may be separated from the others by trapping with tetracyanoethylene in a Diels-Alder reaction). In the field of insect sexpheromones, only a few active substances present the E,E conjugated systems, and E,Z (or Z,E) isomers are much more common. Let us remind the most useful approaches described so far :

The Wittig olefination reaction of α-alkenals (prepared as pure E isomers), if performed under "salt free" conditions (1), leads to a Z,E diene

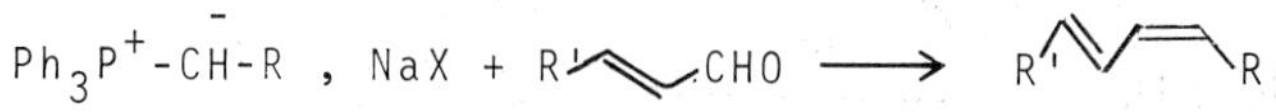

$Ph_3P^+-\overset{-}{C}H-R$, NaX + $R'\diagup\!\!\!\diagdown CHO \longrightarrow R'\diagup\!\!=\!\!\diagdown R$

and has been widely exemplified in the field of pheromones (2).

<u>Preparation of a Z-enyne</u> followed by anti-hydrogenation of the triple bond (3)

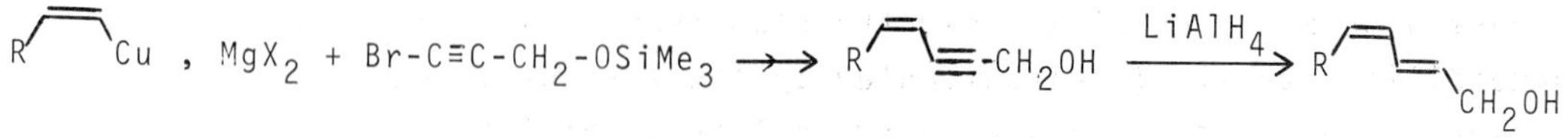

$R\diagup\!\!=\!\!\diagdown Cu$, MgX_2 + $Br-C\equiv C-CH_2-OSiMe_3 \longrightarrow R\diagup\!\!=\!\!\diagdown\!\!-\!\!\equiv\!\!-CH_2OH \xrightarrow{LiAlH_4} R\diagup\!\!=\!\!\diagdown\!\!=\!\!\diagdown_{CH_2OH}$

or preparation of a E-enyne, for example :

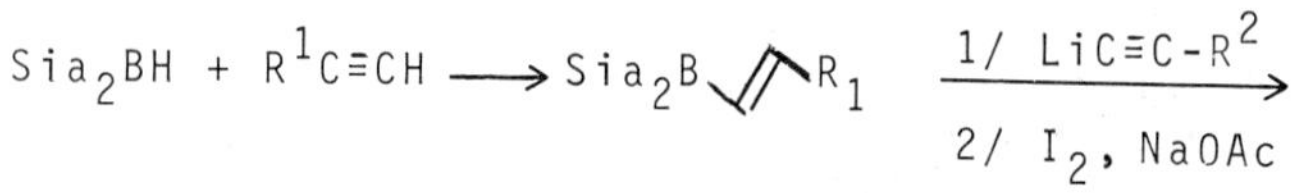

Sia_2BH + $R^1C\equiv CH \longrightarrow Sia_2B\diagdown\!\!\diagup\!\!\diagdown R_1 \xrightarrow[\text{2/ }I_2,\text{ NaOAc}]{\text{1/ LiC}\equiv\text{C-}R^2} R^2-C\equiv C\diagdown\!\!\diagup\!\!\diagdown R^1$ (4)

$Bu\diagdown\!\!\diagup\!\!\diagdown^I$ + $ClZn-C\equiv C-R \xrightarrow{Pd°(PPh_3)_4 \text{ cat.}} R-C\equiv C\diagdown\!\!\diagup\!\!\diagdown Bu$ (5)

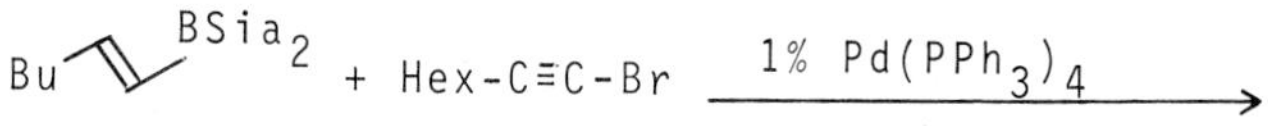

$Bu\diagdown\!\!\diagup\!\!\diagdown^{BSia_2}$ + $Hex-C\equiv C-Br \xrightarrow{1\% \text{ Pd}(PPh_3)_4} Hex-C\equiv C\diagdown\!\!\diagup\!\!\diagdown Bu$ (6)

followed by syn-hydrogenation of the triple bond by Lindlar's semi hydrogenation (7) or, better, by bicyclohexyl borane (8), disiamyl borane (4,9) or zinc in propanol in the presence of potassium cyanide (10) - Also conjugated diynes may be partly reduced to Z,Z-dienes by bicyclohexyl boranes (11).

Once the dienic skeletton is set up, an allylic function may be substituted

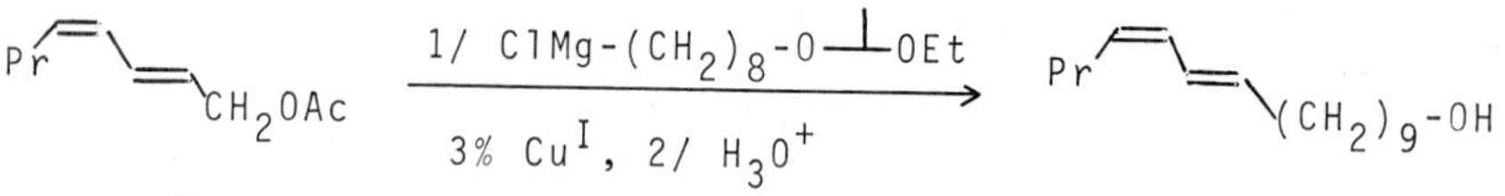

$Pr\diagup\!\!=\!\!\diagdown\!\!\diagdown_{CH_2OAc} \xrightarrow[\text{3\% Cu}^I, \text{ 2/ }H_3O^+]{\text{1/ ClMg-}(CH_2)_8\text{-O-}\!\!\perp\!\!\text{-OEt}} Pr\diagup\!\!=\!\!\diagdown\!\!\diagdown(CH_2)_9\text{-OH}$ (3)

$\underset{R^1}{\diagup}\!\!\diagdown\!\!\overset{+}{N}\!\!\diagdown \xrightarrow{OH^-} R^1\diagup\!\!=\!\!\diagdown\!\!=\!\!\diagdown N\!\!< \xrightarrow[\text{2/ RMgX, 5\% Cu}^I]{\text{1/ MeI}} R^1\diagup\!\!=\!\!\diagdown\!\!=\!\!\diagdown R$ (12)

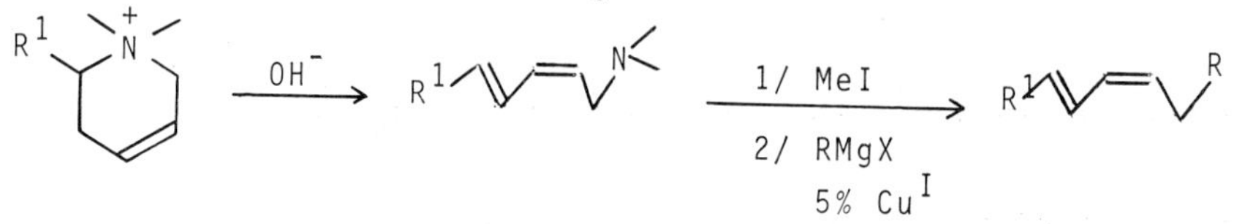

However the copper catalyzed substitution (by a Grignard reagent) of the ester, or ammonium moiety, leads to substantial amounts of undesired isomers.

<u>Direct coupling of two vinylic synthons</u> is a straight forward method. Dang and Linstrumelle developped a palladium catalyzed Grignard reaction (13)

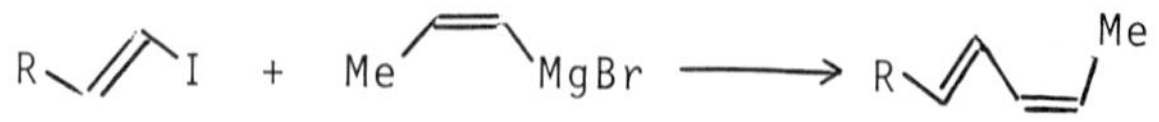

$R\diagdown\!\!\diagup\!\!\diagdown^I$ + $Me\diagup\!\!=\!\!\diagdown MgBr \longrightarrow R\diagdown\!\!\diagup\!\!\diagdown\!\!=\!\!\diagup^{Me}$

leading to a pure diene when the iodo alkene is of E type, but when Z, a mixture of isomers is obtained.

The general access to E-vinyl zirconium (aluminium) derivatives by hydrozirconation (14), hydro alumination (15) or by methyl alumination of 1-alkynes (16) led Negishi et al to couple these organometallics with iodo alkenes under palladium or Nickel catalysis. Dienes of high purity are obtained (16).

$$\text{Pent}-\equiv \xrightarrow{R_2AlH} \text{Pent}\diagdown\!\!=\!\!\diagup AlR_2 \xrightarrow[\text{Pd}°(PPh_3)_4 \text{Cat.}]{I\diagdown\!\!=\!\!\diagup Bu} \text{Pent}\diagdown\!\!=\!\!\diagup\diagdown\!\!=\!\!\diagup Bu \qquad 55\%$$
(>99% E,Z)

The starting organometallics is of E configuration. If the hydrometallation is performed on an internal alkyne, the sluggishness of the corresponding organometallics (Zr,Al) is circumvented by their transformation into a zinc derivative, more reactive towards vinyl palladium halides (16)

$$\underset{\text{Al}\diagdown}{\overset{\text{Et}\quad\text{Et}}{\diagdown\!=\!\diagup}} \xrightarrow[\substack{2/\ Pd°(PPh_3)_4\ (10\%)\\3/\ Br\diagdown\!=\!\diagup\underset{\text{COOMe}}{\text{Me}}}]{1/\ ZnCl_2\ 20\%} \underset{\diagup\!=\!\diagdown\underset{\text{COOMe}}{\text{Me}}}{\overset{\text{Et}\quad\text{Et}}{\diagdown\!=\!\diagup}} \qquad 72\%$$

The palladium mediated substitution of bromoalkenes by vinyl boranes, according to Suzuki et al., is a versatile method for the preparation of E-E, E-Z and Z-Z dienes. Better yields are obtained, however, from the E-organometallic reagents (80-90%) as compared to the Z ones (40-50%) (6, 17)

$$\text{Bu}\diagup\!\!=\!\!\diagdown B\diagup^O_O\diagdown\!\!\bigcirc + \text{Ph}\diagdown\!\!=\!\!\diagup Br \xrightarrow{Pd°L_4\ cat} \overset{Bu}{\diagdown\!\!=\!\!\diagup}\diagdown\!\!=\!\!\diagup Ph \qquad 86\%$$
(99%, Z,E)

This scheme has been developped for the synthesis of pheromones (9).

<u>Rearrangement of unsaturated skelettons</u> to dienes of high purity has been achieved, starting from β-allenic esters (18)

$$\text{Pent}\diagup\!\!=\!\!\bullet\!\!=\!\!\diagdown\!\!\diagup COOEt \xrightarrow{Al_2O_3} \text{Pent}\diagup\!\!=\!\!\diagdown\diagup\!\!=\!\!\diagdown COOR \qquad 82\%$$
(100% E,Z)

or acetates of bis allylic carbinols (19)

$$\underset{\diagdown\!=\!\diagup\diagup\overset{OAc}{|}\diagdown\diagup\!=\!\diagdown}{} \xrightarrow[cat.]{Pd^{II}} \underset{\diagdown\!=\!\diagup\diagdown\diagup\overset{OAc}{|}\diagdown\!=\!\diagdown}{} \qquad 100\%$$

or metallation-isomerisation of skipped terminal dienes (20)

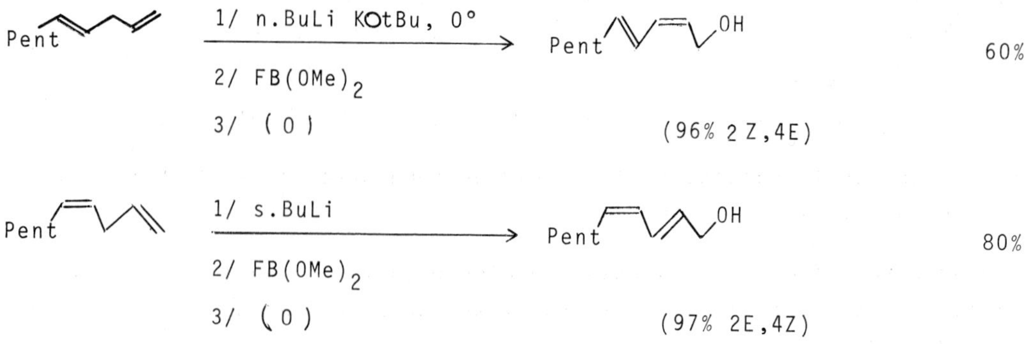

Pent—CH=CH—CH=CH₂
1/ n.BuLi KOtBu, 0°
2/ FB(OMe)₂
3/ (O)
→ Pent-CH=CH-CH=CH-OH   60%

(96% 2Z,4E)

Pent-CH=CH-CH₂-CH=CH₂
1/ s.BuLi
2/ FB(OMe)₂
3/ (O)
→ Pent-CH=CH-CH=CH-CH₂OH   80%

(97% 2E,4Z)

The palladium catalyzed decomposition of β-acetoxy carboxylic acids, creates an E ethylenic unit, but isomers are formed in 10-20% amounts (21)

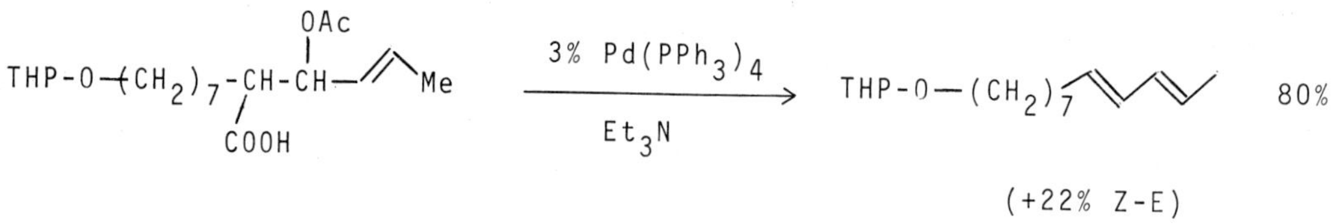

$$THP-O-(CH_2)_7-CH(OAc)-CH(COOH)-CH=CH-Me \xrightarrow[Et_3N]{3\% \ Pd(PPh_3)_4} THP-O-(CH_2)_7-CH=CH-CH=CH-Me \quad 80\%$$

(+22% Z-E)

Several reviews may be consulted (2,9,22-24).

REACTION OF VINYL-COPPER REAGENTS WITH VINYLIC HALIDES

Our first approach to conjugated dienes stemmed from the study of vinyl copper species, obtained by carbocupration of terminal alkynes according to the following two pathways (25) :

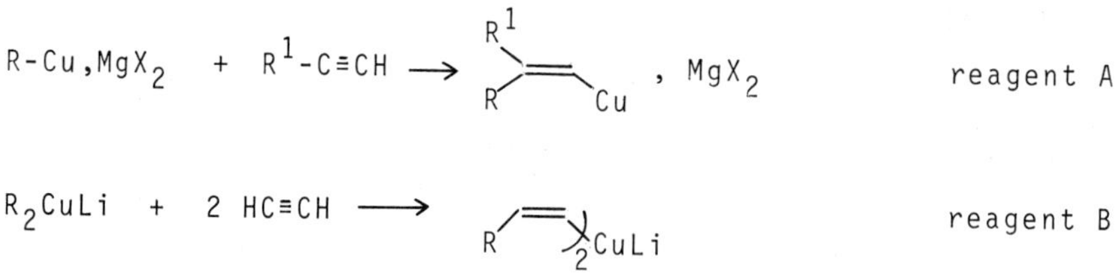

$$R-Cu, MgX_2 \ + \ R^1-C\equiv CH \ \rightarrow \ \underset{R}{\overset{R^1}{>}}\!\!=\!\!\underset{Cu}{<} \ , \ MgX_2 \qquad \text{reagent A}$$

$$R_2CuLi \ + \ 2 \ HC\equiv CH \ \rightarrow \ \left(\underset{R}{>}\!\!=\!\!<\right)_2 CuLi \qquad \text{reagent B}$$

The first (disubstituted) vinylic reagents are derived from a Grignard reagent. The second ones (monosubstituted) are made from the corresponding lithium derivatives. Both are able to react, under appropriate conditions, with saturated organic halides (Csp_3-X), the first half of the lithio cuprate reacting more rapidly than the second one, which has a reactivity rather similar to that of reagent A. Hence a way to trisubstituted olefins or Z disubstituted ones.

Reagent A, or modified reagent B, also react cleanly with haloacetylenes (25) (Csp-X) leading to enynes of general formula

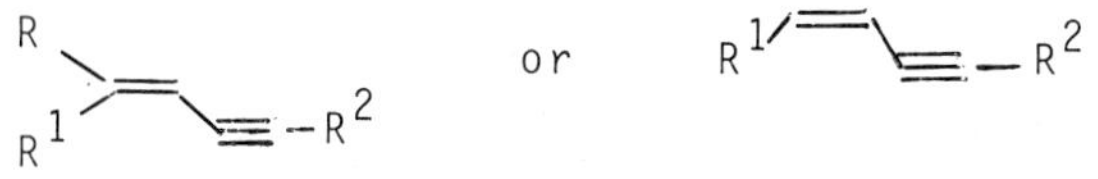

$$\underset{R^1}{\overset{R}{>}}=\kern-1em =-R^2 \quad \text{or} \quad R^{1}\!/\!\!=\kern-1em =-R^2$$

We anticipated that reagents A or B would be good precursors of dienes by reaction with haloalkenes (Csp_2-X).

However, the reactions of saturated cuprates with vinylic halides described so far (26) necessitate a large excess of the cuprate reagent; and vinylic cuprates gave disappointingly low yields of dienes.

Whatsoever, the high nucleophilicity of cuprate reagents might be well suited for their reaction with vinyl palladium (or Nickel) halides.

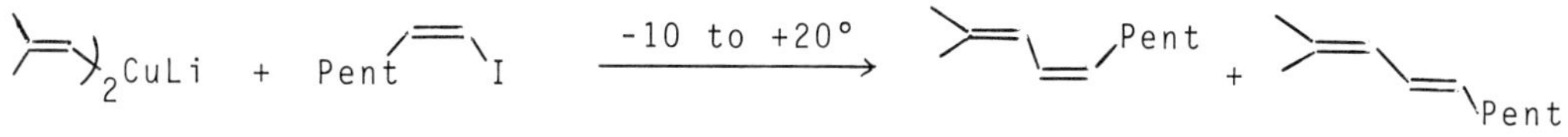

$$\left(\!\!>\!=\!\!\right)_2\!CuLi \;+\; Pent\!\!\diagup\!\!=\!\!\diagdown\!I \quad\xrightarrow{-10 \text{ to } +20°}\quad >\!\!=\!\!\diagdown\!\!\diagup^{Pent} \;+\; >\!\!=\!\!\diagdown\!\!\diagdown_{Pent}$$

Catalyst	time (h)	Yield	Z/E
none	72	5	—
$NiBr_2$, 2 PPh_3 (3%)	4	51	80/20
$Pd(PPh_3)_4$ (5%)	24	50	91/ 9

A severe limitation has to be taken into account, since, contrary to the corresponding Nickel catalyzed Grignard reactions, particularly developped by Kumada et al. (27), the copper reagents decompose, upon heating above 0°C, to symmetrical dienes

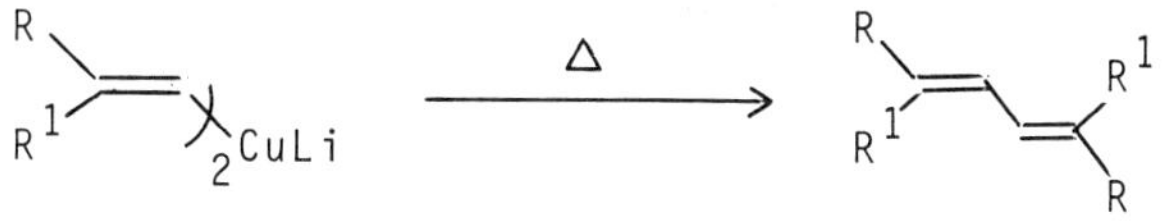

$$\left(\underset{R^1}{\overset{R}{>}}\!\!=\!\!\right)_2\!CuLi \quad\xrightarrow{\Delta}\quad \underset{R^1}{\overset{R}{>}}\!\!=\!\!\diagdown\!\!\underset{R}{\overset{R^1}{>}}$$

so that the ordinary conditions for such coupling reactions (reflux of ethereal solvents) must be avoided.

We have seen above that vinylic zinc reagents are prone to give a palladium catalyzed condensation with haloalkenes, and one may question wether a metal-metal exchange

$$=\!\!\underset{|}{C}\!-\!Cu \;\longrightarrow\; =\!\!\underset{|}{C}\!-\!Zn\!-$$

might be possible. We had previously shown that vinyl copper reagents may be transformed into their mercury or tin analogs by action of mercury dihalides or tin tetrahalides (28).

In the case of zinc, preliminary experiments show that exchange is indeed possible, from a cuprate, but not from an organocopper reagent.

$$Vi_2CuLi \xrightleftharpoons{ZnCl_2} ViZnCl + ViCu, LiCl \xrightarrow{ZnCl_2} \!\!\!\!\!\!/\!\!\!\!\!\!\to 2\ ViZnCl$$

Vice versa, dimethyl zinc, when treated by Cu^{II} or Cu^{I} halides, precipitates methyl copper (29)

$$4\ Me_2Zn + CuX_2 \longrightarrow \underline{MeCu}\!\!\downarrow + MeH + Me\text{-}Me + ZnX_2$$

The reaction of a symmetrical vinyl cuprate with a vinylic iodide was therefore attempted, in the presence of a zinc halide, and a catalytic amount of $Pd(PPh_3)_4$ (30)

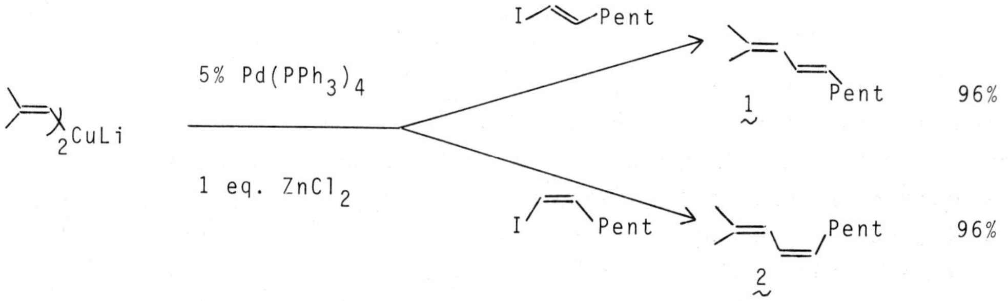

1 is 99.5% E, **2** is 99.6% Z. It turns out that the nucleophilic attack, and the reductive elimination from the Pd^{II} intermediate, all proceed with total retention of configuration. To ascertain the retention of configuration in the vinyl-copper/vinyl Zinc exchange one can check with E or Z vinylic cuprates :

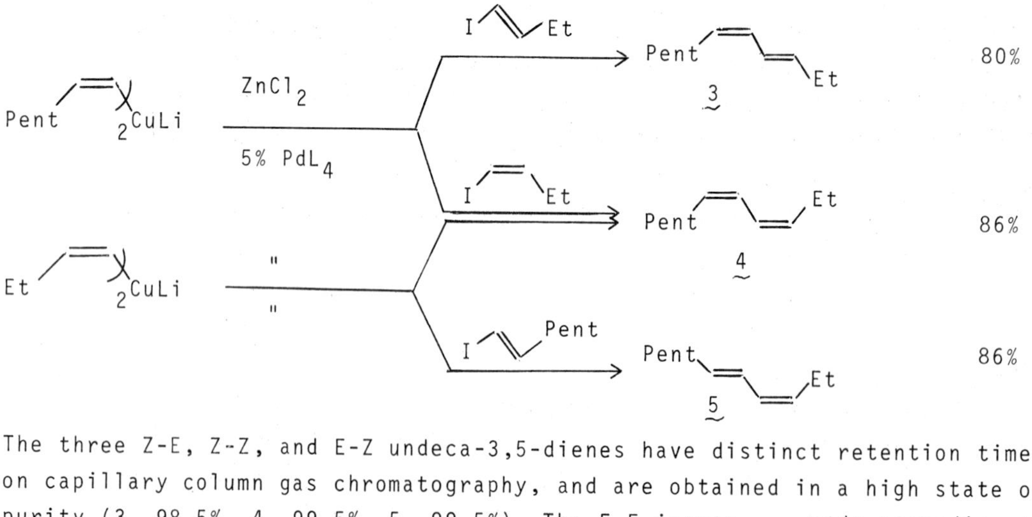

The three Z-E, Z-Z, and E-Z undeca-3,5-dienes have distinct retention times on capillary column gas chromatography, and are obtained in a high state of purity (**3** 98,5%, **4** 99.5%, **5** 99.5%). The E,E isomer, made according to Negishi (31), has been compared, and is totally absent.

This highly selective method has been successfully applied to the reaction of aryl iodides (32) or heterocyclic iodides, with simple vinylic cuprates, as well as functionnalized ones

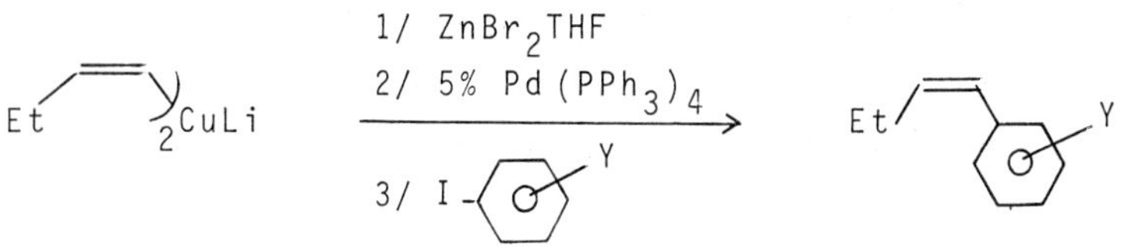

Y :	H	p.Br	p.OMe	p.NO$_2$	p.COOMe	o.COOMe
Yield%:	67	71	65	80	81	72

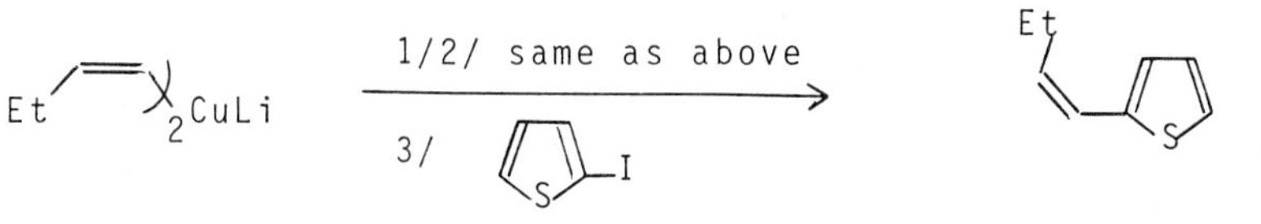

In the preceeding schemes, only one vinyl group of the lithium cuprate reagent is used ; this fact represents a limitation when the organic moiety is an elaborated one. We therefore looked for conditions where the only vinyl group of reagent A (from RMgX) or both vinyl groups of reagent B (from RLi) would be consumed. The first trials were performed with an other reagent, namely a *magnesium* symmetrical vinyl cuprate, and showed that, not only zinc halides were necessary, but also that both vinyl groups were consumed in their presence (in opposition with lithium reagents B)

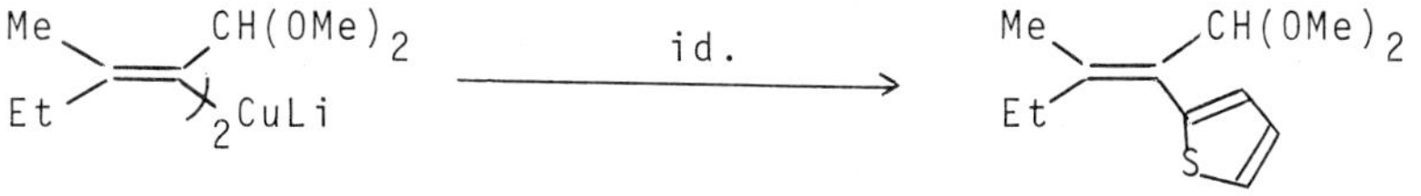

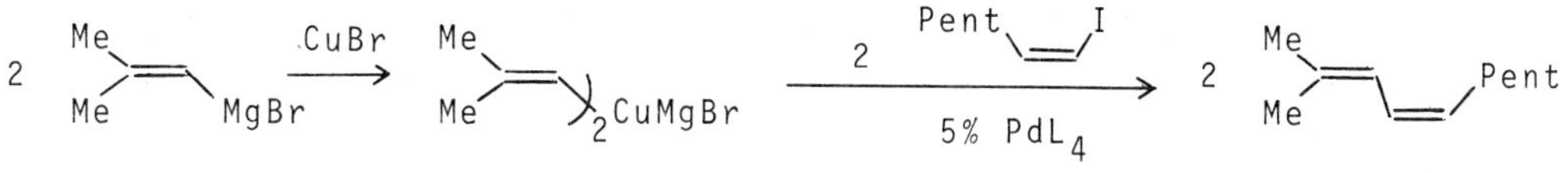

with ZnX$_2$: yield 94%, without ZnX$_2$: 24%. On the other hand, unsymmetrical homo-(or hetero-) magnesium cuprates behave abnormally in the sense that R groups, known to be "non transferable" in copper chemistry, tend to be now easily transferred (see table 1).

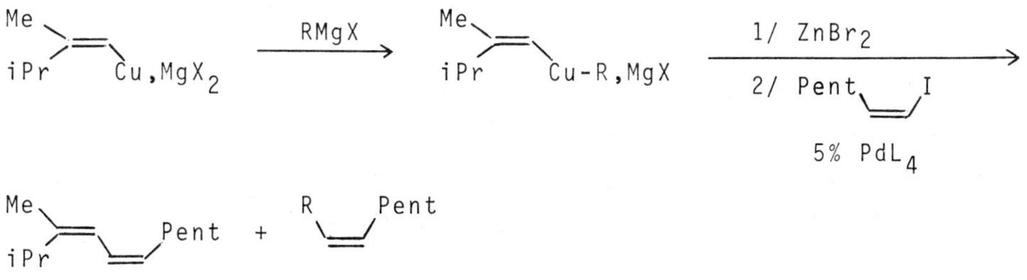

TABLE 1 - Ratio of transfer of the vinyl moiety versus the R group according to the nature of R

R	$\underset{\sim}{6}$ / $\underset{\sim}{7}$
BuC≡C	44/56
iPr	86/14
Me	10/90
tBuO	98/ 2
PhS	94/ 6

This discrepancy must originate from the first step, namely the Cu $\rightarrow$ Zn transmetallation.

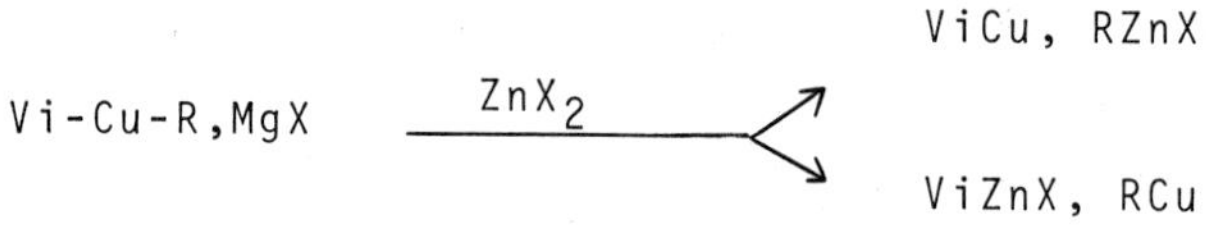

$$\text{Vi-Cu-R,MgX} \xrightarrow{ZnX_2} \begin{array}{c} \text{ViCu, RZnX} \\ \\ \text{ViZnX, RCu} \end{array}$$

The formation of the active vinylzinc species may be thwarted by the kinetic formation of the alkyl (alkynyl) zinc derivative. However terbutoxide, or phenylthio groups are not transferred, and may be used successfully. These heterogroups have been introduced by G. Posner (33) as ate-complexing reagents for the preparation of RCuZLi complexes in THF. Their magnesium counterparts, considered here, have a less stabilizing effect, and we observed some analogies between Vi-Cu, and Vi-Cu-OtBuMgX species, indicating that the former reagents might be used in lieu of the latter. Even more, the vinyl copper reagents A (in the presence of Magnesium halides) need not the presence of added zinc halides (but the magnesium divinylcuprate does !)

It follows therefore, that without added magnesium terbutoxide, and without zinc salts, the simplest reagents (A) may be used according to :

Scheme 1

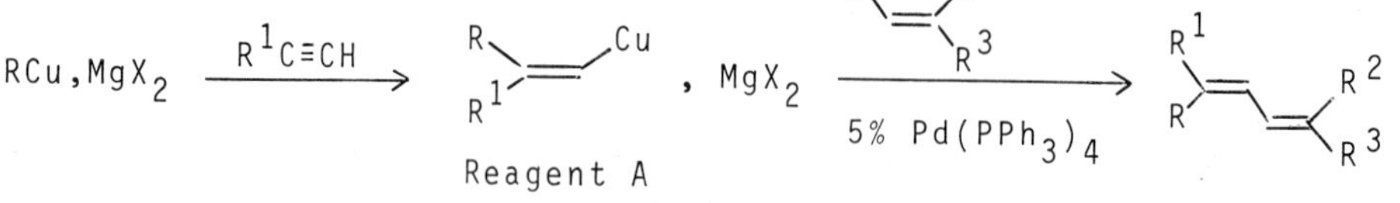

$$RCu, MgX_2 \xrightarrow{R^1C\equiv CH} \underset{R^1}{\overset{R}{\diagdown}}\!\!=\!\!\underset{}{\overset{Cu}{\diagup}} , MgX_2 \xrightarrow[5\% \text{ Pd(PPh}_3)_4]{I\diagdown\!\!=\!\!\underset{R^3}{\overset{R^2}{\diagup}}} \underset{R^1}{\overset{R^1}{\diagdown}}\!\!=\!\!\diagup\!\!\diagdown\!\!=\!\!\underset{R^3}{\overset{R^2}{\diagup}}$$

Reagent A

This scheme allows the easy preparation of 1-3 conjugated dienes with variable substituents on carbon number 1 or/and 4.

TABLE 2 - Some examples of dienes prepared according to scheme 1, with reagent A (R^1=Me, R=Et) (34)

1-iodo alkene	Product	Yield	Purity
 Pent-CH=CH-I	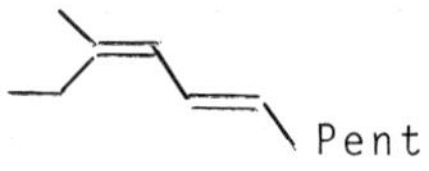 diene with Pent	78	99.5 Z,E
 Pent-CH=CH-I (Z)	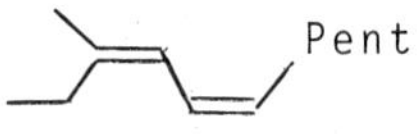 diene with Pent	70	99.8 Z,Z
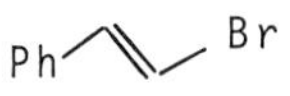 Ph-CH=CH-Br	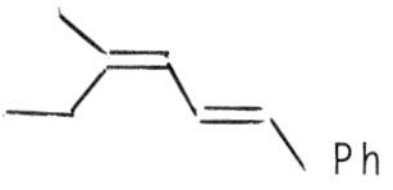 diene with Ph	74	99 Z,E
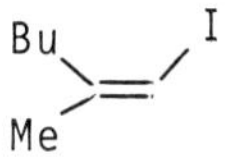 Bu(Me)C=CH-I	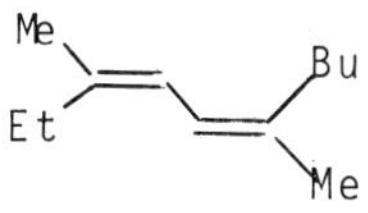 Me,Et,Bu,Me tetrasubstituted diene	55	99.8 Z,Z

This reaction shows that vinyl copper reagents, in the presence of magnesium salts, give the palladium catalyzed coupling reaction (whereas the vinyl copper reagents in the presence of lithium salts do not react). We were thus led to anticipate that *both* vinyl groups of a lithium divinyl cuprate should react if magnesium salts are added : the experiment corroborates this assumption

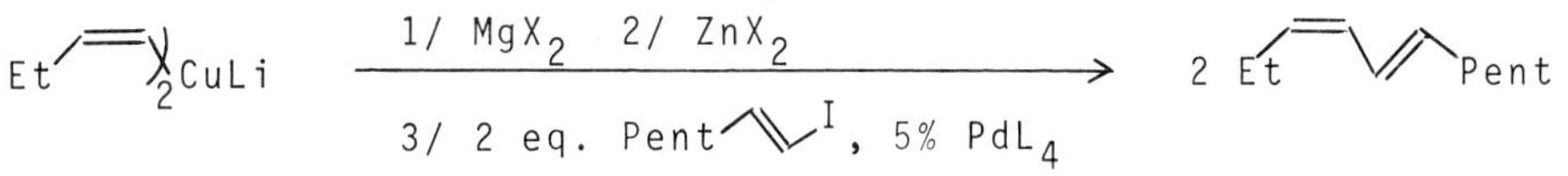

$$\text{Et-}\underset{2}{\diagup\!\!=\!\!\diagdown}\text{CuLi} \xrightarrow[\text{3/ 2 eq. Pent}\diagup\!\!=\!\!\diagdown\text{I, 5\% PdL}_4]{\text{1/ MgX}_2 \quad \text{2/ ZnX}_2} 2 \text{ Et}\diagup\!\!=\!\!\diagdown\!\!\diagup\!\!=\!\!\diagdown\text{Pent} \qquad 84\%$$

(a 41% yield is obtained in the absence of MgX_2)

DOUBLE CARBOCUPRATION TECHNIQUES

The success of the carbocupration of terminal acetylenes is due to the inertness of the vinyl copper species thus formed, towards the starting acetylenes, so that no polymerisation occurs

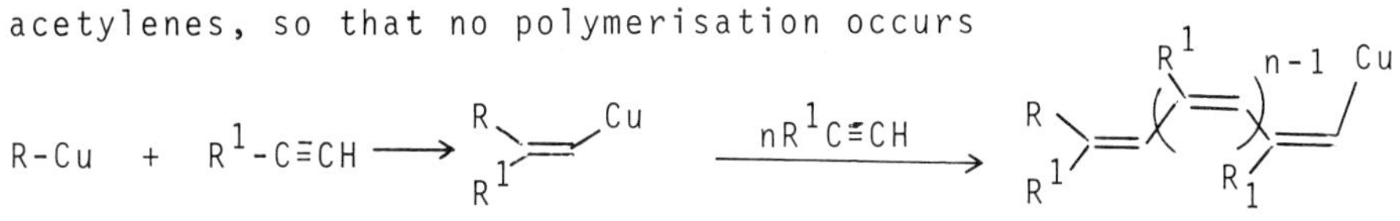

$$\text{R-Cu} + R^1\text{-C}\equiv\text{CH} \longrightarrow \underset{R^1}{\overset{R}{\diagup}}\!\!=\!\!\underset{}{\overset{Cu}{\diagdown}} \xrightarrow{nR^1C\equiv CH} \underset{R^1}{\overset{R}{\diagup}}\!\!=\!\!\underset{R^1}{\overset{}{\diagdown}}\!\!\left(\!\!=\!\!\right)_{n-1}\!\!\overset{Cu}{\diagup}$$

However, vinyl copper reagents may add to functionnalized alkynes : alkyl propiolates may be used, and we found that a lithium divinyl cuprate inserts two equivalents of ethyl propiolate (35), leading to 2-E,4-Z ethyl decadienoate

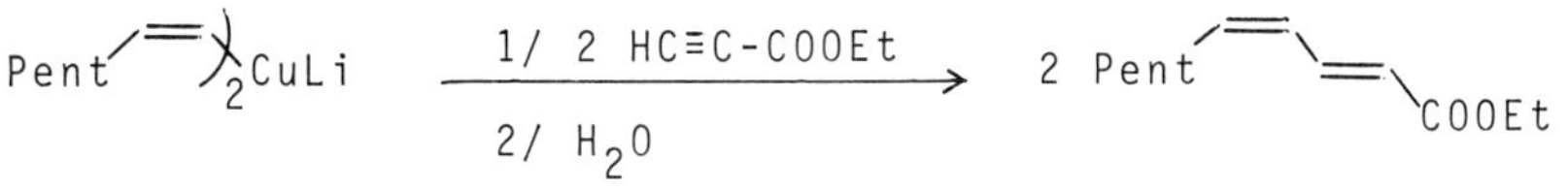

$$\text{Pent-}\underset{2}{\diagup\!\!=\!\!\diagdown}\text{CuLi} \xrightarrow[\text{2/ H}_2\text{O}]{\text{1/ 2 HC}\equiv\text{C-COOEt}} 2 \text{ Pent}\diagup\!\!=\!\!\diagdown\!\!\diagup\!\!=\!\!\diagdown\text{COOEt} \qquad 78\%$$

(Bartletts'pear) (95% 2-E,4-Z). The corresponding acetal is also sufficiently activated to insert one vinyl group (only) (32)

$$\text{Pent} \diagup\!\!\!=\!\!\!\diagdown_2 \text{CuLi} \xrightarrow{HC\equiv C-CH(OEt)_2} \text{Pent} \diagup\!\!\!=\!\!\!\diagdown\text{CH(OEt)}_2 \quad 76\%$$

Also heterosubstituted acetylenes, may enter the carbocupration twice, for example :

$$Bu_2CuLi \xrightarrow[2/\ 2\ HC\equiv C-SEt]{1/\ 2\ HC\equiv CH} Bu\diagup\!\!=\!\!\diagdown\!\!\diagup\!\!=\!\!\diagdown_{SEt}\!\!{}_2 CuLi \longrightarrow Bu\diagup\!\!=\!\!\diagdown\!\!\diagup\!\!=\!\!\diagdown_{SEt} \quad 86\%$$

$$Bu_2CuLi \xrightarrow[2/\ 2\ HC\equiv C-OEt]{1/\ 2HC\equiv CH} Bu\diagup\!\!=\!\!\diagdown\!\!\diagup\!\!=\!\!\diagdown_{OEt}\!\!{}_2 CuLi \xrightarrow{H_3O^+} Bu\diagdown\!\!\diagup\!\!=\!\!\diagdown\!\!\diagup\!\!\diagdown_{O} \quad 75\%$$

or only once, with a disubstituted aryl thio acetylene

$$Bu_2CuLi \xrightarrow[\substack{2/\ 1\ MeC\equiv C-SPh \\ 3/\ H_2O}]{1/\ 2\ HC\equiv CH} Bu\diagup\!\!=\!\!\diagdown\!\!\diagup\!\!=\!\!\diagdown_{Me\ SPh} \quad 98\%$$

The opposite regio selectivity observed, according to the hetero atom positionned on the starting acetylene, leads to the formation of a vinylic carbenoïd species (case of sulfur) or an E-β alkoxy vinyl cuprate (in the case of oxygen), which are respectively equivalents of acyl anions, or enolates of ethylenic ketones

$$R\diagup\!\!=\!\!\diagdown\!\!{}^M_{SEt} \sim R\diagup\!\!=\!\!\diagdown\!\!\diagdown\!\!{}^\ominus_O \quad \text{and} \quad R\diagup\!\!=\!\!\diagdown\!\!{}^M_{OEt} \sim R\diagup\!\!=\!\!\diagdown\!\!\diagdown\!\!{}^\ominus_O$$

This double carbocupration represents a more challenging process, in the synthesis of conjugated dienes, since one ends ups with a dienyl copper reagent, which may be further used for chain elongation (reminding "living polymers"). However, the reaction should be applicable to non functionnalized alkynes, in order to be of general interest.

We indeed observed, that under particular conditions, the following bis carbocupration

$$R_2CuLi \xrightarrow{4\ HC\equiv CH} R\diagup\!\!=\!\!\diagdown\!\!\diagup\!\!=\!\!\diagdown{}_2 CuLi$$

proceeded to some limited extent. We were thus led to a systematic study of the addition of vinyl cuprates to acetylene, according to their structural features, namely the degree of substitution of the sp$_2$ carbon atoms

$$\underset{R^2}{\overset{R^1}{\diagdown}}\!\!=\!\!\underset{{}_2CuLi}{\overset{R^3}{\diagup}} + HC\equiv CH \xrightarrow{?} \underset{R^2}{\overset{R^1}{\diagdown}}\!\!=\!\!\overset{R^3}{\diagup}\!\!\diagdown\!\!=\!\!\diagup{}_2CuLi$$

Since the temperature of addition is now higher than that which was required in the case of di*alkyl* cuprates, the results are highly depending upon the thermal stability of the starting vinyl cuprate, and the resulting dienyl cuprate. It turns out that R^3 groups, geminated to copper are much more stabilizing than R^1 or R^2 alkyl groups, so that the reagents of type

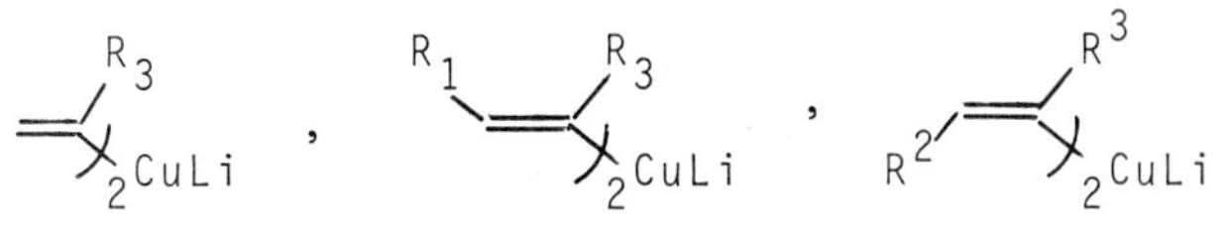

do insert smoothly two equivalents of acetylene, leading respectively to reagents of type :

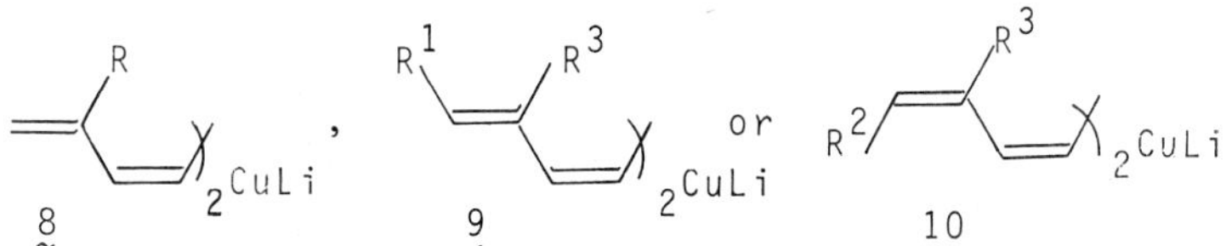

8, **9** or **10**

it is worthy of note that no further addition is observed, this fact may be tentatively assigned to an internal chelation of the copper atom by the former C=C bond.

It should be emphasized, that all these three dienylcuprates are very useful synthons for the elaboration of insect pheromones, and natural products in general. Reagent **9** presents an interesting wide versatility since groups R^1 and R^3 may be part of a cycle, and further condensation with an electrophilic reagent may lead to interesting cyclisations such as the following :

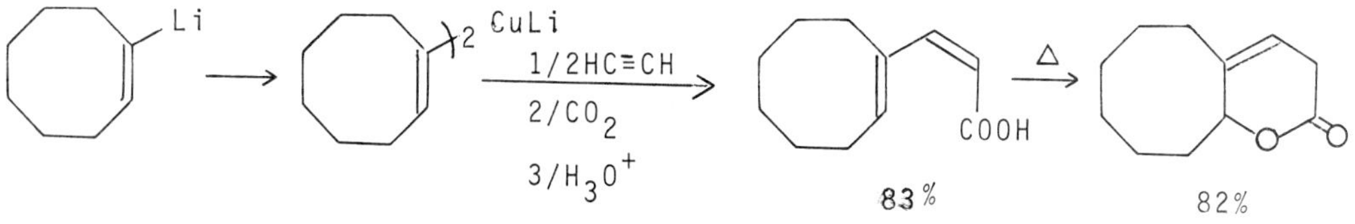

In summary, vinyl copper derivatives, derived from carbo-metallation of acetylene or alkynes are good precursors of conjugated dienes, via the palladium mediated coupling reaction with iodo alkenes. Even more suited for synthesis are the dienyl cuprates of given geometry, which may introduce directly the dienyl synthon in more sophisticated structures.

AKNOWLEDGEMENTS

The authors are particularly indebted to N. Jabri who is involved in the palladium catalyzed reactions and Mrs J. Mathe for typing the manuscript. They thank the C.N.R.S. for financial support.

REFERENCES

1. M. Schlosser, K.F. Christmann, *Liebigs Ann. Chem.* 708, 1 (1967).
2. H.J. Bestmann, O. Vostrowsky, *Chemistry and Physics of lipids* 24, 335 (1979)
3. J.F. Normant, A. Commerçon, J. Villieras, *Tetrahedron Lett.*, 1465 (1975)
4. E.I. Negishi, G. Lew, T. Yoshida, *J.C.S. Chem. Comm.* 874 (1973)
5. A.O. King, N. Okukado, E. Negishi, *J.C.S. Chem. Comm.*, 683 (1977)
6. N. Miyaura, K. Yamada, A. Suzuki, *Tetrahedron Lett.* 3437 (1979)
7. D. Holme, E.R.H. Jones, M.C. Whiting, *Chem. and Ind.* 928 (1956)
8. G. Zweifel, N.L. Polston, *J. Amer. Chem. Soc.* 92, 4068 (1970)
9. R. Rossi, A. Carpita, M. Grazia Quirici, *Tetrahedron* 37, 2616 (1981)
10. F. Näf, R. Decorzant, W.T. Thommen, B. Willhalm, G. Ohloff, *Helv. Chim. Acta.* 58, 1016 (1975)
11. P.I. Svirskaya, C.C. Leznoff, W.L. Roelofs, *Synthetic Comm.* 10, 391 (1980)
12. G. Decodts, G. Dressaire, Y. Langlois, *Synthesis* 510 (1979)
13. H.P. Dang, G. Linstrumelle, *Tetrahedron Lett.* 191 (1978)
14. J. Schwartz, *J. Organometal. Chem. Library* 1, 461 (1976)
15. G. Wilke, H. Müller, *Chem. Ber.* 89, 444 (1956)
16. E.I. Negishi, *Aspects of Mechanism and organometallic Chemistry*, J.H. Brewster, Ed., Plenum Press, New York, 285 (1978), and *Pure and Appl. Chem.* 53, 2333 (1981)
17. N. Miyaura, H. Suginome, A. Suzuki, *Tetrahedron Lett.* 22, 127 (1981)
18. S. Tsuboi, T. Masuda, H. Makino, A. Takeda, *Tetrahedron Lett.* 23, 209 (1982)
19. B.T. Golding, G. Pierpoint, R. Aneja, *J.C.S. Chem. Comm.* 1030 (1981)
20. H. Bosshardt, M. Schlosser, *Helv. Chim. Acta* 63, 2393 (1980)
21. B.M. Trost, J.M. Fortunak, *J. Amer. Chem. Soc.* 102, 2841 (1980)
22. C.A. Henrick, *Tetrahedron* 33, 1845 (1977)
23. R. Rossi, *Synthesis* 817 (1977)
24. K. Mori, *Aspects of Mechanism and organometallic Chemistry*, J.H. Brewster, Ed., Plenum Press, New York, 285 (1978), and *Pure and Appl. Chem.* 53, 2333 (1981)
25. J.F. Normant, A. Alexakis, *Synthesis* 841 (1981)
26. G.H. Posner, *Organic reactions* 22, 253 (1975)
27. M. Kumada, *Pure Appl. Chem.* 52, 669 (1980)
28. J.F. Normant, C. Chuit, G. Cahiez, J. Villieras, *Synthesis* 803 (1974)
29. K.H. Thiele, J. Köhler, *J. Organometal. Chem.* 12, 225 (1968)
30. N. Jabri, A. Alexakis, J.F. Normant, *Tetrahedron Lett.* 22, 959 (1981)
31. E.I. Negishi, N. Okukado, A.O. King, D.E. Horn, B.I. Spiegel, *J. Amer. Chem. Soc.* 100, 2254 (1978)
32. N. Jabri, A. Alexakis, J.F. Normant, *Tetrahedron Lett.* 22, 3851 (1981)
33. G.H. Posner, C.E. Whitten, J.J. Sterling, *J. Amer. Chem. Soc.* 95, 7788 (1973)
34. N. Jabri, A. Alexakis, J.F. Normant, *Tetrahedron Lett.* 23, 1589 (1982)
35. A. Alexakis, G. Cahiez, J.F. Normant, *Tetrahedron* 36, 1961 (1980)

QUINONE SYNTHESIS WITH ORGANOMETALLIC REAGENTS

M. F. Semmelhack, Tadahisa Sato, J. Bozell, L. Keller, W. Wulff, A. Zask and E. Spiess

Department of Chemistry, Princeton University, Princeton, N.J. 08544, USA

Abstract - Efforts directed toward the naphthoquinone antibiotics have focussed on transition metal-promoted reactions. Initial targets are frenolicin and nanaomycin A, closely related compounds with a substituted tetrahydropyran ring fused to a naphthoquinone unit. Two general approaches have been completed. In the first case, the simple naphthoquinone, juglone, is elaborated through nickel-promoted conjugate addition of an acyl anion and trapping of the resulting enolate with allyl iodide. Alternatively, the intramolecular cycloaddition of an alkyne unit with an arylcarbene-chromium complex (derived from anisole) leads to precisely the same intermediate. Then palladium-catalyzed intramolecular alkoxy-carbonylation of the double bond in the allyl side chain gives the pyran ring. Simple protecting group manipulations provide the natural products. The strategy involving the carbene-chromium chemistry is then applied to the more complicated target, granaticin. An elaborate analog of anisole is required for the key step, and it is prepared in a short series of stereospecific operations. The conversion to granaticin follows the same lines as the frenolicin.

INTRODUCTION

Deoxyfrenolicin (<u>1</u>)(1,2) and nanomycin (<u>2</u>) are structurally closely related naphthoquinones with moderate antibiotic and antifungal activity.(3,4) The more complex analog, granaticin (<u>3</u>)(5) shows much more potent biological activity, including inhibition of nucleic acid synthesis, (6) P-388 leukemia,(7) and some mouse tumors.(8) The common structural feature is the isochromanoquinone unit (<u>4</u>). The interesting biological activity may be responsible for substantial recent synthesis activity, including the total synthesis of frenolicin,(8) nanaomycin A,(8,9,10) and kalafugin.(10)

1: R = nPr
2: R = Me

3

4

We are pursuing the synthesis of granaticin (<u>3</u>), and have focussed our attention on the construction of the quinone and pyran rings by developing routes to deoxyfrenolicin and nanaomycin A. More recently, we have prepared the appropriately functionalized bicyclo[2.2.2] ring system important in the left-hand portion of granaticin, and have begun to follow a strategy for elaborating it to the natural product. In this paper, we describe two approaches to the simpler naphthoquinones <u>1</u> and <u>2</u>, and summarize the progress toward granaticin. A third successful approach to deoxyfrenolicin, employing the same key intermediate, was reported last year. (11)

CONJUGATE ADDITION OF NICKEL ACYLATE COMPLEXES

The key intermediate has the form <u>5</u>, from which intramolecular alkoxycarbonylation (eq. 1) can produce the proper pyran ring.(11) One new strategy is outlined in eq. 1. Naphthoquinone monoketals such as <u>6</u> are readily available and useful intermediates known to undergo conjugate addition of hetero atom and stabilized carbanion nucleophiles. (12)

However, conjugate addition of reactive carbanions has not been successful; dialkylcuprates have been observed to initiate reductive cleavage of benzoquinone monoketal rather than addition. (13) In the early stages of this work, we investigated the use of HMPA to promote 1,4-addition of more reactive nucleophiles, (14) as well as addition of cuprates. In general, the procedures were not useful, with the exception of reactions with 2-propyl-2-lithio-1,3-dithiane. In addition of this anion to 6b, only the product of 1,4-addition was obtained (7, 86%). Trapping of the intermediate enolate anion was best achieved by addition of a five-fold excess of allyl bromide. The product (8) was obtained as a single isomer (presumably trans) in 60% yield.

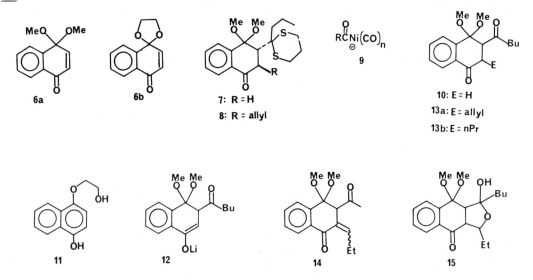

Corey and Hegedus reported that acyl-nickel carbonylate anions (9), generated from nickel tetracarbonyl and alkyllithium reagents, will react with α,β-unsaturated ketones via 1,4-addition of the acyl unit.(15) No quinone or quinone ketal examples were reported, the yields with cyclic enones were only moderate, and no attempt to trap the presumed enolate anion intermediate was reported. In spite of vigorous development of other carbonyl anion equivalents capable of conjugate addition, little further investigation of this reagent has appeared. We have examined the reaction of 9 with monoketals of naphthoquinones and find, with minor modifications of the published procedure, that 1,4-adducts are generally obtained in excellent yield. For example, the dimethylketal (6a) of naphthoquinone reacts at -50° in THF with the acyl complex from n-butyllithium and Ni(CO)$_4$ to give the adduct (10) in 91% yield. Yields are also good with n-butylmagnesium chloride (10, 75%). Reductive cleavage is not observed with 6a during the addition, but structure 11 is detected (12% yield) from 6b. The lithium enolate (12) assumed to be present in these reactions can be trapped by addition of excess allyl iodide. In this way, 13a was obtained in 85% yield from addition of the nBuLi/Ni(CO)$_4$ combination to 6a, followed by addition of a solution of allyl iodide (5 mol-equiv) in HMPA. Attempted alkylation (nPrI) of the enolate led to a mixture of products from O-alkylation (20%), C-alkylation (21%), and multiple alkylation (45%). Similarly, reaction of the enolate 12b with propionaldehyde was complicated by elimination (14, 40%) and hemi-ketal formation (15, 34%).

Deoxyfrenolin (1) and nanaomycin A (2) were prepared in seven steps from the known (16) juglone monoketal methyl ether,16 (Scheme 1 . Addition of n-propyllithium to nickel carbonyl (-50°, 0.5 hr) gave a brown solution in ether to which was added 16 at -50°. After 1.5 hr at -50°, excess allyl iodide (2 mol-equiv) and HMPA were added, and the mixture was stirred at 25° for 15 hr. The adduct 17 was isolated in 81% yield (based on 16) as a yellow solid, mp 113.5-114°. Hydrolysis of the ketal, reduction of the side chain carbonyl, and oxidation converted 17 to the hydroxyquinone 18 (79% yield overall) as an orange oil. A similar sequence starting from methyllithium produced 19 (54%) and then 20 (69%; mp 76-78°).

Reaction of 18 under alkoxy carbonylation conditions (11) involved a mixture of $PdCl_2(CH_3CN)_2$ (0.1 mol-equiv) and $CuCl_2$ (3.0 mol-equiv) in methyl alcohol under a positive pressure of CO (1.1 atm). After 3.3 hr at 25°, the crude product was obtained and chromatographed to provide a mixture of 21a and 21b (70% yield). Analytical HPLC indicated a ratio of 21a:21b=3:1. The major component (21a, trans) was obtained by crystallization as orange needles, mp 134-136°. The minor product (21b, cis) was also obtained by crystallization from the mother liquor, mp 144-148.5° and the isomers were identified by NMR spectral analysis. Treatment of the phenol ethers with BBr_3 causes demethylation to the phenol for both 21a and 21b and complete isomerization of the cis arrangement in 21b into the natural trans series, 22 (84% yield of 22). The ester 22 has been converted to (±)-deoxyfrenolicin (1) and correlated with a sample of (+)-frenolicin derived from nature. (17)

Scheme 1

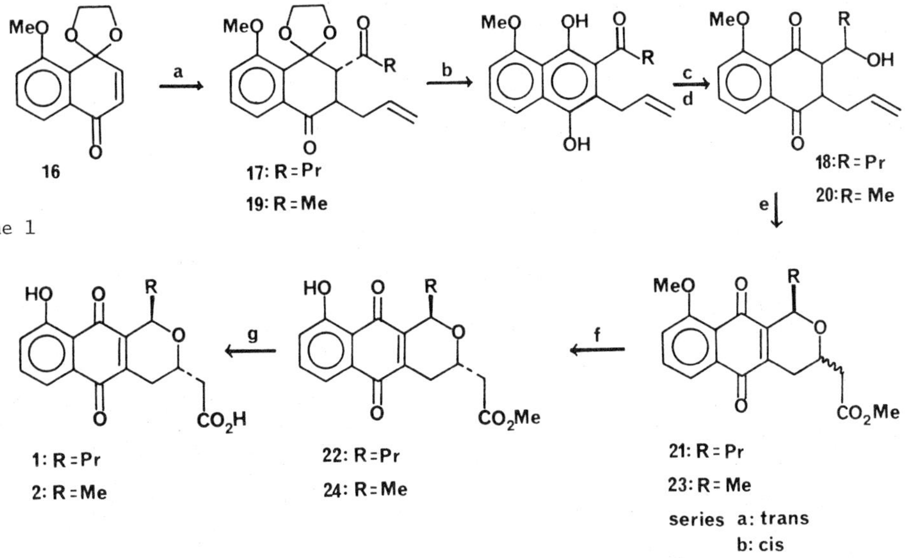

series a: trans
b: cis

a. $nPrLi/Ni(CO)_4/Et_2O$, -50°, 1.5 hr; allyl iodide/HMPA, 25°, 15 hr.
b. 2:1 dioxane: 6 N HCl, 25°, 2 days. (c) $NaBH_4$/THF. (d) DDQ/CH_3OH, 0°, 0.5 hr. (e) $PdCl_2(CH_3CN)_2$ (0.1 mol-eq)/$CuCl_2$ (3.0 mol-eq)/CH_3OH/ CO (1.1 atm), 25°, 3.3 hr. (f) from 21: 10 mol-eq of BBr_3/CH_2Cl_2, 0°, 10 min.

Reaction of 20 under the same alkoxy-carbonylation conditions (25°/6 hr) produced pyrano-ester isomers 23a and 23b in 89% yield and a ratio trans:cis = 3:2/ The major isomer (23a) was isolated by crystallization from hexane-ethyl acetate, mp 132.5-135°. The minor isomer (23b) was crystallized from the mother liquor, mp 144.5-145°; it can be be equilibrated with 23a in sulfuric acid. (8) The phenol methyl ether is cleaved with AlCl$_3$ and the methyl ester is hydrolyzed with dilute aqueous base to give nanaomysin A (2).

QUINONES FROM INTRAMOLECULAR ALKYNE-CARBENE CYCLOADDITION

The second general approach is based on the cycloaddition of an alkyne with an arylcarbene-chromium complex (18) as outlined in eq. 2. The regioselectivity of this reaction is beginning to be defined (18,19) and suggests that steric effects are dominant. The larger of the two groups R_1 and R_2 ends up nearest the free OH in the product naphthol complex, as indicated in eq. 2. The most direct application of this reaction in the synthesis of our key intermediate 18 would require the regioselectivity shown in eq. 2, counter

to the established trend. While it might be possible to change the functional groups on the alkyne substituents in 24 and thereby influence the regioselectivity in a favorable way, we have chosen instead to develop intramolecular versions of the alkyne/carbene cycloaddition reaction which will allow high predictability in the orientation of the alkyne during the cycloaddition. An important simplification in implementing this strategy is the anhydride-like reactivity of acetoxy-carbene complexes. Stoichiometric

replacement of acetate with alkoxy, amino, and thioalkoxy nucleophiles has been reported. (20). We prepared the known (21) acylate salt, 25, precipitated from water with tetra-methylammonium ion. The salt can be prepared in large quantity and stored indefinitely as the solid. It is activated for exchange by acylation at -20° in dichloromethane to produce the delicate, deep red acetate, 26. Without purification, 26 is allowed to react with a series of alkynols (27), chosen to test for ring size preferences in the cycload-dition reaction. The exchanged alkoxy-carbene complexes (represented by 28) are red oils and have been characterized by NMR spectroscopy but not highly purified. They are unstable above 25°; after 20-30 hr at 35°, complete conversion to a new complex (29) is observed. This product was particularly sensitive to donor solvents such as acetone, in which the Cr(CO)$_3$ unit moves to the less substituted ring (30) and then is detached completely. For ease of isolation, the Cr(CO)$_3$ unit was rapidly detached from 29 with triphenylphosphine in the presence of triethylamine and acetic anhydride to give the acetyl derivatives, 31. The Table displays the results of the cyclization reactions. From the reaction conditions required for complete conversion, it is clear that these intramolecular examples proceed considerably faster than related intermolecular cases (60°/10-20 hr). All three ring sizes (5,6, and 7) can form efficiently, but terminal alkynes are distinctly less effective than the disubstituted alkynes.

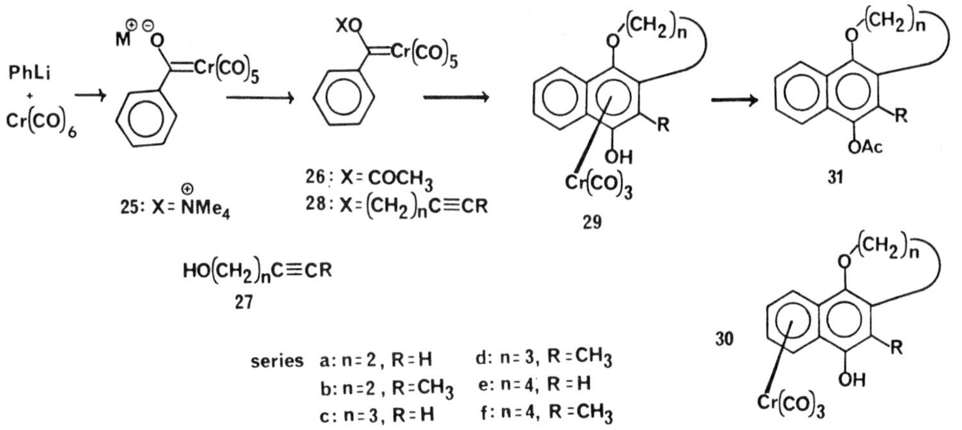

Table. Intramolecular Alkyne-Carbene Reaction

Alkyne	Reaction Conditions[a]	Naphthol Acetate, 31	Isolated Yield[b]
27a	26hr/25°	31a	16%
27b	20hr/35°	31b	81%
27c	44hr/25°	31c	18%
27d	44hr/35°	31d	62%
27e	20hr/35°	31e	38%
27f	46hr/35°	31f	62%

(a) In ether solution with complexes 28 used without purification.
(b) The yields are calculated overall from 25 as starting material and are based on chromatographically pure materials.

A specific example shows how the method is applied to simple regiochemical control. The alkyne 32 is prepared by alkylation of hex-3-yne-2-ol with ethyl bromoacetate and reduction of the ester with lithium aluminum hydride. The following sequence is carried out under argon. Complex 25 (1.1 mmol) is dissolved in dichloromethane (20 ml) and the yellow solution is cooled to -20°. Acetyl chloride (0.87 ml, 1.2 mmol) is added as a solution in dichloromethane over 5 min. The resulting deep red solution is warmed to -10° and stirred for 40 min. The alkynol 32 (1.1 mmol) is added and the solution is stirred at 25° for 1 hr. After the solvent is removed at aspirator vacuum, the residue is triturated with pentane and filtered. From the filtrate, the pentane is removed to leave a deep red oil (33, 86% yield). It is dissolved in ether and heated at 35° for 20 hr. The solvent is removed to leave an orange-yellow powder, which is dissolved in acetone (10 ml) and treated by sequential addition of triphenylphosphine (2.9 mmol), acetic anhydride (1.5 ml), and triethylamine (0.5 ml). After the solution has been stirred for 17.5 hr, the solvent is removed and the residue is purified by MPLC (silica gel; hexane followed by

to yield 34 as a colorless solid in 65% yield. Cleavage of the acetate, oxidation with DDQ, and detachment of the hydroxyethyl unit produces the disubstituted quinone, 35.

Using intramolecular reaction to control regioselectivity, we can report a short, convergent pathway for the synthesis of deoxyfrenolicin (1) as outlined in Scheme 1, starting again from o-bromoanisole. Reaction of o-lithioanisole with chromium hexacarbonyl gave a lithium salt which could be exchanged with tetramethylammonium ion and precipitated from water as 36. Formation of the acetate and immediate displacement of acetate by the primary hydroxyl group of alcohol 37 gave carbene complex 38 as a red oil (88% yield from 36). When heated at reflux in ether, complex 38 began to cyclize; after 64 hr, the volatile material was removed at reduced pressure to leave a crude product expected to be complex 39. The free ligand (40) was obtained by treatment with excess triphenylphosphine in acetone at 25°, but, more efficiently, treatment of the crude complex 39 with 2,3-dichloro-5,6-dicyanoquinone in acetonitrile caused removal of chromium and conversion to quinone 41. Chromatography of the product from oxidation provided 41 as a yellow oil in 51% yield. By simple treatment with aqueous acid, the hydroxyethyl side chain was lost and the keto-hydroquinone (42) was obtained in 95% yield. Presumably enolization produced the ortho-quinomethide 43. Internal addition of the hydroxyl group would give ketal 44, which is hydrolyzed. The quinone ketal 45 is obtained from 41 simply on standing in air for several weeks, presumably by a similar process and air oxidation. As before, hydride reduction of 42 followed by oxidation back to the quinone provides 18, the key common intermediate in our syntheses. Conversion of 18 to (±)-deoxyfrenolicin (1) was discussed above.

Scheme . Synthesis of Deoxyfrenolicin, 1.

(a) nBuLi, ether, -78°; Cr(CO), ether, 25°: $Me_4N^{\oplus} Br^{\ominus}$, H_2O. (b) AcCl, -20°, CH_2Cl_2; alcohol 37 (1.0 mol-eq), CH_2Cl_2, 25°, 6 hr. (c) 35°, 64 hr, ether, DDQ, CH_3CN, 1 hr. (d) 5 N H_2SO_4, MeOH, 6 days; (e) $NaBH_4$, THF, 25°, 24 hr; DDQ, MeOH, 0°, 1 hr.

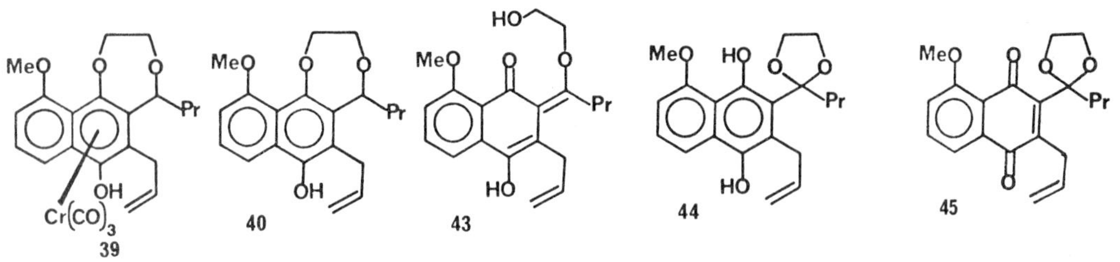

Application of this approach to granaticin (3) requires the somewhat more complicated anisole derivative, 46. We have prepared 46 from 1,4-dimethoxybenzene in a highly stereoselective manner, and have converted 46 via complex 47 into the key intermediate (48). Alkoxycarbonylation is expected to give 49 on the way to granaticin, 3.

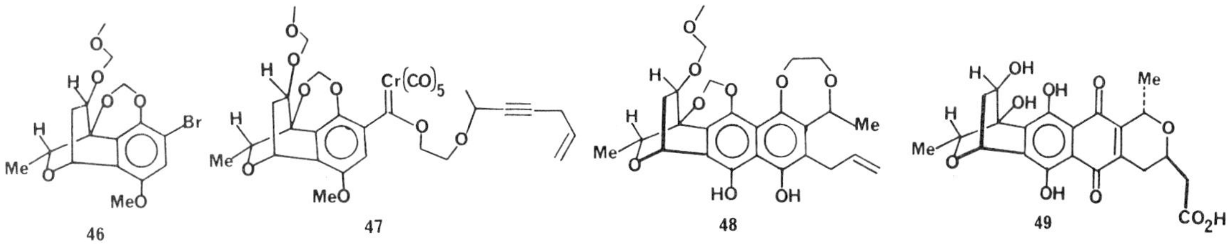

ACKNOWLEDGEMENTS

We are pleased to acknowledge generous financial support from the National Institutes of Health through research grants (AI-25926 and CA 26727) and postdoctoral fellowships (to WW for 1978-1980 and to ES for 1979-1981).

REFERENCES

1. G.A. Ellestad, M.P. Kunstmann, H.A. Whaley, and E.L. Patterson, J. Am. Chem. Soc., 90, 1325 (1968).
2. Y. Iwai, A. Kora, Y. Takahashi, T. Hayashi, J. Awaya, R. Masuma, R. Ōiwa, and S. Ōmura, J. Antiobiot. 31, 959 (1978).
3. (a) S. Ōmura, H. Tanaka, Y. Koyana, R. Ōiwa, and M. Katagiri, J. Antiobiotics, 25, 363, (1974); (b) H. Tanaka, Y. Koyama, H. Marumo, R. Ōiwa, M. Katagiri, T. Nagai, and S. Ōmura, ibid., 28, 860 (1975); and (c) H. Tanaka, Y. Koyama, T. Nagai, H. Marumo, and S. Ōmura, ibid., 78, 868 (1975).
4. Deoxyfrenolicin shows antibacterial activity in vitro and anti-fungal activity against a variety of fungi in vivo: (a) J.C. van Meter, M. Cann, and N. Bohonos, "Antibacterial Agents Annual, 1960," Plenum Press, New York, 1961; and (b) Y. Iwai, A. Kora, Y. Takahashi, T. Awaya, R. Masuma, R. Ōiwa, and S. Ōmura, J. Antibiot. 31, 959 (1978).
5. (a) V. Prelog, et al., Helv. Chem. Acta, 40, 1262 (1957); and (b) S. Barcza, M. Brufani, et al., ibid., 49, 1736 (1966).
6. (a) W. Kersten and A. Ogilvie, Z. Klin. Chem. Klin. Biochem., 13, 371 (1975); and (b) Cf. M.L. Sethi, J. Pharm. Sci., 66, 130 (1977).
7. C-J. Chang, H.G. Floss, P. Soong, and C-t Chang, J. Antibiot., 28, 156 (1975).
8. (a) A. Ichihara, M. Ubukata, H. Oikawa, K. Murakami, and S. Sakamura, Tetrahedron Lett. 4463 (1980); and (b) Y. Naruta, H. Uno, and K. Maruyama, Chem. Lett., 609-612 (1982).
9. T. Kometani, Y. Takeuchi, and E. Yoshii, J. Chem. Soc. Perkin I, 1197 (1981).
10. T. Li and R.H. Ellison, J. Am. Chem. Soc., 100, 6263 (1978).
11. (a) M.F. Semmelhack, Pure Appl. Chem., 53, 2379 (1981); (b) A. Zask, PhD Thesis, Princeton University, 1982; and (d) M.F. Semmelhack and A. Zask, submitted for publication.
12. For examples with benzoquinone monoketal and leading references, see: (a) L.H. Foster and D.A. Payne, J. Am. Chem. Soc., 100, 2384-2344 (1978); and (b) K.A. Parker and Suck-ku Kang, J. Org. Chem., 45, 1218-1224 (1980).
13. For example with lithiodemethylcuprate, see: A. Nilsson and A. Ronlan, Tetrahedron Lett., 1107-1111 (1975).
14. For examples of HMPA promoting 1,4-addition of dithioacetal anions, see: (a) C.A. Brown and A. Yamaichi, Chem. Comm., 100-102 (1979); and (b) J. Lucchetti, W. Dumont, and A. Krief, Tetrahedron Lett., 2695-2699 (1979).
15. E.J. Corey and L.S. Hegedus, J. Am. Chem. Soc., 91, 4926-4927 (1969).
16. D.M.S. Wheeler, and D.J. Crouse, J. Org. Chem., 46, 1814 (1981).
17. We are grateful to Dr. D.B. Borders, Lederle Division of American Cyanamid Co., for a generous sample of (-)-frenolicin which we converted to (+)-deoxyfrenolicin(1) following the earlier report.[1]
18. (a) K.H. Dötz, Angew. Chem. Int. Ed., 14, 644-645 (1975); (b) K.H. Dötz and B. Fügen-Köster, Chem. Ber., 113, 1449-1457 (1980); and (c) K.H. Dötz and I. Pruskil, J. Organomet. Chem., 209, C4-C6 and references therein (1981).
19. W.D. Wulff, C. Tang, and J.S. McCallum, J. Am. Chem. Soc., 103, 7677-7678 (1981).
20. (a) J.A. Connor and E.M. Jones, J. Chem. Soc. A, 3368-3372 (1971); and (b) J.A. Connor and E.M. Jones, J. Chem. Soc. Chem. Comm., 570-571 (1971).
21. E.O. Fischer and A. Maasböl, Chem. Ber., 100, 2445-2456 (1967).

ORGANO-SILICON MEDIATED SYNTHETIC REACTIONS

Isao Kuwajima

Department of Chemistry, Tokyo Institute of Technology, Meguro-ku, Tokyo 152, Japan

Abstract. 1. 1,2-Bis(trimethylsiloxy)-1-cyclobutene has proved to be an exceptionally useful reagent for cyclopentane annulation and formation of 1,4-dicarbonyl compounds. Stereochemical features of its addition reaction to cyclohexanone derivatives and the spiroannulation reaction of the adducts have been established in several cases. By using this reagent, a short synthesis of showdomycin has also been described. 2. Stereo- and regio-selective synthesis of silyl enol ethers has been achieved through butyllithium induced rearrangement of 1-trimethylsilylallylic alcohols. Selective conversion of 1-trimethylsilylpropargyl alcohols to siloxyallenes has also been effected under the similar reaction conditions.

1. Cyclopentane Annulation and 1,4-Dicarbonyl Compounds.

Silyl enol ethers behave as excellent nucleophiles toward various types of electrophiles (ref. 1). Among them, 1,2-bis(trimethylsiloxy)-1-cyclobutene $\underline{1}$, easily available from a succinic ester, is a unique reagent to offer a useful cyclobutanone source. It reacts with aldehydes or acetals to give the corresponding adducts $\underline{2}$ or $\underline{3}$ in high yields. Several useful transformations from these adducts $\underline{3}$ have been achieved with high efficacies.

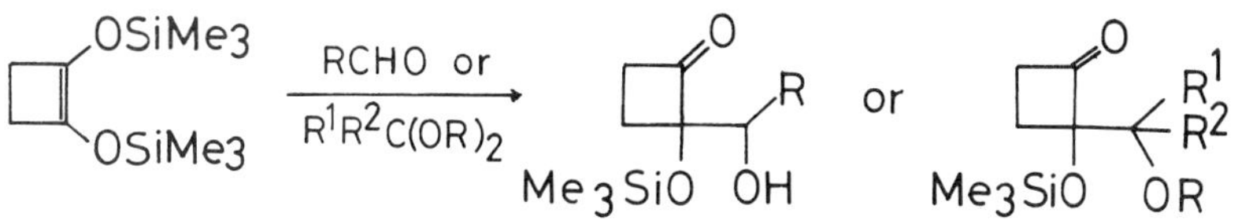

On treatment with trifluoroacetic acid (TFA), the adducts $\underline{3}$ undergo rearrangement to yield 2-mono- or 2,2-disubstituted cyclopentane-1,3-diones ($\underline{4}$ or $\underline{5}$) (ref. 2), useful intermediates for construction of steroid CD ring or trichothecane skeleton. Selective migration of acyl group should be remarkable and its regio-isomer $\underline{6}$ has never been formed. This reaction appears to be effected mainly by the ring strain of cyclobutanone; it has been reported by other group that such ring enlargement does not occur with similar cyclopentanone adducts (ref. 3).

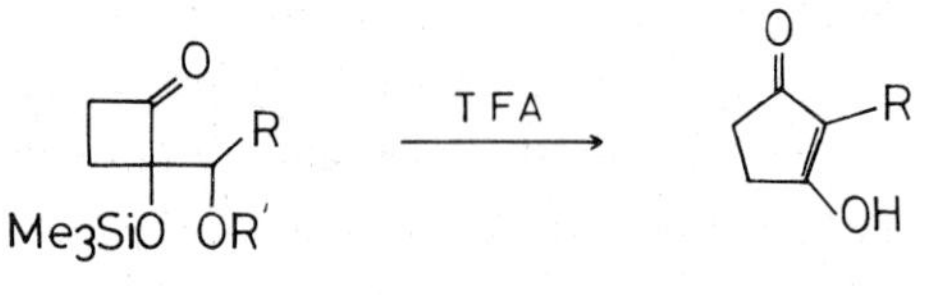

(Eq. 2)

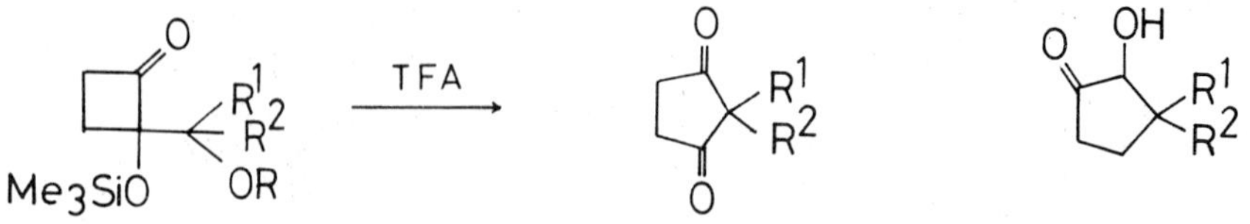

(Eq. 3)

Under the influence of stannic chloride, the adducts 3 can be transformed selectively to the γ-keto esters 7. One-pot preparation of 7 has also been achieved more conveniently: treatment of a ketal with cyclobutene 1 in the presence of stannic chloride (0.5 eq) leads directly to the formation of the same product 7 through an initial aldol type reaction followed by ring opening of the adduct (Reductive Succinoylation) (ref. 4). In contrary to the formation of 4, a range of application has been limited to the adducts bearing tert-alkyl ether moieties presumably due to stability of sec-alkyl ether linkages under such conditions. If desired, the product can be isolated as its silyl enol ether 8 through non-aqueous work-up of the reaction mixture. Several examples of synthetic applications are shown in the following equations.

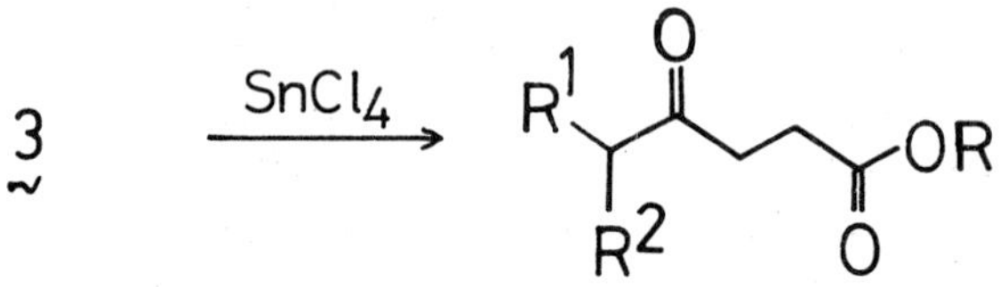

(Eq. 4)

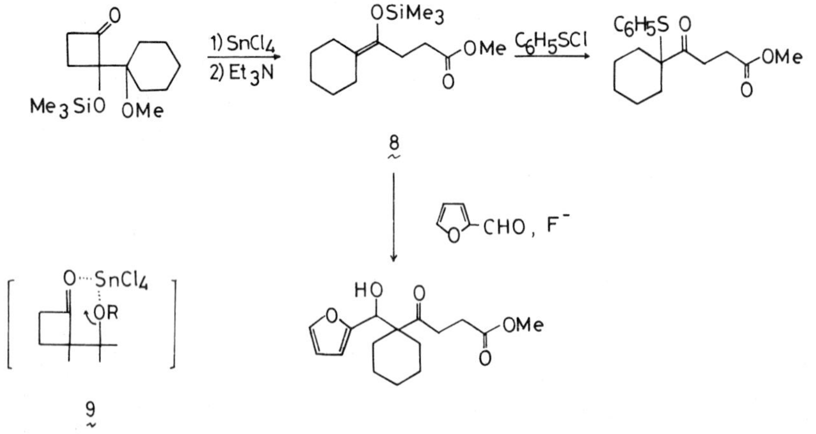

(Eq. 5)

On the basis of a crossover experiment, a complex 9 has been suggested to break down directly to silyl enol ether. The completely different effect of proton and Lewis acid on cyclobutanone 3 is remarkable, and probably indicates that 3 behaves as bidentate ligand of the Lewis acid as in 9.

Steric congestion around carbonyl prevents a general introduction of various nucleophiles on the adducts 3, but methylmagnesium iodide or hydride (from LAH) adds to their carbonyl to afford mixtures of cis- and trans-cyclobutanediol derivatives 10 or 11. Lead tetraacetate oxidation of 10 constitutes a good methodology for 5-alkoxy 1,4-diketones 12 (ref. 5). The procedure appears to be applicable to a general synthesis of alkoxy 1,n-diketones although this has not been attested.

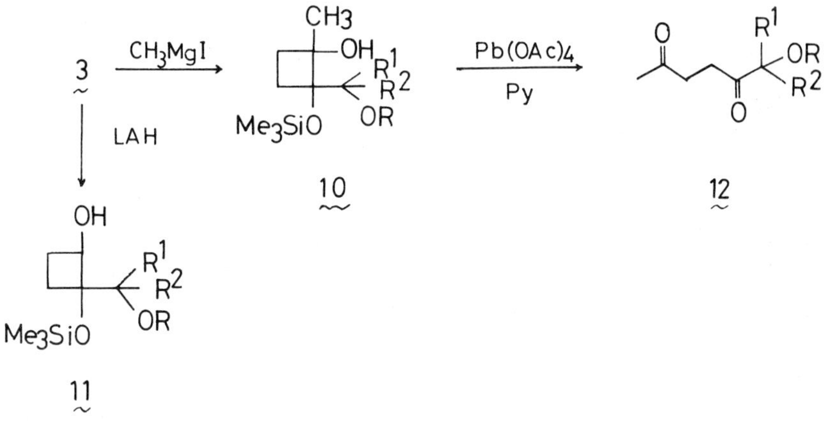

(Eq. 6)

Cyclobutanol derivatives 11 also undergo ring enlargement in the presence of TFA or Lewis acid. Stannic chloride induces such type of reaction most effectively to give the β-hydroxycyclopentanones 13 in good yields (ref. 6). It is reasonable to assume a similar mechanism with reductive succinoylation, which suggests an initial ring opening to yield silyl enol ethers including aldehyde group. Internal aldol type reaction may afford 13. However, formation of α-hydroxycyclopentanones 14 as minor products (3-9%) may suggest a partial participation of direct rearrangement.

When the alcohol 10 is used, the ring opening 1,4-diketone is also obtained together with the β-hydroxycyclopentanone. This may be attributable to lower reactivity of ketone carbonyl to electrophilic attack of enol ethers.

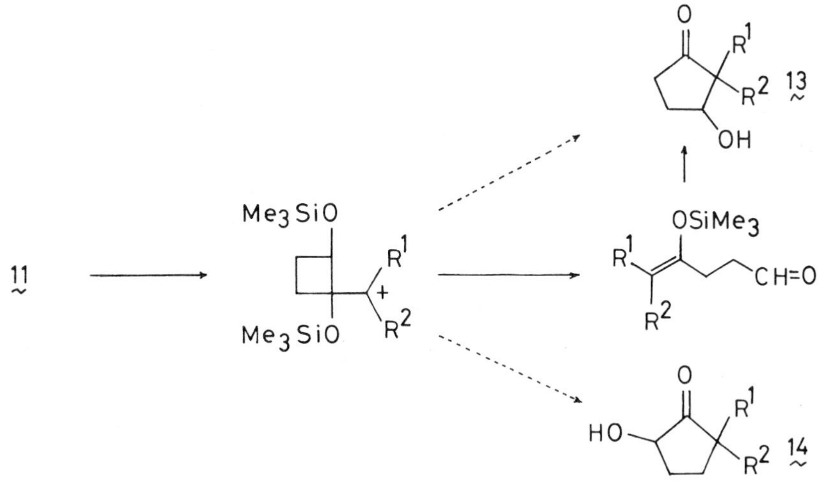

(Eq. 7)

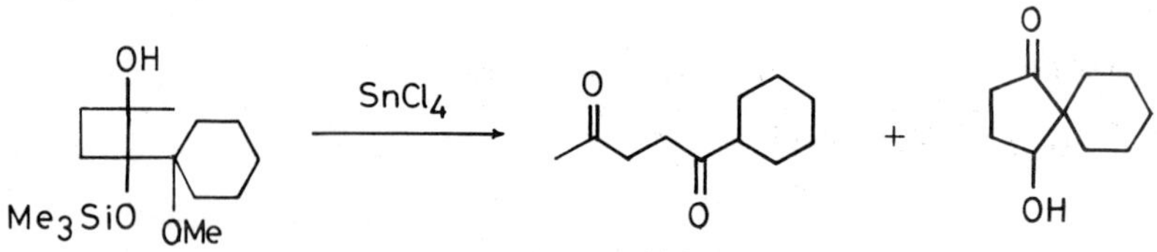

(Eq. 8)

It is worthy to note that, under similar reaction conditions, use of 3, 10, or 11 enables to generate the silyl enol ether bearing an ester, a ketone, or an aldehyde group, respectively, on its 4-position. The final product has been determined in each case by nature of the included carbonyl group, which is parallel to the known reactivity to electrophilic attack of silyl enol ethers (ref. 7).

It has hitherto been discussed on the directing effects of carbonyl and alkoxy groups in the ring opening or rearrangement reaction of cyclobutane rings. In addition to these groups, a large participation of a carbon-carbon double bond is also anticipated in this type of reactions.

Wittig olefination ($Ph_3PCH_3^+Br^-$ and KH in THF) of 3 affords the methylenecyclobutane 15 in high yield, irrespective of steric congestion around ketone function. Methylenecyclobutanes 15 have exhibited a similar tendency to rearrangement under the influence of TFA. Although formation of a methylenecyclopentanone has been detected in an initial stage of the reaction, 2-substituted 2-cyclopentenone 16 is usually obtained as a final product. A remarkable acceleration observed may reflect a greater participation of a C=C bond (compare the half-life period with that of 3). Methylenecyclobutanes derived from aromatic or α,β-unsaturated acetals usually give 16 in good yields, whereas those from saturated aliphatic acetals do not. Lability of silyl tert-allylic ether moieties may be responsible to the failure of the latter cases.

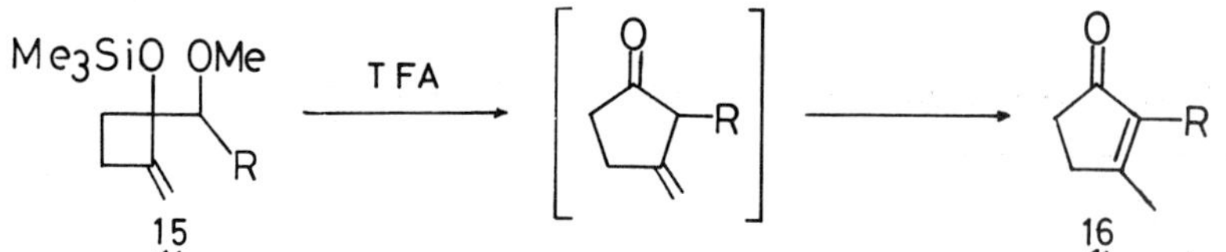

R	% yield	$t_{1/2}$ (min)	$t_{1/2}^*$ at 35°C
C_6H_5-	81	2.5	2 h
$p-ClC_6H_4-$	83	7	
$p-MeC_6H_4-$	90	<1	
~~	80	<2	
C_6H_5~~	97	<1	

(Eq. 9)

* $t_{1/2}$ of the rearrangement of Me₃SiO OMe cyclobutanone-R

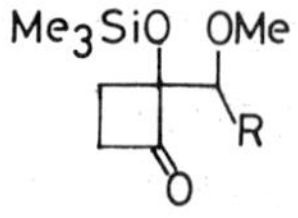

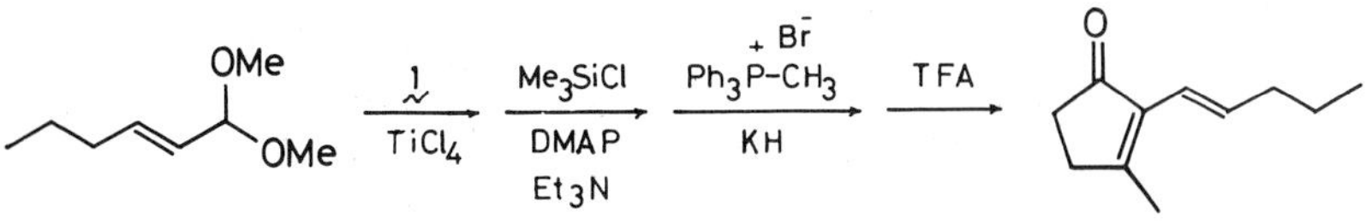

(Eq. 10)

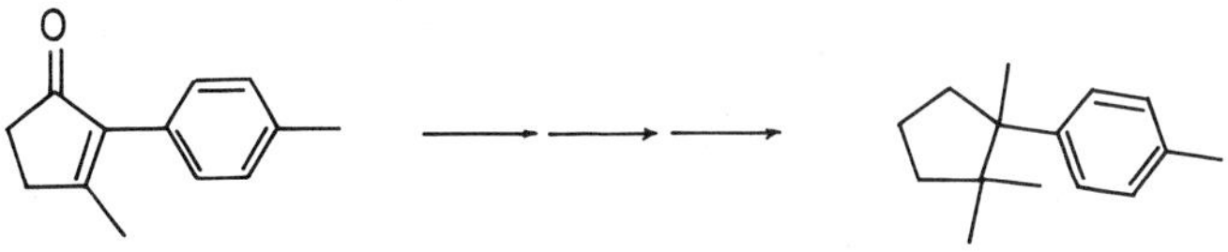

(Eq. 11)

Several difficulties are usually encountered on arylation or vinylation of ketone enolates. This method may be employed as a good alternative as shown in the synthesis of cuparene.

Presumably because of their preferable cleavage of tert-alkyl ethers, methylene- or ethylidene-cyclobutanes bearing such moieties rearrange smoothly to the corresponding cyclopentanone derivatives 17. In the latter case, the stereochemical integrity of C=C bond has proved not to be disturbed during the rearrangement process. For example, the ethylidene substrate 18 affords 19; stereochemical outcome has been established by comparison with its isomer prepared as shown below. By way of this rearrangement, the ethylidenecyclobutane 20 can be effectively transformed to the bicyclic product 21, a useful intermediate for steroid synthesis.

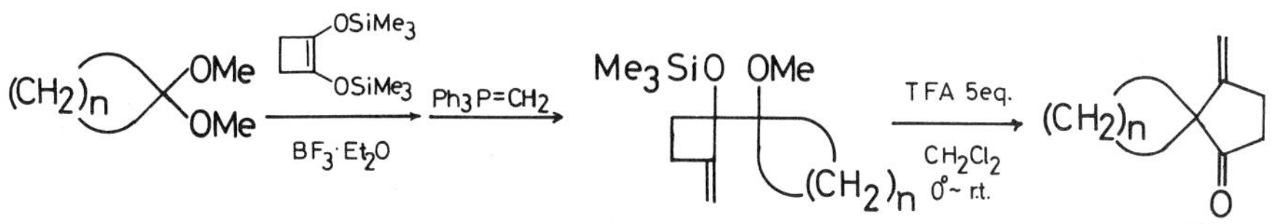

Ketal	Aldol,% yield	Wittig,% yield		Rearr.,% yield
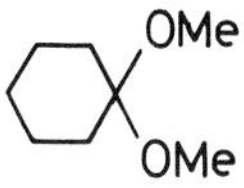 (cyclohexane ketal) OMe/OMe	90	92[a]	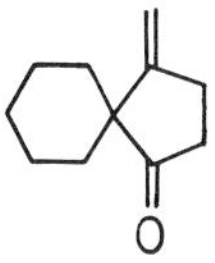	74
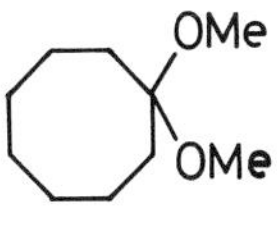 (cyclooctane ketal) OMe/OMe	70	90[b]	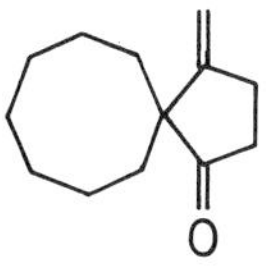	82
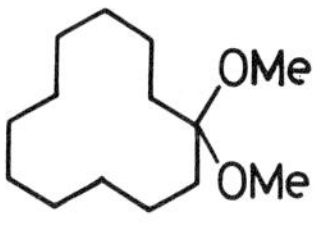 (cyclododecane ketal) OMe/OMe	92	67[c]	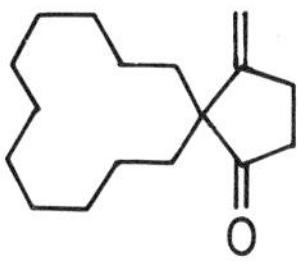	72

(Eq. 12)

a: $(Ph_3PCH_3)^+ Br^-$/KH/r.t., 3h.
b: $(Ph_3PCH_3)^+ Br^-$/t-AmOK/toluene, refl. 3h.
c: $(Ph_3PCH_3)^+ BF_4^-$/KH/r.t., 3h.

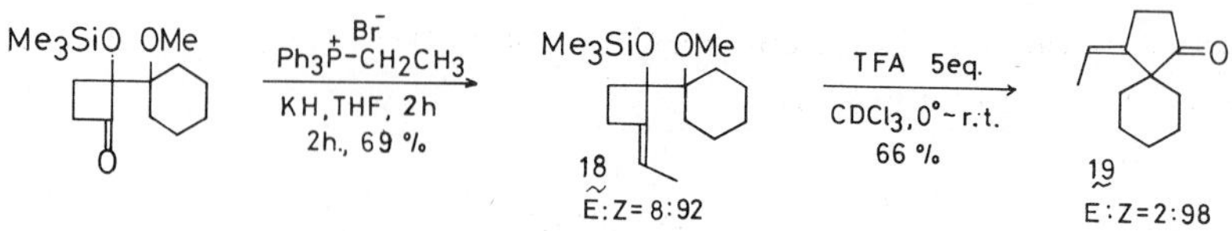

(Eq. 13)

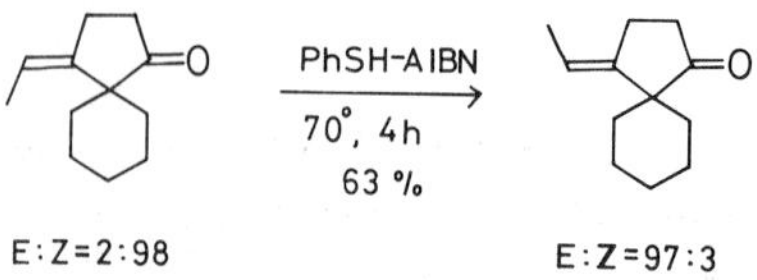

(Eq. 14)

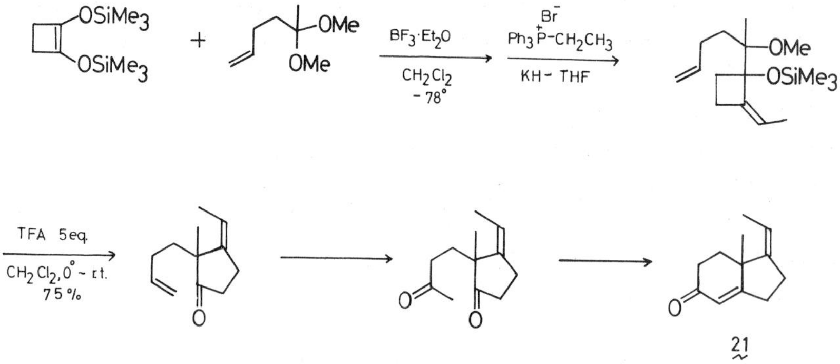

(Eq. 15)

In addition, three sequences of the spiroannulation procedure have revealed remarkable features on stereochemical aspects. Both the initial aldol reaction and the rearrangement proceed highly stereoselectively. The results of 4-t-butylcyclohexanone dimethylketal are shown in the following scheme. Exclusive equatorial addition in the first step and complete conversion of the last one should be remarkable.

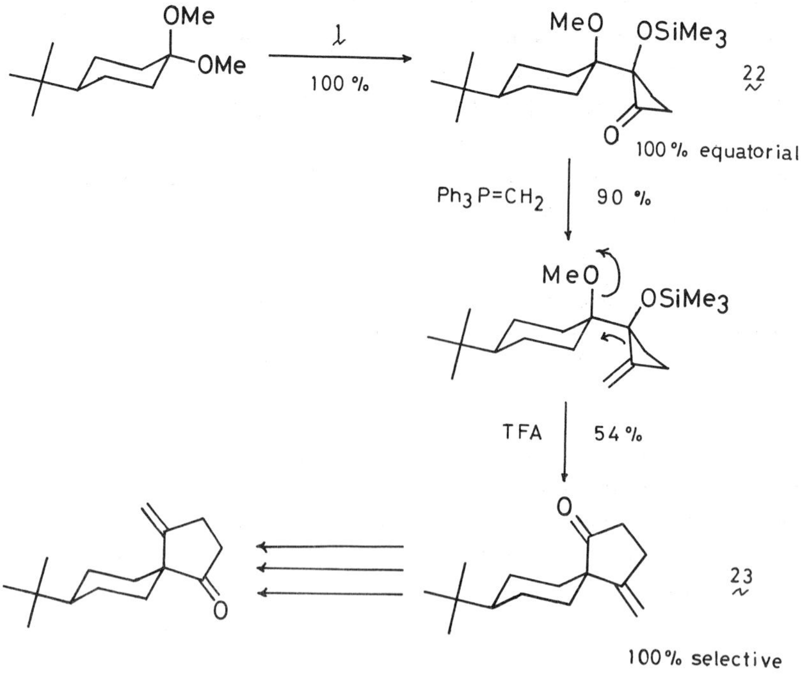

(Eq. 16)

The coupling of the dibenzylketal with **1** followed by oxidative ring cleavage of cyclobutanone and removal of benzyl group gives selectively the axial alcohol **24**, whose stereochemistry has been confirmed by comparison with a mixture of alcohols prepared by reductive succinoylation and MCPBA oxidation.

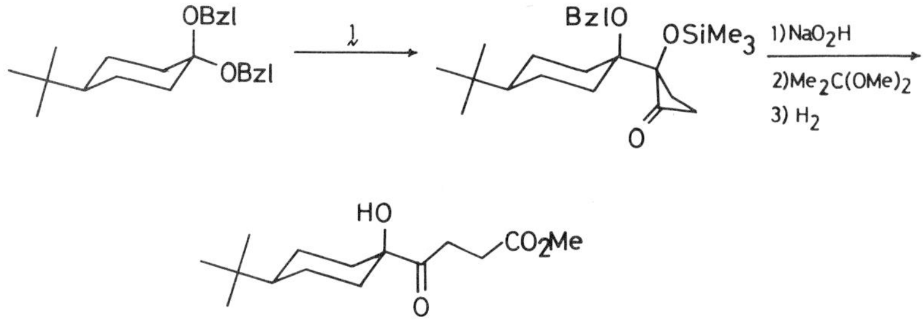

(Eq. 17)

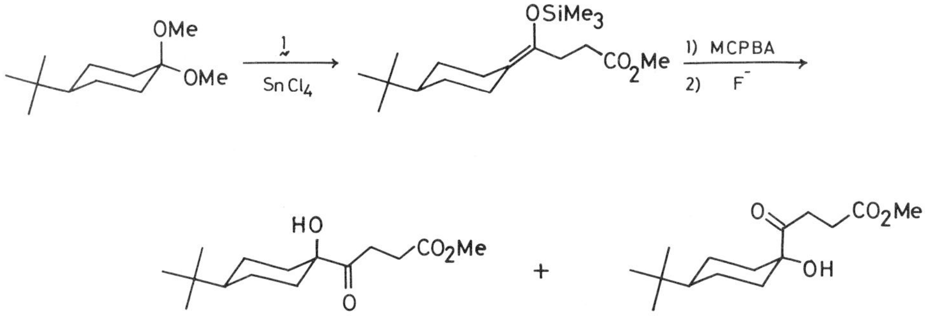

(Eq. 18)

The aldol reaction of 3-methylcyclohexanone ketal is again highly stereoselective irrespective of an expected flexibility of its conformation. Only two diastereomers have been detected and afford a single keto acid **27** on oxidative treatment. The subsequent two procedures gives the spiro ketone **26** as a single stereoisomer.

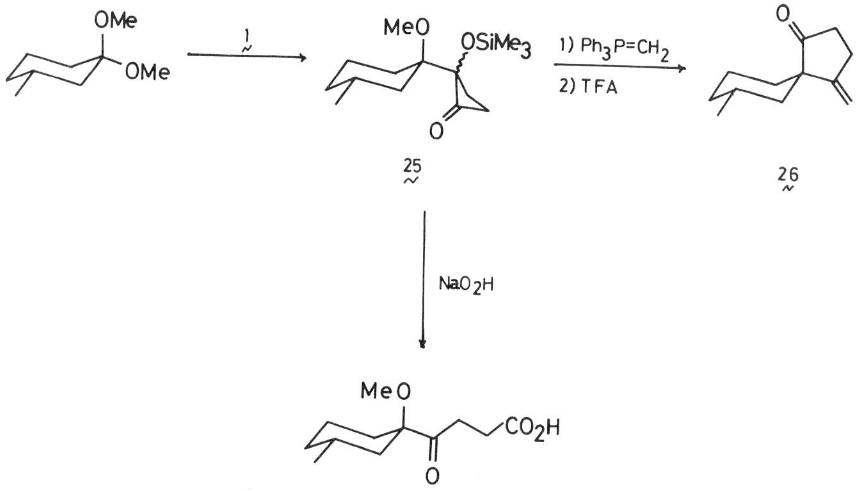

(Eq. 19)

The reaction of 2-methylcyclohexanone ketal has disclosed more of the mechanistic information about the rearrangement, since the initial aldol reaction gives a mixture of stereoisomers, 28 and 29. The stereochemical integrity of each isomer has been confirmed by removing the chiral center irrelevant to the axial-equatorial problem via oxidative cleavage. The remaining two reactions performed on 28 and 29 afford the spiro ketones 30 and 31, The absence of crossover has ascertained the full inversion at the migration turminus during the rearrangement.

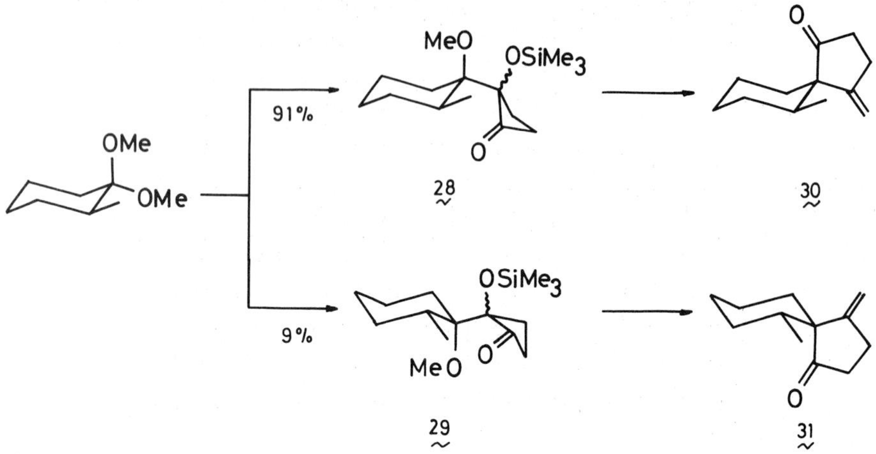

(Eq. 20)

Much attention has been focused on synthetic approaches to nucleosides having carbon-glycosyl linkages because of their important antibiotic properties and potent anticancer and antiviral activities. Some of them, e.g. showdomycin and pyrazomycin, possess four carbon units on C-1 position of ribose. The cyclobutene 1 has turned out to be an efficient four carbon source for the synthesis of showdomycin. Synthetic scheme is shown below. The initial condensation of 2,3,5-tri-O-benzoyl-D-ribofuranosyl acetate 32 with 1 proceeds quite efficiently in the presence of stannic chloride to give the desired β-anomer 33 in high yield. Synthesis of showdomycin 34 has been carried out through the following 3 steps in 70% overall yield;(i) conversion of 33 into its silyl enol ether with lithium hexamethyldisilazide and trimethylchlorosilane followed by in situ oximination and ring cleavage with nitrosyl chloride, (ii) removal of benzoyl group with methanolic ammonia, and

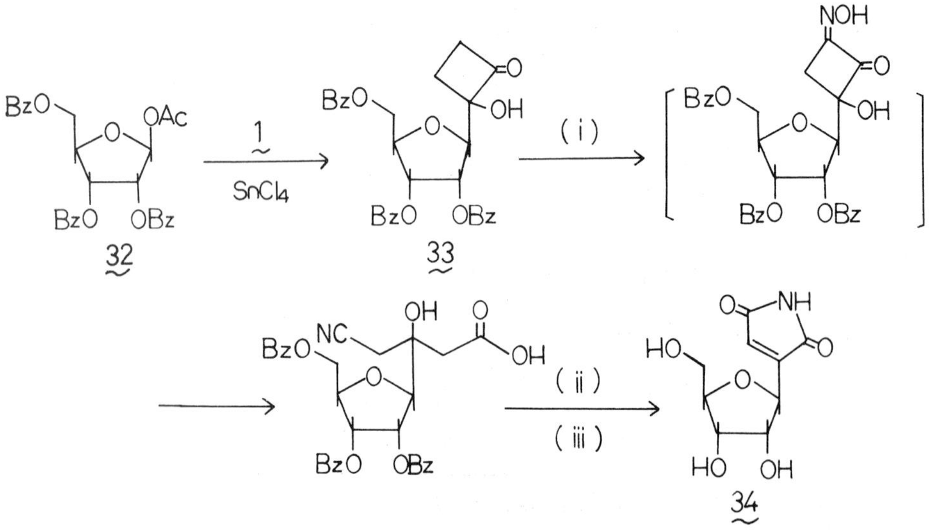

(Eq. 21)

(iii) cyclization and dehydration with trifluoroacetic anhydride (ref. 8).

In contrast with selective condensation of 1 on C-1 site, use of other acyclic silyl enol ethers is quite problematic due to an ambident electrophilic character of 32; the silyl enol ether has a remarkable influence to decide the reaction site. As shown in the Table, regiochemistry of the reaction is definitly determined and no regio-isomer has been detected in every case. Enol ethers of usual acyclic ketones 35a-b and of esters 35c-d give the products arising from attack on C-2 benzoxyl, whereas the reaction takes place selectively on C-1 site with enols of ketones 35e-g having α-hetero substituents to afford the desired 37e-g (ref. 9).

These regiochemical aspects may be in part attributable to the difference of actual nucleophilic species. ^{1}H NMR studies have revealed that 35a reacts with stannic chloride to yield the trichlorostannyl ketone, but 35e is quite stable under such conditions. These observations have suggested that α-hetero substituent may prevent silyl/stannyl group exchange and 35e-g react with well precedented cationic intermediate in a similar manner with olefinic or aromatic substrates, whereas, with regard to 35a-d, actual nucleophiles seem to be α-stannyl carbonyl compounds to react on C-2 benzoxyl group selectively.

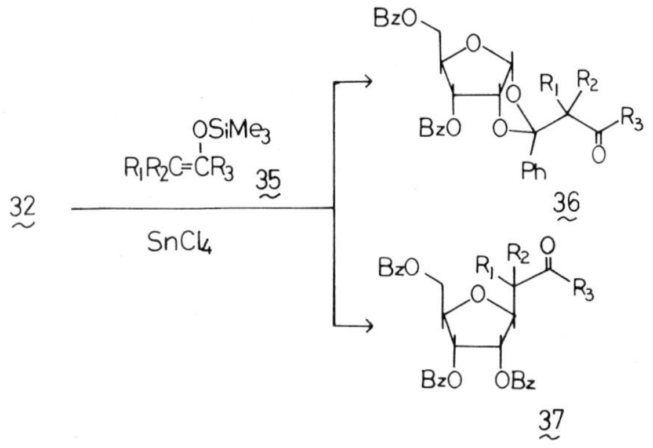

$$32 \xrightarrow[SnCl_4]{\underset{R_1R_2C=CR_3}{OSiMe_3} \; 35} 36 \; / \; 37 \quad \text{(Eq. 22)}$$

35	R_1	R_2	R_3	36(%)	37(%)
a)	CH_3	H	C_2H_5	88	--
b)	H	H	$C(CH_3)_3$	55	--
c)	H	H	OC_2H_5	32	--
d)	SC_6H_5	H	OC_2H_5	82	--
e)	SC_6H_5	H	CH_3	--	83
f)	SC_6H_5	H	$C(CH_3)_3$	--	45
g)	$OSiMe_3$	$-(CH_2)_2-$		--	92

Since phenylthio group is easily removed by reduction, the use of such silyl enol ether as 35e is generally feasible for introduction of 2-ketoalkyl group on C-1 site of the ribose derivative.

2. Stereo- and Regio-selective Synthesis of Silyl Enol Ethers.

Various procedures have recently been developed for the preparation of silyl enol ethers. In most of them, silylation of ketones has been achieved with moderate or sufficient regioselectivity by generating the corresponding enolates. Further appropriate choice of reagents or reaction conditions has sometimes enabled their stereoselective preparation. However, two types of methylene groups neighboring to carbonyl group can not be differentiated satisfactorily by the procedures reported so far.

Use of 1-trimethylsilylallylic alcohols 38 appears to bring a solution to this problem. The alcohols 38 have easily been prepared by treating acyl-trimethylsilanes (ref. 10) with vinylmagnesium bromide. On treating with an equimolar amount of butyllithium, the resulting alkoxides 39 exist in equilibrium with allyllithium (homoenolates) 40 via rearrangement of silyl group. Although this equilibrium appears to be greatly favored to the alkoxides 39, addition of an alkyl iodide shifts it to the right side through selective alkylation of the homoenolates 40. Generation of lithium alkoxides 39 in hexane followed by addition of THF and alkyl iodide to the resulting solution gives the corresponding (Z)-silyl enol ethers 41 in good yields (ref. 11).

$$\text{R-CO-SiMe}_3 \xrightarrow{\text{MgBr}} \text{R-C(OH)(SiMe}_3\text{)-CH=CH}_2 \quad \textbf{38} \tag{Eq. 23}$$

$$\textbf{38} \xrightarrow{\text{BuLi}} \left[\text{R-C(OLi)(SiMe}_3\text{)-CH=CH}_2 \; \textbf{39} \rightleftharpoons \text{Me}_3\text{SiO-Li, R-CH=CH-CH}_2 \; \textbf{40} \right] \xrightarrow{\text{R'-Hal.}} \text{R-C(OSiMe}_3\text{)=CH-CH}_2\text{-R'} \quad \textbf{41} \tag{Eq. 24}$$

Further, treatment of the alcohol 38 with a catalytic amount (ca.5 mol%) of butyllithium induces a facile conversion to the corresponding silyl enol ether 42 with high stereoselectivity (ref. 12). The reaction totally proceeds through a proton abstraction of the homoenolate 40 from the starting alcohol 38.

The reaction rate as well as the stereoselectivity has been influenced by substituents in the following two ways. The rearrangement usually completes within 30 min at -20°C with high selectivity (Z:E = ca.95:5).

$$\underset{\textbf{38}}{\text{R}^1\text{C(OH)(R}^2\text{SiMe}_3\text{)-CH=CH-R}^3} \xrightarrow{\text{cat.BuLi}} \left[\underset{\textbf{ }}{\text{R}^1\text{C(OLi)(R}^2\text{SiMe}_3\text{)-CH=CH-R}^3} \rightleftharpoons \underset{\textbf{40}}{\text{Me}_3\text{SiO-Li, R}^1\text{(R}^2\text{)CH-CH=CH-R}^3} \right] \xrightarrow{\textbf{38}}$$

$$\underset{\textbf{42}}{\text{Me}_3\text{SiO-C(R}^1\text{)(R}^2\text{)-CH=CH-R}^3} \;+\; \text{R}^1\text{C(OLi)(R}^2\text{SiMe}_3\text{)-CH=CH-R}^3 \tag{Eq. 25}$$

Substituents at β'-position facilitate the reaction to enhance the selectivity (Z:E = ca.99:1), whereas those on γ-position retard it and sometimes lower the selectivity (ca.90:10--95:5).

Stereo-control of tetra-substituted silyl enol ethers is also effected by the present procedure. A catalytic amount of butyllithium induces the rearrangement of the alcohol 43 or 44 to the corresponding silyl enol ether 45 or 46 in good yield with at least 98% stereoselectivity in each case. Requisite alcohols 43 and 44 have been prepared by the modified method of Cohen (ref. 13). It is an interesting problem to controll the stereochemistry of quaternary carbon atoms via aldol reaction. These stereochemically pure enol ethers may be useful for such purpose (ref. 14).

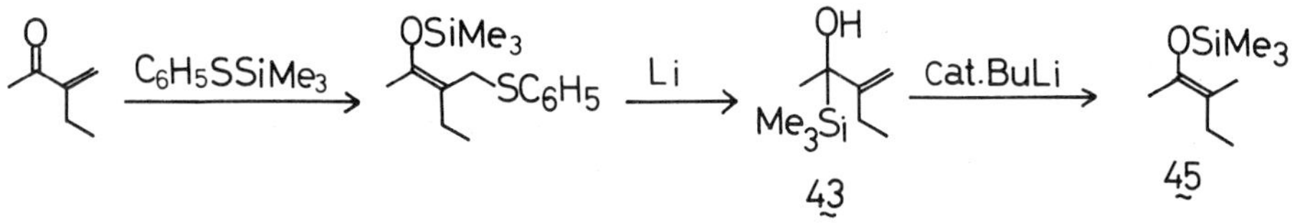

(Eq. 26)

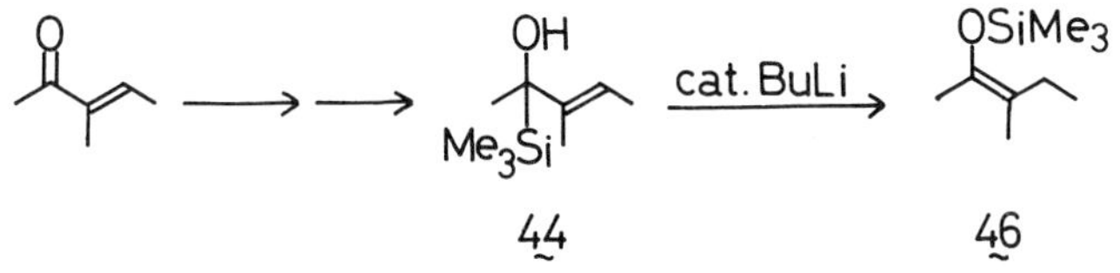

(Eq. 27)

Application of these procedures to 1-trimethylsilylpropargyl alcohols 47 affords 1-trimethylsiloxyallenes 48 or 49 as shown below (ref. 15).

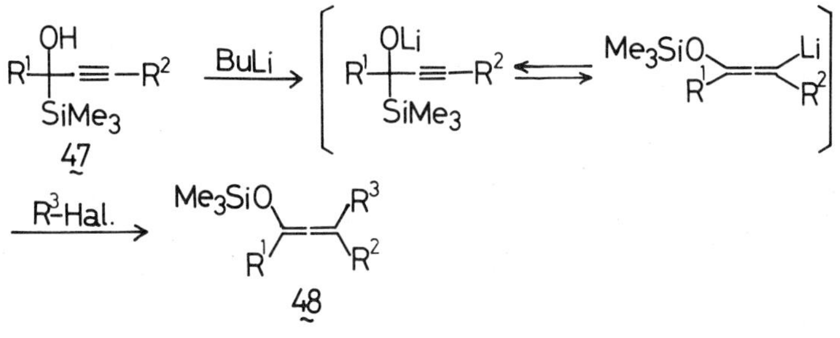

(Eq. 28)

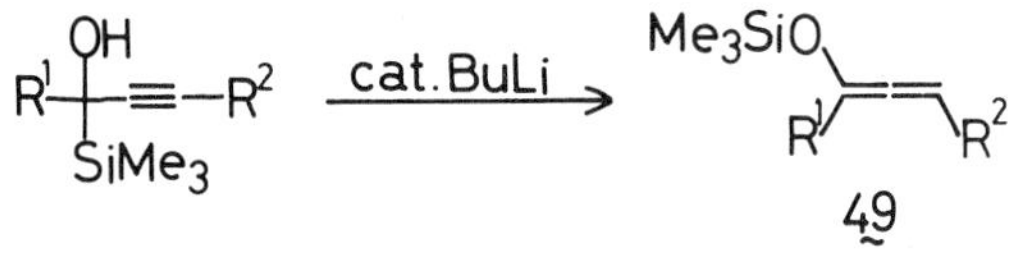

(Eq. 29)

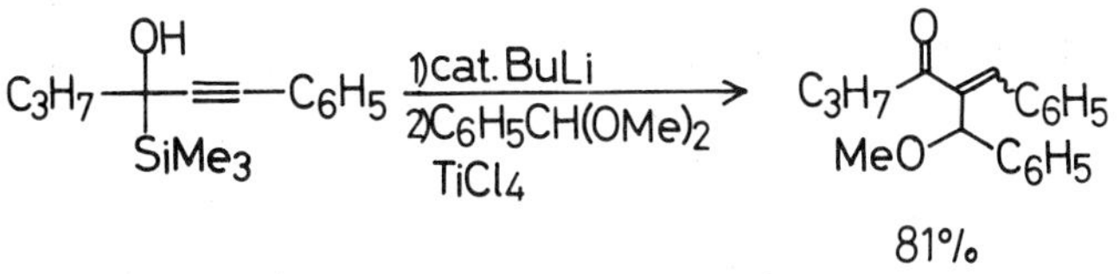

(Eq. 30)

References

1. E. W. Colvin, Silicon in Organic Synthesis, p.198, Butterworths, London (1980).
2. E. Nakamura and I. Kuwajima, J. Am. Chem. Soc., **99**, 961-963 (1977).
3. G. Pattenden and S. Teague, Tetrahedron Lett., **23**, 1403-1404 (1982).
4. E. Nakamura, K. Hashimoto, and I. Kuwajima, J. Org. Chem., **42**, 4166-4167 (1977).
5. I. Kuwajima, I. Azegami, and E. Nakamura, Chem. Lett., 1431-1434 (1978).
6. I. Kuwajima and I. Azegami, Tetrahedron Lett., 2369-2372 (1979).
7. T. Mukaiyama, Angew. Chem. Int. Ed. Engl., **16**, 817-826 (1977).
8. T. Inoue and I. Kuwajima, J. Chem. Soc., Chem. Commun., 251-253 (1980).
9. Y. S. Yokoyama, M. R. H. Elmoghayar, and I. Kuwajima, Tetrahedron Lett., **23**, 2673-2676 (1982).
10. I. Kuwajima, M. Arai, and T. Sato, J. Am. Chem. Soc., **99**, 4181-4182 (1977); I. Kuwajima, A. Mori, and M. Kato, Bull. Chem. Soc. Jpn., **53**, 2634-2638 (1980).
11. I. Kuwajima and M. Kato, J. Chem. Soc., Chem. Commun., 708-709 (1979).
12. I. Kuwajima, M. Kato, and A. Mori, Tetrahedron Lett., **21**, 2745-2748 (1980).
13. T. Cohen and J. R. Matz, J. Am. Chem. Soc., **102**, 6900-6902 (1980).
14. I. Kuwajima, M. Kato, and A. Mori, Tetrahedron Lett., **21**, 4291-4294 (1980).
15. I. Kuwajima and M. Kato, Tetrahedron Lett., **21**, 623-626 (1980).

TRIMETHYLSILANOL AS A LEAVING GROUP IN PREPARATIVE ORGANIC CHEMISTRY

Helmut Vorbrüggen

Research Laboratories, Schering AG, Berlin-Bergkamen, 1000 Berlin 65, Federal Republic of Germany

Abstract - Trimethylsilylation not only protects reactive hydroxyl groups but leads also to activation of amide, lactame and aromatic N-Oxide moieties permitting addition-elimination as well as substitution reactions. Silylation of ethyl(or methyl)-4-chloroacetate gives readily the reactive ethyl(or methyl)-3-trimethylsilyloxy-4-chloro-crotonates, which can be used for the synthesis of new 1,2 substituted (4,5)-imidazole acetic acid esters and the synthesis of annelated 5-membered rings.

Silylation of organic compounds is commonly used to protect oxygen functions during chemical reactions (Ref. 1-3). However silylation can also activate oxygen functions, e.g. in amide, lactam - especially heterocyclic lactam- or N-oxide groups - to permit new synthetic reactions with elimination of trimethylsilanol.

PROPERTIES OF TRIMETHYLSILYLOXY GROUPS

Due to the high affinity of silicon to oxygen (Ref. 3) and the high mobility of trimethylsilyl groups (Ref. 4), there is usually obtained the thermodynamically controlled silylation product.

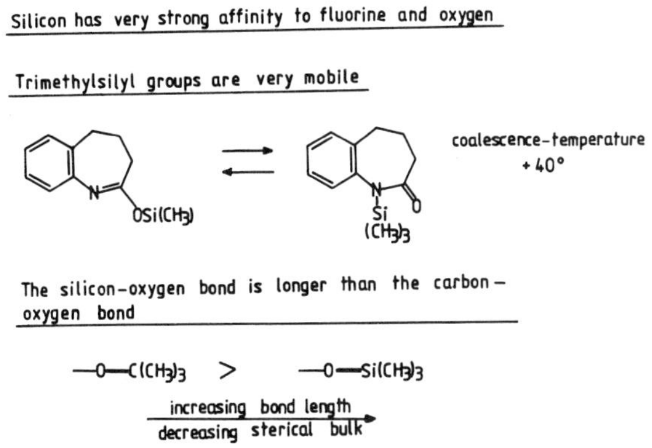

Compared to the oxygen-carbon bond, the oxygen-silicon bond is much longer resulting in a considerable decreased steric requirement of a trimethylsilyloxy group compared to a t-butoxy group (Ref. 5).

Reaction of an activated silylated synthon with a nucleophile produces trimethylsilanol bp 98.8° as leaving group which dimerizes to hexamethyldisiloxane bp 99-101° and water. This dimerization is accelerated by the presence of acidic or basic catalysts (Ref. 6).

Properties of trimethylsilanol

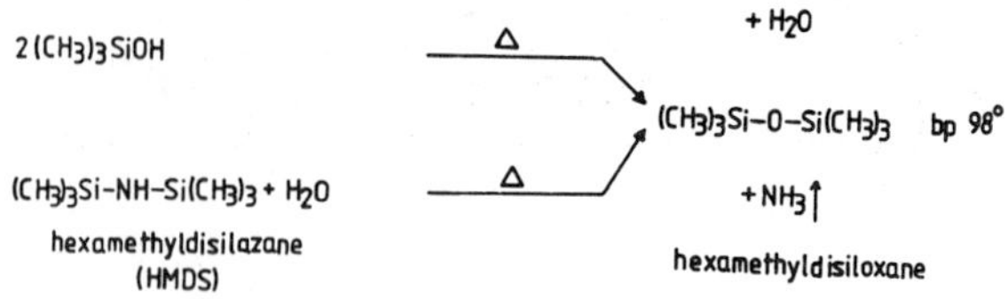

$$2(CH_3)_3SiOH \xrightarrow{\Delta} (CH_3)_3Si-O-Si(CH_3)_3 \quad bp\ 98° \quad + H_2O$$

$$(CH_3)_3Si-NH-Si(CH_3)_3 + H_2O \xrightarrow{\Delta} (CH_3)_3Si-O-Si(CH_3)_3 + NH_3\uparrow$$

hexamethyldisilazane (HMDS) → hexamethyldisiloxane

Transsilylation of silylated hydroxyl groups

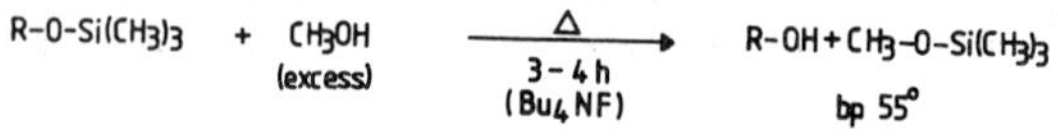

$$R-O-Si(CH_3)_3 + CH_3OH\ (excess) \xrightarrow[\text{(Bu}_4\text{NF)}]{\Delta,\ 3-4\ h} R-OH + CH_3-O-Si(CH_3)_3 \quad bp\ 55°$$

The water which is formed during the dimerization has to be removed by a silylating agent, usually hexamethyldisilazane (HMDS) bp 126° or trimethylchlorosilane (TCS)/triethylamine, whereupon hexamethyldisiloxane and ammonia or triethylamine-hydrochloride are obtained.

Finally, after completion of the chemical reaction, the remaining protecting O-trimethylsilyl groups have to be removed by transsilylation with excess boiling methanol.

SILYLATION-AMINATION OF HYDROXY-N-HETEROCYCLES

In their classical synthesis of cytidine, Todd, et al. (Ref. 7) treated 4-0-ethyl uridine with methanolic ammonia at 80° in an autoclave to afford cytidine and ethanol as a leaving group.

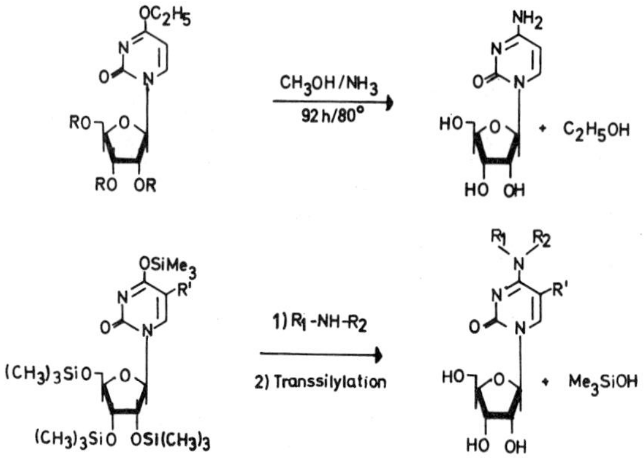

Since 4-0-ethyl uridine is only accessable via several synthetic steps, we wondered whether the corresponding persilylated uridine containing an activated 4-iminosilyl ether would not react analogously with primary and secondary amine to give N^4-substituted cytidines and trimethylsilanol as a leaving group.

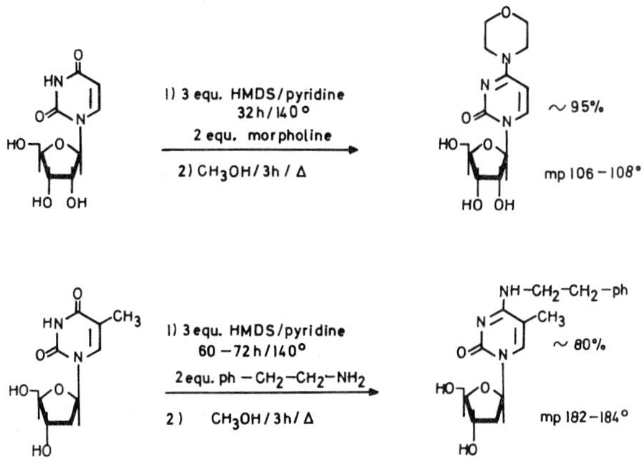

Using excess hexamethyldisilazane (HMDS) for the persilylation of uridine or thymidine lead in a one step-one pot reaction in high yields to the corresponding persilylated N^4-substituted cytidines, which crystallized on transsilylation with methanol (Ref. 8).

Whereas the silylated pyrimidine-nucleosides can be aminated without acidic catalysts, the addition-elimination reaction of primary or secondary amines to the silylated aromatic purine-nucleosides proceeds only in the presence of Lewis acids to yield the desired N^6-substituted adenosines.

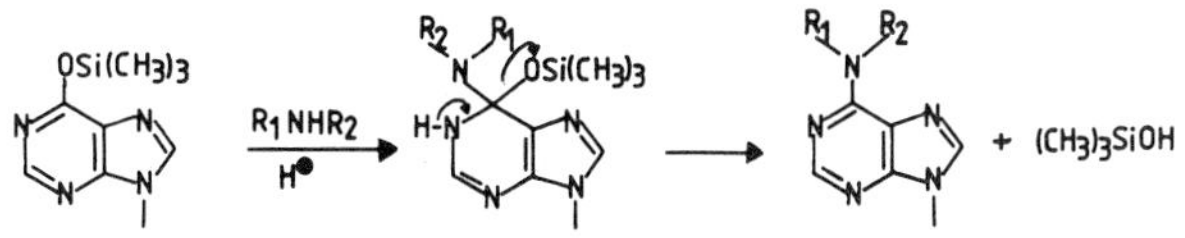

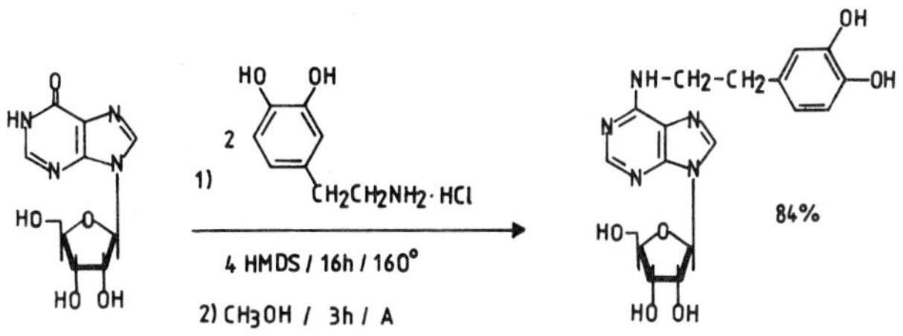

Thus on heating inosine and dopamine hydrochloride with excess HMDS afforded, after transsilylation, an 84 % yield of the corresponding N^6-substituted adenosine. During this reaction, the sensitive catechol-moieties in dopamine hydrochloride and the end product as well alcoholic hydroxyl groups of the ribose-moiety are protected as trimethylsilyl ethers (Ref. 9).

To demonstrate that this silylation is generally applicable, a whole range of different aromatic hydroxy-N-heterocycles was aminated.

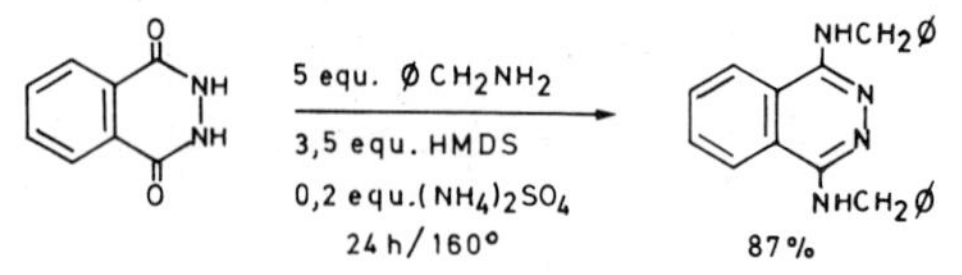

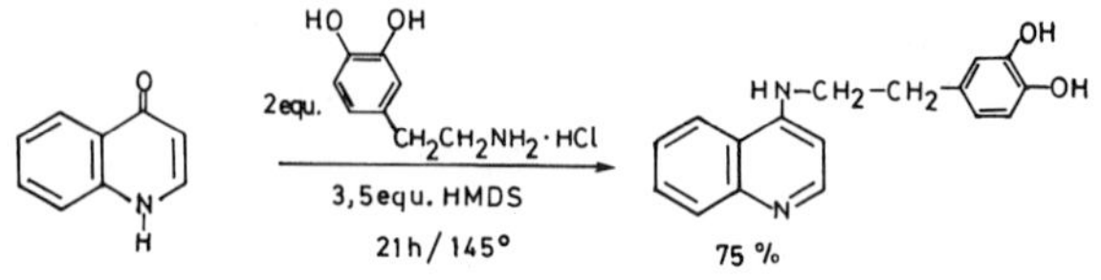

The silylation-amination of 2,3-dihydro-phthalazine-1,4-dione as well as of 1,4-dihydro-quinoline-4-one with benzylamine and dopamine hydrochloride my serve as examples (Ref. 10).

SILYLATION-CYANATION OF PYRIDINE-N-OXIDES

As described previously in the chemical literature, pyridine- or quinoline-N-oxides can be converted in 40 - 70 % yield into the corresponding 2- or 4-cyano-pyridines or -quinolines by alkylation or acylation of the oxide-function followed by treatment with aqueous alkali cyanide (Ref. 11).

Activation of the pyridine-N-oxide function and cyanation can be combined to a simple and efficient one step procedure by treatment of pyridine-, quinoline- or isoquinoline-N-oxides with trimethylsilylcyanide/triethylamine in acetonitrile or N,N-dimethylformamide (DMF). The strong affinity of the "hard" trimethylsilyl group for the "hard" N-oxide moiety and the "soft" cyano group for the "soft" α-carbon atom leads to a selective 1,2-addition to give 2-cyanopyridines, 2-cyanoquinolines or 1-cyanoisoquinolines in high yields.

Since trimethylsilylcyanide is expensive and gradually decomposes on heating, it is much more convenient to prepare trimethylsilylcyanide in situ (Ref. 12). Addition of excess trimethylchlorosilane (TCS) to 3-hydroxypyridine-N-oxide, excess sodium cyanide and triethylamine in abs. DMF and subsequent heating to 100 - 110° afforded, after transsilylation with methanol 90 % of 2-cyano-5-hydroxypyridine.

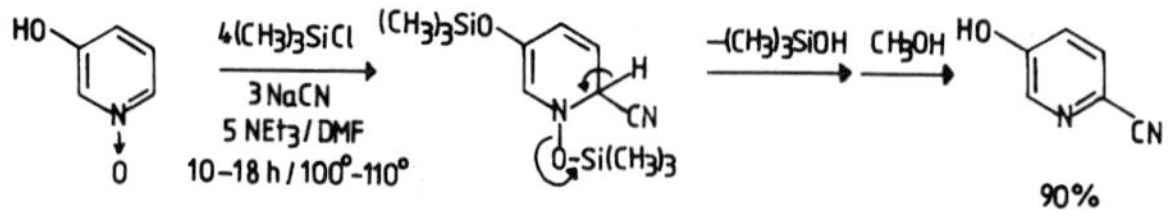

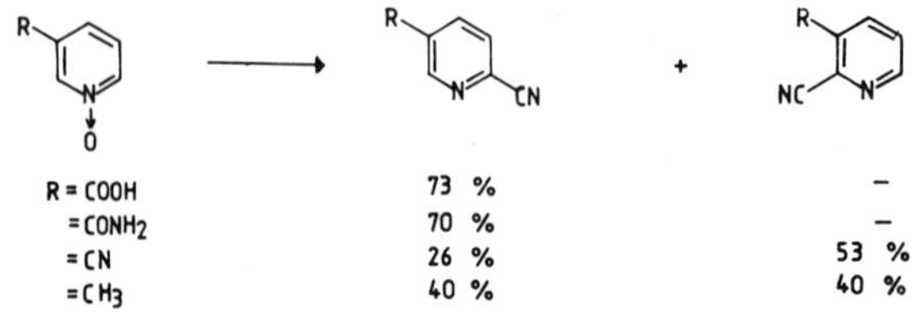

R =		
COOH	73 %	—
CONH$_2$	70 %	—
CN	26 %	53 %
CH$_3$	40 %	40 %

Analogously, silylated nicotinic acid - as well as nicotinic acid amide-N-oxide afforded the corresponding 2,5-disubstituted derivatives in high yields. 3-Methyl- and especially 3-cyanopyridine-N-oxide yielded, in addition to the 2,5-disubstituted derivatives, increasing amounts of 2,3-disubstituted derivatives. This effect is presumably due to decreased sterical hindrance of the 3-methyl- and the 3-cyano-substituent to nucleophilic attack at the 2-position. Other pyridine-N-oxides react analogously (Ref. 13). Further applications to other N-oxide systems as well as nitrones are being investigated.

SILYLATION-ALLYLATION OF PYRIDINE-N-OXIDES

As described by Sakurai (Ref. 14), allyltrimethylsilane forms on treatment with tetra-N-butylammonium fluoride (TBAF) the "soft" allylic anion and "hard" trimethylsilylfluoride.

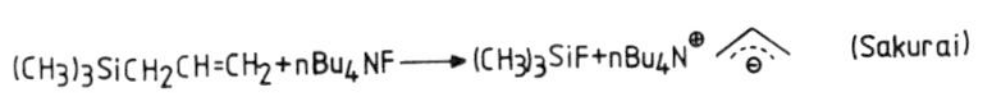

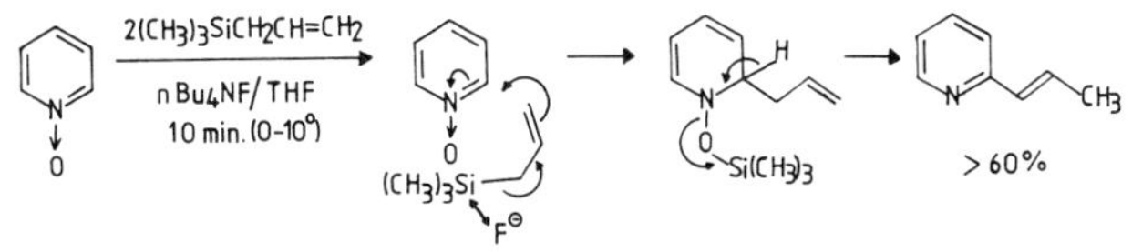

$$(CH_3)_3SiCH_2CH=CH_2 + nBu_4NF \longrightarrow (CH_3)_3SiF + nBu_4N^{\oplus} \overset{\ominus}{\diagdown} \quad (Sakurai)$$

$$(CH_3)_3SiOH + (CH_3)_3SiF \rightleftharpoons (CH_3)_3Si-O-Si(CH_3)_3 + HF$$
$$(CH_3)_3SiCH_2CH=CH_2 + HF \longrightarrow (CH_3)_3SiF + CH_3CH=CH_2$$

Reaction of pyridine-N-oxide in tetrahydrofuran (THF) with 2 equivalents of allyltrimethylsilane and 0.1 equivalent of abs. TBAF gave a 60 % yield of 2-propenylpyridine, which is probably formed by base catalysed rearrangement of initially generated 2-allyl-pyridine (Ref. 15). Analogous reactions of substituted pyridine-N-oxides and further applications to other heterocyclic N-oxides will be published shortly.

REACTIONS WITH O-SILYLATED ETHYL(OR METHYL)-4-CHLOROACETOACETATES

Because we need large amounts of imidazole(4,5)-acetic acid, we became interested in the reaction between the commercially available ethyl-4-chloroacetoacetate and amidines. However, due to the rather unreactive chloromethyl group, ethyl-4-chloroacetoacetate reacts with amidines to afford exclusively 6-chloromethyl-pyrimidine-4-one in high yield (Ref. 16).

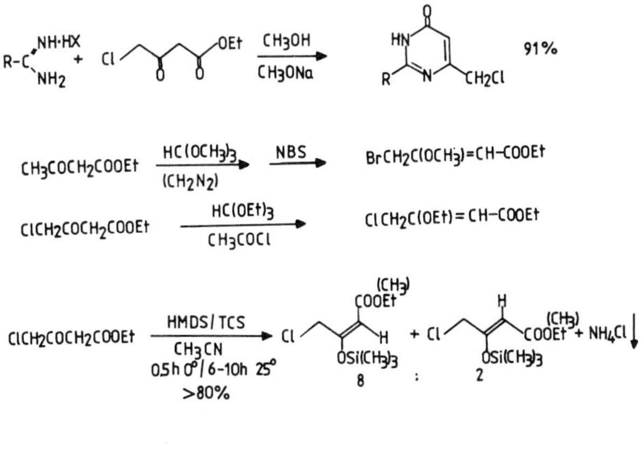

Since the halogen in enol ethers of ethyl-4-chloro(or 4-bromo)acetoacetates is allylic and thus much more reactive than the halogen in the parent ketone, these enol ethers have been widely used for synthetic purposes (Ref. 17). However, the preparation of these derivatives requires several reaction steps and proceeds often in low yields (Ref. 18).

We therefore prepared the corresponding enolsilyl ethers of ethyl(or methyl)-4-chloroacetoacetates and investigated their reactivity in alkylation reactions followed by subsequent cyclizations with elimination of trimethylsilanol.

Ethyl(or methyl)-4-chloroacetate, hexamethyldisilazane (HMDS) and trimethylchlorosilane (TCS) gave in acetonitrile at ambient temperature a 8 : 2 mixture of (E) and (Z) ethyl(or methyl)-3-trimethylsilyloxy-4-chloro-crotonate in more than 80 % yield as determined by NMR.

Reaction of different amidines or their salts with ethyl(or methyl)-3-trimethylsilyloxy-4-chloro-crotonate afforded readily a whole range of substituted imidazole-5-acetic acid esters in good yields (Ref. 19).

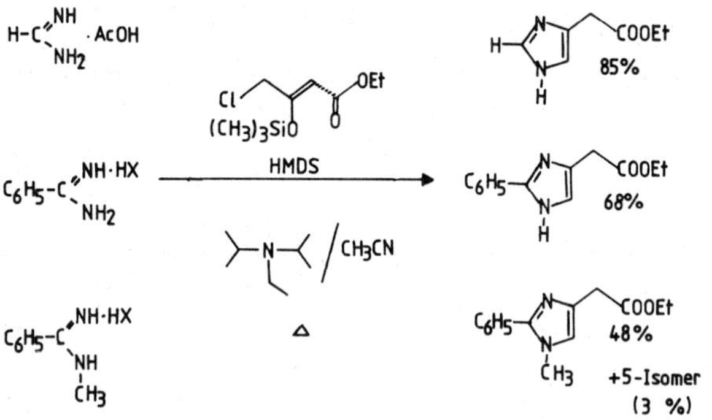

In these reactions, alkylation of the free or N-silylated amidines is probably followed by addition of the second nitrogen moiety to the unsaturated ester system, elimination of trimethylsilanol and rearrangement to the imidazole.

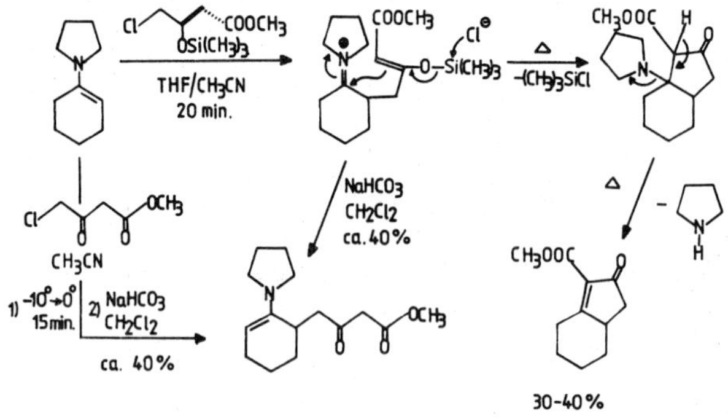

Further applications of these reagents to the synthesis of other nitrogen heterocycles are being investigated.

Reaction of the pyrrolidine enamine of cyclohexanone with methyl-3-trimethylsilyloxy-4-chloro-crotonate gave after several hours an isolable intermediate, which cyclizes slowly at ambient temperature in ca. 30 - 40 % yield to the 9-methoxycarbonyl-bicyclo[4.3.0]non-1,9-ene-8-one. The enamine intermediate, the structure of which has not as yet been unequivocally determined, can also be obtained by alkylation of the pyrrolidine enamine of cyclohexanone with methyl-4-chloroacetoacetate. Analogous reactions, e.g. with pyrrolidine enamine of cyclopentanone are presently being studied (Ref. 20).

Finally, I would like to emphasize that independent from our own experimental results discussed above, other groups have investigated further silylation-activation reactions. Thus for example aliphatic sulfoxides (Ref. 21), nitro compounds (Ref. 22) and nitrones (Ref. 23) give rise to interesting rearrangements and additions.

Acknowledgement - I thank my two very able and dedicated technicians Mr. Konrad Krolikiewicz and Miss Bärbel Bennua for their active and intelligent collaboration and Dr. D. Rosenberg for the physical data.

REFERENCES

1. L. Birkofer and A. Ritter, Angew. Chemie 77, 414 (1965), Angew. Chemie Intern. Ed. 4, 417 (1965).
2. B.E. Cooper, Chem. Ind. 794 (1978)
3. E. Colvin, Silicon in Organic Synthesis, Butterworths, London (1981).
4. J.F. Klebe, Acc. Chem. Res. 3 299 (1970).
5. H. Förster and F. Vögtle, Angew. Chemie 89, 443 (1977), Angew. Chemie Intern. Ed. 16 429 (1977).
6. W.T. Grubb, J. Amer. Chem. Soc. 76, 3408 (1954).
7. G.A. Howard, B. Lythgoe and A.R. Todd, J. Chem. Soc. (London) 1052 (1947).
8. H. Vorbrüggen, K. Krolikiewicz and U. Niedballa, Liebigs Ann. Chem. 988 (1975).
9. H. Vorbrüggen and K. Krolikiewicz, Liebigs Ann. Chem. 745 (1976)
10. Preliminary publication: H. Vorbrüggen, Angew. Chemie 84, 348 (1972), Angew. Chemie Intern. Ed. 11, 305 (1972).
    H. Vorbrüggen and K. Krolikiewicz to be published.
11. M. Henze, Ber. 69, 1566 (1936.
    W.E. Feely and E.M. Beavers, J. Amer. Chem. Soc. 81, 4004 (1959).
    T. Okamoto and H. Tani, Chem. Pharm. Bull. 7, 925 (1959).
    H. Tani, Chem. Pharm. Bull. 7, 930 (1959).
12. J.K. Rasmussen and S.M. Heilmann, Synthesis 219 (1978).
13. H. Vorbrüggen and K. Krolikiewicz to be published.
14. H. Sakurai, Pure & Appl. Chem. 54, 1 (1982).
15. H. Vorbrüggen and K. Krolikiewicz to be published.
16. German Patent Application DE 2167194, C. A. 76, 72539 (1972).
17. K. Koshima and K. Sakai, Tet. Letters 3333 (1972).
    S.M. Weinreb and J. Auerbach, J. Amer. Chem. Soc. 97, 2503 (1975).
    G. Stork, R.K. Boeckmann, D.F. Taber, W.C. Still and J. Singh, J. Amer. Chem. Soc. 101, 7101 (1979).
18. D.G.F.R. Klostermans, Rec. 70, 79 (1951).
    E.B. Reid and W.R. Ruby, J. Amer. Chem. Soc. 73, 1054 (1951)
    R. Finding, G. Zimmermann and U. Schmidt, Monatsh. Chem. 102, 214 (1971).
19. N. Schwarz and H. Vorbrüggen to be published.
20. H. Vorbrüggen and B. Bennua to be published.
21. T.S. Chou, Tet. Letters 725 (1974).
    J.J. de Koning, H.J. Kooreman, H.S. Tan and J. Verweij, J. Org. Chem. 40, 1346 (1975).
    B.S. Bal and H.W. Pinnick, J. Org. Chem. 44, 3727 (1979).
    A.F. Janzen, G.N. Lypka and R.E. Wasylishen, Can. J. Chem. 58, 60 (1980).

22. K. Torssell and O. Zeuthen, Acta Chem. Scand. B 32, 118 (1978).
    S.C. Sharma and K. Torssell, Acta Chem. Scand. B 33, 379 (1979).
    S.K. Mukerji and K. Torssell, Acta Chem. Scand. B 35, 643 (1981).
    M. Asaoka, T. Mukuta and H. Takei, Tet. Letters 735 (1981)
23. O. Tsuge, S. Urano and T. Iwasaki, Bull. Chem. Soc. Jpn. 53, 485 (1980).

REACTION OF SILYL ENOL ETHERS WITH CARBENES AND CARBENOIDS. NEW SYNTHESES OF VARIOUS CARBONYL COMPOUNDS

Jean-Marie Conia and Luis Blanco

Laboratoire des Carbocycles, Université de Paris-Sud, 91405 Orsay, France

Abstract – The reaction of carbenes or carbenoid species (CH_2, CBr_2, CClMe) with silyl enol ethers leads to silyloxycyclopropanes and easy ring opening of these products yields various interesting carbonyl compounds. An account of the recent literature in the field is given and the main results obtained in the Laboratoire des Carbocycles at Orsay are reported. The halocarbenes CBr_2 and CClMe lead to unsaturated carbonyl compounds because ring opening always occurs with the elimination of $XSiMe_3$. Ring enlargement of cycloalkanones into higher 2-methyl-cycloalkenones and the direct preparation of $\underline{E}$ and (or) $\underline{Z}$ αβ-ethylenic ketones and aldehydes from the appropriate halosilyloxycyclopropane isomers, without $\underline{Z} \longrightarrow \underline{E}$ isomerisation, are discussed in greater detail.

INTRODUCTION

The importance of organosilicon chemistry has greatly increased in recent years. Silyl enol ethers $\underline{1}$ in particular became highly valued synthetic tools (for some reviews, (see Ref. 1-6). The purpose of the present lecture is to show the considerable interest of the reaction of silyl enol ethers with carbenes or carbenoid species followed by silyloxycyclopropane ring opening, for the synthesis of many types of carbonyl compounds (2) (7).

The reactions of various common silyl enol ethers with three carbenoid species (CH_2, CBr_2 and CClMe), investigated in particular in various Japanese and American laboratories and in the Laboratoire des Carbocycles at Orsay, are reported and discussed. The results show the important potentiality of such reactions because of the two possible bond cleavages (a and b) occurring in the cyclopropane rings. The direction of cleavage depends on the nature and the stereochemistry of the substituents and on the reagent and the reaction conditions. E.g.

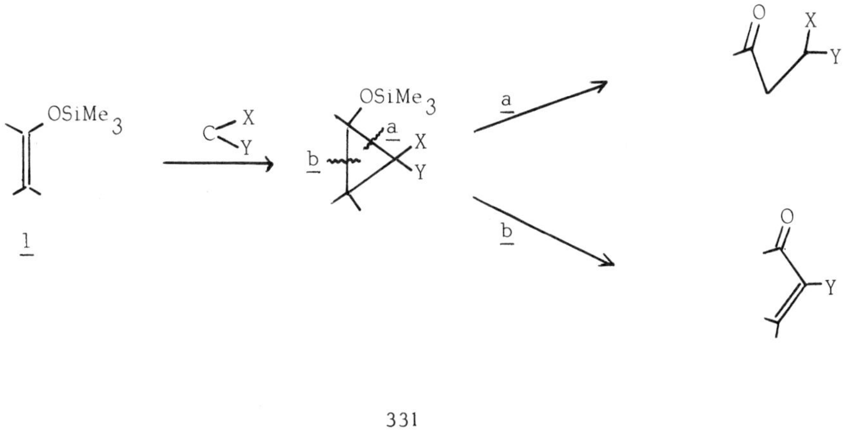

Many methods of preparation of silyl enol ethers are now known, but the selective preparation of the **α**- or **α'**-isomer from an unsymmetrical ketone, or of the Z or E stereoisomer from an open-chain ketone, or of the 1- or 2-silyloxy-1,3-diene from an **αβ**-ethylenic ketone can sometimes still represent a real problem. Except in certain cases, however, we did not encounter this problem and it even appeared that, with a good spinning band column, it is possible to separate many E and Z silyl enol ether isomers. Concerning the preparation and the determination of structure of silyl enol ethers in general, see for instance (8-10).

REACTION OF SILYL ENOL ETHERS WITH A METHYLENE CARBENOID

Silyl enol ethers **1** undergo ready reaction with the Simmons-Smith reagent to give silyloxycyclopropane **2**. In this field, a first useful improvement has been made to the original conditions (11) : the use of a zinc-silver couple (2) (12) (13) which is more reactive than the classical zinc-copper couple (Note a). A second improvement has been made by the use of pyridine instead of an acid aqueous isolation procedure ; pyridine precipitates ZnI_2 and its use avoids hydrolysis when the desired product is unstable in acid aqueous medium.

$$1 \xrightarrow[2°) \text{Pyridine}]{1°) CH_2I_2, Zn-Ag} 2$$

An advantage of Simmons-Smith reagents and other carbenoid species is that they preferentially attack the relatively electron-rich enol ether double bond in the cases of silyl enol ethers derived from ethylenic carbonyl compounds ; the double cyclopropanation of such silyl enol ethers occurs when an excess of the reagent is used (13) (17). In this lecture we describe the preparation of cyclopropanols by the conversion of the silyloxy groups of silyloxycyclopropanes **2** into hydroxyl groups without opening of the cyclopropane ring, and the conversion of **2** by the reaction with various reagents, generally *via* a bond cleavage **a**, into various carbonyl compounds ; finally we report two special rearrangements of 1-vinyl-1-silyloxycyclopropanes, their ring enlargements into silyl enol ethers of cyclopentanones and (*via* the cyclopropanols) into cyclobutanones.

Preparation of cyclopropanols

The synthesis of silyloxycyclopropanes **2** from silyl enol ethers **1**, followed by hydrolysis of the $OSiMe_3$ group without opening of the cyclopropane ring, allows easy access to cyclopropanols. The first conversion of trimethylsilyloxycyclopropanes into

---

Note a. Concerning the use of a zinc-silver couple for the cyclopropanation of silyl enol ethers in particular we are surprised to read in the literature either the claim of spectacular results or the confession of obvious failures ; we think that the former are normal but the latter are always due to failures in the preparation of the zinc-silver couple itself. Actually silyl enol ethers very easily undergo the Simmons-Smith reaction and in our hands none has failed to do so. For an efficient preparation of the zinc-silver couple, see (13).

cyclopropanols by methanolysis was observed in 1972 (14), and this technique has been used in many cases (15) (16) (21).

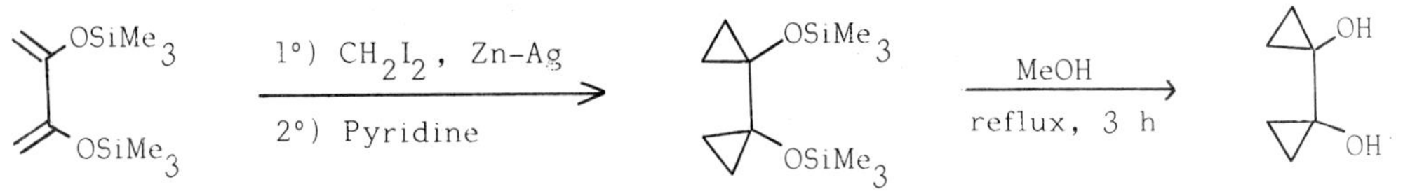

Methanolysis is accelerated in the presence of NaOH ; for example 30 minutes at room temperature are sufficient for a total conversion with MeOH containing 10% of a 0,1 molar aqueous solution of NaOH (18). Acidic hydrolysis seems less useful (17) although it has been claimed that toluenesulphonic acid is also a good catalyst (20). In any case the sequence silyl enol ethers ⟶ silyloxycyclopropanes ⟶ cyclopropanols constitutes by far the simplest and best method of preparation of cyclopropanols.

α-Methylation of aldehydes and ketones

It is well known that cyclopropanols are opened in alcaline or acid medium leading to carbonyl compounds. Silyloxycyclopropanes 2 thus lead to α-methylated aldehydes and ketones 3 in very good yields (15). Here the best method is to reflux 2 with NaOH in dilute MeOH + H_2O solution (15), or to pyrolyze 2 at 130° (18) See Scheme 1. For ring opening in acid or basic media, see also (17) (see note a).

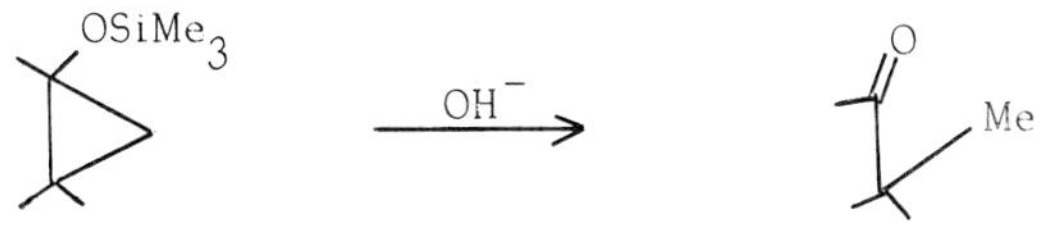

All these ring openings involve bond cleavages a. But we must draw attention to the special case of the reaction cyclopentanone ⟶ 2-methyl-cyclopentanone. It is the one case in this field in which bond cleavage b also occurs, for 2-methyl-cyclopentanone is accompanied by a small amount of the ring enlargement product cyclohexanone. This method possesses two main advantages. Firstly it is specific : only a monomethylation product is obtained. Secondly it is regiospecific : monomethylation can be oriented either to the α or to the α' position in the case of an unsymmetrical ketone. The now classical example is the methylation in the α or α' position of 2-methyl-cyclohexanone, as required (15).

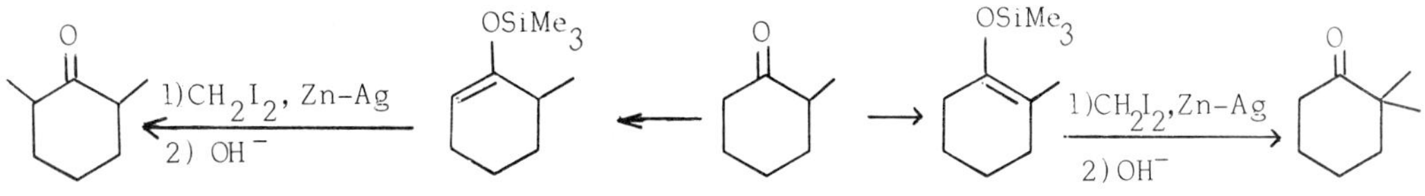

Nota a. Contrary to what has been reported by some authors (5) (22) we never recommended the use of strong bases such as t-BuOK to effect ring opening. In fact the use of t-BuOK leads to good results in the α-methylation of ketones which are relatively stable in alkaline medium (17), but is unsuitable for aldehydes, or unstable ketones like cyclopentanone.

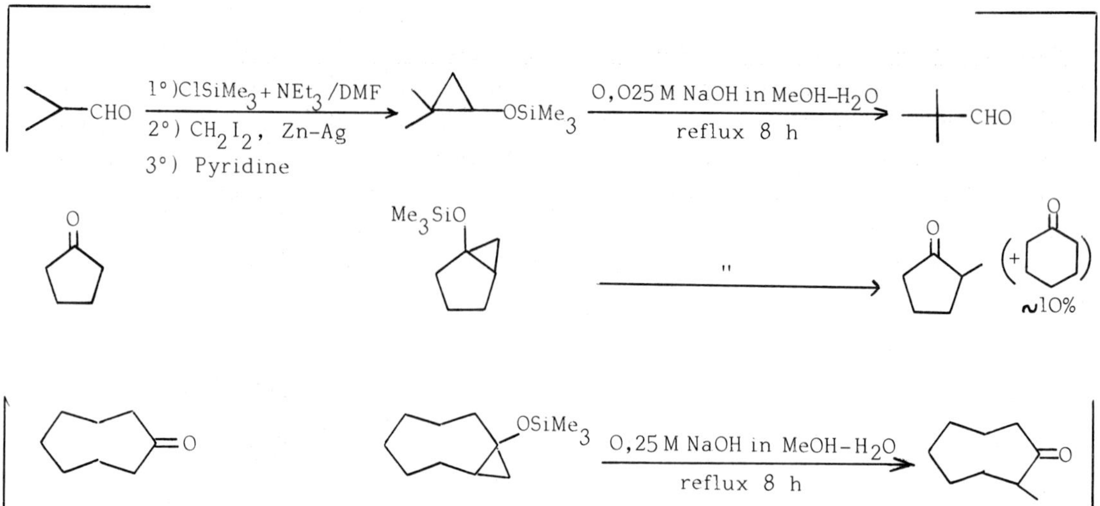

Scheme 1. Examples of **α**-monomethylation of carbonyl compounds (15).

The same sequence can be applied to ethylenic ketones (19). Example : $\underline{4} \to \underline{5} \to \underline{6} \to \underline{7}$

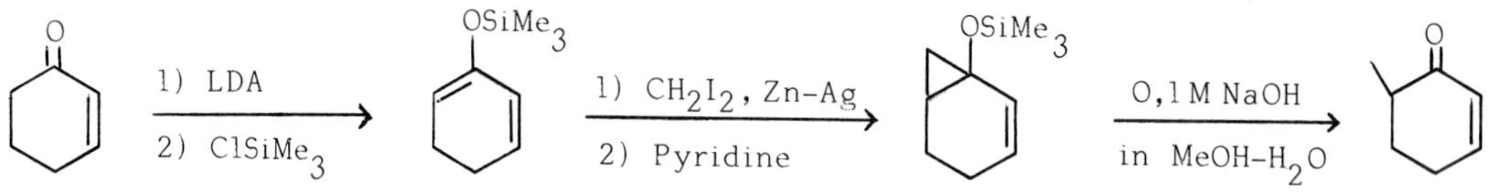

It has been also applied in the steroid field. For example it is possible to prepare the 4- and 2-methyl derivatives $\underline{9}$ and $\underline{10}$ of testosterone $\underline{8}$ via the corresponding silyloxydienes and their monocyclopropanation products.

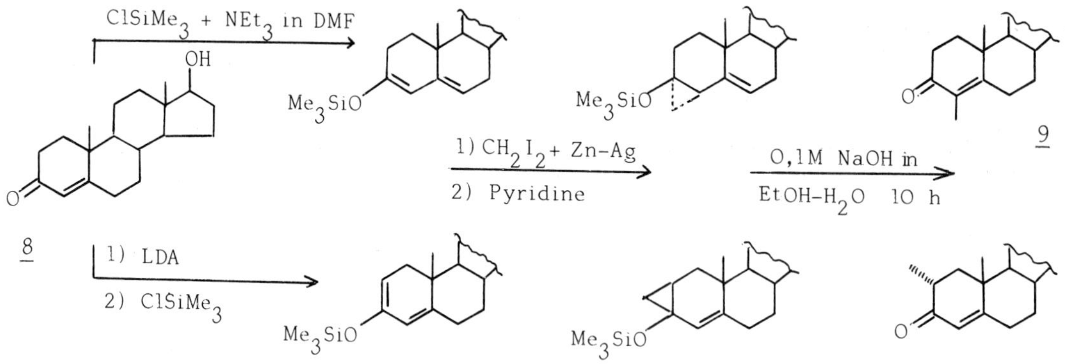

Preparation of **α**-bromomethyl-ketones

Whereas the bromination of silyl enol ethers themselves leads to **α**-bromo-aldehydes and -ketones and allows the preparation of **α**- or **α'**-bromo-ketones from unsymmetrical ketones as required, and even of **α'**-bromo-**αβ**-ethylenic ketones from **αβ**-ethylenic ketones (25) (26), the bromination of silyloxycyclopropanes $\underline{2}$ gives the derived **α**-bromomethyl-aldehydes or ketones $\underline{11}$ (27).

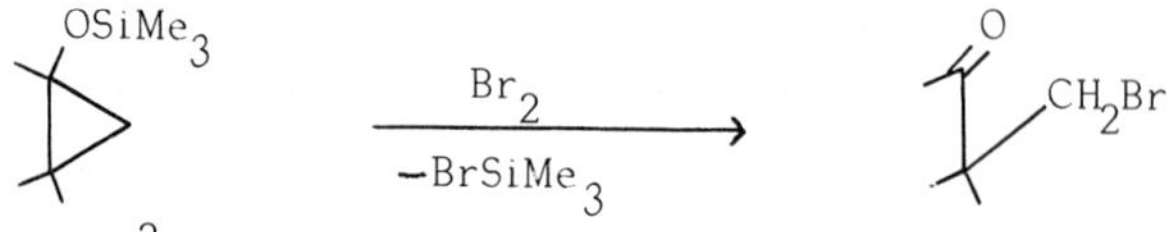

Preparation of 2-methylene-cycloalkanols and spiro [n.2] alkanols and alkanones

When the Simmons-Smith cyclopropanation of cyclic silyl enol ethers is carried out under the original conditions and in concentrated solution, the silyloxycyclopropanes 12 so formed undergo zinc iodide-induced isomerisation to silyl enol ethers of 2-methylene-cycloalkanols 13 (28). The use of an excess of CH_2I_2 and of diethylzinc as a co-reagent leads to further cyclopropanation giving silyloxy-spiro [n.2] alkanes 14 which can be hydrolysed or oxidized to the corresponding alcohols or ketones (29). The yields are generally excellent.

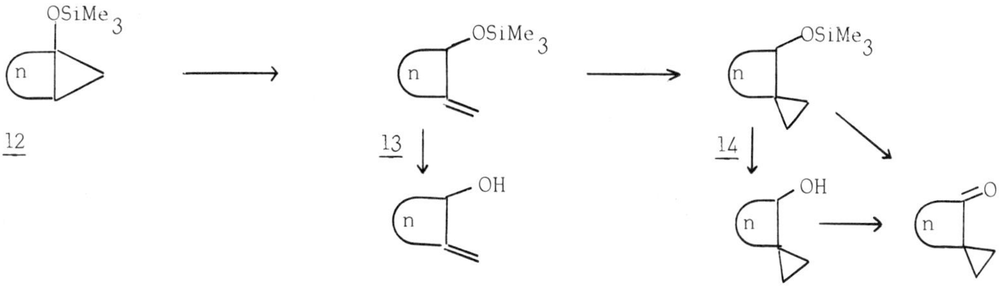

Preparation of α-methylene-ketones and γ-ketoesters

Silyloxycyclopropanes are quantitatively converted into β-acetoxymercuri-ketones 15 by the reaction with mercuric acetate. Successive treatment with $PdCl_2$ or $PdCl_2$ + CO gives α-methylene-ketones 16 or γ-ketoesters 17 respectively in good yields (30). The ring cleavage takes place highly selectively at the least substituted cyclopropane carbon atom. The sequence is particularly useful in the open-chain series, but less so in the cyclic series because of the possible polymerisation or isomerisation of the α-methylene-cycloalkanones.

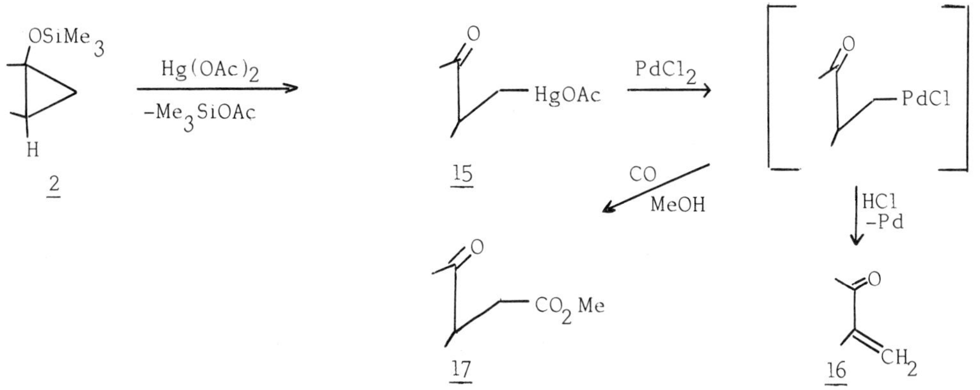

For another preparation of α-methylene-ketones and for the preparation of β-ketoesters, see below.

Preparation of ω-unsaturated acids

The treatment of silyloxycyclopropanes derived from silyl enol ethers of cycloalkanones 12 with lead (IV) acetate (LTA) in acetic acid leads to ω-unsaturated acids 18.

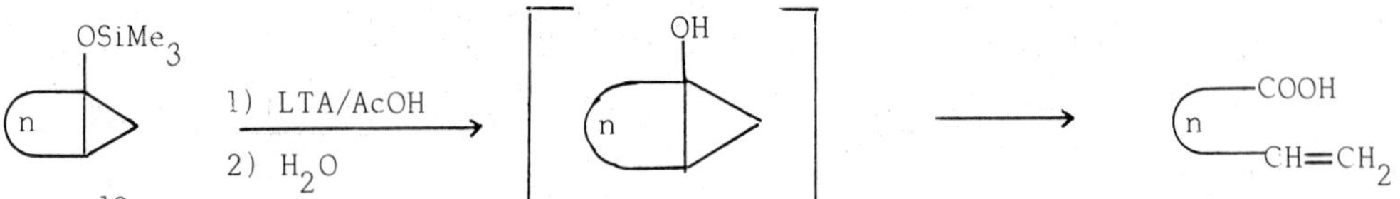

The fragmentation process involves oxidation of the cyclopropanol intermediate.

Synthesis of 2-cycloalkenones via ring enlargement of cycloalkanones

The reaction of the silyloxycyclopropanes derived from the silyl enol ethers of cycloalkanones 12 with iron (III) chloride, followed by treatment with sodium acetate in methanol gives 2-cycloalkenones 19 in high yield (22) (23).

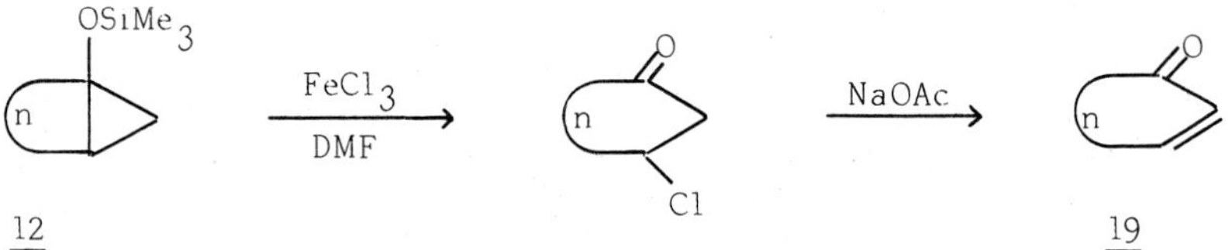

Such a one-carbone ring homologation is well explained by the authors by a mechanism involving an alkoxy radical intermediate which undergoes homolytic bond cleavage b.

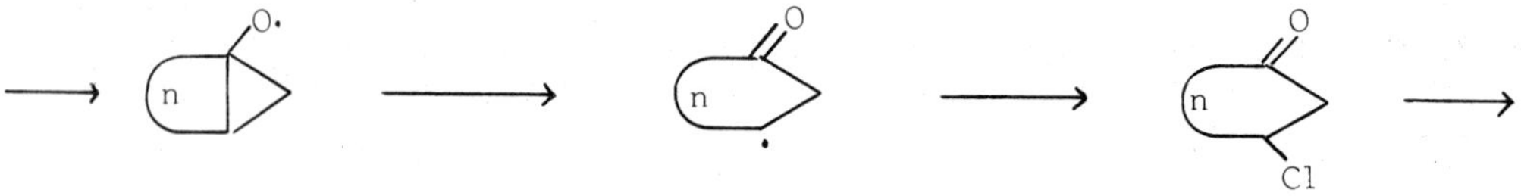

From 1-vinyl-1-silyloxycyclopropanes to cyclobutanones or cyclopentanones

It was mentioned above that the silyloxycyclopropane derivatives of 2-cyclohexenones undergo ready cyclopropane ring opening into 6-methyl-2-cyclohexenones : $\underline{4} \rightarrow \underline{5} \rightarrow \underline{6} \rightarrow \underline{7}$. But when a cisoid or labile enone is the precursor, a different course of cyclopropane ring opening takes place, either by a thermal or by an acid-catalysed rearrangement, via the corresponding cyclopropanol, into a cyclobutanone (24)(32)(33)(34). This is a common rearrangement, one of the numerous and now well known ring enlargements ($C_3 \longrightarrow C_4$) (35). See Scheme 2.

On the other hand, merely on heating, 1-vinyl-1-silyloxycyclopropanes undergo smooth thermal rearrangement into cyclopentanone silyl enol ethers (36).

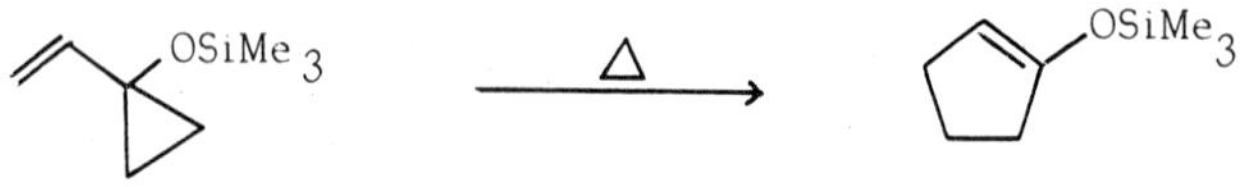

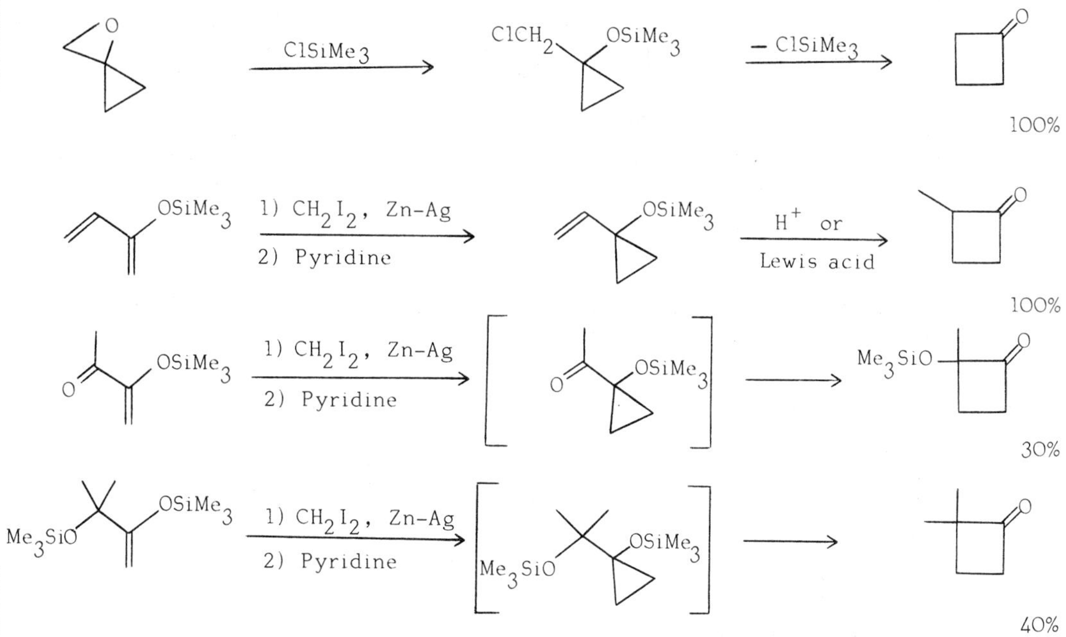

Scheme 2. Examples of silyloxycyclopropane ring expansions into cyclobutanones

Scheme 3 illustrates the synthetic interest of both of these ring expansions (37).

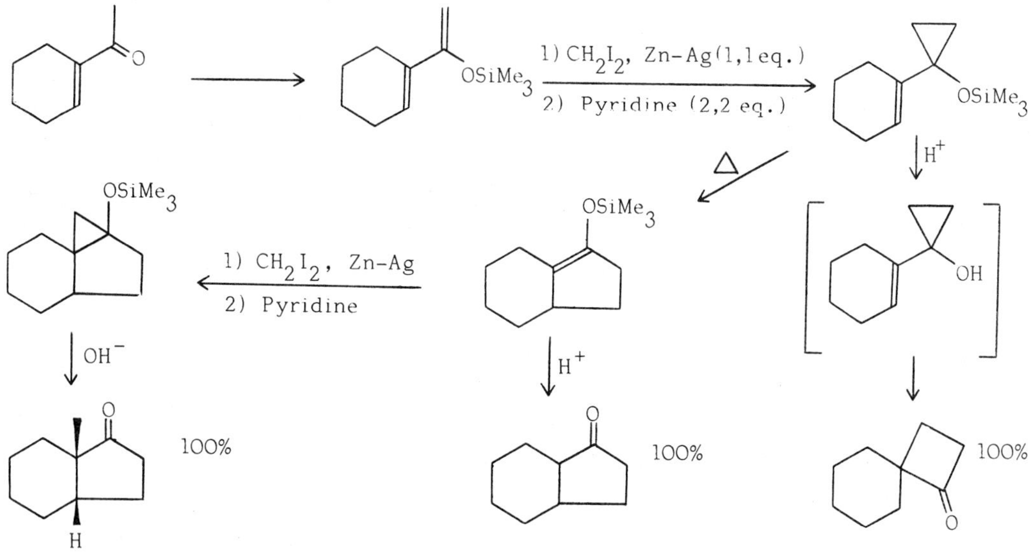

Scheme 3. Applications of both ring expansions of 1-vinyl-1-silyloxycyclopropanes

Various 1-silyloxy-1-vinylcyclopropanes have recently been prepared and converted into 1-silyloxy-cyclopentenes, and advantage has been taken of the specificity of the thermal rearrangement (38). For example, when two vinylcyclopropane moieties are involved, as in 20, which a priori leads to a 1-silyloxy-3-cyclopropyl-cyclopentene

21 or to a 1-cyclopentenyl-1-silyloxycyclopropane 22, only the former is formed illustrating the effect of the OSiMe$_3$ group in the cyclopropane ring (33).

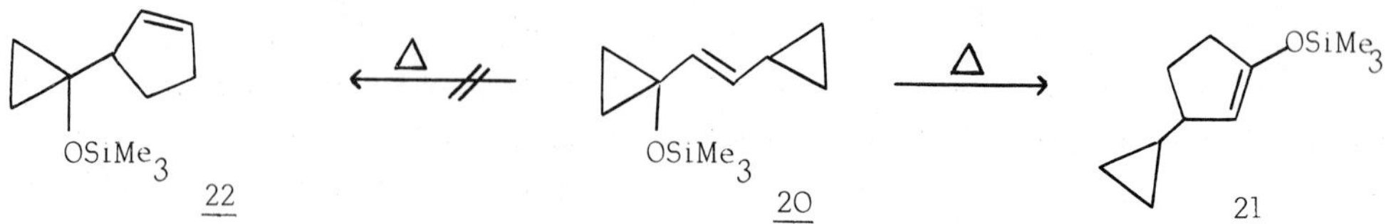

On the other hand it is known that the position of this group on the double bond of the vinylcyclopropane system, can on the contrary, hinder the rearrangement (39).

REACTION OF SILYL ENOL ETHERS WITH DIHALOCARBENES

The addition of mono- or dihalocarbenes to silyl enol ethers yields 2-mono- or 2,2-dihalo-1-silyloxycyclopropanes. A priori, because of the presence of one or, better, two leaving groups in the 2-position, cyclopropane ring opening, which involves a disrotatory bond cleavage b, assists the separation of the leaving group.

Synthesis of α-halo-αβ-unsaturated carbonyl compounds

2,2-Dihalo-1-silyloxycyclopropanes 23, easily obtained by dihalocarbene addition to silyl enol ethers of aldehydes and cyclic or acyclic ketones, do, indeed, undergo smooth thermal or acid-catalysed rearrangement (41) to a new class of products: α-halo-αβ-unsaturated carbonyl compounds 24.

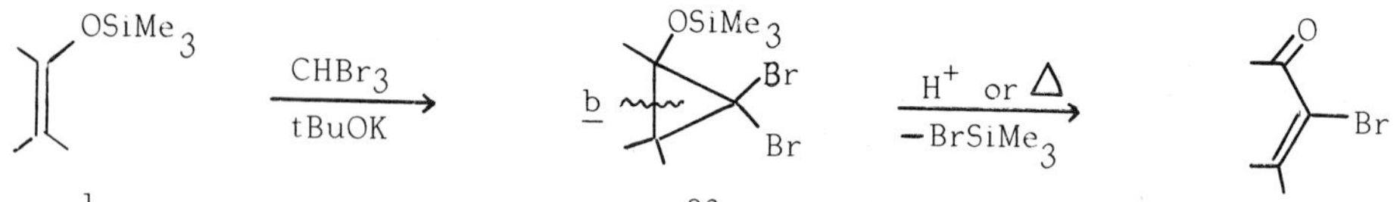

The reaction is general and advantage can be taken of regioselective silyl enol ether generation. Example :

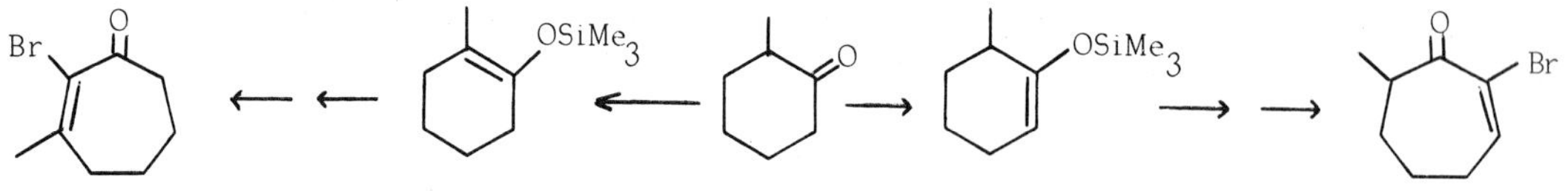

Preparation of β-methyl-αβ-unsaturated ketones 26

A one-carbon homologation method uses the treatment of 2,2-dichloro-1-ethoxycyclopropanes 25 by two equivalents of methyllithium to afford β-methyl-αβ-unsaturated ketones (here we mention the reaction starting from alkyl enol ethers, which has been applied to the synthesis of dl-muscone ; presumably this route would be even more satisfactory starting from silyl enol ethers).

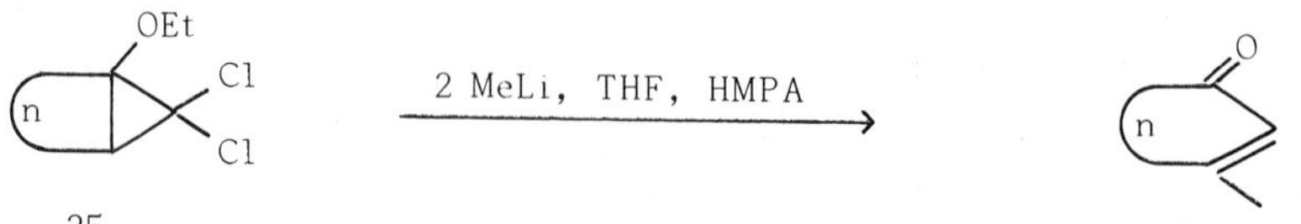

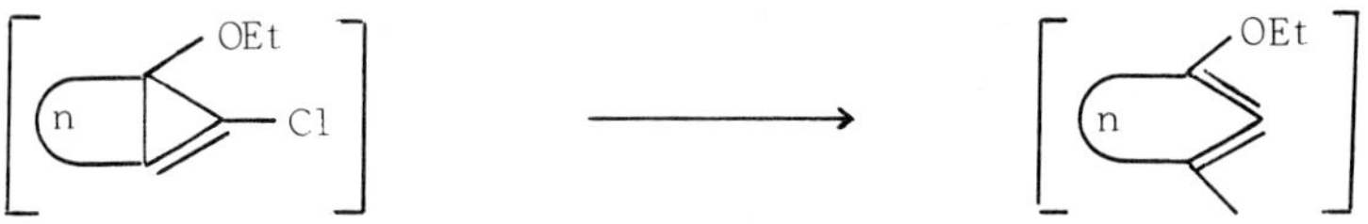

The intermediates in this reaction are assumed to be a chlorocyclopropene derivative, formed by HCl elimination by one equivalent of MeLi, and an ethoxyallene derivative formed by a subsequent addition of another equivalent of MeLi inducing the ring enlargement of the resulting carbenoid species.

Formation of β-ketoesters 28

Electrolysis, in alcohols in the presence of iron (III) nitrate, of dichloro-1-trimethyl-silyloxybicyclo [n.1.0] alkanes 27 leads to anodic bond cleavage a ; it is a convenient method for the introduction of an alkoxycarbonyl group at the α-position of some cycloalkanones (43).

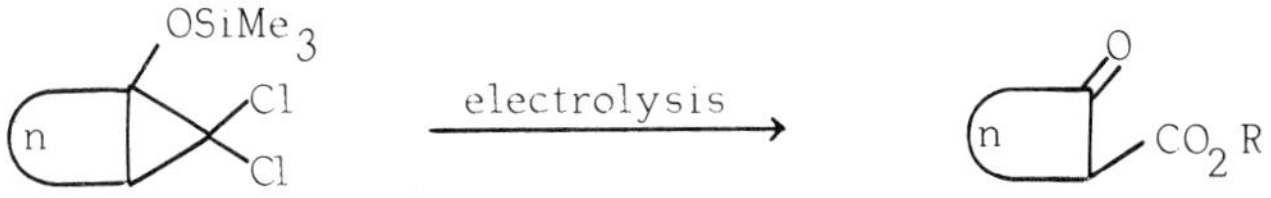

In contrast anodic oxidation of the corresponding ethyl ethers 25 affords acyclic ω-dichlorovinylidene-esters 29 preferentially but in low yields (44).

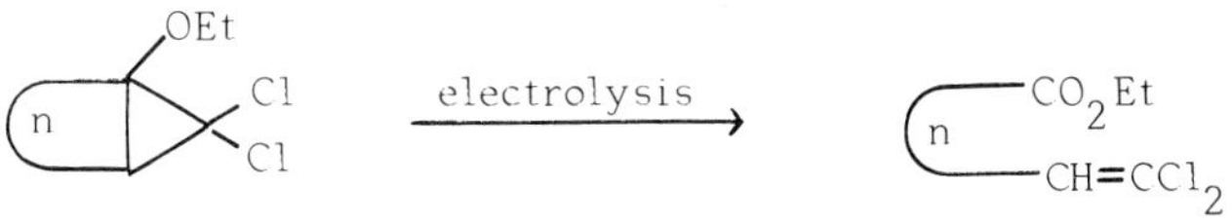

Lastly, it has been shown that the above-mentioned preparation of ω-unsaturated acids by the treatment of 1-silyloxybicyclo [n.1.0] alkanes with lead (IV) acetate seems general. It has been applied, for instance, to the preparation of ω-dichloro unsaturated acids possessing a cis double bond 31, from the appropriate dichlorocarenols 30 (45). E.g.

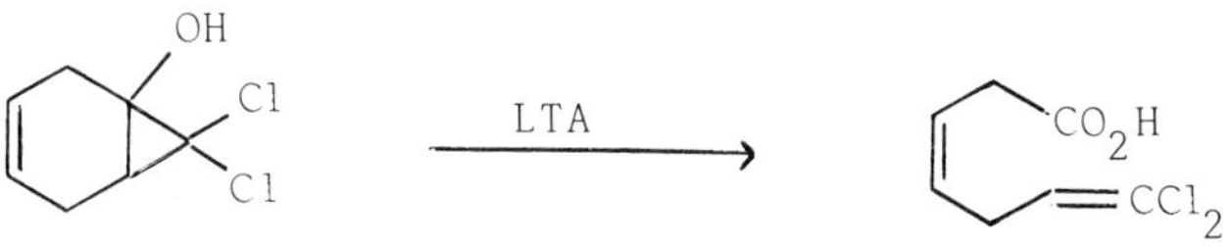

REACTION OF SILYL ENOL ETHERS WITH CHLOROMETHYLCARBENE

As mentioned above, saturated aldehydes and ketones are readily converted into higher α-halo- αβ-ethylenic homologues 24 on addition of dihalocarbene CX_2 to trimethylsilyl enol ethers, followed by opening of the cyclopropane ring. Likewise the cycloaddition of chloromethylcarbene, easily prepared from 1,1-dichloroethane and butyllithium according to (50), to trimethylsilyl enol ethers, followed by the elimination of chlorotrimethylsilane from the mixture of Z and E chloro-silyloxy-methyl-cyclopropane isomers 32, leads to the higher α-methyl- αβ -ethylenic carbonyl compounds 33 (46) (47).

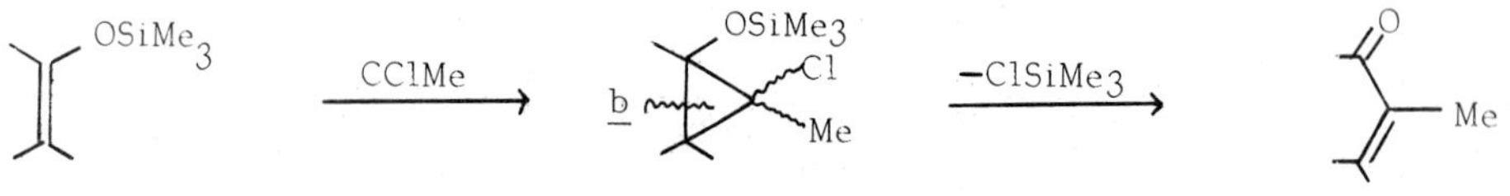

Conversion of cycloalkanones into higher α-methylcycloalkenones

This short sequence is obviously of interest from the point of view of synthesis, and, when the crude mixture of Z and E chloro-silyloxy-methylcyclopropane isomers 34 is heated either in toluene (or xylene) or in MeOH + NEt$_3$ under reflux, it leads to the corresponding higher ring 2-methyl-cycloalkenone 35 (46). But the rate of the rearrangement of the E isomer, favoured by the trans configuration of OSiMe$_3$ and Cl groups,

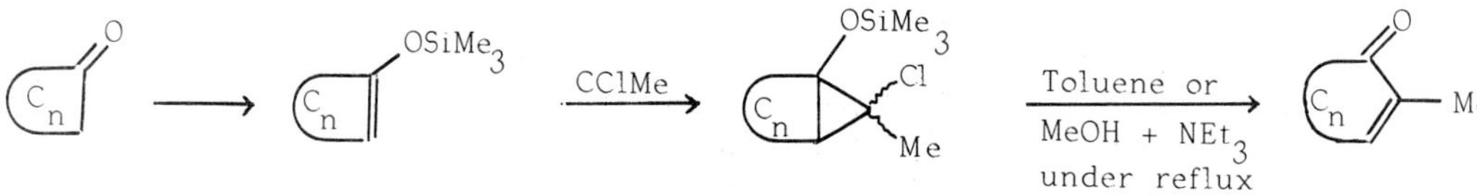

is generally greater than that of the Z isomer. Thus, a careful investigation of the reaction applied to the conversion of C_5, C_6, C_7 cycloalkanones into C_6, C_7, C_8 2-methyl-cycloalkenones 35a, b, c shows that both of E and Z chloro-silyloxy-methyl-bicyclo [n.1.0] alkane intermediates 34a, b, c lead to the same higher Z 2-methyl-cycloalkenone, the E isomers reacting more rapidly than the Z isomers.

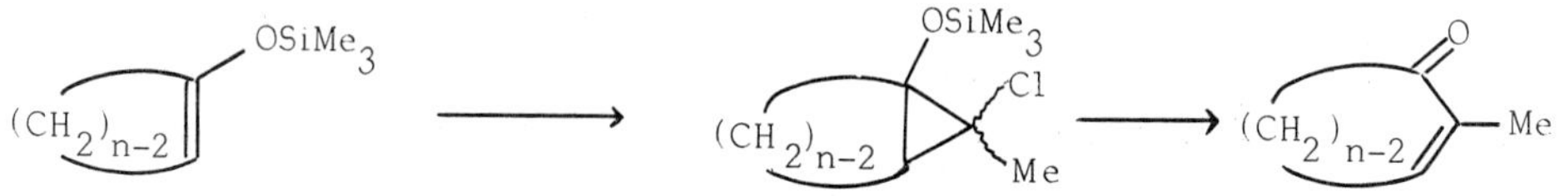

n = 5   cyclopentanone   34a   35a
    6   cyclohexanone   b   b
    7   cycloheptanone   c   c
    8   cyclooctanone   d   d

This is particularly clear for E and Z chloro-silyloxy-methyl-bicyclo [3.1.0] hexane isomers 34a, the former being very unstable even at room temperature and quickly rearranged, and the latter being very stable and recovered unchanged after several days in boiling toluene, or converted into the very stable chlorhydrin 36 after 15 h in boiling methanol. (However, by treatment with silver acetate in CH$_3$CN, 34aZ can be rearranged into 2-methyl-cyclohexenone 35a).

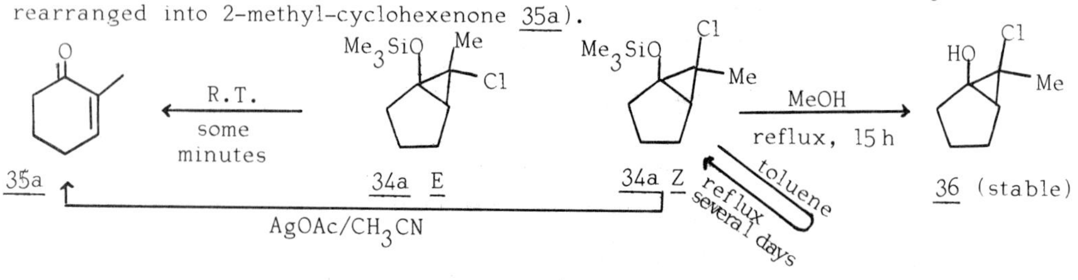

In fact, in the conversion of C_5, C_6, C_7, C_8 cycloalkanones, via E and Z chloro-silyloxy-methyl-cyclopropanes 34a, b, c, d, into 2-methyl-cycloalkenones 35a, b, c, d, the rates of opening of the E silyloxycyclopropane intermediates decrease with increasing sizes, whereas these rates increase for the corresponding Z isomers and Z-34d reacts faster than E-34d. The reaction in toluene does not involve a true thermal concerted mechanism, for $ClSiMe_3$, thus HCl, are also formed. Presumably the reaction of the E isomers, but not the Z isomers, in MeOH + NEt_3 follows a concerted process of the type A.

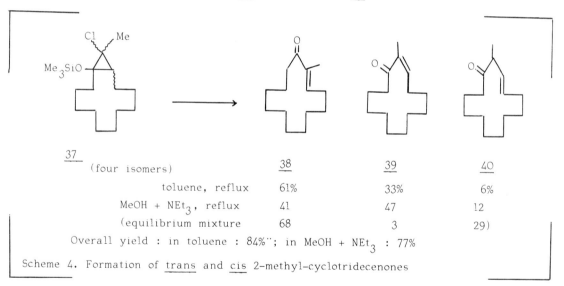

The medium ring ketone cyclododecanone can be converted into trans and cis 2-methyl-cyclotridecenones 38 and 39 (61 and 33% respectively) accompanied by a small amount (6%) of the trans deconjugated isomer 40. The reaction occurs via both stereoisomeric cyclododecanone silyl enol ethers and by heating in toluene of the four chloro-silyloxy-methyl-cyclopropane derivatives 37. When the same reaction is carried out in MeOH + NEt_3 a still greater proportion (47%) of the least stable cis isomer 39 is formed, and this isomer can thus be obtained in ca 35% yield from 37. See Scheme 4.

	37 (four isomers)	38	39	40
toluene, reflux		61%	33%	6%
MeOH + NEt_3, reflux		41	47	12
(equilibrium mixture		68	3	29)

Overall yield : in toluene : 84%"; in MeOH + NEt_3 : 77%

Scheme 4. Formation of trans and cis 2-methyl-cyclotridecenones

Such a ring enlargement has been applied to the silyl enol ether of 5,5-dimethyl-cyclohexenone and thus a new synthesis of eucarvone 41, with an overall yield of 30%, has been achieved (49).

Conversion of open-chain carbonyl compounds into α-methylene-ketones and α-methyl-αβ-unsaturated ketones and aldehydes

This conversion can be readily carried out via the reaction sequence 1 ⟶ 32 ⟶ 33. Thus, on heating the crude mixture of E and Z chloro-silyloxy-methyl-cyclopropane isomers 42, obtained from CClMe and the silyl enol ethers of acetophenone, methyl-cyclopropylketone and methylpentylketone, for some hours in toluene or better in MeOH + NEt$_3$ under reflux, the corresponding α-methylene-ketones 43 are obtained in good yields (70-90% from the silyl enol ethers).

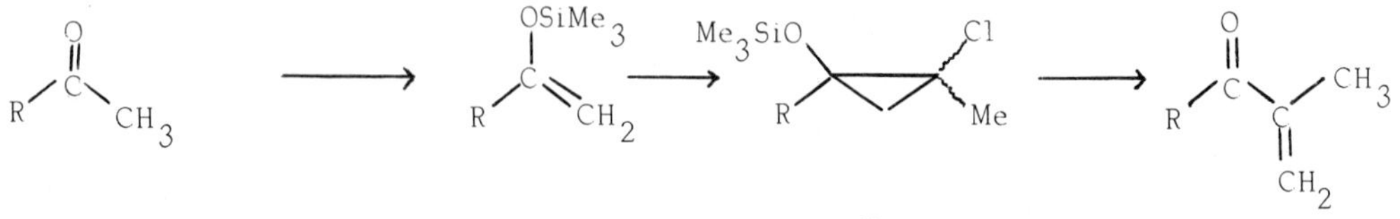

R = phenyl, cyclopropyl or pentyl

Likewise from isobutanal and from methylisopropylketone it is easy to prepare the corresponding α-methyl-αβ-ethylenic aldehyde and ketone 44 in 68 and 78% yield, respectively, from the silyl enol ethers (47).

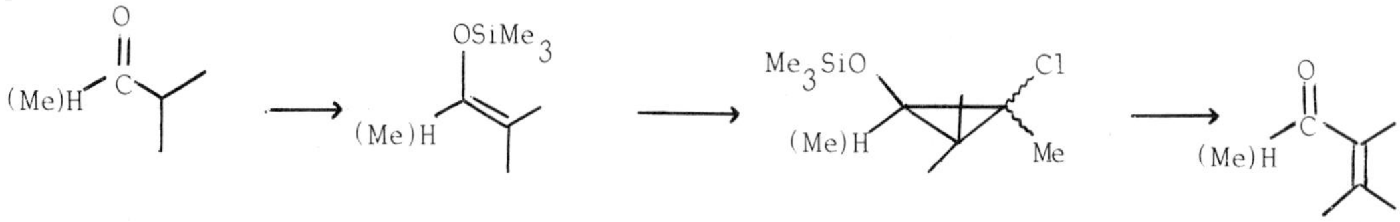

The reaction has been applied to the synthesis of natural products such as (±)-nuciferal 45 and (±) manicone 46 (49). In both cases, the trans product only is obtained (see below).

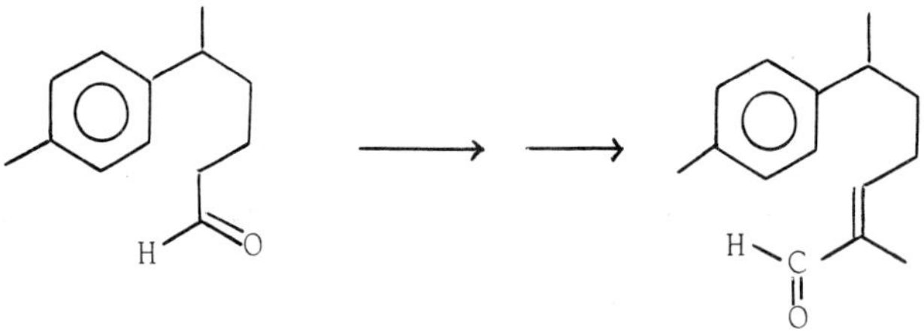

Preparation of Z and (or) E enals and enones from the appropriate silyloxy-chloro-cyclopropanes

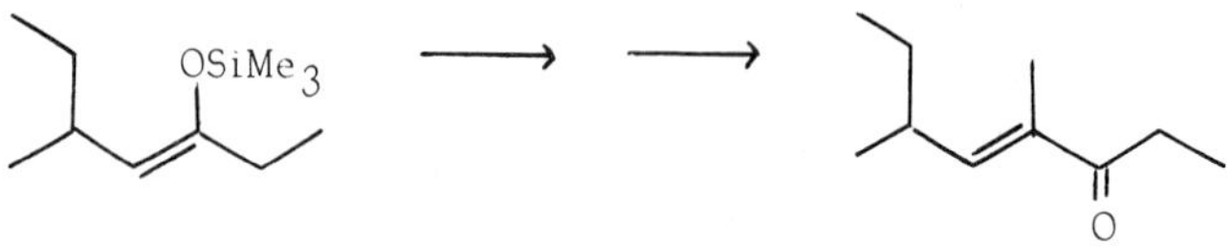

It appears clearly now that it must be possible to predict the possibility of direct stereospecific synthesis of cis and (or) trans αβ-ethylenic carbonyl compounds R—CH=CH—CO—R(H), and of Z and E isomers of more substituted carbonyl compounds, by using such a ring opening reaction carried out from the appropriate chloro-silyloxy-cyclopropane derivatives. The interest of this reaction is considerable, in particular because of the small number of the known syntheses of cis-enones and cis-enals, all the less numerous as these compounds undergo very facile cis ⟶ trans isomerisation.

One reaction has been investigated in detail : the conversion of an open-chain aldehyde n-heptanal, via the trans and cis silyl enol ethers 47 and 48, isolated by careful distillation, and then the four chloro-silyloxy-methylcyclopropane derivatives 49 (E and Z) and 50 (E and Z) into 2-methyl-oct-2-enals 51Z and 51E (48).

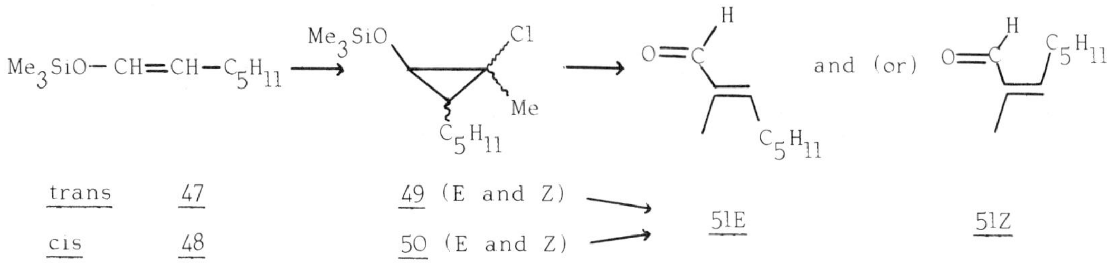

trans	47	49 (E and Z)		
cis	48	50 (E and Z)	51E	51Z

With CClMe each of the enol ethers 47 and 48 gave two chloro-silyloxy-cyclopropane derivatives 49E and 49Z (70/30) from 47 and 50E and 50Z (50/50) from 48 (see Scheme 6).

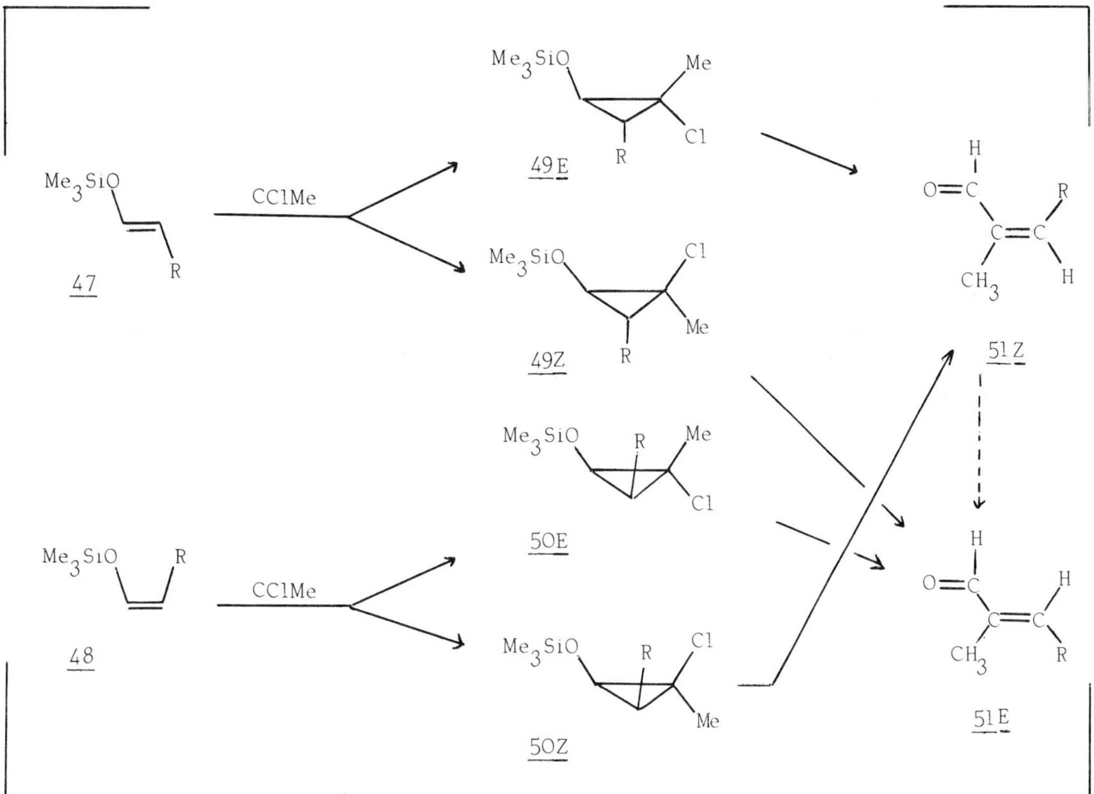

Scheme 6. (R = C_5H_{11}) Preparation of E and (or) Z 2-methyl-oct-2-enals 51

All four, isolated by chromatography on silica-gel and then heated separately or together in refluxing xylene for one day, were entirely converted into 2-methyl-oct-2-enal 51E (yield ∼ 85%). But the behaviour in various solvents at various temperatures of each of them separately was also investigated. It was easy to establish that 51Z is the kinetic product from 49E and 50Z and that 51E is the kinetic product from 49Z and 50E and so to perfect a good preparation of 2-methyl-oct-2-enal 51Z and to avoid the very facile isomerisation 51Z ⟶ 51E. The best way appears to be to operate via the corresponding chlorocyclopropanols of 49E and 50Z, for example to dissolve 49E and 50Z in slightly acidified methanol and after ten minutes to add NEt_3. Two hours and 9 hours respectively in this solvent at 20° are then sufficient to obtain a mixture of 51Z and 51E (overall yield : ∼ 90%) very rich in the Z isomer (95/5 and 80/11) because in this case practically no Z ⟶ E isomerisation occurs. The reaction is highly stereoselective and probably stereospecific and it is a good illustration of the rules of Woodward, Hoffmann and De Puy.

REFERENCES

1. S.S. Washburne, J. Organometal. Chem., a 83, 155-211 (1974) ; b 123, 1-73 (1976).
2. J.M. Conia, Pure and Applied Chemistry, 43, 317-321 (1975).
3. R. Calas and J. Dunogues, J. Organometal. Chem. Library, 2, 277-404 (1976).
4. J.K. Rasmussen, Synthesis, 91-110 (1977).
5. L. Birkofer and O. Stuhl, Topics in Current Chemistry - Springer Verlag, 88, 33-81 (1980).
6. E. Colvin, Silicon in Organic Syntheses, Butterworths London, 1981.
7. Ref (1a), p. 164-167 ; Ref (1b), p. 19-22 ; Ref (3) ; Ref (4), p. 96-110 ; Ref (5), p. 53-55 ; Ref (6), p. 243-250.
8. Ref (6), p. 198-213.
9. L. Blanco, Thesis, Orsay 1982, p. 10-18 and a paper to be published.
10. H.O. House, L.J. Czuba, M. Gall and H.D. Olmstead, J. Org. Chem., 34, 2324-2336 (1969).
11. H.E. Simmons and R.D. Smith, J. Amer. Chem. Soc., 80, 5323-5324 (1958) and 81, 4256-4264 (1959).
12. J.M. Denis, C. Girard and J.M. Conia, Synthesis, 549-551 (1972) ; Fieser and Fieser, Reagents for Organic Synthesis, Wiley, New-York, 4, 430 (1974).
13. G. Rousseau and J.M. Conia, Tetrahedron Letters, 649-652 (1981).
14. J.M. Denis and J.M. Conia, Tetrahedron Letters, 4593-4596 (1972).
15. J.M. Conia and C. Girard, Tetrahedron Letters, 2767-2770 (1973).
16. C. Girard and J.M. Conia, Tetrahedron Letters, 3333-3334 (1974).
17. G.M. Rubottom and M.J. Lopez, J. Org. Chem., 38, 2097-2099 (1973).
18. S. Murai, T. Aya and N. Sonoda, J. Org. Chem., 38, 4354-4356 (1973).
19. C. Girard and J.M. Conia, Tetrahedron Letters, 3327-3328 (1974).
20. R. Le Goaller and J.L. Pierre, Bull. Soc. Chim Fr., 1531-1532 (1973).
21. C. Girard and J.M. Conia, J. Chem. Res., S 182 and M 2348-2385 (1978).
22. Y. Ito, S. Fujii and T. Saegusa, J. Org. Chem., 41, 2073-2074 (1976).
23. Y. Ito, S. Fujii, M. Nakastsuka, F. Kawamoto and T. Saegusa, Organic Synthese, 59, 289-290 (1973).
24. B.M. Trost and M.J. Bogdanowicz, J. Amer. Chem. Soc., 95, 289-290 (1973).
25. R.H. Reuss and A. Hassner, J. Org. Chem., 39, 1785-1787 (1974).
26. L. Blanco, P. Amice and J.M. Conia, Synthesis, 3, 194-196 (1976).
27. S. Murai, Y. Seki and N. Sonoda, J.C.S. Chem. Comm., 1032-1033 (1974).
28. S. Murai, T. Aya, T. Renge, I. Ryu and N. Sonoda, J. Org. Chem., 39, 858-859 (1974). I. Ryu, S. Murai, S. Otani and N. Sonoda, Tetrahedron Letters, 1995-1998 (1977).
29. I. Ryu, S. Murai and N. Sonoda, Tetrahedron Letters, 4611-4614 (1977).
30. I. Ryu, K. Matsumoto, M. Ando, S. Murai and N. Sonoda, Tetrahedron Letters, 4283-4286 (1980).

31. G.M. Rubottom, R. Marrero, D.S. Krueger and J.L. Schreiner, Tetrahedron Letters 4013-4015 (1977).
32. J. Salaün and J.M. Conia, Tetrahedron Letters, 2849-2852 (1972).
33. J.P. Barnier, B. Garnier, C. Girard, J.M. Denis, J. Salaün and J.M. Conia, Tetrahedron Letters, 1747-1750 (1973) ; J. Salaün, B. Garnier and J.M. Conia, Tetrahedron, 30, 1413-1421 (1974).
34. H.H. Wasserman, R.E. Cochoy and M.S. Baird, J. Amer. Chem. Soc., 91, 2375-2376 and 4943 (1969).
35. J.M. Conia and M.J. Robson, Angew. Chem., 87, 505-516 (1975) ; Internat. Ed., 14, 473-485 (1975).
36. B.M. Trost and M.J. Bogdanowicz, J. Amer. Chem. Soc., 95, 289 and 5311-5321 (1973) ; B.M. Trost, Acc. Chem. Res., 7, 85-92 (1974) ; B.M. Trost, Y. Nishimura, K. Yamamoto and S.S. Mc Elvain, J. Amer. Chem. Soc., 101, 1328-1330 (1979).
37. C. Girard, P. Amice, J.P. Barnier and J.M. Conia, Tetrahedron Letters, 3329-3332 (1974).
38. J. Salaün and J. Ollivier, Nouv. J. Chim., 5, 587-594 (1981).
39. S.A. Monti, F.G. Cowherd and T.W. Mc Aninch, J. Org. Chem., 40, 858-862 (1975) and references cited therein ; B. Trost and P.H. Scudder, J. Org. Chem., 46, 506-509 (1981).
40. J.P. Barnier, B. Garnier, C. Girard, J.M. Denis, J. Salaün and J.M. Conia, Tetrahedron Letters, 1747 (1973).
41. P. Amice, L. Blanco and J.M. Conia, Synthesis, 196-197 (1976).
42. T. Hiyama, T. Mishima, K. Kitatani and H. Nozaki, Tetrahedron Letters, 3297-3300 (1974).
43. S. Torii, T. Okamoto and N. Ueno, J.C.S. Chem. Comm., 293-294 (1978).
44. M. Klehr and H.J. Schäfer, Angew. Chem. Internat. Ed., 14, 247-248 (1975).
45. T.L. Macdonald, Tetrahedron Letters, 4201-4204 (1978).
46. L. Blanco, P. Amice and J.M. Conia, Synthesis, 289-291 (1981).
47. L. Blanco, P. Amice and J.M. Conia, Synthesis, 291-293 (1981).
48. J.M. Conia and L. Blanco, to be published.
49. L. Blanco, N. Slougui, G. Rousseau and J.M. Conia, Tetrahedron Letters, 645-648 (1981).
50. S. Arora and P. Binger, Synthesis, 801-803 (1974).

A NEW SYNTHETIC APPROACH TO 1,4-DICARBONYL SYSTEMS AND FUNCTIONALIZED CYCLOPENTENONES BASED ON THE HORNER-WITTIG REACTION OF PHOSPHONATES CONTAINING SULFUR

Marian Mikolajczyk

Centre of Molecular and Macromolecular Studies, Polish Academy of Sciences, Department of Organic Sulfur Compounds, 90-362 Lodz, Boczna 5, Poland

<u>Abstract</u> - A general synthesis of 1,4-dicarbonyl compounds employing the Horner-Wittig reaction of the properly substituted α-phosphoryl sulfides with mono-carbonyl or the half-protected 1,3-dicarbonyl compounds has been developed. The utility of this method is demonstrated by the syntheses of dihydrojasmone, methylenomycin B and 4-hydroxy-2-cyclopentenones. The synthetic applications of the addition of elemental sulfur and selenium to phosphonate carbanions are also presented.

INTRODUCTION

The application of organic sulfur and phosphorus compounds in organic synthesis (Refs.1 to 3) and particularly in reversal (Umpolung) of normal reactivity of nucleophilic and electrophilic centers (Refs. 4 and 5) is a subject of considerable current interest. For the past few years, work in this Laboratory has centered on the preparation and reactions of α-phosphoryl-substituted organosulfur compounds of general formula (1). They were found to be useful reagents for effecting a variety of synthetic transformations.

$(RO)_2\overset{\underset{\parallel}{O}}{P}-\overset{\underset{|}{Y}}{C}H-X$

(1)

(1a), X=SR, Y=H, alkyl, aryl
(1b), X=S(O)R, Y=H, alkyl, aryl
(1c), X=S(O)$_2$R, Y=H, alkyl, aryl
(1d), X=SR, Y=OR
(1e), X=Y=SR

Thus, α-phosphoryl sulfoxides (1b), which recently became readily available in racemic (Refs.6 to 8) and optically active forms (Ref.9), undergo the Pummerer-type reactions (Ref.10), halogenation, alkylation, oxidation and reduction (Ref.11). Owing to the presence of the phosphonate moiety, α-phosphoryl sulfoxides (1b) are key substrates in the synthesis of α,β-unsaturated sulfoxides based on the Horner-Wittig reaction (Refs.9 and 12). The reactions of α-phosphoryl sulfoxides (1b) mentioned above are summarized in Scheme I.

S,S-Thioacetals of formylphosphonates (1e) are the reagents of choice for the synthesis of ketene S,S-thioacetals (Ref.13). The Horner-Wittig reaction of (1e) with carbonyl compounds was found to be a general reaction which affords ketene S,S-thioacetals in high yields. It is interesting to note that the Horner-Wittig reaction of (1e) with aromatic aldehydes can also be carried out under two-phase transfer catalytic conditions.

$(RO)_2\overset{\underset{\parallel}{O}}{P}CH\genfrac{}{}{0pt}{}{\diagup SR}{\diagdown SR} \xrightarrow[\text{3.H}^+]{\text{1.Base} \atop \text{2.R}^1\text{R}^2\text{CO}} \genfrac{}{}{0pt}{}{R^1}{R^2}\!\!\diagdown\!\!\diagup C=C\genfrac{}{}{0pt}{}{\diagup SR}{\diagdown SR} + (RO)_2PO_2H$

(1e)

Scheme I.

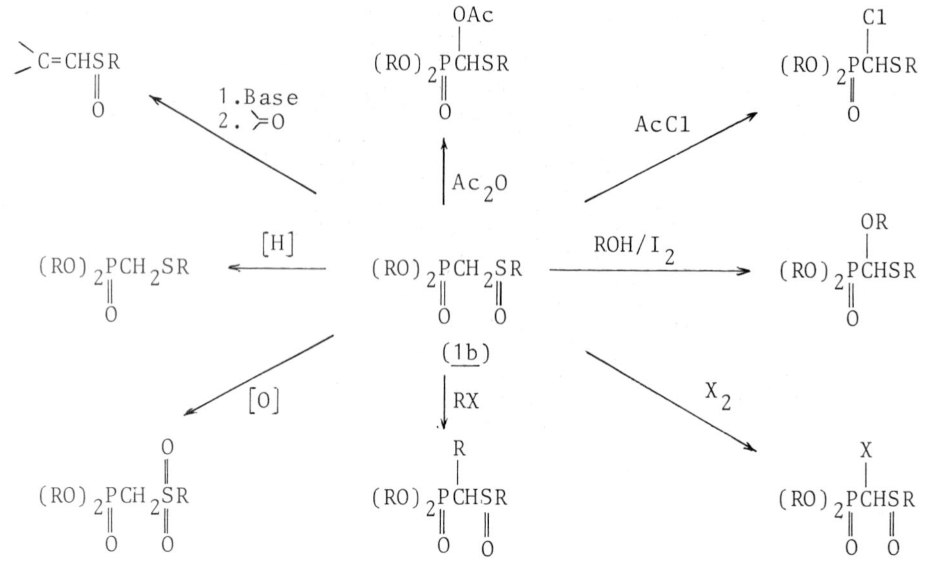

[Scheme showing transformations of (RO)$_2$P(O)CH$_2$SR (1b) with various reagents:
- 1.Base, 2. >=O → >C=CHSR with P(O) group
- Ac$_2$O → (RO)$_2$PCH(OAc)SR
- AcCl → (RO)$_2$PCH(Cl)SR
- [H] → (RO)$_2$PCH$_2$SR
- ROH/I$_2$ → (RO)$_2$PCH(OR)SR
- [O] → (RO)$_2$P(O)CH$_2$S(O)R
- RX → (RO)$_2$PCH(R)SR
- X$_2$ → (RO)$_2$PCH(X)SR]

Our interest in the chemistry of α-phosphoryl sulfides (1a) was stimulated by the fact that they are key reagents in the synthesis of ketones. This method, developed by Corey and Shulman (Ref.14), involves the Horner-Wittig reaction of an α-phosphoryl sulfide (2) bearing an alkyl group on the α-carbon atom with a carbonyl component followed by hydrolysis of the α,β-unsaturated sulfide formed.

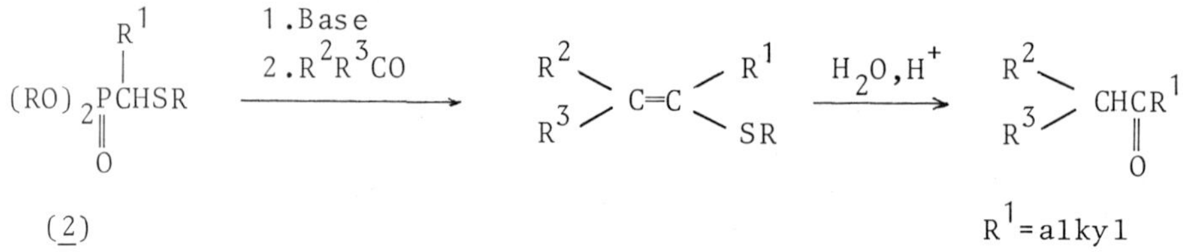

$$(RO)_2P(O)CHR^1SR \xrightarrow{1.\text{Base},\ 2.R^2R^3CO} R^2R^3C=C(R^1)SR \xrightarrow{H_2O,H^+} R^2R^3CHC(O)R^1$$

(2)   R^1 = alkyl

Since the original carbonyl carbon atom of an electrophile (R^{2}R^{3}CO) is reduced to a saturated carbon atom and the α-carbon atom of a phosphonate is transformed into a carbonyl carbon atom in a final product, the synthesis of ketones shown above may be in general classified as a nucleophilic acylation with elaboration at the electrophilic center or reductive nucleophilic acylation. It should be noted, however, that the main limitation of this approach to the synthesis of ketones involved the general synthesis of the α-phosphoryl sulfide (2) containing the alkyl or aryl α-substituent R. Finding such a synthetic method was the starting point of the study presented in this account. However, the major objective of our study was the application of α-phosphoryl sulfides (2) as an acyl anion equivalents for the synthesis of 1,4-dicarbonyl compounds which have proven to be valuable intermediates for the elaboration of cyclopentenone-systems. 2-Cyclopentenones are structural units contained in many important natural products such as prostanoids, jasmonoids, methylenomycins and pentenomycins.
Before reporting the results on the synthesis of 1,4-dicarbonyl compounds and functionalized cyclopentenones it seems desirable to describe briefly the addition of elemental sulfur and selenium to phosphonate carbanions and its synthetic utility.

SYNTHETIC APPLICATIONS OF ADDITION OF ELEMENTAL SULFUR AND SELENIUM TO PHOSPHONATE CARBANIONS

Although the metallated phosphonates may by sulfenylated directly by means of dialkyl or diaryl disulfides (Ref.15) it was found that more convenient and general is the two-step procedure which involves addition of elemental sulfur to the metallated phosphonates and subsequent alkylation (Ref.16). Addition of elemental sulfur to the lithium salts of both unsubstituted and variously substituted phosphonates takes place readily at and above -20°C in THF solution affording the corresponding lithium mercaptides which can be alkylated in situ or acidified and then alkylated under phase-transfer catalytic conditions. According to this procedure we were able to synthesize a variety of substituted or unsubstituted α-phosphoryl sulfides (1a), symmetrical or unsymmetrical S,S-thioacetals of formylphosphonates (1e) and O,S-thioacetals of formylphosphonates (1d) in high yields.

$$(RO)_2\overset{\text{O}}{\underset{\|}{P}}CH_2R^1 \xrightarrow[\text{3.MeI}]{\substack{1.\text{n-BuLi} \\ 2.\overline{S}_8}} (RO)_2\overset{\text{O}}{\underset{\|}{P}}CHR^1 \text{ (SMe)}$$

R¹ = H, alkyl, aryl, SR, OR

Alkylation of the thiols resulting from the sulfur addition to phosphonate carbanions with ethylene ketal of 5-chloro-pentanone-2 followed by oxidation and deblocking of the carbonyl group affords sulfoxides (3a,b) and sulfones (4a,b). Their internal Horner-Wittig reaction results in the formation of the cyclic, unsaturated six-membered sulfoxides (5a,b) and sulfones (6a,b) (Ref.17).

Scheme II

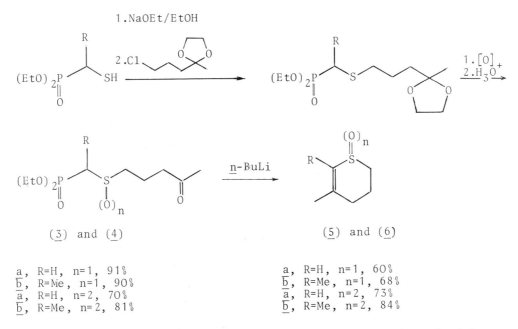

(3) and (4)

a, R=H, n=1, 91%
b, R=Me, n=1, 90%
a, R=H, n=2, 70%
b, R=Me, n=2, 81%

(5) and (6)

a, R=H, n=1, 60%
b, R=Me, n=1, 68%
a, R=H, n=2, 73%
b, R=Me, n=2, 84%

From the viewpoint of the synthesis of ketones according to the principle discussed above the most important product, which can be obtained via sulfur addition, is α-phosphoryl-α-phenyl-methyl methyl sulfide (7). It was found to react with aldehydes giving vinyl sulfides which upon hydrolysis were converted into the corresponding aromatic ketones (8).

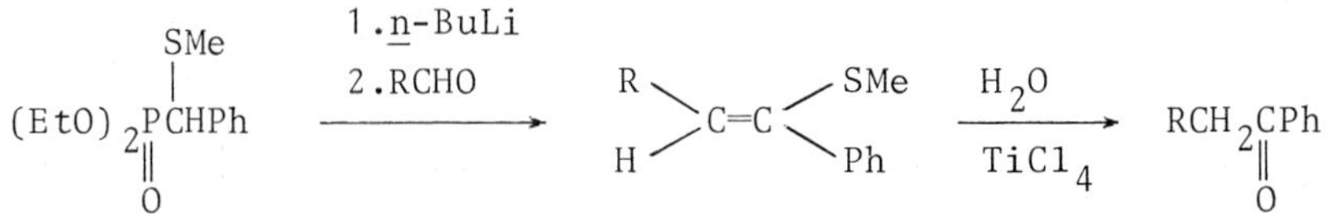

$$(EtO)_2\underset{\underset{O}{\|}}{P}CHPh \xrightarrow[2.RCHO]{1.\underline{n}-BuLi} \underset{H}{\overset{R}{>}}C=C\underset{Ph}{\overset{SMe}{<}} \xrightarrow[TiCl_4]{H_2O} RCH_2\underset{\underset{O}{\|}}{C}Ph$$
$$(\underline{7}) \hspace{6cm} (\underline{8})$$

with SMe on the phosphonate carbon in compound (7).

In consequence, the reaction sequence shown above may be considered as a new method for the conversion of any aldehyde into an aromatic ketone. Moreover, this method in combination with the synthesis of unsubstituted phosphonates (Arbusov or Michaelis-Becker reaction) used for the preparation of sulfides (1a) makes it possible to convert an alkyl halide into a ketone as schematically shown below.

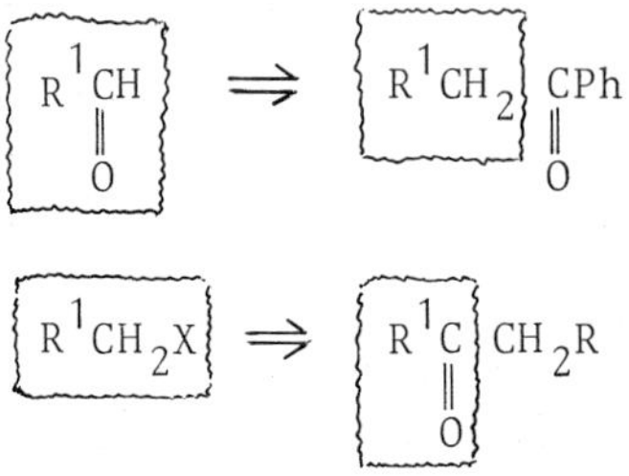

$$\{R^1\underset{\underset{O}{\|}}{C}H\} \Rightarrow \{R^1CH_2\}\underset{\underset{O}{\|}}{C}Ph$$

$$\{R^1CH_2X\} \Rightarrow \{R^1\underset{\underset{O}{\|}}{C}\}CH_2R$$

In an extension of our work on addition of sulfur to phosphonate carbanions, we found (Ref.18) that elemental selenium reacts also readily with lithium phosphonates to give α-phosphoryl lithioselenols which, without isolation, are converted into α-phosphoryl selenides (9) on treatment with methyl iodide.

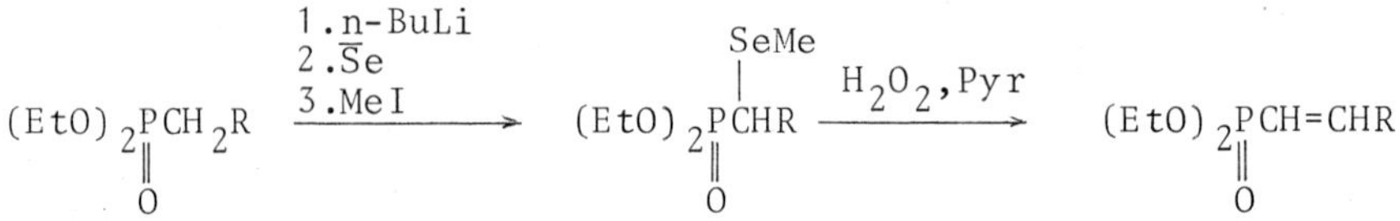

$$(EtO)_2\underset{\underset{O}{\|}}{P}CH_2R \xrightarrow[3.MeI]{\substack{1.\underline{n}-BuLi \\ 2.\overline{S}e}} (EtO)_2\underset{\underset{O}{\|}}{P}\overset{SeMe}{\underset{|}{C}}HR \xrightarrow{H_2O_2, Pyr} (EtO)_2\underset{\underset{O}{\|}}{P}CH=CHR$$
$$(\underline{9}) \hspace{5cm} (\underline{10})$$

A convenient synthesis of α-phosphoryl selenides (9) in combination with the well-known selenoxide elimination represents a new way for the preparation of vinylphosphonates (10). The utility of this approach is exemplified by the synthesis of diethyl α-methylthiovinylphosphonate (11) as shown in Scheme III.

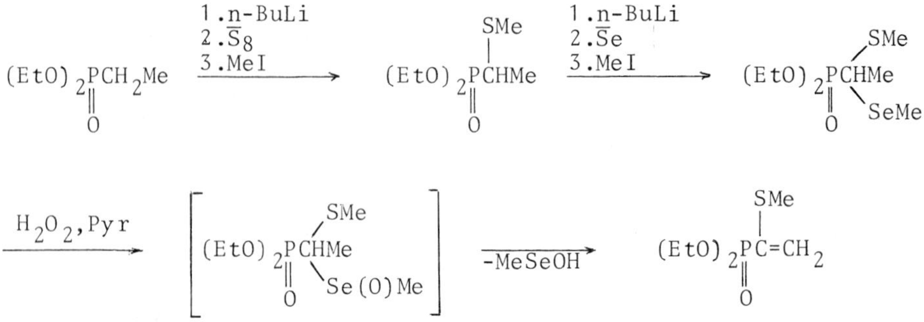

$$(EtO)_2\underset{\underset{O}{\|}}{P}CH_2Me \xrightarrow[3.MeI]{\substack{1.\underline{n}-BuLi \\ 2.\overline{S}_8}} (EtO)_2\underset{\underset{O}{\|}}{P}\overset{SMe}{\underset{|}{C}}HMe \xrightarrow[3.MeI]{\substack{1.\underline{n}-BuLi \\ 2.\overline{S}e}} (EtO)_2\underset{\underset{O}{\|}}{P}CHMe\underset{SeMe}{\overset{SMe}{<}}$$

$$\xrightarrow{H_2O_2, Pyr} \left[(EtO)_2\underset{\underset{O}{\|}}{P}CHMe\underset{Se(O)Me}{\overset{SMe}{<}}\right] \xrightarrow{-MeSeOH} (EtO)_2\underset{\underset{O}{\|}}{P}\overset{SMe}{\underset{|}{C}}=CH_2$$
$$(\underline{11})$$

Another synthesis of vinylphosphonate (11) involves the Peterson-type reaction of α-phosphoryl-α-trimethylsilyl-methyl methyl sulfide (12) with formaldehyde (Ref.19).

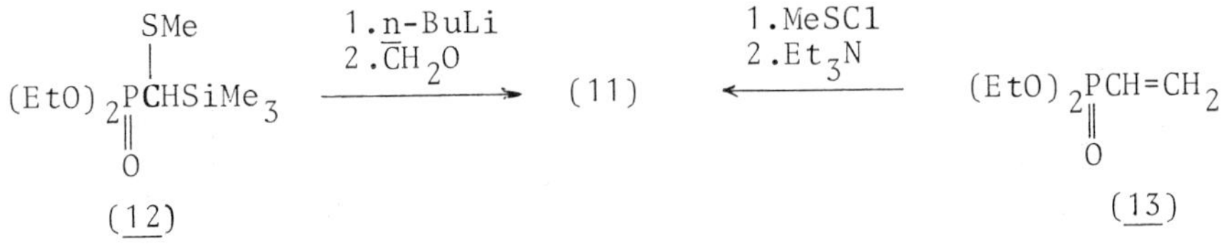

$$(EtO)_2\underset{O}{\overset{\|}{P}}CH(SMe)SiMe_3 \quad \xrightarrow[2.\overline{CH}_2O]{1.\text{n-BuLi}} \quad (11) \quad \xleftarrow[2.Et_3N]{1.MeSCl} \quad (EtO)_2\underset{O}{\overset{\|}{P}}CH=CH_2$$

(12) (13)

It was found, however, that for a large scale preparation of (11) a different synthetic approach is more convenient. It consists of the addition of metanesulfenyl chloride to diethyl vinylphosphonate (13) followed by dehydrochlorination (Ref.20). The latter procedure is based on the results of Ivin and coworkers (Ref.21) on the addition of sulfenyl chlorides to vinylphosphonates.

It should be noted that vinylphosphonate (11) represents a new masked reagent of the type shown below and its reactivity may be compared with that of ketene S,S-thioacetal monooxide.

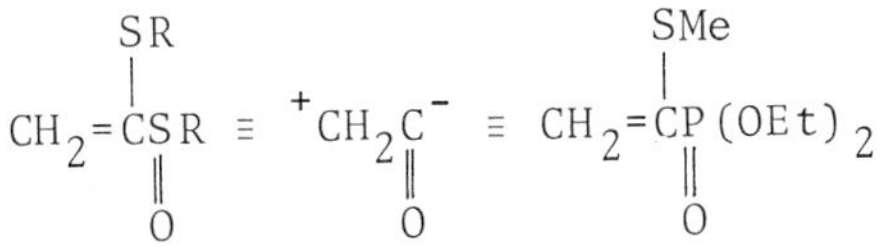

$$CH_2=\underset{O}{\overset{SR}{\underset{\|}{C}}}SR \equiv \,^+CH_2\underset{O}{\overset{SR}{\underset{\|}{C}}}^- \equiv CH_2=\underset{O}{\overset{SMe}{\underset{\|}{C}}}P(OEt)_2$$

Therefore, (11) behaves as a Michael acceptor and reacts with a variety of anionic nucleophiles to produce the corresponding addition products shown in Scheme IV.

Scheme IV.

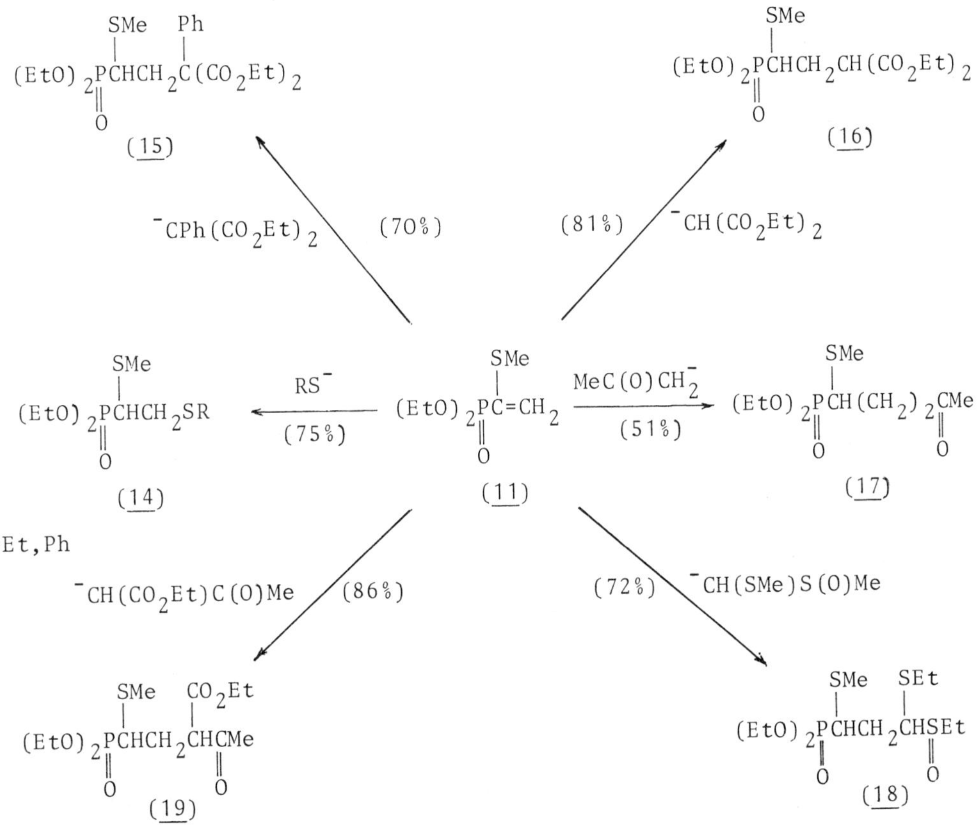

$$(EtO)_2\underset{O}{\overset{SMe}{\underset{\|}{P}}}CHCH_2\underset{}{\overset{Ph}{C}}(CO_2Et)_2 \qquad\qquad (EtO)_2\underset{O}{\overset{SMe}{\underset{\|}{P}}}CHCH_2CH(CO_2Et)_2$$

(15) (16)

$$^-CPh(CO_2Et)_2 \quad (70\%) \qquad (81\%) \quad ^-CH(CO_2Et)_2$$

$$(EtO)_2\underset{O}{\overset{SMe}{\underset{\|}{P}}}CHCH_2SR \xleftarrow[(75\%)]{RS^-} (EtO)_2\underset{O}{\overset{SMe}{\underset{\|}{P}}}C=CH_2 \xrightarrow[(51\%)]{MeC(O)CH_2^-} (EtO)_2\underset{O}{\overset{SMe}{\underset{\|}{P}}}CH(CH_2)_2\underset{O}{\overset{\|}{C}}Me$$

(14) (11) (17)

R=Et,Ph

$$^-CH(CO_2Et)C(O)Me \quad (86\%) \qquad (72\%) \quad ^-CH(SMe)S(O)Me$$

$$(EtO)_2\underset{O}{\overset{SMe}{\underset{\|}{P}}}CHCH_2\underset{O}{\overset{CO_2Et}{\underset{\|}{C}H}}CMe \qquad\qquad (EtO)_2\underset{O}{\overset{SMe}{\underset{\|}{P}}}CHCH_2\underset{O}{\overset{SEt}{\underset{\|}{C}H}}SEt$$

(19) (18)

Since all the addition products (14) to (19) contain the α-phosphoryl sulfide moiety, which according to the Umpolung concept may be regarded as a masked acyl anion, they can be further applied for the synthesis of functionalized ketones and diketones.

Thus, for example, the Horner-Wittig reaction of (14) with aromatic aldehydes followed by a mild hydrolysis of vinyl sulfides formed enabled us to obtain one of the two regioisomers of β-ketosulfide (20) in a selective manner (Ref.20). In this context, it is interesting to note that this isomer of (20) would be formed by sulfenylation of the less thermodynamically stable enolate anion derived from the ketone (21).

$$(EtO)_2\underset{O}{\underset{\|}{P}}CH(SMe)CH_2SR \xrightarrow[\text{2.ArCHO}]{\text{1.NaH}} \underset{H}{\overset{Ar}{>}}C=C\underset{CH_2SR}{\overset{SMe}{<}} \xrightarrow[\text{MeCN}]{H_2O, TiCl_4} ArCH_2\underset{O}{\underset{\|}{C}}CH_2SR$$

(14)　　　　　　　　　　　(E+Z)　　　　　　　　　　　(20a)

$$ArCH_2\underset{O}{\underset{\|}{C}}CH_3$$
(21)

Base ↙　　↘ Base

ArCH$_2$C̄CH$_2$　　　　　ArC̄HCCH$_3$
 ‖　　　　　　　　　　　　 ‖
 O　　　　　　　　　　　　 O

↓ RSSR　　　　　　　　　↓ RSSR

ArCH$_2$CCH$_2$SR　　　　ArCHCCH$_3$ (with SR)
 ‖　　　　　　　　　　　　 ‖
 O　　　　　　　　　　　　 O

(20a)　　　　　　　　　　　(20b)

Especially interesting are the products (17) and (19) because they can be applied as intermediates for the synthesis of 1,4-diketones. These results will be discussed in the next section of this account.

SYNTHESIS OF 1,4-DICARBONYL COMPOUNDS AND FUNCTIONALIZED 2-CYCLOPENTENONES BY MEANS OF α-PHOSPHORYL SULFIDES

General considerations

The most versatile synthesis of 2-cyclopentenones involves the initial preparation of acyclic 1,4-dicarbonyl compounds and their subsequent intramolecular base-catalyzed aldol condensation (Ref.22). Therefore, the synthesis of 1,4-dicarbonyl compounds is still a subject of an extensive study. Our approach to the synthesis of 1,4-dicarbonyl compounds employing the Horner-Wittig reaction of the properly substituted α-phosphoryl sulfides was deduced from a retrosynthetic analysis (routes A and B) shown in Scheme V.

According to the route A the precursor of the 1,4-dicarbonyl system is the vinyl sulfide (22) which can be obtained by the Horner-Wittig reaction of the already mentioned α-phosphoryl sulfide (17) with a carbonyl component. The same 1,4-dicarbonyl compound may alternatively be prepared from the isomeric vinyl sulfide (23) in which the double bond is located between the carbon atoms 3 and 4. The synthesis of the latter can be accomplished by reacting α-phosphoryl sulfide (1a) with 1,3-dicarbonyl compound under the Horner-Wittig reaction conditions.

Scheme V

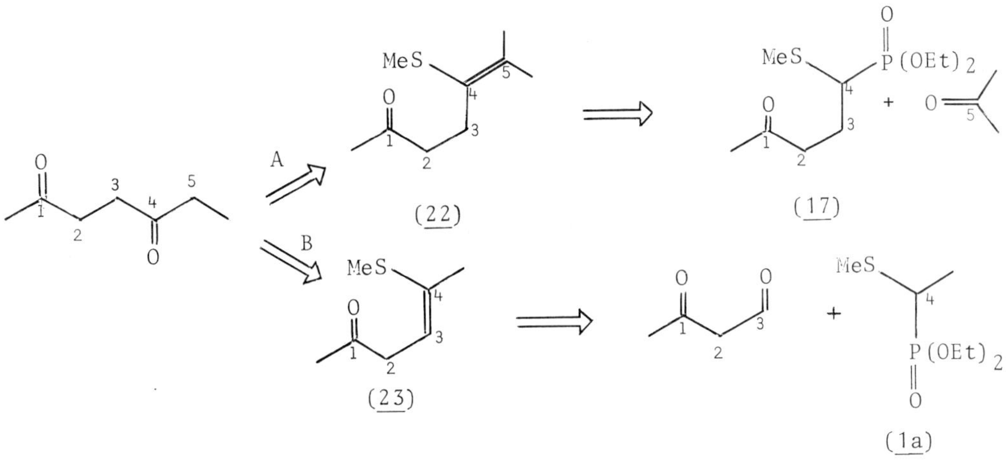

In practical realization of the synthesis of 1,4-dicarbonyl compounds indicated by the route A it was necessary to use instead of α-phosphoryl sulfide (17) its derivative (24) with the protected carbonyl group in order to avoid the generation of the enolate anion on treatment with a strong base like n-butyllithium. Although the acid-catalyzed reaction of (17) with ethylene glycol leads to the requisite product (24) in high yield we found that it may be prepared in a shorter way by the Arbusov reaction of triethyl phosphite with 5-chloro-2-pentanone ethylene ketal followed by the sulfur addition and methylation.

Scheme VI

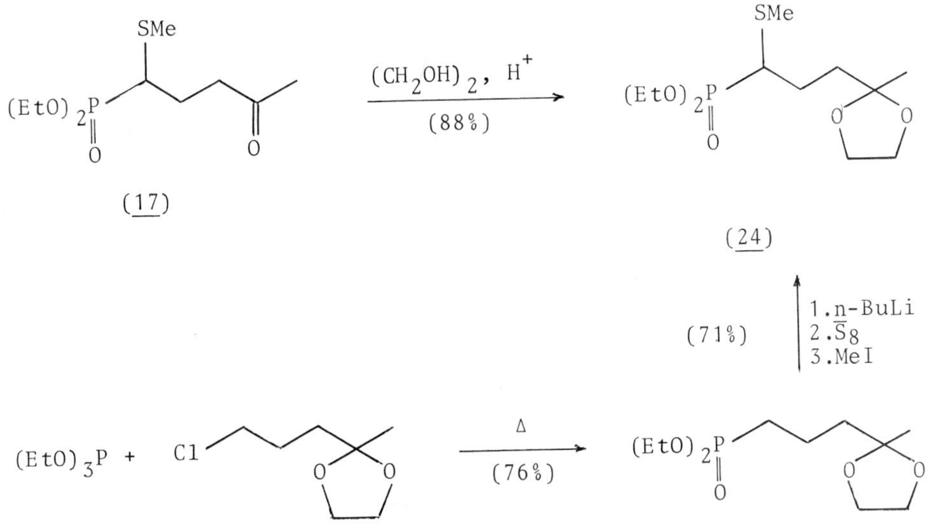

Similarly, in order to accomplish the conversion of 1,3- into 1,4-dicarbonyl compounds suggested by retrosynthetic analysis (route B) it was necessary to take the half-protected 1,3-dicarbonyl compounds for the Horner-Wittig reaction. Since the aldehyde (25) can be easily prepared from ethyl acetoacetate this compounds was chosen in our studies as convenient starting material.

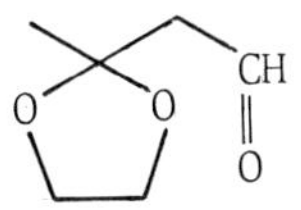

The synthesis of some selected 1,4-dicarbonyl compounds and the corresponding 2-cyclopentenones derived from them based on the principles discussed herein is described below.

Synthesis of dihydrojasmone

A great number of compounds have been identified in the essential oil of Jasminum grandiflorum (L), most of which are responsible for the characteristic jasmine fragrance. Of particular importance are cis-jasmone (26) and methyljasmonate (27). Since both cis-jasmone (26) and its derivative (28) with a saturated side-chain are components of many perfumes, their syntheses have recently attracted a great attention (Ref.24). Moreover, the synthesis of dihydrojasmone (28) is usually carried out to test the quality of a new synthesis of 1,4-dicarbonyl system.

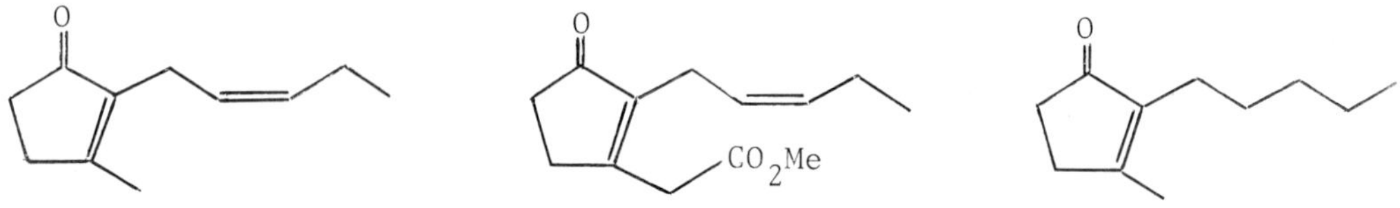

(26)   (27)   (28)

For this reason we decided to synthesize (Refs.18 and 25) dihydrojasmone (28) to demonstrate the usefulness of our novel strategy for the construction of the 1,4-dicarbonyl and cyclopentenone skeleton. Treatment of the lithium salt of (24) with n-hexanal results in the formation of the corresponding addition product which was rather stable and could not be decomposed to the Horner-Wittig reaction products even on a prolonged reflux in THF solution. It was found, however, that the isolated β-hydroxy phosphonate (mixture of two diastereomers) undergoes a facile decomposition at room temperature on treatment with sodium hydride in the presence of catalytic amount of 18-crown-6. The vinyl sulfide formed was hydrolysed under acidic conditions to give undecane-2,5-dione (29) - a precursor of (28). Its cyclization carried out in a standard way afforded dihydrojasmone (28). The overall yield of (28) prepared in this way was 66% from (24).

Scheme VII

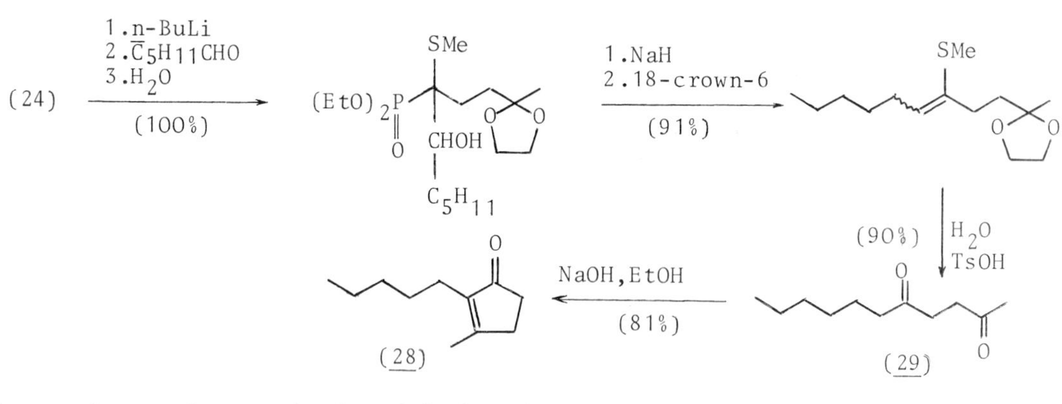

In an alternative synthesis of (28) aldehyde (25) and diethyl α-methylthio-n-heptanephosphonate (30) were used as the Horner-Wittig reaction components. The latter compound was prepared in two ways as shown in Scheme VII. Also in this case the addition product formed from the lithium salt of (30) and aldehyde (25) had to be isolated and decomposed under the conditions discussed above. The resulting vinyl sulfide was hydrolyzed to diketone (29) which, in turn, was transformed into dihydrojasmone (28). The overall yield of this process was 47% from (25).

Scheme VIII

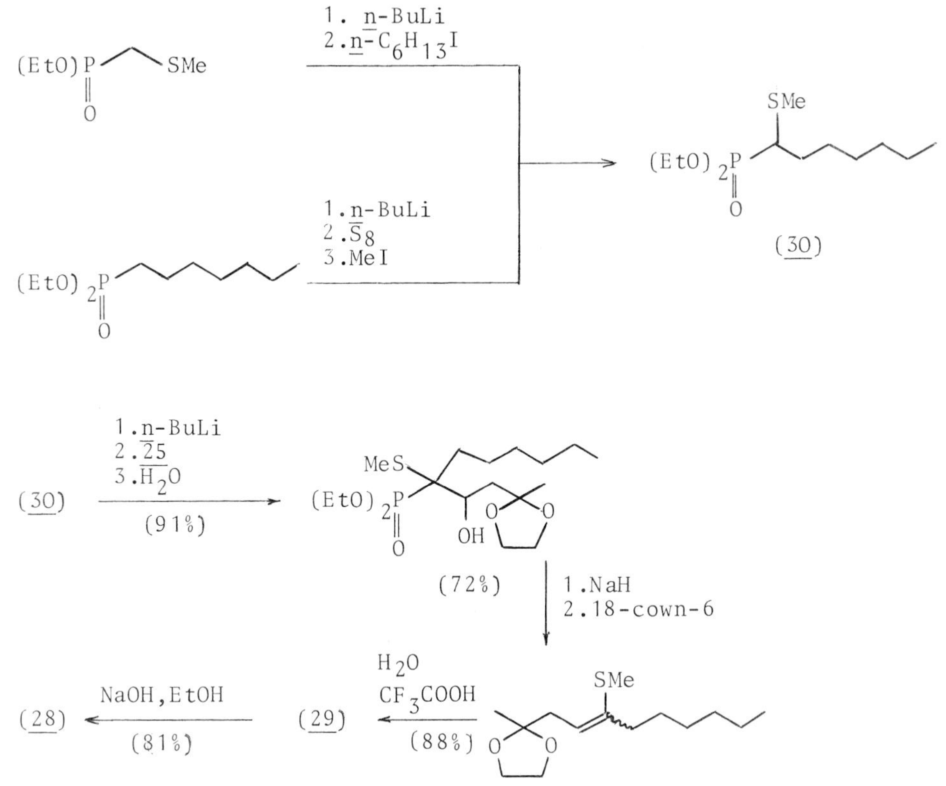

Synthesis of methylenomycin B

Methylenomycin A (31a), epimethylenomycin (31b), desepoxy-4,5-didehydromethylenomycin A (32) and methylenomycin B (33) have recently been isolated from the culture filtrate of a streptomycete strain (Streptomyces violaceoruber) and belong to a family of cyclopentenoid antibiotics. For the past few years, several groups have reported syntheses of methylenomycins (Refs.26 and 27)

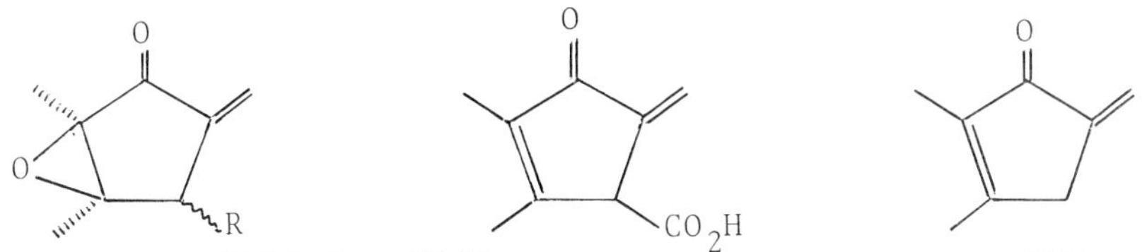

(31a), R=α-CO_2H; (31b), R=β-CO_2H (32)   (33)

In the course of our study it was found that our new approach to 1,4-dicarbonyl systems may also be utilized in the total synthesis of methylenomycin B (33) (Ref.25). A sub-target in this synthesis, n-heptane-2,5-dione (34) was prepared in two ways. The first involved the Horner-Wittig reaction of (24) with acetaldehyde and subsequent hydrolysis of the vinyl sulfide (35a) formed. The second route comprised the synthesis of the isomeric vinyl sulfide (35b) via the Horner-Wittig reaction of (25) with diethyl α-methyl-thio--n-propanephosphonate and hydrolysis of (35b) to (34). The cyclization of (34) under basic conditions gave 2,3-dimethylcyclopenten-2-one (36). The introduction of the exocyclic α-methylene function to (36) was accomplished according to Jernow et al. (Ref.27) via the hydroxymethylation followed by dehydration. After column chromatography methylenomycin B(33) was isolated in an overall yield 16% from (25) and showed spectral properties identical with those reported in the literature. The reactions discussed above are shown in Scheme IX.

Scheme IX.

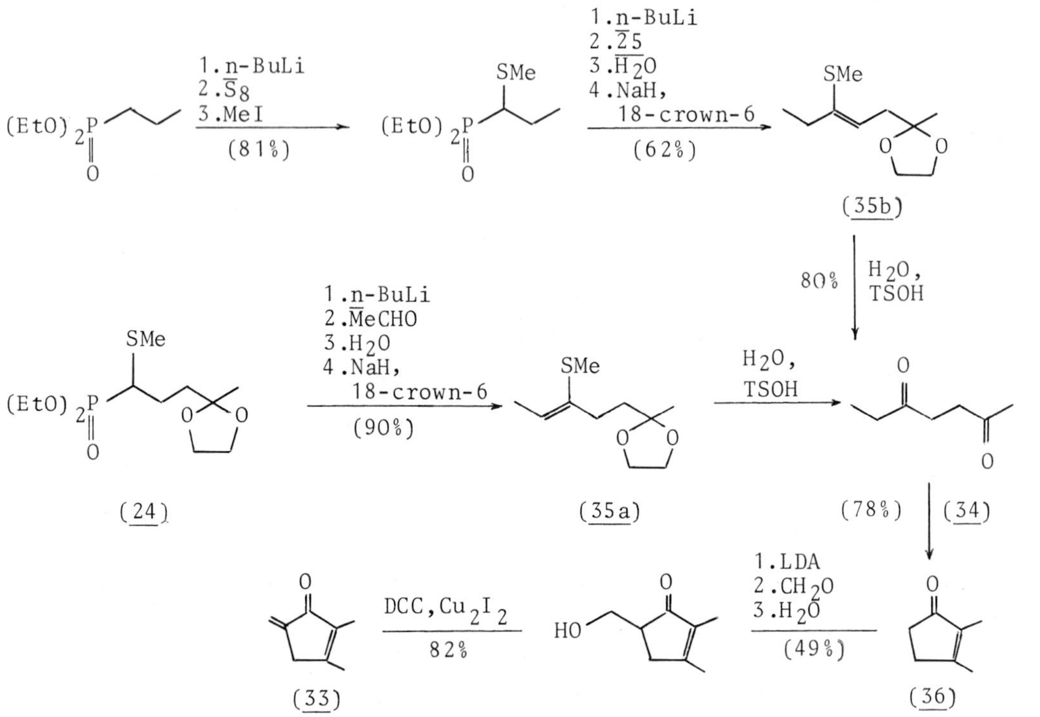

The synthesis of desepoxy-4,5-didehydromethylenomycin A (32) starting from (19) is being investigated.

Synthesis of 4-hydroxy-cyclopentenones

In an extension of our work on the synthesis of functionalized cyclopentenones using α-phosphoryl sulfides as key reagents, we examined the possibility of the synthesis of 4-hydroxy-cyclopentenones. Many natural products such as jasmonoids and prostanoids incorporate this structural unit.

Our synthetic strategy was based on retrosynthetic analysis shown in Scheme X.

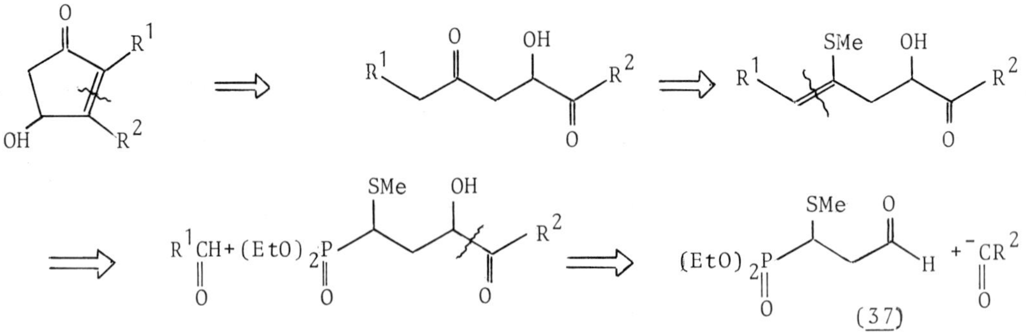

It indicated that phosphono-aldehyde (37) would be a key reagent. Having two reactive centers (an electrophilic center at the carbonyl carbon atom and a nucleophilic center at the phosphonate α-carbon atom) this compound should be able to react with an acyl anion equivalent as well as to undergo the Horner-Wittig reaction with a carbonyl component affording, in consequence, the desired 4-hydroxy-cyclopentenone system via the corresponding 2-hydroxy--1,4-diketone.

It was found that (37) can be easily and efficiently obtained in three steps starting from triethyl phosphite and diethyl acetal of 3-chloro-propanol-1 (Scheme XI). As expected, the addition of the acetyl anion equivalent |MeC̄(SEt)S(O)Et| to (37) occurs readily and gives, after protection of the hydroxy group, the corresponding addition product (39). However, less satisfactory were the results of the Horner-Wittig reaction of (39) with benzaldehyde. Although we were able to obtain the corresponding vinyl sulfide (40) - a potential source of 2-phenyl-3-methyl-cyclopent-2-en-1-one - its chemical yield was rather low. This can be explained by the steric hindrance occuring in the adduct and its good solvation properties.

Scheme XI

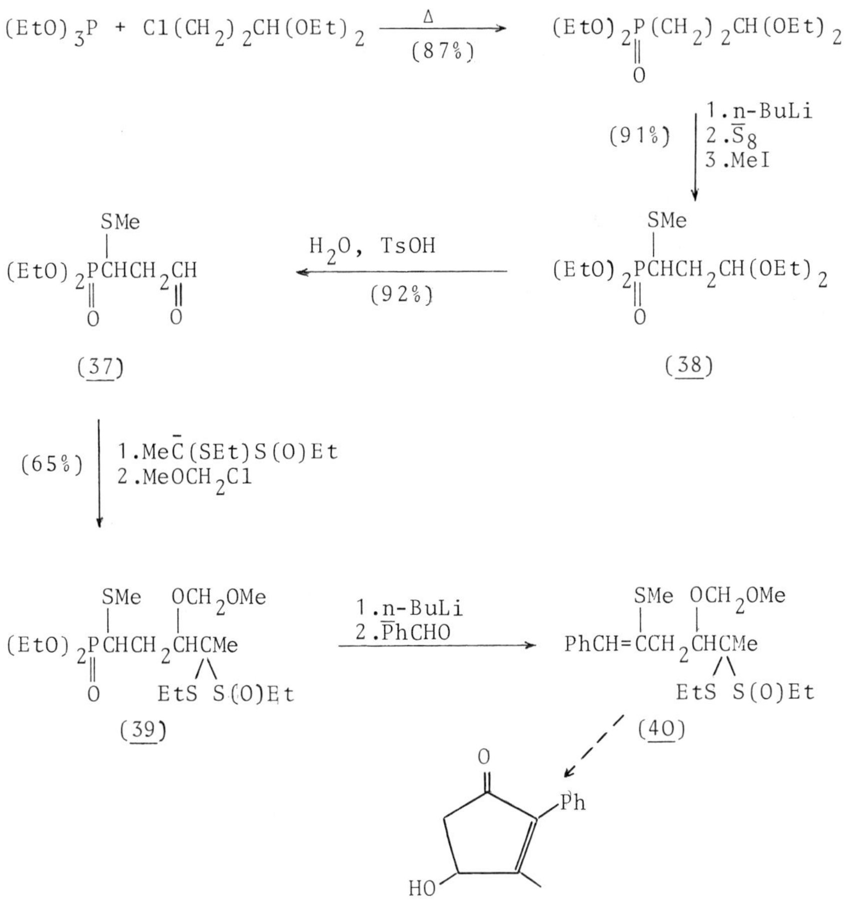

$$(EtO)_3P + Cl(CH_2)_2CH(OEt)_2 \xrightarrow[(87\%)]{\Delta} (EtO)_2P(O)(CH_2)_2CH(OEt)_2$$

$$\xrightarrow[(91\%)]{\substack{1.\,n\text{-BuLi}\\2.\,\bar{S}_8\\3.\,MeI}}$$

(38): $(EtO)_2P(O)CH(SMe)CH_2CH(OEt)_2$

$$\xrightarrow[(92\%)]{H_2O,\,TsOH}$$

(37): $(EtO)_2P(O)CH(SMe)CH_2CHO$

(65%) | 1. MeC̄(SEt)S(O)Et
      | 2. MeOCH_2Cl

(39): $(EtO)_2P(O)CH(SMe)CH_2CH(OCH_2OMe)C(Me)(SEt)(S(O)Et)$

$$\xrightarrow{\substack{1.\,n\text{-BuLi}\\2.\,\overline{Ph}CHO}}$$

(40): $PhCH=C(SMe)CH_2CH(OCH_2OMe)C(Me)(SEt)(S(O)Et)$

↓ (dashed) 3-hydroxy-2-phenyl-3-methyl-cyclopent-... (HO, Ph on cyclopentenone)

In order to overcome the difficulties encountered it was necessary to modify our original strategy for the synthesis of the title compounds. Therefore, the protected phosphono-aldehyde (38) was treated at first with benzaldehyde. The vinyl sulfide (41) formed was selectively converted into the corresponding aldehyde (42) which on treatment with an acetyl anion equivalent and with chloromethyl methyl ether gave (40) in a better yield.

Scheme XII

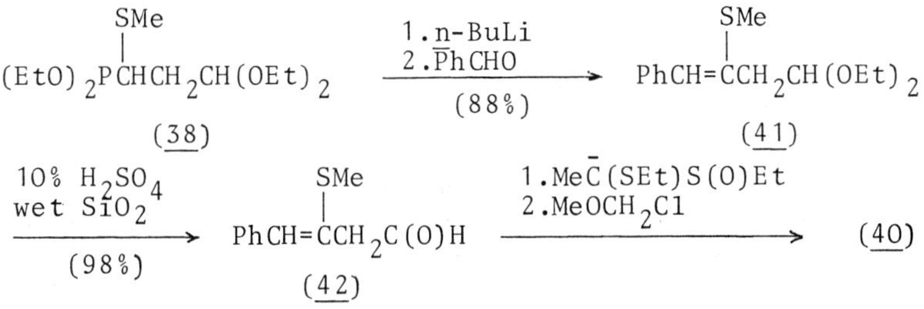

(38): $(EtO)_2P(O)CH(SMe)CH_2CH(OEt)_2$ $\xrightarrow[(88\%)]{\substack{1.\,n\text{-BuLi}\\2.\,\overline{Ph}CHO}}$ (41): $PhCH=C(SMe)CH_2CH(OEt)_2$

$\xrightarrow[(98\%)]{\substack{10\%\,H_2SO_4\\ \text{wet }SiO_2}}$ (42): $PhCH=C(SMe)CH_2C(O)H$ $\xrightarrow{\substack{1.\,MeC̄(SEt)S(O)Et\\2.\,MeOCH_2Cl}}$ (40)

Although the synthetic approach to 4-hydroxy-cyclopentenones is, at present, at the stage of model experiments we hope to apply it for the synthesis of some natural products.

Acknowledgements - I am very indebted to my coworkers, particularly to Dr.S.Grzejszczak, whose names appear in the references, for their skilful contributions, withour their efforts this review could not have been written. I also thank the Polish Academy of Sciences for a financial support of this work within the project MR-I-12.

REFERENCES

1. B.M.Trost and L.S.Melvin, Sulfur Ylides, Academic Press, New York (1975).
2. E.Block, Reactions of Organosulfur Compounds, Academic Press, New York (1978).
3. Organophosphorus Reagents in Organic Synthesis, Ed. by J.I.G.Cadogan, Academic Press, London (1979).
4. D.Seebach, Angew.Chem., 91, 259 (1979).
5. S.F.Martin, Synthesis, 633 (1979).
6. M.Mikolajczyk and A.Zatorski, Synthesis, 669 (1973).
7. J.Drabowicz and M.Mikolajczyk, Synthesis, 758 (1978).
8. J.Drabowicz, W.Midura and M.Mikolajczyk, Synthesis, 39 (1979).
9. M.Mikolajczyk, W.Midura, S.Grzejszczak, A.Zatorski and A.Chefczynska, J.Org.Chem., 43, 473 (1978).
10. M.Mikolajczyk, B.Costisella, S.Grzejszczak and A.Zatorski, Tetrahedron Letters, 477 (1976); M.Mikolajczyk, A.Zatorski, S.Grzejszczak, B.Costisella and W.Midura, J.Org.Chem., 43, 2518 (1978).
11. M.Mikolajczyk, S.Grzejszczak and W.Midura, unpublished results.
12. M.Mikolajczyk, S.Grzejszczak and A.Zatorski, J.Org.Chem., 40, 1979 (1975).
13. M.Mikolajczyk, S.Grzejszczak, A.Zatorski and B.Mlotkowska, Tetrahedron Letters, 2731 (1976); M.Mikolajczyk, S.Grzejszczak, A.Zatorski, B.Mlotkowska, H.Gross and B.Costisella, Tetrahedron, 34, 3081 (1978).
14. E.J.Corey and J.I.Shulman, J.Org.Chem., 35, 777 (1970).
15. M.Mikolajczyk, P.Balczewski and S.Grzejszczak, Synthesis, 127 (1980).
16. M.Mikolajczyk, S.Grzejszczak, A.Chefczynska and A.Zatorski, J.Org.Chem., 44, 2967 (1979).
17. M.Mikolajczyk, S.Grzejszczak, W.Midura, M.Popielarczyk and J.Omelanczuk, Phosphorus Chemistry, p.55, ACS Symposium Series 171 (1981).
18. M.Mikolajczyk, S.Grzejszczak and K.Korbacz, Tetrahedron Letters, 22, 3097 (1981).
19. M.Mikolajczyk, P.Balczewski and S.Grzejszczak, unpublished results.
20. M.Mikolajczyk, P.Kielbasinski and S.Grzejszczak, unpublished results.
21. S.Z.Ivin, W.K.Promonienkov,and B.I.Tetelbaum, Zh.Obshch.Khim., 37, 486 (1967).
22. R.A.Ellison, Synthesis, 397 (1973).
23. T.Oishi, M.Nagai and Y.Ban, Tetrahedron Letters, 491 (1968); T.R.Kelly and W.G.Tsang, Tetrahedron Letters, 4457 (1978).
24. See for example: Natural Products Chemistry, Volume 2, p.21-27, Ed. K.Nakanishi et.al.Academic Press, New York, 1975; T.L.Ho, Synth.Commun., 4, 265 (1977).
25. M.Mikolajczyk, S.Grzejszczak and P.Lyzwa, Tetrahedron Letters, 23, 2237 (1982),
26. R.M.Scarborough, Jr. and A.B.Smith,III, J.Amer.Chem.Soc.,99. 7085 (1977); R.M.Scarborough, Jr., B.H.Toder and A.B.Smith, III, J.Amer.Chem.Soc.,102 3904 (1980); D.Boschelli, R.M.Scarborough, Jr. and A.B.Smith III, Tetrahedron Letters, 22, 19 (1981); M.Koreeda and Y.P.Liang Chen, Tetrahedron Letters, 22, 15 (1981);Y.Takahashi, K.Isobe,H.Hagiwara, H.Kosugi and H.Uda, J.Chem.Soc., Chem.Commun., 714 (1981); Y.Takahashi, H.Kosugi and H.Uda, J.Chem.Soc., Chem.Commun., 496 (1982);R.F.Newton, D.P.Reynolds and T.Eyre, Synth.Commun., 11, 527 (1981); H.J.Altenbach and R.Korff, Angew.Chem.Int.Ed.Engl., 21, 371 (1982).
27. J.Jernow, W.Tautz, P.Rosen and T.H.Williams, J.Org.Chem., 44, 4212 (1979).

VINYL AND BETA-ALKOXY RADICALS IN ORGANIC SYNTHESIS

Gilbert Stork

Chemistry Department, Columbia University, New York, NY 10027, USA

Abstract. The heretofore not used cyclization of vinyl radicals, usually generated by abstraction of halogen from a vinyl bromide or iodide, into a suitably situated double bond is shown to be a very valuable construction method. The process can be extended to the initial formation of the required vinyl radical by internal addition of an intramolecularly held radical center to an appositely placed triple bond. A particularly valuable version of this latter process is one in which the initiating radical is derived from a mixed acetal with a beta bromine atom. Indeed, the cyclization of beta bromo mixed acetals of propargylic and allylic alcohols appears to be a very promising synthetic operation.

A number of years ago our attention was drawn to the report by Ohloff (1) that the expected transformation of the epoxide of alpha ionone to the transposed allylic alcohol upon reaction with hydrazine in methanol was only partially realized. In addition to the normal Wharton product, a bicyclic compound was also obtained, as shown in Fig. 1. The mechanism of this

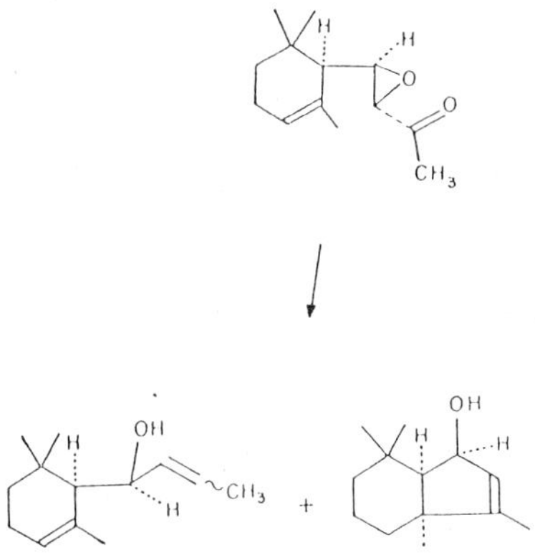

Figure 1

intriguing reaction was rather mysterious at the time. We decided to study it and considered two possibilities: the process might involve some more or less concerted decomposition of an alkyl diimide intermediate; or there might be induced decomposition of the diimide, with formation of a vinyl radical which could then undergo cyclization (Fig. 2). Together with

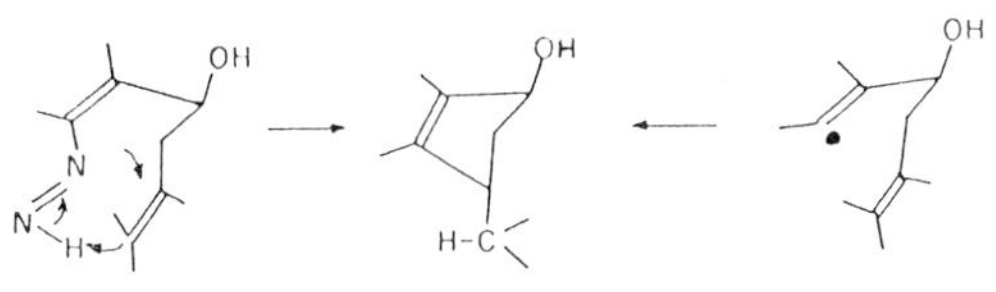

Figure 2

Paul Williard, we set about studying this problem. It seemed expedient to use a stereo-chemical criterion: The concerted process implies a cis relationship between the new carbon-carbon and carbon-hydrogen bonds while the intermediacy of a vinyl radical would produce a mixture in which the new bonds would be trans as well as cis to each other. The problem was simpler to formulate than to solve and it is only recently that we were able to show (with Neil Baine) that the formation of the two new bonds is indeed stereorandom. The construction of the system which served to establish that fact is shown in Fig. 3 while we

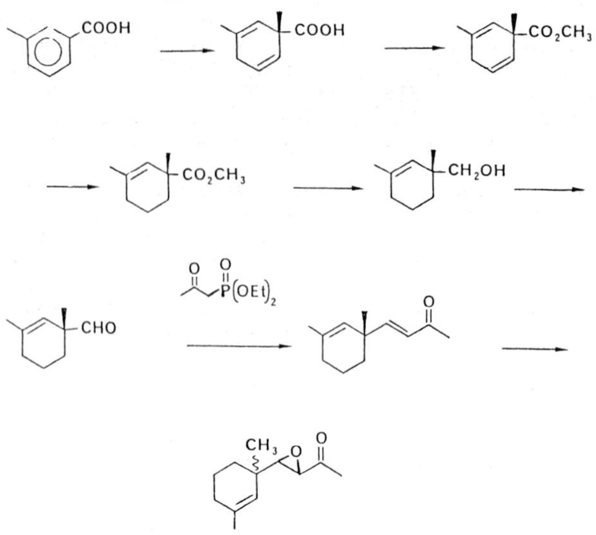

Figure 3

indicate in Fig. 4 the two possible outcomes of the cyclization step. In fact, this particular cyclization only occurred in very poor yield (10%) but the important point is that both possible isomers of the product were formed in roughly equal amounts (see Fig. 5). Al-

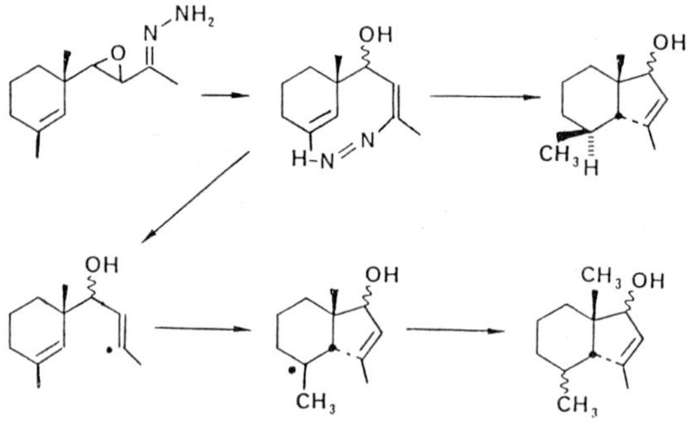

Figure 4

though this result rules out the concerted diimide mechanism it falls short of establishing the radical mechanism rigorously. Nevertheless, we now believe that the existing data are best accomodated by the assumption that treatment of alpha, beta-epoxy ketones with hydrazine is a route to vinyl radicals.

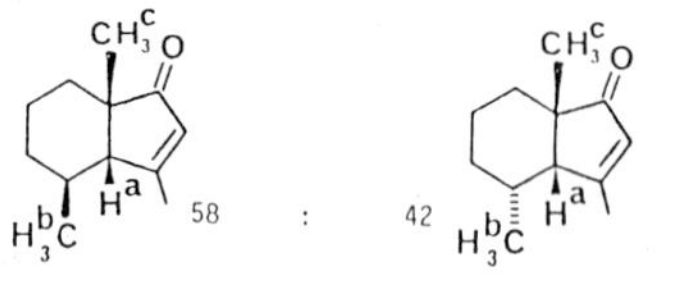

58 : 42

	δ (ppm)	δ (ppm)
H_a	2.07, J=1.5 Hz	2.46, J=4.2 Hz
H_3^bC	1.18, J=7 Hz	1.08, J=6.5 Hz
H_3^cC	1.15	1.08

Figure 5

In any event, we were struck by the fact that cyclizations involving vinyl radicals appeared largely unprecedented and we decided to investigate the possible synthetic use of such cyclizations, a process which can be schematized as in Fig. 6, whether the vinyl radical arises via an epoxyketone or, e.g., by the reduction of a vinyl halide with a tin hydride.

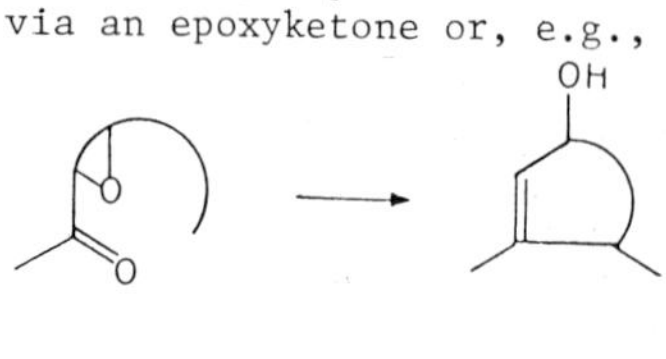

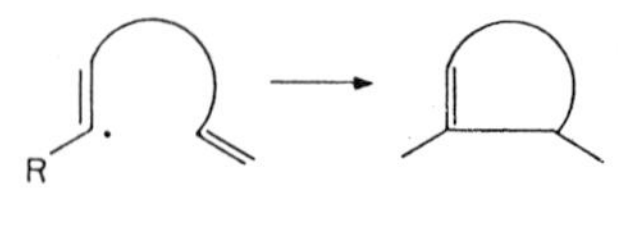

Figure 6

Vinyl and Beta-Alkoxy Radicals in Organic Synthesis

We were immediately struck by the fact that the cyclization of a vinyl radical into an apposite double bond would have a very considerable advantage over the widely studied cyclizations involving alkyl radicals(2). In the latter process one starts, as with 6-bromo-1-hexene, with two synthetically valuable functionalities: a primary bromide and a double bond. The product of cyclization, however, is a hydrocarbon, with the loss of both functions. Were the vinyl radical cyclization a feasible reaction, one would retain in the product a double bond in a completely defined position, independent of the relative stability of various double bond isomers, since no plausible path would be available for the isomerization of the cyclization product.

The possibility of achieving a useful cyclization would depend critically, as it does with the alkyl radical analog, on the relative rates of the cyclization process and that of hydrogen abstraction by the radical prior to cyclization. In the archetypal example of an alkyl radical, the 5-hexenyl radical, it is well-known principally from the work of Walling (3) and of Ingold (4) that at Molar concentration of tin hydride the rate of cyclization is only 1/10 that of hydrogen abstraction without cyclization. The important-and observed-corollary is that the synthetically useful cyclization predominates at low tin hydride concentration. The various processes are illustrated in Fig. 7.

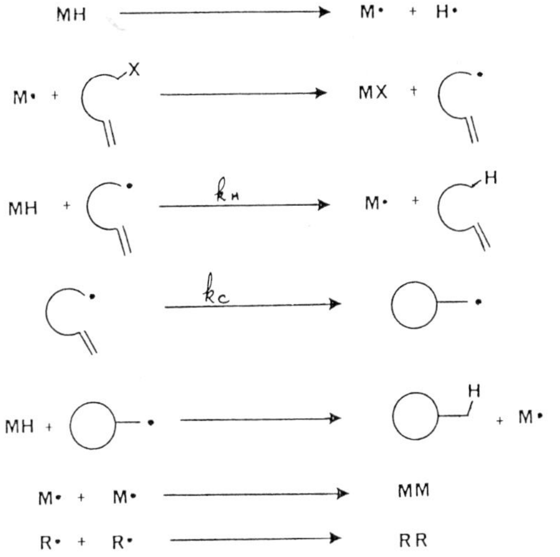

Figure 7

An initial concern was obviously that the anticipated higher energy of a vinyl radical would speed up the rate, k_H, of hydrogen abstraction, but one could hope- as appears to be true- that the rate of cyclization would be similarly affected so that the crucial k_c/k_H ratio would not necessarily be unfavorably affected.

As is now known, vinyl radical cyclizations form a useful method for the construction of functionalized rings. We now illustrate some of its features. Figure 8 shows the first cyclization involving the vinyl radical from a vinyl halide which we studied.

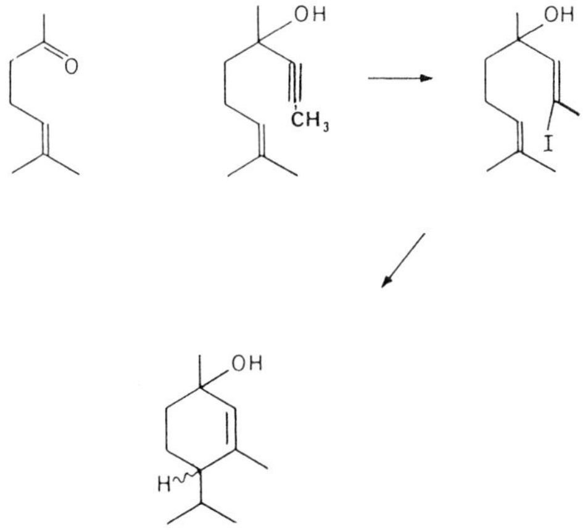

Figure 8

It proceeded in high yield under the typical conditions we used with all vinyl bromides and iodides: refluxing in benzene with tributyl tin hydride in the presence of azobisisobutyronitrile. The ring formed is, as expected, six- rather than seven-membered. More interesting is that a free hydroxyl is not deleterious to the reaction and needs not be protected. This is, of course, expected from the great homolytic bond strength of the hydroxyl group.

A very important point emerges from consideration of Fig. 9. We see here the demonstration that the stereochemistry of the vinyl halide is inconsequential, as a result of the known

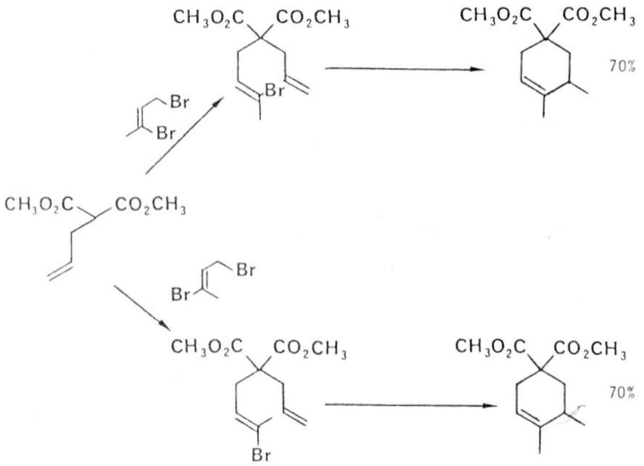

Figure 9

very low inversion barrier of (alkyl-substituted) vinyl radicals (5). This enlarges the range of possible methods which might be used to make the starting vinyl halides. The example also illustrates the compatibility of an ester function with the radical cyclization process.

The formation of a six- rather than a seven-membered ring in the examples of Figs. 8 and 9 is as expected. On the other hand, in the a priori not obvious case of five- vs. six-membered ring formation, the process schematized in Fig. 10 is subject to constraints such

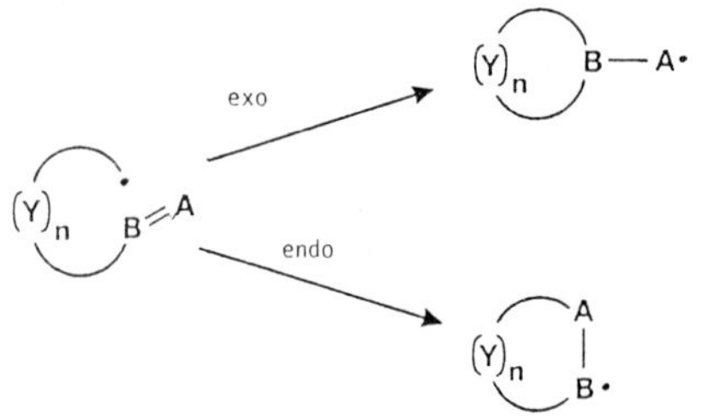

Figure 10

that when n=4, the exo process is favored. This is of course well-known with the saturated radical case shown in Fig. 11 (6).

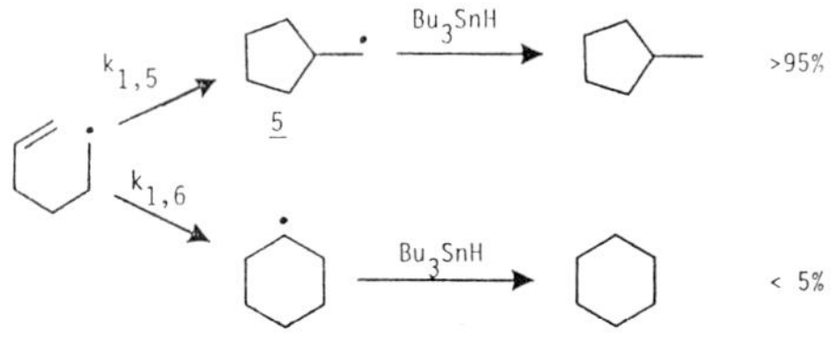

Figure 11

The constraint which leads to favoring a cyclopentane rather than a cyclohexane ring, even though this results in a somewhat less stable primary radical, is geometric in nature. Recent calculations (7) have led to a picture shown in Fig. 12 for the approach of a radical

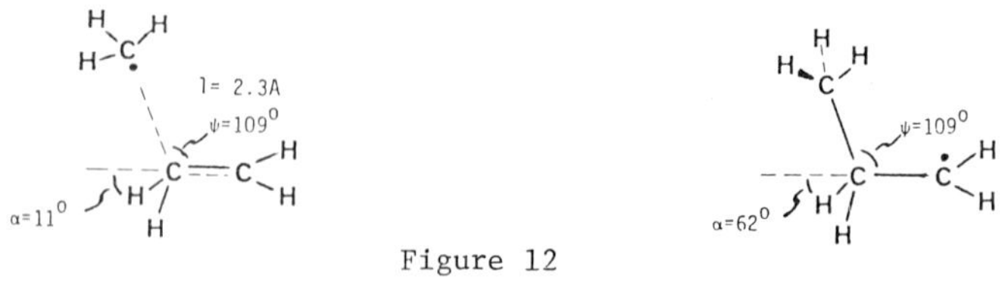

Figure 12

center to an olefin: it would thus take more energy to reach the required position to form a six-membered ring, and that energy is clearly more than that which favors a secondary over a primary alkyl radical. The energy cost (a guess might be 5-6 kcal in favor of the five-geometry) is not very high, however. This is illustrated in Fig. 13 which shows a 2:1 advantage for the formation of the interesting spirosystem at the top of the figure. That case illustrates another remarkable feature of vinyl radical cyclization to which we shall have occasion to return later: the ability to form bonds at completely substituted centers. On the other hand, the lower part of Fig. 13 shows that in a competition between cyclopentane formation leading to a primary radical and cyclohexane formation via a tertiary one, the sole product here is the six-membered ring. Further examples in which the (presumed)

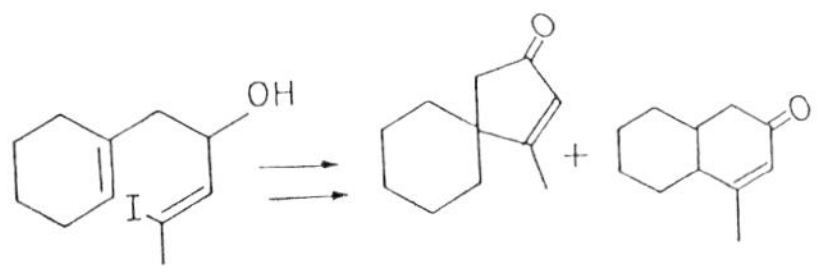

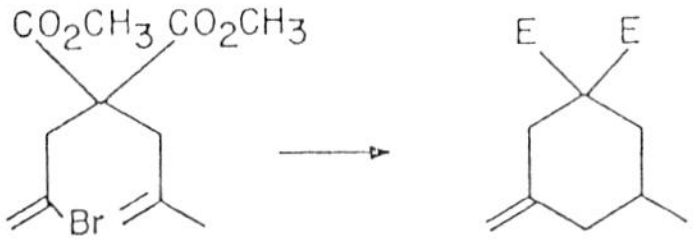

Figure 13

vinyl radicals are produced by the epoxyketone-hydrazine route are shown in Figs. 14 and 15. Since strain in the transition state is involved in favoring cyclopentane formation,

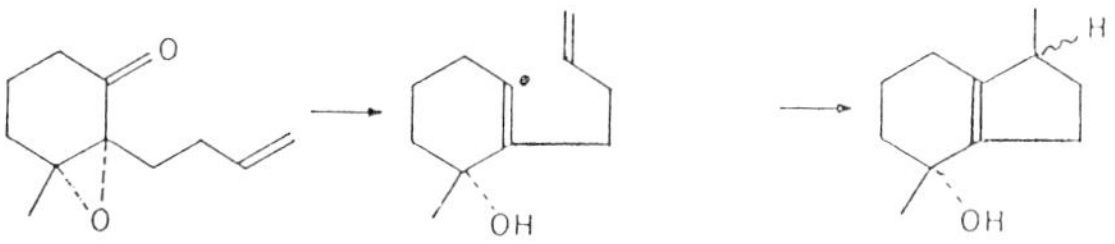

Figure 14

strain rather than substitution at the newly formed radical center is certainly sufficient to explain the result of Fig. 16: The alternative to the observed *trans* hydrindane formation

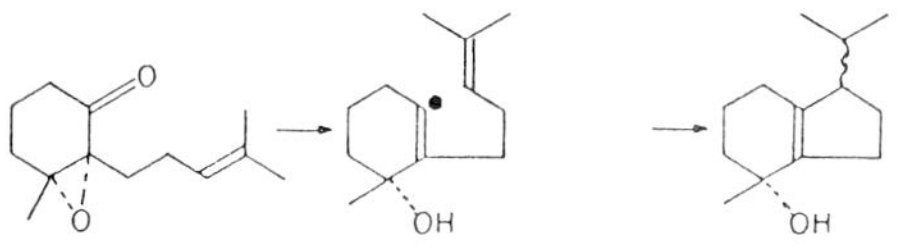

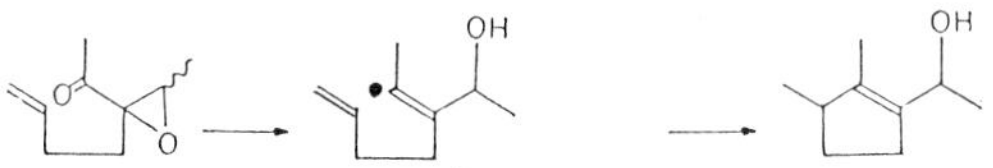

Figure 15

would have been a *trans* five-five system which would be several kilocalories higher in strain energy. We draw attention again to the important feature of these cyclizations which

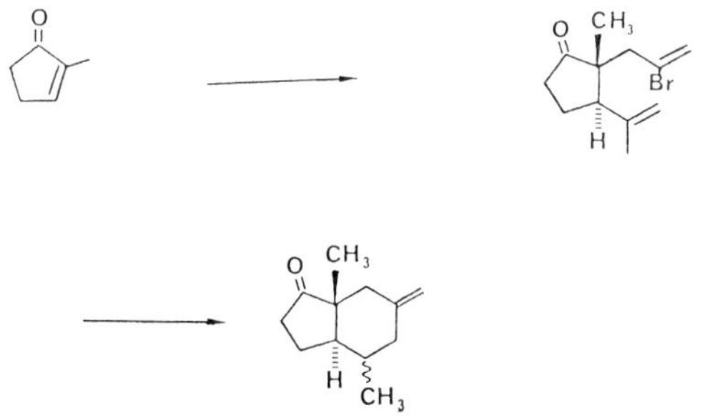

Figure 16

is that the surviving double bond remains at its anticipated position.

We draw attention in the following two Figures (17 and 18) to polycyclic constructions which take advantage of the vinyl radical process to produce a double bond at a position useful for the further control of stereochemistry. The construction of the starting material for the high yield radical cyclization process is worthy of special notice because we found that only one of the two possible products of alkylation was obtained with

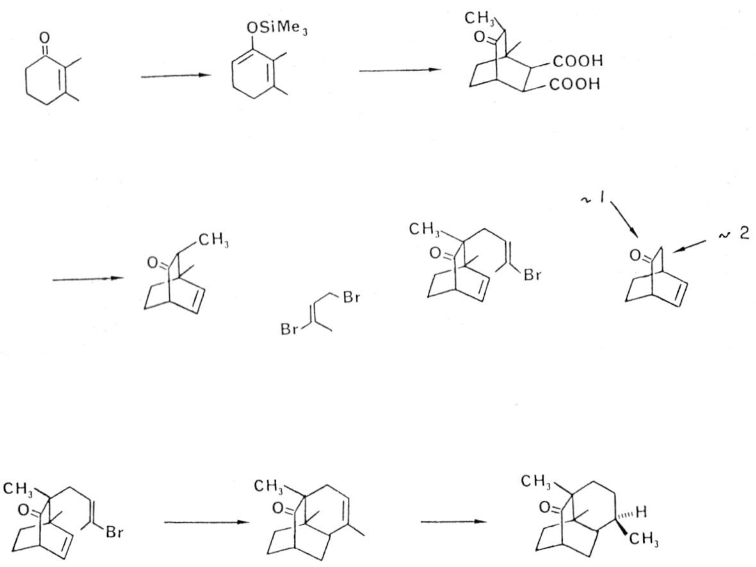

Figure 17

1,3-dibromo-2-butene, i.e. that in which alkylation has taken place on the double bond side. A similar result is shown in the alkylation of Fig. 18. These stereochemical results are especially useful for the construction of 2.2.2. bicyclooctanes possessing additional rings and are to be contrasted with the results of Grignard addition to a similar system: approach from the olefinic side is favored, but only by about 2:1, as indicated at the right of Fig. 17.

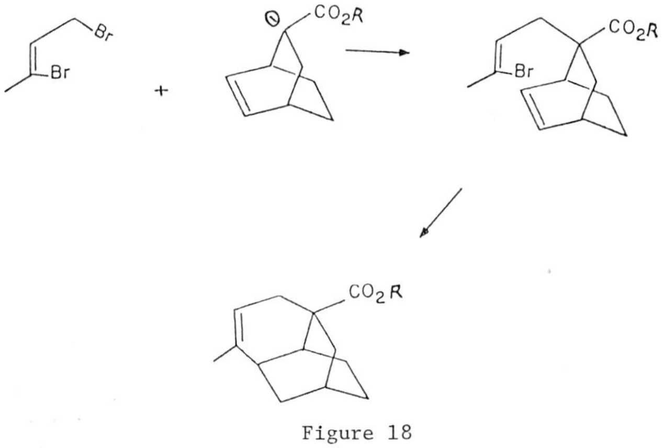

Figure 18

One last example will lead us to what will occupy the rest of this paper: cyclizations initiated by radicals in the beta-position to an acetal center. The cyclization shown in Fig. 19 which proceeds in high yield was conceived as a possible model for a construction of

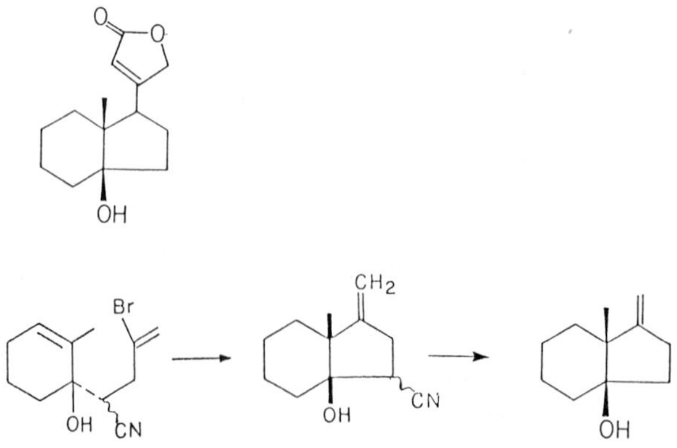

Figure 19

the C/D ring system of the cardiac aglycones. Indeed, the cyclization shown in Fig. 19 illustrates very well the great potential of vinyl radical cyclizations in synthesis: a double bond is produced at a known position for possible further elaboration, a free hydroxyl survives without protection, as does a cyano group and, most significantly, the process is not prevented by the steric congestion implicit in the formation of a quaternary center in the product. It would, of course, be possible to devise schemes for the transformation of the methylene group in the hydrindane product to the unsaturated lactone characteristic of the cardiac aglycones. We had, however, been considering other possible processes that might lead to vinyl radicals in addition to those starting with vinyl halides or epoxyketones. One possibility would obviously be external addition of a radical species to the acetylene function of a suitable acetylenic olefin. We concluded that the difficulty of ensuring addition of the external radical specifically to the acetylene rather than to the olefin could be surmounted if one could make the addition sterically biased in favor of the acetylene. This could in principle be achieved by using an internal radical within easy bonding distance from the acetylene, but not from the olefin. One useful possibility might involve the production of a radical such as is shown in Fig. 20.

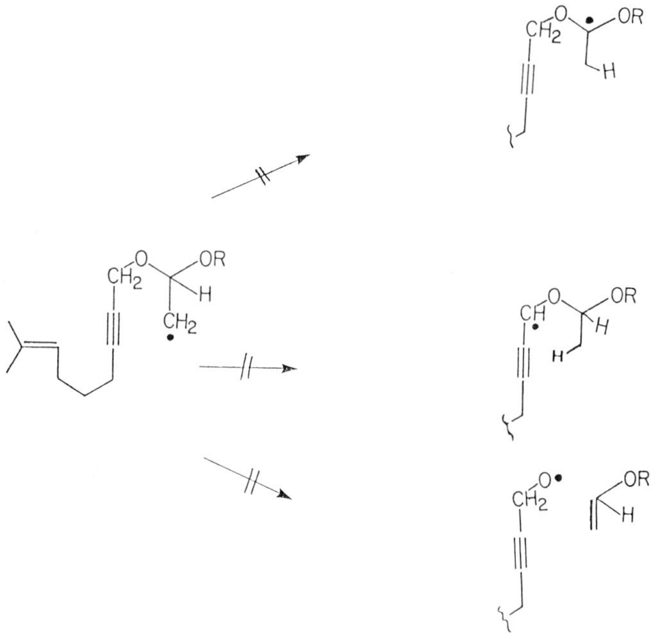

Figure 20

We note immediately an extraordinary advantage of radical processes over ionic counterparts. It is hard to think of a situation with so many pathways which might possibly compete with the otherwise reasonable cyclization into the triple bond. They are shown in Fig. 20. Reading from top to bottom: 1. Were there an electronically plausible path permitting 1,2 hydrogen migration to a radical center, it would obviously be a very favorable process, resulting in considerable stabilization of the new radical. There appears to be no such path. 2. If the geometry involving hydrogen migration via a five-membered ring were not rather unfavorable, the transformation leading from an unstabilized primary radical to one stabilized by both an acetylenic and an alpha-alkoxy one would be thermodynamically very favorable. More about this below. 3. Beta-scission, one possible result of which is shown in Fig. 20, is not particularly favorable (contrast the ionic process!) since an alkoxy radical is a rather high energy species.

With respect to the perfectly possible but difficult process of hydrogen transfer through a five atom transition state (cf. 2. above) it is interesting that the intuitively acceptable implication that the low energy path for hydrogen transfer must be collinear has been buttressed recently by theoretical work of Walch and Dunning (8) on the potential energy surface for the ·OH + H_2 reaction (Fig. 21). The lowest energy point corresponds actually to a 15° departure from colinearity in this particular case (top right of Fig. 21) while it is linear for the O + H_2 reaction (top left of Fig.). The important point is that the energy cost goes up steeply as one goes away from essential linearity. The net result is that the transition state for hydrogen transfer involving a so-called six-membered ring is actually geometrically much more like a five-ring while transfer via a so-called five-membered ring would actually imply a strain close to that of a four-ring.

In accordance with these considerations, we found (with Bob Mook) that the bromoacetal readily obtainable from cyclohexenol and the relevant alpha,beta-dibromoether cyclizes cleanly to the bicyclic acetal as shown in Fig. 22. The obligatory cis relationship

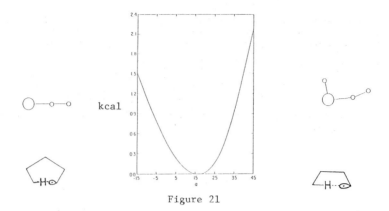

Figure 21

stereochemistry of the resulting bicyclic system is confirmed by oxidation to the known cis bicyclic ketone.

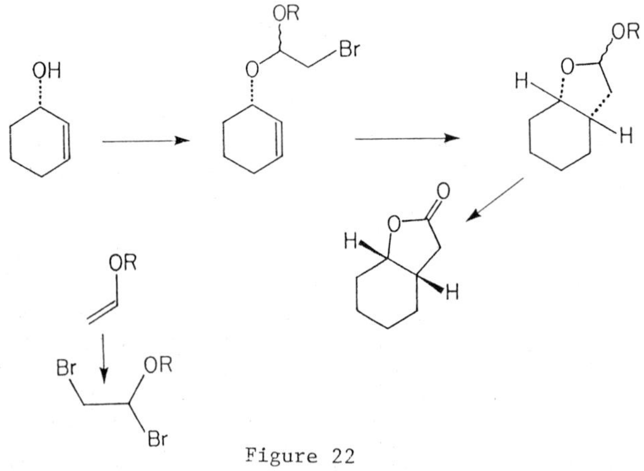

Figure 22

On the other hand, and significantly, the system shown in Fig. 23 is now capable of strain-free hydrogen transfer with the formation of the desirable tertiary allylic system. That process and the more desirable cyclization have been found (with Scott Biller) to take

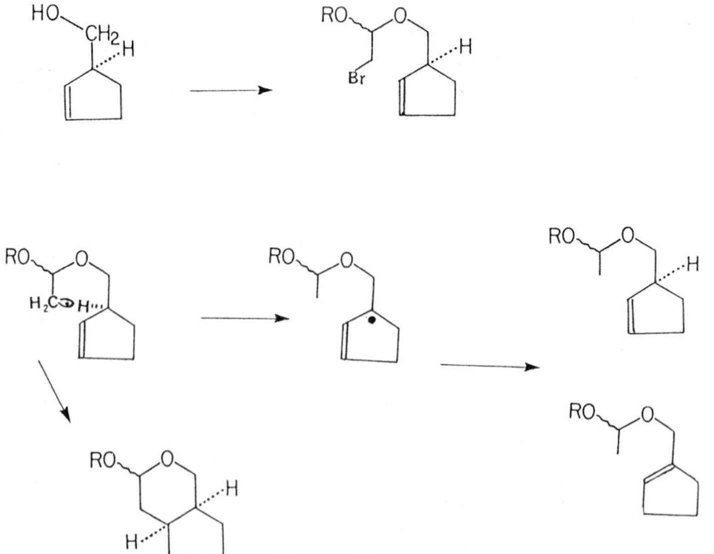

Figure 23

place with about equal ease. The fact that, even here, cyclization competes successfully suggests that judiciously placed structural features would give exclusive cyclization: this is indeed the case as we showed with the system of Fig. 24. The major isomer of the high yield cyclization is shown here, presumably the result of facilitated hydrogen donation by the stannane from the convex side of the initially formed bicyclic radical.

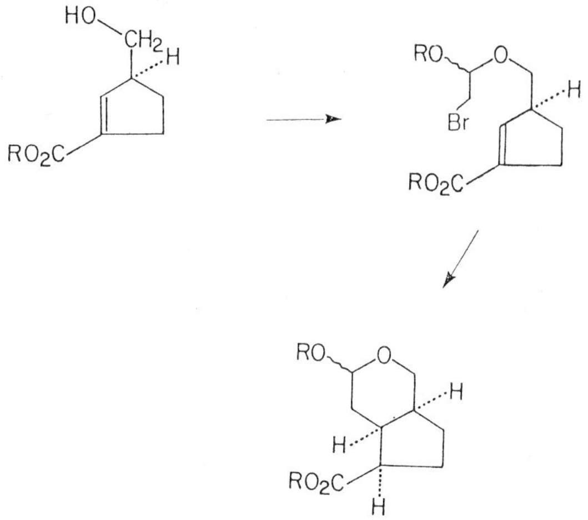

Figure 24

We now return to the possible formation of a vinyl radical capable of cyclization into an olefin, via the internal addition of an acetal radical such as the ones just described, to a triple bond. In the event, we were gratified to see that the process shown in Fig. 25 does indeed take place in good yield. This particular model is obviously considerably less

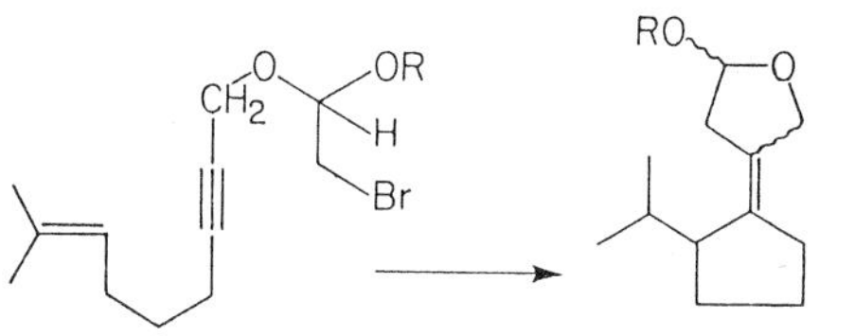

Figure 25

crowded than the tetrasubstituted olefin which would result from the possible cyclization shown at the bottom of Fig. 26. That cyclization, not so incidentally perhaps, reveals the

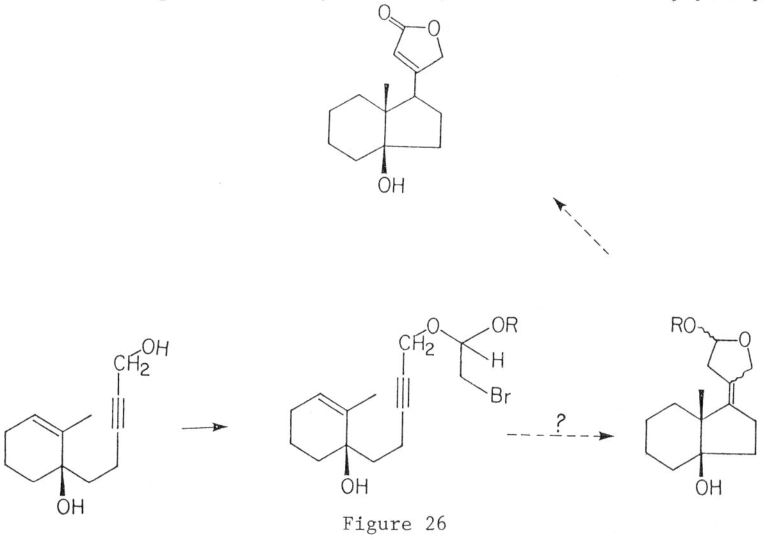

Figure 26

relationship to the temporarily forgotten cardiac aglycones. We have now (with Bob Mook) examined the closely related case shown on Fig. 27: cyclization takes place in good yield even here. Fig. 28 shows the use of the mixed acetal obtained from the dibromide of 2-chloroethyl vinyl ether. This was chosen because of the instability of the tetrasubstituted olefins to the acid conditions needed to hydrolyze the cyclized acetals: Treatment of the 2-chloro products with lithium and ammonia gives, quantitatively, the free hemiacetal. This can then be oxidized to a beta, gamma unsaturated lactone which now isomerizes to a single compound with the lactone system of the cardiac aglycones. Whether the cardiac aglycones themselves can be made by such a process is an interesting question which is now

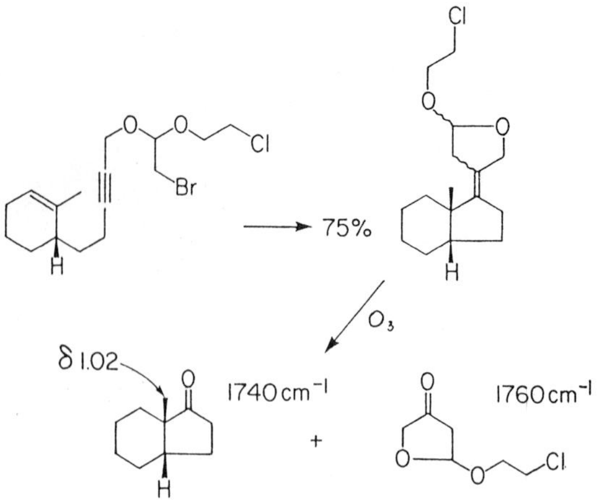

Figure 27

being addressed (with Jeff Stryker). This will require the solution of an unrelated problem, that of the control of the stereochemistry in the addition of a carbanion to the enone

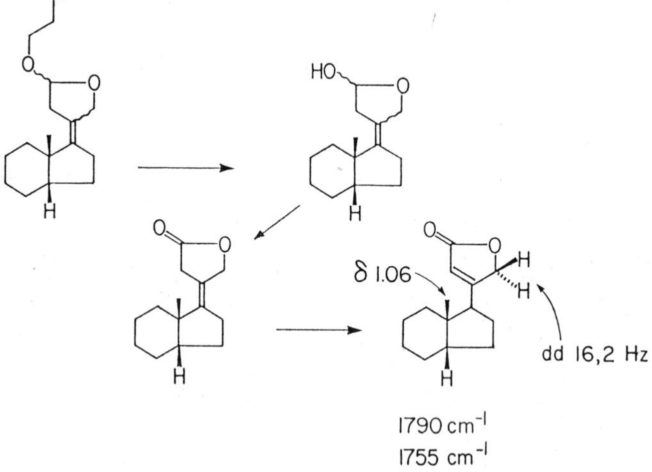

Figure 28

at the top of Figure 29.

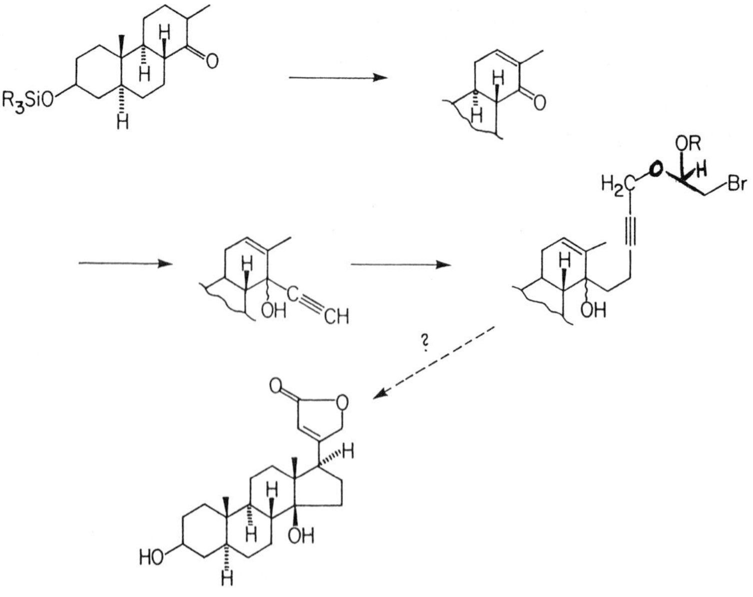

Figure 29

Vinyl and Beta-Alkoxy Radicals in Organic Synthesis   369

It was pointed out at the beginning of this paper that the big advantage of vinyl radical cyclization is the retention of functionality in the form of a specifically located double bond in the product. The loss of functionality when an alkyl radical is used explains the rarity of the process as a step in the construction of complex molecules. It is obvious, however, that pre-existing functionality which is not involved in the cyclization process will still be present after the new ring is formed. This process is especially valuable when, as with our beta-dialkoxy radicals, the survival of the particular functionality is compatible only with a radical process. It will be appreciated, for instance, that (cf. Fig. 22) even the simplest application of the haloacetal cyclization should have relevance to prostaglandin synthesis.

It is worth emphasizing, finally, that the benefit which accrues from vinyl radical cyclization, namely a product with a double bond at a known place can also be obtained by the cyclization of an alkyl radical into an acetylene. This is of course involved in the double cyclizations leading to alpha,beta-unsaturated lactones. We would like to end this report by illustrating this point with a simple mono-cyclization situation: We have been studying a number of construction processes for the stereoselective construction of chains of chiral centers bearing alternating methyl and hydroxyl groups. For instance, we show in Fig. 30 how stereospecific hydroxylation and stereospecific inversion lead from a cis-4-

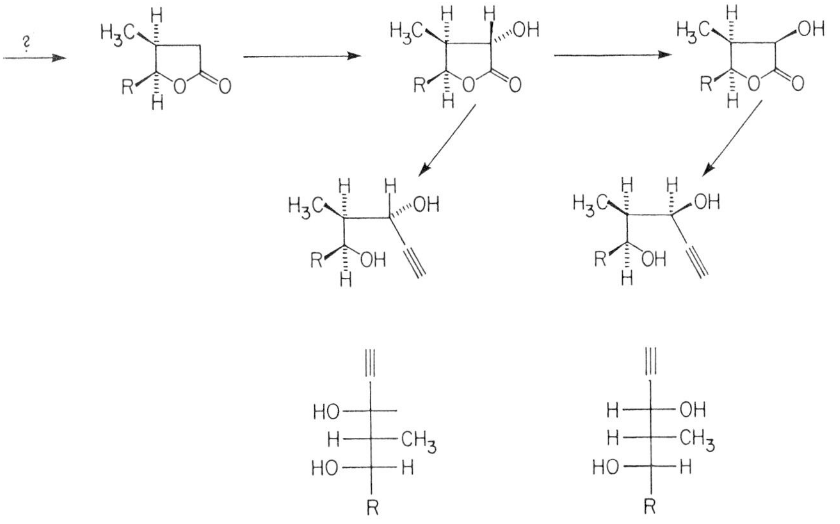

Figure 30

methyl-5-substituted 2-furanone to the production of the two Fischer projections shown at the bottom of that Figure. The problem was how to make the initial cis disubstituted lactone from an ethynyl carbinol, so the process could be repetitive. The problem and the solution (with Scott Rychnovsky) are shown in Fig. 31. The problem is that catalytic

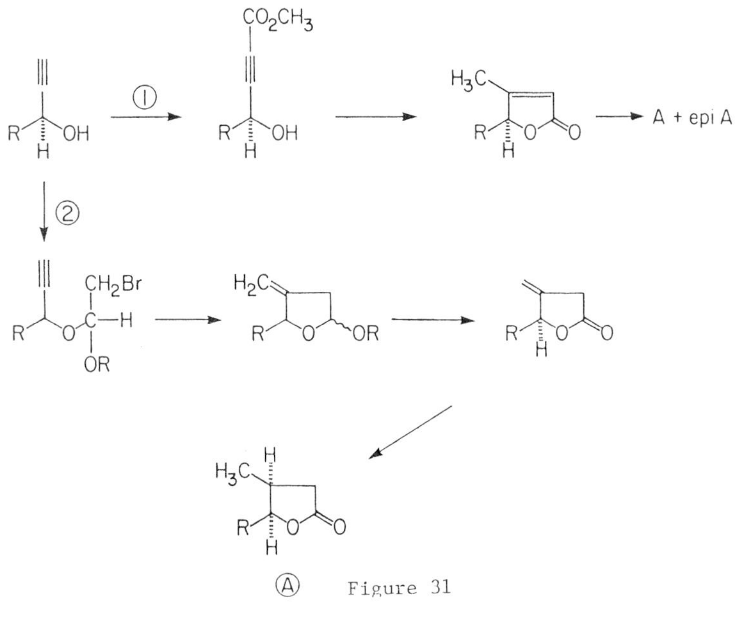

Ⓐ   Figure 31

hydrogenation of the alpha,beta unsaturated lactone (ex. path 1) by a whole variety of means gives at best only a 7:1 ratio in favor of the expected cis dihydro product.

Hydrogenation should be much more favorable to the cis dihydro product, starting with the exocyclic methylene tautomer. This could not be made by deconjugation of the alpha,beta unsaturated lactone. It was, however, readily made via bromoacetal radical cyclization of path 2 which indeed led via homogeneous hydrogenation of the 4-methylene lactone, to the desired A, in an acceptable ratio (greater than 60:1) to its undesired isomer.

We are confident that many applications of the cyclization of vinyl radicals and of radicals from haloacetal or other entities with special features will come forth from further work in our laboratories, as well as in those of others.

Acknowledgment - This work was supported by grants from the National Science Foundation and the National Institutes of Health.

REFERENCES

1. G. Ohloff and G. Uhde, Helv. Chim. Acta 53, 531 (1970). G. Stork and P. G. Williard, J. Am. Chem. Soc. 99, 7067 (1977).
2. For an excellent review, see A. L. J. Beckwith and K. U. Ingold, Rearrangements in Ground and Excited States, Paul de Mayo, Ed. pp. 182-220, Academic Press, New York (1980).
3. C. Walling, J. H. Cooley, A. A. Ponaras and E. J. Racah, J. Am. Chem. Soc. 88, 5361 (1966); C. Walling and A. Cioffari, ibid. 94, 6059 (1972); M. Julia and his associates have disclosed a number of reactions which lead to preferential formation of cyclohexane rings by a process involving thermodynamic equilibrium between the possible five- and six-membered rings. For examples of this synthetically useful process, cf. M. Julia, Pure and Applied Chemistry, 40, 553 (1974).
4. D. J. Carlson and K. U. Ingold, J. Am. Chem. Soc. 90, 7047 (1968).
5. R. W. Fessenden and R. H. Schuler, J. Chem. Phys. 39, 2147 (1963).
6. cf. reference 3 and A. L. J. Beckwith and G. Moad, Chem. Commun. 472 (1974).
7. M. J. S. Dewar and S. Olivella, J. Am. Chem. Soc. 100, 5290 (1978).
8. S. P. Walch and T. H. Dunning, Jr., J. Chem. Phys. 72, 1303 (1980).

SELECTIVE HYDROGENOLYSIS OF C-C BONDS IN SMALL RING COMPOUNDS

Hans Musso

Institut für Organische Chemie, Universität Karlsruhe, D-7500 Karlsruhe, Federal Republic of Germany

Abstract - Two factors have been shown to direct hydrogenolysis of C-C bonds in small cycloalkanes on Pd/C katalysts. In polycyclic hydrocarbons like basketane snoutane and cubane the most prestressed bond and therefore, the longest bond is openend first. The product giving greatest relief of strain is formed predominantly. In 1.1-disubstituted cyclopropanes with electron donating groups (R=alkyl, OR and NR_2) the opposite bond is hydrogenolysed only; on the other hand electron withdrawing substituents (R=CO_2R, COR, CN, C_6H_5, CH=CHR) favour cleavage of the adjacent bond exclusively. Recent results on bicyclo[n.1.0]alkanes, vinylcyclopropanes and 1-aminocyclopropane carboxylic acid will be presented.

INTRODUCTION

Carbon carbon single bonds are resistant to hydrogenolysis by molecular hydrogen activated with a heterogeneous metal catalyst like Palladium, Platinum or Nickel under normal conditions; this means: atmospheric pressure and room temperature. Although this reaction is exothermic with about 15 kcal/mol, it has

$$-CH_2-CH_2- \xrightarrow{H_2} -CH_3 \quad H_3C- \quad \Delta H = -15 \text{ kcal/mol}$$

a high activation energy and this is the reason why temperatures of 300°C are needed. Recently, P.v.R.Schleyer (1) has shown 190°C to be enough for C-C bond hydrogenolysis in 1-ethyladamantene but 280°C for the next step over Nickel on alumina.

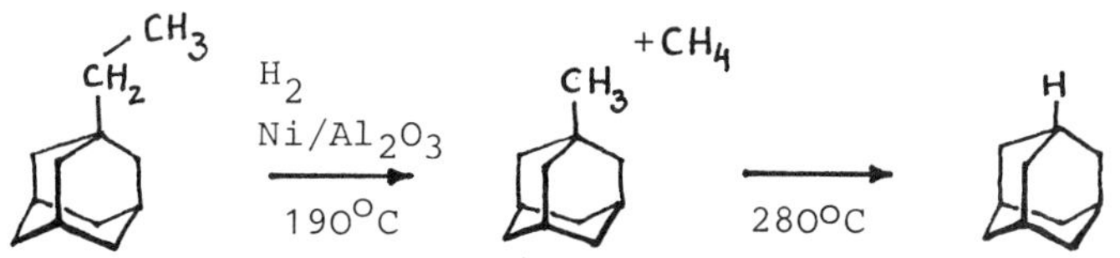

The ring opening of strained cyclopropane and cyclobutane has been observed first by Willstätter and Bruce (2) in 1907 over Nickel catalysts at 80 and 180°C. Ring strain energy (SE) of 28 and 26 kcal/mol lowers the actication energy and the temperature needed. This situation is illustrated in Fig.1

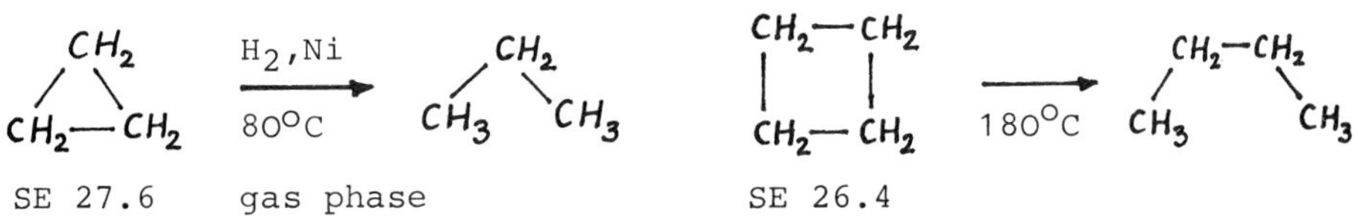

and we may conclude from this that the most strained bond in a system of small rings is the longest bond and this will be opened first or faster than the other bonds.

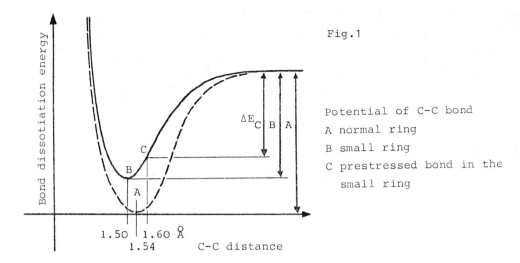

Fig.1

Potential of C-C bond
A normal ring
B small ring
C prestressed bond in the small ring

Apart from many more or less systematic studies on alkyl substituted cyclopropanes in the past which have been summerised by Newham (3) two preparative routes have been developed which use this reaction. The transformation of ketones to geminal dimethyl derivatives and construction of tert. butyl groups from ester groups. The last step in these synthetic sequences proceeds specific, in the substituted cyclopropane the adjacent bond b is not attacked the ring is cleaved at the opposite bond a only.

POLYCYCLIC HYDROCARBONS

We became interested in these hydrogenations when we were not able to reproduce the hydrogenation of bascetane 2. In the literature (Ref.4) the symmetrical compound 1 was claimed to be the product. Hydrogenation of 2 is fast indeed at room temperature without pressure over Palladium on carbon (Pd/C) and highly selectiv. Not a trace of compound 1 could be detected. We isolated the dihydro derivative 3 and twistane 4 as the final product (Ref.5). The reason for this has been elucidated by force field or molecular mechanics calculations (Ref.6) and X-ray structure analysis (Ref.7). On cleavage of the longer side bonds b 28 kcal/mol more strain is released as in the case of the shorter symmetrical bond a. The same is found for secocubane 6. None of compound 5 could be detected. The formation of 7 is energetically favoured by 34 kcal/mol and bicyclo[2.2.2]octane 8 is the final product (Ref.8). Again the difference of bond lenght a and b is significant 0.04 Å (Ref.9).

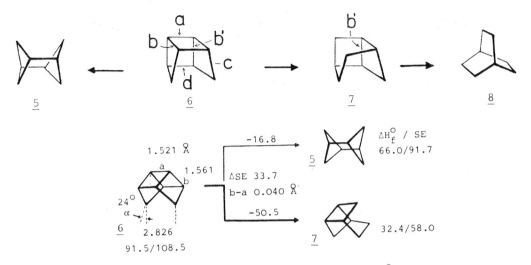

When these two carbon atoms with a distance of 2.825 Å are pulled together this angle α is decreased from 24° to 13° in homocubane 9. The distance is now 2.277 Å and by mechanical strain bond b gets longer. Bond difference b-a is decreased to 0.018 Å only and the energy difference in release of strain

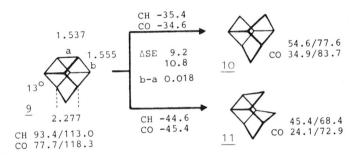

goes down to about 10 kcal/mol, here both products 10 and 11 are observed. Over Pd/C in the hydrocarbon 15 % of bond a is cleaved, in the ketone 9 especially in the dimethyl acetal over Rhodium on alumina up to 40 % of 10 can be isolated (Ref.9 & 10).

In all these cases the bonds b and b' are opened from different sides of the molecule. In the hydrogenation of snoutane 12 the three membered rings are cleaved both from the same side of the molecule. Isotwistane 14 is the main product (5c). From the X-ray structure analysis of the dihydro compound 13 - actually the dimethyl dicarboxylate but the same is found for the hydrocarbon by mm - calculations - we see that the bond a which is cleaved is the longest bond in the cyclopropane ring indeed. The agreement of the bond lenght found and calculated is not very high but the order is reproduced by the calculation correctly (Ref.11), see Table 1.

Tab.1 Found and calculated bond lenght in Å

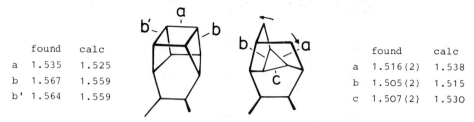

	found	calc
a	1.535	1.525
b	1.567	1.559
b'	1.564	1.559

	found	calc
a	1.516(2)	1.538
b	1.505(2)	1.515
c	1.507(2)	1.530

There are many more reports in the literature about the selectiv hydrogenation of small rings in polycyclic hydrocarbons. Two representativ examples are given here. Barettane 15 and bishomocubane 17 both yield the hydrocarbon 16 exclusively as shown by de Meijere (12) and Yonemitsu and there collaborators (Ref.13).

15   154.9/176.5   16   8.3/41.8   17   48.3/75.9

BICYCLO[n.1.0]ALKANES

In the larger rings with n=6 and 5 substituent effects direct ring opening to the methyl cycloalkane of the same ring size (Tab.2). In cases of n=4 and 3 strain effects begin to compete and increasing amounts of enlarged rings are formed 7 and 20 %. In bicyclo[2.1.0]pentane cyclopentane is the only product detected (Ref.14) as reported by Criegee (15) in 1957. At that time small amounts of methyl cyclobutane could not be excludet. Therefore, in all the experiments reported here from our laboratory special care was taken with the detection of small amounts of isomeric products by direct comparison with authentic samples of all the possible products by v.p.c. and n.m.r. In most cases 0.1 % is identified with certainty, ±1 % is the utmost error.

Tab.2 Hydrogenation Products from Bicyclo[n.1.0]alkanes

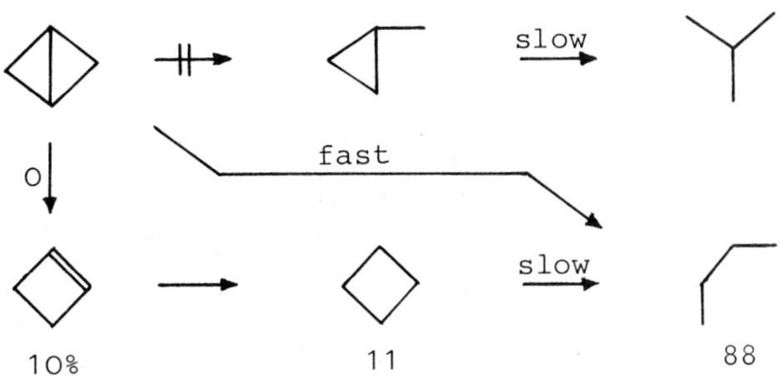

Bicyclo[1.1.0]butane reacts very fast and n-butane is the main product. Most of the cyclobutane is formed by rearrangement of the starting material on the catalyst to form cyclobutene and hydrogenation of that. Cyclobutane and methyl cyclopropane are hydrogenated much slower and this could not be detected as an intermediate. Corresponding results have been obtained on bicyclo[1.1.0]butane 1- and 2-ethyl carboxylates, 2-methylbutyric acid ester is the principal product. In case of the 1-carboxylate after two minutes the three olefinic isomeric dihydro intermediates can be detected (Ref.14). These are interconvertible under hydrogenation and react very fast to the final product.

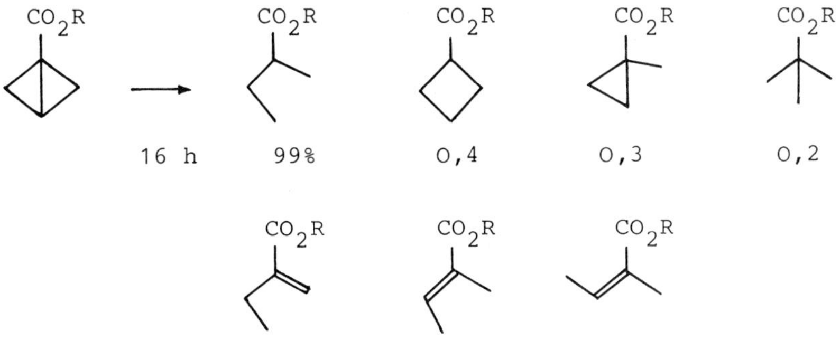

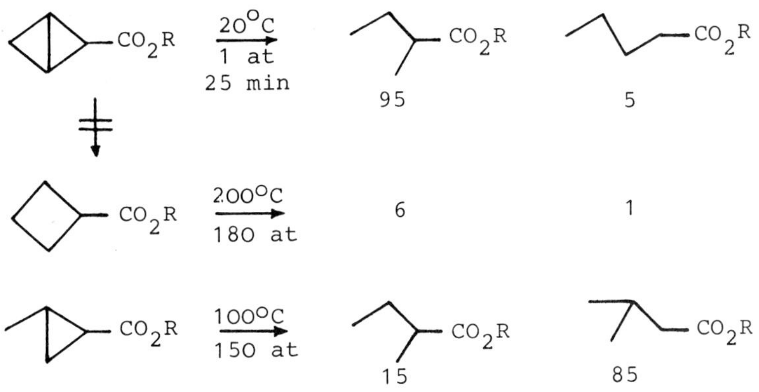

In bicyclo[2.1.0]pentane the longest most prestressed bond is the central one; in bicyclo[1.1.0]butane the central bond is stronger and shorter as the periferic ones. In bicyclobutane both bonds are broken simultaneously after the first attack of hydrogen atoms. An olefin type intermediate is bond as a metal complex on the surface of the heterogeneous katalyst befor an olefinic dihydro product can be liberated into the solution.

SUBSTITUENT EFFECTS

In 1.1-disubstituted cyclopropanes pure substituent effects are operating. Two different types of products are possible (Tab.3). In compounds with electron donating groups like alkyl, hydroxymethyl, alkoxy, alkylamines the opposite bond a is hydrogenolysed only and isopropane derivatives are formed. With electron acceptors like ester groups and other derivatives of the carboxy group as well as acyl, vinyl and phenyl the adjacent bond b is opened and n-propyl derivatives are isolated with high selectivity (Ref.16). These effects can be interpreted by the theory of Hoffmann, Stohrer and Günther (17) about the influence of substituents on the equilibrium of cycloheptatriene norcaradiene by electronic interaction with the Walsh orbitals of the cyclopropane ring (Fig.2). Electron donating groups fill into the lowest unoccupied orbital with antibonding character of the opposite bond a; acceptors take electrons from the highest occupied orbital which has bonding character in the adjacent bond b.

Table 3 Hydrogenation Products of 1.1-disubstituted cyclopropanes

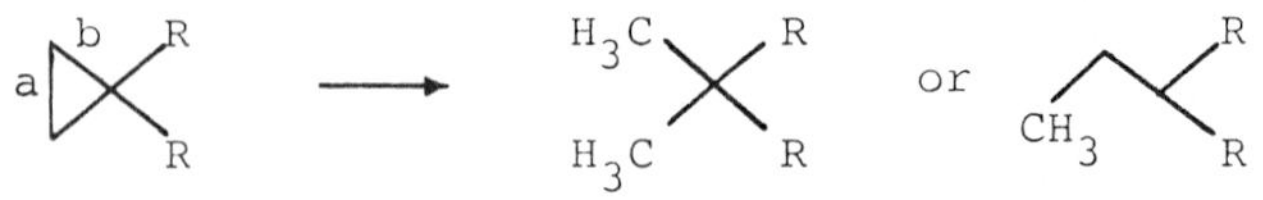

R =

a CH_3
   CH_2-OH
   OH
   OCH_3
   OH, OCH_3
   $OCOCH_3$
   $OH, OCOCH_3$

b CO_2H
   CO_2CH_3
   $CO_2C_2H_5$
   $CONH_2$
   CN
   $COCH_3$
   $CH=CH_2$
   C_6H_5

Fig.2

R = $-CO_2R$
    $-CN$
    $-NO_2$
    $-CF_3$

R = $-CH_3$
    $-CH_2-OH$
    $-OR$
    $-NR_2$

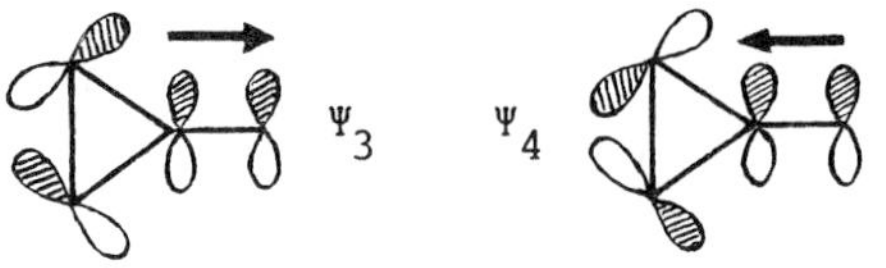

More sophisticated calculations as well as X-ray analysis and micro wave spectroscopy have shown the bonds which are hydrogenolysed to have longer C-C distance than the other bond in many examples (Ref.18 & 19) as indicated in some of the formula.

Vinyl derivatives have beeen studied recently in more detail because it appears as an interesting question to what extent the unsaturated group can be hydrogenated without attacking the cyclopropane ring (Ref.20 & 21). Unhindered double bonds in general are saturated much faster.

Divinylcyclopropane 18 takes up two moles of H_2 in about 10 min and 3-ethylpentane 20 is the only product. No 3.3-dimethyl pentane 22 could be detected. The hydrogenation of 1.1-diethyl cyclopropane to 21 is much slower. When interrupted after 20 or 30 s 3-ethyl-2-pentene 19 can be detected as an intermediate in up to 48 % in the reaction mixture. The spiro-diene 23 yields 98 % ethyl cyclopentane 26. The saturated spirane 27 and 1.1-dimethyl cyclopentane 28 in some runs are seen by v.p.c. in traces (less than 0.1 %). Diolefin 24 and the monoolefin 25 again are intermediates.

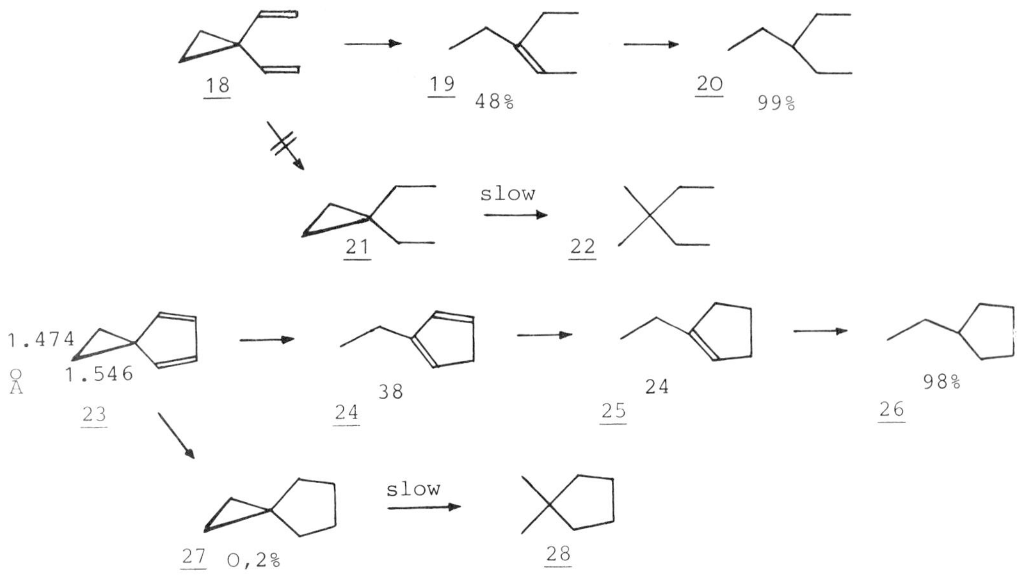

This may be valued as a visible sign of 1.4-addition of hydrogen and substantical conjugation between cyclopropane rings and unsaturated groups. A metal type complex like 29 or 30 (Fig.3) may be used to interpret these results and the formation of more than 90 % of p-diethyl benzene 32 from the dispiro-diolefin 31. The saturated compounds 33, 34 and 35 can be detected in trace amounts only (Ref.21)

Fig.3 Intermediates on the katalyst

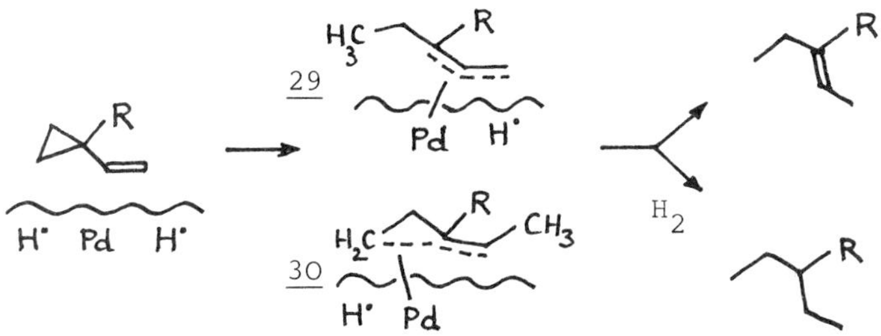

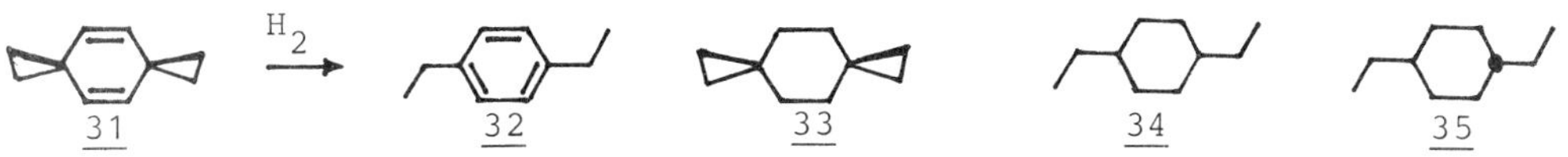

Very many combinations of different donating groups and electron acceptors are possible and it can become boring to study these competing effects. 1-Aminocyclopropane carboylic acid 36 should be the last compound presented today. As axpected different product rations of iso- and n-aminobutric acid 37 and 38 are obtained depending on the pH of the solution (Ref.21).

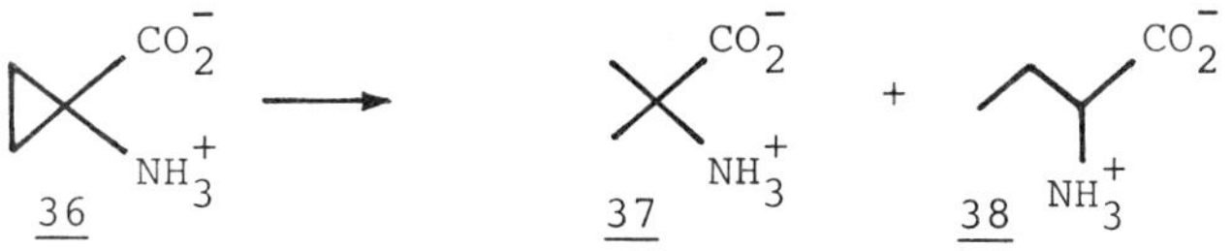

		37	38
Pd/C	CH_3OH	90%	10%
Pd/C	CH_3OH, NH_3	98,5	1,5
Pt	CH_3CO_2H	23	77

Acknowledgement - I have to thank my collaborators, who did this work, Doris Büchle-Kallfass, Martin Gagel, Claus Gröger, Helmuth Guth, Christa Hauschild, Winfried Hertzsch, Annelie Kuiper, Sigrid Maurer, Said Pirsch, Susanne Pleyer, Ingrid Roßnagel, Andy N.Sasaki, Wolfgang Schwarz, Karl J.Stahl, Reinhard Stober, Claudia Witte, Karin Woisetschläger, Reinhard Zunker, especially Prof.Eiji Ōsawa from Sapporo for all the molecular mechanics calculations and Dr.Berthold Deppisch in Karlsruhe for X-ray structure determinations. This work was supported by the Deutsche Forschungsgemeinschaft and the Fonds der Chemischen Industrie.

REFERENCES

1. P.Grubmüller, P.v.R.Schleyer and M.A.McKervey, Tetrahedron Letters 1979, 181
   P.Grubmüller, W.F.Maier, P.v.R.Schleyer, M.A.McKervey and J.R.Rooney, Chem.Ber. 113, 1989 (1980)
2. R.Willstätter and J.Bruce, Ber.Dtsch.Chem.Ges. 40, 3979, 4456 (1907)
3. J.Newham, Chem.Reviews 63, 123 (1963)
4. S.Masamune, H.Cuts and M.Hogben, Tetrahedron Letters 1966, 1017
5. a) E.Ōsawa, P.v.R.Schleyer, L.Chang and V.V.Cane, Tetrahedron Letters 1974, 4189
   b) A.N.Sasaki, R.Zunker and H.Musso, Chem.Ber. 106, 2992 (1973)
   c) H.Musso, Chem.Ber. 108, 337 (1975)
6. N.L.Allinger, Advances in Physical Organic Chemistry, Vol. 13, S.1-82 (1976); E.Ōsawa and H.Musso, Topics in Stereochemistry, Vol.13, 117-193 (1982)
7. J.P.Schaffer and K.K.Walthers, Tetrahedron 27, 5281 (1971)
8. R.Stober and H.Musso, Angew.Chem. 89, 430 (1977)
9. E.Ōsawa, I.Roßnagel, K.J.Toyne and H.Musso, to be published
10. K.J.Toyne, J.Chem.Soc. Perkin I, 1976, 1346
11. B.Deppisch, H.Guth, H.Musso and E.Ōsawa, Chem.Ber. 109, 2956 (1976)
12. D.Bosse and A.de Meijere, Chem.Ber. 111, 2223 (1978)
13. K.Hirao, T.Iwakuma, M.Taniguchi, E.Abe, O.Yonemitsu, T.Date and K.Kotera, J.Chem.Soc. Chem.Comm. 1974, 691, J.Chem.Soc. Perkin I 1980, 163
14. W.Hertzsch, K.J.Stahl and H.Musso, to be published
15. R.Criegee and A.Rimmelin, Chem.Ber. 90, 414 (1957)
16. C.Gröger and H.Musso, Angew.Chem. 88, 415 (1976)
    C.Gröger, H.Musso and I.Roßnagel, Chem.Ber. 113, 3621 (1980)
17. R.Hoffmann, Tetrahedron Letters 1970, 2907
    R.Hoffmann and W.D.Stohrer, J.Amer.Chem.Soc. 93, 6941 (1971)
    H.Günther, Tetrahedron Letters 1970, 5173
    S.Durmaz and H.Kollmar, J.Amer.Chem.Soc. 102, 6943 (1980)
18. F.H.Allen, Acta Cryst.Sect. B 36, 81 (1980); Tetrahedron 38, 645 (1982)
19. M.D.Harmony, S.N.Mathur, J.-I.Choe, M.Kattija-Ari, A.E.Howard and S.W.Staley, J.Am.Chem.Soc. 103, 2961 (1981)
20. D.Büchle-Kallfass, M.Gagel, Ch.Hauschild, H.Musso, I.Roßnagel and C.Schreiber, Israel J.Chem. 21, 190 (1981)
21. I.Roßnagel, W.Schwarz and H.Musso, to be published

RECENT SYNTHETIC DEVELOPMENTS IN ANNULENE CHEMISTRY

Emanuel Vogel

Institut für Organische Chemie der Universität zu Köln, Greinstraße 4, D-5000 Köln 41, Federal Republic of Germany

Abstract — Recent progress in synthetic methodology has not only provided the basis for the construction of natural products of high structural sophistication but has also been an incentive for the synthesis of complex, non-natural organic molecules which are of interest for theoretical, structural, topological, or other reasons. Molecules prominent within the latter category are the annulenes of various types. The main synthetic efforts in this domain currently focus on Hückel-type bridged annulenes (and dehydroannulenes) since such compounds usually qualify as aromatic, in both physical and chemical respects, and thus promise to add a new dimension to aromatic chemistry. It is the purpose of this lecture to outline the recent contributions from the author's laboratory to these developments.

INTRODUCTION

About a decade after their discovery, Sondheimer's [18]annulene 1 and 1,6-methano[10]annulene 2 from our laboratory were introduced into Organic Syntheses: Vol. 54, 1, 11 (1974). In his editorial preface to this volume, Robert E. Ireland expressed the opinion that "the syntheses of both compounds represent landmarks in organic chemistry, and that their availability through Organic Syntheses procedures may spur further work on these interesting structures".

At that time, we thus felt encouraged to persist in our endeavours to develop the branch of annulene chemistry that had emerged from the Hückel-type aromatic compounds 2, 1,6-oxido[10]annulene, and 1,6-imino[10]annulene (1a-f).

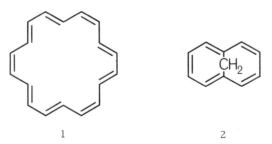

1   2

The structural relationship existing between the 1,6-bridged [10]annulenes and naphthalene had opened the interesting perspective that the potentially aromatic bridged [4n+2]annulenes 3, 4, 5, etc., corresponding to the classical aromatic hydrocarbons, the acenes, could likewise be prepared. In general, these rather rigid bridged [4n+2]annulenes give rise to geometrical isomerism (syn/anti-isomers) since bridge inversion, which could occur by conformational flip, is effectively inhibited by steric constraints.

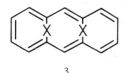 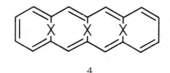 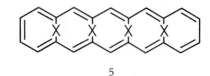

      3   4   5

X=CH$_2$, O, NH

As a more detailed inspection of molecular models of 3 and its homologues reveals, only the isomers with the bridges in an all-syn-arrangement — and with bridges not too spatially demanding — possess C$_{(4n+2)}$-perimeters flattened out sufficiently to allow delocalization of the π-electron system and hence aromaticity of the respective molecules. The anti-isomers, exhibiting strongly puckered perimeters, are expected to be devoid of aromaticity. That these steric considerations measure up to reality became manifest when syn-1,6:8,13-bis-oxido[14]annulene 6 and anti-1,6:8,13-bismethano[14]annulene 7, the first bridged [14]annulenes with an anthracene perimeter to be synthesized, were found to be aromatic and olefinic molecules, respectively (1c,d).

                            6   7

In our quest to make bridged [10]annulenes with a naphthalene perimeter "common" aromatic molecules, and to probe the scope and limitations of the synthesis of bridged [4n+2]-annulenes with an acene perimeter, we have been able up to now to devise two expedient and complementary approaches to bridged annulenes of the present type. The first one, embracing the synthesis of a great variety of bridged [10]- and [14]annulenes is based on the Birch reduction products of naphthalene (1,4,5,8-tetrahydronaphthalene) and anthracene (1,4,5,8,9,10-hexahydroanthracene), respectively, as starting materials. The second, more recent one, takes advantage of the synthon cycloheptatriene-1,6-dicarboxaldehyde 11. It is the virtue of 11 that it provides access to homologous series of CH$_2$-bridged [10]-, [14]-, [18]-, and even [22]annulenes by a "building-block" strategy. The dialdehyde 11, moreover, promises to open up an entry into the realm of bridged aza[4n+2]annulenes through appropriate methods of annelation.

The fact that the bridged [10]annulenes and many of the syn-bridged [14]annulenes are distinguished by remarkable stability, and can now be prepared in sizable quantities, has rendered these annulenes rewarding objects for detailed chemical exploration.

    BRIDGED [10]ANNULENES

In the first part of this lecture I would like to focus your attention on some new aspects of the synthesis and chemistry of bridged [10]annulenes and aza[10]annulenes, specifically of the parent 1,6-methano[10]annulene 2 and of the heteroannulene 2,7-methanoaza[10]annulene 20.

1,6-Methano[10]annulenes from cycloheptatriene-1,6-dicarboxaldehyde 11
───

In the last few years the synthesis of 1,6-methano[10]annulenes has received added momentum by the introduction of a new synthon, namely the afore-mentioned cycloheptatriene-1,6-dicarboxaldehyde 11. This dialdehyde has recently become accessible on a preparative scale due to the discovery that cycloheptatriene 8, under appropriate conditions, can be acetylated at the termini of the double bond system to give 1,6-diacetylcycloheptatriene 9. The subsequent conversion of the diacetyl compound into the dialdehyde, by way of cycloheptatriene-1,6-dicarboxylic acid 10, is achieved effectively by conventional procedures (2).

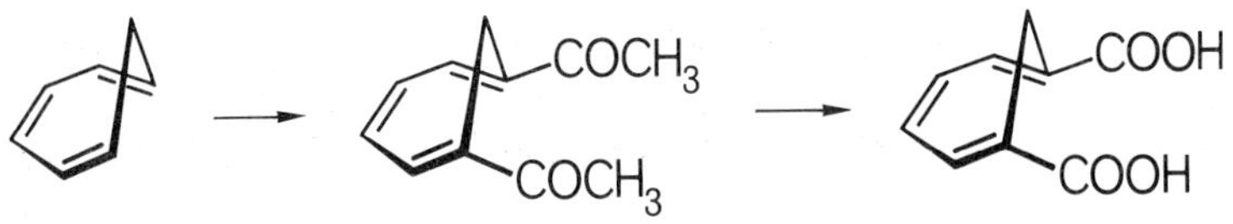

The synthetic usefulness of cycloheptatriene-1,6-dicarboxaldehyde 11 becomes best apparent if we regard it as "homo-phthalaldehyde". Similar to the classical phthalaldehyde, 11 especially invites chemical transformations leading to annelations.

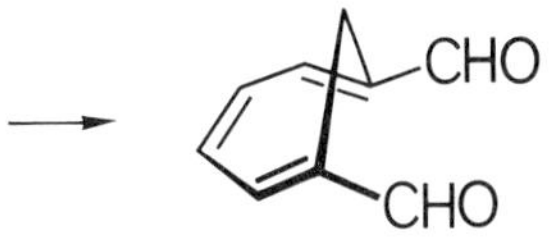

As regards the conversion of 11 to 1,6-methano[10]annulene 2, we have been able to devise no fewer than three methods of annelation, all of which are based on 10π-electrocyclic processes. The method of choice consists of the following steps: i) successive olefination of 11 with the Wittig reagents methylene(triphenyl)phosphorane and chloromethylene(triphenyl)-phosphorane to give 12 and ii) heating of 12 in dimethylformamide to yield 2 by "cyclodehydrohalogenation", i.e., by a 10π-electrocyclic process combined with elimination of hydrogen chloride from the labile 3,4-dihydro-1,6-methano[10]annulene intermediate 13 (1f).

This scheme has since been extended into a general procedure for the synthesis of 3- and 3,4-disubstituted 1,6-methano[10]annulenes by appropriate variation of the Wittig reagents in the initial two-stage olefination step.

As representative examples I would like to mention the smooth preparation of 3-bromo-1,6-methano[10]annulene 15 and of 3,4-dicarbomethoxy-1,6-methano[10]annulene 18 (the phthalic ester analogue in the 1,6-methano[10]annulene series) from the respective olefination products of 11, namely 14 and 17.

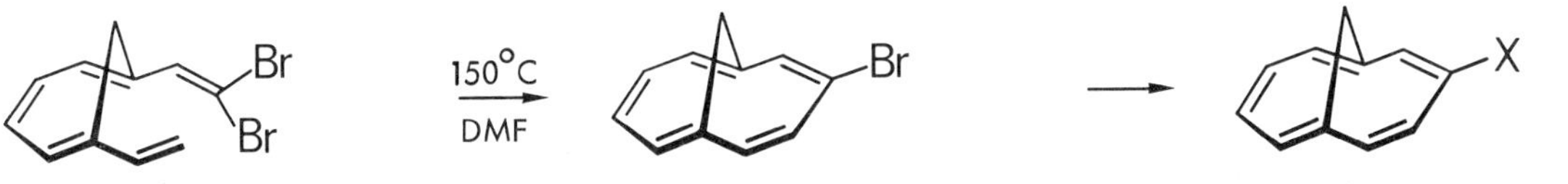

X=CN, OCH$_3$, N(CH$_3$)$_2$, Si(CH$_3$)$_3$

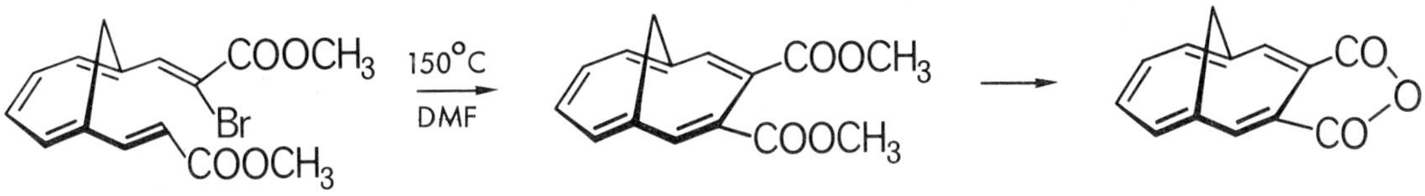

The bromide <u>15</u> is to be regarded as a key compound in the field since it can be converted into numerous other 3-substituted 1,6-methano[10]annulenes (<u>16</u> with X = CN, OCH$_3$, N(CH$_3$)$_2$, Si(CH$_3$)$_3$, etc.), some of which not readily available by other routes. Attempts to hydrolyze the diester <u>18</u>, interestingly, afford the anhydride <u>19</u> rather than the corresponding diacid as the product isolated. This finding seems to indicate that on going from 2,3-disubstituted naphthalenes to 3,4-disubstituted 1,6-methano[10]annulenes the distance between the substituents is reduced, as suggested by steric considerations.

In view of the fact that 1,6-methano[10]annulene preferentially furnishes 2-substituted derivatives on treatment with electrophilic reagents, the synthesis of 3- and 3,4-disubstituted 1,6-methano[10]annulenes from <u>11</u> through the above annelation scheme has filled a hitherto existing gap in 1,6-methano[10]annulene chemistry (1f).

2,7-Methanoaza[10]annulene <u>20</u>
___

A virtue of cycloheptatriene-1,6-dicarboxaldehyde <u>11</u> is that it also should provide an entry to the essentially untouched field of 1,6-methano[10]annulenes containing one or two nitrogen atoms in the peripheral ring, i.e., methanoaza- and methanodiaza[10]annulenes, by annelation procedures. Such novel heterocycles, constituting 10π-analogues of the classical heteroaromatic compounds pyridine, pyrimidine, pyridazine, and pyrazine, respectively, unquestionably attract great interest both for theoretical and for practical reasons.

Naturally, it has been one of our main goals in this field to construct the two possible bridged aza[10]annulenes derived from 1,6-methano[10]annulene, i.e., 2,7-methanoaza-[10]annulene <u>20</u> and 3,8-methanoaza[10]annulene <u>21</u>. While <u>21</u> has so far escaped synthesis, <u>20</u> has been prepared from <u>11</u> by an annelation scheme that features the isocyanate <u>22</u> as the key intermediate. The supposition that <u>22</u> would isomerize thermally by a 10π-electrocyclic process – involving the isocyanate carbon nitrogen double bond – to the azaenone <u>23</u>, and that <u>23</u> would spontaneously undergo a prototropic shift to give the lactam <u>24</u> was borne out by experiment. On heating <u>22</u> in toluene, <u>24</u> is obtained as the main product. The subsequent transformation of <u>24</u> into <u>20</u> was best effected via the O-tosylate of <u>24</u>. 2,7-Methanoaza[10]annulene thus prepared proved to be a stable aromatic compound (3).

Interestingly, 2,7-methanoaza[10]annulene <u>20</u>, having a pk$_a$-value of 3.20, is more weakly basic than pyridine or quinoline (pk$_a$ = 5.23 and 4.94, respectively). Despite its low basicity, <u>20</u> affords a stable hydrochloride with dry hydrogen chloride in ether, reacts with suitable alkylating agents to give the corresponding quarternary salts, and is capable of forming an N-oxide – a 10π-analogue of pyridine oxide – on treatment with peracids.

Electrophiles, such as bromine or acetyl chloride, preferentially attack 20 at the "nitrogen-free" segment of the aza[10]annulene ring to give either adducts or substitution products.

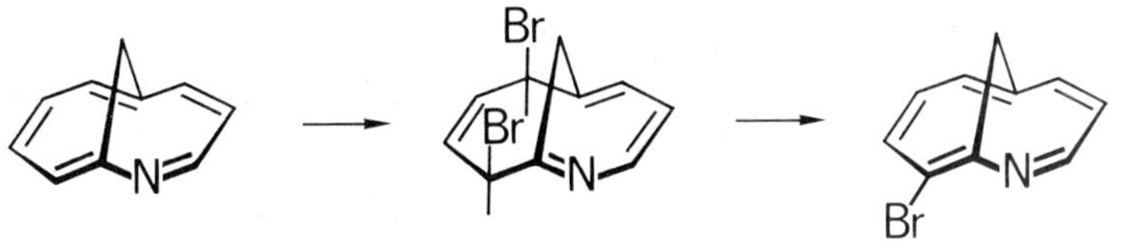

20   25   26

Bromination of 20 parallels that of 1,6-methano[10]annulene in that bromine stereoselectively approaches the molecule from the syn-position with formation of the adduct 25. Although 25 is a labile compound, it has been possible to isolate it and to determine its stereochemistry by an X-ray study. Subsequent treatment of 25 with DBN affords 3-bromo-2,7-methanoaza[10]annulene 26 as the only product (4). An X-ray investigation performed on 26 has served to adduce structural evidence in favor of the aromatic nature of the 2,7-methanoaza[10]annulene system (5).

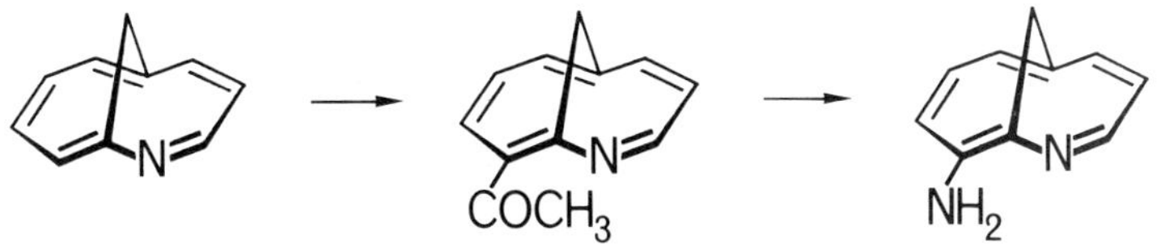

20   27   28

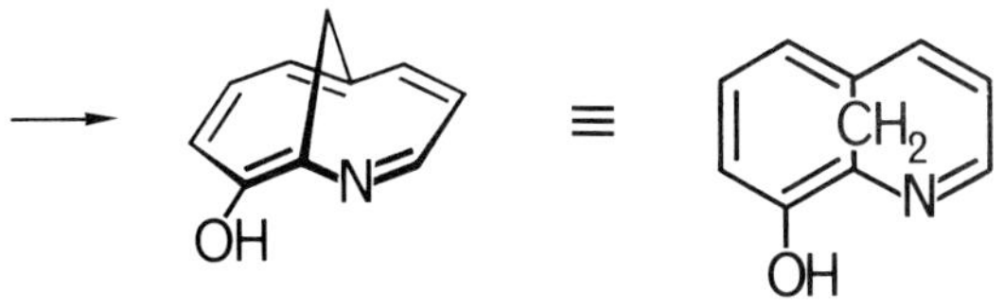

29

In the reaction of 20 with acetyl chloride the nitrogen atom appears to exert a directional effect since 3-acetyl-2,7-methanoaza[10]annulene 27 is formed as the by far predominant product of substitution (6-acetyl-2,7-methanoaza[10]annulene can be traced). While attempts to hydrolyze 26 to 3-hydroxy-2,7-methanoaza[10]annulene 29 have as yet failed, this interesting hydroxyl compound - to be regarded as "homo-8-hydroxyquinoline" - is obtained when 27 is degraded to the amine 28 and the latter treated with dilute hydrochloric acid. The intriguing question as to whether 29 shares the propensity of 8-hydroxyquinoline to form metal salts stabilized by chelation is currently being examined (6).

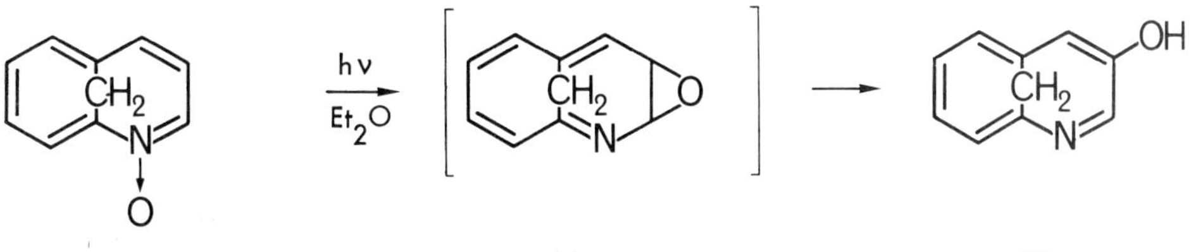

30   31   32

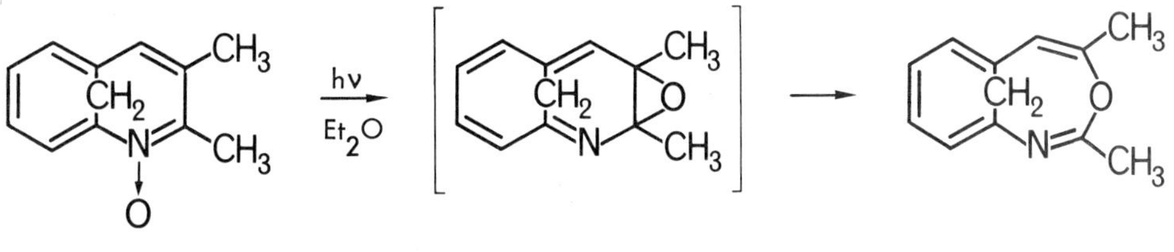

33   34   35

The afore-mentioned 2,7-methanoaza[10]annulene N-oxide 30 undergoes a remarkable photochemical isomerization providing access to 9-hydroxy-2,7-methanoaza[10]annulene 32 which matches 29 with regard to the position of the hydroxyl group relative to the nitrogen atom. In all likelihood, the conversion of 30 to 32 involves the oxaziridine isomer of 30 and the heteroarene oxide 31 as reactive intermediates. This assumption derives support from the observation that the photolysis of 33 affords the vinylogous 1,3-oxazepine 35 which must originate from 34 by a thermally allowed 10π-electrocyclic process (7).

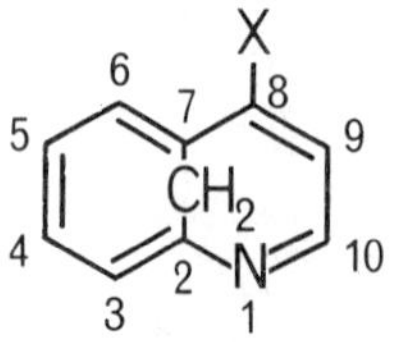

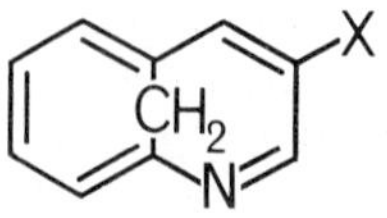

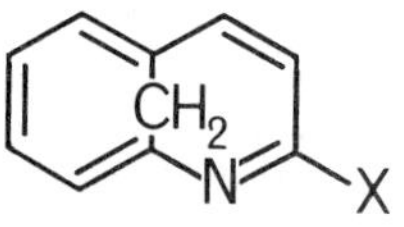

The relatively limited availability of substituted 2,7-methanoaza[10]annulenes by functionalization of the parent 20 made it imperative to develop independent routes to such compounds, if one of our main objects in the field, the study of substituent effects in the 2,7-methanoaza[10]annulene system, were to be realized. Thus, 8-, 9-, and 10-substituted 2,7-methanoaza[10]annulenes (36, 37, and 38, respectively), which are of special interest as they invite comparisons with the corresponding pyridines, became targets of synthesis.

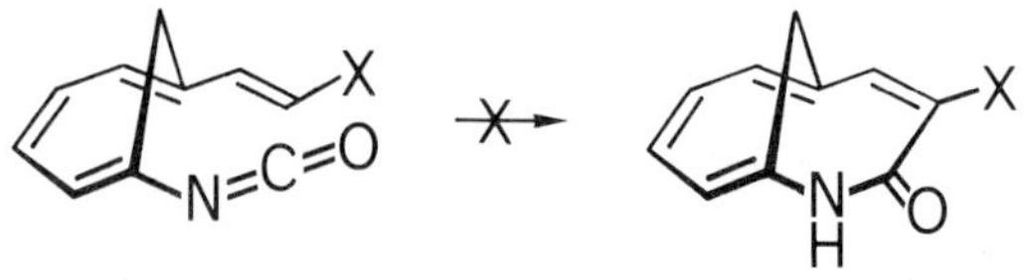

As anticipated, the scheme leading to 20 readily lent itself to the preparation of 8-methyl-2,7-methanoaza[10]annulene 53 as well as of some other 8-substituted 2,7-methanoaza-[10]annulenes. Contrary to expectation, however, this scheme does not extend to the synthesis of 9-substituted 2,7-methanoaza[10]annulenes 37 because the respective isocyanate

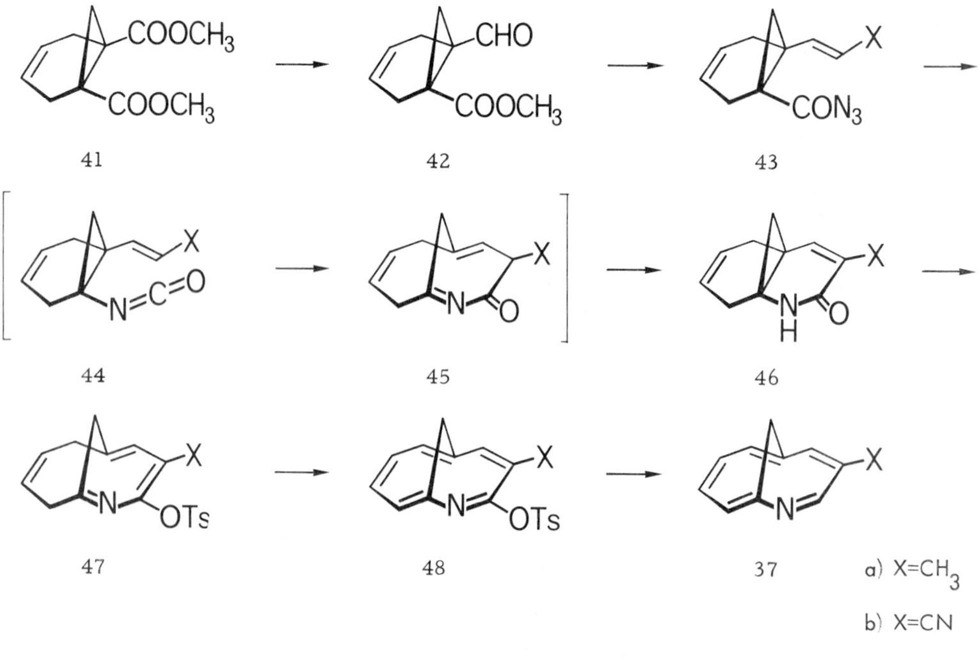

a) X=CH₃
b) X=CN

intermediates 39 no longer possess the necessary driving force to undergo the 10π-electrocyclic rearrangement to the corresponding α-pyridone analogues 40 at temperatures low enough to exclude competitive reactions effectively.

Failure of 39 to rearrange to 40 prompted attempts to prepare the aza[10]annulenes 37 from the suitably substituted cis-2-vinylcyclopropylisocyanates 44 (available by conventional methods from 41 via 42 and 43). Due to the release of ring strain of the cyclopropane ring, these isocyanates were expected to exhibit a pronounced tendency to isomerize by a process akin to a Cope rearrangement to give the azaenones 45 which, for thermodynamic reasons, should isomerize further by a prototropic shift of the azomethine π-bond to afford the respective lactams with unchanged ring skeleton (2-azabicyclo[4.4.1]undeca-5,8,10-trien-3-ones). In fact, the isocyanates – on generation from the azides 43 – experienced the Cope-type rearrangement, but the azaenones 45 thus formed stabilized themselves by a homo-1,5-sigmatropic shift (alternative mechanisms conceivable) to give the lactams 46. This devious course of the gross isomerization of the cis-2-vinylcyclopropylisocyanates 44, fortunately, did not jeopardize our synthetic efforts directed at 37, since a smooth pathway leading from the lactams 46 to the desired products (37a, 37b) has been devised (intermediates 47 and 48) (8).

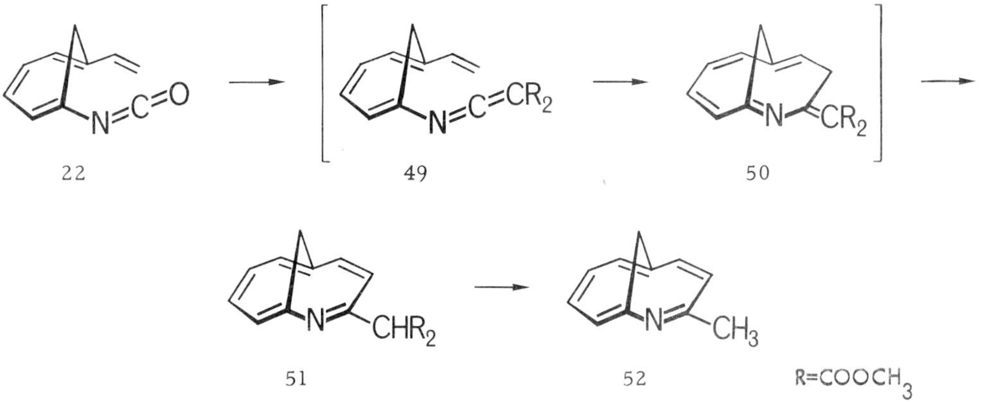

The methodology, based on electrocyclic or Cope-type rearrangements, that proved so useful in the synthesis of 20 and 37 also opened an attractive route to 10-alkyl-substituted 2,7-methanoaza[10]annulenes, specifically to the 10-methyl compound 52. As found by Ch.S. Kim most recently (9), 6-vinyl-1-cycloheptatrienylisocyanate 22 when treated with dicarbomethoxymethylene(triphenyl)phosphorane directly yields 51. Obviously, this reaction is best explained by assuming that the ketene imine 49 and the dicarbomethoxymethylene compound 50 resulting from the latter by electrocyclic ring closure occur as transient reaction intermediates. Conversion of the diester 51 to 10-methyl-2,7-methanoaza[10]annulene 52 is smoothly effected by alkaline hydrolysis and heating of the mono-acid thus produced in tetrahydrofuran.

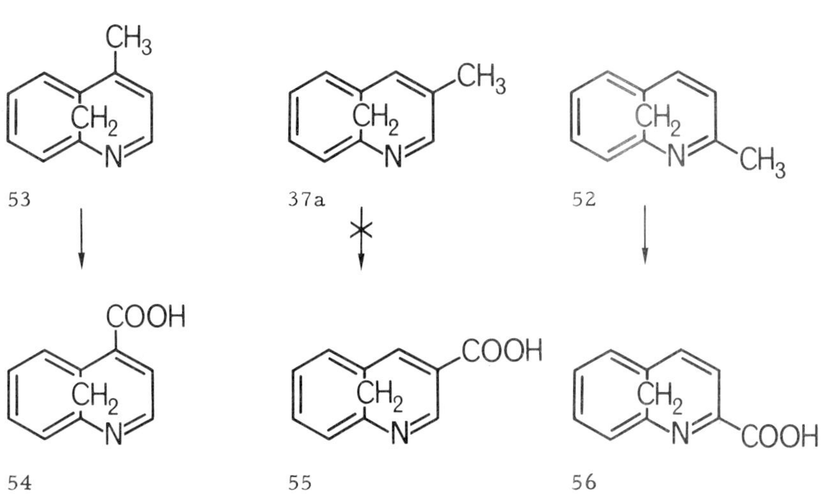

Attesting to the astonishing chemical stability of the 2,7-methanoaza[10]annulene system, 53 and 52 are oxidized to the acids 54 and 56, respectively, by sequential treatment with selenium dioxide and silver oxide, whereas 37a remains essentially unchanged under these conditions (1f). This, incidentally, is the same reactivity pattern observed in the oxidation of the three methylpyridines.

The efforts to convert the methyl compound 37a to the acid 55 were motivated to a large extent by the prospect that 55 would make accessible the amide 57 which, constituting a 10 π-analogue of nicotinamide, is of obvious interest as a pyridine nucleotide model. As the route to 57 from 37a was blocked because of the failure of the oxidation experiments, efforts were made to obtain 57 from the nitrile 37b. While the attempted alkaline hydrolysis of 37b led only to ill-defined products, treatment of 37b with 95 % sulphuric acid at room temperature efficiently yielded 57. This on reaction with 75 % sulphuric acid at 130°C finally gave 55. That 55 survives these drastic conditions suggests that it is capable of forming stable pyridinium-type salts with the strong mineral acid. A collaboration with biochemists has now been initiated in order to find out if 57 and related molecules exhibit biological activity.

The as yet preliminary studies on the synthesis of transition metal complexes of 20 may serve to illustrate that the chemical parallels between 20 and pyridine, emphasized in the preceding discussion, clearly have their limits. Thus, 20, featuring a relatively weakly basic nitrogen atom, reacts with triamminetricarbonylchromium to give the π-complex 58 rather than a σ-complex with the metal bound to nitrogen, whereas pyridine and also quinoline have been reported to yield only σ-complexes when treated with metal carbonyls. As evidenced by its ^{1}H NMR spectrum and by an X-ray investigation, 58 bears a close structural relationship to tricarbonyl-1,6-methano[10]annulenechromium 59 described previously (10). Most significantly, the annulene ligand in 58 still harbours a delocalized 10π-electron system, as is the case for that in 59 (11).

The ramifications of the fact that 20 and its derivatives are chiral have yet to be explored.

BRIDGED [14]ANNULENES

General remarks

Following this report on some current trends in the synthesis and chemistry of bridged [10]annulenes and aza[10]annulenes, I would like to proceed by discussing our recent synthetic pursuits in the area of the syn/anti-stereoisomeric bridged [14]annulenes with an anthracene perimeter. It should be recalled here that, with the exception of syn-1,6:8,13-bis-methano[14]annulene 79, to be referred to later, all of the numerous presently known bridged [14]annulenes of this type, be they of syn- or anti-stereochemistry, have been prepared from 1,4,5,8,9,10-hexahydroanthracene. While some of these annulenes, such as the aromatic syn-1,6:8,13-bisoxido[14]annulene 6 and the olefinic anti-1,6:8,13-bismethano-[14]annulene 7, are accessible by fairly short reaction pathways others require laborious routes to be made.

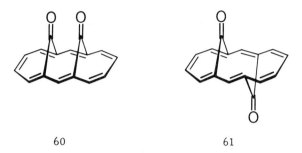

60   61

Thus, quite an effort had to be spent in order to arrive at the syn- and anti-15,16-dioxo-1,6:8,13-bismethano[14]annulenes (60 and 61, respectively) embarking from hexahydroanthracene (12,13). Although the recent syntheses of 60 and 61 are not part of this lecture, a brief comment on these rather unusual structures would seem to be appropriate. The syn-isomer 60 is particularly noteworthy, not so much because it qualifies as aromatic (as expected), but because it exhibits truly astounding thermal properties for a structure of this type. It seems hard to believe that 60 – formally to be regarded as a "carbonyl complex" of anthracene – remains unchanged when it is subjected to a flash vacuum pyrolysis at 500°C. By contrast to 60, its anti-isomer 61 is an olefinic compound assumed to possess fluctuating π-bonds.

As has been discussed elsewhere (1c,d,e), the various syn-bridged and anti-bridged [14]-annulenes with an anthracene perimeter have proved to be ideal substrates for investigations relating the molecular geometry of a (4n+2)π-electron system to π-electron delocalization. In fact, it was largely through such studies that it became apparent that a cyclically conjugated (4n+2)π-electron system is capable of tolerating rather striking deviations from planarity without π-electron delocalization being significantly impaired.

syn-1,6-Imino-8,13-methano[14]annulene 62 and syn-1,6:8,13-bisimino[14]annulene 63
―――

The potentially aromatic NH-bridged [14]annulenes 62 and 63 have been synthetic goals in our laboratory for quite some time since these molecules possess structural features making them equally interesting both as annulenes and as novel heterocycles. The annulene 63 promised to be one of the most intriguing representatives among the bridged [14]-annulenes with an anthracene perimeter because it bears some relationship to the porphyrins. Although various approaches to 62 and 63 can be conceived it occurred to us that "hexahydroanthracene-routes" would offer the best possible chances for success.

62   63

syn-1,6-Imino-8,13-methano[14]annulene 62 came within reach when it was found, that the hydrocarbon 64, available by Simmons-Smith cyclopropanation of hexahydroanthracene, could be converted regio- and stereoselectively into the aziridine 66 by way of 65 by taking advantage of the method of A. Krief. Subsequent transformation of 66 to 62 was routine except for the fact that 67, representing the first 1H-azepine ever isolated and identified unequivocally, turned out to be an intermediate hardly less interesting than the target molecule. While dehydrogenation of 67 to the annulene 62 by means of DDQ was complicated by side reactions, this step could finally be realized by treating 67 with oxygen in the presence of potassium-t-butoxide, a dehydrogenating reagent which had also come to the rescue in the synthesis of syn-1,6:8,13-bisoxido[14]annulene 6.

syn-1,6-Imino-8,13-methano[14]annulene 62 matches 60 and the other syn-bridged [14]annulenes with an anthracene perimeter in that it is a stable compound which, as indicated by its ^{1}H NMR spectrum and by structural data, is aromatic (14).

Inevitably, a description of the molecular geometry of 62 must include an answer to the subtle question of whether the NH-bridge proton is to be assigned the exo- or the endo-configuration with respect to the neighboring CH$_2$-bridge (62a or 62b, respectively). Since the steric arrangement of the NH-proton in 62 is of great topical interest, in view of the much discussed problem of the relative sizes of the free electron pair and the H-atom on the trigonal imino nitrogen, this assignment was given special attention. Confirming earlier tentative conclusions based on ^{1}H NMR findings, a detailed X-ray structure determination of 62 has now firmly established that the NH-proton possesses the exo-configuration as depicted in 62a (15). The NH-proton of 62 thus avoids the congested endo-position suggesting that it is sterically more demanding than the free electron pair on the nitrogen atom in question.

The synthetic studies on 62 had some unexpected chemical ramifications outside annulene chemistry in that they had provided the clue for the preparation of the hitherto elusive 1H-azepine 71, the parent of the well-known N-substituted 1H-azepines developed by K. Hafner and others. The properties of the intermediate 67 (a 2,6-disubstituted 1H-azepine) suggested to us that 71 might be stable enough to be characterized or even isolated if the compound were generated under strictly neutral conditions at low temperature. Accordingly, 1-carbomethoxy-1H-azepine 68 was treated with trimethylsilyl iodide and the trimethylsilyl ester 69 thus produced subjected to hydrolysis by methanol in the cold (-78°C). Contrary to expectation, the product of the latter reaction was not 71 but the carbamic acid 70, which actually could be isolated below -30°C. Heating 70 in chloroform briefly to room temperature and then cooling again afforded NMR spectroscopically pure 71 in solution. Not too surprisingly in view of its previous record, the molecule is extremely reactive, polymerizing fairly rapidly in chloroform and other solvents even at -30°C. Significantly, 71 could be shown by NMR spectroscopic investigations to exist as the uniform seven-membered ring valence tautomer, the concentration of the bicyclic benzene imine valence tautomer 72 being below 1 % (16).

Attempts at the synthesis of syn-1,6:8,13-bisimino[14]annulene 63 have been stagnating for years, as the bisaziridine 76, the key intermediate of the envisaged route to 63, proved to be unexpectedly difficult to come by. The problem was that the seemingly ideal precursor for 76, the diazidodiol 74 (readily formed regioselectively by nucleophilic opening of the epoxide rings of the hexahydroanthracene bisepoxide 73 by azide ion in the presence of magnesium chloride) behaved unusually as far as the reactivity of its hydroxyl groups was concerned. Although virtually all pertinent acylating agents including triflic anhydride and sulphuryl chloride (entropy favored formation of the cyclic sulphate 75) were tried, the hydroxyl functions of 74 – presumably because of strong intramolecular hydrogen bonding suggested by their 1,3-diaxial position – could not be converted into leaving groups as was required for the subsequent aziridine ring closures.

It was only by the recent observation that 74 on reaction with sulphur trioxide/sulphuric acid in ether does afford 75 as the sole isolable product, that this stalemate has been overcome. As anticipated, reduction of 75 with lithium aluminum hydride to give 76 occurred smoothly. When 76 was subjected to the familiar bromination-dehydrobromination reaction sequence, the respective bis-(1H-azepine), i.e., 7,14-dihydro-syn-1,6:8,13-bisimino-[14]annulene, was obtained, but the dehydrogenation of this chemically very capricious molecule has as yet remained inconclusive with regard to the formation of 63. Efforts to avoid the intermediacy of the 7,14-dihydro compound on the route to 63 by employing protection techniques have led, with remarkable ease, to the preparation of the syn-1,6:8,13-bisimino[14]annulene derivative 78 from 77, the latter being available on reaction of 76 with thionyl chloride in the presence of triethylamine. Utilization of thionyl chloride as a protecting agent has since been found to serve our purpose exceedingly well, for treatment of 78 with lithium aluminum hydride readily yields the desired 63. This new bridged [14]-annulene is a scarlet-red stable aromatic compound. Work towards full spectroscopic and structural characterization of 63 was still in progress at the deadline for submission of the manuscript (17).

CH$_2$-BRIDGED [4n+2]ANNULENES BY A BUILDING-BLOCK APPROACH

I now would like to return to CH$_2$-bridged [4n+2]annulenes and give you a brief account of our recent successful efforts to synthesize – by a building-block approach – the all-syn-homologues of 1,6-methano[10]annulene, i.e., syn-1,6:8,13-bismethano[14]annulene 79, syn,syn-1,6:8,17:10,15-trismethano[18]annulene 80, and lately even syn,syn,syn-1,6:8,21-10,19:12,17-tetrakismethano[22]annulene 81 (1f).

Molecular models indicate that 79, 80, and 81 must possess bent annulene rings due to the crowding of the juxtaposed inner bridge hydrogen atoms. Nevertheless, these annulenes were anticipated to qualify as aromatic molecules, as we had shown previously that the steric conditions for aromaticity are not very stringent.

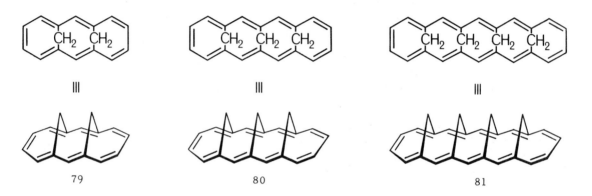

79   80   81

syn-1,6:8,13-Bismethano[14]annulene 79 has been a synthetic goal for us from the outset of our studies on bridged annulenes, but all attempts to prepare the hydrocarbon from hexahydroanthracene have proved futile. In order to overcome this impasse it was necessary to devise an entirely new synthetic strategy. An approach deserving this description emerged from the discovery that bicyclo[5.4.1]dodeca-2,5,7,9,11-pentaene-3,5-dicarboxaldehyde 82, which should exist as an equilibrating mixture of a syn- and an anti-conformer, is present solely as the syn-conformer. Clearly, this rather unexpected stereochemical finding makes the dialdehyde 82 a starting material "par excellence" for the preparation of 79.

82

The dialdehyde 82, first encountered in chemical studies on 1H-3,8-methanocyclopropa-[10]annulene, has now become readily accessible by a three-step route starting from cycloheptatriene-1,6-dicarboxaldehyde 11 (2).

The crucial first step of the route to 82 is the Wittig-Horner reaction of 11 with the new bisphosphonate 83, derived from α,α'-dibromoglutaric diethyldicarboxylate. Although earlier reports on ring closure reactions utilizing bifunctional phosphoranes were not very encouraging, 84 is obtained in good yield. The subsequent conversion of 84 to 82 is then effected routinely by the reaction sequence of diisobutylaluminum hydride reduction and DDQ oxidation.

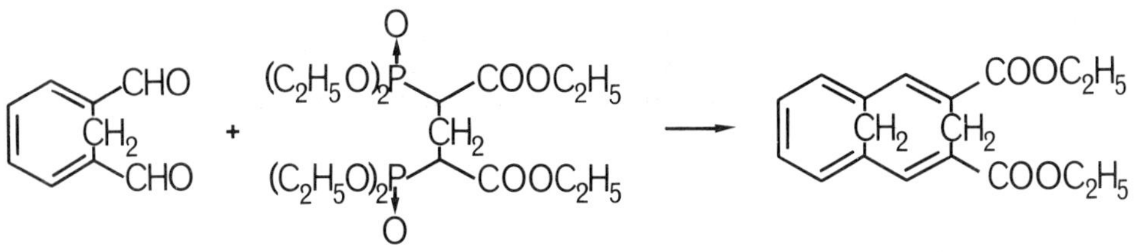

In the search for an expedient pathway from 82 to 79 our attention focused on the annelation scheme that had proved to be useful in the synthesis of 2 from 11. In fact, all the various versions of this scheme could be applied successfully to 82. However, best results were achieved when 85, readily prepared from 82 by successive treatment with methylene(triphenyl)phosphorane and chloromethylene(triphenyl)phosphorane, was chosen as the key intermediate. As anticipated, 85 experiences a 14π-electrocyclic process on heating in dimethylformamide and the cyclization product 86 thus formed spontaneously undergoes elimination of hydrogen chloride furnishing 79 in good yield.

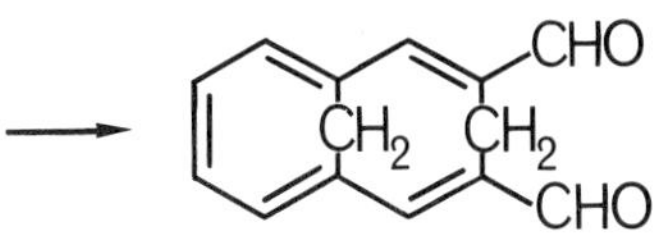

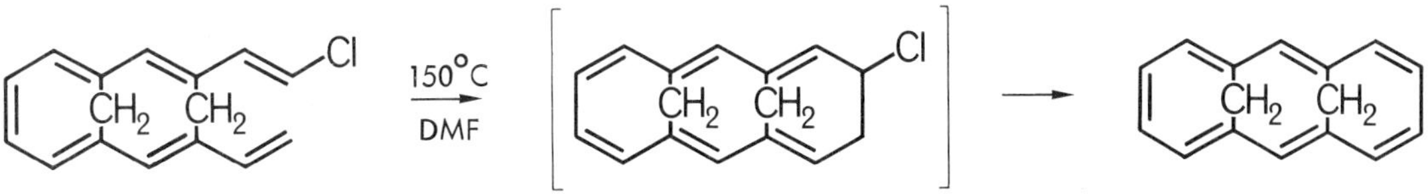

From the mechanistic point of view it is interesting to note that, by analogy with 85, the acetylenic hydrocarbon 87 is subject to a thermal ring closure by a 14π-electrocyclic process. This ring closure of 87 leads to the strained allene 88 which can be envisaged to give the product isolated, the annulene 79, by a thermally allowed 1,13-sigmatropic shift. Although the conversion of 87 to 79 proceeds very smoothly, the preparation of 79 via 87 is not economical because of the relative inaccessibility of 87 (18).

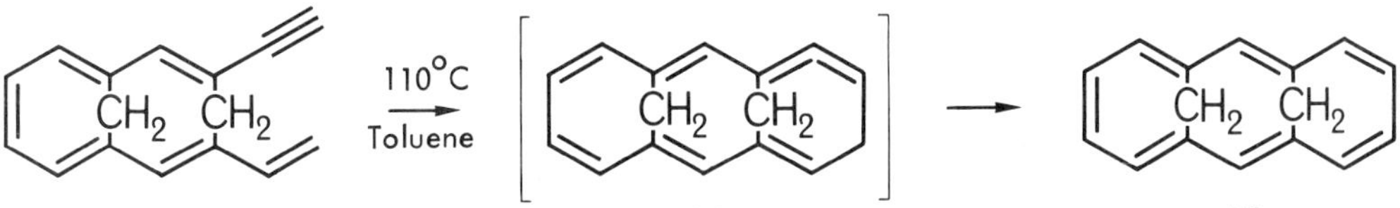

syn-1,6:8,13-Bismethano[14]annulene 79 obtained by these routes is a stable orange compound whose physical properties show it to be aromatic. In contrast to 2 and 89, however, the conformationally strained 79 behaves chemically as an olefin in that it affords addition rather than substitution products on treatment with electrophilic reagents.

The ^{1}H NMR spectrum of 79 (Fig. 3) indicates that this [14]annulene bears an especially close relationship to 1,6:8,13-propanediylidene[14]annulene 89 which is formally derived from 79 by replacement of its inner bridge hydrogen atoms by a CH_2-group. Thus, the resonances of the annulene protons, showing the familiar pattern of an AA'XX'-system and a singlet, occur in the range of $\delta = 7.2$-7.9, whereas the resonances of the outer bridge protons appear as the X-part of the AX-system of the CH_2-protons at $\delta = -1.13$.

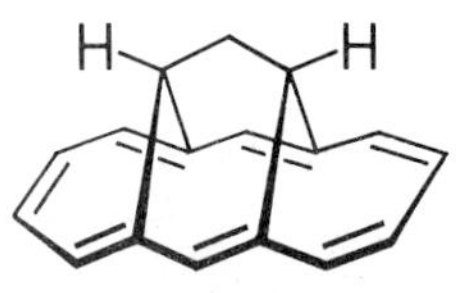

89

These values are virtually the same as those observed for the resonances of the corresponding protons of 89. The resonances of the sterically interacting inner bridge protons, constituting the A-part of the AX-system, appear at $\delta = 0.95$, i.e., more than 2 ppm downfield from those of the outer bridge protons. The occurrence of these resonances at strikingly low field must be attributed in large measure to a deshielding effect arising from the proximity of the inner bridge protons. In parallel with these NMR findings, the electronic spectra of 79 and 89 are also in excellent agreement.

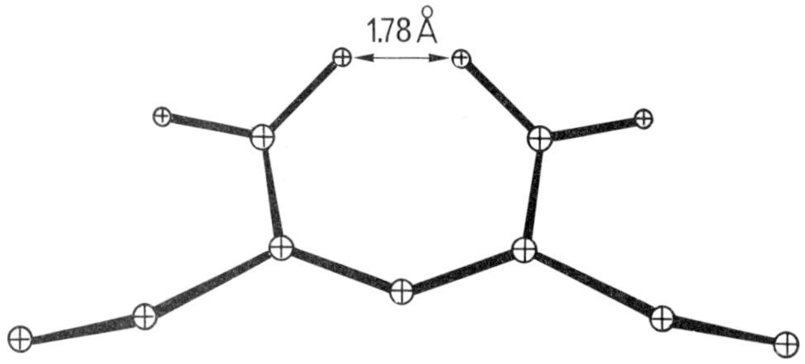

Fig. 1. X-ray structure of syn-1,6:8,13-bismethano[14]annulene 79.

The steric compression of the inner bridge hydrogen atoms of 79, as deduced from molecular models and confirmed by the NMR spectrum, manifests itself most clearly in the X-ray structure determination of the hydrocarbon (Fig. 1). These hydrogen atoms are found to be only 1.78 Å apart.

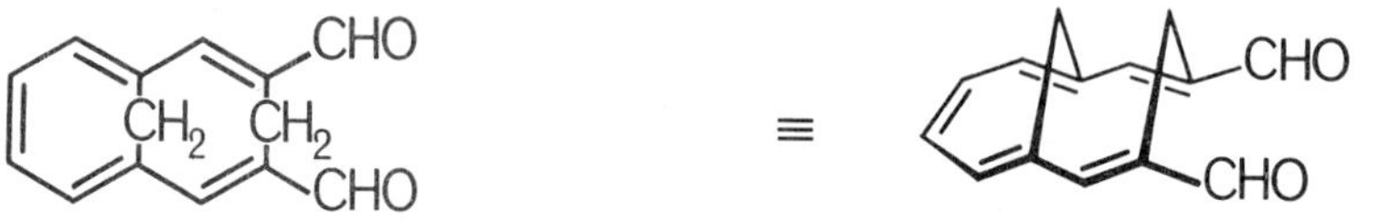

82

90

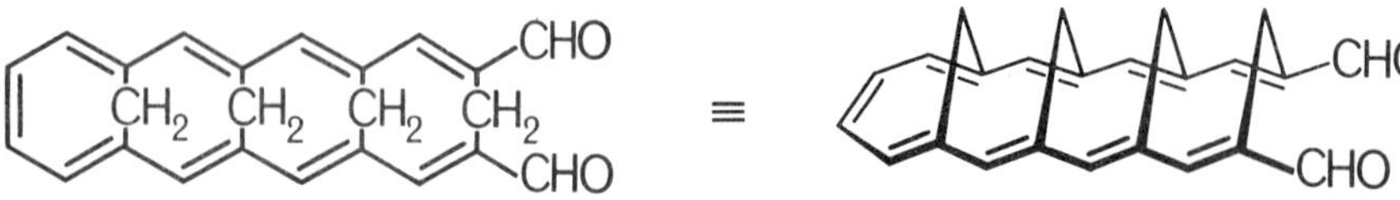

91

The unexpected existence of bicyclo[5.4.1]dodeca-2,5,7,9,11-pentaene-3,5-dicarboxaldehyde 82 as the syn-conformer justified the assumption that its homologues 90 and 91 (syn- and syn,syn-configurational isomers, respectively) would likewise be present as the respective syn-conformers. A conceivable way to build up the α,ω-polyene dialdehydes 90 and 91 was to start from 82 and to reapply to this compound – and subsequently to 90 – the same homologation methodology, i.e., the three-step sequence of: i) Wittig-Horner olefination by means of 83, ii) diisobutylaluminum hydride reduction, and iii) DDQ oxidation, that has served to convert 11 into 82. Attesting to the usefulness of this methodology, both homologations could be realized straightforwardly, selectively affording the syn- and syn,syn-configurational isomers, respectively. Verifying the above-mentioned assumption regarding conformation, X-ray structure determinations of these two next higher homologues of 82 showed them to be the respective syn-conformers. That the existence of 90 and 91 as the syn-conformers is not limited to the crystalline state, but, most importantly for the current project, also extends to the state in solution, is indicated by spectroscopic investigations.

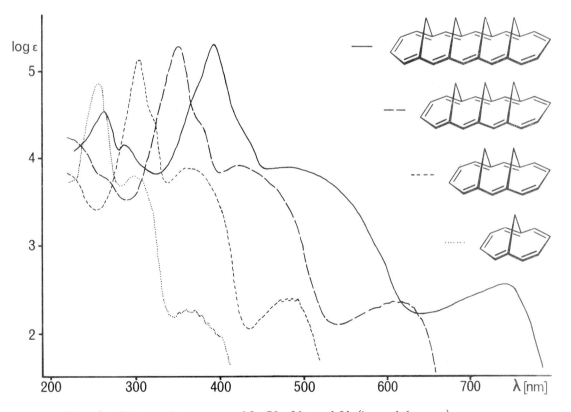

Possession of the dialdehyde 90 in reasonable quantity promised a ready completion of the synthesis of 80 by applying to 90 the annelation scheme that had led to 2 and 79 from the corresponding dialdehyde precursors. In fact, successive treatment of 90 with methylene(triphenyl)phosphorane and chloromethylene(triphenyl)phosphorane afforded 92, and when this compound was subjected to thermolysis in boiling dimethylformamide, 80 was produced in good yield by the anticipated sequence of an 18π-electrocyclic process and elimination of hydrogen chloride. Other versions of the annelation scheme have also been tried but were found to be less satisfactory.

syn,syn-1,6:8,17:10,15-Trismethano[18]annulene 80 is a bronze-coloured compound which, in striking contrast to its lower homologues 2 and 79, shows a great tendency to polymerize. The olefinic reactivity of 80 seems to be caused, however, by Baeyer- and Pitzer-strain rather than by loss of resonance, for according to its spectroscopic properties the hydrocarbon exhibits distinct aromatic character (19).

Fig. 2. Electronic spectra of 2, 79, 80, and 81 (in cyclohexane).

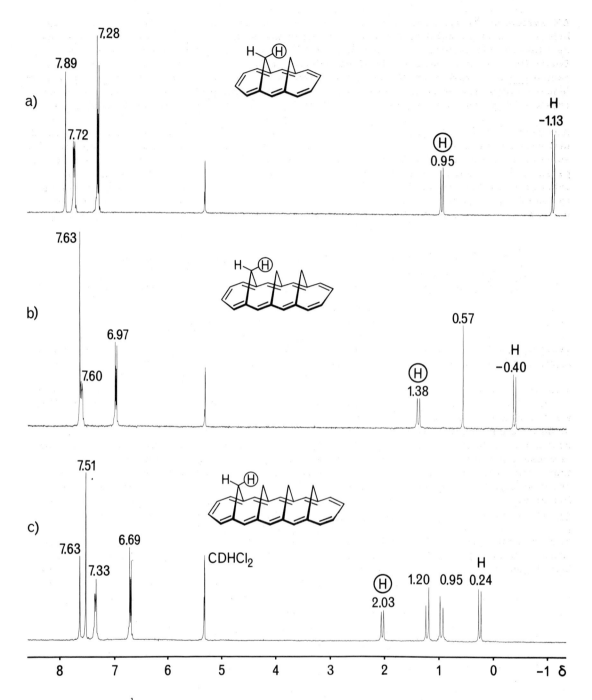

Fig. 3. ^{1}H NMR spectra (300 MHz-FT) of a) syn-1,6:8,13-bismethano[14]annulene 79, b) syn,syn-1,6:8,17:10,15-trismethano[18]annulene 80, and c) syn,-syn,syn-1,6:8,21:10,19:12,17-tetrakismethano[22]annulene 81 (in each case in CS_2/CD_2Cl_2, lock: CD_2Cl_2).

Impressive evidence for the aromaticity of 80 is provided by the relationship of its electronic spectrum to those of 2 and 79. As seen from Fig. 2, the spectra of the three annulenes have remarkably similar shape, but on going from the [10]- to the [14]- and to the [18]annulene all of the bands undergo bathochromic shifts, as predicted by theory.

In parallel with these findings, the ^{1}H NMR spectrum of 80 (Fig. 3) shows that the molecule sustains a diamagnetic ring current. Thus, the annulene protons give rise to an AA'XX'-system at δ = 7.30 and a singlet at δ = 7.63, and the bridge protons occur as an

AX-system at $\delta = 1.38$ and -0.40 (inner and outer protons, respectively, of the two outer bridges) and as a singlet at $\delta = 0.57$ (protons of the central bridge). However, it is apparent that the diamagnetic ring current in 80 has decreased by comparison with that in 79. Due to its absorption pattern, the NMR spectrum also furnishes clear evidence that the expected syn, syn-stereoisomer having C_{2v}-symmetry is present, and not the syn, anti-isomer (equally compatible with the configuration of the dialdehyde precursor). The nature of the crystals of 80 has thus far precluded an X-ray structure determination of the compound.

Attempts to convert the dialdehyde 91 into 81 by the proven annelation scheme, i.e., via 93, met with difficulties which arose from the poor solubility properties and the instability of some of the compounds involved. It was only after a laborious study of the various reaction parameters that satisfactory conditions for the two successive Wittig olefinations and the subsequent cyclodehydrohalogenation could be found. Interestingly, the cyclodehydrohalogenation of 93, featuring a rate-determining 22π-electrocyclic process, requires slightly more rigorous conditions than those needed for the corresponding reactions of 85 and 92.

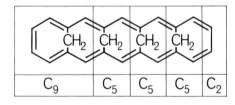

syn, syn, syn-1,6:8,21:10,19:12,17-Tetrakismethano[22]annulene 81, thus finally brought to light, is a brownish-black compound that is even more prone to polymerize than 80 (20).

The ^{1}H NMR spectrum of 81 shows the molecule to possess C_{2v}-symmetry (delocalized or rapidly fluctuating double bonds), since the annulene protons appear as an AA'XX'-system and two singlets (for two and four protons, respectively), whereas, correspondingly, the bridge protons occur as two AX-systems (Fig. 3). As inferred from the location of the resonances of both types of protons (indicated in the spectrum) a diamagnetic ring current is present in the molecule, but has decreased further with respect to that in 80. The electronic spectrum of 81 is still of the same type as those of its lower homologues; not too surprisingly, however, all of the bands in this spectrum have progressed further towards longer wave lengths (Fig. 2). A more detailed description of the π-electron structure of 81 must await the outcome of additional investigations currently in progress.

The syntheses presented in the last section of this lecture constitute the experimental realization of a concept which we have pursued for a long time, namely the development of a homologous series of CH_2-bridged Hückel-type [4n+2]annulenes by a building-block approach.

Fig. 4. Building-block approach to CH_2-bridged [4n+2]annulenes.

Let me recapitulate briefly the essentials of this approach by the schematic representation shown in Fig. 4. All of our syntheses start from the major building-block, the C_9-unit provided by the synthon cycloheptatriene-1,6-dicarboxaldehyde 11. Employing the homologation sequence, we attach one, two, or three C_5-units to the C_9-moiety and thus arrive at the respective α,ω-polyene dialdehydes possessing syn-stereochemistry with regard to both configuration (refers to 90 and 91) and conformation. The desired CH_2-bridged annulenes with 14-, 18-, and 22π-electrons are then obtained by adding the terminal C_2-unit by the three step sequence of Wittig reaction, $(4n+2)\pi$-electrocyclic process, and elimination. By omitting the C_5-units this building-block approach also covers the synthesis of the parent 1,6-methano[10]annulene 2.

The synthetic strategy portrayed in Fig. 4 has the added virtue of being rather flexible in that the terminating step, for example, can be varied. This raises the prospect that other series of CH$_2$-bridged (4n+2)π-electron systems, such as series of carbenium ions, carbanions, and azaannulenes will materialize. Provided that appropriate building-blocks become available there is even a chance that series of bridged (4n+2)π-electron systems with bridges other than CH$_2$ can be developed.

OTHER BRIDGED [4n+2]ANNULENES

For almost two decades research on bridged [4n+2]annulenes has concentrated on the 1,6-bridged [10]annulenes, on V. Boekelheide's 15,16-dihydropyrenes (21), such as 97 (actually, the first bridged [14]annulene to become known), and the bridged [14]annulenes with an anthracene perimeter as substrates. In recent years, however, the material basis of the field has broadened considerably as the continued efforts to develop Hückel-aromatic compounds has brought to light quite a number of novel bridged [10]-, [14]-, and [18]annulenes. The major new additions may be referred to here briefly.

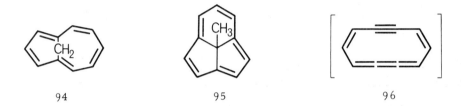

94   95   96

The existence and aromaticity of 2 have inspired both S. Masamune (22) and L.T. Scott (23) to synthesize the isomeric 1,5-methano[10]annulene 94, a molecule closely related structurally to azulene. Although 94 has been predicted by molecular mechanics calculations to possess a rather distorted C$_{10}$-perimeter and hence to be olefinic, this [10]annulene also qualifies as an aromatic compound. Varying the geometry of the C$_{10}$-perimeter even further, C.W. Rees (24) has been able to prepare the no less intriguing hydrocarbon 7b-methyl-7bH-cyclopent[cd]indene 95 and to demonstrate its aromaticity. By contrast, efforts to stabilize [10]annulenes by removal of hydrogens, i.e., by introduction of acetylene-cumulene bonds, have as yet been futile. The didehydro[10]annulene 96, parent of the acetylene-cumulene didehydroannulenes developed so successfully by M. Nakagawa (25), has remained a challenge to the organic chemist.

An elegant access to bridged [14]annulenes with a dicyclopenta[ef,kl]heptalene perimeter, such as anti-15,16-dimethyl-1,4:8,11-ethanediylidene[14]annulene 98, has been provided by the observation of K. Müllen (26) that the dianion of dicyclopenta[ef,kl]heptalene experiences alkylation at the two central carbon atoms. Representatives of both syn- and anti-isomers of these [14]annulenes are available through this approach.

The long-standing efforts in the author's laboratory to translate, in addition to anthracene, also phenanthrene into bridged [14]annulenes have finally been brought to fruition by the synthesis of one of the syn/anti-isomeric 1,6:7,12-bismethano[14]annulenes.

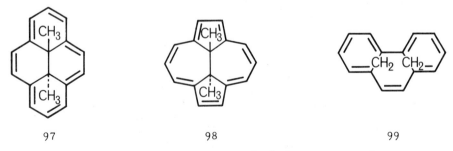

97   98   99

Our concept to build-up syn-1,6:7,12-bismethano[14]annulene 99 and/or its stereoisomer depended on the availability of 1,6-diiodocycloheptatriene 101 as starting material. Although 101 had been obtained by us in the course of our studies on the chemistry of benzocyclopropene 100 as early as 1965 (27) (100 had then to be prepared laboriously from 1,6-methano[10]annulene 2), it was only through the discovery of R. Okazaki from Tokyo University (28) that the conversion of 100 to 101 is a photochemical process that 101 could be made available on a preparative scale. Exchange of one of the iodine atoms of 101 by a cyanide group to give 102 can be achieved satisfactorily if the diiodide is heated with cuprous cyanide in a 1:1 molar ratio. Treatment of 102 with tetrakis(triphenylphosphine)nickel(0) furnishes the coupling product 103 which can be converted smoothly into 104 by means of diisobutylaluminum hydride. While attempts to subject 104 to intramolecular reductive coupling using low valent titanium species or phosphines led only to traces of hydrocarbon material, one of the syn/anti-isomeric 1,6:7,12-bismethano[14]annulenes was produced in 15 % yield when the system WCl_6/n-butyllithium was employed as the reducing agent (method of Sharpless). Similar to 79, this new [14]annulene is an air-stable orange-coloured compound (29).

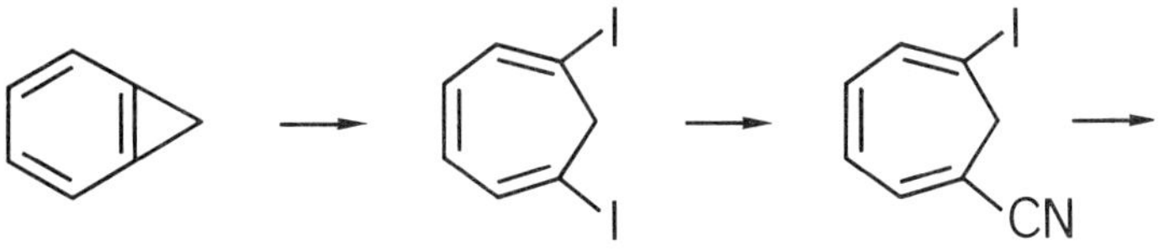

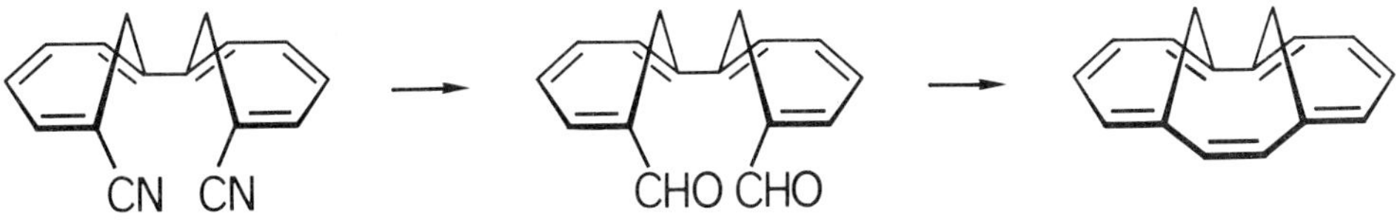

The ^{1}H NMR spectrum of the 1,6:7,12-bismethano[14]annulene isomer in question, apart from proving the gross structure of the molecule, is indicative of the presence of a diamagnetic ring current and therefore suggests that one is dealing with the syn-isomer 99. Additional evidence in favor of 99 comes from the electronic spectrum for this is remarkably similar to that of 79. An X-ray structure investigation on the compound is still in progress but from preliminary findings it is already evident that the syn-isomer 99 is indeed present.

OCTALENE-[14]ANNULENE CONVERSION

A presentation of our recent synthetic studies in the annulene field would not be balanced without mentioning some new developments in the chemistry of the olefinic 14π-electron system octalene 105, a molecule which had been the subject of intensive theoretical discussion prior to its synthesis a few years ago (30). Having the double bonds arranged around the periphery of the two eight-membered rings, octalene can be derived formally from Sondheimer's [14]annulene 106 by connecting carbon atoms 1 and 8 and thus, in a sense, constitutes a perturbed [14]annulene (zero-bridged [14]annulene).

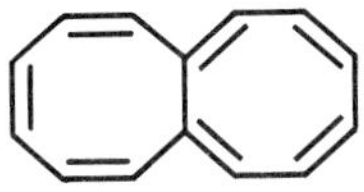

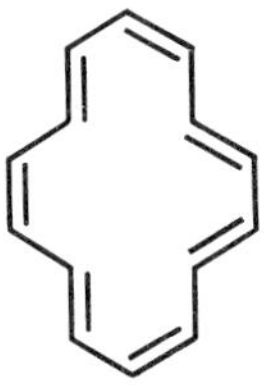

The structural relationship existing between 105 and 106 invited attempts at a novel approach to [14]annulenes by cleavage of the central octalene carbon carbon bond (C_{13}-C_{14}). Conceptually, this unusual kind of cleavage might be initiated by addition of appropriate reagents to the quarternary carbon atoms of 105 with formation of 13,14-dihydrooctalenes. There is strong evidence to support the assumption that such dihydrooctalenes are prone to isomerize thermally to the corresponding [14]annulenes by electrocyclic pathways. Whereas the chance of realizing the crucial addition step of this sequence seems remote with octalene itself, such a step is predicted to be more facile with its ionic derivative, the diamagnetic octalene dianion 107. As borne out by spectroscopic findings, in particular by ^{13}C NMR chemical shift data, as well as by MO-calculations, 107 possesses the highest local π-charge density at the quarternary carbon atoms. Accordingly, 107 should be susceptible to attack by electrophilic reagents preferentially at these positions.

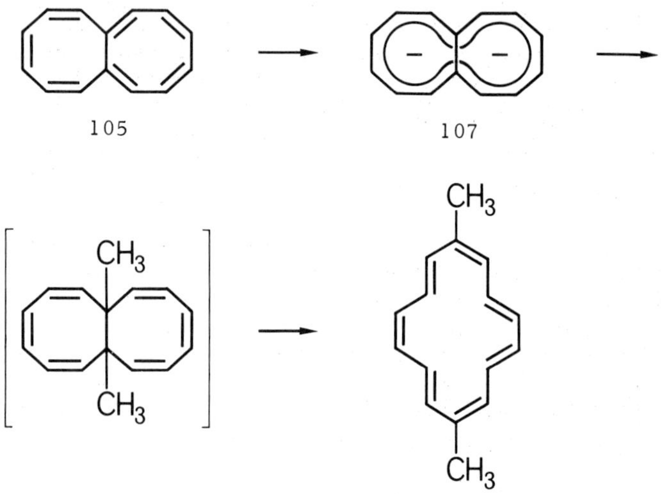

Experimental realization of this concept proved to be remarkably straightforward in that treatment of 107 in tetrahydrofuran at -80°C with dimethylsulphate afforded 1,8-dimethyl-[14]annulene 109a as the sole isolable product. Similar to the parent [14]annulene, 109a exhibits a temperature dependent NMR spectrum and is distinguished by a pronounced diamagnetic ring current. In further analogy to [14]annulene, 109a must have a pyrene type configuration as the NMR spectrum is indicative of four inner protons (31).

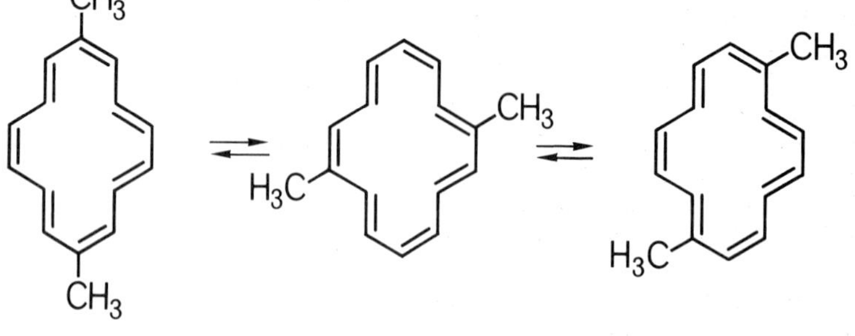

The dynamic process responsible for the dependence of the NMR spectrum is a rotation of the trans double bonds around the neighboring single bonds which causes the inner and outer ring protons to be exchanged. From this conformational interconversion it is to be concluded that 1,8-dimethyl[14]annulene, in solution, exists as an equilibrating mixture of 109a, 109b, and 109c. An X-ray structural analysis performed on 1,8-dimethyl[14]annulene shows that in its crystalline state the compound is solely present as the conformer 109a. In accord with predictions based on molecular models, 109a avoids the severe crowding of the inner hydrogen atoms that would obtain in the planar molecule. The carbon skeleton of 109a, featuring torsional angles of up to 20°, adopts the shape of a puckered loop. Nevertheless, the carbon carbon bond lengths are found to be typical of benzenoid aromatic bonds (1.364-1.407 Å) (31). These structural findings on 1,8-dimethyl[14]annulene attest to our previous conclusions, derived from a study of bridged [14]annulenes, that the steric condition for aromaticity, i.e., the planarity of the carbon skeleton, is far from being rigorous.

From the preparative point of view, the conversion of octalene to 1,8-dimethyl[14]annulene nicely complements Sondheimer's synthesis of the parent [14]annulene since treatment of the latter with electrophilic reagents leads to polymerization rather than to substitution products (32). Further work on this new approach to the [14]annulene system is currently in progress.

We were about to communicate these results to Professor Sondheimer privately, since we were sure he would have enjoyed them, when we learned about his tragic death. It is therefore appropriate to dedicate this publication to his memory, and thus to express our deep appreciation of the great stimulation he has given to us and to many others by opening one of the most fascinating chapters of modern organic chemistry: Annulene Chemistry.

Acknowledgements – It is a pleasure to express my deep gratitude to my coworkers – names cited in the references – whose efforts, persistence, and ability has made possible the development of this research program.

The support of our work by the Ministerium für Wissenschaft und Forschung des Landes Nordrhein-Westfalen, by the Deutsche Forschungsgemeinschaft, and by a NATO Research Grant is gratefully acknowledged.

REFERENCES

1. For previous reviews, see a) E. Vogel, Chem. Soc. Spec. Publ., 21, 113 (1967); b) Chimia, 22, 21 (1968); c) Proc. Robert A. Welch Found. Conf. Chem. Res., 12, 215 (1968); d) Pure Appl. Chem., 28, 355 (1971); e) Isr. J. Chem., 20, 215 (1980); f) Pure Appl. Chem., 54, 1015 (1982).
2. E. Vogel, H.M. Deger, J. Sombroek, J. Palm, A. Wagner, and J. Lex, Angew. Chem., 92, 43 (1980).
3. M. Schäfer-Ridder, A. Wagner, M. Schwamborn, H. Schreiner, E. Devrout, and E. Vogel, Angew. Chem., 90, 894 (1978).
4. E. Vogel, M. Schwamborn, and T. Kinkel, unpublished results.
5. R. Destro, M. Simonetta, and E. Vogel, J. Am. Chem. Soc., 103, 2863 (1981).
6. E. Vogel, T. Kinkel, G. Hilken, and M. Schwamborn, unpublished results.
7. E. Vogel and G. Hilken, unpublished results.
8. E. Vogel, M. Schwamborn, and B. Glinka, unpublished results
9. Ch.S. Kim and E. Vogel, unpublished results.
10. E.O. Fischer, H. Rühle, E. Vogel, and W. Grimme, Angew. Chem., 78, 548 (1966).
11. G. Hilken, T. Kinkel, M. Schwamborn, J. Lex, H. Schmickler, and E. Vogel, Angew. Chem., submitted for publication.
12. M. Balci, R. Schalenbach, and E. Vogel, Angew. Chem., 93, 816 (1981).
13. E. Vogel, R. Nitsche, and H.-U. Krieg, Angew. Chem., 93, 818 (1981).
14. E. Vogel, U. Brocker, and H. Junglas, Angew. Chem., 92, 1051 (1980).
15. M. Simonetta, private communication.
16. E. Vogel, H.-J. Altenbach, J.-M. Drossard, H. Schmickler, and H. Stegelmeier, Angew. Chem., 92, 1053 (1980).
17. E. Vogel, F. Kuebart, J.A. Marco, and R. Andree, unpublished results.
18. E. Vogel and W. Püttmann, unpublished results.
19. W. Wagemann, M. Iyoda, H.-M. Deger, J. Sombroek, and E. Vogel, Angew. Chem., 90, 988 (1978).
20. E. Vogel, H.-W. Engels, M. Hanelt, and J. Palm, unpublished results.

21. V. Boekelheide, Proc.Robert A. Welch Found.Conf.Chem.Res., 12, 83 (1968); Pure Appl.Chem., 44, 751 (1975); T. Otsubo, R. Gray, and V. Boekelheide, J.Am.Chem. Soc., 100, 2449 (1978).
22. S. Masamune, D.W. Brooks, K. Morio, and R.L. Sobczak, J.Am.Chem.Soc., 98, 8277 (1976); S. Masamune and D.W. Brooks, Tetrahedron Lett., 1977, 3239.
23. L.T. Scott and W.R. Brunsvold, J.Am.Chem.Soc., 100, 4320 (1978); L.T. Scott, W.R. Brunsvold, M.A. Kirms, and I. Erden, Angew.Chem., 93, 282 (1981); L.T. Scott, W.R. Brunsvold, M.A. Kirms, and I. Erden, J.Am.Chem.Soc., 103, 5216 (1981).
24. T.L. Gilchrist, C.W. Rees, D. Tuddenham, and D.J. Williams, Chem.Commun., 1980, 691; T.L. Gilchrist, D. Tuddenham, R. McCague, C.J. Moody, and C.W. Rees, Chem.Commun., 1981, 657; Z. Lidert and C.W. Rees, Chem.Commun., 1982, 499.
25. M. Nakagawa, Pure Appl.Chem., 44, 885 (1975).
26. W. Huber, J. Lex, T. Meul, and K. Müllen, Angew.Chem., 93, 401 (1981).
27. E. Vogel, W. Grimme, and S. Korte, Tetrahedron Lett., 1965, 3625.
28. R. Okazaki, M. O-oka, N. Tokitoh, Y. Shishido, and N. Inamoto, Angew.Chem., 93, 833 (1981).
29. E. Vogel, W. Püttmann, W. Duchatsch, and Th. Schieb, unpublished results.
30. E. Vogel, H.-V. Runzheimer, F. Hogrefe, B. Baasner, and J. Lex, Angew.Chem., 89, 909 (1977); J.F.M. Oth, K. Müllen, H.-V. Runzheimer, P. Mues, and E. Vogel, Angew.Chem., 89, 910 (1977); K. Müllen, J.F.M. Oth, H.-W. Engels, and E. Vogel, Angew.Chem., 91, 251 (1979); E. Vogel, H.-W. Engels, S. Matsumoto, and J. Lex, Tetrahedron Lett., 1982, 1797; H.-W. Engels, J. Lex, and E.Vogel, Tetrahedron Lett., 1982, 1801.
31. E. Vogel, H.-W. Engels, W. Huber, J. Lex, and K. Müllen, J.Am.Chem.Soc., in press.
32. F. Sondheimer, Acc.Chem.Res., 5, 81 (1972).

VICARIOUS NUCLEOPHILIC SU[BSTITUTION OF] HYDROGEN. A NEW MET[HOD OF] NUCLEOPHILIC ALKYLATION OF [...]

Mieczyslaw Makosza

*Institute of Organic Chemistry, Polish Academy of Sciences[...]
Poland*

Abstract - Carbanions containing leaving groups at the carba[nion center] react with variety of nitroarenes replacing hydrogen in the o- [or p-] positions to the nitro group by the carbanion moiety, the substitu[ent] acts as a vicarious leaving group.

This new type of the nucleophilic substitution of hydrogen offers a ge[neral] method for introduction of functionalized alkyl substituents into o- or p- positions of nitroarenes and also some aromatic heterocycles. This reaction is therefore a process complementary to the Friedel-Crafts reaction for those arenes which are resistant toward electrophilic attack. The scope, limitations, some specific features and the mechanism of this reaction as well as some of its synthetic consequences will be discussed.

Introduction of alkyl or functionalized alkyl substituents into aromatic ring via electroph[ilic] substitution of hydrogen, known as the Friedel-Crafts reaction is a process of enormou[s] importance and embraces really a vast field of application (Ref.1). It has however a su[bstan]tial limitation - aromatic compounds containing electronwithdrawing substituents, particu[lar]ly nitro group are reluctant toward electrophilic attack and, as a rule, do not enter the Friedel-Crafts reaction.

On the other hand nitroarenes are susceptible toward nucleophilic attack, so a variet[y of] nucleophiles including carbanions, can replace a number of nucleofugal substituents [(Hal,] OR, NO_2 etc.) located in the ortho- or para- positions to the nitro group (Ref.2).
Nucleophilic substitution of hydrogen, contrary to the electrophilic substitution o[f ...] takes place only in very few cases because hydride anion H^- is, unlike to proton[, a] poor leaving group.

This lecture will present a new general process of the nucleophilic substitutio[n of hydrogen] in nitroarenes and some aromatic heterocycles by carbanions, and as such a [method for] the direct introduction of α-functionalized alkyl substituents into the aroma[tic ring.] Nucleophilic substitution of halogen in p- or o- chloronitrobenzenes with c[arbanions proceeds] via addition of the carbanion to the nitroarene to form adduct A (known as [...]) which subsequently eliminates chloride anion yielding the product (schem[e ...]).
The ability of p-chloronitrobenzene to form adducts with nucleophiles is [a consequence] of a powerful electronwithdrawing nitro group, therefore the isomeric

that in its crystalline state the compound is solely present as the conformer 109a. In accord with predictions based on molecular models, 109a avoids the severe crowding of the inner hydrogen atoms that would obtain in the planar molecule. The carbon skeleton of 109a, featuring torsional angles of up to 20°, adopts the shape of a puckered loop. Nevertheless, the carbon carbon bond lengths are found to be typical of benzenoid aromatic bonds (1.364-1.407 Å) (31). These structural findings on 1,8-dimethyl[14]annulene attest to our previous conclusions, derived from a study of bridged [14]annulenes, that the steric condition for aromaticity, i.e., the planarity of the carbon skeleton, is far from being rigorous.

From the preparative point of view, the conversion of octalene to 1,8-dimethyl[14]annulene nicely complements Sondheimer's synthesis of the parent [14]annulene since treatment of the latter with electrophilic reagents leads to polymerization rather than to substitution products (32). Further work on this new approach to the [14]annulene system is currently in progress.

We were about to communicate these results to Professor Sondheimer privately, since we were sure he would have enjoyed them, when we learned about his tragic death. It is therefore appropriate to dedicate this publication to his memory, and thus to express our deep appreciation of the great stimulation he has given to us and to many others by opening one of the most fascinating chapters of modern organic chemistry: Annulene Chemistry.

Acknowledgements – It is a pleasure to express my deep gratitude to my coworkers – names cited in the references – whose efforts, persistence, and ability has made possible the development of this research program.

The support of our work by the Ministerium für Wissenschaft und Forschung des Landes Nordrhein-Westfalen, by the Deutsche Forschungsgemeinschaft, and by a NATO Research Grant is gratefully acknowledged.

REFERENCES

1. For previous reviews, see a) E. Vogel, Chem.Soc.Spec.Publ., 21, 113 (1967); b) Chimia, 22, 21 (1968); c) Proc.Robert A. Welch Found.Conf.Chem.Res., 12, 215 (1968); d) Pure Appl.Chem., 28, 355 (1971); e) Isr.J.Chem., 20, 215 (1980); f) Pure Appl.Chem., 54, 1015 (1982).
2. E. Vogel, H.M. Deger, J. Sombroek, J. Palm, A. Wagner, and J. Lex, Angew. Chem., 92, 43 (1980).
3. M. Schäfer-Ridder, A. Wagner, M. Schwamborn, H. Schreiner, E. Devrout, and E. Vogel, Angew.Chem., 90, 894 (1978).
4. E. Vogel, M. Schwamborn, and T. Kinkel, unpublished results.
5. R. Destro, M. Simonetta, and E. Vogel, J.Am.Chem.Soc., 103, 2863 (1981).
6. E. Vogel, T. Kinkel, G. Hilken, and M. Schwamborn, unpublished results.
7. E. Vogel and G. Hilken, unpublished results.
8. E. Vogel, M. Schwamborn, and B. Glinka, unpublished results
9. Ch.S. Kim and E. Vogel, unpublished results.
10. E.O. Fischer, H. Rühle, E. Vogel, and W. Grimme, Angew.Chem., 78, 548 (1966).
11. G. Hilken, T. Kinkel, M. Schwamborn, J. Lex, H. Schmickler, and E. Vogel, Angew. Chem., submitted for publication.
12. M. Balci, R. Schalenbach, and E. Vogel, Angew.Chem., 93, 816 (1981).
13. E. Vogel, R. Nitsche, and H.-U. Krieg, Angew.Chem., 93, 818 (1981).
14. E. Vogel, U. Brocker, and H. Junglas, Angew.Chem., 92, 1051 (1980).
15. M. Simonetta, private communication.
16. E. Vogel, H.-J. Altenbach, J.-M. Drossard, H. Schmickler, and H. Stegelmeier, Angew.Chem., 92, 1053 (1980).
17. E. Vogel, F. Kuebart, J.A. Marco, and R. Andree, unpublished results.
18. E. Vogel and W. Püttmann, unpublished results.
19. W. Wagemann, M. Iyoda, H.-M. Deger, J. Sombroek, and E. Vogel, Angew.Chem., 90, 988 (1978).
20. E. Vogel, H.-W. Engels, M. Hanelt, and J. Palm, unpublished results.

21. V. Boekelheide, Proc.Robert A. Welch Found.Conf.Chem.Res., 12, 83 (1968); Pure Appl.Chem., 44, 751 (1975); T. Otsubo, R. Gray, and V. Boekelheide, J.Am.Chem. Soc., 100, 2449 (1978).
22. S. Masamune, D.W. Brooks, K. Morio, and R.L. Sobczak, J.Am.Chem.Soc., 98, 8277 (1976); S. Masamune and D.W. Brooks, Tetrahedron Lett., 1977, 3239.
23. L.T. Scott and W.R. Brunsvold, J.Am.Chem.Soc., 100, 4320 (1978); L.T. Scott, W.R. Brunsvold, M.A. Kirms, and I. Erden, Angew.Chem., 93, 282 (1981); L.T. Scott, W.R. Brunsvold, M.A. Kirms, and I. Erden, J.Am.Chem.Soc., 103, 5216 (1981).
24. T.L. Gilchrist, C.W. Rees, D. Tuddenham, and D.J. Williams, Chem.Commun., 1980, 691; T.L. Gilchrist, D. Tuddenham, R. McCague, C.J. Moody, and C.W. Rees, Chem.Commun., 1981, 657; Z. Lidert and C.W. Rees, Chem.Commun., 1982, 499.
25. M. Nakagawa, Pure Appl.Chem., 44, 885 (1975).
26. W. Huber, J. Lex, T. Meul, and K. Müllen, Angew.Chem., 93, 401 (1981).
27. E. Vogel, W. Grimme, and S. Korte, Tetrahedron Lett., 1965, 3625.
28. R. Okazaki, M. O-oka, N. Tokitoh, Y. Shishido, and N. Inamoto, Angew.Chem., 93, 833 (1981).
29. E. Vogel, W. Püttmann, W. Duchatsch, and Th. Schieb, unpublished results.
30. E. Vogel, H.-V. Runzheimer, F. Hogrefe, B. Baasner, and J. Lex, Angew.Chem., 89, 909 (1977); J.F.M. Oth, K. Müllen, H.-V. Runzheimer, P. Mues, and E. Vogel, Angew.Chem., 89, 910 (1977); K. Müllen, J.F.M. Oth, H.-W. Engels, and E. Vogel, Angew.Chem., 91, 251 (1979); E. Vogel, H.-W. Engels, S. Matsumoto, and J. Lex, Tetrahedron Lett., 1982, 1797; H.-W. Engels, J. Lex, and E.Vogel, Tetrahedron Lett., 1982, 1801.
31. E. Vogel, H.-W. Engels, W. Huber, J. Lex, and K. Müllen, J.Am.Chem.Soc., in press.
32. F. Sondheimer, Acc.Chem.Res., 5, 81 (1972).

VICARIOUS NUCLEOPHILIC SUBSTITUTION OF HYDROGEN. A NEW METHOD OF NUCLEOPHILIC ALKYLATION OF NITROARENES

Mieczyslaw Makosza

Institute of Organic Chemistry, Polish Academy of Sciences, 01-224 Warsaw, Poland

Abstract - Carbanions containing leaving groups at the carbanionic center react with variety of nitroarenes replacing hydrogen in the o- and p-positions to the nitro group by the carbanion moiety, the substituent X acts as a vicarious leaving group.

This new type of the nucleophilic substitution of hydrogen offers a general method for introduction of functionalized alkyl substituents into o- or p-positions of nitroarenes and also some aromatic heterocycles. This reaction is therefore a process complementary to the Friedel-Crafts reaction for those arenes which are resistant toward electrophilic attack. The scope, limitations, some specific features and the mechanism of this reaction as well as some of its synthetic consequences will be discussed.

Introduction of alkyl or functionalized alkyl substituents into aromatic ring via electrophilic substitution of hydrogen, known as the Friedel-Crafts reaction is a process of enormous importance and embraces really a vast field of application (Ref.1). It has however a substantial limitation - aromatic compounds containing electronwithdrawing substituents, particularly nitro group are reluctant toward electrophilic attack and, as a rule, do not enter the Friedel-Crafts reaction.

On the other hand nitroarenes are susceptible toward nucleophilic attack, so a variety of nucleophiles including carbanions, can replace a number of nucleofugal substituents (Cl, OR, NO_2 etc.) located in the ortho- or para- positions to the nitro group (Ref.2). Nucleophilic substitution of hydrogen, contrary to the electrophilic substitution of hydrogen takes place only in very few cases because hydride anion H^- is, unlike to proton H^+, a very poor leaving group.

This lecture will present a new general process of the nucleophilic substitution of hydrogen in nitroarenes and some aromatic heterocycles by carbanions, and as such a new method of the direct introduction of α-functionalized alkyl substituents into the aromatic rings.

Nucleophilic substitution of halogen in p- or o- chloronitrobenzenes with carbanions occurs via addition of the carbanion to the nitroarene to form adduct A (known as σ-complex) which subsequently eliminates chloride anion yielding the product (scheme 1).

The ability of p-chloronitrobenzene to form adducts with nucleophiles is due to the presence of a powerful electronwithdrawing nitro group, therefore the isomeric adduct B should be formed as well (Note a).

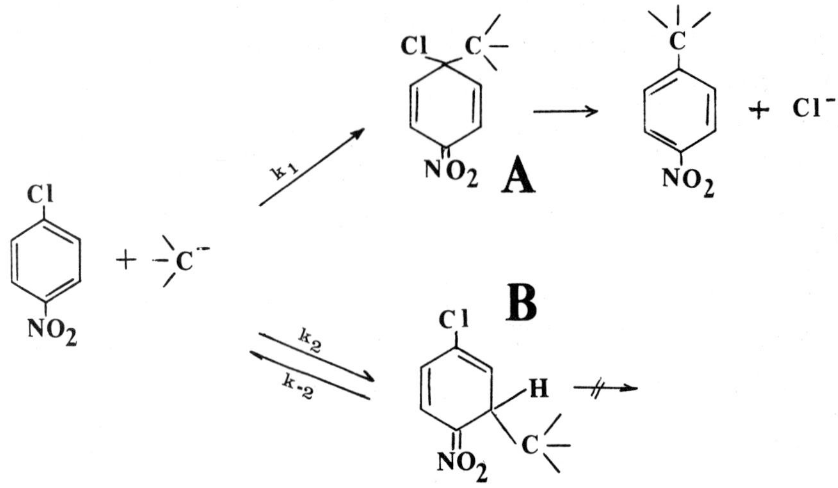

$$\text{(1)}$$

However, the formation of adduct B does not lead to the nucleophilic replacement of hydrogen since it would require the elimination of the hydride anion - a poor leaving group. Nevertheless there are some examples of further transformations of complexes B in which overall results can be considered as the removal of the hydride anion via redox process involving the nitro group (scheme 2).

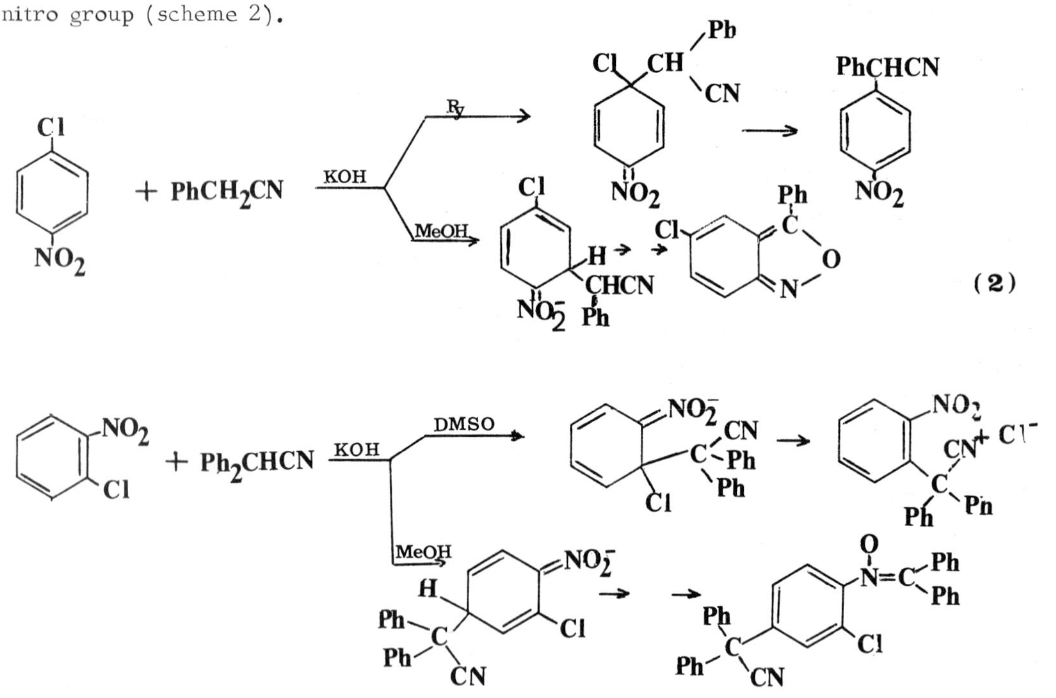

$$\text{(2)}$$

Note a. In the further discussion letters A and B will be used correspondingly for 6-complexes formed via addition of nucleophiles to nitroarenes in positions bearing a nucleofugal group and hydrogen.